OPTICAL PROPERTIES OF SURFACES

Second Edition

OPTICAL PROPERTIES OF SURFACES

Second Edition

Dick Bedeaux

Institute of Chemistry, Norwegian University of Science and Technology, Norway

Jan Vlieger

Leiderdorp, The Netherlands (retired)

ICP Imperial College Press

Published by

Imperial College Press
57 Shelton Street
Covent Garden
London WC2H 9HE

Distributed by

World Scientific Publishing Co. Pte. Ltd.
5 Toh Tuck Link, Singapore 596224
USA office: Suite 202, 1060 Main Street, River Edge, NJ 07661
UK office: 57 Shelton Street, Covent Garden, London WC2H 9HE

British Library Cataloguing-in-Publication Data
A catalogue record for this book is available from the British Library.

OPTICAL PROPERTIES OF SURFACES, SECOND EDITION

ISBN 1-86094-450-7

Printed in Singapore by World Scientific Printers (S) Pte Ltd

PREFACE TO THE SECOND EDITION

The first edition of this monograph appeared two years ago. In this second edition we have added a chapter on the reflection of light by a gyrotropic medium. In the literature there are two methods to describe homogeneous gyrotropic media. These methods lead to different reflection amplitudes of such a medium using the so-called standard boundary conditions (Fresnel). The theory developed in the first edition of the book is shown to be eminently useful to elucidate the origin of this difference. Gyrotropic contributions due to the interfacial layer are also discussed.

Furthermore we added a section on self-affine surfaces in the chapter on rough surfaces. Such surfaces are simultaneously flat and rough over the same range of lengths. They have a well-defined roughness exponent. Reflection studies may give information about this exponent.

We are grateful to Professor M. Osipov from the Strathclyde University, Glasgow, Scotland, who drew our attention to the problems regarding the reflection amplitudes for gyrotropic media and helped us to address these questions. We are also indebted to Dr. S. Gheorghiu from the Technical University in Delft, The Netherlands, for his help with the section on self-affine surfaces and to Dr. C. Chassagne from the Norwegian University of Science and Technology, Trondheim, for her help with the figure in that section.

We also want to thank Caroline C. Vlieger for designing the cover of both editions of this book.

February, 2004

Dick Bedeaux (Trondheim, Norway)
Jan Vlieger (Leiderdorp, The Netherlands)

Address for correspondence:
Dick Bedeaux, Institute of Chemistry,
Norwegian University of Science and Technology,
7491 Trondheim, Norway.
Email: bedeaux@phys.chem.ntnu.no

PREFACE TO THE FIRST EDITION

The aim of the book is to present in a systematic manner the exact results the authors obtained over the years for the description of the optical properties of thin films and rough surfaces. This work found its origin 30 years ago in a discussion of one of us (J.V.), during a sabbatical, with the late Professor Dr. G.D. Scott of the University of Toronto, Canada, about the Maxwell Garnett theory. Many questions arose which resulted in a paper by Vlieger on the reflection and transmission of light by a two-dimensional square lattice of polarizable dipoles (Physica 64, page 63, 1973). When the other author (D.B.) read this paper he was surprised by the efforts to describe the results in terms of a layer of a finite thickness, using Maxwell Garnett, while the original model had a polarizability, which was so clearly restricted to the plane of the surface. A choice of the optical thickness in terms of the distance between the dipoles seemed to be constructed. It appeared much more reasonable to replace a thin layer, compared to the wavelength of the incident light, by an infinitesimally thin layer, than the other way around. Of course the formulation of a new theory in terms of an infinitesimally thin polarizable dipole layer, which was compatible with Maxwell's equations, is easier said than done. It turned out to be necessary to introduce singularities in the electric and magnetic field at the surface in addition to the occurrence of such singularities in the sources of these fields (Physica A 67, page 55, 1973). A general description was developed, using constitutive coefficients, to describe the electromagnetic response of the surface. Having thus convinced ourselves that this approach was feasible, we started to apply these ideas to thin island films and rough surfaces.

In the treatment of surfaces a confusing element is "where to choose the precise location of the surface". For instance, for an island film one has two possible choices, one through the average centre of the islands and the other on the surface of the usually flat substrate. The constitutive coefficients depend on this choice. As the choice of this dividing surface is only a matter of convenience in the mathematical description, it is clear that the relevant observable properties, like for instance the ellipsometric angles, are independent of this choice. To make this independence clear so-called "invariants" were introduced. These invariants are appropriately chosen combinations of the constitutive coefficients independent of the location of the dividing surface. The introduction of such invariants was first done by Lekner, for the special case of thin stratified layers, and described in 1987 in his monograph on the "Theory of Reflection". One of the chapters in this book considers the case of thin stratified layers in detail and compares with Lekner's work.

When one moves a dipole from one choice of the dividing surface to another the dipole moment remains the same. The displacement leads to a contribution to the quadrupole moment, however. In order to properly describe the equivalence of different choices of the dividing surface it is therefore necessary to describe the surface to quadrupolar order. The relative importance of these quadrupolar terms is very dependent on the nature of the surface. For highly absorbing metal island films they are not important. If the island material is dielectric or when the surface is rough, the quadrupolar terms are found to be very important. In the comparison with Lekner's results for thin stratified layers, these quadrupolar terms are found to be essential.

For island films, thin compared to the wavelength of the incident light, two aspects are found to be of importance. The first is the interaction of the islands with their image in the substrate. The second is the interaction with the other islands, and with their images. For the first problem the shape of the island is very important. Over the years we were able to construct explicit solutions for spheres, truncated spheres, spheroids and truncated spheroids. For different shapes, the same amount of island material is found to lead to very different optical properties. This was also the reason to construct these explicit solutions, as they give insight into which precise aspect of the shape might be responsible for certain observed behavior. For the interaction along the surface one would expect the correlations in the distribution to be important. This, however, turned out not to be the case. Only for coverages larger than 50% this starts to be an issue. A square and a triangular array lead to essentially the same properties as a random distribution for coverages below 50%. Nevertheless the interaction with the other islands, though not dependent on the details of the distribution, changes the polarizability of the islands considerably.

For rough surfaces the correlations along the surface play a more essential role. The quadrupolar contributions are crucial in this case. It was in fact in the study of the contribution of capillary waves on fluid surfaces to the ellipsometric coefficient, that we discovered the relevance of these quadrupolar contributions.

Over the years we had many stimulating contacts. Over a period of more than 25 years Professor Dr. O. Hunderi from the Norwegian University of Science and Technology, Trondheim, Norway, has been a source of inspiration. His knowledge of the properties of island films has been a great help. We are also indebted to Professors Dr. C.G. Grandqvist and Dr. G.A. Niklasson from the University of Uppsala, Sweden, for many discussions about island films. For the foundation of the use of singular fields, charge and current densities we are grateful to Professor Dr. A.M. Albano from Bryn Mawr College, Penn., USA. We had a very rewarding collaboration with Dr. R. Greef from the University of Southampton, UK, on the optical properties of films sparsely seeded with spherical islands with a size comparable to the wavelength. Though this subject is not covered in this book, it added much to our understanding of the subject.

In the past decade we had an active collaboration with the group of Professor Dr. P. Schaaf and Dr. E.K. Mann from the Institute Charles Sadron, Strasbourg, France, and with Dr. G.J.M. Koper of our own institution. Our insight in the use and the practical relevance of invariants gained immensely due to this work.

Over the years we had many graduate students who contributed to the contents of this book. In chronological order we have: Dr. B.J.A. Zielinska, Dr. M.M. Wind, Dr. P.A. Bobbert, Dr. E.M. Blokhuis, Dr. M. Haarmans, Dr. E.A. van der Zeeuw and R. van Duijvenbode. In particular the theoretical work of Wind and Bobbert led to significant progress for the foundation of the whole methodology.

Recently I. Simonsen from the Norwegian University of Science and Technology, Trondheim, Norway, (Ingve.Simonsen@phys.ntnu.no), wrote software to perform the calculations outlined in chapters 4-10. These programs have been put into operation together with R. Lazzari from the CEA Grenoble, France (Lazzari@drfmc.ceng.cea.fr) and it is now possible to make practical use of the analytical results in these chapters, see http://www.phys.ntnu.no/~ingves/Software/GranularFilm/. We are very grateful to both of them for this and for many clarifying discussions. About 60 % of the figures in this book were made by Lazzari who thereby contributed greatly to the usefulness and the clarity of the sections on island films with applications. The software is available for use. For this purpose one should consult the above mentioned web site and in case of difficulties contact either Simonsen or Lazzari. For the other figures we are grateful to J. van der Ploeg from our group and to L. Nummedal from the Norwegian University of Science and Technology, Trondheim, Norway. Nummedal was also, on many occasions, a great help solving the various software problems.

Leiden, February, 2001

Dick Bedeaux
Jan Vlieger

Address for correspondence:
Dick Bedeaux, Leiden Institute of Chemistry, Leiden University
P.O. Box 9502, 2300 RA Leiden, The Netherlands
Email: bedeaux@chem.leidenuniv.nl

CONTENTS

Chapter 1
INTRODUCTION

It is the aim of this book to describe the optical properties of surfaces with a thickness small compared to the wavelength of the incident light. The emphasis will be on two kinds of surfaces. The first kind consists of a film of discrete islands, small compared to the wavelength, attached to a flat substrate. An important example of such films are metallic films. The second kind is the rough surface. In that case the surface resembles a landscape with hills and valleys. The height is again small compared to the wavelength of the light. In the second edition a chapter has been added on the reflection from a gyrotropic medium.

Historically the first description of metallic films was given by Maxwell Garnett in 1904[1]. He developed a theory for metallic glasses, assuming the metal to be distributed in the form of small spherical islands. The polarizability of such an island may be shown to be equal to

$$\alpha = 4\pi\epsilon_a R^3 \frac{\epsilon - \epsilon_a}{\epsilon + 2\epsilon_a} \tag{1.1}$$

where ϵ_a is the dielectric constant of the glass, ϵ the complex frequency dependent dielectric "constant" of the metal, and R the radius of the sphere. Using the Lorentz–Lorenz formula one then finds for the effective complex frequency dependent dielectric constant of the metallic glass the following formula

$$\frac{\epsilon_{eff} - \epsilon_a}{\epsilon_{eff} + 2\epsilon_a} = \phi \frac{\epsilon - \epsilon_a}{\epsilon + 2\epsilon_a} \tag{1.2}$$

where ϕ is the volume fraction of the spheres. This formula is successful, in that it explains for instance the striking colors of the metallic glass and their dependence on the volume fraction. In the same paper Maxwell Garnett also applied his theory to metallic films. In that case the islands are on the surface of the glass and surrounded by the ambient with a dielectric constant ϵ_a. The volume fraction is a parameter which is not systematically defined. In practice one fits it to the experimental data, like for instance to the minimum in the transmission, and interprets it as the weight thickness divided by the so-called optical thickness. The experiment in this way measures the optical thickness of the film.

Even though the Maxwell Garnett theory is very useful to describe the qualitative behavior of thin metallic films, the quantitative agreement is not very satisfactory. One has tried to improve this along various lines. One observation is that Lorentz–Lorenz is not adequate. An alternative popular choice for the effective dielectric constant is given, for instance, by the symmetric Bruggeman formula[2], [3]

$$\phi\frac{\epsilon_{eff}-\epsilon}{\epsilon_{eff}+2\epsilon}+(1-\phi)\frac{\epsilon_{eff}-\epsilon_a}{\epsilon_{eff}+2\epsilon_a}=0 \tag{1.3}$$

One may show that this expression reduces to Maxwell Garnett for small volume fractions. For a history and a description of alternative effective medium theories one is referred to Landauer [4]. It is not our aim to discuss the various methods to make effective medium theories work. For this we refer to various reviews, [5], [6], [7], [8] and [9].

There are two major reasons why effective medium theories are only qualitatively correct for surfaces. The first reason is that the direct electromagnetic interaction between the islands along the surface is taken into account using some local field argument. The choice of this local field is appropriate for a three dimensional distribution of islands, as for instance in a metallic glass, but not for a two dimensional array. The second reason is that all these theories neglect the electromagnetic interaction with the substrate. The electric field due to the images of the spheres is not taken into account. These images cause the polarizability of the spheres to be different in the directions along and normal to the surface. The surface breaks the symmetry. A dipolar model for this effect was first given by Yamaguchi, Yoshida and Kinbara [10].

In this book a theory for thin island films and rough surfaces is given, which describes both the direct electromagnetic interaction along the surface and the interaction with the substrate. The electromagnetic properties of the surface are described in terms of four susceptibilities, γ, β, τ and δ. The first coefficient γ gives the integrated surface polarization parallel to the surface in terms of the electric field along the surface. The second coefficient β gives the integrated surface polarization normal to the surface in terms of the electric displacement field normal to the surface. The third and the fourth coefficients τ and δ are of quadrupolar order. They are not very important for the description of metallic films, where γ and β dominate the behavior. For rough surfaces, but also for films of latex spheres on a glass substrate, these quadrupolar terms are found to be needed, however. The book discusses the general case for which also the integrated surface magnetization is taken along. For the details of this aspect, which requires the introduction of magnetic analogs of the above susceptibilities, we refer to the main text. The work described was done over many years in our group in Leiden and will be referred to when used in the text.

For thin island films the analysis in this book is based on the calculation of the polarizabilities of the islands. The surface is assumed to be isotropic for translation along the surface and rotation around a normal. All islands are therefore (statistically) equivalent. Effects due to electromagnetic interaction between islands are calculated assuming the islands to be identical. Both regular arrays and random arrays of islands are considered. The analysis to dipolar order gives the (average) polarizabilities parallel, $\alpha_{\|}$, and normal, $\alpha_{\perp}$, to the surface per island. The resulting dipolar susceptibilities are

$$\gamma=\rho\alpha_{\|} \text{ and } \beta=\rho\alpha_{\perp}/\epsilon_a^2 \tag{1.4}$$

where ρ is the number of islands per unit of surface area. The (average) quadrupole polarizabilities parallel, $\alpha_{\parallel}^{10}$, and normal, $\alpha_{\perp}^{10}$, to the surface per island give the dipole moment of the island in the direction parallel to the surface in terms of the parallel derivative of the electric field along the surface, and the dipole moment of the island in the direction normal to the surface in terms of the normal derivative of the electric field normal to the surface, respectively. The resulting susceptibilities are

$$\tau = -\rho\alpha_{\parallel}^{10} \text{ and } \delta = -\rho[\alpha_{\perp}^{10} + \alpha_{\parallel}^{10}]/\epsilon_a \tag{1.5}$$

See chapter 5 for a discussion of the definition of the polarizabilities and their relation to the susceptibilities in the small ρ case. Chapters 5–10 give the explicit calculation of dipole and quadrupole polarizabilities for spheres, spheroids, truncated spheres and truncated spheroids on a substrate, both for small and finite ρ.

One may substitute the polarizabilities of the sphere surrounded by the ambient into eq.(1.4). This gives

$$\gamma = \epsilon_a^2\beta = \rho\alpha = 4\pi\epsilon_a R^3\rho\frac{\epsilon - \epsilon_a}{\epsilon + 2\epsilon_a} = 3t_w\epsilon_a\frac{\epsilon - \epsilon_a}{\epsilon + 2\epsilon_a} \tag{1.6}$$

where the weight thickness was identified with $t_w = 4\pi R^3\rho/3$. In the calculation of the integrated excess quadrupole moments one must specify the location of the surface, on which one locates this dipole moment. The natural choice for this dividing surface is the surface of the substrate. Shifting the dipoles from the center of the spheres to the surface of the substrate leads to quadrupole polarizabilities given by

$$\alpha_{\parallel}^{10} = \alpha_{\perp}^{10} = -R\alpha \tag{1.7}$$

Substituting this equation, together with eq.(1.1) for the polarizability, into eq.(1.5) then gives, with eq.(1.6),

$$\tau = \frac{\epsilon_a}{2}\delta = 3t_w R\epsilon_a\frac{\epsilon - \epsilon_a}{\epsilon + 2\epsilon_a} = R\gamma = R\epsilon_a^2\beta \tag{1.8}$$

It should be noted that the quadrupole moment, due to a constant field, can be identified with the dipole moment, due to a gradient field, on the basis of symmetry considerations, cf. chapter 5. All interactions between the spheres have been neglected. The above expressions are therefore only correct in the low coverage, i.e. low weight thickness regime. Also they assume that the interaction with the image charges in the substrate is unimportant. This is only correct if the difference between the dielectric constants of the ambient and the substrate is negligible.

In order to compare with the Maxwell Garnett or the Bruggeman theory, one must calculate the susceptibilities for a thin layer with a dielectric constant ϵ_{eff} and a thickness t_{opt}. In chapter 11 the expressions for the susceptibilities of a stratified medium are derived. Applying these expressions to a thin layer one finds

$$\begin{aligned} \gamma &= t_{opt}(\epsilon_{eff} - \epsilon_a), \qquad \beta = t_{opt}(\epsilon_{eff} - \epsilon_a)/\epsilon_{eff}\epsilon_a \\ \tau &= \frac{1}{2}t_{opt}^2(\epsilon_{eff} - \epsilon_a), \quad \delta = \frac{1}{2}t_{opt}^2(\epsilon_{eff}^2 - \epsilon_a^2)/\epsilon_{eff}\epsilon_a \end{aligned} \tag{1.9}$$

Substituting the Maxwell Garnett equation, (1.2) with $\phi = t_w/t_{opt}$, one finds for γ and β

$$
\begin{aligned}
\gamma &= t_w(\epsilon - \epsilon_a)\left[1 + \frac{1}{3\epsilon_a}\left(1 - \frac{t_w}{t_{opt}}\right)(\epsilon - \epsilon_a)\right]^{-1} \\
\beta &= t_w\epsilon_a^{-2}(\epsilon - \epsilon_a)\left[1 + \frac{1}{3\epsilon_a}\left(1 + 2\frac{t_w}{t_{opt}}\right)(\epsilon - \epsilon_a)\right]^{-1}
\end{aligned}
\qquad (1.10)
$$

For a small weight thickness these expressions reduce to eq.(1.6) as expected. For larger weight thicknesses the usual depolarization factor of $1/3$ for a sphere is replaced by $\frac{1}{3}(1 - t_w/t_{opt})$ along the surface and by $\frac{1}{3}(1 + 2t_w/t_{opt})$ normal to the surface. The sum of the depolarization factors over three directions remains one. This modification is due to the electromagnetic interaction between the spheres along the surface. In the Maxwell Garnett (Lorentz-Lorenz) theory this is accounted for by distinguishing between the incident field and the local field in which the dipole is placed. As he uses a local field appropriate for a three-dimensional distribution of spheres the result is inadequate. The formulae for τ and δ analogous to those given in eq.(1.10) for γ and β may similarly be given.

The way to improve these effective medium theories is to calculate the polarizability of the islands with the full electromagnetic interaction with other islands and their images. This is the subject of chapters 5 through 10 in this book.

If one describes a surface as a 2-dimensional transition layer between the ambient and the substrate, as is done in this book, one needs to position this surface at some convenient location. For the island films there are 2 natural choices. The first is the surface of the substrate. The second is a plane through the (average) center of the islands. The first choice, used above, makes it necessary to shift the dipole in the center of the islands to the surface of the substrate and as a consequence one needs to introduce multipoles, to the order of approximation in size over wavelength one wants to describe. The second choice makes it necessary to extend the substrate material to the center of the islands. Both choices lead to different surface susceptibilities. In view of the fact that the experimental results are independent of the choice of this dividing surface, they can only depend on combinations of these susceptibilities, which do not depend on this choice. Such combinations are called *invariants*. This book contains an extensive discussion of these invariants. All measurable quantities, discussed in particular in chapter 4 in this book, are furthermore given in terms of these invariants. For the special case of stratified surface layers, such invariants were introduced by Lekner [11]. This case, and an extension thereof to magnetic stratification, is discussed in chapter 11.

The second chapter is meant as a introduction to the use of excess electromagnetic fields, electric current densities and charge densities. This is done using a number of simple examples. The aim of this chapter is to prepare the reader for the third chapter, where the validity of the Maxwell equations is extended to electromagnetic fields, electric current densities and charge densities, which are generalized functions. For the last three chapters on the wave equation and its general solution, general linear response theory and surface roughness, this extension is crucial.

Much of the results in these chapters could not easily be obtained without the use of generalized functions. The general solution of the wave equation, including the surface, is used in the chapter about linear response theory to verify that the constitutive relations, given in chapter 3, are the only correct choice in view of the source observer symmetry [12]. The general solution is used in the chapter 14 on surface roughness to obtain contributions, due to correlations along the surface, to the surface susceptibilities.

At the end of the second edition a chapter on the reflection of a gyrotropic medium has been added. There are two generally accepted methods to describe the gyrotropic nature of a homogeneous phase. Surprisingly enough these two methods lead to different reflection amplitudes, when one uses the so-called standard boundary conditions (Fresnel). The origin of this difference is discussed from the point of view developed in this book. Additional modifications of the reflection amplitudes, due to the possible gyrotropic nature of the interfacial layer, are discussed.

The book uses the somewhat old fashioned c.g.s. system of units rather than the now generally accepted SI units. The reason for this is that in the c.g.s. unit system, for instance, the electric fields, the displacement fields and the polarization densities all have the same dimension, while they do not have this in SI units. In the algebra using the fields one must work with similarly dimensioned fields. While this is natural in c.g.s. units it is not in SI units. Of course one may use SI units and correct this by adding the appropriate power of the dielectric and magnetic permittivities of vacuum everywhere. We choose not to do this. In the experimentally relevant formulae we have always used forms such that the unit system does not matter. Nevertheless one should be careful regarding this point when applying the needed formulae.

Chapter 2
EXCESS CURRENTS, CHARGE DENSITIES AND FIELDS

2.1 Introduction

It is the aim of this chapter to explain how the concept of an excess of the electric current, the charge density, the electric and magnetic fields as well as excesses of the electric displacement field and the magnetic induction may be used to describe the electro-magnetic properties of a boundary layer, or as Gibbs [13] calls it a surface of discontinuity, between two homogeneous phases. Generally speaking there are two aspects of such a boundary layer which must be considered in this context. One is the so-called surface roughness of the interface and the other is the presence of a different material in the boundary layer. The general reason to introduce excess quantities is that for many considerations the detailed structure of the surface is not really needed. If one considers for instance a surface which conducts an electric current, due to a thin layer of metal, it is for many purposes sufficient to know the total current along the surface rather than its precise distribution on length scales comparable to the thickness of the interfacial layer. An extremely important difference between an interface and the bulk of the adjacent phases is the inherent asymmetry of the interface in its response to fields normal to the surface or parallel to the interface. This is eminently clear for a thin metal film between two dielectric media. If one applies an electric field along the film this results in a current while an electric field normal to the film does not lead to a current. In the following section this example of a plane parallel conducting film will be discussed in some detail in order to clarify the introduction of the various excess quantities. In the third section the same discussion is given for a plane parallel dielectric film with the purpose to show how an excess dielectric displacement field can be introduced for that example.

The examples in the second and the third section are of course extremely well-known simple cases which do not really need the introduction of excess quantities. They are in fact the simplest examples of a stratified layer, and one may solve the Maxwell equations for this case explicitly for an arbitrary incident field [14], [11]. The real purpose of this book is to consider an interfacial region which may not only be rough but may in addition also contain islands of different materials. The only real restriction on the complexity of the interfacial structure is that the thickness of the layer is sufficiently small compared to the wavelength of the incident light. In the fourth section of this chapter the general definition of excess quantities will be discussed in detail. It will then also become clear how the boundary conditions for the extrapolated bulk fields are affected by these excess quantities. In particular it is found, which excess quantities are important in these boundary conditions. Excess quantities, which do not affect the boundary conditions, have as a consequence no relevance for the reflection and transmission of light by the interface. To further clarify this matter the role of higher order moments of the excess currents, charge densities and fields is shortly discussed in section 2.5. In the last section a definition is given of the interfacial polarization and magnetization densities.

2.2 Excess electric current, charge density and electric field in a conducting layer

Consider two homogeneous non-conducting media separated by a stratified boundary layer of conducting material. For a general discussion of the procedure explained in this section, the reader is referred to Albano et al [15], [16]. The analysis in this section is meant to clarify the concepts and not to be complete. The conductivity σ in such a stratified layer varies only in the direction orthogonal to the layer. The z-axis is chosen orthogonal to this layer so that as a consequence σ depends only on z. It is now possible to apply a constant electric field directed parallel to the conducting layer

$$\mathbf{E}(\mathbf{r}) = (E_x, E_y, 0) \tag{2.1}$$

where $\mathbf{r} = (x, y, z)$ indicates the position. The resulting electric current density in the layer is then given by

$$\mathbf{I}(\mathbf{r}) = \mathbf{I}(z) = (I_x, I_y, 0) = (\sigma(z)E_x, \sigma(z)E_y, 0) \tag{2.2}$$

A quantity of practical interest is the total current per unit of length along the surface which is defined by

$$\mathbf{I}^s \equiv \int_{-\infty}^{\infty} \mathbf{I}(z)dz \tag{2.3}$$

Upon substitution of eq.(2.2) into this definition one obtains the following expression for this total current in terms of the applied electric field

$$\mathbf{I}^s = (I_x^s, I_y^s, 0) = (\sigma^s E_x, \sigma^s E_y, 0) = \sigma^s(E_x, E_y, 0) \tag{2.4}$$

Here σ^s is defined by

$$\sigma^s \equiv \int_{-\infty}^{\infty} \sigma(z)dz \tag{2.5}$$

and figures as the total conductivity of the boundary layer. The formula for this total conductivity shows in other words that the 'conductances of the sub-layers' are put in parallel so that they add up. If one measures the total current along the surface one finds direct experimental information about this total conductivity. In the description of this phenomenon one needs only two properties. One of these properties is the boundary condition for the electric field along the layer, which states that this field is continuous across the layer. The other property one needs is the total conductivity σ^s. Details of $\sigma(z)$ do not manifest themselves in this context.

One may also consider a system in which the two homogeneous media outside the conducting boundary layer are also conducting, with conductivities σ^+ and σ^-. The conductivity $\sigma(z)$ then approaches these values outside the layer rather than going to zero. Usually (but not necessarily) σ^+ and σ^- are much smaller than the value of $\sigma(z)$ in the layer so that the main contribution to the current is still in this layer. In this case one would like to know how much extra current flows along the layer compared to what one would expect on the basis of the conductivities away from it. For this purpose it is convenient to introduce a so-called excess electric current density by

$$\mathbf{I}_{ex}(z) \equiv \mathbf{I}(z) - \mathbf{I}^-\theta(-z) - \mathbf{I}^+\theta(z) = [\sigma(z) - \sigma^-\theta(-z) - \sigma^+\theta(z)](E_x, E_y, 0) \quad (2.6)$$

where $\mathbf{I}^-$ and $\mathbf{I}^+$ are the constant current densities in the homogeneous media. Furthermore $\theta(s)$ is the so-called Heaviside function ($\theta(s) \equiv 0$ for $s < 0$ and $\theta(s) \equiv 1$ for $s > 0$). The excess current density now becomes equal to zero outside the layer and the total excess current density may be defined as

$$\mathbf{I}^s \equiv \int_{-\infty}^{\infty} \mathbf{I}_{ex}(z)dz \quad (2.7)$$

One may for this case define the total excess conductivity of the layer as

$$\sigma^s \equiv \int_{-\infty}^{\infty} [\sigma(z) - \sigma^-\theta(-z) - \sigma^+\theta(z)]dz \quad (2.8)$$

and one again has

$$\mathbf{I}^s = (I_x^s, I_y^s, 0) = (\sigma^s E_x, \sigma^s E_y, 0) = \sigma^s(E_x, E_y, 0) \quad (2.9)$$

There is one matter which has led to considerable discussion in the introduction of such excess quantities for the description of the behavior of boundary layers [13], this in particular in the context of thermodynamics: How should one choose the precise location of the $z = 0$ dividing surface in the boundary layer? It is clear that there is some freedom in this choice and the above definitions show that the total excess current as well as the total excess conductivity depend upon it. Of course the physics of the system must depend on combinations of these coefficients which are invariant for the choice of the dividing surface. The construction of such invariant combinations will be discussed in great detail in chapter 3.

One may also analyze the system described above under the influence of a time-independent electric field orthogonal to the layer. In this case one has an electric current in the z-direction, which will be constant as a consequence of charge conservation:

$$\mathbf{I}(\mathbf{r}) = \mathbf{I}(z) = (0, 0, I_z) \quad (2.10)$$

It follows that, if the electric field is normal to the interface, there is no excess current. It follows using Ohm's law, $I_z = \sigma(z)E_z(z)$, that the electric field is given by

$$\mathbf{E}(\mathbf{r}) = (0, 0, E_z(z)) = (0, 0, \frac{1}{\sigma(z)}I_z) \tag{2.11}$$

Taking the dielectric constant equal to unity in this example, the divergence of this field gives the charge density and one then has

$$\rho(\mathbf{r}) = \rho(z) = \operatorname{div}\mathbf{E}(\mathbf{r}) = \frac{d}{dz}E_z(z) = [\frac{d}{dz}\frac{1}{\sigma(z)}]I_z \tag{2.12}$$

Note that the charge density drops to zero outside the interfacial region. The total (excess) charge density can therefore be defined as

$$\rho^s \equiv \int_{-\infty}^{\infty}[\frac{d}{dz}\frac{1}{\sigma(z)}]I_z dz = (\frac{1}{\sigma^+} - \frac{1}{\sigma^-})I_z \tag{2.13}$$

It is worth noting that this quantity is independent of $\sigma(z)$. Thus the details of the stratification do not affect the excess charge.

One may also define an excess electric field for this case by

$$\mathbf{E}_{ex}(z) \equiv \mathbf{E}(z) - \mathbf{E}^-\theta(-z) - \mathbf{E}^+\theta(z) = (0, 0, [\frac{1}{\sigma(z)} - \frac{1}{\sigma^-}\theta(-z) - \frac{1}{\sigma^+}\theta(z)]I_z) \tag{2.14}$$

where $\mathbf{E}^-$ and $\mathbf{E}^+$ are the electric fields in the homogeneous media. The total excess of the electric field may now be defined as

$$\mathbf{E}^s \equiv \int_{-\infty}^{\infty}\mathbf{E}_{ex}(z)dz = (0, 0, E_z^s) \tag{2.15}$$

Substituting eq.(2.14) it follows that

$$\mathbf{E}^s = (0, 0, E_z^s) = (0, 0, R^s I_z) = R^s\mathbf{I} \tag{2.16}$$

where the total excess resistivity is defined by

$$R^s \equiv \int_{-\infty}^{\infty} dz[\frac{1}{\sigma(z)} - \frac{1}{\sigma^-}\theta(-z) - \frac{1}{\sigma^+}\theta(z)] \tag{2.17}$$

The formula for this total resistivity shows that the 'resistances of the sub-layers' are put in series, so that they add up. This is thus very different from the case described above for an electric field parallel to the layer where the conductivities of the sub-layers were put in parallel. If the electric field is parallel to the layer one finds an important contribution from the layer if its conductivity is large compared to the surrounding medium. For a field which is orthogonal to the layer, however, one finds an important contribution if the conductivity of the layer is much smaller than the conductivity of the surrounding medium. It is this very characteristic difference in the response of the layer to fields orthogonal and parallel to the layer which is the origin of many of the interesting electromagnetic properties of the layer.

If one measures the potential difference from one side of the layer to the other one has

$$\Phi(d) - \Phi(-d) = \int_{-d}^{d} E_z(z)dz = E_z^s + d(E_z^+ + E_z^-) = [R^s + d(\frac{1}{\sigma^+} + \frac{1}{\sigma^-})]I_z \quad (2.18)$$

This expression shows that both E_z^s and R^s may be determined by measuring the potential difference as a function of d and extrapolation to $d = 0$. If the conductivity of the layer is sufficiently small compared to the conductivities outside the layer one may even neglect the dependence on d. It then follows that the potential difference over the layer directly gives the total excess of the normal component of the electric field and the resistivity of the layer

$$\Phi(d) - \Phi(-d) \cong E_z^s = R^s I_z \quad (2.19)$$

In the description of current flow across the layer one therefore needs only two properties. One of these properties is the boundary condition for the electric current in the direction orthogonal to the layer, which states that this current is continuous across the layer. The other property one needs is the total resistivity R^s. Details of $\sigma(z)$ do not manifest themselves in this context.

The above analysis shows how total excess currents, charge densities and electric fields are defined and explains the possible usefulness of these quantities. It was found that the electric current density may have an excess which is directed along the interfacial layer whereas the electric field may have an excess in the direction orthogonal to the layer. One may wonder if the current density may in principle also have an excess in the direction normal to the layer or similarly whether the electric field may have an excess in the direction along the layer. An excess of the electric current in the normal direction would be important in a description of the redistribution of charge in the layer. This is a process which will be described, in the context of this book, in terms of a time dependent contribution to the normal component of an excess polarization density. An excess of the electric field along the layer would imply that one has a total excess electric potential which upon differentiation along the layer would give the excess electric field along this layer. This is clearly unphysical. In the description of the electro-magnetic properties of surfaces using excess current densities and fields given in this book certain components of these excess quantities will be chosen to be zero. As will be clear from the above discussion the reason for such a choice is not always the same. Sometimes such an excess contribution is simply unphysical and sometimes it is more convenient to describe the corresponding physical effect in a different manner. A final more pragmatic reason to take an excess quantity zero is that this excess is only relevant in the description of higher order moments in the normal direction of certain variables in the layer, cf. section 2.4 below. The description of such higher order moments of the electromagnetic fields is not the aim of this book.

2.3 Excess electric and displacement fields in a dielectric layer

Consider two homogeneous non-conducting dielectric media separated by a stratified boundary layer of non-conducting dielectric material [16]. The position dependent dielectric constant $\epsilon(z)$ varies only in the direction orthogonal to the layer and approaches the values ϵ^+ and ϵ^- outside the layer. A constant electric field is applied parallel to the dielectric layer

$$\mathbf{E}(\mathbf{r}) = (E_x, E_y, 0) \tag{2.20}$$

The resulting electric displacement field is then given by

$$\mathbf{D}(\mathbf{r}) = \mathbf{D}(z) = (D_x, D_y, 0) = (\epsilon(z)E_x, \epsilon(z)E_y, 0) \tag{2.21}$$

Analogous to the procedure in the previous section it is convenient to define an excess displacement field by

$$\mathbf{D}_{ex}(z) \equiv \mathbf{D}(z) - \mathbf{D}^-\theta(-z) - \mathbf{D}^+\theta(z) = [\epsilon(z) - \epsilon^-\theta(-z) - \epsilon^+\theta(z)](E_x, E_y, 0) \tag{2.22}$$

where $\mathbf{D}^-$ and $\mathbf{D}^+$ are the constant electric displacement fields away from the layer in the homogeneous media. The excess displacement field now becomes equal to zero outside the layer and the total excess displacement field may be defined as

$$\mathbf{D}^s \equiv \int_{-\infty}^{\infty} \mathbf{D}_{ex}(z)dz \tag{2.23}$$

One may for this case define the total excess dielectric constant of the layer as

$$\gamma_e \equiv \int_{-\infty}^{\infty} [\epsilon(z) - \epsilon^-\theta(-z) - \epsilon^+\theta(z)]dz \tag{2.24}$$

and one obtains [17]

$$\mathbf{D}^s = (D_x^s, D_y^s, 0) = (\gamma_e E_x, \gamma_e E_y, 0) = \gamma_e(E_x, E_y, 0) \tag{2.25}$$

One may also analyze the system described above under the influence of a time-independent electric field orthogonal to the layer. In this case one has a displacement field in the z-direction, which is constant as a consequence of the fact that the charge density is zero

$$\mathbf{D}(\mathbf{r}) = \mathbf{D}(z) = (0, 0, D_z) \tag{2.26}$$

It follows that, if the electric field is normal to the interface, there is no excess displacement field. It furthermore follows using $D_z = \epsilon(z)E_z(z)$, that the electric field is given by

$$\mathbf{E}(\mathbf{r}) = (0, 0, E_z(z)) = (0, 0, \frac{1}{\epsilon(z)}D_z) \tag{2.27}$$

One may now define an excess electric field for this case by

$$\mathbf{E}_{ex}(z) \equiv \mathbf{E}(z) - \mathbf{E}^-\theta(-z) - \mathbf{E}^+\theta(z) = (0, 0, [\frac{1}{\epsilon(z)} - \frac{1}{\epsilon^-}\theta(-z) - \frac{1}{\epsilon^+}\theta(z)]D_z) \tag{2.28}$$

The total excess of the electric field, cf. eq. (2.15), is therefore given by [17]

$$\mathbf{E}^s = (0, 0, E_z^s) = (0, 0, -\beta_e D_z) = -\beta_e \mathbf{D} \tag{2.29}$$

where β_e is defined by

$$\beta_e \equiv -\int_{-\infty}^{\infty} dz \left[\frac{1}{\epsilon(z)} - \frac{1}{\epsilon^-}\theta(-z) - \frac{1}{\epsilon^+}\theta(z)\right] \tag{2.30}$$

This coefficient is minus the excess of the inverse dielectric constant.

The analogy with the conducting system in the previous section is clear. If the electric field is parallel to the layer the dielectric constants of the sub-layers are so to say put in parallel. If the electric field is normal to the layer, however, they are put in series. Furthermore one finds that, if the dielectric constant of the layer is much larger than the dielectric constants of the surrounding homogeneous media, the constitutive coefficient γ_e is much smaller than the coefficient β_e and dominates the behavior of the layer. If the dielectric constant of the layer is much smaller than the dielectric constants of the surrounding homogeneous media, the constitutive coefficient β_e is much larger than the coefficient γ_e and dominates the behavior of the layer. It is again this very characteristic difference in the response of the layer in the directions orthogonal and parallel to the layer, which is the origin of many of the interesting electromagnetic properties of the layer.

2.4 Excess currents, charge density and fields and the boundary conditions

The starting point of this analysis are the Maxwell equations which are given by

$$\begin{aligned} \operatorname{rot}\mathbf{E}(\mathbf{r},t) &= -\frac{1}{c}\frac{\partial}{\partial t}\mathbf{B}(\mathbf{r},t)\ , \qquad \operatorname{div}\mathbf{D}(\mathbf{r},t) = \rho(\mathbf{r},t) \\ \operatorname{rot}\mathbf{H}(\mathbf{r},t) &= \frac{1}{c}\frac{\partial}{\partial t}\mathbf{D}(\mathbf{r},t) + \frac{1}{c}\mathbf{I}(\mathbf{r},t)\ , \quad \operatorname{div}\mathbf{B}(\mathbf{r},t) = 0 \end{aligned} \tag{2.31}$$

where $\mathbf{B}$ is the magnetic induction, $\mathbf{H}$ the magnetic field and c the velocity of light. The system is now separated in two regions by the choice of a dividing surface. This dividing surface must be chosen such that it is located in the transition layer between the two bulk phases. In principle this dividing surface may move and have a time dependent curvature. Though this general situation is very interesting, cf. [18], most experiments are done for flat non-moving surfaces. This book is therefore restricted to this case. It should be emphasized that much of the fundamental aspects of the introduction of excess quantities have also been formulated for dividing surfaces, which are both moving and curved [15], [16]. The restriction to flat non-moving surfaces is merely a pragmatic restriction to the case of most practical interest. The $x - y$ plane is chosen to coincide with the dividing surface.

Away from the interface the behavior of the solution of the Maxwell equations is governed by the bulk constitutive coefficients. Using these bulk constitutive coefficients one may extrapolate the solution in the bulk phases back to the surface. The difference between the real fields and the extrapolated fields is what will be called the position dependent excess field. The extrapolated fields satisfy the Maxwell equations given above up to and including the dividing surface but with constitutive relations containing the position independent bulk constitutive coefficients. The position dependent excess electric field is defined by

$$\mathbf{E}_{ex}(\mathbf{r},t) \equiv \mathbf{E}(\mathbf{r},t) - \mathbf{E}^{+}(\mathbf{r},t)\theta(z) - \mathbf{E}^{-}(\mathbf{r},t)\theta(-z) \tag{2.32}$$

where the super indices + and − indicate the extrapolated electric fields in the $z > 0$ and the $z < 0$ region, respectively. Furthermore $\theta(z)$ is the Heaviside function; $\theta(z) \equiv 1$ if $z > 0$ and $\theta(z) \equiv 0$ if $z < 0$. In a similar way one may define the position dependent excess displacement, magnetic and the magnetic induction fields as well as the position dependent excess current and charge density. Using the Maxwell equations for the real fields and for the extrapolated fields one finds the following equations for the excess fields

$$\text{rot}\, \mathbf{E}_{ex}(\mathbf{r},t) + \hat{\mathbf{z}} \times (\mathbf{E}_{\|}^{+}(\mathbf{r}_{\|},0,t) - \mathbf{E}_{\|}^{-}(\mathbf{r}_{\|},0,t))\delta(z) = -\frac{1}{c}\frac{\partial}{\partial t}\mathbf{B}_{ex}(\mathbf{r},t)$$

$$\text{div}\, \mathbf{D}_{ex}(\mathbf{r},t) + (D_{z}^{+}(\mathbf{r}_{\|},0,t) - D_{z}^{-}(\mathbf{r}_{\|},0,t))\delta(z) = \rho_{ex}(\mathbf{r},t)$$

$$\text{rot}\, \mathbf{H}_{ex}(\mathbf{r},t) + \hat{\mathbf{z}} \times (\mathbf{H}_{\|}^{+}(\mathbf{r}_{\|},0,t) - \mathbf{H}_{\|}^{-}(\mathbf{r}_{\|},0,t))\delta(z) = \frac{1}{c}\frac{\partial}{\partial t}\mathbf{D}_{ex}(\mathbf{r},t) + \frac{1}{c}\mathbf{I}_{ex}(\mathbf{r},t)$$

$$\text{div}\, \mathbf{B}_{ex}(\mathbf{r},t) + (B_{z}^{+}(\mathbf{r}_{\|},0,t) - B_{z}^{-}(\mathbf{r}_{\|},0,t))\delta(z) = 0 \tag{2.33}$$

where $\hat{\mathbf{z}} \equiv (0,0,1)$ is the normal to the dividing surface, the subindex $\|$ indicates the projection of the corresponding vector on the $x - y$ plane, e.g. $\mathbf{r}_{\|} \equiv (x,y)$, and the subindex z indicates the normal component of the corresponding vector. The contributions in the above equation proportional to the δ-function find there origin in the derivative of the Heaviside function with respect to z, $\partial\theta(z)/\partial z = \delta(z)$ and $\partial\theta(-z)/\partial z = -\delta(z)$, which appear if one calculates the rotation or the divergence of the extrapolated fields as defined in eq.(2.32). One may easily convince oneself that these terms are related to the boundary condition by considering what shall be called a Fresnel interface for which the dividing surface is a sharp surface at which the constitutive coefficients change discontinuously from one bulk value to the other bulk value. In that case the real fields and the extrapolated fields are identical all the way up to the dividing surface. As a consequence the excess field is zero. This implies that the δ-function contributions must also be zero so that

$$\begin{aligned} \mathbf{E}_{\|}^{+}(\mathbf{r}_{\|},0,t) - \mathbf{E}_{\|}^{-}(\mathbf{r}_{\|},0,t) &= 0\ ,\ D_{z}^{+}(\mathbf{r}_{\|},0,t) - D_{z}^{-}(\mathbf{r}_{\|},0,t) = 0 \\ \mathbf{H}_{\|}^{+}(\mathbf{r}_{\|},0,t) - \mathbf{H}_{\|}^{-}(\mathbf{r}_{\|},0,t) &= 0\ ,\ B_{z}^{+}(\mathbf{r}_{\|},0,t) - B_{z}^{-}(\mathbf{r}_{\|},0,t) = 0 \end{aligned} \tag{2.34}$$

These are the usual boundary conditions for a Fresnel surface.

In order to see how the excess fields modify these boundary conditions it is convenient to Fourier transform the excess fields with respect to the $z-$coordinate. For the excess electric field one has for instance

$$\mathbf{E}_{ex}(\mathbf{r}_{\|},k_z,t) \equiv \int_{-\infty}^{\infty} dz\ \exp(-ik_z z)\mathbf{E}_{ex}(\mathbf{r},t) \tag{2.35}$$

and similarly for the other excess fields, current and charge density. One may then Fourier transform eq.(2.33) which gives

$$\begin{aligned}&\nabla_{\|}\times\mathbf{E}_{ex}(\mathbf{r}_{\|},k_z,t)+ik_z\hat{\mathbf{z}}\times\mathbf{E}_{ex,\|}(\mathbf{r}_{\|},k_z,t)\\&+\hat{\mathbf{z}}\times[\mathbf{E}_{\|}^{+}(\mathbf{r}_{\|},z=0,t)-\mathbf{E}_{\|}^{-}(\mathbf{r}_{\|},z=0,t)]=-\frac{1}{c}\frac{\partial}{\partial t}\mathbf{B}_{ex}(\mathbf{r}_{\|},k_z,t)\end{aligned}$$

$$\begin{aligned}\nabla_{\|}.\mathbf{D}_{ex,\|}(\mathbf{r}_{\|},k_z,t)+ik_zD_{ex,z}(\mathbf{r}_{\|},k_z,t)+[D_z^{+}(\mathbf{r}_{\|},z&=0,t)-D_z^{-}(\mathbf{r}_{\|},z=0,t)]\\&=\rho_{ex}(\mathbf{r}_{\|},k_z,t)\end{aligned}$$

$$\begin{aligned}&\nabla_{\|}\times\mathbf{H}_{ex}(\mathbf{r}_{\|},k_z,t)+ik_z\hat{\mathbf{z}}\times\mathbf{H}_{ex,\|}(\mathbf{r}_{\|},k_z,t)\\&+\hat{\mathbf{z}}\times[\mathbf{H}_{\|}^{+}(\mathbf{r}_{\|},z=0,t)-\mathbf{H}_{\|}^{-}(\mathbf{r}_{\|},z=0,t)]=\frac{1}{c}\frac{\partial}{\partial t}\mathbf{D}_{ex}(\mathbf{r}_{\|},k_z,t)+\frac{1}{c}\mathbf{I}_{ex}(\mathbf{r}_{\|},k_z,t)\end{aligned}$$

$$\nabla_{\|}.\mathbf{B}_{ex,\|}(\mathbf{r}_{\|},k_z,t)+ik_zB_{ex,z}(\mathbf{r}_{\|},k_z,t)+[B_z^{+}(\mathbf{r}_{\|},z=0,t)-B_z^{-}(\mathbf{r}_{\|},z=0,t)]=0 \tag{2.36}$$

where $\nabla_{\|}\equiv(\partial/\partial x,\partial/\partial y)$. These equations are valid for all values of k_z. In order to obtain formulae for the jumps in the extrapolated bulk fields in their simplest form we now set k_z equal to zero

$$\begin{aligned}\hat{\mathbf{z}}\times[\mathbf{E}_{\|}^{+}(\mathbf{r}_{\|},z&=0,t)-\mathbf{E}_{\|}^{-}(\mathbf{r}_{\|},z=0,t)]\\&=-\nabla_{\|}\times\mathbf{E}_{ex}(\mathbf{r}_{\|},k_z=0,t)-\frac{1}{c}\frac{\partial}{\partial t}\mathbf{B}_{ex}(\mathbf{r}_{\|},k_z=0,t)\end{aligned}$$

$$D_z^{+}(\mathbf{r}_{\|},z=0,t)-D_z^{-}(\mathbf{r}_{\|},z=0,t)=-\nabla_{\|}.\mathbf{D}_{ex,\|}(\mathbf{r}_{\|},k_z=0,t)+\rho_{ex}(\mathbf{r}_{\|},k_z=0,t)$$

$$\begin{aligned}&\hat{\mathbf{z}}\times[\mathbf{H}_{\|}^{+}(\mathbf{r}_{\|},z=0,t)-\mathbf{H}_{\|}^{-}(\mathbf{r}_{\|},z=0,t)]\\=&-\nabla_{\|}\times\mathbf{H}_{ex}(\mathbf{r}_{\|},k_z=0,t)+\frac{1}{c}\frac{\partial}{\partial t}\mathbf{D}_{ex}(\mathbf{r}_{\|},k_z=0,t)+\frac{1}{c}\mathbf{I}_{ex}(\mathbf{r}_{\|},k_z=0,t)\end{aligned}$$

$$B_z^{+}(\mathbf{r}_{\|},z=0,t)-B_z^{-}(\mathbf{r}_{\|},z=0,t)=-\nabla_{\|}.\mathbf{B}_{ex,\|}(\mathbf{r}_{\|},k_z=0,t) \tag{2.37}$$

The boundary conditions are now obtained by taking the parallel parts of the first and the third equation together with the second and the fourth equation. This gives

$$E_x^{+}(\mathbf{r}_{\|},z=0,t)-E_x^{-}(\mathbf{r}_{\|},z=0,t)=\frac{\partial}{\partial x}E_{ex,z}(\mathbf{r}_{\|},k_z=0,t)-\frac{1}{c}\frac{\partial}{\partial t}B_{ex,y}(\mathbf{r}_{\|},k_z=0,t)$$

$$E_y^{+}(\mathbf{r}_{\|},z=0,t)-E_y^{-}(\mathbf{r}_{\|},z=0,t)=\frac{\partial}{\partial y}E_{ex,z}(\mathbf{r}_{\|},k_z=0,t)+\frac{1}{c}\frac{\partial}{\partial t}B_{ex,x}(\mathbf{r}_{\|},k_z=0,t)$$

$$D_z^{+}(\mathbf{r}_{\|},z=0,t)-D_z^{-}(\mathbf{r}_{\|},z=0,t)=-\nabla_{\|}.\mathbf{D}_{ex,\|}(\mathbf{r}_{\|},k_z=0,t)+\rho_{ex}(\mathbf{r}_{\|},k_z=0,t)$$

$$\begin{aligned} H_x^+(\mathbf{r}_\|, z &= 0,t) - H_x^-(\mathbf{r}_\|, z=0,t) \\ &= \frac{\partial}{\partial x} H_{ex,z}(\mathbf{r}_\|, k_z = 0, t) + \frac{1}{c}\frac{\partial}{\partial t} D_{ex,y}(\mathbf{r}_\|, k_z = 0, t) + \frac{1}{c} I_{ex,y}(\mathbf{r}_\|, k_z = 0, t) \end{aligned}$$

$$\begin{aligned} H_y^+(\mathbf{r}_\|, z &= 0,t) - H_y^-(\mathbf{r}_\|, z=0,t) \\ &= \frac{\partial}{\partial y} H_{ex,z}(\mathbf{r}_\|, k_z = 0, t) - \frac{1}{c}\frac{\partial}{\partial t} D_{ex,x}(\mathbf{r}_\|, k_z = 0, t) - \frac{1}{c} I_{ex,x}(\mathbf{r}_\|, k_z = 0, t) \end{aligned}$$

$$B_z^+(\mathbf{r}_\|, z=0,t) - B_z^-(\mathbf{r}_\|, z=0,t) = -\nabla_\|.\mathbf{B}_{ex,\|}(\mathbf{r}_\|, k_z = 0, t) \tag{2.38}$$

It follows from these boundary conditions that the jumps in the extrapolated fields are given in terms of the total excess of $\mathbf{D}_{ex,\|}$, $E_{ex,z}$, $\mathbf{B}_{ex,\|}$, $H_{ex,z}$, $\mathbf{I}_{ex,\|}$ and ρ_{ex}. Notice in this context that the $k_z = 0$ value of a field is given by the integral of the z-dependent excess field and is therefore the total excess of this field. It is important to realize that the jumps in the extrapolated fields are not affected by the total excesses of $D_{ex,z}$, $\mathbf{E}_{ex,\|}$, $B_{ex,z}$, $\mathbf{H}_{ex,\|}$, $I_{ex,z}$; these total excesses therefore have no effect on the reflection and transmission amplitudes and may as such be neglected in the description of the optical properties of the interface. This fact will be used in the last section of this chapter in the definition of the interfacial polarization and magnetization densities.

In order to make the notation more compact the total excesses, which do play a role in the boundary conditions, will be written in the following form [16]

$$\mathbf{D}_\|^s(\mathbf{r}_\|, t) \equiv \mathbf{D}_{ex,\|}(\mathbf{r}_\|, k_z = 0, t) = \int_{-\infty}^{\infty} dz \mathbf{D}_{ex,\|}(\mathbf{r}_\|, z, t)$$

$$E_z^s(\mathbf{r}_\|, t) \equiv E_{ex,z}(\mathbf{r}_\|, k_z = 0, t) = \int_{-\infty}^{\infty} dz\ E_{ex,z}(\mathbf{r}_\|, z, t)$$

$$\mathbf{B}_\|^s(\mathbf{r}_\|, t) \equiv \mathbf{B}_{ex,\|}(\mathbf{r}_\|, k_z = 0, t) = \int_{-\infty}^{\infty} dz \mathbf{B}_{ex,\|}(\mathbf{r}_\|, z, t)$$

$$H_z^s(\mathbf{r}_\|, t) \equiv H_{ex,z}(\mathbf{r}_\|, k_z = 0, t) = \int_{-\infty}^{\infty} dz\ H_{ex,z}(\mathbf{r}_\|, z, t)$$

$$\mathbf{I}_\|^s(\mathbf{r}_\|, t) \equiv \mathbf{I}_{ex,\|}(\mathbf{r}_\|, k_z = 0, t) = \int_{-\infty}^{\infty} dz \mathbf{I}_{ex,\|}(\mathbf{r}_\|, z, t)$$

$$\rho^s(\mathbf{r}_\|, t) \equiv \rho_{ex}(\mathbf{r}_\|, k_z = 0, t) = \int_{-\infty}^{\infty} dz\ \rho_{ex}(\mathbf{r}_\|, z, t) \tag{2.39}$$

Using these lowest order moments of the excess quantities the boundary conditions become [17]

$$E_x^+(\mathbf{r}_\parallel, z=0,t) - E_x^-(\mathbf{r}_\parallel, z=0,t) = \frac{\partial}{\partial x}E_z^s(\mathbf{r}_\parallel,t) - \frac{1}{c}\frac{\partial}{\partial t}B_y^s(\mathbf{r}_\parallel,t)$$

$$E_y^+(\mathbf{r}_\parallel, z=0,t) - E_y^-(\mathbf{r}_\parallel, z=0,t) = \frac{\partial}{\partial y}E_z^s(\mathbf{r}_\parallel,t) + \frac{1}{c}\frac{\partial}{\partial t}B_x^s(\mathbf{r}_\parallel,t)$$

$$D_z^+(\mathbf{r}_\parallel, z=0,t) - D_z^-(\mathbf{r}_\parallel, z=0,t) = -\nabla_\parallel.\mathbf{D}_\parallel^s(\mathbf{r}_\parallel,t) + \rho^s(\mathbf{r}_\parallel,t)$$

$$H_x^+(\mathbf{r}_\parallel, z=0,t) - H_x^-(\mathbf{r}_\parallel, z=0,t) = \frac{\partial}{\partial x}H_z^s(\mathbf{r}_\parallel,t) + \frac{1}{c}\frac{\partial}{\partial t}D_y^s(\mathbf{r}_\parallel,t) + \frac{1}{c}I_y^s(\mathbf{r}_\parallel,t)$$

$$H_y^+(\mathbf{r}_\parallel, z=0,t) - H_y^-(\mathbf{r}_\parallel, z=0,t) = \frac{\partial}{\partial y}H_z^s(\mathbf{r}_\parallel,t) - \frac{1}{c}\frac{\partial}{\partial t}D_x^s(\mathbf{r}_\parallel,t) - \frac{1}{c}I_x^s(\mathbf{r}_\parallel,t)$$

$$B_z^+(\mathbf{r}_\parallel, z=0,t) - B_z^-(\mathbf{r}_\parallel, z=0,t) = -\nabla_\parallel.\mathbf{B}_\parallel^s(\mathbf{r}_\parallel,t) \tag{2.40}$$

where the super index s from surface indicates the total excess of the corresponding quantity.

2.5 Higher order moments of the excess currents, charge density and fields

In deriving the above boundary conditions from the Maxwell equations it was sufficient to use the parallel parts of the first and the third identity in eq.(2.37). The normal parts give the following two equations

$$\begin{aligned} 0 &= \nabla_\parallel \times \mathbf{E}_{ex,\parallel}(\mathbf{r}_\parallel, k_z=0,t) + \frac{1}{c}\frac{\partial}{\partial t}B_{ex,z}(\mathbf{r}_\parallel, k_z=0,t) \\ 0 &= \nabla_\parallel \times \mathbf{H}_{ex,\parallel}(\mathbf{r}_\parallel, k_z=0,t) - \frac{1}{c}\frac{\partial}{\partial t}D_{ex,z}(\mathbf{r}_\parallel, k_z=0,t) \\ & \quad -\frac{1}{c}I_{ex,z}(\mathbf{r}_\parallel, k_z=0,t) \end{aligned} \tag{2.41}$$

These two expressions, which follow directly from the Maxwell equations, relate lowest order moments of excess quantities which are unimportant for the jump in the extrapolated fields. In order to see the role of these moments one can expand the identities in eq.(2.36) to linear order in k_z and consider the linear term. The discussion will be restricted to the second relation as this is sufficient to illustrate this point

$$\begin{aligned} D_{ex,z}(\mathbf{r}_\parallel, k_z &= 0,t) = -i[\frac{\partial}{\partial k_z}\rho_{ex}(\mathbf{r}_\parallel,k_z,t)]_{k_z=0} + i\nabla_\parallel.[\frac{\partial}{\partial k_z}\mathbf{D}_{ex,\parallel}(\mathbf{r}_\parallel,k_z,t)]_{k_z=0} \\ &= -\int_{-\infty}^{\infty} dz\ z[\rho_{ex}(\mathbf{r},t) - \nabla_\parallel.\mathbf{D}_{ex,\parallel}(\mathbf{r},t)] \end{aligned} \tag{2.42}$$

This equation shows that $D_{ex,z}(\mathbf{r}_{\|}, k_z = 0, t)$ is important if one wants to describe the first order moment of the excess of the difference of the charge distribution with the divergence of the displacement field. Similarly the lowest order moments of $\mathbf{E}_{ex,\|}$, $B_{ex,z}$, $\mathbf{H}_{ex,\|}$ and $I_{ex,z}$ are important only if one wants to describe first order moments of the excess quantities. *The boundary conditions derived and given above show rigorously that the first order moment of the excess currents, charge density and fields have no effect on the jump of the extrapolated fields at the dividing surface* [15], [16]. These first order moments are therefore not important for the subject of this book. In the description of the optical properties of surfaces given in this book it is therefore possible to set these first order, and similarly the higher order, moments of the excess quantities equal to zero without affecting the rigor of the description. As follows from the above equation one must then also set the zeroth order moments of $D_{ex,z}$, $\mathbf{E}_{ex,\|}$, $B_{ex,z}$, $\mathbf{H}_{ex,\|}$ and $I_{ex,z}$, which are given in terms of the first order moments, equal to zero. It should be emphasized that, in view of the fact that the boundary conditions are independent of D_z^s, $\mathbf{E}_{\|}^s$, B_z^s, $\mathbf{H}_{\|}^s$ and I_z^s, setting them equal to zero, as is done in this book, is merely a matter of mathematical convenience. It does not imply that these quantities are in fact equal to zero.

2.6 Boundary conditions in terms of the surface polarization and magnetization

Similar to the description in bulk phases it is convenient to introduce polarization and magnetization densities for the interface. The choice of the definition of these quantities is not quite straightforward in the sense that, unlike in the bulk, it is not possible to just use as definition the difference between the interfacial electric (or magnetic) displacement fields and the interfacial electric (or magnetic) fields. As discussed above half of these fields do not affect the boundary conditions and can therefore not possibly be used in such a definition. The proper definition in fact is

$$\mathbf{P}^s(\mathbf{r}_{\|}, t) \equiv (\mathbf{D}_{\|}^s(\mathbf{r}_{\|}, t), -E_z^s(\mathbf{r}_{\|}, t)) \text{ and } \mathbf{M}^s(\mathbf{r}_{\|}, t) \equiv (\mathbf{B}_{\|}^s(\mathbf{r}_{\|}, t), -H_z^s(\mathbf{r}_{\|}, t)) \quad (2.43)$$

Using eq.(2.39) one therefore has

$$\begin{aligned} \mathbf{P}^s(\mathbf{r}_{\|}, t) &= \int_{-\infty}^{\infty} dz(\mathbf{D}_{ex,\|}(\mathbf{r}_{\|}, z, t), -E_{ex,z}(\mathbf{r}_{\|}, z, t)) \\ \mathbf{M}^s(\mathbf{r}_{\|}, t) &= \int_{-\infty}^{\infty} dz(\mathbf{B}_{ex,\|}(\mathbf{r}_{\|}, z, t), -H_{ex,z}(\mathbf{r}_{\|}, z, t)) \end{aligned} \quad (2.44)$$

It should again be emphasized that these are the proper expressions for the calculation of the interfacial polarization and magnetization densities in every particular example like for instance island films or rough surfaces. It is not correct to replace the integrands by the excess polarization and magnetization densities. *As such a replacement is intuitively appealing it can not be warned against sufficiently!*

The boundary conditions for the jumps in the fields, cf. eq.(2.40), may now also be given in terms of the interfacial polarization and magnetization densities. Using the definition of the polarization and magnetization density given in eq.(2.43) one finds

$$\begin{aligned}
E_x^+(\mathbf{r}_\parallel, z &= 0,t) - E_x^-(\mathbf{r}_\parallel, z=0,t) = -\frac{\partial}{\partial x}P_z^s(\mathbf{r}_\parallel,t) - \frac{1}{c}\frac{\partial}{\partial t}M_y^s(\mathbf{r}_\parallel,t) \\
E_y^+(\mathbf{r}_\parallel, z &= 0,t) - E_y^-(\mathbf{r}_\parallel, z=0,t) = -\frac{\partial}{\partial y}P_z^s(\mathbf{r}_\parallel,t) + \frac{1}{c}\frac{\partial}{\partial t}M_x^s(\mathbf{r}_\parallel,t) \\
D_z^+(\mathbf{r}_\parallel, z &= 0,t) - D_z^-(\mathbf{r}_\parallel, z=0,t) = -\nabla_\parallel.\mathbf{P}_\parallel^s(\mathbf{r}_\parallel,t) + \rho^s(\mathbf{r}_\parallel,t) \\
H_x^+(\mathbf{r}_\parallel, z &= 0,t) - H_x^-(\mathbf{r}_\parallel, z=0,t) = -\frac{\partial}{\partial x}M_z^s(\mathbf{r}_\parallel,t) + \frac{1}{c}\frac{\partial}{\partial t}P_y^s(\mathbf{r}_\parallel,t) + \frac{1}{c}I_y^s(\mathbf{r}_\parallel,t) \\
H_y^+(\mathbf{r}_\parallel, z &= 0,t) - H_y^-(\mathbf{r}_\parallel, z=0,t) = -\frac{\partial}{\partial y}M_z^s(\mathbf{r}_\parallel,t) - \frac{1}{c}\frac{\partial}{\partial t}P_x^s(\mathbf{r}_\parallel,t) - \frac{1}{c}I_x^s(\mathbf{r}_\parallel,t) \\
B_z^+(\mathbf{r}_\parallel, z &= 0,t) - B_z^-(\mathbf{r}_\parallel, z=0,t) = -\nabla_\parallel.\mathbf{M}_\parallel^s(\mathbf{r}_\parallel,t) \qquad (2.45)
\end{aligned}$$

for the jumps.

As was said above, it is in the mathematical analysis often convenient to introduce the following zero fields

$$\mathbf{E}_\parallel^s(\mathbf{r}_\parallel,t) \equiv D_z^s(\mathbf{r}_\parallel,t) \equiv \mathbf{H}_\parallel^s(\mathbf{r}_\parallel,t) \equiv B_z^s(\mathbf{r}_\parallel,t) \equiv I_z^s(\mathbf{r}_\parallel,t) \equiv 0 \qquad (2.46)$$

One of the obvious advantages of the introduction of these fields is that it follows together with the definition of the interfacial polarization and magnetization densities, eq.(2.43), that

$$\mathbf{P}^s(\mathbf{r}_\parallel,t) = \mathbf{D}^s(\mathbf{r}_\parallel,t) - \mathbf{E}^s(\mathbf{r}_\parallel,t) \text{ and } \mathbf{M}^s(\mathbf{r}_\parallel,t) = \mathbf{B}^s(\mathbf{r}_\parallel,t) - \mathbf{H}^s(\mathbf{r}_\parallel,t) \qquad (2.47)$$

As said before, the validity of this relation is a matter of mathematical convenience only. The introduction of the zero fields in eq.(2.46) does not represent either a restrictive assumption in the description of the boundary conditions or some profound truth! To further underline this fact it is good to realize that in general

$$\begin{aligned}
\mathbf{E}_\parallel^s(\mathbf{r}_\parallel,t) &\neq \mathbf{E}_{ex,\parallel}(\mathbf{r}_\parallel,k_z=0,t), \quad D_z^s(\mathbf{r}_\parallel,t) \neq D_{ex,z}(\mathbf{r}_\parallel,k_z=0,t), \\
\mathbf{H}_\parallel^s(\mathbf{r}_\parallel,t) &\neq \mathbf{H}_{ex,\parallel}(\mathbf{r}_\parallel,k_z=0,t), \quad B_z^s(\mathbf{r}_\parallel,t) \neq B_{ex,z}(\mathbf{r}_\parallel,k_z=0,t), \\
I_z^s(\mathbf{r}_\parallel,t) &\neq I_{ex,z}(\mathbf{r}_\parallel,k_z=0,t) \qquad (2.48)
\end{aligned}$$

As explained above, the left-hand side is equal to zero by definition, whereas the right-hand side may be unequal to zero.

Chapter 3
MAXWELL'S EQUATIONS WITH SINGULAR FIELDS

3.1 Introduction

In the previous chapter it was shown how one may introduce the concept of excess current and charge densities and excess fields as a way to describe properties of a thin boundary layer without considering the detailed behavior of these quantities as a function of position in the direction normal to this layer. In particular the discontinuity of the extrapolated bulk field at the dividing surface could be expressed in terms of these excesses. In this chapter it will be shown how these excess quantities may be described as singular contributions to the corresponding field [16]. The first to introduce singularities in the polarization and magnetization along the surface in order to obtain discontinuities in the extrapolated fields, were Stahl and Wolters [19]. They did not consider the effects of singularities in the normal direction, however. In section 3.2 explicit expressions for the current and charge densities and the electromagnetic fields, containing such singular contributions, are given and in section 3.3 the validity of the Maxwell equations is then discussed, for these singular fields. It is found in this context that at the dividing surface the Maxwell equations reduce to the boundary conditions, given in section 2.4, for the fields. It is precisely this fact, which makes the extension of the validity of the Maxwell equations to the singular fields possible. In section 3.4 charge conservation is treated. The boundary conditions for the fields are used to obtain a charge conservation relation at the surface. In a general analysis of time dependent electromagnetic phenomena it is convenient to absorb the induced electric current in the displacement field. This simplification is discussed in section 3.5.

The constitutive relations are given in section 3.6 for a surface in the absence of spatial dispersion [17]. In section 3.7 the constitutive relations are given for the more general case that spatial dispersion is taken into account. The reason to do this finds its origin in the choice of the location of the dividing surface. The excess polarization and magnetization are placed as equivalent polarization and magnetization densities on the dividing surface. It will be clear that they are in fact coupled to the field at their original position rather then to the field value at the dividing surface. This implies that the coefficients characterizing spatial dispersion will sensitively depend on the choice of the dividing surface. As the optical properties of the interface do not depend on the choice of the dividing surface it must be possible to construct invariant combinations of the constitutive coefficients. For the special case of stratified surfaces such "invariants" were first introduced by Lekner [11]. In order to get a proper understanding of the nature of these invariants in the general case the introduction of coefficients characterizing spatial dispersion is essential. In this respect the description of a surface differs in a crucial way from the description in the bulk. In the bulk one may place the dipole moments in the "center" of the molecules involved. The possible quadrupole moment relative to this center, though interesting, has only a small influence on the response of the bulk material. For a surface all dipole moments are placed on the dividing surface. The size of the quadrupole moment relative to the position of the dividing surface is of the order of the thickness of the layer which is much larger than the size of the molecules. This leads to the fact that the quadrupole density on the dividing surface has a much greater impact on the optical properties of the surface than the analogy with the bulk would suggest. Due to the lack of symmetry of the surface in the direction normal to the surface one will generally find that a homogeneous electric field will yield a quadrupole moment density. In view of time reversal invariance, or as it is sometimes called source observer symmetry, [12], this also implies that a gradient field will yield a dipole moment density at the surface. In the subsequent chapters this will be discussed, for instance for island films, in great detail. In section 3.8 the dependence of the constitutive coefficients on the choice of the dividing surface is discussed. Using these results the invariants are constructed and discussed in section 3.9. The last two sections discuss how the constitutive coefficients change if the dividing surface is shifted through a homogeneous background and how to add two adjacent but distinct films to one single film. The formulae describing such a shift and the addition of adjacent films will turn out to be extremely useful in subsequent applications.

3.2 Singularities in the fields, currents and charge densities

Consider again a system, consisting of two different homogeneous media separated by a thin boundary layer. It is assumed that a dividing surface may be chosen within this boundary layer which is both flat and non-moving. In this context it should be noted that this assumption does not imply that the boundary layer as a whole is time independent. In principle the properties of the boundary layer may depend on the time as long as the layer remains close to the time independent dividing surface. It is possible to extend the formalism to interfaces for which a dividing surface must be chosen, which is both curved and moving. This complicates matters considerably, however, and is of less practical use. This more general case will therefore not be considered in this book. As dividing surface the $x-y$ plane is chosen. The explicit form of the charge density and the electric current density as a function of the position $\mathbf{r}$ and the time t including these singular contributions is [16]

$$\rho(\mathbf{r},t) = \rho^{-}(\mathbf{r},t)\theta(-z) + \rho^{s}(\mathbf{r}_{\|},t)\delta(z) + \rho^{+}(\mathbf{r},t)\theta(z) \tag{3.1}$$

$$\mathbf{I}(\mathbf{r},t) = \mathbf{I}^{-}(\mathbf{r},t)\theta(-z) + \mathbf{I}^{s}(\mathbf{r}_{\|},t)\delta(z) + \mathbf{I}^{+}(\mathbf{r},t)\theta(z) \tag{3.2}$$

The superscript $-$ indicates that the corresponding density refers to the medium in the half-space $z<0$, whereas the superscript $+$ in a similar way refers to the medium in the half-space $z>0$. Furthermore $\delta(z)$ is the Dirac delta-function. Finally $\rho^{s}(\mathbf{r}_{\|},t)$

and $\mathbf{I}^s(\mathbf{r}_\parallel, t)$ are the so-called (total) excess charge and electric current densities introduced in the previous section. It is important to realize that ρ^s is a density per unit of surface area, while $\mathbf{I}^s$ is a current density per unit of length. In contrast to ρ^s and $\mathbf{I}^s$, $\rho^\pm$ are charge densities per unit of volume and $\mathbf{I}^\pm$ current densities per unit of surface. These excess densities only depend on the position $\mathbf{r}_\parallel = (x, y)$ along the dividing surface. As has been discussed extensively in the last two sections of the previous chapter the excess current density may be chosen to be directed along this surface so that

$$I_z^s(\mathbf{r}_\parallel, t) = 0 \tag{3.3}$$

without loss of generality regarding the nature of the boundary conditions for the fields.

In the previous chapter it was similarly found that the electromagnetic fields also have excesses. These excesses are again described as singular contributions to the fields at the dividing surface in $z = 0$. Thus one has for the electric field $\mathbf{E}$, the magnetic field $\mathbf{H}$, the electric displacement field $\mathbf{D}$ and the magnetic induction $\mathbf{B}$ [17], [16]:

$$\begin{aligned}
\mathbf{E}(\mathbf{r}, t) &= \mathbf{E}^-(\mathbf{r}, t)\theta(-z) + \mathbf{E}^s(\mathbf{r}_\parallel, t)\delta(z) + \mathbf{E}^+(\mathbf{r}, t)\theta(z) \\
\mathbf{H}(\mathbf{r}, t) &= \mathbf{H}^-(\mathbf{r}, t)\theta(-z) + \mathbf{H}^s(\mathbf{r}_\parallel, t)\delta(z) + \mathbf{H}^+(\mathbf{r}, t)\theta(z) \\
\mathbf{D}(\mathbf{r}, t) &= \mathbf{D}^-(\mathbf{r}, t)\theta(-z) + \mathbf{D}^s(\mathbf{r}_\parallel, t)\delta(z) + \mathbf{D}^+(\mathbf{r}, t)\theta(z) \\
\mathbf{B}(\mathbf{r}, t) &= \mathbf{B}^-(\mathbf{r}, t)\theta(-z) + \mathbf{B}^s(\mathbf{r}_\parallel, t)\delta(z) + \mathbf{B}^+(\mathbf{r}, t)\theta(z)
\end{aligned} \tag{3.4}$$

As discussed in the last two sections of the previous chapter the following contributions to the singular fields are chosen equal to zero

$$\begin{aligned}
\mathbf{E}_\parallel^s(\mathbf{r}_\parallel, t) &\equiv (E_x^s(\mathbf{r}_\parallel, t), E_y^s(\mathbf{r}_\parallel, t)) = 0, \quad D_z^s(\mathbf{r}_\parallel, t) = 0 \\
\mathbf{H}_\parallel^s(\mathbf{r}_\parallel, t) &\equiv (H_x^s(\mathbf{r}_\parallel, t), H_y^s(\mathbf{r}_\parallel, t)) = 0, \quad B_z^s(\mathbf{r}_\parallel, t) = 0
\end{aligned} \tag{3.5}$$

It will become clear below why it is convenient to set these singular contributions, which do not effect the boundary conditions for the fields, equal to zero.

3.3 Maxwell's equations for singular fields and the boundary conditions

The fundamental assumption on which much of the analysis in this book is founded is that the Maxwell equations

$$\begin{aligned}
\operatorname{rot} \mathbf{E}(\mathbf{r}, t) &= -\frac{1}{c}\frac{\partial}{\partial t}\mathbf{B}(\mathbf{r}, t), & \operatorname{div} \mathbf{D}(\mathbf{r}, t) &= \rho(\mathbf{r}, t) \\
\operatorname{rot} \mathbf{H}(\mathbf{r}, t) &= \frac{1}{c}\mathbf{I}(\mathbf{r}, t) + \frac{1}{c}\frac{\partial}{\partial t}\mathbf{D}(\mathbf{r}, t), & \operatorname{div} \mathbf{B}(\mathbf{r}, t) &= 0
\end{aligned} \tag{3.6}$$

are valid for the singular form of the fields and densities given in the previous section. c is the speed of light. Away from the dividing surface these equations reduce to their usual form. At the dividing surface the Maxwell equations for the singular fields and densities yield boundary conditions for the asymptotic value of the bulk fields at the dividing surface. In order to see how these boundary conditions are found the second Maxwell equation will now be analyzed in detail. Upon substitution of the singular form of the fields given in eqs. (3.1) and (3.4) into the second Maxwell equation $\text{div}\mathbf{D} = \rho$ one obtains

$$\begin{aligned} & \text{div}[\mathbf{D}^-(\mathbf{r},t)\theta(-z) + \mathbf{D}^s(\mathbf{r}_\|,t)\delta(z) + \mathbf{D}^+(\mathbf{r},t)\theta(z)] \\ = \; & [\text{div}\,\mathbf{D}^-(\mathbf{r},t)]\theta(-z) + [\text{div}\,\mathbf{D}^s(\mathbf{r}_\|,t)]\delta(z) + [\text{div}\,\mathbf{D}^+(\mathbf{r},t)]\theta(z) \\ & +\mathbf{D}^-(\mathbf{r},t).\,\text{grad}\,\theta(-z) + \mathbf{D}^s(\mathbf{r}_\|,t).\,\text{grad}\,\delta(z) + \mathbf{D}^+(\mathbf{r},t).\,\text{grad}\,\theta(z) \\ = \; & \rho^-(\mathbf{r},t)\theta(-z) + \rho^s(\mathbf{r}_\|,t)\delta(z) + \rho^+(\mathbf{r},t)\theta(z) \end{aligned} \tag{3.7}$$

The dot indicates a contraction of two vectors, i.e. $\mathbf{a}.\mathbf{b} \equiv a_x b_x + a_y b_y + a_z b_z$. In order to simplify the above expression one uses

$$\text{grad}\,\theta(z) = \hat{\mathbf{z}}\delta(z), \quad \text{grad}\,\theta(-z) = -\hat{\mathbf{z}}\delta(z) \text{ and } \text{grad}\,\delta(z) = \hat{\mathbf{z}}\frac{d\delta(z)}{dz} \tag{3.8}$$

where $\hat{\mathbf{z}} \equiv (0, 0, 1)$ is the unit vector normal to the dividing surface. In the further analysis it is convenient to introduce a three- and a two-dimensional gradient operator

$$\nabla \equiv (\frac{\partial}{\partial x}, \frac{\partial}{\partial y}, \frac{\partial}{\partial z}) \quad \text{and} \quad \nabla_\| \equiv (\frac{\partial}{\partial x}, \frac{\partial}{\partial y}) \tag{3.9}$$

Substituting eq.(3.8) into eq.(3.7) and using the fact that the normal component of the electric displacement field is zero one obtains

$$\begin{aligned} & [\text{div}\,\mathbf{D}^-(\mathbf{r},t) - \rho^-(\mathbf{r},t)]\theta(-z) \\ & +[D_z^+(\mathbf{r}_\|,0,t) - D_z^-(\mathbf{r}_\|,0,t) + \nabla_\|.\mathbf{D}_\|^s(\mathbf{r}_\|,t) - \rho^s(\mathbf{r}_\|,t)]\delta(z) \\ & +[\text{div}\,\mathbf{D}^+(\mathbf{r},t) - \rho^+(\mathbf{r},t)]\theta(z) = 0 \end{aligned} \tag{3.10}$$

The term proportional to the δ-function contains the values of the normal component of the bulk displacement fields extrapolated to the dividing surface. An important point is that no contribution proportional to the normal derivative of the δ-function appears in the above equation. This is a consequence of the fact that the singular contribution to the displacement field was chosen along the surface. Notice furthermore that the dot is also used to indicate a contraction of 2-dimensional vectors.

The above analysis shows that it is only possible to extend the validity of the second Maxwell equation to fields with singularities at the dividing surface if D_z^s is zero. One could in principle further extend the singular nature of the fields at the dividing surface by also allowing contributions proportional to derivatives of δ-functions. In that case it is no longer necessary in the analysis to take D_z^s equal to zero. As discussed in the last two sections of the previous chapter such a generalization in no way affects the boundary conditions and is therefore of no use. In this book it is therefore appropriate to use that the singular fields given in eq.(3.5) as well as the normal component of the singular current density may be set equal to zero without loss of generality. This also assures the possibility to extend the validity of the other Maxwell equations to fields with singularities.

From the above equation it first of all follows that the electric displacement field in the bulk phases satisfies the second Maxwell equation in its usual form

$$\begin{aligned} \operatorname{div} \mathbf{D}^-(\mathbf{r},t) &= \rho^-(\mathbf{r},t) \quad \text{for } z<0 \\ \operatorname{div} \mathbf{D}^+(\mathbf{r},t) &= \rho^+(\mathbf{r},t) \quad \text{for } z>0 \end{aligned} \tag{3.11}$$

Furthermore one finds at the dividing surface the following boundary condition for the normal component of the electric displacement field

$$D_z^+(\mathbf{r}_\parallel,0,t) - D_z^-(\mathbf{r}_\parallel,0,t) = \rho^s(\mathbf{r}_\parallel,t) - \nabla_\parallel.\mathbf{D}_\parallel^s(\mathbf{r}_\parallel,t) \tag{3.12}$$

This is precisely the boundary condition derived also in section 2.4.

In a completely analogous manner one may verify that also the other Maxwell equations are satisfied in their usual form in the bulk phases. These equations in fact read like those given in eq.(3.6) but with superscripts + or − to restrict their validity to the regions occupied by the two phases, where $z>0$ or $z<0$ respectively. As boundary conditions at the dividing surface one obtains for the fields:

$$\begin{aligned} \mathbf{E}_\parallel^+(\mathbf{r}_\parallel,0,t) - \mathbf{E}_\parallel^-(\mathbf{r}_\parallel,0,t) &= \hat{\mathbf{z}} \times \frac{\partial}{c\partial t}\mathbf{B}_\parallel^s(\mathbf{r}_\parallel,t) + \nabla_\parallel E_z^s(\mathbf{r}_\parallel,t) \\ \mathbf{H}_\parallel^+(\mathbf{r}_\parallel,0,t) - \mathbf{H}_\parallel^-(\mathbf{r}_\parallel,0,t) &= -\hat{\mathbf{z}} \times \frac{1}{c}\mathbf{I}_\parallel^s(\mathbf{r}_\parallel,t) - \hat{\mathbf{z}} \times \frac{\partial}{c\partial t}\mathbf{D}_\parallel^s(\mathbf{r}_\parallel,t) + \nabla_\parallel H_z^s(\mathbf{r}_\parallel,t) \\ B_z^+(\mathbf{r}_\parallel,0,t) - B_z^-(\mathbf{r}_\parallel,0,t) &= -\nabla_\parallel.\mathbf{B}_\parallel^s(\mathbf{r}_\parallel,t) \end{aligned} \tag{3.13}$$

Eqs.(3.12) and (3.13) are the complete set of boundary conditions already derived for the fields in section 2.4.

The analysis above shows not only that the Maxwell equations may be extended to be valid for the singular fields introduced above, but also that the boundary conditions are now in a simple way contained in the Maxwell equations for the singular fields. The use of the Maxwell equations for these singular fields thus assures that the boundary conditions are in a natural way accounted for.

For the further analysis in the book it is convenient to express the jumps in the bulk fields at the dividing surface in terms of the interfacial polarization and magnetization densities defined in the last section of the previous chapter. One then obtains

$$\begin{aligned} \mathbf{E}_\parallel^+(\mathbf{r}_\parallel,0,t) - \mathbf{E}_\parallel^-(\mathbf{r}_\parallel,0,t) &= \hat{\mathbf{z}} \times \frac{\partial}{c\partial t}\mathbf{M}_\parallel^s(\mathbf{r}_\parallel,t) - \nabla_\parallel P_z^s(\mathbf{r}_\parallel,t) \\ D_z^+(\mathbf{r}_\parallel,0,t) - D_z^-(\mathbf{r}_\parallel,0,t) &= \rho^s(\mathbf{r}_\parallel,t) - \nabla_\parallel.\mathbf{P}_\parallel^s(\mathbf{r}_\parallel,t) \\ \mathbf{H}_\parallel^+(\mathbf{r}_\parallel,0,t) - \mathbf{H}_\parallel^-(\mathbf{r}_\parallel,0,t) &= -\hat{\mathbf{z}} \times \frac{1}{c}\mathbf{I}_\parallel^s(\mathbf{r}_\parallel,t) - \hat{\mathbf{z}} \times \frac{\partial}{c\partial t}\mathbf{P}_\parallel^s(\mathbf{r}_\parallel,t) - \nabla_\parallel M_z^s(\mathbf{r}_\parallel,t) \\ B_z^+(\mathbf{r}_\parallel,0,t) - B_z^-(\mathbf{r}_\parallel,0,t) &= -\nabla_\parallel.\mathbf{M}_\parallel^s(\mathbf{r}_\parallel,t) \end{aligned} \tag{3.14}$$

which is identical to eq.(2.45). Notice the fact that, if all the interfacial current and charge densities as well as the interfacial polarization and magnetization densities are zero, these boundary conditions reduce to the usual continuity conditions for the fields. In that case the interface is merely a sharp transition from one bulk phase to the other. It serves as a mathematical boundary and has no distinct physical characteristics. An interfacial charge density, which typically occurs on the surface of a conductor, leads to a discontinuity in the normal component of the electric displacement field. This is also a well-known phenomenon. An interfacial electric current density leads to a discontinuity of the parallel component of the magnetic field. As we discussed in the previous chapter such an interfacial current will typically occur in a thin conducting layer. An other well-known case where such an interfacial electric current occurs is at the surface of a superconductor. In this case the interfacial current is needed, because the parallel component of the magnetic field jumps from a finite value to zero. It follows from the above equations that not only the excesses of the charge and current densities lead to discontinuities in the fields. Also the interfacial polarization and magnetization densities, which are the results of excesses of the fields, cf. eq.(2.44), themselves lead to such discontinuities. It are, in particular, the occurrence of such excesses and their consequences, which form the central issue in this book.

3.4 Charge conservation

If one takes the time derivative of the second Maxwell equation and combines the result with the third Maxwell equation one obtains

$$\frac{\partial}{\partial t}\rho(\mathbf{r},t) = -\operatorname{div}\mathbf{I}(\mathbf{r},t) \tag{3.15}$$

This is the usual differential form of the law of charge conservation. Now, however, both the charge density and the electric current density may have a singular contribution on the dividing surface due to the occurrence of excesses in these quantities. Substituting the singular form of these densities, given in eqs.(3.1) and (3.2), into the above equation one obtains, using a derivation which is completely analogous to the one given in the previous section for the second Maxwell equation, in the bulk phases the following expressions

$$\frac{\partial}{\partial t}\rho^{\pm}(\mathbf{r},t) = -\operatorname{div}\mathbf{I}^{\pm}(\mathbf{r},t) \quad \text{for } 0 < \pm z \tag{3.16}$$

which has the usual form. At the dividing surface one obtains in a similar way

$$\frac{\partial}{\partial t}\rho^{s}(\mathbf{r}_{\parallel},t) = -\nabla_{\parallel}.\mathbf{I}^{s}(\mathbf{r}_{\parallel},t) - I_z^{+}(\mathbf{r}_{\parallel},0,t) + I_z^{-}(\mathbf{r}_{\parallel},0,t) \tag{3.17}$$

This equation shows that the excess of the charge density not only changes due to a possible two-dimensional divergence of the excess current density along the dividing surface, but also due to electric currents which flow from the bulk phases to the interfacial layer or vice versa. In the form of eq. (3.17) given above it is clear that the equation may be interpreted as a balance equation for the excess charge density. One may alternatively write this equation as follows

$$I_z^{+}(\mathbf{r}_{\parallel},0,t) - I_z^{-}(\mathbf{r}_{\parallel},0,t) = -\frac{\partial}{\partial t}\rho^{s}(\mathbf{r}_{\parallel},t) - \nabla_{\parallel}.\mathbf{I}^{s}(\mathbf{r}_{\parallel},t) \tag{3.18}$$

if one wants to emphasize the fact that this equation also serves as a boundary condition for the normal component of the bulk currents in the two phases. As has already been discussed above, the excess current has no component in the direction normal to the dividing surface, cf. eq.(3.3).

In order to simplify expressions, a short hand notation for the jump of an arbitrary field $a(\mathbf{r}, t)$ at the dividing surface is introduced:

$$a_-(\mathbf{r}_\parallel, t) \equiv a^+(\mathbf{r}_\parallel, 0, t) - a^-(\mathbf{r}_\parallel, 0, t) \tag{3.19}$$

Using this short hand notation eq.(3.17) may be written in the form

$$\frac{\partial}{\partial t}\rho^s(\mathbf{r}_\parallel, t) = -\nabla_\parallel . \mathbf{I}^s(\mathbf{r}_\parallel, t) - I_{z,-}(\mathbf{r}_\parallel, t) \tag{3.20}$$

The jump in the electric current at the dividing surface $I_{z,-}$ thus gives the net flow of charge away from the interfacial layer. If $I_{z,-}$ is negative this implies that there is a net flow from the bulk phases into the interfacial layer.

3.5 Generalized electric displacement field

In time dependent problems it is convenient to define a generalized electric displacement field. This generalized displacement field is an appropriately chosen combination of the usual electric displacement field and the electric current density [20]. Not only is this procedure commonly accepted as convenient for time dependent problems but it has an additional advantage in the description of surfaces. The reason is that in the equations given above it is necessary to take the normal components of the excesses of both the electric displacement field and the electric current density equal to zero. If one introduces the generalized electric displacement field it is only necessary to take the normal component of the excess of this field equal to zero. This is less restrictive. An important example of a system for which this difference is crucial is the electric double layer in an electrolyte. In that case the excess of the normal component of the generalized electric displacement field is zero whereas the excesses of both the normal component of the electric current and of the electric displacement field may be unequal to zero.

In order to define the generalized electric displacement field the Fourier transform of the various quantities with respect to the time is introduced. For the displacement field this Fourier transform is for instance defined by

$$\mathbf{D}(\mathbf{r}, \omega) \equiv \int_{-\infty}^{\infty} dt \; e^{i\omega t} \mathbf{D}(\mathbf{r}, t) \tag{3.21}$$

The definition for the other quantities is similar.

The Maxwell equations for the position and frequency dependent fields are found upon Fourier transformation of the equations given in eq.(3.6). This gives

$$\begin{aligned} \operatorname{rot} \mathbf{E}(\mathbf{r}, \omega) &= i\frac{\omega}{c}\mathbf{B}(\mathbf{r}, \omega) \,, & \operatorname{div} \mathbf{D}(\mathbf{r}, \omega) = \rho(\mathbf{r}, \omega) \\ \operatorname{rot} \mathbf{H}(\mathbf{r}, \omega) &= \frac{1}{c}\mathbf{I}(\mathbf{r}, \omega) - i\frac{\omega}{c}\mathbf{D}(\mathbf{r}, \omega) \,, & \operatorname{div} \mathbf{B}(\mathbf{r}, \omega) = 0 \end{aligned} \tag{3.22}$$

The generalized electric displacement field is now defined by

$$\mathbf{D}'(\mathbf{r},\omega) \equiv \mathbf{D}(\mathbf{r},\omega) + \frac{i}{\omega}\mathbf{I}(\mathbf{r},\omega) \tag{3.23}$$

In terms of this generalized displacement field the Maxwell equations (3.22) become

$$\begin{aligned} \text{rot}\,\mathbf{E}(\mathbf{r},\omega) &= i\frac{\omega}{c}\mathbf{B}(\mathbf{r},\omega)\ , \quad \text{div}\,\mathbf{D}'(\mathbf{r},\omega) = 0 \\ \text{rot}\,\mathbf{H}(\mathbf{r},\omega) &= -i\frac{\omega}{c}\mathbf{D}'(\mathbf{r},\omega)\ , \quad \text{div}\,\mathbf{B}(\mathbf{r},\omega) = 0 \end{aligned} \tag{3.24}$$

The second Maxwell equation was obtained in this form by using conservation of charge which is given in the position and frequency representation by

$$i\omega\ \rho(\mathbf{r},\omega) = \text{div}\,\mathbf{I}(\mathbf{r},\omega) \tag{3.25}$$

In the rest of the book the electric current will always be absorbed in this manner in the electric displacement field. This is in practice always convenient with the notable exception of time independent problems. As this book primarily considers optical problems this is no limitation. It should be noted that one may also absorb the magnetization density in the polarization density. This would result in a description in which the magnetic permeability is part of the dielectric constant. This will not be done in this book.

The restrictions on the singular behavior of the fields now become

$$\begin{aligned} \mathbf{E}^s_{\parallel}(\mathbf{r}_{\parallel},\omega) &= 0, \quad D'^s_z(\mathbf{r}_{\parallel},\omega) = D^s_z(\mathbf{r}_{\parallel},\omega) + \frac{i}{\omega}I^s_z(\mathbf{r}_{\parallel},\omega) = 0 \\ \mathbf{H}^s_{\parallel}(\mathbf{r}_{\parallel},\omega) &= 0, \quad B^s_z(\mathbf{r}_{\parallel},\omega) = 0 \end{aligned} \tag{3.26}$$

It should be stressed again that, if one uses the generalized electric displacement field, it is sufficient to take the normal component D'^s_z of its excess equal to zero. It is no longer necessary to take both D^s_z and I^s_z separately equal to zero!

The definition of the interfacial polarization and magnetization densities now becomes

$$\mathbf{P}'^s(\mathbf{r}_{\parallel},\omega) \equiv (\mathbf{D}'^s_{\parallel}(\mathbf{r}_{\parallel},\omega), -E^s_z(\mathbf{r}_{\parallel},\omega))\ ,\ \mathbf{M}^s(\mathbf{r}_{\parallel},\omega) \equiv (\mathbf{B}^s_{\parallel}(\mathbf{r}_{\parallel},\omega), -H^s_z(\mathbf{r}_{\parallel},\omega)) \tag{3.27}$$

Using eq.(3.26) it follows that the interfacial polarization and magnetization densities are again also given by

$$\mathbf{P}'^s(\mathbf{r}_{\parallel},\omega) \equiv \mathbf{D}'^s(\mathbf{r}_{\parallel},\omega) - \mathbf{E}^s(\mathbf{r}_{\parallel},\omega), \quad \mathbf{M}^s(\mathbf{r}_{\parallel},\omega) \equiv \mathbf{B}^s(\mathbf{r}_{\parallel},\omega) - \mathbf{H}^s(\mathbf{r}_{\parallel},\omega) \tag{3.28}$$

It should again be emphasized that eq.(3.27) is really the proper definition of the interfacial polarization and magnetization densities rather than eq.(3.28). See the last section of the previous chapter for a more extended discussion of this point.

As in section 3.3 one may again substitute the singular form of the fields into the Maxwell equations given in eq.(3.24). One then obtains for the fields in the bulk regions

$$\begin{aligned} \text{rot}\,\mathbf{E}^{\pm}(\mathbf{r},\omega) &= i\frac{\omega}{c}\mathbf{B}^{\pm}(\mathbf{r},\omega), \quad \text{div}\,\mathbf{D}'^{\pm}(\mathbf{r},\omega) = 0 \\ \text{rot}\,\mathbf{H}^{\pm}(\mathbf{r},\omega) &= -i\frac{\omega}{c}\mathbf{D}'^{\pm}(\mathbf{r},\omega), \quad \text{div}\,\mathbf{B}^{\pm}(\mathbf{r},\omega) = 0 \ \text{ for }\ 0 < \pm z \end{aligned} \tag{3.29}$$

On the dividing surface one obtains in this way the following boundary conditions for the fields

$$\begin{aligned}
\mathbf{E}_{\|,-}(\mathbf{r}_\|,0,\omega) &\equiv \mathbf{E}_\|^+(\mathbf{r}_\|,0,\omega)-\mathbf{E}_\|^-(\mathbf{r}_\|,0,\omega)=-i\frac{\omega}{c}\hat{\mathbf{z}}\times\mathbf{M}_\|^s(\mathbf{r}_\|,\omega)-\nabla_\| P_z^{\prime s}(\mathbf{r}_\|,\omega)\\
D'_{z,-}(\mathbf{r}_\|,0,\omega) &\equiv D_z^{\prime +}(\mathbf{r}_\|,0,\omega)-D_z^{\prime -}(\mathbf{r}_\|,0,\omega)=-\nabla_\|.\mathbf{P}_\|^{\prime s}(\mathbf{r}_\|,\omega)\\
\mathbf{H}_{\|,-}(\mathbf{r}_\|,0,\omega) &\equiv \mathbf{H}_\|^+(\mathbf{r}_\|,0,\omega)-\mathbf{H}_\|^-(\mathbf{r}_\|,0,\omega)=i\frac{\omega}{c}\hat{\mathbf{z}}\times\mathbf{P}_\|^{\prime s}(\mathbf{r}_\|,\omega)-\nabla_\| M_z^s(\mathbf{r}_\|,\omega)\\
B_{z,-}(\mathbf{r}_\|,0,\omega) &\equiv B_z^+(\mathbf{r}_\|,0,\omega)-B_z^-(\mathbf{r}_\|,0,\omega)=-\nabla_\|.\mathbf{M}_\|^s(\mathbf{r}_\|,\omega) \qquad (3.30)
\end{aligned}$$

These equations, containing the generalized electric displacement field, are most simply obtained from the corresponding equations given in section 3.3 by replacing $\mathbf{D}$ by $\mathbf{D}'$, $\mathbf{P}^s$ by $\mathbf{P}'^s$ and setting ρ and $\mathbf{I}$ equal to zero. From this point on the description will always be in terms of the generalized electric displacement field. To simplify matters this generalized electric displacement field will from now on simply be called the electric displacement field and the prime as a superscript of this field, and of the corresponding polarization density, will be further dropped.

In the bulk regions the constitutive relations for the generalized displacement field and for the magnetic induction are

$$\mathbf{D}^\pm(\mathbf{r},\omega)=\epsilon^\pm(\omega)\mathbf{E}^\pm(\mathbf{r},\omega) \quad \text{and} \quad \mathbf{B}^\pm(\mathbf{r},\omega)=\mu^\pm(\omega)\mathbf{H}^\pm(\mathbf{r},\omega) \qquad (3.31)$$

It should again be emphasized that $\mathbf{D}^\pm$ and $\epsilon^\pm$ contain the contributions due to possible conductivities $\sigma^\pm$ of the bulk media even though this is no longer indicated explicitly by a prime. The coefficients $\epsilon^\pm(\omega)$ and $\mu^\pm(\omega)$ are the frequency dependent dielectric permeability and the magnetic permeability respectively. Using these constitutive equations one may derive the wave equation for the electric field by taking the rotation of the first Maxwell equation. This results in

$$\Delta\mathbf{E}^\pm(\mathbf{r},\omega)=(\frac{\partial^2}{\partial x^2}+\frac{\partial^2}{\partial y^2}+\frac{\partial^2}{\partial z^2})\mathbf{E}^\pm(\mathbf{r},\omega)=-(\frac{\omega}{c})^2\epsilon^\pm(\omega)\mu^\pm(\omega)\mathbf{E}^\pm(\mathbf{r},\omega) \qquad (3.32)$$

where Δ is the Laplace operator. Notice the fact the other bulk fields satisfy the same wave equation.

3.6 Constitutive relations for isotropic interfaces without dispersion

In order to make practical use of the boundary conditions for the fields one needs constitutive equations which express the interfacial polarization and magnetization densities in terms of the extrapolated fields in the adjacent bulk phases. It is noted that the electric and the magnetic fields have a different symmetry for time reversal. This implies that, in the absence of spatial dispersion, the polarization may be chosen in such a way that it couples only to the electric field. Similarly the magnetization may be chosen in such a way that it only couples to the magnetic field. First consider the polarization. There are three electric field components on both sides of the surface and one could in principle express the excess polarization in all these six fields. One could in fact also use the bulk displacement fields rather than the electric fields. Since they are related to the bulk electric fields by the constitutive relations in the bulk this does not represent an independent choice. It is possible to convince oneself that only three independent fields are necessary. The reason for this is that the jump of the parallel components of the bulk electric fields and of the normal component of the bulk displacement fields can themselves be expressed, using the boundary conditions, in combinations of the excess fields. One may therefore always express the excess polarization in terms of the averages of the bulk fields at the surface. It is most convenient in this context to use the averages of the parallel components of the bulk electric fields and of the normal components of the displacement fields. This was discussed in detail in section 2.3, cf. also eq.(2.43). The most general constitutive relation [21] in the absence of spatial dispersion is therefore given by

$$\mathbf{P}^s(\mathbf{r}_\parallel,\omega) = \boldsymbol{\xi}_e^s(\omega).(\mathbf{E}_{\parallel,+}(\mathbf{r}_\parallel,\omega), D_{z,+}(\mathbf{r}_\parallel,\omega)) \tag{3.33}$$

The subindex + denotes the average of the extrapolated values of the corresponding bulk field at the dividing surface. For an arbitrary field a this average is defined by

$$a_+(\mathbf{r}_\parallel,\omega) \equiv \frac{1}{2}[a^-(\mathbf{r}_\parallel, z=0,\omega) + a^+(\mathbf{r}_\parallel, z=0,\omega)] \tag{3.34}$$

The constitutive relations for the excesses of the magnetic fields may in the absence of spatial dispersion similarly be given by

$$\mathbf{M}^s(\mathbf{r}_\parallel,\omega) = \boldsymbol{\xi}_m^s(\omega).(\mathbf{H}_{\parallel,+}(\mathbf{r}_\parallel,\omega), B_{z,+}(\mathbf{r}_\parallel,\omega)) \tag{3.35}$$

In the above constitutive equations no assumption was made regarding the nature of the surface. The constitutive coefficients for the interface will, in view of the source observer symmetry, [12], be symmetric matrices. The constitutive relations should be used in the boundary conditions for the fields at the surface. For an interface between gyrotropic media one needs off-diagonal elements in the above interfacial constitutive matrices for a proper description of the possible boundary conditions. This extension will be discussed in chapter 15.

In this book the surface will always be assumed to be isotropic. This isotropy is a two-dimensional invariance of the interface with respect to translation, rotation and reflection in the $x-y$ plane. This isotropy implies for instance that two-dimensional vectors like $\mathbf{D}_\parallel^s$ and $\mathbf{B}_\parallel^s$ do not couple to two-dimensional scalars like $D_{z,+}$ and $B_{z,+}$. In that case the constitutive matrices reduce to the following form

$$\begin{aligned} \boldsymbol{\xi}_e^s(\omega) &= \gamma_e(\omega)(\hat{\mathbf{x}}\hat{\mathbf{x}} + \hat{\mathbf{y}}\hat{\mathbf{y}}) + \beta_e(\omega)\hat{\mathbf{z}}\hat{\mathbf{z}} && (3.36)\\ \boldsymbol{\xi}_m^s(\omega) &= \gamma_m(\omega)(\hat{\mathbf{x}}\hat{\mathbf{x}} + \hat{\mathbf{y}}\hat{\mathbf{y}}) + \beta_m(\omega)\hat{\mathbf{z}}\hat{\mathbf{z}} && (3.37)\end{aligned}$$

where $\hat{\mathbf{x}}$, $\hat{\mathbf{y}}$ and $\hat{\mathbf{z}}$ are the unit vectors in the x, y and z directions respectively. Substitution in the above constitutive relations results in

$$\mathbf{P}_\parallel^s(\mathbf{r}_\parallel,\omega) = \gamma_e(\omega)\mathbf{E}_{\parallel,+}(\mathbf{r}_\parallel,\omega) \tag{3.38}$$

$$P_z^s(\mathbf{r}_\parallel,\omega) = \beta_e(\omega)D_{z,+}(\mathbf{r}_\parallel,\omega) \tag{3.39}$$

for the polarization, and

$$\mathbf{M}_{\parallel}^{s}(\mathbf{r}_{\parallel},\omega)=\gamma_{m}(\omega)\mathbf{H}_{\parallel,+}(\mathbf{r}_{\parallel},\omega) \tag{3.40}$$

$$M_{z}^{s}(\mathbf{r}_{\parallel},\omega)=\beta_{m}(\omega)B_{z,+}(\mathbf{r}_{\parallel},\omega) \tag{3.41}$$

for the magnetization. In section 2.3 the electric constitutive coefficients γ_e and β_e where calculated for the special case of a thin plan parallel layer. While this is, of course, a very simple example it does illustrate one general property of these coefficients. They have the dimensionality of a length. This length could be interpreted in terms of an effective optical thickness as measured, for instance, using ellipsometry. How this effective optical thickness is related to the size of the islands and the distance between the islands for an island film, or to the mean square height for a rough surface, or whatever length scale involved, will be discussed in the following chapters.

3.7 Constitutive relations for isotropic interfaces with spatial dispersion

If a surface is spatially dispersive the polarization and the magnetization are no longer related to the fields in a local manner. The reason for this non-local nature of the response finds its origin in the heterogeneous nature of the interface. Polarization and magnetization are distributed over a thin layer which may contain discrete islands and may be rough. In the description using an equivalent polarization and magnetization density on a sharply defined dividing surface the whole distribution of polarization and magnetization is shifted to the dividing surface. The shifted polarization and magnetization densities are, however, coupled to the field at the original position. This naturally leads to a non-local dependence on the field. In view of the fact that the shift is over a distance of the order of the thickness of the layer, and of the fact that the bulk fields vary over a distance of the order of a wavelength, the size of the non-local contribution is of a relative order of the thickness divided by the wavelength.

One of the reasons to take the effect of the non-local nature of the interfacial coefficients into account is related to the choice of the location of the dividing surface, [22], [23], [24], [25]. As the coefficients, which will characterize this effect, find their origin in shifting polarization and magnetization densities to the dividing surface, it is evident that they will sensitively depend on this choice. As the optical properties do not depend on this choice they can be expressed in invariant combinations of these coefficients. In order to fully understand the nature of these invariants the introduction of coefficients describing the non-local nature of the response functions is crucial. In the following sections the dependence of the various constitutive coefficients on the location of the dividing surface and the choice of these invariants will be discussed in detail.

The discussion will be restricted to isotropic surfaces. Consider first the polarization. The parallel component of the excess polarization may for that case also have a contributions proportional to the parallel derivative of the normal component of the average bulk displacement field and a contribution proportional to the normal derivative of the average parallel electric field. For an isotropic surface these spatial derivatives are the only ones with the proper symmetry. As it is most convenient to have an interfacial response function which only depends on the wave vector along the surface it is preferable to rewrite the term with the normal derivative. Using the first Maxwell equation, cf. eq.(3.24), and the constitutive relations in the bulk for the bulk fields one may write this term as a sum of a term proportional to the parallel derivative of the average of the normal displacement field which can be combined with the other term, and of a term proportional to the time derivative of the average of the parallel component of the magnetic field. Notice that terms containing jumps of the fields at the surface can as usual be eliminated using the boundary conditions. The normal component of the excess polarization may similarly have a contribution proportional to the normal derivative of the average of the normal component of the bulk displacement field which has the right symmetry. This normal derivative may be rewritten as the parallel divergence of the average of the parallel component of the bulk electric field. For the magnetization one may go through the analogous argumentation. As a result one finds the following constitutive equations to first order in the non-locality

$$\mathbf{P}_{\parallel}^{s}(\mathbf{r}_{\parallel},\omega)=\gamma_{e}(\omega)\mathbf{E}_{\parallel,+}(\mathbf{r}_{\parallel},\omega)-\delta_{e}(\omega)\nabla_{\parallel}D_{z,+}(\mathbf{r}_{\parallel},\omega)+i\frac{\omega}{c}\tau(\omega)\hat{\mathbf{z}}\times\mathbf{H}_{\parallel,+}(\mathbf{r}_{\parallel},\omega) \quad (3.42)$$

$$P_{z}^{s}(\mathbf{r}_{\parallel},\omega)=\beta_{e}(\omega)D_{z,+}(\mathbf{r}_{\parallel},\omega)+\delta_{e}(\omega)\nabla_{\parallel}.\mathbf{E}_{\parallel,+}(\mathbf{r}_{\parallel},\omega) \quad (3.43)$$

$$\mathbf{M}_{\parallel}^{s}(\mathbf{r}_{\parallel},\omega)=\gamma_{m}(\omega)\mathbf{H}_{\parallel,+}(\mathbf{r}_{\parallel},\omega)-\delta_{m}(\omega)\nabla_{\parallel}B_{z,+}(\mathbf{r}_{\parallel},\omega)+i\frac{\omega}{c}\tau(\omega)\hat{\mathbf{z}}\times\mathbf{E}_{\parallel,+}(\mathbf{r}_{\parallel},\omega) \quad (3.44)$$

$$M_{z}^{s}(\mathbf{r}_{\parallel},\omega)=\beta_{m}(\omega)B_{z,+}(\mathbf{r}_{\parallel},\omega)+\delta_{m}(\omega)\nabla_{\parallel}.\mathbf{H}_{\parallel,+}(\mathbf{r}_{\parallel},\omega) \quad (3.45)$$

In addition to the above motivation of the constitutive equations one may consider the dependence of the constitutive coefficients on the location of the dividing surface. This is done in the following section and in particular also in appendix A. This analysis further clarifies the origin of the dispersive constitutive coefficients δ_e , δ_m and τ. In the rest of the book the frequency dependence of the constitutive coefficients will usually not be indicated explicitly. Unless indicated otherwise this dependence is always implicitly present.

In the above constitutive equations there appear only seven independent constitutive coefficients. Three of these coefficients, γ_e , β_e and δ_e, are of electric origin, three of these coefficients, γ_m , β_m and δ_m, are of magnetic origin and one coefficient τ couples the electric to the magnetic field and vice versa. The reason that there are only seven independent coefficients is that the system is isotropic and on the microscopic level time reversal invariant. The fact that the same coefficient δ_e appears in eqs. (3.42) and (3.43) is a consequence of the source observer symmetry. The same is true for δ_m in eqs. (3.44) and (3.45). The fact that only one τ appears in the constitutive equations is also due to the source observer symmetry. An in depth discussion of these source observer symmetry relations, [12], is given in chapter 13.

The coefficients γ_e, β_e, γ_m and β_m give contributions to the excess electric and the magnetic dipole polarizations in terms of homogeneous fields. They have the dimensionality of a length (in c.g.s. units). The coefficients describing the dispersion δ_e, δ_m and τ give contributions to the excess electric and the magnetic dipole polarizations in terms of derivatives of the fields. They have the dimensionality of a length squared. Their influence on the optical properties will therefore usually be a factor optical thickness divided by the wavelength smaller than the other coefficients. As will become clear in the chapters on island films and also in the chapter on rough surfaces, these coefficients play, for certain optical properties, a more important role than one would expect on the basis of this estimate. In the rest of the book it will be often used that if a dipole moment density can be caused by field gradients that, as a consequence of source observer symmetry, a quadrupole moment density will be caused by a homogeneous field. The constitutive coefficients of the two phenomena can be expressed in each other using source observer symmetry. This will turn out to be very convenient in the evaluation of these coefficients.

If one considers a given surface, one has two possibilities to choose the direction of the z-axis. If one knows the constitutive coefficients for one choice, one finds the coefficients for the other choice by taking the same values for the non-dispersive coefficients γ_e , β_e , γ_m and β_m and by changing the sign of the dispersive coefficients δ_e , δ_m and τ. This may be verified, using the above constitutive equations, by inverting the direction of the x- and the z-axis. In order to keep the coordinate system right-handed, one must invert the direction of two axes.

3.8 Dependence of the constitutive coefficients on the location of the dividing surface

In the definition of excess quantities discussed in the previous section one introduces the concept of a dividing surface. The precise choice of its location in the interfacial region is somewhat arbitrary and has no physical significance. Measurable quantities will not depend on the special choice of the dividing surface. To verify this independence explicitly, the dependence of the constitutive coefficients on this choice will now be given [22], [23]. This will make it possible to define invariant combinations of the constitutive coefficients which are independent of the choice of the dividing surface. This will be done in the next section. The experimental results may always be expressed in terms of these invariants alone and consequently these invariants contain the real physical characteristics of the surface.

Consider, in order to clarify this matter in some more detail, the stratified dielectric boundary layer discussed in section 2.3. In this section the dividing surface was chosen in $z = 0$, cf. in this context in particular eq.(2.24). If the dividing surface is chosen in $z = d$ this expression for the total excess dielectric constant becomes

$$\gamma_e(d) \equiv \int_{-\infty}^{\infty} [\epsilon(z) - \epsilon^{-}\theta(d - z) - \epsilon^{+}\theta(z - d)]dz \tag{3.46}$$

It follows from this definition that

$$\gamma_e(d) = \gamma_e(0) + (\epsilon^{+} - \epsilon^{-})d \tag{3.47}$$

It may similarly be shown that, cf. eq.(2.30),

$$\beta_e(d) = \beta_e(0) - [(\epsilon^{+})^{-1} - (\epsilon^{-})^{-1}]d \tag{3.48}$$

For this example it is now also possible to define the following invariant

$$I_e \equiv \gamma_e(d) - \epsilon^-\epsilon^+\beta_e(d) \tag{3.49}$$

For the stratified dielectric medium discussed in section 2.3 this invariant becomes

$$I_e = \int_{-\infty}^{\infty} [\epsilon(z) - \epsilon^-][\epsilon(z) - \epsilon^+]/\epsilon(z)dz \tag{3.50}$$

As will be discussed in detail in section 4.8 one finds this invariant if one measures the ellipsometric coefficient.

For the general case it is also possible to give the explicit dependence of the interfacial constitutive coefficients on the position of the dividing surface. In appendix A it is shown that if the dividing surface is chosen in $z = d$, where d may be positive or negative, the dependence is given, to second order in d, by:

$$\begin{aligned}
\gamma_e(d) &= \gamma_e(0) + (\epsilon^+ - \epsilon^-)d = \gamma_e(0) + \epsilon_- d \\
\gamma_m(d) &= \gamma_m(0) + (\mu^+ - \mu^-)d = \gamma_m(0) + \mu_- d \\
\beta_e(d) &= \beta_e(0) - [(\epsilon^+)^{-1} - (\epsilon^-)^{-1}]d = \beta_e(0) - (1/\epsilon)_- d \\
\beta_m(d) &= \beta_m(0) - [(\mu^+)^{-1} - (\mu^-)^{-1}]d = \beta_m(0) - (1/\mu)_- d \\
\delta_e(d) &= \delta_e(0) + \gamma_e(0)(1/\epsilon)_+ d + \beta_e(0)\epsilon_+ d - \epsilon_+(1/\epsilon)_- d^2 \\
\delta_m(d) &= \delta_m(0) + \gamma_m(0)(1/\mu)_+ d + \beta_m(0)\mu_+ d - \mu_+(1/\mu)_- d^2 \\
\tau(d) &= \tau(0) + \gamma_e(0)\mu_+ d - \gamma_m(0)\epsilon_+ d + \frac{1}{2}(\mu_+\epsilon_- - \mu_-\epsilon_+)d^2
\end{aligned} \tag{3.51}$$

The proof of these relations is given for the general case where all constitutive coefficients for the surface as well as for the bulk regions may be complex functions of the frequency. The reason to give these relation only to second order in d is the fact that the only natural choice is near or within the interfacial region between the two bulk phases. As the constitutive coefficients themselves are at most of the second order in the interfacial thickness, it would not be consistent to go to a higher order in d. Inspection of the above equations shows that they are (anti)symmetric for the interchange of the electric and the magnetic quantities. This property is a consequence of the symmetry of the Maxwell equations (3.24) and the resulting symmetry of the boundary conditions (3.30) for the interchange of the electric and the magnetic fields. They remain unchanged, if one interchanges the electric with the magnetic field and the electric displacement field with minus the magnetic induction. Together with the usual constitutive equations in the bulk and the constitutive equations, eqs.(3.42) through (3.45), at the interface this interchange is equivalent with simultaneously interchanging $\epsilon^\pm$ with $-\mu^\pm$, γ_e with $-\gamma_m$, β_e with $-\beta_m$, δ_e with δ_m and replacing τ by $-\tau$. As one may easily verify this interchange, when performed in the expressions given in eq.(3.51), yields the same set of formulae.

In the applications it is sometimes convenient to shift the location of the dividing surface over a distance d, where d may again be positive or negative, through a homogeneous medium. As an example one could for instance consider a soap film, where one could alternatively take the dividing surface in the center of the film, or at the outer surfaces. In that case one has $\epsilon^- = \epsilon^+ = \epsilon$ and $\mu^- = \mu^+ = \mu$. The formulae in equation (3.51) then reduce to

$$\begin{aligned}
\gamma_e(d) &= \gamma_e(0), \quad \gamma_m(d) = \gamma_m(0) \\
\beta_e(d) &= \beta_e(0), \quad \beta_m(d) = \beta_m(0) \\
\delta_e(d) &= \delta_e(0) + \gamma_e(0)d/\epsilon + \beta_e(0)d\ \epsilon \\
\delta_m(d) &= \delta_m(0) + \gamma_m(0)d/\mu + \beta_m(0)d\ \mu \\
\tau(d) &= \tau(0) + \gamma_e(0)d\ \mu - \gamma_m(0)d\ \epsilon
\end{aligned} \tag{3.52}$$

The above formulae show clearly that the introduction of the second order coefficients δ_e, δ_m and τ is intimately related to the choice of the position of the dividing surface.

3.9 Invariants

Using the dependence of the constitutive coefficients on the choice of the location of the dividing surface given in the previous section, it is now possible to define combinations of these constitutive coefficients independent of this choice, [22], [23], [24], [25]. It should be noted in this context, that the frequency dependence of the constitutive coefficients implies that the invariant combinations also depend on the frequency. For ease of notation this will not be explicitly indicated in their definitions. In the special example discussed there such an invariant combination of γ_e and β_e was in fact already given, cf. eq.(3.49). Experimental consequences derived, using the general method discussed in this book, will only depend on these invariant combinations. The following invariant combinations of the linear coefficients are found

$$\begin{aligned}
I_e &\equiv \gamma_e(d) - \epsilon^+\epsilon^-\beta_e(d) \\
I_m &\equiv \gamma_m(d) - \mu^+\mu^-\beta_m(d) \\
I_{em} &\equiv (\mu^+ - \mu^-)\gamma_e(d) - (\epsilon^+ - \epsilon^-)\gamma_m(d)
\end{aligned} \tag{3.53}$$

The first of these invariants is the one given already for the example in the previous chapter. It is a combination of the electric constitutive coefficients alone. The second one is the analogous magnetic combination. The third is of mixed electric and magnetic origin. Note that one could also define a mixed invariant in terms of β_e and β_m. This is, however, not an additional invariant combination since it may be written as $I_{em} - \mu_- I_e + \epsilon_- I_m$. For a non-magnetic system only the first invariant I_e is unequal to zero. As a consequence of the symmetry of the Maxwell equations (3.24) for the interchange of the electric and the magnetic fields, the invariants can, and have been chosen such, that they transform into plus or minus each other or themselves. If one interchanges $\epsilon^\pm$ with $-\mu^\pm$, γ_e with $-\gamma_m$ and β_e with $-\beta_m$, I_e interchanges with $-I_m$ and I_{em} with $-I_{em}$.

Invariant combinations containing the second order constitutive coefficients are

$$I_{\delta,e} \equiv \delta_e(d) - \frac{1}{2}\left[\frac{\epsilon^+ + \epsilon^-}{\epsilon^+ - \epsilon^-}\right]\gamma_e(d)\beta_e(d)$$

$$I_{\delta,m} \equiv \delta_m(d) - \frac{1}{2}\left[\frac{\mu^+ + \mu^-}{\mu^+ - \mu^-}\right]\gamma_m(d)\beta_m(d)$$

$$I_\tau \equiv \tau(d) - \frac{1}{4}\left[\frac{\mu^+ + \mu^-}{\epsilon^+ - \epsilon^-}\right]\gamma_e^2(d) + \frac{1}{4}\left[\frac{\epsilon^+ + \epsilon^-}{\mu^+ - \mu^-}\right]\gamma_m^2(d) \tag{3.54}$$

It is important to realize that no second order invariants can be defined purely in terms of the second order constitutive coefficients. As on the linear level, one has three invariants, which are of an electric, a magnetic and a mixed origin respectively. In this case, however, both $I_{\delta,e}$ and I_τ remain unequal to zero in a non-magnetic system.

As the 7 interfacial constitutive coefficients are in general complex functions of the frequency, one has 14 real functions of the frequency, which describe a surface. The 6 invariants given above are also complex functions of the frequency, and the real and imaginary parts therefore give 12 real functions of the frequency, which represent measurable quantities, independent of the choice of the dividing surface, and which may be found experimentally. A common practice in choosing the location of the dividing surface is to take one of the constitutive coefficients zero. If the constitutive coefficients are complex one may take either the real or the imaginary part zero. It must therefore be possible to give one more invariant which is a combination of the real and the imaginary part of the constitutive coefficients. The following combination gives such an invariant

$$I_c \equiv \mathrm{Im}\left(\frac{(\mu^+ + \mu^-)\gamma_e(d) + (\epsilon^+ + \epsilon^-)\gamma_m(d)}{2(\epsilon^+\mu^+ - \epsilon^-\mu^-)}\right) \tag{3.55}$$

Im indicates the imaginary part of the corresponding quantity. Notice the fact that the first order constitutive coefficient I_c is by definition always a real function of the frequency unlike the other invariants which are complex functions of the frequency. It should be realized that other choices of the invariant combinations are possible. These alternative choices are, however, always combinations of the above invariants. The particular choices made above reflect the symmetry of the Maxwell equations (3.24) for the interchange of the electric and the magnetic fields. They have been chosen such that they transform into plus or minus each other or themselves if one interchanges $\epsilon^\pm$ with $-\mu^\pm$, γ_e with $-\gamma_m$ and β_e with $-\beta_m$, δ_e with δ_m and τ with $-\tau$. It follows from the above equations that $I_{\delta,e}$ interchanges with $I_{\delta,m}$, I_τ with $-I_\tau$ and I_c with I_c.

For any given interface one may choose the direction of the z-axis in two different ways. If one goes from one choice to the other one finds that I_e and I_m remain the same while all the other invariants change sign. This follows from the above expressions for the invariants and using the fact that δ_e , δ_m and τ as well as $(\epsilon^+ - \epsilon^-)$ and $(\mu^+ - \mu^-)$ also change sign.

A case which deserves special attention is where not only the ambient and the substrate, but also the interface are non-absorbing. In that case the bulk as well as the interfacial constitutive coefficients are real. Consequently the imaginary parts of the invariants I_e, I_m, I_{em}, $I_{\delta,e}$, $I_{\delta,m}$ and I_τ, as well as I_c, are zero, so that there are only 6 real invariant combinations of the 7 real interfacial constitutive coefficients in that case.

In this book the emphasis is on systems which are non-magnetic, i.e. $\mu^{\pm} = 1$ and $\gamma_m = \beta_m = \delta_m = 0$. In that case the number of non-zero complex interfacial constitutive coefficients reduces to 4 and there are 3 non-zero complex invariants. The complete set of invariants is in this case therefore given by

$$\begin{aligned}
I_e &\equiv \gamma_e(d) - \epsilon^+\epsilon^-\beta_e(d) \\
I_{\delta,e} &\equiv \delta_e(d) - \frac{1}{2}\left[\frac{\epsilon^+ + \epsilon^-}{\epsilon^+ - \epsilon^-}\right]\gamma_e(d)\beta_e(d) \\
I_\tau &\equiv \tau(d) - \frac{1}{2}(\epsilon^+ - \epsilon^-)^{-1}\gamma_e^2(d) \\
I_c &\equiv \mathrm{Im}[\gamma_e(d)/(\epsilon^+ - \epsilon^-)]
\end{aligned} \tag{3.56}$$

In the non-magnetic case one thus has 7 invariant real combinations of the 8 real and imaginary parts of the interfacial constitutive coefficients. If one considers again the case that both the ambient, the substrate and the interface are non-absorbing, only the real parts of I_e, $I_{\delta,e}$ and I_τ are unequal to zero, so that there are only 3 invariant real combinations of the 4 non-zero real constitutive coefficients.

There is one other case which deserves special attention. If one matches the dielectric constants and the magnetic permeabilities of the bulk phases, $\epsilon^- = \epsilon^+ \equiv \epsilon$ and $\mu^- = \mu^+ \equiv \mu$, the invariants $I_{\delta,e}$, $I_{\delta,m}$, I_τ and I_c all become infinite. It is clear that the invariants given above are not appropriate. For this case the center of the islands is a natural choice of the dividing surface and the introduction of invariants is unnecessary.

3.10 Shifting the dividing surface for a film embedded in a homogeneous medium

For most boundary layers the bulk media away from the interface are different. In the calculation of the interfacial constitutive coefficients the location of the dividing surface is then chosen in a somewhat arbitrary manner in the transition layer between these two bulk media. As discussed in the previous section the resulting constitutive coefficients depend on this choice and it becomes convenient to introduce combinations which are invariant for this choice. For more complex transition layers one often has more than one "natural" choice of the location of the dividing surface. An example, which will be discussed in great detail in the next section, is an island film. In that case one obvious choice is the surface of the substrate. The islands are lying on the surface of the substrate surrounded by the ambient medium. As the polarization of the islands is most logically located in their center a second natural choice of the location of the dividing surface would be at a distance equal to the average distance of these island to the substrate. Of course both choices yield the same invariants and as such there is no real problem.

The reason to address this point, however, is of a more practical nature. In the evaluation of these invariants it is often convenient to calculate them in a number of steps. In the first step one introduces two dividing surfaces rather then one. One is the surface of the substrate where there are no excess fields, so that for that surface the constitutive coefficients are zero. The second dividing surface is located at the average distance of the islands to the substrate. The polarization and magnetization of the islands then give the constitutive coefficients for the second dividing surface. This dividing surface is embedded in the homogeneous ambient medium. Between the two dividing surfaces one has the ambient medium. One could in fact even consider a film with several layers of islands. This would result in one dividing surface for the substrate and one for each layer. Between all these dividing surfaces one has ambient medium. Of course it is not the intention to finally analyze the properties of this system using this sequence of dividing surfaces. The aim is to find the combined effect of the whole sequence. The method to obtain the constitutive coefficients describing this combined effect is to first shift the location of the dividing surfaces in the direction of the surface of the substrate until they are separated by not more than an infinitesimal layer of ambient. In this section it will be explained how this should be done. In the next section it will then be explained, how one superposes the constitutive coefficients of dividing surfaces separated by an infinitesimal distance.

In eq.(3.52) formulae were given for the change of the constitutive coefficients due to a shift of the location of the dividing surface over a distance d through a homogeneous medium. These formulae may be used to shift the position of the dividing surface from the position d_1 relative to the substrate to a distance d_2. In this case one has $\epsilon^- = \epsilon^+ = \epsilon$ and $\mu^- = \mu^+ = \mu$. The formulae in equation (3.52) then give

$$
\begin{aligned}
\gamma_e(d_2) &= \gamma_e(d_1), \quad \gamma_m(d_2) = \gamma_m(d_1) \\
\beta_e(d_2) &= \beta_e(d_1), \quad \beta_m(d_2) = \beta_m(d_1) \\
\delta_e(d_2) &= \delta_e(d_1) + \gamma_e(d_1)(d_2 - d_1)/\epsilon + \beta_e(d_1)(d_2 - d_1)\epsilon \\
\delta_m(d_2) &= \delta_m(d_1) + \gamma_m(d_1)(d_2 - d_1)/\mu + \beta_m(d_1)(d_2 - d_1)\mu \\
\tau(d_2) &= \tau(d_1) + \gamma_e(d_1)(d_2 - d_1)\mu - \gamma_m(d_1)(d_2 - d_1)\epsilon
\end{aligned}
\tag{3.57}
$$

If one chooses the $x - y$ plane along the surface of the substrate and takes d_2 infinitesimally small one may use the superposition formulae given in the next section to calculate the constitutive coefficients of the combined surface. If one has more dividing surfaces one may shift them all to the surface of the substrate using the above formulae. It is important to keep the order in the sequence intact. Using the superposition formulae one again obtains the constitutive coefficients for the combined surface. The above formulae again show clearly that the introduction of the second order coefficients δ_e, δ_m and τ is intimately related to the shift of the polarization and magnetization densities to the dividing surface of the combined film.

3.11 Superposition of adjacent films

As discussed above the distribution of material in the interface is in many cases such that one may actually distinguish more than one separate film placed on top of each other. A trivial example is the case of two plane parallel layers above a substrate. A more sophisticated example, which will be discussed in chapter 11, is a stratified layer which can be considered to be a superposition of a whole stack of very thin plane parallel layers. Also if one places an island film above a rough substrate one may superimpose the effects of the roughness and the effects of the film if they are uncorrelated.

Consider two films placed in $-\ell$ and ℓ where ℓ is infinitesimally small. The film placed in $-\ell$ has the constitutive coefficients $\gamma_{e,1}$, $\gamma_{m,1}$, $\beta_{e,1}$, $\beta_{m,1}$, $\delta_{e,1}$, $\delta_{m,1}$, τ_1 and the film placed in ℓ has the constitutive coefficients $\gamma_{e,2}$, $\gamma_{m,2}$, $\beta_{e,2}$, $\beta_{m,2}$, $\delta_{e,2}$, $\delta_{m,2}$, τ_2. At both films the fields have jumps given by eq.(3.30) in terms of the interfacial polarization and magnetization densities of the corresponding film. The total jump across both films is found by simply adding the two separate jumps together and one may then conclude using eq.(3.30) that this total jump is again given by this equation but now containing the total interfacial polarization and magnetization densities, i.e. the sum of the densities of the separate films. The interfacial polarization and magnetization densities are for both films given by the constitutive equations (3.42)–(3.45). The problem which now arises is that these relations contain the field averages at both films which both contain the field between the two films. This field between the two films must be eliminated in order to obtain the constitutive coefficients of the superimposed film. Consider for instance

$$\mathbf{E}_{\|,+,1} = \frac{1}{2}[\mathbf{E}_{\|}^{-} + \mathbf{E}_{\|}(z=0)] = \frac{1}{2}[\mathbf{E}_{\|}^{-} + \mathbf{E}_{\|}^{+}] - \frac{1}{2}[\mathbf{E}_{\|}^{+} - \mathbf{E}_{\|}(z=0)]$$

$$= \mathbf{E}_{\|,+} + \frac{i\omega}{2c}\hat{\mathbf{z}} \times \mathbf{M}_{\|,2}^{s}(z=0) + \frac{1}{2}\nabla_{\|}P_{z,2}^{s}$$

$$= \mathbf{E}_{\|,+} + \frac{i\omega}{2c}\gamma_{m,2}\hat{\mathbf{z}} \times \mathbf{H}_{\|,+,2}(z=0) + \frac{1}{2}\beta_{e,2}\nabla_{\|}D_{z,+,2} \tag{3.58}$$

In the last identity eqs.(3.44) and (3.43) were used for the second film and terms of second order in the thickness were neglected as they would lead to contributions of the third order in the constitutive coefficients for the superimposed film. For the same reason the average fields at the second film in eq.(3.58) may be replaced by the averages over the superimposed film, so that

$$\mathbf{E}_{\|,+,1} = \mathbf{E}_{\|,+} + \frac{i\omega}{2c}\gamma_{m,2}\hat{\mathbf{z}} \times \mathbf{H}_{\|,+}(z=0) + \frac{1}{2}\beta_{e,2}\nabla_{\|}D_{z,+} \tag{3.59}$$

Similarly one finds

$$\mathbf{E}_{\|,+,2} = \mathbf{E}_{\|,+} - \frac{i\omega}{2c}\gamma_{m,1}\hat{\mathbf{z}} \times \mathbf{H}_{\|,+}(z=0) - \frac{1}{2}\beta_{e,1}\nabla_{\|}D_{z,+}$$

$$D_{z,+,1} = D_{z,+} + \frac{1}{2}\gamma_{e,2}\nabla_{\|}.\mathbf{E}_{\|,+} \text{ and } D_{z,+,2} = D_{z,+} - \frac{1}{2}\gamma_{e,1}\nabla_{\|}.\mathbf{E}_{\|,+} \tag{3.60}$$

For the magnetic fields the corresponding relations are

$$\mathbf{H}_{\|,+,1} = \mathbf{H}_{\|,+} - \frac{i\omega}{2c}\gamma_{e,2}\hat{\mathbf{z}} \times \mathbf{E}_{\|,+}(z=0) + \frac{1}{2}\beta_{m,2}\nabla_{\|}B_{z,+}$$

$$\mathbf{H}_{\|,+,2} = \mathbf{H}_{\|,+} + \frac{i\omega}{2c}\gamma_{e,1}\hat{\mathbf{z}} \times \mathbf{E}_{\|,+}(z=0) - \frac{1}{2}\beta_{m,1}\nabla_{\|}B_{z,+}$$

$$B_{z,+,1} = B_{z,+} + \frac{1}{2}\gamma_{m,2}\nabla_{\|}.\mathbf{H}_{\|,+} \text{ and } B_{z,+,2} = B_{z,+} - \frac{1}{2}\gamma_{m,1}\nabla_{\|}.\mathbf{H}_{\|,+} \tag{3.61}$$

Using the above formulae one may now express the total excesses in terms of the average across the superimposed film.

The resulting constitutive coefficients for the superimposed film are then found to be given in terms of the constitutive coefficients of the separate films by

$$\gamma_e = \gamma_{e,1} + \gamma_{e,2}\,, \quad \beta_e = \beta_{e,1} + \beta_{e,2}$$

$$\delta_e = \delta_{e,1} + \delta_{e,2} - \frac{1}{2}\gamma_{e,1}\beta_{e,2} + \frac{1}{2}\gamma_{e,2}\beta_{e,1}$$

$$\tau = \tau_1 + \tau_2 + \frac{1}{2}\gamma_{e,1}\gamma_{m,2} - \frac{1}{2}\gamma_{m,1}\gamma_{e,2}$$

$$\gamma_m = \gamma_{m,1} + \gamma_{m,2}, \quad \beta_m = \beta_{m,1} + \beta_{m,2}$$

$$\delta_m = \delta_{m,1} + \delta_{m,2} - \frac{1}{2}\gamma_{m,1}\beta_{m,2} + \frac{1}{2}\gamma_{m,2}\beta_{m,1} \tag{3.62}$$

In the case that there are no magnetic terms the above relations reduce to

$$\gamma_e = \gamma_{e,1} + \gamma_{e,2}\,, \quad \beta_e = \beta_{e,1} + \beta_{e,2}$$

$$\delta_e = \delta_{e,1} + \delta_{e,2} - \frac{1}{2}\gamma_{e,1}\beta_{e,2} + \frac{1}{2}\gamma_{e,2}\beta_{e,1}$$

$$\tau = \tau_1 + \tau_2 \tag{3.63}$$

In fact only the expression for τ which couples the excess of the electric displacement field to the magnetic field, and vice versa, simplifies.

From the above formulae it follows, that the linear constitutive coefficients of adjacent films can simply be added if one calculates the coefficients for the superimposed film. The quadratic coefficients, however, contain additional cross effects of the linear terms of the separate films. If the film is non-magnetic these cross effects only occur in δ_e.

3.12 Appendix A

If one chooses the dividing surface in $z = d$ the boundary condition for the discontinuity of the extrapolated x-component of the electric field is given by, cf. eq.(3.30),

$$E_{x,-}(\mathbf{r}_\|, d, \omega) = E_x^+(\mathbf{r}_\|, d, \omega) - E_x^-(\mathbf{r}_\|, d, \omega) = i\frac{\omega}{c}M_y^s(\mathbf{r}_\|, d, \omega) - \frac{\partial}{\partial x}P_z^s(\mathbf{r}_\|, d, \omega) \tag{3.64}$$

As the analysis for the jumps in the other fields follows the same lines, most of the explicit calculation in this appendix will be for this case alone. For $d = 0$ the above expression reduces to eq.(3.30). As the explicit dependence on $\mathbf{r}_\|$ and ω is not pertinent to the analysis in this appendix it will further be suppressed. The constitutive relations become if the dividing surface is chosen in $z = d$

$$P_z^s(d) = \beta_e(d)D_{z,+}(d) + \delta_e(d)\nabla_\|.\mathbf{E}_{\|,+}(d)$$

$$M_y^s(d) = \gamma_m(d)H_{y,+}(d) - \delta_m(d)\frac{\partial}{\partial y}B_{z,+}(d) + i\frac{\omega}{c}\tau(d)E_{x,+}(d) \tag{3.65}$$

In order to derive the desired relations the above equations are expanded to second order in d. For the jumps this gives

$$E_{x,-}(d) = E_{x,-}(0) + d[\frac{\partial}{\partial z}E_{x,-}(z)]_{z=0} + \frac{1}{2}d^2[\frac{\partial^2}{\partial z^2}E_{x,-}(z)]_{z=0} \tag{3.66}$$

The normal derivatives of the field must be eliminated using the Maxwell equations in the bulk regions, eq.(3.29). The first derivative gives

$$[\frac{\partial}{\partial z}E_{x,-}(z)]_{z=0} = [\text{rot}\,\mathbf{E}(0)]_{y,-} + \frac{\partial}{\partial x}E_{z,-}(0) = i\frac{\omega}{c}B_{y,-}(0) + \frac{\partial}{\partial x}E_{z,-}(0) \tag{3.67}$$

The second derivative is most conveniently eliminated using the wave equation (3.32)

$$[\frac{\partial^2}{\partial z^2}E_{x,-}(z)]_{z=0} = -\{[(\frac{\omega}{c})^2\epsilon\mu + \Delta_\|]E_x(0)\}_-$$

$$= -(\frac{\omega}{c})^2(\epsilon\mu)_-E_{x,+}(0) - [(\frac{\omega}{c})^2(\epsilon\mu)_+ + \Delta_\|]E_{x,-}(0) \tag{3.68}$$

where it was used that the jump of a product satisfies $(ab)_- = a_-b_+ + a_+b_-$. Furthermore $\Delta_\| \equiv \partial^2/\partial x^2 + \partial^2/\partial y^2$. The method to eliminate the jumps on the right-hand side of eq.(3.67) is

$$B_{y,-}(0) = (\mu H_y(0))_- = \mu_-H_{y,+}(0) + \mu_+H_{y,-}(0)$$

$$= \mu_-H_{y,+}(0) + i\,\mu_+\frac{\omega}{c}P_x^s(0) - \mu_+\frac{\partial}{\partial y}M_z^s(0)$$

$$= \mu_- H_{y,+}(0) + i\ \mu_+ \frac{\omega}{c}\gamma_e(0)E_{x,+}(0) - \mu_+\beta_m(0)\frac{\partial}{\partial y}B_{z,+}(0) \tag{3.69}$$

In the last identity only contributions linear in the thickness of the interface where taken along, which made it possible to use eqs. (3.38) and (3.41). Quadratic terms would lead to contributions of the third order upon substitution into eq.(3.66). Notice in this context that d will be of the order of the thickness of the interface for any reasonable choice of the dividing surface. Similarly one finds

$$E_{z,-}(0) = (\frac{1}{\epsilon}D_z(0))_- = (\frac{1}{\epsilon})_- D_{z,+}(0) + (\frac{1}{\epsilon})_+ D_{z,-}(0) = (\frac{1}{\epsilon})_- D_{z,+}(0) - (\frac{1}{\epsilon})_+ \nabla_\parallel.\mathbf{P}^s_\parallel(0)$$

$$= (\frac{1}{\epsilon})_- D_{z,+}(0) - (\frac{1}{\epsilon})_+ \gamma_e(0)\nabla_\parallel.\mathbf{E}_{\parallel,+}(0) \tag{3.70}$$

In eq.(3.68) one may take $E_{x,-}(0)$ to zeroth order in the thickness of the interface as this contribution is multiplied by d^2. To zeroth order this jump is zero. Combining eqs.(3.66)–(3.70) one obtains

$$E_{x,-}(d) = E_{x,-}(0) + d[i\frac{\omega}{c}\mu_- H_{y,+}(0) - \mu_+(\frac{\omega}{c})^2\gamma_e(0)E_{x,+}(0) - i\frac{\omega}{c}\mu_+\beta_m(0)\frac{\partial}{\partial y}B_{z,+}(0)$$

$$+(\frac{1}{\epsilon})_-\frac{\partial}{\partial x}D_{z,+}(0) - (\frac{1}{\epsilon})_+\gamma_e(0)\frac{\partial}{\partial x}\nabla_\parallel.\mathbf{E}_{\parallel,+}(0)] - \frac{1}{2}d^2(\frac{\omega}{c})^2(\epsilon\mu)_- E_{x,+}(0) \tag{3.71}$$

There remains the problem to expand the excess fields on the right-hand side of eq.(3.64). Using the constitutive relations (3.65) one has

$$P^s_z(d) = P^s_z(0) + d[\beta^{(1)}_e D_{z,+}(0) + \beta_e(0)\frac{\partial}{\partial z}D_{z,+}(0) + \delta^{(1)}_e\nabla_\parallel.\mathbf{E}_{\parallel,+}(0)$$

$$+\delta_e(0)\frac{\partial}{\partial z}\nabla_\parallel.\mathbf{E}_{\parallel,+}(0)] + \frac{1}{2}d^2[\beta^{(2)}_e D_{z,+}(0) + 2\beta^{(1)}_e\frac{\partial}{\partial z}D_{z,+}(0) + \beta_e(0)\frac{\partial^2}{\partial z^2}D_{z,+}(0)$$

$$+\delta^{(2)}_e\nabla_\parallel.\mathbf{E}_{\parallel,+}(0) + 2\delta^{(1)}_e\frac{\partial}{\partial z}\nabla_\parallel.\mathbf{E}_{\parallel,+}(0) + \delta_e(0)\frac{\partial^2}{\partial z^2}\nabla_\parallel.\mathbf{E}_{\parallel,+}(0)] \tag{3.72}$$

where the superindices (1) and (2) indicate the first and the second derivatives in $z = 0$, respectively. Notice the fact that $\beta^{(1)}_e$, $\beta^{(2)}_e$ and $\delta^{(2)}_e$ are of the zeroth order in the thickness while $\delta^{(1)}_e$ is of the first order. Neglecting terms of higher than second order and eliminating the z-derivatives of the field in eq.(3.72) gives

$$P^s_z(d) = P^s_z(0) + d[\beta^{(1)}_e D_{z,+}(0) - \epsilon_+\beta_e(0)\nabla_\parallel.\mathbf{E}_{\parallel,+}(0) + \delta^{(1)}_e\nabla_\parallel.\mathbf{E}_{\parallel,+}(0)]$$

$$+\frac{1}{2}d^2[\beta^{(2)}_e D_{z,+}(0) - 2\ \epsilon_+\beta^{(1)}_e\nabla_\parallel.\mathbf{E}_{\parallel,+}(0) + \delta^{(2)}_e\nabla_\parallel.\mathbf{E}_{\parallel,+}(0)] \tag{3.73}$$

Similarly one finds

$$M_y^s(d) = M_y^s(0) + d[\gamma_m^{(1)} H_{y,+}(0) + (\frac{1}{\mu})_+ \gamma_m(0) \frac{\partial}{\partial y} B_{z,+}(0) + i\frac{\omega}{c}\gamma_m(0)\epsilon_+ E_{x,+}(0)$$

$$-\delta_m^{(1)} \frac{\partial}{\partial y} B_{z,+}(0) + i\frac{\omega}{c}\tau^{(1)} E_{x,+}(0)] + \frac{1}{2}d^2[\gamma_m^{(2)} H_{y,+}(0) + 2(\frac{1}{\mu})_+ \gamma_m^{(1)} \frac{\partial}{\partial y} B_{z,+}(0)$$

$$+2i\frac{\omega}{c}\gamma_m^{(1)}\epsilon_+ E_{x,+}(0) - \delta_m^{(2)} \frac{\partial}{\partial y} B_{z,+}(0) + i\frac{\omega}{c}\tau^{(2)} E_{x,+}(0)] \qquad (3.74)$$

to second order in the thickness.

Substitution of eqs.(3.71), (3.73) and (3.74) into eq.(3.64) gives after some rearrangement of the terms

$$-i\frac{\omega}{c}[\frac{1}{2}d\ \gamma_m^{(2)} - \mu_- + \gamma_m^{(1)}] H_{y,+}(0)$$

$$-(\frac{\omega}{c})^2[\mu_+\gamma_e(0) + \frac{1}{2}d(\epsilon\mu)_- - \gamma_m(0)\epsilon_+ - \tau^{(1)} - d\ \gamma_m^{(1)}\epsilon_+ - \frac{1}{2}d\ \tau^{(2)}] E_{x,+}(0)$$

$$-i\frac{\omega}{c}[\mu_+\beta_m(0) + (\frac{1}{\mu})_+\gamma_m(0) - \delta_m^{(1)} + d(\frac{1}{\mu})_+\gamma_m^{(1)} - \frac{1}{2}d\ \delta_m^{(2)}] \frac{\partial}{\partial y} B_{z,+}(0)$$

$$+[\beta_e^{(1)} + (\frac{1}{\epsilon})_- + \frac{1}{2}d\ \beta_e^{(2)}] \frac{\partial}{\partial x} D_{z,+}(0)$$

$$-[(\frac{1}{\epsilon})_+\gamma_e(0) + \epsilon_+\beta_e(0) - \delta_e^{(1)} + d\ \epsilon_+\beta_e^{(1)} - \frac{1}{2}d\delta_e^{(2)}] \frac{\partial}{\partial x} \nabla_\parallel . \mathbf{E}_{\parallel,+}(0) = 0 \qquad (3.75)$$

In view of the fact that the electromagnetic fields may be chosen arbitrarily all the expressions between square brackets in this equation must be zero. This gives

$$\frac{1}{2}d\ \gamma_m^{(2)} - \mu_- + \gamma_m^{(1)} = 0$$

$$\mu_+\gamma_e(0) + \frac{1}{2}d(\epsilon\mu)_- + \gamma_m(0)\epsilon_+ - \tau^{(1)} - d\ \gamma_m^{(1)}\epsilon_+ - \frac{1}{2}d\ \tau^{(2)} = 0$$

$$\mu_+\beta_m(0) + (\frac{1}{\mu})_+\gamma_m(0) - \delta_m^{(1)} + d(\frac{1}{\mu})_+\gamma_m^{(1)} - \frac{1}{2}d\ \delta_m^{(2)} = 0$$

$$\beta_e^{(1)} + (\frac{1}{\epsilon})_- + \frac{1}{2}d\ \beta_e^{(2)} = 0$$

$$(\frac{1}{\epsilon})_+\gamma_e(0) + \epsilon_+\beta_e(0) - \delta_e^{(1)} + d\ \epsilon_+\beta_e^{(1)} - \frac{1}{2}d\delta_e^{(2)} = 0 \tag{3.76}$$

As these equations are true for arbitrary values of d it follows that

$$\gamma_m^{(1)} = \mu_- \text{ and } \gamma_m^{(2)} = 0$$

$$\tau^{(1)} = \mu_+\gamma_e(0) - \epsilon_+\gamma_m(0) \text{ and } \tau^{(2)} = (\epsilon\mu)_- - 2\ \gamma_m^{(1)}\epsilon_+ = \mu_+\epsilon_- - \mu_-\epsilon_+$$

$$\delta_m^{(1)} = \mu_+\beta_m(0) + (\frac{1}{\mu})_+\gamma_m(0) \text{ and } \delta_m^{(2)} = 2(\frac{1}{\mu})_+\gamma_m^{(1)} = 2(\frac{1}{\mu})_+\mu_- = -2(\frac{1}{\mu})_-\mu_+$$

$$\beta_e^{(1)} = -(\frac{1}{\epsilon})_- \text{ and } \beta_e^{(2)} = 0$$

$$\delta_e^{(1)} = (\frac{1}{\epsilon})_+\gamma_e(0) + \epsilon_+\beta_e(0) \text{ and } \delta_e^{(2)} = 2\ \epsilon_+\beta_e^{(1)} = -2\ \epsilon_+(\frac{1}{\epsilon})_- \tag{3.77}$$

Together with the analogous results for γ_e and β_m, which one finds if one performs the same analysis for the parallel component of the magnetic field, the relations given in eq.(3.51) now follow. The other boundary conditions lead to exactly the same relations.

Chapter 4
REFLECTION AND TRANSMISSION

4.1 Introduction

In this chapter expressions will be derived for the angle of incidence dependent reflection and transmission amplitudes in terms of the seven electromagnetic interfacial constitutive coefficients introduced in the previous chapter. For non-magnetic systems the resulting reflectances, transmittances as well as ellipsometric coefficients are given. Contrary to the amplitudes these quantities are measurable and therefore independent of the choice of the dividing surface. They can therefore be expressed in terms of the invariants introduced in the previous chapter. The expressions in terms of these invariants are given.

The media on both sides of the surface are chosen to be homogeneous and isotropic. The radiation field in these media is a solution of the Maxwell equations, cf. eq.(3.24),

$$\mathrm{rot}\,\mathbf{E}(\mathbf{r},\omega) = i\frac{\omega}{c}\mathbf{B}(\mathbf{r},\omega), \qquad \mathrm{div}\,\mathbf{D}(\mathbf{r},\omega) = 0$$

$$\mathrm{rot}\,\mathbf{H}(\mathbf{r},\omega) = -i\frac{\omega}{c}\mathbf{D}(\mathbf{r},\omega), \quad \mathrm{div}\,\mathbf{B}(\mathbf{r},\omega) = 0 \tag{4.1}$$

The constitutive equations are

$$\mathbf{D}(\mathbf{r},\omega) = \epsilon(\omega)\mathbf{E}(\mathbf{r},\omega) \text{ and } \mathbf{B}(\mathbf{r},\omega) = \mu(\omega)\mathbf{H}(\mathbf{r},\omega) \tag{4.2}$$

where ϵ and μ are the frequency dependent dielectric constant and magnetic permeability. If one calculates the rotation of the first Maxwell equation, and uses the second and the third Maxwell equation together with the constitutive relations one obtains the wave equation

$$\Delta\mathbf{E}(\mathbf{r},\omega) = -(\frac{\omega}{c})^2\epsilon(\omega)\mu(\omega)\mathbf{E}(\mathbf{r},\omega) = -(\frac{\omega}{c})^2 n^2(\omega)\mathbf{E}(\mathbf{r},\omega) \tag{4.3}$$

Here Δ is the Laplace operator

$$\Delta \equiv \frac{\partial^2}{\partial x^2} + \frac{\partial^2}{\partial y^2} + \frac{\partial^2}{\partial z^2} \tag{4.4}$$

Furthermore the refractive index of the medium is identified with

$$n(\omega) \equiv \sqrt{\epsilon(\omega)\mu(\omega)} \text{ with } \mathrm{Im}\,n(\omega) \geq 0 \tag{4.5}$$

If there is absorption in the medium $n(\omega)$ will be a complex function of ω. This is the consequence of the fact that either the dielectric constant or the magnetic permeability are complex functions of ω. In that case the plane wave solution will decay exponentially in the direction of propagation. Its source must be at a finite distance of the observer.

As discussed in all textbooks on electromagnetism [26], [14] the above equations for the field in a homogeneous isotropic medium have so-called plane wave solutions of the following form

$$\mathbf{E}(\mathbf{r},\omega) = \mathbf{E}_0 \exp[i\mathbf{k}_0.\mathbf{r}]2\pi\ \delta(\omega - \omega_0)$$

$$\mathbf{B}(\mathbf{r},\omega) = (c/\omega_0)\mathbf{k}_0 \times \mathbf{E}_0 \exp[i\mathbf{k}_0.\mathbf{r}]2\pi\ \delta(\omega - \omega_0) \tag{4.6}$$

with $\mathbf{k}_0.\mathbf{k}_0 = [(\omega_0/c)n(\omega_0)]^2$ and $\mathbf{E}_0.\mathbf{k}_0 = 0$. As a function of position and time the plane wave is given by

$$\mathbf{E}(\mathbf{r},t) = \mathbf{E}_0 \exp[i\mathbf{k}_0.\mathbf{r} - i\omega_0 t]$$

$$\mathbf{B}(\mathbf{r},t) = (c/\omega_0)\mathbf{k}_0 \times \mathbf{E}_0 \exp[i\mathbf{k}_0.\mathbf{r} - i\omega_0 t] \tag{4.7}$$

It should be noted that in a non-absorbing medium the wave vector is real valued and the directions of both the electric and the magnetic field are orthogonal to the direction of propagation of the wave $\hat{\mathbf{k}}_0 \equiv \mathbf{k}_0/|\mathbf{k}_0|$. Furthermore the directions of the electric and the magnetic field are orthogonal to each other. The direction of the electric field is called the direction of the polarization of the wave. There are therefore two independent polarization directions which are both normal to the direction of propagation.

In an absorbing medium the refractive coefficient has an imaginary part. In that case one usually takes $\mathbf{k}_0 = (\omega_0/c)n(\omega_0)\hat{\mathbf{k}}_0$. The amplitude of the field decreases exponentially in the direction of propagation $\hat{\mathbf{k}}_0$. The electric and the magnetic field are then again orthogonal to each other and to the direction of propagation. In particular the incident and the reflected fields will be of this nature. It is important to realize, however, that the transmitted light may have a somewhat different nature. It follows from the wave equation that the wave vector and the frequency of a solution of the form given in eq.(4.7) are related by $\mathbf{k}_0.\mathbf{k}_0 = [(\omega_0/c)n(\omega_0)]^2$. This gives the freedom to choose two components of the wave vector and the frequency and to determine the third component of the wave vector using this relation. This third component of the wave vector is then usually the component normal to the surface. This matter will be discussed in some more detail in the next section. If, as is often the case, the medium is only weakly absorbing it is sufficient to neglect the imaginary part of the refractive index in most of the analysis except for an overall attenuation factor. In the further analysis in this chapter it will be assumed that the medium, through which the light is incident, is weakly absorbing. The dielectric constant and the magnetic permeability will therefore be chosen to be real (for $z < 0$).

If a plane wave is incident on a flat interface one may distinguish two different cases. One case has a direction of the polarization which is orthogonal to the plane of incidence. One calls this alternatively an s-polarized plane wave or a TE-wave. The s is from the German word for orthogonal (senkrecht) whereas TE stands for transverse electric. The other case has a direction of the polarization which is in the plane of incidence. One calls this alternatively a p-polarized plane wave or a TM-wave. Here p stands for parallel and TM for transverse magnetic. In principle one must calculate the reflection and transmission amplitudes of the TE- and the TM-waves independently. It is possible, however, to use again the symmetry of the Maxwell equations for the interchange of the electric field with the magnetic field and the electric displacement field with minus the magnetic induction, cf. section 3.8. As discussed in that section this is equivalent with simultaneously interchanging $\epsilon^{\pm}$ with $-\mu^{\pm}$, γ_e with $-\gamma_m$, β_e with $-\beta_m$, δ_e with δ_m and replacing τ by $-\tau$. It is therefore sufficient to solve only one of the two problems. The other follows immediately using this symmetry.

In the following sections the amplitudes of the transmitted and reflected fields will be calculated for the general case. In the calculation of transmittances, reflectances and ellipsometric coefficients the analysis will be restricted to the non-magnetic case in order to avoid undue complexity. For the general case one may use the general amplitudes given to calculate the desired property along similar lines. All final expressions for measurable quantities are written using invariants. This makes them visibly independent of the somewhat arbitrary choice of the dividing surface. Except for simple phase factors the same is done for the amplitudes of the reflected and transmitted light. All experimental results can therefore be interpreted directly in terms of these invariant, which are characteristic for the surface considered.

4.2 TE-waves

The medium through which the plane TE-wave is incident is assumed to fill the $z < 0$ half-space. The solution of the wave equation in the $z < 0$ region is a superposition of the incident wave and the reflected wave:

$$\begin{aligned} \mathbf{E}(\mathbf{r},\omega) &= (0,1,0)[r_i \exp(i\mathbf{k}_i.\mathbf{r}) + r_s \exp(i\mathbf{k}_r.\mathbf{r})]2\pi\ \delta(\omega-\omega_0) \\ \mathbf{B}(\mathbf{r},\omega) &= n^-[r_i(-\cos\theta_i,0,\sin\theta_i)\exp(i\mathbf{k}_i.\mathbf{r}) \\ &\quad +r_s(\cos\theta_r,0,\sin\theta_r)\exp(i\mathbf{k}_r.\mathbf{r})]2\pi\ \delta(\omega-\omega_0) \end{aligned} \tag{4.8}$$

It should be noted that r_i and r_s give the amplitudes of the incident and the reflected waves at the $z = 0$ surface. n^- is the real refractive index in the $z < 0$ half-space. The wave vectors of the incident and reflected light are given by

$$\mathbf{k}_i \equiv n^-\frac{\omega}{c}(\sin\theta_i, 0, \cos\theta_i) \text{ and } \mathbf{k}_r \equiv n^-\frac{\omega}{c}(\sin\theta_r, 0, -\cos\theta_r) \tag{4.9}$$

Here θ_i is the angle of incidence and θ_r the angle of reflection, which will as usual be found to be equal. The coordinates have been chosen such that the plane of incidence is the $x-z$ plane. Not only the incident wave but also the reflected wave is a TE-wave.

The transmitted light is given by

$$\begin{aligned} \mathbf{E}(\mathbf{r},\omega) &= (0,1,0)t_s \exp(i\mathbf{k}_t.\mathbf{r})2\pi\ \delta(\omega-\omega_0) \\ \mathbf{B}(\mathbf{r},\omega) &= n^+t_s(-\cos\theta_t,0,\sin\theta_t)\exp(i\mathbf{k}_t.\mathbf{r})2\pi\ \delta(\omega-\omega_0) \end{aligned} \tag{4.10}$$

for $z > 0$, where n^+ is the possibly complex refractive index in the $z > 0$ half-space. The wave vector of the transmitted light is given by

$$\mathbf{k}_t \equiv n^+ \frac{\omega}{c}(\sin\theta_t, 0, \cos\theta_t) \tag{4.11}$$

Here θ_t is the angle of transmission. The transmitted wave is also a TE-wave and t_s is the amplitude for $z = 0$. If the medium through which the light is transmitted is absorbing the refractive index, and therefore the wave vector of the transmitted light, will have an imaginary part. This leads to damping in the direction of propagation.

In order to calculate the reflection and transmission amplitudes r_s and t_s introduced above one must first evaluate the interfacial polarization and magnetization densities. As follows from the constitutive relations, eqs.(3.42) – (3.45), one needs for this purpose the averages, cf. eq.(3.34), of the asymptotic bulk fields at the interface. Using eqs.(4.8)–(4.11) and the constitutive relations in the bulk regions one obtains

$$\begin{aligned}
\mathbf{E}_{\|,+}(\mathbf{r}_\|, \omega) &= \frac{1}{2}(0,1)[r_i \exp(i\mathbf{k}_{i,\|}.\mathbf{r}_\|) + r_s \exp(i\mathbf{k}_{r,\|}.\mathbf{r}_\|) + t_s \exp(i\mathbf{k}_{t,\|}.\mathbf{r}_\|)]2\pi\delta(\omega - \omega_0) \\
D_{z,+}(\mathbf{r}_\|, \omega) &= 0 \\
\mathbf{H}_{\|,+}(\mathbf{r}_\|, \omega) &= -\frac{1}{2}(1,0)\{(n^-/\mu^-)[r_i \cos\theta_i \exp(i\mathbf{k}_{i,\|}.\mathbf{r}_\|) - r_s \cos\theta_r \exp(i\mathbf{k}_{r,\|}.\mathbf{r}_\|)] \\
&\quad + (n^+/\mu^+) t_s \cos\theta_t \exp(i\mathbf{k}_{t,\|}.\mathbf{r}_\|)\} 2\pi\ \delta(\omega - \omega_0) \\
B_{z,+}(\mathbf{r}_\|, \omega) &= \frac{1}{2}\{n^-[r_i \sin\theta_i \exp(i\mathbf{k}_{i,\|}.\mathbf{r}_\|) + r_s \sin\theta_r \exp(i\mathbf{k}_{r,\|}.\mathbf{r}_\|)] \\
&\quad + n^+ t_s \sin\theta_t \exp(i\mathbf{k}_{t,\|}.\mathbf{r}_\|)\} 2\pi\ \delta(\omega - \omega_0)
\end{aligned} \tag{4.12}$$

It is only possible to satisfy the boundary conditions (3.30) if the components of the wave vectors of the incident, reflected and transmitted light parallel to the interface are equal:

$$\mathbf{k}_{i,\|} = \mathbf{k}_{r,\|} = \mathbf{k}_{t,\|} \equiv \mathbf{k}_\| \tag{4.13}$$

In fact this property may also be derived on the basis of translational invariance along the interface, which implies that the above solution must be an eigenfunction of the generator $\partial/\partial\mathbf{r}_\|$ of infinitesimal translations. An immediate consequence of this property is that the plane of incidence is equal to the "plane of reflection" and to the "plane of transmission", a fact which was first established by Snell. Substituting the explicit expressions (4.9) and (4.11) for the wave vectors one furthermore finds the well-known results

$$\theta_i = \theta_r \equiv \theta \text{ and } n^+ \sin\theta_t = n^- \sin\theta \tag{4.14}$$

It is thus found that, independent of possible surface excesses in the fields, the reflection and transmission angles are given by the usual law of reflection and by Snell's law respectively. If n^+ is a complex number, the angle of transmission θ_t is also a complex number. Of course a complex angle of transmission does not have a physical meaning as an angle. It is more transparent in that case to use the wave vector instead. According to Snell's law, the wave vector along the surface $\mathbf{k}_\parallel$ is the same on both sides of the interface. In view of the fact that the ambient has been chosen to be transparent, $\mathbf{k}_\parallel$ is therefore real. The component of the wave vector normal to the surface $k_{t,z}$ is given by $k_{t,z} = \sqrt{[(\omega/c)n^+(\omega)]^2 - k_\parallel^2}$, and has a positive imaginary part. The transmitted wave is attenuated away from the surface. If n^- and n^+ are both real and $n^- > n^+$ one finds that $k_{t,z} = (\omega/c)\sqrt{[n^+]^2 - [n^- \sin\theta]^2}$ is purely imaginary, if one chooses θ larger than the angle of total reflection, $\arcsin(n^+/n^-)$. In that case the transmitted wave is overdamped.

Using the equality of the parallel components of the wave vectors, the expression (4.12) for the averages of the bulk fields simplify to

$$\begin{aligned}
\mathbf{E}_{\parallel,+}(\mathbf{r}_\parallel,\omega) &= \frac{1}{2}(0,1)(r_i + r_s + t_s)\exp(i\mathbf{k}_\parallel.\mathbf{r}_\parallel)2\pi\ \delta(\omega-\omega_0) \\
D_{z,+}(\mathbf{r}_\parallel,\omega) &= 0 \\
\mathbf{H}_{\parallel,+}(\mathbf{r}_\parallel,\omega) &= -\frac{1}{2}(1,0)[(n^-/\mu^-)(r_i - r_s)\cos\theta \\
&\quad +(n^+/\mu^+)t_s\cos\theta_t]\exp(i\mathbf{k}_\parallel.\mathbf{r}_\parallel)2\pi\ \delta(\omega-\omega_0) \\
B_{z,+}(\mathbf{r}_\parallel,\omega) &= \frac{1}{2}n^-\sin\theta(r_i + r_s + t_s)\exp(i\mathbf{k}_\parallel.\mathbf{r}_\parallel)2\pi\ \delta(\omega-\omega_0) \qquad (4.15)
\end{aligned}$$

where in the last equality Snell's law has been used.

The next step is the calculation of the interfacial polarization and magnetization densities, defined by eq.(3.27), (the primes should be dropped in this equation, see remark after eq.(3.30)). Substitution of the averages of the bulk fields (4.15) at the interface into the constitutive relations (3.42)–(3.45) then gives

$$\begin{aligned}
\mathbf{P}^s_\parallel(\mathbf{r}_\parallel,\omega) &= \frac{1}{2}(0,1)\ \exp(i\mathbf{k}_\parallel.\mathbf{r}_\parallel)2\pi\ \delta(\omega-\omega_0) \\
&\quad \times\{\gamma_e(r_i + r_s + t_s) - i\frac{\omega}{c}\tau[(n^-/\mu^-)(r_i - r_s)\cos\theta + (n^+/\mu^+)t_s\cos\theta_t]\} \\
P^s_z(\mathbf{r}_\parallel,\omega) &= 0 \\
\mathbf{M}^s_\parallel(\mathbf{r}_\parallel,\omega) &= -\frac{1}{2}(1,0)\exp(i\mathbf{k}_\parallel.\mathbf{r}_\parallel)2\pi\ \delta(\omega-\omega_0)\{\gamma_m[(n^-/\mu^-)(r_i - r_s)\cos\theta \\
&\quad +(n^+/\mu^+)t_s\cos\theta_t] + i\frac{\omega}{c}[\tau + \delta_m(n^-\sin\theta)^2](r_i + r_s + t_s)\} \\
M^s_z(\mathbf{r}_\parallel,\omega) &= \frac{1}{2}n^-\sin\theta\ \exp(i\mathbf{k}_\parallel.\mathbf{r}_\parallel)2\pi\ \delta(\omega-\omega_0)\{\beta_m(r_i + r_s + t_s) \\
&\quad -i\frac{\omega}{c}\delta_m[(n^-/\mu^-)(r_i - r_s)\cos\theta + (n^+/\mu^+)t_s\cos\theta_t]\} \qquad (4.16)
\end{aligned}$$

Finally one finds two independent linear equations for the reflection and transmission amplitudes, if the above results for the interfacial polarization and magnetization densities are substituted into the boundary conditions (3.30) for the jumps in the bulk fields at $z = 0$. Introducing the abbreviations

$$\gamma_{e,m}(\theta) \equiv \gamma_e + \beta_m(n^-\sin\theta)^2$$

$$\tau_m(\theta) \equiv \tau + \delta_m (n^- \sin\theta)^2 \tag{4.17}$$

one finds, after some rather elaborate and not very illuminating algebra, that the expressions for the reflection and transmission amplitudes can be written in the following form:

$$\begin{aligned} r_s(\theta) \;&=\; r_i\{(n^-/\mu^-)\cos\theta[1+(\frac{\omega}{c})^2\tau_m(\theta)] - (n^+/\mu^+)\cos\theta_t[1-(\frac{\omega}{c})^2\tau_m(\theta)] \\ &+i\frac{\omega}{c}[\gamma_{e,m}(\theta) - \gamma_m(n^-n^+/\mu^-\mu^+)\cos\theta\ \cos\theta_t] \\ &-\frac{1}{2}(\frac{\omega}{c})^2\gamma_m\gamma_{e,m}(\theta)[(n^-/\mu^-)\cos\theta - (n^+/\mu^+)\cos\theta_t]\}/\ D_s(\theta) \end{aligned} \tag{4.18}$$

and

$$t_s(\theta) = 2\ r_i(n^-/\mu^-)\cos\theta\ /\ D_s(\theta) \tag{4.19}$$

The denominator is given by

$$\begin{aligned} D_s(\theta) \;&\equiv\; (n^-/\mu^-)\cos\theta[1+(\frac{\omega}{c})^2\tau_m(\theta)] + (n^+/\mu^+)\cos\theta_t[1-(\frac{\omega}{c})^2\tau_m(\theta)] \\ &-i\frac{\omega}{c}[\gamma_{e,m}(\theta) + \gamma_m(n^-n^+/\mu^-\mu^+)\cos\theta\ \cos\theta_t] \\ &-\frac{1}{2}(\frac{\omega}{c})^2\gamma_m\gamma_{e,m}(\theta)[(n^-/\mu^-)\cos\theta + (n^+/\mu^+)\cos\theta_t] \end{aligned} \tag{4.20}$$

and is the same in both amplitudes.

It should be emphasized that the reflection and transmission amplitudes are not invariant if the position of the dividing surface is changed. It is therefore not possible to write these amplitudes in a form containing only the invariants introduced in section 3.9. The reason for this is the occurrence of phase factors which depend on the choice of the dividing surface. It can, however, be shown that, apart from such phase factors, the above amplitudes $r_s(\theta)$ and $t_s(\theta)$ can be expressed in terms of the complex invariants I_m, I_{em}, $I_{\delta,m}$ and I_τ, defined in eqs.(3.53) and (3.54). One finds from eqs.(4.18)–(4.20), together with eqs.(4.17), (3.53) and (3.54), after lengthy but straightforward calculations, which will not be given here, the following results for the reflection and transmission amplitudes:

$$\begin{aligned} r_s(\theta) \;&=\; r_i \exp\{-i\frac{\omega}{c}\frac{\mu_+\gamma_e + \epsilon_+\gamma_m}{(\epsilon\mu)_-}(n^-\cos\theta + n^+\cos\theta_t)\}\{\frac{n^-}{\mu^-}\cos\theta - \frac{n^+}{\mu^+}\cos\theta_t \\ &+i\frac{\omega}{c}\left[-\frac{I_m}{\mu^-\mu^+}(n^-\sin\theta)^2 + \frac{I_{em}}{(\epsilon\mu)_-}\left(\epsilon_+ + \mu_+\frac{n^-n^+}{\mu^-\mu^+}\cos(\theta+\theta_t)\right)\right] \\ &+\left(\frac{\omega}{c}\right)^2\frac{\mu_+ I_m I_{em}}{(\epsilon\mu)_-\mu_-\mu^-\mu^+}(n^-\sin\theta)^2(n^-\cos\theta + n^+\cos\theta_t) \\ &+\frac{1}{2}\left(\frac{\omega}{c}\right)^2\frac{\mu_+ I_{em}^2}{(\epsilon\mu)_-\epsilon_-\mu_-}[\frac{\epsilon_+}{(\epsilon\mu)_-}\left(\frac{n^-}{\mu^-}\cos\theta - \frac{n^+}{\mu^+}\cos\theta_t\right)(n^-\cos\theta + n^+\cos\theta_t) \\ &+\frac{2}{\mu^-\mu^+}(n^-\sin\theta)^2](n^-\cos\theta + n^+\cos\theta_t) \\ &+\left(\frac{\omega}{c}\right)^2[I_\tau + I_{\delta,m}(n^-\sin\theta)^2]\left(\frac{n^-}{\mu^-}\cos\theta + \frac{n^+}{\mu^+}\cos\theta_t\right)\}/\ D_s(\theta) \end{aligned} \tag{4.21}$$

and

$$t_s(\theta) = 2\ r_i(n^-/\mu^-)\cos\theta\ /\ D_s(\theta) \tag{4.22}$$

The denominator in these expressions is given by

$$\begin{aligned}
D_s(\theta) \;=\; & \exp\{i\frac{\omega}{c}\frac{\mu_+\gamma_e+\epsilon_+\gamma_m}{(\epsilon\mu)_-}(n^-\cos\theta - n^+\cos\theta_t)\}\{\frac{n^-}{\mu^-}\cos\theta + \frac{n^+}{\mu^+}\cos\theta_t \\
& -i\frac{\omega}{c}\left[-\frac{I_m}{\mu^-\mu^+}(n^-\sin\theta)^2 + \frac{I_{em}}{(\epsilon\mu)_-}\left(\epsilon_+ - \mu_+\frac{n^-n^+}{\mu^-\mu^+}\cos(\theta-\theta_t)\right)\right] \\
& +\left(\frac{\omega}{c}\right)^2\frac{\mu_+ I_m I_{em}}{(\epsilon\mu)_-\mu_-\mu^-\mu^+}(n^-\sin\theta)^2(n^-\cos\theta - n^+\cos\theta_t) \\
& +\frac{1}{2}\left(\frac{\omega}{c}\right)^2\frac{\mu_+ I_{em}^2}{(\epsilon\mu)_-\epsilon_-\mu_-}[\frac{\epsilon_+}{(\epsilon\mu)_-}\left(\frac{n^-}{\mu^-}\cos\theta + \frac{n^+}{\mu^+}\cos\theta_t\right)(n^-\cos\theta - n^+\cos\theta_t) \\
& +\frac{2}{\mu^-\mu^+}(n^-\sin\theta)^2](n^-\cos\theta - n^+\cos\theta_t) \\
& +\left(\frac{\omega}{c}\right)^2[I_\tau + I_{\delta,m}(n^-\sin\theta)^2]\left(\frac{n^-}{\mu^-}\cos\theta - \frac{n^+}{\mu^+}\cos\theta_t\right)\}
\end{aligned} \tag{4.23}$$

Here again the notation $a_- \equiv a^+ - a^-$ and $a_+ \equiv \frac{1}{2}(a^+ + a^-)$ has been used and the relation

$$\pm\sin\theta\ \sin\theta_t - \cos\theta\ \cos\theta_t = -\cos(\theta \pm \theta_t) \tag{4.24}$$

which is also valid for complex angles θ_t.

Formula (4.21) together with (4.23) for the reflection amplitude using invariants is rather complicated compared to the original formula, (4.18) together with (4.20). As both formulae are equivalent one may equally well use the simpler expression in terms of the constitutive coefficients. An additional reason to be careful with the formula using the invariants, is the exponential prefactor. The imaginary part of γ_e for metal films can be rather large compared to the wavelength. This leads to unphysical small values of the exponent. It is then more in line with the derivation of eq.(4.18) to expand the exponent again and to return in this manner to the original formula.

In view of the fact that the description of the interface has been restricted to constitutive coefficients of the first and the second order in the interfacial thickness, contributions to the numerator and the denominator of $r_s(\theta)$ and $t_s(\theta)$ which are of third and higher order have been neglected in the above expressions. One may wonder why the denominator is not eliminated by expanding it into a power series in the interfacial coefficients. One could argue that the description of the film is only given to second order in the thickness of the film, so that one should in the further analysis systematically expand the whole expression to second order in the thickness. Though this argument is straightforward and appears to be appealing, it is physically incorrect. The exponential factors are simple phase factors. The rest of the numerator and denominator one could rewrite by dividing out the Fresnel contribution and subsequently expanding the denominator. It is then seen that the resulting expression has the typical structure, in which one sums over multiple internal reflections inside the film. The resulting expression is in fact rather more complicated than the present expression already is, even if one neglects terms of higher than second order. The present expression takes all these multiple reflections into account and is therefore more physical. The reason to hold on to the above form is therefore based on physical considerations rather than on puristic mathematical ones.

The complex exponential phase factors, appearing in the numerator of $r_s(\theta)$ and in the denominator $D_s(\theta)$ of both $r_s(\theta)$ and $t_s(\theta)$, are also correct up to this second order. The reason to prefer the exponential form is that this factor gives the phase factor needed to go from one choice of the dividing surface to another. This follows if one calculates

$$\frac{\mu_+\gamma_e(d)+\epsilon_+\gamma_m(d)}{(\epsilon\mu)_-}-\frac{\mu_+\gamma_e(0)+\epsilon_+\gamma_m(0)}{(\epsilon\mu)_-}=d \tag{4.25}$$

using eq.(3.51). Notice that this equality also implies that the imaginary part of both quotients on the left hand side are equal and therefore independent of d. This is the reason for the choice of I_c as invariant, cf. eq. (3.55).

It is seen by inspection of eqs.(4.21) and (4.23), that the exponential factor in the numerator of $r_s(\theta)$ partly cancels against the exponential factor in the denominator $\Delta_s(\theta)$, so that finally the complex exponential factor

$$f\equiv\exp\left\{-2i\frac{\omega}{c}n^-\cos\theta\frac{\mu_+\gamma_e+\epsilon_+\gamma_m}{(\epsilon\mu)_-}\right\} \tag{4.26}$$

is left in the expression for $r_s(\theta)$. Since n^- has been assumed to be real, so that $n^-\cos\theta$ is also real, the factor f can be written as

$$\begin{aligned} f &\equiv \exp\{-2i\frac{\omega}{c}n^-\cos\theta\,\mathrm{Re}\left[\frac{\mu_+\gamma_e+\epsilon_+\gamma_m}{(\epsilon\mu)_-}\right]\}.\exp\{2\frac{\omega}{c}n^-\cos\theta\,\mathrm{Im}\left[\frac{\mu_+\gamma_e+\epsilon_+\gamma_m}{(\epsilon\mu)_-}\right]\} \\ &= \exp\{-2i\frac{\omega}{c}n^-\cos\theta\,\mathrm{Re}\left[\frac{\mu_+\gamma_e+\epsilon_+\gamma_m}{(\epsilon\mu)_-}\right]\}.\exp\left(2\frac{\omega}{c}I_cn^-\cos\theta\right) \end{aligned} \tag{4.27}$$

where the definition eq.(3.55) of the real invariant I_c has been used. Therefore f is the product of an exponential phase factor and a damping factor due to absorption by the interface. The latter is an expression in terms of the invariant I_c. It is evidently not possible to express the phase factor in terms of invariants as follows from eq.(4.25). This is, of course, related to the fact, that the amplitude $r_s(\theta)$ is not a measurable quantity. In the expression for the reflectance this phase factor will drop out (see section 4.4).

In the expression eq.(4.22) for the transmission amplitude a complex exponential factor appears only in the denominator $D_s(\theta)$, cf. eq.(4.23). It does not cancel partly against such factor in the numerator, as was the case for the reflection amplitude. Therefore $t_s(\theta)$ contains the complex exponential factor

$$F\equiv\exp\left\{-i\frac{\omega}{c}\frac{\mu_+\gamma_e+\epsilon_+\gamma_m}{(\epsilon\mu)_-}\left(n^-\cos\theta-n^+\cos\theta_t\right)\right\} \tag{4.28}$$

It will be clear that, in order to write this expression again as the product of a phase factor and a damping factor, containing I_c, now both n^- and n^+ must be real, so that $(n^- \cos\theta - n^+ \cos\theta_t)$ is real. In this case one finds, using eq.(3.55),

$$F = \exp\{-i\frac{\omega}{c}\,\mathrm{Re}\left[\frac{\mu_+\gamma_e + \epsilon_+\gamma_m}{(\epsilon\mu)_-}\right](n^- \cos\theta - n^+ \cos\theta_t)\} \times \exp\{\frac{\omega}{c} I_c\,(n^- \cos\theta - n^+ \cos\theta_t)\} \tag{4.29}$$

The first factor at the right-hand side of this equation is again the phase factor, which is not expressible in terms of the invariants, but which will drop out in the transmittance (see section 4.4).

If the substrate is absorbing, so that n^+ is complex, it is no longer possible to express the transmittance merely in invariants. In that case there is not only absorption by the interface but also absorption in the substrate over the distance between two possible choices of the dividing surface. As a consequence the transmittance, which is defined upon passing the dividing surface, becomes dependent on the choice of its location. In the following the transmission of light will only be studied in the special case that the medium, through which it is transmitted, is transparent. Then the transmittance is constant in space and expressible in invariants (see section 4.4). If the substrate is weakly absorbing one may still use the formulae given in this book provided that one accounts for the attenuation between the surface and the observer.

In the following section the expressions for $r_s(\theta)$ and $t_s(\theta)$ and the symmetry for the interchange of the electric and the magnetic fields, discussed in section 4.1, will be used in order to calculate the reflection and transmission amplitudes for TM-waves, $r_p(\theta)$ and $t_p(\theta)$ respectively.

In this book the interest is mainly in non-magnetic systems. In such systems one has

$$\mu^{\pm} = 1 \text{ and } \gamma_m = \beta_m = \delta_m = 0 \tag{4.30}$$

This leads to a considerable simplification of the expressions for the amplitudes. In the first place one has $\gamma_{e,m}(\theta) = \gamma_e$ and $\tau_m(\theta) \equiv \tau$, cf. eq.(4.17). Furthermore the complex invariants I_m , I_{em} and $I_{\delta,m}$ are zero. The amplitudes then become

$$r_s(\theta) = r_i \exp\{-i\frac{\omega}{c}(\gamma_e/\,\epsilon_-)(n^- \cos\theta + n^+ \cos\theta_t)\} \times\{n^- \cos\theta - n^+ \cos\theta_t + \left(\frac{\omega}{c}\right)^2 I_\tau\,(n^- \cos\theta + n^+ \cos\theta_t)\}/\; D_s(\theta) \tag{4.31}$$

for the reflection amplitude, and

$$t_s(\theta) = 2\; r_i n^- \cos\theta \;/\; D_s(\theta) \tag{4.32}$$

for the transmission amplitude. The denominator in these expressions is given by

$$D_s(\theta) = \exp\{i\frac{\omega}{c}(\gamma_e/\epsilon_-)(n^- \cos\theta - n^+ \cos\theta_t)\} \times\{n^- \cos\theta + n^+ \cos\theta_t + \left(\frac{\omega}{c}\right)^2 I_\tau\,(n^- \cos\theta - n^+ \cos\theta_t)\} \tag{4.33}$$

In this non-magnetic case the refractive indices are given by $n^{\pm} = \sqrt{\varepsilon^{\pm}}$. Note the fact that the reflection and transmission amplitudes only contain γ_e and I_τ, where the invariant is now given by eq.(3.56). The reason for this is the fact that the electric field of the TE-waves is directed along the interface. An excess of the normal component of the electric field therefore does not occur. The exponential factors in numerators and denominators may again be written as a product of a phase factor and a damping factor. The damping factor now contains the invariant $I_c = \mathrm{Im}(\gamma_e/\epsilon_-)$, cf. eq.(3.56).

The reflectance and transmittance for non-magnetic systems, which can be calculated from the corresponding amplitudes, are given for an arbitrary angle of incidence in section 4.4. Special choices, which are convenient to do experiments, are for instance: normal incidence, 45 degrees and near to the Brewster angle and the angle of total reflection. These are discussed in subsequent sections.

To zeroth order in the thickness of the interface the reflection and transmission amplitudes reduce to

$$\begin{aligned} r_s^0(\theta) &= r_i(n^- \cos\theta - n^+ \cos\theta_t)/(n^- \cos\theta + n^+ \cos\theta_t) \\ t_s^0(\theta) &= 2\, r_i n^- \cos\theta \;/(n^- \cos\theta + n^+ \cos\theta_t) \end{aligned} \tag{4.34}$$

which are the well-known Fresnel amplitudes.

4.3 TM-waves

The medium through which the plane TM-wave is incident is, as in the preceding section, assumed to fill the $z < 0$ half-space. The solution of the wave equation in the $z < 0$ region is a superposition of the incident wave and the reflected wave:

$$\begin{aligned} \mathbf{E}(\mathbf{r},\omega) &= -[r_i(-\cos\theta_i, 0, \sin\theta_i)\exp(i\mathbf{k}_i.\mathbf{r}) \\ &\quad + r_p(\cos\theta_r, 0, \sin\theta_r)\exp(i\mathbf{k}_r.\mathbf{r})]2\pi\delta(\omega-\omega_0) \\ \mathbf{B}(\mathbf{r},\omega) &= n^-(0,1,0)[r_i\exp(i\mathbf{k}_i.\mathbf{r}) + r_p\exp(i\mathbf{k}_r.\mathbf{r})]2\pi\;\delta(\omega-\omega_0) \end{aligned} \tag{4.35}$$

Here r_i is the amplitude of the incident wave. The wave vectors of the incident and reflected light are again given by eq.(4.9). In the present case the direction of the polarization is in the plane of incidence, i.e. the $x - z$ plane. Not only the incident wave but also the reflected wave is a TM-wave. The amplitude of the reflected light is r_p.

The transmitted light is given by

$$\begin{aligned} \mathbf{E}(\mathbf{r},\omega) &= t_p(\cos\theta_t, 0, -\sin\theta_t)\exp(i\mathbf{k}_t.\mathbf{r})2\pi\;\delta(\omega-\omega_0) \\ \mathbf{B}(\mathbf{r},\omega) &= n^+(0,1,0)t_p\exp(i\mathbf{k}_t.\mathbf{r})2\pi\;\delta(\omega-\omega_0) \end{aligned} \tag{4.36}$$

for $z > 0$, where the wave vector of the transmitted light is given by eq.(4.11). The transmitted wave is also a TM-wave. t_p is the amplitude of the transmitted light.

It is now possible to obtain the expressions for r_p and t_p from the expressions for r_s and t_s given in eqs.(4.21)–(4.23) using the symmetry of the Maxwell equations for the interchange of the electric with the magnetic field and the electric displacement field with minus the magnetic induction. As discussed in section 4.1 this is equivalent to the interchange of $\epsilon^{\pm}$ with $-\mu^{\pm}$, γ_e with $-\gamma_m$, β_e with $-\beta_m$, δ_e with δ_m and replacing τ by $-\tau$. For the invariants this implies that one must interchange I_e with $-I_m$, I_{em} with $-I_{em}$, $I_{\delta,e}$ with $I_{\delta,m}$ and I_τ with $-I_\tau$. In this way one finds from eqs.(4.21)–(4.23), after multiplication of numerator and denominator with -1, for the reflection amplitude

$$\begin{aligned} r_p(\theta) &= r_i \exp\{-i\frac{\omega}{c}\frac{\mu_+\gamma_e+\epsilon_+\gamma_m}{(\epsilon\mu)_-}(n^-\cos\theta+n^+\cos\theta_t)\}\{\frac{n^-}{\epsilon^-}\cos\theta-\frac{n^+}{\epsilon^+}\cos\theta_t \\ &-i\frac{\omega}{c}\left[\frac{I_e}{\epsilon^-\epsilon^+}(n^-\sin\theta)^2+\frac{I_{em}}{(\epsilon\mu)_-}\left(\mu_++\epsilon_+\frac{n^-n^+}{\epsilon^-\epsilon^+}\cos(\theta+\theta_t)\right)\right] \\ &-\left(\frac{\omega}{c}\right)^2\frac{\epsilon_+I_eI_{em}}{(\epsilon\mu)_-\epsilon_-\epsilon^-\epsilon^+}(n^-\sin\theta)^2(n^-\cos\theta+n^+\cos\theta_t) \\ &+\frac{1}{2}\left(\frac{\omega}{c}\right)^2\frac{\epsilon_+I_{em}^2}{(\epsilon\mu)_-\epsilon_-\mu_-}[\frac{\mu_+}{(\epsilon\mu)_-}\left(\frac{n^-}{\epsilon^-}\cos\theta-\frac{n^+}{\epsilon^+}\cos\theta_t\right)(n^-\cos\theta+n^+\cos\theta_t) \\ &+\frac{2}{\epsilon^-\epsilon^+}(n^-\sin\theta)^2](n^-\cos\theta+n^+\cos\theta_t) \\ &-\left(\frac{\omega}{c}\right)^2\left[I_\tau-I_{\delta,e}(n^-\sin\theta)^2\right]\left(\frac{n^-}{\epsilon^-}\cos\theta+\frac{n^+}{\epsilon^+}\cos\theta_t\right)\}/\ D_p(\theta) \end{aligned} \tag{4.37}$$

and for the transmission amplitude

$$t_p(\theta) = 2\ r_i(n^+/\epsilon^+)\cos\theta\ /\ D_p(\theta) \tag{4.38}$$

Note the fact that t_p has been multiplied with an additional factor $n^+\epsilon^-/n^-\epsilon^+$, which originates from the fact that **E** and **B** map into **H** and $-$**D** rather than into **B** and $-$**E**. The denominator in these expressions is given by

$$\begin{aligned} D_p(\theta) &= \exp\{i\frac{\omega}{c}\frac{\mu_+\gamma_e+\epsilon_+\gamma_m}{(\epsilon\mu)_-}(n^-\cos\theta-n^+\cos\theta_t)\}\{\frac{n^-}{\epsilon^-}\cos\theta+\frac{n^+}{\epsilon^+}\cos\theta_t \\ &+i\frac{\omega}{c}\left[\frac{I_e}{\epsilon^-\epsilon^+}(n^-\sin\theta)^2+\frac{I_{em}}{(\epsilon\mu)_-}\left(\mu_+-\epsilon_+\frac{n^-n^+}{\epsilon^-\epsilon^+}\cos(\theta-\theta_t)\right)\right] \\ &-\left(\frac{\omega}{c}\right)^2\frac{\epsilon_+I_eI_{em}}{(\epsilon\mu)_-\epsilon_-\epsilon^-\epsilon^+}(n^-\sin\ \theta)^2(n^-\cos\theta-n^+\cos\theta_t) \\ &+\frac{1}{2}\left(\frac{\omega}{c}\right)^2\frac{\epsilon_+I_{em}^2}{(\epsilon\mu)_-\epsilon_-\mu_-}[\frac{\mu_+}{(\epsilon\mu)_-}\left(\frac{n^-}{\epsilon^-}\cos\theta+\frac{n^+}{\epsilon^+}\cos\theta_t\right)(n^-\cos\theta-n^+\cos\theta_t) \\ &+\frac{2}{\epsilon^-\epsilon^+}(n^-\sin\theta)^2](n^-\cos\theta-n^+\cos\theta_t) \\ &-\left(\frac{\omega}{c}\right)^2\left[I_\tau-I_{\delta,e}(n^-sin\theta)^2\right]\left(\frac{n^-}{\epsilon^-}\cos\theta-\frac{n^+}{\epsilon^+}\cos\theta_t\right)\} \end{aligned} \tag{4.39}$$

In the above equations again the notation $a_-\equiv a^+-a^-$ and $a_+\equiv\frac{1}{2}(a^++a^-)$ has been used, as well as the relation eq.(4.24).

The expressions for the reflection and transmission amplitudes for p-polarized light, found above are, of course, just as those for s-polarized light in the previous section, valid up to second order in the interfacial thickness. For the same reason one should, however, not expand the denominator in order to combine the resulting terms to second order with those of the numerator.

Notice the important fact that the complex exponential phase factors, appearing in $r_p(\theta)$ and $r_s(\theta)$ are the same. Consequently these phase factors drop out in the quotient of $r_p(\theta)$ and $r_s(\theta)$, which is the quantity $\rho(\theta)$, measured in ellipsometry (see section 4.8). As it should be, this is therefore again an expression in terms of merely invariants. As will be clear, this is also the case for other measurable quantities, like for instance the reflectance and transmittance (see section 4.4).

For non-magnetic systems, which is the main topic in this book, one may set $\mu^{\pm} = 1$ and $\gamma_m = \beta_m = \delta_m = 0$, cf. eq.(4.30). The complex invariant I_{em} is therefore also zero. The above expressions for the reflection and transmission amplitudes then simplify and are, after multiplication of numerators and denominators with the factor $n^- n^+$, given by

$$\begin{aligned} r_p(\theta) &= r_i \exp\{-i\frac{\omega}{c}(\gamma_e/\ \epsilon_-)(n^- \cos\theta + n^+ \cos\theta_t)\} \\ &\quad \times\{n^+ \cos\theta - n^- \cos\theta_t - i\frac{\omega}{c}\frac{n^-}{n^+} I_e \sin^2\theta \\ &\quad - \left(\frac{\omega}{c}\right)^2 (I_\tau - I_{\delta,e}\epsilon^- \sin^2\theta)(n^+ \cos\theta + n^- \cos\theta_t)\}/\hat{D}_p(\theta) \end{aligned} \tag{4.40}$$

and

$$t_p(\theta) = 2\ r_i n^- \cos\theta\ /\hat{D}_p(\theta) \tag{4.41}$$

The denominator $\hat{D}_p(\theta)$ in these expressions is given by

$$\begin{aligned} \hat{D}_p(\theta) &\equiv n^- n^+ D_p(\theta) \\ &= \exp\{i\frac{\omega}{c}(\gamma_e/\ \epsilon_-)(n^- \cos\theta - n^+ \cos\theta_t)\} \times \{n^+ \cos\theta + n^- \cos\theta_t + \\ &\quad i\frac{\omega}{c}\frac{n^-}{n^+} I_e \sin^2\theta - \left(\frac{\omega}{c}\right)^2 (I_\tau - I_{\delta,e}\epsilon^- \sin^2\theta)(n^+ \cos\theta - n^- \cos\theta_t)\} \end{aligned} \tag{4.42}$$

It is clear from these formulae that the explicit expressions for the amplitudes of p-polarized light in the non-magnetic case do not simplify as much as the amplitudes for s-polarized light. The reason for this more complicated behavior of TM-waves is the fact that the electric field has components both in the direction normal to the surface and in the direction parallel to the surface. As a consequence the interfacial constitutive coefficient γ_e and the invariants I_e, $I_{\delta,e}$ and I_τ appear in the reflection and transmission amplitudes. TE-waves have an electric field which is parallel to the surface and in this case only γ_e and I_τ appear in the amplitudes.

Lekner [11] was the first to introduce invariants in the description of the optical properties of non-magnetic stratified layers. A detailed comparison with his formulae is given in chapter 11 on stratified layers. Here it is only noted that his definition of the reflection amplitude for p-polarized light differs a sign with the definition used in this book.

To zeroth order in the thickness of the interface the reflection and transmission amplitudes again reduce to

$$\begin{aligned} r_p^0(\theta) &= r_i(n^+\cos\theta - n^-\cos\theta_t)/(n^+\cos\theta + n^-\cos\theta_t) \\ t_p^0(\theta) &= 2\,r_i n^-\cos\theta\;/(n^+\cos\theta + n^-\cos\theta_t) \end{aligned} \tag{4.43}$$

which are the well-known Fresnel amplitudes.

4.4 Reflectance and transmittance in non-magnetic systems

The average energy flow is given by the time average of the Poynting vector which, for a plane wave, may be written in the following form ([14], page 33)

$$\mathbf{S} = \frac{1}{2}c\,\mathrm{Re}[\mathbf{E}\times\mathbf{H}^*] \equiv S\hat{\mathbf{k}}_0 \tag{4.44}$$

where $\hat{\mathbf{k}}_0$ is the unit vector in the direction of the wave vector. The absolute value S is the light intensity, while the direction of the energy flow is for a plane wave in the direction of the wave vector. The asterisk denotes complex conjugation.

The reflectances are defined as the ratios of the intensity of the reflected and the incident light. Consequently one finds, using the above expression for the incident and reflected plane waves,

$$R_s \equiv |r_s/r_i|^2 \text{ and } R_p \equiv |r_p/r_i|^2 \tag{4.45}$$

Analogously one finds for the transmittances

$$T_s \equiv (t_s/r_i)^2\frac{(n^+/\mu^+)\cos\theta_t}{(n^-/\mu^-)\cos\theta} \text{ and } T_p \equiv (t_p/r_i)^2\frac{(n^+/\mu^+)\cos\theta_t}{(n^-/\mu^-)\cos\theta} \tag{4.46}$$

In the last equation both n^-, μ^- and n^+, μ^+ are real. This is in agreement with the assumption, made in the previous sections, that the transmission of light will only be studied in the special case that also the medium, through which the light is transmitted, is transparent. (For reflection only n^- and μ^- are assumed to be real). The angle dependent factors originate from the fact that the component of the energy flow orthogonal to the interface is used, rather than the energy flow in the direction of propagation. For the reflected light this makes no difference. Straightforward substitution of the amplitudes given in the previous sections leads to explicit expressions for the reflectances and transmittances.

Before giving the general expressions for the reflectances and transmittances for the non-magnetic case, it is convenient to discuss the so-called Fresnel values. These are found, if one considers a surface where the refractive index changes discontinuously from one constant value to another constant value. For this purpose one simply substitutes the Fresnel amplitudes (4.34) and (4.43) into the above equations for the non-magnetic case. In the study of interfaces the deviations of the reflectances and transmittances from their Fresnel values contain information about the interfacial constitutive coefficients and therefore about the structure of the surface. One therefore usually divides the experimental values of the reflectances and transmittances by these Fresnel values which are given by

$$\begin{aligned} R_s^0(\theta) &= |(n^- \cos\theta - n^+ \cos\theta_t)/(n^- \cos\theta + n^+ \cos\theta_t)|^2 \\ T_s^0(\theta) &= 4n^- n^+ \cos\theta \ \cos\theta_t/(n^- \cos\theta + n^+ \cos\theta_t)^2 \end{aligned} \tag{4.47}$$

and

$$\begin{aligned} R_p^0(\theta) &= |(n^+ \cos\theta - n^- \cos\theta_t)/(n^+ \cos\theta + n^- \cos\theta_t)|^2 \\ T_p^0(\theta) &= 4n^- n^+ \cos\theta \ \cos\theta_t/(n^+ \cos\theta + n^- \cos\theta_t)^2 \end{aligned} \tag{4.48}$$

where the dependence of θ_t on θ is given by Snell's law, eq.(4.14). In the above expressions for the transmittances, both n^- and n^+ are assumed to be real, in those for the reflectances only n^-. This is the reason that absolute value symbols have been used in the latter expressions. Notice the fact that for a transparent substrate the sum of the Fresnel reflectance and transmittance is equal to unity

$$R_s^0(\theta) + T_s^0(\theta) = 1 \text{ and } R_p^0(\theta) + T_p^0(\theta) = 1 \tag{4.49}$$

Using Snell's law one may rewrite the reflectances in the following form

$$R_s^0(\theta) = |\sin(\theta - \theta_t)/\sin(\theta + \theta_t)|^2 \tag{4.50}$$

$$R_p^0(\theta) = |\operatorname{tg}(\theta - \theta_t)/\operatorname{tg}(\theta + \theta_t)|^2 \tag{4.51}$$

These expressions are valid both for a transparent and an absorbing substrate. There are two angles of incidence for which the denominator is not finite. The first possibility is at normal incidence when both denominators and numerators are zero. For that case it is more convenient to use eq.(4.48) which gives

$$R_s^0(0) = R_p^0(0) = |(n^- - n^+)/(n^- + n^+)|^2 \tag{4.52}$$

The other possibility for a transparent substrate is an infinite denominator, which only occurs in R_p^0 at the so-called Brewster angle θ_B^0, when the sum of the angle of incidence θ_B^0 and the angle of transmission is equal to 90 degrees:

$$\theta_B^0 + \theta_t(\theta_B^0) = \pi/2 \tag{4.53}$$

At the Brewster angle there is no reflection of p-polarized light:

$$R_p^0(\theta_B^0) = 0 \tag{4.54}$$

and reflected light is as a consequence fully polarized normal to the plane of incidence. This implies, using eq.(4.48), that

$$\begin{aligned} n^+ \cos\theta_B^0 &= n^- \cos\ \theta_t(\theta_B^0) = n^- \cos(\frac{\pi}{2} - \theta_B^0) = n^- \sin\ \theta_B^0 \\ &\Rightarrow \ \theta_B^0 = \arctan(n^+/n^-) \end{aligned} \tag{4.55}$$

For an absorbing substrate one also introduces a Brewster angle. The definition is then, however, less clear. Possible definitions for that case will be discussed in section 4.6 on the reflection of p-polarized light, and in section 4.8 on ellipsometry. The reason to study reflection near the Brewster angle is the great sensitivity to how much the surface differs from the step function Fresnel interface. In this way one may, for instance, estimate the amount of oxide on a metal surface, study surface roughness and other such surface properties. This will be discussed in detail in the following sections.

Substitution of the reflection amplitude, eq.(4.31) with eq.(4.33), for s-polarized light into eq.(4.45) gives after some straightforward algebra, using eq.(3.56), up to second order in the interfacial thickness,

$$\begin{aligned} R_s(\theta) &= \exp(4\frac{\omega}{c}I_c n^- \cos\theta)\{|n^- \cos\theta - n^+ \cos\theta_t|^2 + 2\left(\frac{\omega}{c}\right)^2 \{I'_\tau\,[(n^- \cos\theta)^2 \\ &\quad -|n^+ \cos\theta_t|^2] - 2I''_\tau n^- \cos\theta\ \mathrm{Im}(n^+ \cos\theta_t)\}\}/\ \Delta_s(\theta) \end{aligned} \tag{4.56}$$

for the reflectance of a surface. The denominator is given by

$$\begin{aligned} \Delta_s(\theta) &\equiv |n^- \cos\theta + n^+ \cos\theta_t|^2 + 2\left(\frac{\omega}{c}\right)^2 \{I'_\tau\,[(n^- \cos\theta)^2 - |n^+ \cos\theta_t|^2] \\ &\quad +2I''_\tau n^- \cos\theta\ \mathrm{Im}(n^+ \cos\theta_t)\} \end{aligned} \tag{4.57}$$

Here and in the following the prime and double prime indicate the real part and the imaginary part of the corresponding quantity, respectively.

All formulae above are for the general case that the ambient, through which the light is incident, is transparent. The substrate may either be transparent or absorbing. If the substrate is absorbing the angle of transmission is always a complex number. If the substrate is transparent and its refractive index larger than the refractive index of the ambient, the angle of transmission is real and can therefore be interpreted as the physical direction of the transmitted light. If the substrate is transparent and its refractive index smaller than the refractive index of the ambient, one has the so-called phenomenon of total reflection. In that case the angle of transmission is real for an angle of transmission smaller than the angle of total reflection, which is given by $\arcsin(n^+/n^-)$. See in this connection the text below eq.(4.14). When the light is incident at an angle larger than the angle of total reflection, the angle of transmission is given by $\cos\theta_t = i[(\epsilon^-/\epsilon^+)\sin^2\theta - 1]^{1/2}$ and is therefore a complex number. In this respect this case becomes similar to the case with an absorbing substrate and one should use the expression for the reflectance given above.

In the case of a transparent substrate, and provided that the angle of incidence is smaller than the angle of total reflection in the case that $n^- > n^+$, the above equations reduce to

$$R_s(\theta) = \exp(4\frac{\omega}{c}I_c n^- \cos\theta)[(n^- \cos\theta - n^+ \cos\theta_t)^2 - 2\left(\frac{\omega}{c}\right)^2 I'_\tau\,(\epsilon^+ - \epsilon^-)]/\ \Delta_s(\theta) \tag{4.58}$$

with

$$\Delta_s(\theta) \equiv (n^- \cos\theta + n^+ \cos\theta_t)^2 - 2\left(\frac{\omega}{c}\right)^2 I'_\tau\,(\epsilon^+ - \epsilon^-) \tag{4.59}$$

where also Snell's law, eq.(4.14), has been used.

For a transparent substrate, provided that the angle of incidence is smaller than the angle of total reflection in the case that $n^- > n^+$, one finds similarly for the transmittance, upon substitution of eqs.(4.32) and (4.33) into eq.(4.46) (for $\mu^\pm = 1$) and using eq.(3.56) and Snell's law,

$$T_s(\theta) = 4\, n^- n^+ \cos\theta\, \cos\theta_t \exp[2\frac{\omega}{c} I_c (n^- \cos\theta - n^+ \cos\theta_t)] / \Delta_s(\theta) \tag{4.60}$$

where the denominator $\Delta_s(\theta)$ is given by eq.(4.59).

As discussed in the previous chapter, both I_τ and I_c are invariants. The above expressions are therefore explicitly independent of the choice of the dividing surface. Another quantity of interest is the energy adsorption by the interface. If the substrate is absorbing, it is not possible to distinguish between absorption by the interface versus the absorption just below the interface in the substrate. For a non-absorbing substrate one finds, up to second order in the interfacial thickness, for the absorption by the interface

$$\begin{aligned} Q_s(\theta) &\equiv 1 - R_s(\theta) - T_s(\theta) \\ &= 4\frac{\omega}{c} n^- \cos\theta\, I_c(\epsilon^+ - \epsilon^-) \exp[2\frac{\omega}{c} I_c (n^- \cos\theta - n^+ \cos\theta_t)] / \Delta_s(\theta) \end{aligned} \tag{4.61}$$

where eqs.(4.58) and (4.60) have been used and where $\Delta_s(\theta)$ is given by eq.(4.59). It follows, that the interface does not absorb energy from the s-polarized light wave, if I_c is equal to zero. In view of the fact that I_c is proportional to γ_e'' it also follows that there is no absorption of energy by the interface, if the imaginary part of the constitutive coefficient γ_e is equal to zero. For a transparent substrate γ_e'' is an invariant: this in accordance with eq.(3.51).

If one measures the reflectance, transmittance or the absorption of s-polarized light by an interface between two different transparent media, one obtains information about two invariants I_c and I_τ'. In the case that the substrate is absorbing, a reflection experiment with s-polarized light will, in addition, give information about I_τ''.

Similarly one finds for p-polarized light by substitution of eqs.(4.40) and eq.(4.42) into eqs.(4.45) and using eq.(3.56), up to second order in the interfacial thickness,

$$\begin{aligned} R_p(\theta) = \; & \exp(4\frac{\omega}{c} I_c n^- \cos\theta) \{ |n^+ \cos\theta - n^- \cos\theta_t - i\frac{\omega}{c}\frac{n^-}{n^+} I_e \sin^2\theta|^2 \\ & -2\left(\frac{\omega}{c}\right)^2 [I_\tau' - I_{\delta,e}'(n^- \sin\theta)^2](|n^+|^2 \cos^2\theta - (n^-)^2 |\cos\theta_t|^2) \\ & +4\left(\frac{\omega}{c}\right)^2 [I_\tau'' - I_{\delta,e}''(n^- \sin\theta)^2] \operatorname{Im}(n^- n^{+*} \cos\theta\, \cos\theta_t)\} / \Delta_p(\theta) \end{aligned} \tag{4.62}$$

for the reflectance of a surface. The denominator is given by

$$\begin{aligned} \Delta_p(\theta) \equiv \; & |n^+ \cos\theta + n^- \cos\theta_t + i\frac{\omega}{c}\frac{n^-}{n^+} I_e \sin^2\theta|^2 \\ & -2\left(\frac{\omega}{c}\right)^2 [I_\tau' - I_{\delta,e}'(n^- \sin\theta)^2](|n^+|^2 \cos^2\theta - (n^-)^2 |\cos\theta_t|^2) \\ & -4\left(\frac{\omega}{c}\right)^2 [I_\tau'' - I_{\delta,e}''(n^- \sin\theta)^2] \operatorname{Im}(n^- n^{+*} \cos\theta\, \cos\theta_t) \end{aligned} \tag{4.63}$$

In the case of a transparent substrate, and in the absence of total reflection, the above equations reduce to

$$\begin{aligned} R_p(\theta) &= \exp(4\frac{\omega}{c}I_c n^- \cos\theta)\{(n^+\cos\theta - n^-\cos\theta_t)^2 \\ &+2\frac{\omega}{c}\frac{n^-}{n^+}I_e'' \sin^2\theta(n^+\cos\theta - n^-\cos\theta_t) + \left(\frac{\omega}{c}\right)^2\frac{\epsilon^-}{\epsilon^+}|I_e|^2\sin^4\theta \\ &-2\left(\frac{\omega}{c}\right)^2(I_\tau' - I_{\delta,e}'\epsilon^-\sin^2\theta)(\epsilon^+\cos^2\theta - \epsilon^-\cos^2\theta_t)\}/\ \Delta_p(\theta) \end{aligned} \quad (4.64)$$

with

$$\begin{aligned} \Delta_p(\theta) &= (n^+\cos\theta + n^-\cos\theta_t)^2 - 2\frac{\omega}{c}\frac{n^-}{n^+}I_e''\sin^2\theta(n^+\cos\theta + n^-\cos\theta_t) \\ &+\left(\frac{\omega}{c}\right)^2\{\frac{\epsilon^-}{\epsilon^+}|I_e|^2\sin^4\theta \\ &-2(I_\tau' - I_{\delta,e}'\epsilon^-\sin^2\theta)(\epsilon^+\cos^2\theta - \epsilon^-\cos^2\theta_t)\} \end{aligned} \quad (4.65)$$

For the transmittance through the transparent substrate one finds in an analogous way, by substituting eqs.(4.41) and (4.42) into eq.(4.46) (for $\mu^\pm = 1$), and using eq.(3.56),

$$T_p(\theta) = 4\,n^-n^+\cos\theta\ \cos\theta_t\exp[2\frac{\omega}{c}I_c(n^-\cos\theta - n^+\cos\theta_t)]/\ \Delta_p(\theta) \quad (4.66)$$

where the denominator $\Delta_p(\theta)$ is given by eq.(4.65).

The energy absorption by the interface is given, again for the transparent substrate, by

$$\begin{aligned} Q_p(\theta) &\equiv 1 - R_p(\theta) - T_p(\theta) = 4\frac{\omega}{c}n^-\cos\theta\ \exp[2\frac{\omega}{c}I_c(n^-\cos\theta - n^+\cos\theta_t)] \\ &\times\{I_c(\epsilon^+ - \epsilon^-)(\sin^2\theta + \cos^2\theta_t) - I_e''\sin^2\theta \\ &+2\frac{\omega}{c}\frac{1}{n^+}\sin^2\theta\ \cos\theta_t[I_c^2(\epsilon^+ - \epsilon^-)^2 - I_cI_e''(\epsilon^+ - \epsilon^-)]\}/\ \Delta_p(\theta) \end{aligned} \quad (4.67)$$

where eqs.(4.64) and (4.66) have been used and where $\Delta_p(\theta)$ is given by eq.(4.65). It now follows that the interface does not absorb energy from the p-polarized light, if both I_c and I_e'' are zero. It follows from the definition of these invariants, eq.(3.56) with ϵ^+ and ϵ^- real, that $\gamma_e'' = \beta_e'' = 0$. Consequently there is no absorption of energy by the interface, if the imaginary parts of the constitutive coefficients γ_e and β_e are equal to zero.

If one measures the reflectance, transmittance or the absorption of p-polarized light by an interface between two different transparent media, one obtains information about five invariants I_c, I_e', I_e'', I_τ' and $I_{\delta,e}'$. Two of these invariants (I_c and I_τ') may alternatively be obtained using s-polarized light. The other three then subsequently follow from an experiment with p-polarized light. In the case that the substrate is absorbing, a reflection experiment with p-polarized light will, in addition, give information about I_τ'' and $I_{\delta,e}''$, of which the former may also be obtained using s-polarized light.

As one may verify in eqs.(4.64)–(4.67), the only deviations from Fresnel, which are linear in the thickness of the interface, are due to the invariants I_c and I_e'' or, alternatively, to the imaginary parts γ_e'' and β_e'', as follows from eq.(3.56) with ϵ^+ and ϵ^- real. If the surface is non-absorbing, i.e. if $Q_p = Q_s = 0$, these imaginary parts are zero. Consequently the deviation from the Fresnel values of the reflectance and transmittance are for each angle of incidence at least quadratic in the thickness of the interface. This shows the importance of the constitutive coefficients τ and δ_e, and therefore of the invariants I_τ and $I_{\delta,e}$, for the proper understanding of the optical properties of in particular non-absorbing interfaces. For absorbing surfaces the deviations from the Fresnel values will be dominated by β_e'' and γ_e'', and therefore by I_c and I_e''.

4.5 Reflectance and transmittance in non-magnetic systems at normal incidence

In view of the fact that many reflection and transmission experiments are performed at normal incidence it is useful to discuss this special case separately. At normal incidence the expressions for the reflection and transmission amplitudes simplify considerably. Since furthermore one often measures the reflectance and transmittance at normal incidence, it is convenient to give these formulae explicitly. For normally incident light there is no longer a difference between p- and s-polarized light. From eqs.(4.56) and (4.57), or eqs.(4.62) and (4.63), it follows that the reflectance is given by

$$\begin{aligned} R_s(0) &= R_p(0) = \exp(4\frac{\omega}{c}n^- I_c)\{|n^+ - n^-|^2 \\ &-2\left(\frac{\omega}{c}\right)^2 I_\tau' [|n^+|^2 - (n^-)^2] - 4\left(\frac{\omega}{c}\right)^2 I_\tau'' n^- \operatorname{Im}(n^+)\}/\ \Delta_s(\theta) \end{aligned} \quad (4.68)$$

where

$$\Delta_s(0) = \Delta_p(0) = |n^+ + n^-|^2 - 2\left(\frac{\omega}{c}\right)^2 I_\tau' [|n^+|^2 - (n^-)^2] + 4\left(\frac{\omega}{c}\right)^2 I_\tau'' n^- \operatorname{Im}(n^+) \quad (4.69)$$

In the case of a transparent substrate this reduces to

$$R_s(0) = R_p(0) = \exp(4\frac{\omega}{c}n^- I_c)\{(n^+ - n^-)^2 - 2\left(\frac{\omega}{c}\right)^2 I_\tau' (\epsilon^+ - \epsilon^-)\}/\ \Delta_s(\theta) \quad (4.70)$$

with

$$\Delta_s(0) = \Delta_p(0) = (n^+ + n^-)^2 - 2\left(\frac{\omega}{c}\right)^2 I_\tau' (\epsilon^+ - \epsilon^-) \quad (4.71)$$

The transmittance follows from eqs.(4.60) and (4.59), or eqs.(4.66) and (4.65),

$$T_s(0) = T_p(0) = 4\, n^- n^+ \exp[2\frac{\omega}{c}(n^- - n^+)I_c]/\ \Delta_s(0) \quad (4.72)$$

where $\Delta_s(0)$ is given by eq.(4.71).

The energy absorption by the interface is given by

$$Q_s(0) = Q_p(0) = 4\, n^- \frac{\omega}{c} I_c(\epsilon^+ - \epsilon^-) \exp[2\frac{\omega}{c}(n^- - n^+)I_c] / \,\Delta_s(0) \qquad (4.73)$$

as follows from eqs.(4.61) or (4.67).

From the above equations it is found that a reflection experiment with normally incident light on a surface gives at most information about the 3 invariants I_c, I'_τ and I''_τ. This is the case, when the substrate below this surface is absorbing. In the case of a transparent substrate only the 2 invariants I_c and I'_τ can be obtained, both from the reflectance, the transmittance and the absorption of energy. In view of the fact that for normally incident light there is no difference between *p*- and *s*-polarized light, this is only natural.

For a non-absorbing surface on a transparent substrate $I_c = 0$, so that only the invariant I'_τ appears in the reflection and transmission intensities. Experiments at normal incidence are therefore a convenient method to obtain this invariant.

4.6 Reflectance of *p*-polarized light in non-magnetic systems near the Brewster angle

Another interesting special case is the reflectance of *p*-polarized light around the Brewster angle. As the case of a transparent substrate is much simpler, this case is considered first. In section 4.4 it was shown that the Fresnel value of this reflectance is zero in the Brewster angle $\theta_B^0 = \text{arctg}(n^+/n^-)$. Due to the finite value of the interfacial constitutive coefficients the reflectance remains finite for all angles of incidence. Near the Fresnel value of the Brewster angle the reflectance still has a minimum which is, however, no longer equal to zero. The location of this minimum of R_p is also shifted away from the original Brewster angle to a new angle, which is also often called the Brewster angle.

In a reflection experiment one may measure the shift of the Brewster angle, $\delta\theta_p$, the value of the reflectance at the Brewster angle θ_B^0 and the value $R_p(\theta_B^0 + \delta\theta_p)$ in the new minimum without doing a complete analysis of the angular dependence of $R_p(\theta)$. As will become clear below this has certain practical advantages. One may calculate these quantities in a tedious but straightforward manner from the general expression (4.64) together with eq.(4.65). One then finds for the minimum value of the reflectance

$$R_p(\theta_B^0 + \delta\theta_p) = \frac{1}{4}(\frac{\omega}{c})^2(\epsilon^+ + \epsilon^-)^{-1}(I'_e)^2 \qquad (4.74)$$

The value of the reflectance in the original Brewster angle is found to be

$$R_p(\theta_B^0) = \frac{1}{4}\left(\frac{\omega}{c}\right)^2 (\epsilon^+ + \epsilon^-)^{-1}|I_e|^2 \qquad (4.75)$$

The shift of the Brewster angle is given by

$$\begin{aligned}\delta\theta_p \;=\; & \frac{n^- n^+ \epsilon^+}{(\epsilon^+ - \epsilon^-)^2(\epsilon^+ + \epsilon^-)^3}[(\frac{\omega}{c})(\epsilon^+ - \epsilon^-)(\epsilon^+ + \epsilon^-)^{3/2} I''_e \\ & + \frac{1}{2}(\frac{\omega}{c})^2\{\epsilon^-(3\epsilon^+ - \epsilon^-)(I''_e)^2 - ((\epsilon^+)^2 + 4\,\epsilon^+\epsilon^- + (\epsilon^-)^2)(I'_e)^2 \\ & -4((\epsilon^+)^2 - (\epsilon^-)^2)[(\epsilon^+ + \epsilon^-)I'_\tau - \epsilon^-\epsilon^+ I'_{\delta,e}]\}] \end{aligned} \qquad (4.76)$$

All these values have been calculated to second order in the thickness of the interface divided by the wavelength in vacuum. The minimum value is particularly useful in view of the fact that it gives the value of the real part I_e' of one of the invariants directly. A subsequent measurement of the value of the reflectance in the original Brewster angle then also gives the imaginary part of this invariant. The shift of the Brewster angle depends in addition on the real part of two other invariants. The reflectance of p-polarized light for a transparent substrate, eq.(4.64), also depends on I_c. This invariant, which is due to absorption by the interface, only contributes to the asymmetry of the reflectance around the minimum and may as such be measured by scanning reflectometry.

It should be noted that the value of both reflectances above are of second order in the thickness of the interface divided by the wavelength in vacuum, whereas the shift of the Brewster angle contains also a linear term if the surface is absorbing, i.e. I_e'' is unequal to zero. For an absorbing interface the linear term will usually dominate in which case the shift simplifies to

$$\delta\theta_p = \frac{n^- n^+ \epsilon^+}{(\epsilon^+ - \epsilon^-)(\epsilon^+ + \epsilon^-)^{3/2}}(\frac{\omega}{c})I_e'' \qquad (4.77)$$

For an absorbing interface the shift of the Brewster angle and the value of the minimum reflectance of p-polarized light then determine both the real and the imaginary part of the invariant I_e.

If the surface is non-absorbing the shift in the location of the minimum simplifies to

$$\begin{aligned}\delta\theta_p \;=\; & -\frac{n^- n^+ \epsilon^+}{2(\epsilon^+ - \epsilon^-)^2(\epsilon^+ + \epsilon^-)^3}(\frac{\omega}{c})^2\{((\epsilon^+)^2 + 4\,\epsilon^+\epsilon^- + (\epsilon^-)^2)(I_e')^2 \\ & +4((\epsilon^+)^2 - (\epsilon^-)^2)[(\epsilon^+ + \epsilon^-)I_\tau' - \epsilon^-\epsilon^+ I_{\delta,e}']\} \qquad (4.78)\end{aligned}$$

For a non-absorbing surface the value and the location of the minimum thus determine the invariant I_e, which is now real, and a combination of two other invariants which, it should be noted, are also real. If these measurements for a non-absorbing surface are combined with those at normal incidence which yield I_τ one may calculate $I_{\delta,e}$.

In the case of an absorbing substrate the Fresnel reflectance of p-polarized light has a minimum which is larger than zero. The angle for which this minimum occurs, is a natural extension of the Brewster angle for scanning angle reflectometry. The minimum of $R_p^0(\theta)$, given in eq.(4.48), is found to be given by:

$$\mathrm{Re}[(\mathrm{tg}\,^2\theta_B^0 - \epsilon^+/\epsilon^-)\cos\theta_{B,t}^0/n^+] = 0 \qquad (4.79)$$

For a transparent substrate n^+ , and consequently $\cos\theta_{B,t}^0$ and ϵ^+ , are real, so that this expression for the Brewster angle reduces to the usual expression $\mathrm{tg}\,\theta_B^0 = n^+/n^-$. In order to solve the above equation for the case that the substrate is absorbing, it is convenient to eliminate $\cos\theta_{B,t}^0$. Using Snell's law for this purpose one then obtains

$$\begin{aligned}& 1 - 2\sin^2\theta_B^0 + [1 - 3(\epsilon^-/|\epsilon^+|)^2]\sin^4\theta_B^0 \\ & +2(\epsilon^-/|\epsilon^+|)^2[1 + (\epsilon^-/|\epsilon^+|)(\mathrm{Re}\,\epsilon^+/|\epsilon^+|)]\sin^6\theta_B^0 = 0 \qquad (4.80)\end{aligned}$$

This equation has three solutions for $\sin^2\theta_B^0$, only one of which is the correct one. The others were introduced in the process of reformulating the original equation into an algebraic form. One may solve the above equation using standard procedures and one then obtains as solution, introducing $e_1 \equiv \epsilon^-/|\epsilon^+|$ and $e_2 \equiv \mathrm{Re}\,\epsilon^+/|\epsilon^+|$,

$$\theta_B^0 = \arcsin[\{E_1 \cos[\frac{1}{3}(\arcsin E_2 + \pi)] - E_3\}^{1/2}] \tag{4.81}$$

where

$$\begin{aligned} E_1 &\equiv \frac{(1+6e_1^2+9e_1^4+12e_1^3e_2)^{1/2}}{3e_1^2(1+e_1e_2)} \\ E_2 &\equiv \frac{1+9e_1^2+27e_1^4(1-e_1^2)+18e_1^3e_2+54e_1^5e_2(1+e_1e_2)}{(1+6e_1^2+9e_1^4+12e_1^3e_2)^{3/2}} \\ E_3 &\equiv \frac{1-3e_1^2}{6e_1^2(1+e_1e_2)} \end{aligned} \tag{4.82}$$

For a transparent substrate, $e_2 = 1$, it may again be verified that $\mathrm{tg}\,\theta_B^0 = n^+/n^-$. Although this result is obvious the verification from eq.(4.81) is cumbersome.

The minimum Fresnel reflectance for p-polarized light is found to be given by

$$\begin{aligned} R_p^0(\theta_B^0) &= \{\cos^2\theta_B^0 + e_1(1-2e_1e_2\sin^2\theta_B^0 + e_1^2\sin^4\theta_B^0)^{1/2} \\ &\quad -2e_1\cos\theta_B^0(1-2e_1e_2\sin^2\theta_B^0+e_1^2\sin^4\theta_B^0)^{1/4} \\ &\quad \times[\mathrm{Re}(n^+/n^-)\cos\Phi + \mathrm{Im}(n^+/n^-)\sin\Phi]\}/\,\Delta_p^0 \end{aligned} \tag{4.83}$$

where

$$\Phi \equiv \frac{1}{2}\,\mathrm{arctg}[(e_1(1-e_2^2)^{1/2}\sin^2\theta_B^0)/(1-e_1e_2\sin^2\theta_B^0)] \tag{4.84}$$

$$\begin{aligned} \Delta_p^0 &\equiv \cos^2\theta_B^0 + e_1(1-2e_1e_2\sin^2\theta_B^0+e_1^2\sin^4\theta_B^0)^{1/2} \\ &\quad +2e_1\cos\theta_B^0(1-2e_1e_2\sin^2\theta_B^0+e_1^2\sin^4\theta_B^0)^{1/4} \\ &\quad \times[\mathrm{Re}(n^+/n^-)\cos\Phi + \mathrm{Im}(n^+/n^-)\sin\Phi] \end{aligned} \tag{4.85}$$

For the absorbing substrate the location of the minimum of the reflectance of p-polarized light shifts to a somewhat different angle θ_B, while also the value of the minimum changes, due to the interfacial susceptibilities. Due to their complexity it is not useful to give explicit expressions for these modifications in terms of the appropriate invariants, as was done above for the case of a transparent substrate. Using eq.(4.62) one may evaluate the reflectance around the zeroth order Brewster angle. It is most convenient to scan this reflectance around this angle. The invariants may then be obtained by a comparison with the Fresnel reflectance. In this case also the imaginary parts of the invariants I_τ and $I_{\delta,e}$ affect the value of the reflectance.

4.7 Reflectometry around an angle of incidence of 45 degrees

The reason to do reflectometry around an angle of incidence of 45 degrees is the following identity for the Fresnel values of the reflectances

$$R_p^0(\pi/4) - [R_s^0(\pi/4)]^2 = 0 \tag{4.86}$$

Remarkable is the fact that this equality is valid for both transparent and absorbing substrates. The Fresnel value of $R_p - R_s^2$ is positive for $\theta < \pi/4$ and negative for $\theta > \pi/4$. Due to the finite value of the interfacial constitutive coefficients this cross-over is shifted to an angle $\pi/4 + \delta\theta_{ps}$. The two relevant quantities are the shift of the cross-over $\delta\theta_{ps}$ and the value of $R_p(\pi/4) - [R_s(\pi/4)]^2$.

For a transparent substrate the general formulae, eqs.(4.58) and (4.64), give after a tedious but straightforward calculation

$$\begin{aligned} R_p(\pi/4) - R_s^2(\pi/4) = R_p^0(\pi/4)\{2^{3/2}\frac{\omega}{c}\frac{n^-}{(\epsilon^+ - \epsilon^-)^2}[\epsilon^+ I_e'' - (\epsilon^+ - \epsilon^-)^2 I_c] \\ -4(\frac{\omega}{c})^2\frac{n^-\epsilon^-(2\epsilon^+ - \epsilon^-)^{1/2}}{(\epsilon^+ - \epsilon^-)^4}[(\epsilon^+ - \epsilon^-)^2(2\ I_\tau' - \epsilon^+ I_{\delta,e}') - \frac{1}{2}\epsilon^+((I_e')^2 - (I_e'')^2)] \\ +4(\frac{\omega}{c})^2\frac{\epsilon^-}{(\epsilon^+ - \epsilon^-)^4}[\epsilon^+ I_e'' - (\epsilon^+ - \epsilon^-)^2 I_c][\epsilon^+ I_e'' + 3(\epsilon^+ - \epsilon^-)^2 I_c]\} \end{aligned} \tag{4.87}$$

For the shift of the cross-over one finds

$$\begin{aligned} \delta\theta_{ps} &= 2^{-3/2}(\frac{\omega}{c})(\epsilon^+ - \epsilon^-)^{-1}(2\epsilon^+ - \epsilon^-)^{-1/2}[\epsilon^+ I_e'' - (\epsilon^+ - \epsilon^-)^2 I_c] \\ &+\frac{1}{8}(\frac{\omega}{c})^2(\epsilon^+ - \epsilon^-)^{-3}\{2\epsilon^-[\epsilon^+(I_e')^2 - 8(\epsilon^+ - \epsilon^-)^2(I_\tau' - \frac{1}{2}\epsilon^+ I_{\delta,e}')] \\ &-\frac{(\epsilon^+ - \epsilon^-)}{(2\epsilon^+ - \epsilon^-)^2}[(\epsilon^+ - \epsilon^-)^3(4(\epsilon^+)^2 - 3\epsilon^+\epsilon^- + (\epsilon^-)^2)I_c^2 + 2(\epsilon^+ - \epsilon^-)^2\epsilon^-\epsilon^+ I_c I_e'' \\ &-\epsilon^+(4(\epsilon^+)^2 - 5\epsilon^+\epsilon^- + 2(\epsilon^-)^2)(I_e'')^2]\} \end{aligned} \tag{4.88}$$

The last two equations are again given in terms of the invariants to second order in the thickness of the interface.

For an absorbing interface it will usually be sufficient to retain only the first order term and the above equations for a transparent substrate simplify to

$$R_p(\pi/4) - R_s^2(\pi/4) = R_p^0(\pi/4)2^{3/2}\frac{\omega}{c}\frac{n^-}{(\epsilon^+ - \epsilon^-)^2}[\epsilon^+ I_e'' - (\epsilon^+ - \epsilon^-)^2 I_c] \tag{4.89}$$

and for the shift of the cross-over

$$\delta\theta_{ps} = 2^{-3/2}(\frac{\omega}{c})(\epsilon^+ - \epsilon^-)^{-1}(2\epsilon^+ - \epsilon^-)^{-1/2}[\epsilon^+ I_e'' - (\epsilon^+ - \epsilon^-)^2 I_c] \tag{4.90}$$

Notice the fact that both the value of $R_p(\pi/4) - R_s^2(\pi/4)$ and the value of $\delta\theta_{ps}$ are proportional to the same combination of invariants, so that it is sufficient to either measure the value at 45 degrees or the shift of the cross-over.

If the interface is not absorbing, the above expressions, (4.87) and (4.88), for a transparent substrate also simplify considerably. One finds

$$R_p(\pi/4) - R_s^2(\pi/4) = 2\ R_p^0(\pi/4)(\frac{\omega}{c})^2 \frac{n^-(2\epsilon^+ - \epsilon^-)^{1/2}}{(\epsilon^+ - \epsilon^-)^4}$$
$$\times \epsilon^-[\epsilon^+(I_e')^2 - 2(\epsilon^+ - \epsilon^-)^2(2I_\tau' - \epsilon^+ I_{\delta,e}')] \qquad (4.91)$$

and

$$\delta\theta_{ps} = \frac{1}{4}(\frac{\omega}{c})^2(\epsilon^+ - \epsilon^-)^{-3}\epsilon^-[\epsilon^+(I_e')^2 - 2(\epsilon^+ - \epsilon^-)^2(2I_\tau' - \epsilon^+ I_{\delta,e}')] \qquad (4.92)$$

Both the value of $R_p(\pi/4) - R_s^2(\pi/4)$ and the value of $\delta\theta_{ps}$ are again proportional to the same combination of invariants, so that it is sufficient to either measure the value at 45 degrees or the shift of the cross-over.

Concluding this section one finds, though the relation given in eq.(4.86) for transparent and absorbing substrates is remarkable, that scanning the reflectances around 45 degrees gives a relatively limited amount of information about the surface. It is concluded that it is not worthwhile to study, for instance, the case of the absorbing substrate in more detail.

4.8 Ellipsometry

In ellipsometry [27] one studies the ellipsometric function, which is given by the ratio of the reflected amplitudes of p- and s-polarized light

$$\rho(\theta) \equiv \frac{r_p(\theta)}{r_s(\theta)} \equiv \mathrm{tg}(\psi(\theta))e^{i\Delta(\theta)} \qquad (4.93)$$

The so-called ellipsometric angles ψ and Δ are real functions of θ which are chosen such that $0 \leq \psi(\theta) \leq \pi/2$ and $0 \leq \Delta(\theta) < 2\pi$. If one considers a system in which both the ambient and the substrate are non-absorbing, and provided that there is no total reflection, the Fresnel amplitudes $r_p^0(\theta)$ and $r_s^0(\theta)$, cf. eqs.(4.43) and (4.34), give for $\rho^0(\theta) = r_p^0(\theta)/r_s^0(\theta)$ a real function. For these Fresnel amplitudes the Brewster angle is defined by $\rho^0(\theta_B^0) = 0$. This definition implies that $\rho^0(\theta)$ changes sign at the Brewster angle. As a consequence $\Delta^0(\theta)$ jumps from 0 to π at $\theta = \theta_B^0$ while $\psi^0(\theta_B^0) = 0$. It is noted that the resulting Brewster angle is smaller than the angle of total reflection, so that ellipsometry works equally well for a substrate, which has a refractive index smaller, as for a substrate with a refractive index larger than that of the ambient.

The interfacial layer modifies this behavior in the following way: $\psi(\theta)$ now has a minimum which is no longer zero, whereas $\Delta(\theta)$ becomes a continuous function. A convenient definition of the Brewster angle, which accounts for the effects of the interfacial layer, is now given by

$$\rho(\theta_B) \text{ is purely imaginary } \Rightarrow \Delta(\theta_B) = \pi/2 \qquad (4.94)$$

It should be emphasized that the minimum value of $\psi(\theta)$ is not attained precisely at the Brewster angle. Furthermore it should be emphasized that the minimum of $R_p(\theta)$ is neither attained at θ_B nor at the angle where $\psi(\theta)$ is minimal.

Using the angle dependent reflection amplitudes for s- and p-polarized light given in sections 4.2 and 4.3 one finds the following expression

$$\begin{aligned}\rho(\theta) &= (n^- \cos\theta + n^+ \cos\theta_t)(n^+ \cos\theta + n^- \cos\theta_t)^{-1}(n^- \cos\theta - n^+ \cos\theta_t)^{-1} \\ &\quad \{(n^+ \cos\theta - n^- \cos\theta_t) - 2i\frac{\omega}{c} I_e n^- \cos\theta \ \sin^2\theta (n^+ \cos\theta + n^- \cos\theta_t)^{-1} \\ &\quad -2(\frac{\omega}{c})^2 \frac{\epsilon^-}{\epsilon^+} I_e^2 n^+ \cos\theta \sin^4\theta (n^+ \cos\theta + n^- \cos\theta_t)^{-2} \\ &\quad -4\left(\frac{\omega}{c}\right)^2 n^- n^+ \cos\theta \ \cos\theta_t (I_\tau - I_{\delta,e}\epsilon^- \sin^2\theta)(n^+ \cos\theta + n^- \cos\theta_t)^{-1} \\ &\quad +4\left(\frac{\omega}{c}\right)^2 n^- n^+ \cos\theta \ \cos\theta_t I_\tau (n^+ \cos\theta - n^- \cos\theta_t)(\epsilon^+ - \epsilon^-)^{-1}\} \end{aligned} \tag{4.95}$$

from which equation one may in principle derive similar expressions for $\psi(\theta)$ and $\Delta(\theta)$. For $\psi(\theta)$ it is convenient to use the identity

$$\mathrm{tg}(\psi(\theta)) = [R_p(\theta)/R_s(\theta)]^{1/2} \tag{4.96}$$

together with the reflectances given in eqs.(4.58) and (4.64). As the explicit expressions for ψ and Δ are rather complex and not very illuminating, even if one expands in the thickness of the interface to second order, they will not be given explicitly.

There are two quantities which are most commonly measured. One is the ellipsometric coefficient, given by $|\rho(\theta)|$ at the Brewster angle, which is found to be equal to

$$\begin{aligned}|\rho(\theta_B)| &= \mathrm{tg}(\psi(\theta_B)) = (\epsilon^+ - \epsilon^-)^{-1}\{\frac{1}{2}\frac{\omega}{c}(\epsilon^+ + \epsilon^-)^{1/2} I'_e + (\frac{\omega}{c})^2[(\epsilon^+ + \epsilon^-) I''_\tau \\ &\quad -\epsilon^+\epsilon^- I''_{\delta,e} + \frac{1}{2}(\epsilon^+)^2((\epsilon^+)^2 - (\epsilon^-)^2)^{-1} I'_e I''_e]\} \end{aligned} \tag{4.97}$$

Notice that $\psi(\theta_B)$, found from this equation, is small and is in fact to second order given by the same formula. The other quantity measured is the Brewster angle, which is found to be given by

$$\begin{aligned}\theta_B &\equiv \theta_B^0 + \frac{n^- n^+ \epsilon^+}{(\epsilon^+ - \epsilon^-)^2(\epsilon^+ + \epsilon^-)^3}[(\frac{\omega}{c})(\epsilon^+ - \epsilon^-)(\epsilon^+ + \epsilon^-)^{3/2} I''_e \\ &\quad +\frac{1}{2}(\frac{\omega}{c})^2\{\epsilon^-(3\epsilon^+ - \epsilon^-)(I''_e)^2 - ((\epsilon^+)^2 - (\epsilon^-)^2)(I'_e)^2 \\ &\quad -4((\epsilon^+)^2 - (\epsilon^-)^2)[(\epsilon^+ + \epsilon^-) I'_\tau - \epsilon^-\epsilon^+ I'_{\delta,e}]\}] \end{aligned} \tag{4.98}$$

where θ_B^0 is the value found from the Fresnel amplitudes.

If one neglects the second order terms in the ellipsometric coefficient and the Brewster angle the above formulae reduce to

$$|\rho(\theta_B)| = \mathrm{tg}(\psi(\theta_B)) = \frac{1}{2}\frac{\omega}{c}(\epsilon^+ - \epsilon^-)^{-1}(\epsilon^+ + \epsilon^-)^{1/2} I'_e \tag{4.99}$$

and

$$\theta_B \equiv \theta_B^0 + (\frac{\omega}{c}) n^- n^+ \epsilon^+ (\epsilon^+ - \epsilon^-)^{-1}(\epsilon^+ + \epsilon^-)^{-3/2} I''_e \tag{4.100}$$

To linear order one thus measures the real and the imaginary part of the invariant I_e. Comparing the above expression for the ellipsometric definition of the Brewster angle with eq.(4.77) for the reflectometric definition, one sees that to linear order they are the same. To second order this is no longer the case. Comparing eq.(4.99) with eq.(4.74) one sees that measuring $R_p(\theta_B)$ to linear order also gives the real part of the invariant I_e. As such ellipsometry gives exactly the same information as reflectometry using p-polarized light. The only difference is that the ellipsometric coefficient is linear in I_e', whereas $R_p(\theta_B)$ is quadratic in I_e'. As I_e' is usually small it is in practice easier to obtain an accurate value of this invariant using ellipsometry.

If the interface is non-absorbing the shift of the Brewster angle away from the Fresnel value is of second order and is given by

$$\theta_B \equiv \theta_B^0 - \frac{1}{2}(\frac{\omega}{c})^2 \frac{n^- n^+ \epsilon^+}{(\epsilon^+ - \epsilon^-)(\epsilon^+ + \epsilon^-)^2} [(I_e')^2 + 4(\epsilon^+ + \epsilon^-) I_\tau' - 4\epsilon^- \epsilon^+ I_{\delta,e}'] \qquad (4.101)$$

It should be noted that to this order the reflectometric Brewster angle differs from ellipsometric one. The ellipsometric coefficient has no second order contributions for a non-absorbing interface and is therefore still given by eq.(4.99) to second order.

In the case of an absorbing substrate the Fresnel value of the ellipsometric function $\rho^0(\theta)$ is no longer zero at any angle. A proper definition of the Brewster angle for this case is again given by eq.(4.94). Along similar lines as in section 4.6 one now finds

$$1 - 2(1 + e_1 e_2) \sin^2 \theta_B^0 + (1 + 4e_1 e_2 + e_1^2) \sin^4 \theta_B^0 - 2e_1(e_1 + e_2) \sin^6 \theta_B^0 = 0 \qquad (4.102)$$

where e_1 and e_2 have been defined below eq.(4.80). This equation has three solutions for $\sin^2 \theta_B^0$, only one of which is the correct one. The others were introduced in the process of reformulating the original equation into an algebraic form. Analogously to the reflectometric definition one may again solve the above equation using standard procedures. Choosing the correct root is a bit more involved than in the previous case due to a discriminant which changes sign as a function of the value of the dielectric coefficients. As the solution is standard anyway, this will not be further discussed [28]. For a transparent substrate $e_2 = 1$. The solution of the above equation is then given by $\sin^2 \theta_B^0 = (1 + e_1)^{-1}$. This is equivalent to the usual equation, $\mathrm{tg}\, \theta_B^0 = n^+/n^-$. In view of the complexity of the analysis for an absorbing substrate, explicit expressions for $\rho(\theta)$, $\psi(\theta_B)$ and θ_B will not be given to second order in the invariants. One should simply use the expressions for the amplitudes r_s and r_p given in sections 4.2 and 4.3 and evaluate the desired quantity numerically.

4.9 Total reflection

For substrates which are optically less dense than the ambient, $n^+ < n^-$, one has the phenomenon of total reflection. Both the ambient and the substrate are chosen to be transparent. The interface may be absorbing, however. The angle of total reflection θ_T is the angle for which the angle of transmission becomes equal to ninety degrees. Using Snell's law it follows that this angle is given by

$$\theta_T = \arcsin(n^+/n^-) \qquad (4.103)$$

This angle does not depend on the constitutive coefficients of the interface, unlike the Brewster angle. The Brewster angle is always smaller than the angle of total reflection. If the surface is absorbing the name "total reflection" is not really correct. Due to the absorption the reflected intensity is then smaller than the incident intensity. The often used name "evanescent wave spectroscopy" is then more appropriate.

Using the reflectance for s-polarized light given in eq.(4.56), and eliminating the complex angle of transmission using $\cos\theta_t = i(\epsilon^- \sin^2\theta/\epsilon^+ - 1)^{1/2} \equiv iA(\theta)$, one finds

$$R_s(\theta) = 1 + 4\frac{\omega}{c}I_c n^- \cos\theta + 8(\frac{\omega}{c})^2 n^- \cos\theta[I_c^2 n^- \cos\theta - I_\tau'' n^+(\epsilon^- - \epsilon^+)^{-1}A(\theta)] \tag{4.104}$$

One may verify that this reflectance, due to the absorption of the interface, is always smaller than one. Notice the fact that I_c is real and negative, cf. eq.(3.56), in this case. By measuring the reflectance as function of the angle of incidence, $\theta > \theta_T$, one obtains I_c and I_τ''.

Using the reflectance for p-polarized light, given in eq.(4.62), one finds similarly

$$\begin{aligned} R_p(\theta) &= 1 + 4\frac{\omega}{c}n^- \cos\theta[I_c + I_e'' \sin^2\theta\ B(\theta)] + 8(\frac{\omega}{c})^2 I_c^2 \epsilon^- \cos^2\theta \\ &+8(\frac{\omega}{c})^2 n^- \cos\theta\ B(\theta)[2I_c I_e'' n^- \cos\theta\ \sin^2\theta + (I_\tau'' - I_{\delta,e}''\epsilon^- \sin^2\theta)n^+ A(\theta)] \\ &+8(\frac{\omega}{c})^2 n^- \cos\theta\ \sin^4\theta\ B^2(\theta)[(I_e'')^2 n^- \cos\theta - I_e' I_e''(\epsilon^-/n^+)A(\theta)] \end{aligned} \tag{4.105}$$

where $B(\theta) \equiv [\epsilon^+ \cos^2\theta + \epsilon^- A^2(\theta)]^{-1}$. One may verify that the above reflectance, due to the absorption of the interface, is again always smaller than one. By measuring this reflectance as function of the angle of incidence, $\theta > \theta_T$, one may in principle obtain, in addition to I_c and I_τ'', the invariants I_e', I_e'' and $I_{\delta,e}''$.

In general the most important contributions to the reflectances given above are due to the linear invariant. Neglecting second order effects the formulae simplify to

$$R_s(\theta) = 1 + 4\frac{\omega}{c}I_c n^- \cos\theta \tag{4.106}$$

$$R_p(\theta) = 1 + 4\frac{\omega}{c}n^- \cos\theta[I_c + I_e'' \sin^2\theta\ B(\theta)] \tag{4.107}$$

It is again emphasized that both these reflectances are smaller than (or equal to) one, due to the absorption of the interface, i.e. γ_e'' and β_e'' are both positive (or zero).

Also in the case of "total reflection" one may measure the ellipsometric angles. The ellipsometric function for this case becomes, using eqs.(4.93), (4.31) and (4.40),

$$\begin{aligned} \rho(\theta) &= \rho^0(\theta)\{1 - 2i\frac{\omega}{c}I_e n^- \cos\theta B(\theta) \\ &-2(\frac{\omega}{c})^2 \cos\theta[I_e^2 \epsilon^- \sin^4(\theta)(\cos\theta - i(n^-/n^+)A(\theta))B^2(\theta) \\ &+2i\ I_\tau\ n^- n^+ A(\theta)/(\epsilon^- - \epsilon^+) \\ &+2i(I_\tau - I_{\delta,e}\epsilon^- \sin^2\theta)n^- n^+ A(\theta)B(\theta)]\} \end{aligned} \tag{4.108}$$

where $\rho^0(\theta)$ is the Fresnel value. The resulting changes of the ellipsometric angles are to linear order given by

$$\Delta - \Delta^0 = -2\frac{\omega}{c} I_e' n^- \cos\theta \, \sin^2\theta \, B(\theta) \qquad (4.109)$$

$$\psi - \psi^0 = \frac{\omega}{c} I_e'' n^- \cos\theta \, \sin^2\theta \, B(\theta) \qquad (4.110)$$

In this way one may, by measuring the ellipsometric angles for an angle of incidence larger than the angle of total reflection, obtain the complex invariant I_e. It is further noted that beyond this angle $\text{tg}(\psi^0(\theta)) = 1$, so that $\psi^0(\theta) = \pi/4$.

Chapter 5
ISLAND FILMS IN THE LOW COVERAGE LIMIT

5.1 Introduction

Very thin films are in many cases discontinuous, see for instance [29], [30], [4] and references therein. The film material is then distributed in the form of islands on the surface of the substrate. The size of the islands is small compared to the wavelength. As a consequence retardation effects are unimportant for the fields inside and around the islands. The electric field is then described using the Laplace equation for the potential

$$\Delta\ \psi(\mathbf{r}) = 0 \quad \text{with} \quad \mathbf{E}(\mathbf{r}) = -\nabla\psi(\mathbf{r}) \tag{5.1}$$

In this chapter only non-magnetic systems will be considered. If one considers a single island surrounded by the ambient the electro-magnetic response may be characterized by dipole, quadrupole and higher order multipole polarizabilities [26]. If this particle is brought close to the substrate the (multipole) polarizabilities are modified, due to an induced charge distribution on the surface of the substrate, and become dependent on the distance to the substrate. The presence of other islands along the substrate in the neighborhood of the original island also changes the polarizabilities of the island. Both the interaction with the images and the interaction with the other islands result in polarizabilities parallel and orthogonal to the surface which are unequal, even if the islands are spherical. As a consequence the film is anisotropic in its reaction to fields parallel versus fields normal to the surface. If the islands are not symmetric, this may also contribute to the anisotropy, in particular, if the orientation of the island and the direction of the normal are correlated due to, for instance, the deposition process.

Both the interaction with the image charge distribution and with the other islands lead to considerable complications in the calculation of the dipole as well as of the multipole polarizabilities of the islands. The discussion in this chapter will be restricted to the effects due to interaction of one island with its image charge distribution. For a low coverage one may then simply add the contributions of the various islands to obtain the response of the film to an electromagnetic field. The contributions due to interaction with the other islands will be described in the following chapter.

In section 5.2 of this chapter the linear response of an island with an arbitrary shape to a given field is discussed. A definition of the multipole polarizabilities is given. The symmetry properties of these multipole polarizabilities are formulated and discussed. As an extensive use will be made, both in the present and in the following chapters, of expansions of the potential in terms of spherical harmonics the solution of the Laplace equation in spherical coordinates is given. A definition of the multipole polarizabilities is also given in this context. Their relation with the multipole polarizabilities in Cartesian coordinates is discussed in section 5.5 for the dipole and quadrupole case.

In section 5.3 the effects of the image charge distribution will be analyzed in the context of the polarizable dipole model. In this model all the higher order multipole moments of the islands are neglected. The field due to the image charge distribution is in this case also given by a dipole field. The interaction with the original dipole modifies the polarizability of the island. Yamaguchi, Yoshida and Kinbara [10] were the first to introduce such a dipole model to describe the interaction of the islands with the substrate. As will be explained the resulting dipole polarizabilities of the island, parallel and orthogonal to the surface, are directly related to the constitutive coefficients γ_e and β_e, respectively. The dipole model of Yamaguchi, Yoshida and Kinbara [10] was the first explaination of the fact that the experimentally observed sum of the depolarization factors in three directions for island films was less than one [31]. Older theories, [32] and [33], gave different depolarization factors normal and parallel to the surface but the sum over three directions remained one.

In the fourth section it is shown that the susceptibilities δ_e and τ account for the fact that the dipoles are situated at a finite distance from the surface of the substrate. It follows in the context of the polarizable dipole model that β_e and γ_e are proportional to the typical diameter of the islands times the coverage. The coverage is the surface area of the particles as viewed from above divided by the total surface area. For identical spheres the coverage is equal to π times the radius squared times the number of islands per unit of surface area. Furthermore it follows that δ_e and τ are proportional to the diameter times the distance of the center of the islands to the substrate times the coverage. Both the image charge distribution and the quadrupole and the higher order multipoles lead to an additional scaled dependence on the dimensionless ratio of the diameter divided by the distance to the substrate. In general, and not only in the polarizable dipole model, one may in fact use the ratio of the coefficients δ_e and τ with the coefficients γ_e or β_e as a rough measure for the distance of the center of the islands to the substrate. In the fifth section the effects of the image charge distribution will be analyzed in the context of a polarizable quadrupole model. In this model all multipoles higher than second order are neglected. It is shown that the most important effect of the quadrupole polarizabilities are additional contributions to the constitutive coefficients δ_e and τ.

The use of the polarizable dipole and quadrupole models represent the lowest two orders in an expansion in the size of the islands and in their distance to the substrate. When the island is brought close to the surface one must account for the modification of the dipole and quadrupole polarizabilities due to the interactions with higher order multipoles. As is clear from the analysis in sections 5.3, 5.4 and 5.5, it is sufficient to calculate the modification of the dipole and the quadrupole polarizabilities, as these give the coefficients γ_e, β_e, δ_e and τ. The problem to perform such an analysis in practice is the fact that the calculation of the multipole polarizabilities of arbitrary order of an island, surrounded by the ambient medium far from the substrate, is only possible for islands with a simple shape. In section 5.6 this analysis is performed for spherical particles. In section 5.7 this analysis is applied to spherical gold islands on sapphire. Our reason for this choice was an article by Craighead and Niklasson who made such gold on sapphire films with a square array of islands, [34], [35]. Their island were, however, not spherical. In the later chapters this example will be used again and again for other particle shapes.

5.2 Linear response of an island

The most general linear response of an island of arbitrary shape to an incident electric field may be written in the following form

$$\mathbf{P}(\mathbf{r},t) = \int d\mathbf{r}'dt'\boldsymbol{\alpha}(\mathbf{r},t|\mathbf{r}',t').\mathbf{E}(\mathbf{r}',t') \tag{5.2}$$

The source of the incident field is outside the region occupied by the island. The polarizability $\boldsymbol{\alpha}$ satisfies two important conditions. The first of these conditions is a consequence of the fact that the polarization $\mathbf{P}$ is real if the incident field $\mathbf{E}$ is real. This implies that the polarizability is real

$$\boldsymbol{\alpha}(\mathbf{r},t|\mathbf{r}',t') = [\boldsymbol{\alpha}(\mathbf{r},t|\mathbf{r}',t')]^* \tag{5.3}$$

where the asterisk indicates complex conjugation. The second condition follows from the invariance for time reversal and is given by

$$\alpha_{ij}(\mathbf{r},t|\mathbf{r}',t') = \alpha_{ji}(\mathbf{r}',t|\mathbf{r},t') \tag{5.4}$$

where i and $j = 1,2,3$ denote Cartesian components x,y,z. This property is also referred to as the source observer symmetry.

The systems considered in this book are always assumed to be stationary. The polarizability may therefore be written in the following form

$$\boldsymbol{\alpha}(\mathbf{r},t|\mathbf{r}',t') = \boldsymbol{\alpha}(\mathbf{r},t-t'|\mathbf{r}',0) \equiv \boldsymbol{\alpha}(\mathbf{r},t-t'|\mathbf{r}') \tag{5.5}$$

The reality condition is now given by

$$\boldsymbol{\alpha}(\mathbf{r},t|\mathbf{r}') = [\boldsymbol{\alpha}(\mathbf{r},t|\mathbf{r}')]^* \tag{5.6}$$

and the source observer symmetry [12] by

$$\alpha_{ij}(\mathbf{r},t|\mathbf{r}') = \alpha_{ji}(\mathbf{r}',t|\mathbf{r}) \tag{5.7}$$

The stationary nature of the system makes it convenient to Fourier transform the time dependence. Eq.(5.2) then becomes

$$\mathbf{P}(\mathbf{r},\omega) = \int d\mathbf{r}'\boldsymbol{\alpha}(\mathbf{r},\omega|\mathbf{r}').\mathbf{E}(\mathbf{r}',\omega) \tag{5.8}$$

where the Fourier transforms are defined by

$$\boldsymbol{\alpha}(\mathbf{r},\omega|\mathbf{r}') \equiv \int_{-\infty}^{\infty} dt\boldsymbol{\alpha}(\mathbf{r},t|\mathbf{r}')e^{i\omega t} \tag{5.9}$$

for the polarizability, and

$$\mathbf{P}(\mathbf{r},\omega) \equiv \int_{-\infty}^{\infty} dt \mathbf{P}(\mathbf{r},t)e^{i\omega t} \tag{5.10}$$

for the polarization density and a similar relation for the electric field. The reality condition now becomes

$$\boldsymbol{\alpha}(\mathbf{r},\omega|\mathbf{r}') = [\boldsymbol{\alpha}(\mathbf{r},-\omega|\mathbf{r}')]^* \tag{5.11}$$

and the source observer symmetry becomes

$$\alpha_{ij}(\mathbf{r},\omega|\mathbf{r}') = \alpha_{ji}(\mathbf{r}',\omega|\mathbf{r}) \tag{5.12}$$

The island under consideration occupies a finite domain V, in which the frequency dependent dielectric constant differs from the usually (but not necessarily) constant extrapolated dielectric constant of the surrounding medium. The island polarization density is only unequal to zero inside the island and as a consequence one knows that

$$\boldsymbol{\alpha}(\mathbf{r},\omega|\mathbf{r}') = 0 \quad \text{if} \quad \mathbf{r} \notin V \tag{5.13}$$

Using the source observer symmetry it then also follows that

$$\boldsymbol{\alpha}(\mathbf{r},\omega|\mathbf{r}') = 0 \quad \text{if} \quad \mathbf{r}' \notin V \tag{5.14}$$

If one calculates the field due to the polarization distribution in the island at a position far from the island compared to its diameter, it is convenient to expand the polarizability in a multipole expansion. For the field this gives an expansion in powers of the diameter of the island, divided by the distance to the observer. Up to quadrupole order the polarizability may be written as

$$\begin{aligned} \alpha_{ij}(\mathbf{r},\omega|\mathbf{r}') &= \alpha_{ij}(\omega)\delta(\mathbf{r})\delta(\mathbf{r}') - \alpha_{kij}^{10}(\omega)\frac{\partial}{\partial r_k}\delta(\mathbf{r})\delta(\mathbf{r}') - \alpha_{ijk}^{01}(\omega)\delta(\mathbf{r})\frac{\partial}{\partial r_k'}\delta(\mathbf{r}') \\ &\quad + \alpha_{\ell ijk}^{11}(\omega)\frac{\partial}{\partial r_\ell}\delta(\mathbf{r})\frac{\partial}{\partial r_k'}\delta(\mathbf{r}') \end{aligned} \tag{5.15}$$

where the center of the coordinate frame has been chosen inside, or at least close to, the island. If the island has a line of symmetry it is usually most convenient to choose the center of the coordinate system on this line of symmetry. Notice that in the above equation the so-called summation convention is used, i.e. that one sums over equal indices (from 1 to 3). It follows by direct integration from the above expansion that

$$\begin{aligned} \alpha_{ij}(\omega) &= \int d\mathbf{r} \int d\mathbf{r}' \alpha_{ij}(\mathbf{r},\omega|\mathbf{r}') \\ \alpha_{kij}^{10}(\omega) &= -\int d\mathbf{r} \int d\mathbf{r}' r_k \alpha_{ij}(\mathbf{r},\omega|\mathbf{r}') \\ \alpha_{ijk}^{01}(\omega) &= -\int d\mathbf{r} \int d\mathbf{r}' \alpha_{ij}(\mathbf{r},\omega|\mathbf{r}') r_k' \\ \alpha_{\ell ijk}^{11}(\omega) &= \int d\mathbf{r} \int d\mathbf{r}' r_\ell \alpha_{ij}(\mathbf{r},\omega|\mathbf{r}') r_k' \end{aligned} \tag{5.16}$$

Using the source observer symmetry it follows that

$$\alpha_{ij}(\omega) = \alpha_{ji}(\omega), \quad \alpha^{10}_{kij}(\omega) = \alpha^{01}_{jik}(\omega) \text{ and } \alpha^{11}_{\ell ijk}(\omega) = \alpha^{11}_{kji\ell}(\omega) \tag{5.17}$$

These symmetry conditions will be used in the following sections.

Substitution of eq.(5.15) into eq.(5.8) yields

$$\begin{aligned} P_i(\mathbf{r},\omega) &= \alpha_{ij}(\omega)E_j(0,0,0,\omega)\delta(\mathbf{r}) - \alpha^{10}_{kij}(\omega)E_j(0,0,0,\omega)\frac{\partial}{\partial r_k}\delta(\mathbf{r}) \\ &\quad +\alpha^{01}_{ijk}(\omega)[\frac{\partial}{\partial r'_k}E_j(\mathbf{r}',\omega)]_{\mathbf{r}'=(0,0,0)}\delta(\mathbf{r}) \\ &\quad -\alpha^{11}_{\ell ijk}(\omega)[\frac{\partial}{\partial r'_k}E_j(\mathbf{r}',\omega)]_{\mathbf{r}'=(0,0,0)}\frac{\partial}{\partial r_\ell}\delta(\mathbf{r}) \end{aligned} \tag{5.18}$$

The resulting total dipole and quadrupole moments are

$$\begin{aligned} P_i(\omega) &\equiv \int d\mathbf{r}P_i(\mathbf{r},\omega) = \alpha_{ij}(\omega)E_j(0,0,0,\omega) + \alpha^{01}_{ijk}(\omega)[\frac{\partial}{\partial r'_k}E_j(\mathbf{r}',\omega)]_{\mathbf{r}'=(0,0,0)} \\ Q_{ij}(\omega) &\equiv \int d\mathbf{r}r_iP_j(\mathbf{r},\omega) \\ &= \alpha^{10}_{ijk}(\omega)E_k(0,0,0,\omega) + \alpha^{11}_{ijk\ell}(\omega)[\frac{\partial}{\partial r'_\ell}E_k(\mathbf{r}',\omega)]_{\mathbf{r}'=(0,0,0)} \end{aligned} \tag{5.19}$$

These are the relations that will be used in the quadrupole model described in section 5.5. Notice the fact that only the traceless symmetric part of the quadrupole moment contributes to the field. One may therefore replace r_iP_j in the above equation by its symmetric traceless part.

In the theoretical analysis it is often convenient to use the Fourier transform of the polarizability and the fields with respect to the spatial coordinates. The wave vector and frequency dependent polarizability is given by

$$\boldsymbol{\alpha}(\mathbf{k},\omega|\mathbf{k}') \equiv \int_{-\infty}^{\infty} d\mathbf{r}d\mathbf{r}'e^{-i\mathbf{k}.\mathbf{r}}\boldsymbol{\alpha}(\mathbf{r},\omega|\mathbf{r}')e^{i\mathbf{k}'.\mathbf{r}'} \tag{5.20}$$

the polarization by

$$\mathbf{P}(\mathbf{k},\omega) \equiv \int_{-\infty}^{\infty} d\mathbf{r}\mathbf{P}(\mathbf{r},\omega)e^{-i\mathbf{k}.\mathbf{r}} \tag{5.21}$$

and a similar relation for the electric field. The reality condition now becomes

$$\boldsymbol{\alpha}(\mathbf{k},\omega|\mathbf{k}') = [\boldsymbol{\alpha}(-\mathbf{k},-\omega|-\mathbf{k}')]^* \tag{5.22}$$

and the source observer symmetry, [12], becomes

$$\alpha_{ij}(\mathbf{k},\omega|\mathbf{k}') = \alpha_{ji}(-\mathbf{k}',\omega|-\mathbf{k}) \tag{5.23}$$

If one substitutes the multipole expansion, eq.(5.15), into eq.(5.20) one obtains up to quadrupole order

$$\alpha_{ij}(\mathbf{k},\omega|\mathbf{k}') = \alpha_{ij}(\omega) - \alpha^{10}_{\ell ij}(\omega)ik_\ell + \alpha^{01}_{ij\ell}(\omega)ik'_\ell - ik_\ell\alpha^{11}_{\ell ijn}(\omega)ik'_n \tag{5.24}$$

The multipole expansion in wave vector representation is thus simply a Taylor expansion in the wave vector.

One may wonder, whether it is possible to construct the full island polarizability for a given shape. In practice this is only possible for a rather limited number of shapes. A typical example is, of course, a sphere or more generally a particle with a spherical symmetry, which will be treated in section 5.6. A more sophisticated solvable case is that of ellipsoidal particles. In the following sections the interaction with the image charge distribution for these special cases will be discussed in more detail. Even though the shapes are not general, these special cases are useful to test the limits of the validity of the multipole expansion. The general solution for any given shape requires the use of a corresponding complete set of eigenfunctions for the Laplace equation. The only shapes for which such a complete set of eigenfunctions is known and for which the influence of the substrate has been analyzed are prolate and oblate spheroids with the symmetry axis normal to the surface of the substrate. A discussion of the results will be given in the next chapter. The spheroids, in addition to being interesting in themselves, will turn out to be useful to analyze the limitations of the dipole and the quadrupole models. This will also be demonstrated, more explicitly, in the application treated in the next chapter.

For many purposes it is convenient to use the general solution of the Laplace equation in spherical coordinates, $(x, y, z) \equiv r(\sin\theta\ \cos\phi,\ \sin\theta\ \sin\phi,\ \cos\theta)$. This general solution for the electrostatic potential in a region, $a < r < b$, where there are no sources, may be written as

$$\psi(\mathbf{r}) = \sum_{\ell m} A_{\ell m} r^{-\ell-1} Y_\ell^m(\theta,\phi) + \sum_{\ell m} B_{\ell m} r^\ell Y_\ell^m(\theta,\phi) \tag{5.25}$$

where the summation is from $\ell = 0$ to ∞ and $m = -\ell$ through ℓ. Furthermore the spherical harmonics are given by [26], [36]:

$$Y_\ell^m(\theta,\phi) \equiv \left[\frac{2\ell+1}{4\pi}\frac{(\ell-m)!}{(\ell+m)!}\right]^{1/2} P_\ell^m(\cos\theta)(-1)^m e^{im\phi} \tag{5.26}$$

where the associated Legendre functions are given by

$$\begin{aligned} P_\ell^m(x) &= \frac{1}{2^\ell \ell!}(1-x^2)^{m/2}(\frac{d}{dx})^{\ell+m}(x^2-1)^\ell \quad \text{for} \quad m \geq 0 \\ P_\ell^m(x) &= (-1)^m \frac{(\ell+m)!}{(\ell-m)!} P_\ell^{-m}(x) \quad \text{for} \quad m < 0 \end{aligned} \tag{5.27}$$

It should be noted that a sign convention is used for P_ℓ^m in agreement with Morse and Feshbach [37]. It differs by a factor $(-1)^m$ from the one used by Jackson and Rose. This is the reason to add this factor in the definition of Y_ℓ^m used above, so that the above Y_ℓ^m is identical to the one used by Jackson [26] and by Rose [36]. The preference for Morse and Feshbach's definition is motivated by the fact, that this book gives the most systematic treatment of Legendre functions, also in the complex domain. This will be needed in the next chapter about the spheroid.

It follows from the above formulae that

$$Y_\ell^{-m}(\theta,\phi) = (-1)^m [Y_\ell^m(\theta,\phi)]^* \tag{5.28}$$

The normalization and orthogonality condition are

$$\int_0^{2\pi} d\phi \int_0^{\pi} d\theta \, \sin\theta [Y_\ell^m(\theta,\phi)]^* Y_{\ell'}^{m'}(\theta,\phi) = \delta_{\ell\ell'}\delta_{mm'} \tag{5.29}$$

The first term in eq.(5.25) gives the field due to sources located in the region $r < a$ and the second term gives the field due to sources in the region $r > b$. If one uses the above potential to describe the response of an island, one must choose the center of the coordinate frame somewhere inside, or at least close to, the island. The distance a may then be identified with the radius of the smallest sphere which contains the island. The distance b is any distance larger than a. The second term in eq.(5.25) may be identified as the incident field due to external sources, whereas the first term is due to the charge distribution induced in the island. In view of the fact, that the island has no net charge the $\ell = 0$ contribution in the first term is zero. The $\ell = 0$ term in the incident field is a constant contribution to the potential and may be chosen arbitrarily. $A_{\ell m}$ gives a contribution to the potential which decays as $r^{-\ell-1}$ and may therefore, apart from a numerical constant, be identified as an amplitude of the ℓ th order multipole field. The amplitudes of these multipole fields are given in terms of the amplitudes of the incident field, using a polarizability matrix, by

$$A_{\ell m} = -\sum_{\ell' m'}{}' \alpha_{\ell m,\ell' m'} B_{\ell' m'} \quad \text{for } \ell \neq 0 \tag{5.30}$$

where the prime as superindex of the summation indicates that $\ell' \neq 0$. Note that $\alpha_{\ell m,\ell' m'}$ is not symmetric for the interchange of ℓm with $\ell' m'$. In the polarizable dipole model one takes all amplitudes with ℓ and $\ell' \neq 1$ equal to zero. Similarly one takes all amplitudes with ℓ or $\ell' > 2$ equal to zero in the polarizable quadrupole model.

The expansion in spherical harmonics is special in the sense, that each term in the expansion corresponds to only one order in the multipole expansion. If one uses other complete sets of solutions of the Laplace equation, as in particular those most appropriate for an ellipsoidal shaped particle, this is no longer the case. These functions always contain at shorter distances contributions due to higher order multipoles.

5.3 Gamma and beta in the polarizable dipole model

In the polarizable dipole model [10] one neglects the multipole polarizabilities of the islands. The island is in this approximation replaced by a polarizable point dipole which is placed in the origin O of a Cartesian coordinate system in the ambient medium at a distance d from the substrate (see fig.5.1). In the analysis below the positive z-axis is chosen to point into the substrate and $z = d$ is the surface of the substrate. If one uses the formulae from chapter 4 this implies that the light is incident through the ambient medium.

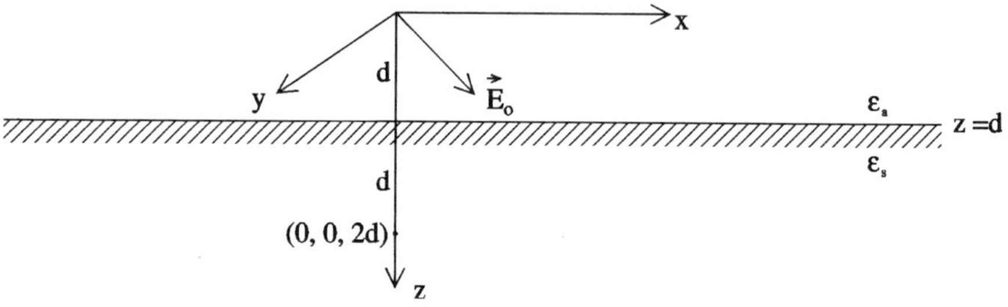

Figure 5.1 Point dipole (quadrupole) above substrate.

As in many experiments the light is incident through the substrate, it will be indicated in the text how the relevant formulae in this section must be modified for that case. The most important change is that the sign of δ_e and τ must be changed, while the signs of β_e and γ_e remain the same. The dielectric constant is ϵ_a in the ambient medium and ϵ_s in the substrate. The polarizability of the island far from the surface and surrounded by the ambient medium is $\boldsymbol{\alpha}$. When placed in a homogeneous electric field $\mathbf{E}_0$ the induced dipole moment $\mathbf{p}$ is given by $\mathbf{p} = \boldsymbol{\alpha}.\mathbf{E}_0$. When the island is close to the surface this dipole moment is modified due to the electric field from an image dipole moment, which is located a distance d on the other side of the surface. The total field in the ambient is given by

$$\begin{aligned}
\mathbf{E}(\mathbf{r},t) &= \mathbf{E}_0 - [4\pi\epsilon_a|(x,y,z)|^3]^{-1}[\mathbf{1} - 3\frac{(x,y,z)(x,y,z)}{|(x,y,z)|^2}].\mathbf{p} \\
&\quad -[4\pi\epsilon_a|(x,y,z-2d)|^3]^{-1}[\mathbf{1} - 3\frac{(x,y,z-2d)(x,y,z-2d)}{|(x,y,z-2d)|^2}].\mathbf{p}_r \\
&\equiv \mathbf{E}_0 + \mathbf{E}_p + \mathbf{E}_{p,r} \qquad (5.31)
\end{aligned}$$

The first term is a constant electric field due to external sources in the ambient. The second term represents the direct field of the polarizable dipole placed in (0,0,0). The last term results from the image of this dipole in the substrate in (0,0,2d). The electric field in the substrate is given by

$$\mathbf{E}(\mathbf{r},t) = (\mathbf{E}_{0,\parallel}, \epsilon_a E_{0,z}/\epsilon_s) - [4\pi\epsilon_s|(x,y,z)|^3]^{-1}[\mathbf{1} - 3\frac{(x,y,z)(x,y,z)}{|(x,y,z)|^2}].\mathbf{p}_t \qquad (5.32)$$

The first term is again the constant field due to external sources, but now in the substrate. The second term represents the "transmitted" field of the dipole. Using the boundary conditions on the surface of the substrate one finds the following relations [26]

$$\mathbf{p}_r = \frac{\epsilon_a - \epsilon_s}{\epsilon_a + \epsilon_s}(\mathbf{p}_\parallel, -p_z), \quad \mathbf{p}_t = \frac{2\epsilon_s}{\epsilon_a + \epsilon_s}\mathbf{p} \qquad (5.33)$$

The dipole in the particle is induced by the sum of the "incident" and the reflected field

$$\mathbf{p} = \boldsymbol{\alpha}.(\mathbf{E}_0 + \mathbf{E}_{p,r}(0,0,0)) \qquad (5.34)$$

Using the above equations one may then calculate the dipole moment of the particle and one finds

$$\mathbf{p} = [\mathbf{1} + (32\pi\epsilon_a d^3)^{-1}\left(\frac{\epsilon_a - \epsilon_s}{\epsilon_a + \epsilon_s}\right)\boldsymbol{\alpha}.\begin{bmatrix} 1 & 0 & 0 \\ 0 & 1 & 0 \\ 0 & 0 & 2 \end{bmatrix}]^{-1}.\boldsymbol{\alpha}.\mathbf{E}_0 \equiv \boldsymbol{\alpha}(0).\mathbf{E}_0 \tag{5.35}$$

This expression gives the polarizability tensor of the particle, modified by the presence of the substrate, as a function of its position $z = 0$ relative to the surface of the substrate, which is positioned at $z = d$.

In this chapter the emphasis is on particles which have a symmetry axis normal to the surface of the substrate. In that case the above formula for the modification of the polarizability close to the substrate can be given for the normal and the parallel components separately:

$$\alpha_{\|}(0) = [1 + A\ \alpha_{\|}]^{-1}\alpha_{\|} \tag{5.36}$$

$$\alpha_z(0) = [1 + 2A\ \alpha_z]^{-1}\alpha_z \tag{5.37}$$

where as a convenient short notation the following factor, which characterizes the strength of the reflected dipole,

$$A \equiv (32\pi\epsilon_a d^3)^{-1}\left(\frac{\epsilon_a - \epsilon_s}{\epsilon_a + \epsilon_s}\right) \tag{5.38}$$

was introduced. If there are two different polarizabilities parallel to the surface eq.(5.36) is valid for both of them. An important aspect of the above equations is the anisotropy in the interaction with the substrate. Even if the particle is spherical, so that $\alpha_{\|} = \alpha_z$, the different responses to fields along the surface and normal to the surface results in $\alpha_{\|}(0) \neq \alpha_z(0)$.

If one covers the substrate with a low density of identical islands, which have a rotational symmetry axis normal to the surface of the substrate, one finds for the first order interfacial susceptibilities

$$\gamma_e(d) = \rho\ \alpha_{\|}(0) = \rho[1 + A\ \alpha_{\|}]^{-1}\alpha_{\|} \tag{5.39}$$

$$\beta_e(d) = \rho\ \epsilon_a^{-2}\alpha_z(0) = \rho\epsilon_a^{-2}[1 + 2A\ \alpha_z]^{-1}\alpha_z \tag{5.40}$$

where ρ is the number of particles per unit of surface area. The argument d of γ_e and β_e indicates that the location of the dividing surface is chosen to be the $z = d$ surface, which in this case coincides with the surface of the substrate. In appendix A the rather technical derivation of the above relation for γ_e and β_e is discussed on the basis of the general method to introduce excess fields, given in section 2.4. As the resulting relations given above are very similar to expressions for the bulk susceptibility in a low density mixture, and therefore have some intuitive appeal, a further elaboration on this derivation in the main text is not given. There is one aspect which is important, however. In writing these formulae the polarization due to the dipoles is in fact taken to be located at the surface of the substrate. As will be discussed in the next section, the origin of the second order interfacial susceptibilities

τ and δ_e is related to a proper choice of this location. Furthermore it is noted that both γ_e and β_e are independent of the choice of the direction of the z-axis. Thus the above expressions remain the same if one substitutes them in the formulae, given in the previous chapter on reflection and transmission, for the case that the light is incident through the substrate.

From the explicit expressions for the polarizabilities, or from those for γ_e and β_e, it follows that, these coefficients approach zero in the $d \to 0$ limit. This is a consequence of the fact, that the distance ($2d$) between the dipole and its image approaches zero in this limit, so that A diverges. This limiting behavior is an indication of the fact that the dipole model becomes unsatisfactory for the description of islands, for which the average distance to the substrate is much smaller than their extension along the surface. It should be emphasized that this behavior is unphysical. As will explicitly be shown in the next chapter for the special case of oblate spheroids, the polarizability remains finite in this limit (see section 6.4).

If one covers the substrate with a low density of islands with different sizes and shapes one may simply add the effects of the different dipoles and one then finds for the first order interfacial susceptibilities

$$\gamma_e(d) = \rho < \alpha_{\|}(0) > \text{ and } \beta_e(d) = \rho\, \epsilon_a^{-2} < \alpha_z(0) > \tag{5.41}$$

where $< ... >$ indicates the average of the particle polarizabilities, i.e. the sum of the polarizabilities of the particles divided by their number.

5.4 Delta and tau in the polarizable dipole model

The origin of the susceptibilities δ_e and τ for island films is related to the choice of the location of the dividing surface. In the calculation of γ_e and β_e in the previous section the location of the dividing surface was chosen to coincide with the surface of the substrate. Such a choice implicitly assumes that the (local or incident) field at the dividing surface is the same as the field in the center of the islands. One may certainly move the induced dipoles to the dividing surface but one should express them in the field in their original center. The response of the film is therefore related in a non-local way to the field. Alternatively one may choose the dividing surface, characterized by the constitutive coefficients γ_e and β_e, to be the plane $z = 0$ in the ambient through the centers of the spheres. In this way one has two dividing surfaces: one is the surface in $z = 0$ where the dipoles are located and the other is the $z = d$ surface of the substrate. In fact, the values of γ_e and β_e given in the previous section are the constitutive coefficients of the surface where the dipoles are located ($\gamma_{e,1}(0) = \gamma_e(d)$, $\beta_{e,1}(0) = \beta_e(d)$, $\delta_{e,1}(0) = \tau_1(0) = 0$), while the surface of the substrate has $\gamma_{e,2}(d) = \beta_{e,2}(d) = \delta_{e,2}(d) = \tau_2(d) = 0$. The validity of the relations $\gamma_{e,1}(0) = \gamma_e(d)$ and $\beta_{e,1}(0) = \beta_e(d)$ is proved in appendix A. For the dipole model (by definition) $\delta_{e,1}(0) = \tau_1(0) = 0$. Of course, one prefers to combine the two films into one film, which is most conveniently placed at the surface of the substrate. For this purpose formulae have been derived in chapter 3 which will now be used. The first step, before adding the effects of the surfaces, is to shift the film describing the effect of the dipoles to a new location at an infinitesimal distance from the substrate surface in the ambient. The constitutive coefficients of the dipolar film for this new location is found, using the formulae given in eq.(3.52), to be given by

$$\begin{aligned} \gamma_{e,1}(d) &= \gamma_e(d), \quad & \beta_{e,1}(d) &= \beta_e(d) \\ \tau_1(d) &= \gamma_e(d)d, \quad & \delta_{e,1}(d) &= \gamma_e(d)d/\epsilon_a + \beta_e(d)d\epsilon_a \end{aligned} \tag{5.42}$$

Now one may superimpose the dipole film with the surface of the substrate, cf. section 3.11, and in view of the fact that the constitutive coefficients of this surface are zero one finds that the above equation also gives the constitutive coefficients of the superimposed film. One therefore finds

$$\tau(d) = \gamma_e(d)d, \quad \delta_e(d) = \gamma_e(d)d/\epsilon_a + \beta_e(d)d\epsilon_a \tag{5.43}$$

while $\gamma_e(d)$ and $\beta_e(d)$ are given in the previous section. For a low density of identical islands with an axis of symmetry orthogonal to the substrate one therefore obtains

$$\begin{aligned} \gamma_e(d) &= \rho\, \alpha_\parallel/(1 + A\alpha_\parallel), \quad & \beta_e(d) &= \rho[\alpha_z/(1 + 2A\alpha_z)]/\epsilon_a^2 \\ \tau(d) &= \rho d\, \alpha_\parallel/(1 + A\alpha_\parallel), \quad & \delta_e(d) &= \rho d[\alpha_\parallel/(1 + A\alpha_\parallel) + \alpha_z/(1 + 2A\alpha_z)]/\epsilon_a \end{aligned} \tag{5.44}$$

If one changes the direction of the z-axis this results in replacing d by $-d$ in these formulae. Due to the fact that furthermore A remains unchanged it follows that γ_e and β_e are independent of the direction of the z-axis, while δ_e and τ change sign if the z-axis changes direction. That this is the case may, of course, also be derived on the basis of more general considerations (see chapter 13). If one uses the above formulae to calculate the reflection and transmission amplitudes, it is important to take the right sign. If, as in the previous chapter, the light is incident through the ambient one may use the above expressions. If, however, the light is incident through the substrate one must, in order to use the formulae given in the previous chapter, interchange ϵ_a and ϵ_s in these formulae in the previous chapter and change the sign of δ_e and τ in the expressions given above, before substituting them into the formulae in the previous chapter.

The reason, that it is also in the dipole model important to take finite values of δ_e and τ, is related to the occurrence of phase factors. Light reflected from the film can either be reflected directly from the island film or be transmitted and subsequently reflected by the surface of the substrate. It is clear that the phase difference between these two contributions can be important. The above analysis and the resulting finite value of δ_e and τ account for the phase differences between the actual location of the islands and the surface of the substrate to linear order.

If one performs an experimental analysis of a film, one obtains information about the interfacial susceptibilities. It is now clear from the above results that the ratios of τ and δ_e over either γ_e or β_e are a measure of the typical size of the islands of the material deposited on the surface of the substrate. As such the coefficients τ and δ_e, though their modification of the optical properties may be relatively small, nevertheless contain interesting new information. The coefficients γ_e and β_e, which usually have a much larger influence, are a measure of the amount of material deposited on the surface of the substrate. They are therefore a measure of what one usually calls the weight thickness.

5.5 Polarizable quadrupole model

In the polarizable quadrupole model one neglects the multipole polarizabilities of the islands of higher than quadrupole order. The island is in this approximation replaced by a polarizable point dipole and a polarizable point quadrupole which have the same position. The choice of this position is not unimportant, in view of the fact that the quadrupole moment depends on this choice. If the island has a center of symmetry it is most convenient to use this as the location of the dipole and the quadrupole moment. If there is no such center of symmetry one could, for instance, use the center of mass. It should be emphasized that the results in this section are not affected by this choice. The reason for this is that the change in the position of the dipole is accounted for by the modification of the quadrupole moment. The change in the location of the quadrupole moment results in a change of the octupole moment. As octupole moments are neglected in the model the location of the quadrupole moments may be chosen arbitrarily inside or very close to the island.

The total field in the ambient is given in this case by

$$\begin{aligned}
\mathbf{E}(\mathbf{r},t) &= \mathbf{E}_0 - [4\pi\epsilon_a|(x,y,z)|^3]^{-1}[\mathbf{1} - 3\frac{(x,y,z)(x,y,z)}{|(x,y,z)|^2}].\mathbf{p} \\
&\quad -3[2\pi\epsilon_a|(x,y,z)|^5]^{-1}[\mathbf{1} - \frac{5}{2}\frac{(x,y,z)(x,y,z)}{|(x,y,z)|^2}].[(x,y,z).\mathbf{Q}] \\
&\quad -[4\pi\epsilon_a|(x,y,z-2d)|^3]^{-1}[\mathbf{1} - 3\frac{(x,y,z-2d)(x,y,z-2d)}{|(x,y,z-2d)|^2}].\mathbf{p}_r \\
&\quad -3[2\pi\epsilon_a|(x,y,z-2d)|^5]^{-1}[\mathbf{1} - \frac{5}{2}\frac{(x,y,z-2d)(x,y,z-2d)}{|(x,y,z-2d)|^2}] \\
&\quad .[(x,y,z-2d).\mathbf{Q}_r] \equiv \mathbf{E}_0 + \mathbf{E}_p + \mathbf{E}_q + \mathbf{E}_{p,r} + \mathbf{E}_{q,r}
\end{aligned} \tag{5.45}$$

where the same coordinate system has been used as in the previous sections. The first term is a constant electric field due to external sources in the ambient. The second and the third terms represent the direct field of the polarizable dipole and the polarizable quadrupole placed in (0,0,0) (see fig.5.1). The last two terms result from the images of the dipole and the quadrupole in the substrate in (0,0,$2d$). The transmitted field may be written down in the same manner. Using the boundary conditions on the surface of the substrate one finds the following relations [26]

$$\mathbf{p}_r = \frac{\epsilon_a - \epsilon_s}{\epsilon_a + \epsilon_s}(\mathbf{p}_\parallel, -p_z), \quad \mathbf{Q}_r = \frac{\epsilon_a - \epsilon_s}{\epsilon_a + \epsilon_s}\begin{bmatrix} \mathbf{Q}_{\parallel,\parallel} & -\mathbf{Q}_{\parallel,z} \\ -\mathbf{Q}_{z,\parallel} & Q_{z,z} \end{bmatrix} \tag{5.46}$$

The field polarizing the dipole and the quadrupole, which will be referred to as the local field, is the sum of the incident field and the reflected fields:

$$\mathbf{E}_{loc} \equiv \mathbf{E}_0 + \mathbf{E}_{p,r} + \mathbf{E}_{q,r} \tag{5.47}$$

The resulting dipole and quadrupole moments are given by

$$\begin{aligned}
p_i &= \alpha_{ij}E_{loc,j}(0,0,0) + \alpha^{01}_{ijk}[\nabla_k E_{loc,j}](0,0,0) \\
Q_{ij} &= \alpha^{10}_{ijk}E_{loc,k}(0,0,0) + \alpha^{11}_{ijk\ell}[\nabla_\ell E_{loc,k}](0,0,0)
\end{aligned} \tag{5.48}$$

where a summation convention over double indices has been used. Notice that the superscript 00 has been omitted in the tensor α_{ij}, in order to use the same notation for this quantity as in the previous section.

As has been discussed in detail in section 2 the island response functions satisfy the following symmetry conditions

$$\alpha_{ij}(\omega) = \alpha_{ji}(\omega), \quad \alpha^{01}_{ijk}(\omega) = \alpha^{10}_{kji}(\omega), \quad \alpha^{11}_{ijk\ell}(\omega) = \alpha^{11}_{\ell kji}(\omega) \tag{5.49}$$

Moreover it follows from the symmetric traceless nature of the quadrupole moment, that $\alpha^{01}_{ijk}(\omega) = \alpha^{10}_{kji}(\omega)$ is symmetric traceless in the indices kj and that $\alpha^{11}_{ijk\ell}(\omega) = \alpha^{11}_{\ell kji}(\omega)$ is symmetric traceless both in the indices ij and $k\ell$.

If the island has a symmetry axis $\mathbf{n}$ it is possible to give the most general form of the island response functions, consistent with the above symmetry relations, in the following form:

$$\begin{aligned}
\alpha_{ij} &= \alpha_{\parallel}(\delta_{ij} - n_i n_j) + \alpha_{\perp} n_i n_j \\
\alpha^{01}_{ijk} &= \alpha^{10}_{kji} = \frac{1}{2}\alpha^{10}_{\parallel}(\delta_{ij}n_k + \delta_{ik}n_j - 2\, n_i n_j n_k) - \frac{1}{3}\alpha^{10}_{\perp} n_i(\delta_{jk} - 3\, n_j n_k) \\
\alpha^{11}_{ijk\ell} &= \frac{1}{2}\alpha^{11}(\delta_{ik}\delta_{j\ell} + \delta_{i\ell}\delta_{jk} - \frac{2}{3}\delta_{ij}\delta_{k\ell}) \\
&\quad + \frac{1}{4}\alpha^{11}_{\parallel}(n_i n_k \delta_{j\ell} + n_i n_\ell \delta_{jk} + n_j n_k \delta_{i\ell} + n_j n_\ell \delta_{ik} - 4\, n_i n_j n_k n_\ell) \\
&\quad + \frac{1}{9}\alpha^{11}_{\perp}(\delta_{ij} - 3\, n_i n_j)(\delta_{k\ell} - 3\, n_k n_\ell)
\end{aligned} \tag{5.50}$$

Using the above equations one may now calculate the dipole and the quadrupole moments of the island in a homogeneous electric field. The further analysis will be restricted to the case that the symmetry axis of the island is orthogonal to the surface of the substrate, so that $\mathbf{n} = \hat{\mathbf{z}}$. The subindex $\perp$ may then also be replaced by z.

If the incident field is orthogonal to the interface, $\mathbf{E}_0 = E_0\hat{\mathbf{z}}$, one finds, using the boundary conditions at the surface of the substrate, that the dipole and the quadrupole moment are given by

$$\mathbf{p} = p_z\hat{\mathbf{z}} \text{ and } \mathbf{Q} = Q_z(\hat{\mathbf{z}}\hat{\mathbf{z}} - \frac{1}{3}\mathbf{1}) \tag{5.51}$$

with

$$\begin{aligned}
p_z &= D_z^{-1}\{\alpha_z[1 + (3A/d^2)(3\alpha^{11} + 2\alpha^{11}_z)] - (6A/d^2)(\alpha^{10}_z)^2\}E_0 \\
Q_z &= D_z^{-1}\{\alpha^{10}_z[1 + 3(A/d)\alpha^{10}_z] - \frac{3}{2}\alpha_z(A/d)(3\alpha^{11} + 2\alpha^{11}_z)\}E_0
\end{aligned} \tag{5.52}$$

where the denominator is given by

$$\begin{aligned}
D_z &= 1 + 2(A/d)(d\alpha_z - 3\alpha^{10}_z) + (3A/2d^2)(2 + A\alpha_z)(3\alpha^{11} + 2\alpha^{11}_z) \\
&\quad -3(A/d)^2(\alpha^{10}_z)^2
\end{aligned} \tag{5.53}$$

and where

$$A \equiv (32\pi\epsilon_a d^3)^{-1}\frac{\epsilon_a - \epsilon_s}{\epsilon_a + \epsilon_s} \tag{5.54}$$

The above formulae show that the relevant dipole and quadrupole polarizabilities of the island are modified by the fact that the island is situated at a distance d to the substrate and become equal to

$$\begin{aligned}
\alpha_z(0) &= D_z^{-1}\{\alpha_z[1+(3A/d^2)(3\alpha^{11}+2\alpha_z^{11})]-(6A/d^2)(\alpha_z^{10})^2\} \\
\alpha_z^{10}(0) &= D_z^{-1}\{\alpha_z^{10}[1+3(A/d)\alpha_z^{10}]-\frac{3}{2}\alpha_z(A/d)(3\alpha^{11}+2\alpha_z^{11})]
\end{aligned} \tag{5.55}$$

The argument 0 denotes the position of the point dipole and quadrupole on the z-axis. If one now compares the dipole polarizability normal to the surface of the substrate with the expression given in eq.(5.37), one sees that the above expression reduces to this simpler form if the quadrupole polarizabilities are taken equal to zero. If the island is far from the surface of the substrate, A is approximately zero and $\alpha_z(0)$ and $\alpha_z^{10}(0)$ reduce to the polarizabilities α_z and α_z^{10} as is to be expected.

If the incident field is parallel to the interface, e.g. $\mathbf{E}_0 = E_0\widehat{\mathbf{x}}$, one finds, using the boundary conditions at the surface of the substrate, that the dipole and the quadrupole moment are given by

$$\mathbf{p} = p_x\widehat{\mathbf{x}} \text{ and } \mathbf{Q} = \frac{1}{2}Q_{zx}(\widehat{\mathbf{x}}\widehat{\mathbf{z}}+\widehat{\mathbf{z}}\widehat{\mathbf{x}}) \tag{5.56}$$

with

$$\begin{aligned}
p_x &= D_\parallel^{-1}\{\alpha_\parallel[1+(3A/d^2)(2\alpha^{11}+\alpha_\parallel^{11})]-(3A/d^2)(\alpha_\parallel^{10})^2\}E_0 \\
Q_{zx} &= D_\parallel^{-1}\{\alpha_\parallel^{10}[1+\frac{3}{2}(A/d)\alpha_\parallel^{10}]-\frac{3}{2}\alpha_\parallel(A/d)(2\alpha^{11}+\alpha_\parallel^{11})\}E_0
\end{aligned} \tag{5.57}$$

where the denominator is given by

$$\begin{aligned}
D_\parallel &= 1+(A/d)(d\alpha_\parallel+3\alpha_\parallel^{10})+(3A/4d^2)(4+A\alpha_\parallel)(2\alpha^{11}+\alpha_\parallel^{11}) \\
&\quad -\frac{3}{4}(A/d)^2(\alpha_\parallel^{10})^2
\end{aligned} \tag{5.58}$$

Notice the important fact that the interaction with the image charge distribution in the substrate is not found to lead to diagonal elements of the quadrupole moment of the island. That this should be the case may in fact, in a more general context, be understood on the basis of the symmetry of the charge distribution. The above formulae show that the relevant dipole and quadrupole polarizabilities of the islands are modified by the fact that the island is situated at a distance d to the substrate and become equal to

$$\begin{aligned}
\alpha_\parallel(0) &= D_\parallel^{-1}\{\alpha_\parallel[1+(3A/d^2)(2\alpha^{11}+\alpha_\parallel^{11})]-(3A/d^2)(\alpha_\parallel^{10})^2\} \\
\alpha_\parallel^{10}(0) &= D_\parallel^{-1}\{\alpha_\parallel^{10}[1+\frac{3}{2}(A/d)\alpha_\parallel^{10}]-\frac{3}{2}\alpha_\parallel(A/d)(2\alpha^{11}+\alpha_\parallel^{11})\}
\end{aligned} \tag{5.59}$$

If one now compares the dipole polarizability parallel to the surface of the substrate with the expression given in eq.(5.36), one sees that the above expression reduces to this simpler form, if the quadrupole polarizabilities are taken equal to zero. Furthermore $\alpha_\parallel(0)$ and $\alpha_\parallel^{10}(0)$ again reduce to $\alpha_\parallel$ and $\alpha_\parallel^{10}$ if the island is far from the surface so that A is small.

In order to identify interfacial constitutive coefficients, it is convenient to shift the position of the dipole and the quadrupole to the dividing surface at $z = d$, leaving them in the ambient. Such a shift leaves the dipole moment unaffected while the quadrupole moment increases with the traceless part of the symmetrized product of the dipole moment with the displacement vector $\hat{\mathbf{z}}d$. This shift therefore leaves the dipole polarizabilities unchanged but gives different quadrupole polarizabilities. In this manner one finds

$$
\begin{aligned}
\alpha_z(d) &= \alpha_z(0) \\
\alpha_\parallel(d) &= \alpha_\parallel(0) \\
\alpha_z^{10}(d) &= \alpha_z^{10}(0) - d\alpha_z(0) \\
&= D_z^{-1}\{\alpha_z^{10} - d\alpha_z + \frac{9}{2}(A/d)[2(\alpha_z^{10})^2 - \alpha_z(3\alpha^{11} + 2\alpha_z^{11})]\} \\
\alpha_\parallel^{10}(d) &= \alpha_\parallel^{10}(0) - d\alpha_\parallel(0) \\
&= D_\parallel^{-1}\{\alpha_\parallel^{10} - d\alpha_\parallel + \frac{9}{2}(A/d)[(\alpha_\parallel^{10})^2 - \alpha_\parallel(2\alpha^{11} + \alpha_\parallel^{11})]\} \qquad (5.60)
\end{aligned}
$$

For the dipole model these expressions reduce to

$$
\alpha_z(d) = \alpha_z/(1 + 2A\alpha_z), \qquad \alpha_\parallel(d) = \alpha_\parallel/(1 + A\alpha_\parallel) \qquad (5.61)
$$

$$
\alpha_z^{10}(d) = -d\alpha_z/(1 + 2A\alpha_z), \quad \alpha_\parallel^{10}(d) = -d\alpha_\parallel/(1 + A\alpha_\parallel) \qquad (5.62)
$$

If these expressions are now compared with eqs.(5.39), (5.40), (5.43) and (5.44) for a low density of identical islands with the symmetry axis orthogonal to the substrate one may identify

$$
\begin{aligned}
\gamma_e(d) &= \rho\, \alpha_\parallel(d) = \rho\alpha_\parallel/(1 + A\alpha_\parallel) \\
\beta_e(d) &= \rho\, \alpha_z(d)/\epsilon_a^2 = \rho[\alpha_z/(1 + 2A\alpha_z)]/\epsilon_a^2 \\
\tau(d) &= -\rho\, \alpha_\parallel^{10}(d) = \rho d[\alpha_\parallel/(1 + A\alpha_\parallel)] \\
\delta_e(d) &= -\rho[\alpha_\parallel^{10}(d) + \alpha_z^{10}(d)]/\epsilon_a \\
&= \rho d[\alpha_\parallel/(1 + A\alpha_\parallel) + \alpha_z/(1 + 2A\alpha_z)]/\epsilon_a \qquad (5.63)
\end{aligned}
$$

In the quadrupole model one should make the same identification and one then obtains

$$
\begin{aligned}
\gamma_e(d) &= \rho\, \alpha_\parallel(d) = \rho\, D_\parallel^{-1}\{\alpha_\parallel[1 + (3A/d^2)(2\alpha^{11} + \alpha_\parallel^{11})] - (3A/d^2)(\alpha_\parallel^{10})^2\} \\
\beta_e(d) &= \rho\, \alpha_z(d)/\epsilon_a^2 = \rho\, D_z^{-1}\{\alpha_z[1 + (3A/d^2)(3\alpha^{11} + 2\alpha_z^{11})] - (6A/d^2)(\alpha_z^{10})^2\}/\epsilon_a^2 \\
\tau(d) &= -\rho\, \alpha_\parallel^{10}(d) = \rho D_\parallel^{-1}\{d\alpha_\parallel - \alpha_\parallel^{10} - \frac{9}{2}(A/d)[(\alpha_\parallel^{10})^2 - \alpha_\parallel(2\alpha^{11} + \alpha_\parallel^{11})]\} \\
\delta_e(d) &= -\rho[\alpha_\parallel^{10}(d) + \alpha_z^{10}(d)]/\epsilon_a \\
&= \rho[D_\parallel^{-1}\{d\alpha_\parallel - \alpha_\parallel^{10} - \frac{9}{2}(A/d)[(\alpha_\parallel^{10})^2 - \alpha_\parallel(2\alpha^{11} + \alpha_\parallel^{11})]\} \\
&\quad + D_z^{-1}\{d\alpha_z - \alpha_z^{10} - \frac{9}{2}(A/d)[2(\alpha_z^{10})^2 - \alpha_z(3\alpha^{11} + 2\alpha_z^{11})]\}]/\epsilon_a \qquad (5.64)
\end{aligned}
$$

for the interfacial constitutive coefficients. As a motivation for the use of the same identification in the dipole and in the quadrupole model the following. The value of the quadrupole moment of a particle depends on the position around which the dipole and the quadrupole moment are calculated. In fact one may always modify or even eliminate the quadrupole moment by choosing a different location. The above identification of the constitutive coefficients makes them independent of the rather arbitrary choice of the center of the particle and is therefore the only correct one. If one changes the direction of the z-axis, one finds again that γ_e and β_e are independent of the direction of the z-axis, while δ_e and τ change sign if the z-axis changes direction. If one uses the above formulae to calculate the reflection and transmission amplitudes, it is important to take the right sign. If, as in the previous chapter, the light is incident through the ambient one may use the above expressions. If the light is incident through the substrate one must, in order to use the formulae given in the previous chapter, interchange ϵ_a and ϵ_s in these formulae in the previous chapter and change the sign of δ_e and τ in the expressions given above, before substituting them into the formulae in the previous chapter.

Also in the polarizable quadrupole model the polarizabilities and the constitutive coefficients approach zero in the $d \rightarrow 0$ limit. This limiting behavior is, as also discussed in the context of the polarizable dipole model, unphysical. The polarizable quadrupole model is therefore unsatisfactory for the description of islands, for which the average distance to the substrate is much smaller than their extension along the surface. In the next chapter it will explicitly be shown, in section 6.3, for the special case of oblate spheroids, that the polarizability remains finite in this limit.

As discussed in the previous section the coefficients δ_e and τ are related to the choice of the location of the dipole moment. While the dipole polarizabilities $\alpha_\parallel$ and α_z are not affected by a change in this location one finds that the quadrupole polarizabilities $\alpha_\parallel^{10}$ and α_z^{10} depend on this location. It was found in the previous section that δ_e and τ can be given in terms of these quadrupole polarizabilities if one chooses the location infinitesimally close to the surface of the substrate (cf. also eq.(5.63)). In the present section this property was used to derive expressions for the constitutive coefficients for the quadrupole model. It should be emphasized in this context that the quadrupole polarizabilities, though being dependent on the chosen location, can not all be eliminated by a different choice of the "electromagnetic" center of the particle. One could, of course, eliminate the real or the imaginary part of either $\alpha_\parallel^{10}$ or α_z^{10} but this is in general only one out of the four contributions. Thus the quadrupole model cannot be reduced to the dipole model by an appropriate choice of the location of the dipole.

If one considers a film with a low density of islands, which have a certain dispersion in size, but which all have their symmetry axis orthogonal to the substrate, one may use the following averages

$$\begin{aligned} &\gamma_e(d) = \rho < \alpha_\parallel(d) >, \qquad \beta_e(d) = \rho < \alpha_z(d) > /\epsilon_a^2 \\ &\tau(d) = -\rho < \alpha_\parallel^{10}(d) >, \quad \delta_e(d) = -\rho[< \alpha_\parallel^{10}(d) > + < \alpha_z^{10}(d) >]/\epsilon_a \end{aligned} \tag{5.65}$$

From the above expressions one may generally conclude that the susceptibilities β_e and γ_e give the average polarization per unit of surface area in terms of the incident field at the interface. Similarly the susceptibilities δ_e and τ give the average quadrupole moment per unit of surface area in terms of the incident field at the surface. Both, but in particular, the quadrupole moment have been calculated using the surface of the substrate as the dividing surface.

If one calculates the invariants one finds up to linear order in the density ρ

$$\begin{array}{ll} I_e = \gamma_e(d) - \epsilon_a\epsilon_s\beta_e(d), & I_c = \mathrm{Im}\left[\gamma_e(\varepsilon_s - \varepsilon_a)^{-1}\right] \\ I_\tau = \tau(d), & I_{\delta,e} = \delta_e(d) \end{array} \qquad (5.66)$$

One may now draw the, in practical applications, convenient conclusion that for low densities the ratios of these invariants are independent of the density. In practical situation the coverage is often a not very well-known function of the time. It is then convenient to compare the experimental ratios with the above predictions, rather than the invariants themselves.

One may relate the polarizabilities introduced in eqs.(5.48) and (5.50) to the polarizabilities introduced in eq.(5.30), if the particle has a symmetry axis. Choosing the z-axis along this symmetry axis one may show that

$$\begin{aligned} \alpha_{10,10} &= \alpha_z/4\pi\epsilon_a \\ \alpha_{11,11} &= \alpha_{1,-1;1,-1} = \alpha_\parallel/4\pi\epsilon_a \\ \alpha_{21,11} &= \frac{3}{5}\alpha_{11,21} = \alpha_{2,-1;1,-1} = \frac{3}{5}\alpha_{1,-1;2,-1} = 3\alpha_\parallel^{10}/4\pi\epsilon_a\sqrt{5} \\ \alpha_{20,10} &= \frac{3}{5}\alpha_{10,20} = \alpha_z^{10}/2\pi\epsilon_a\sqrt{5/3} \\ \alpha_{21,21} &= \alpha_{2,-1;2,-1} = 3(\alpha_\parallel^{11} + 2\alpha^{11})/4\pi\epsilon_a \\ \alpha_{20,20} &= (2\alpha_z^{11} + 3\alpha^{11})/2\pi\epsilon_a \\ \alpha_{22,22} &= \alpha_{2,-2;2,-2} = 3\alpha^{11}/2\pi\epsilon_a \end{aligned} \qquad (5.67)$$

where ϵ_a is the dielectric constant of the medium in which the island is embedded. In the derivation of this equation use is made of linear relations, which can be derived between the amplitudes $A_{\ell m}(\ell = 1,2)$ and the dipole and quadrupole moments $\mathbf{p}$ and $\mathbf{Q}$ of the particle, and between the amplitudes $B_{\ell m}(\ell = 1,2)$ and the incident electric field $\mathbf{E}$ and its gradient $\nabla\mathbf{E}$ in the center of the particle. As the derivation of these relations is not very illuminating, it will not be given explicitly. It should be noted that, due to the symmetry of the island for rotation about the z-axis, all matrix elements which couple different values of m are zero, $\alpha_{\ell m,\ell' m'} = \alpha_{\ell m,\ell' m}\delta_{mm'}$. Furthermore it is clear from the above expressions, that $\alpha_{\ell m,\ell' m'}$ is not symmetric for the interchange of ℓm with $\ell' m'$.

5.6 Spherical islands

A model system which is often used for the description of island films is the case that the islands are spherically symmetric. The main reason for the attractiveness of this model is its relative simplicity. For spherical particles one may calculate the properties of the film using numerical methods to a large extent exactly. One may account in this way for the image charge distribution to all orders in the multipole expansion. Furthermore one may also account for the interaction between different spheres and their image charge distribution to all orders in the multipole expansion for an arbitrary coverage. This makes the island film using spherical islands an ideal testing ground for the usefulness of various approximations. If the islands are reasonably spherical, which is often the case if they are small, it is also good for practical applications. The special case of a homogeneous spherical island above a substrate has been solved, using bispherical coordinates, by Ruppin [38]. The numerical results, solving the linear set of equations given by him and those presented below for this special case, are identical.

Before discussing some of these aspects, first the island in a homogeneous medium far from the surface is considered. Due to the spherical symmetry of the island the polarizability matrix becomes diagonal and depends only on ℓ. One then has

$$\alpha_{\ell m,\ell' m'} = \alpha_\ell \delta_{\ell\ell'} \delta_{mm'} \tag{5.68}$$

The relation between the induced and the incident amplitudes, eq.(5.30), simplifies in that case to

$$A_{\ell m} = -\alpha_\ell B_{\ell m} \tag{5.69}$$

For a homogeneous sphere with a radius R and a dielectric constant ϵ embedded in a medium with dielectric constant ϵ_a , both frequency dependent, the multipole polarizabilities are given by [14]

$$\alpha_\ell = \frac{\ell(\epsilon - \epsilon_a)}{\ell\epsilon + (\ell+1)\epsilon_a} R^{2\ell+1} \tag{5.70}$$

Another interesting case is a sphere coated with a layer of a different material. The multipole polarizabilities of such a sphere are given by [39]

$$\alpha_\ell = \frac{\ell[(\epsilon_c - \epsilon_a)(\ell\epsilon + (\ell+1)\epsilon_c)R_c^{2\ell+1} + (\epsilon - \epsilon_c)(\ell\epsilon_a + (\ell+1)\epsilon_c)R^{2\ell+1}]}{(\ell\epsilon_c + (\ell+1)\epsilon_a)(\ell\epsilon + (\ell+1)\epsilon_c)R_c^{2\ell+1} + \ell(\ell+1)(\epsilon - \epsilon_c)(\epsilon_c - \epsilon_a)R^{2\ell+1}} R_c^{2\ell+1} \tag{5.71}$$

where ϵ and ϵ_c are the complex dielectric constants of the core and the layer coating the surface. Furthermore R is the radius of the core and R_c the radius of the core plus the layer, which coats the surface. The derivation of the above expressions for α_ℓ is relatively straightforward. One uses a potential field of the form given in eq.(5.25) in each layer and chooses the amplitudes such that the potential as well as the dielectric constant times the normal derivative of the potential are continuous at each surface.

The polarizabilities in the above described dipole and quadrupole models, which are useful for a comparison with the exact (numerical) results, are given for a spherically symmetric island by

$$\begin{aligned} \alpha_z &= \alpha_\parallel = 4\pi\epsilon_a\alpha_1 \equiv \alpha_d \\ \alpha^{11} &= 2\pi\epsilon_a\alpha_2/3 \equiv \alpha_q \\ \alpha_z^{10} &= \alpha_\parallel^{10} = \alpha_z^{11} = \alpha_\parallel^{11} = 0 \end{aligned} \tag{5.72}$$

This follows from eqs.(5.67) and (5.68). As discussed in the previous section the results using the dipole model are found from those using the quadrupole model, if one sets the quadrupole polarizability α_q equal to zero. It is therefore sufficient to give the formulae describing the interaction with the substrate only for the quadrupole model. Substitution of the above relations into eq.(5.55) and using eq.(5.53) gives for the polarizabilities, as modified by the proximity of the substrate, for an electric field normal to the substrate

$$\begin{aligned} \alpha_z(0) &= \alpha_d[1 + (9A/d^2)\alpha_q]/\ D_z \\ \alpha_z^{10}(0) &= -\frac{9}{2}\alpha_d\alpha_q A\ /\ d\ D_z \end{aligned} \tag{5.73}$$

where

$$D_z = 1 + 2A\ \alpha_d + (9A/2d^2)(2 + A\alpha_d)\alpha_q \tag{5.74}$$

and where, cf. eq.(5.54),

$$A \equiv (32\pi\epsilon_a d^3)^{-1}\frac{\epsilon_a - \epsilon_s}{\epsilon_a + \epsilon_s} \tag{5.75}$$

In eq.(5.73) the argument 0 denotes the position of the center of the sphere on the z-axis, where the point dipole and quadrupole are located (see fig.5.1). It should be noted that the quadrupole polarizability $\alpha_z^{10}(0)$, which is zero if the sphere is far from the surface of the substrate, becomes finite when the sphere approaches the surface. The resulting quadrupole polarizability is proportional to the product of the dipole and the quadrupole polarizabilities α_d and α_q. In the dipole model the quadrupole polarizability therefore remains zero independent of the distance to the substrate.

If the electric field is directed parallel to the surface of the substrate, one finds in a similar way, by substitution of the polarizabilities given in eq.(5.72) into eq.(5.59) and using eq.(5.58), for the polarizabilities, as modified by the proximity of the substrate, for an electric field parallel to the substrate

$$\begin{aligned} \alpha_\parallel(0) &= \alpha_d[1 + (6A/d^2)\alpha_q]/\ D_\parallel \\ \alpha_\parallel^{10}(0) &= -3\ \alpha_d\alpha_q A\ /\ d\ D_\parallel \end{aligned} \tag{5.76}$$

in terms of α_d and α_q, and where

$$D_\parallel = 1 + A\ \alpha_d + (3A/2d^2)(4 + A\alpha_d)\alpha_q \tag{5.77}$$

Again the quadrupole polarizability, $\alpha_\parallel^{10}(0)$, which is zero if the sphere is far from the surface of the substrate, becomes finite when the sphere approaches the substrate.

The resulting susceptibilities of the film are given by eq.(5.64) which reduces for this case to

$$\begin{aligned} \gamma_e(d) &= \rho\alpha_d[1 + (6A/d^2)\alpha_q]/D_\parallel \\ \beta_e(d) &= \rho\alpha_d[1 + (9A/d^2)\alpha_q]/\epsilon_a^2 D_z \\ \tau(d) &= \rho d\alpha_d[1 + (9A/d^2)\alpha_q]/D_\parallel \\ \delta_e(d) &= \rho d\alpha_d\{[1 + (9A/d^2)\alpha_q]/D_\parallel + [1 + (27A/2d^2)\alpha_q]/D_z\}/\epsilon_a \end{aligned} \tag{5.78}$$

where A, D_z and $D_\parallel$ are given by eqs.(5.75), (5.74) and (5.77) respectively. The argument d of the above coefficients denotes again the position of the dividing surface at $z = d$, which coincides with the surface of the substrate. The invariants can be found by substitution of the above formulae into eq.(5.66).

For the application of the above formulae to the special case of homogeneous spheres one may use the following polarizabilities, cf. eqs.(5.70) and (5.72),

$$\begin{aligned} \alpha_d &= 4\pi\epsilon_a R^3 \frac{\epsilon - \epsilon_a}{\epsilon + 2\epsilon_a} \\ \alpha_q &= \frac{4}{3}\pi\epsilon_a R^5 \frac{\epsilon - \epsilon_a}{2\epsilon + 3\epsilon_a} \end{aligned} \tag{5.79}$$

where R is the radius of the sphere and ϵ the frequency dependent dielectric constant of the island material.

Substitution of these polarizabilities and eq.(5.75) into eqs.(5.73) and (5.74) gives

$$\begin{aligned} \alpha_z(0) &= 4\pi\epsilon_a R^3 \frac{\epsilon - \epsilon_a}{\epsilon + 2\epsilon_a}[1 + \frac{3}{8}(\frac{R}{d})^5 \frac{(\epsilon_a - \epsilon_s)(\epsilon - \epsilon_a)}{(\epsilon_a + \epsilon_s)(2\epsilon + 3\epsilon_a)}] / \; D_z \\ \alpha_z^{10}(0) &= -\frac{3}{4}\pi\epsilon_a R^4 (\frac{R}{d})^4 \frac{(\epsilon_a - \epsilon_s)(\epsilon - \epsilon_a)^2}{(\epsilon_a + \epsilon_s)(\epsilon + 2\epsilon_a)(2\epsilon + 3\epsilon_a)} / D_z \end{aligned} \tag{5.80}$$

where

$$\begin{aligned} D_z = \; & 1 + \frac{1}{4}\frac{(\epsilon_a - \epsilon_s)}{(\epsilon_a + \epsilon_s)}\{(\frac{R}{d})^3 \frac{(\epsilon - \epsilon_a)}{(\epsilon + 2\epsilon_a)} \\ & + \frac{3}{2}(\frac{R}{d})^5 \frac{(\epsilon - \epsilon_a)}{(2\epsilon + 3\epsilon_a)}[1 + \frac{1}{16}(\frac{R}{d})^3 \frac{(\epsilon_a - \epsilon_s)(\epsilon - \epsilon_a)}{(\epsilon_a + \epsilon_s)(\epsilon + 2\epsilon_a)}]\} \end{aligned} \tag{5.81}$$

These formulae show that, as is to be expected, the modification of the polarizabilities is most pronounced when the spheres touch the surface of the substrate.

Substitution of the polarizabilities, given in eq.(5.79), and eq.(5.75) into eqs.(5.76) and (5.77) gives

$$\begin{aligned} \alpha_\parallel(0) &= 4\pi\epsilon_a R^3 \frac{\epsilon - \epsilon_a}{\epsilon + 2\epsilon_a}[1 + \frac{1}{4}(\frac{R}{d})^5 \frac{(\epsilon_a - \epsilon_s)(\epsilon - \epsilon_a)}{(\epsilon_a + \epsilon_s)(2\epsilon + 3\epsilon_a)}] / \; D_\parallel \\ \alpha_\parallel^{10}(0) &= -\frac{1}{2}\pi\epsilon_a R^4 (\frac{R}{d})^4 \frac{(\epsilon_a - \epsilon_s)(\epsilon - \epsilon_a)^2}{(\epsilon_a + \epsilon_s)(\epsilon + 2\epsilon_a)(2\epsilon + 3\epsilon_a)} / \; D_\parallel \end{aligned} \tag{5.82}$$

where

$$\begin{aligned} D_\parallel = \; & 1 + \frac{1}{8}\frac{(\epsilon_a - \epsilon_s)}{(\epsilon_a + \epsilon_s)}\{(\frac{R}{d})^3 \frac{(\epsilon - \epsilon_a)}{(\epsilon + 2\epsilon_a)} \\ & + 2(\frac{R}{d})^5 \frac{(\epsilon - \epsilon_a)}{(2\epsilon + 3\epsilon_a)}[1 + \frac{1}{32}(\frac{R}{d})^3 \frac{(\epsilon_a - \epsilon_s)(\epsilon - \epsilon_a)}{(\epsilon_a + \epsilon_s)(\epsilon + 2\epsilon_a)}]\} \end{aligned} \tag{5.83}$$

Again one finds that the modification of the polarizabilities, due to the presence of the substrate, is most pronounced, when the spheres touch the surface of the substrate.

If one substitutes the polarizabilities of a homogeneous sphere, cf. eq.(5.79), into the interfacial susceptibilities, eq.(5.78), one finds

$$\begin{aligned}
\gamma_e(d) &= 4\pi\ \rho\ \epsilon_a R^3 \frac{\epsilon - \epsilon_a}{\epsilon + 2\epsilon_a}[1 + \frac{1}{4}(\frac{R}{d})^5 \frac{(\epsilon_a - \epsilon_s)(\epsilon - \epsilon_a)}{(\epsilon_a + \epsilon_s)(2\epsilon + 3\epsilon_a)}]/\ D_{\|} \\
\beta_e(d) &= 4\pi\ \rho\ R^3 \frac{\epsilon - \epsilon_a}{\epsilon + 2\epsilon_a}[1 + \frac{3}{8}(\frac{R}{d})^5 \frac{(\epsilon_a - \epsilon_s)(\epsilon - \epsilon_a)}{(\epsilon_a + \epsilon_s)(2\epsilon + 3\epsilon_a)}]/\ \epsilon_a D_z \\
\tau(d) &= 4\pi\ \rho\epsilon_a d\ R^3 \frac{\epsilon - \epsilon_a}{\epsilon + 2\epsilon_a}[1 + \frac{3}{8}(\frac{R}{d})^5 \frac{(\epsilon_a - \epsilon_s)(\epsilon - \epsilon_a)}{(\epsilon_a + \epsilon_s)(2\epsilon + 3\epsilon_a)}]/\ D_{\|} \\
\delta_e(d) &= 4\pi\ \rho\ d\ R^3 \frac{\epsilon - \epsilon_a}{\epsilon + 2\epsilon_a}\{[1 + \frac{3}{8}(\frac{R}{d})^5 \frac{(\epsilon_a - \epsilon_s)(\epsilon - \epsilon_a)}{(\epsilon_a + \epsilon_s)(2\epsilon + 3\epsilon_a)}]/\ D_{\|} \\
&\quad + [1 + \frac{9}{16}(\frac{R}{d})^5 \frac{(\epsilon_a - \epsilon_s)(\epsilon - \epsilon_a)}{(\epsilon_a + \epsilon_s)(2\epsilon + 3\epsilon_a)}]/\ D_z\} \qquad (5.84)
\end{aligned}$$

where D_z and $D_{\|}$ are given by eqs.(5.81) and (5.83) respectively. If one changes the direction of the z-axis, one has again, that γ_e and β_e are independent of the direction of the z-axis, while δ_e and τ change sign if the z-axis changes direction. If one uses the above formulae to calculate the reflection and transmission amplitudes, it is important to take the right sign. If, as in the previous chapter, the light is incident through the ambient one may use the above expressions. If, however, the light is incident through the substrate one must, in order to use the formulae given in the previous chapter, interchange ϵ_a and ϵ_s in these formulae in the previous chapter and change the sign of δ_e and τ in the expressions given above, before substituting them into the formulae in the previous chapter.

Another case worth considering is a sphere coated with a layer of a different material. The polarizabilities of such an island are given by, cf. eq.(5.71),

$$\begin{aligned}
\alpha_d &= 4\pi\epsilon_a R_c^3 \left[\frac{(\epsilon_c - \epsilon_a)(\epsilon + 2\epsilon_c)R_c^3 + (\epsilon - \epsilon_c)(\epsilon_a + 2\epsilon_c)R^3}{(\epsilon_c + 2\epsilon_a)(\epsilon + 2\epsilon_c)R_c^3 + 2(\epsilon - \epsilon_c)(\epsilon_c - \epsilon_a)R^3}\right] \\
\alpha_q &= \frac{4}{3}\pi\epsilon_a R_c^5 \left[\frac{(\epsilon_c - \epsilon_a)(2\epsilon + 3\epsilon_c)R_c^5 + (\epsilon - \epsilon_c)(2\epsilon_a + 3\epsilon_c)R^5}{(2\epsilon_c + 3\epsilon_a)(2\epsilon + 3\epsilon_c)R_c^5 + 6(\epsilon - \epsilon_c)(\epsilon_c - \epsilon_a)R^5}\right] \qquad (5.85)
\end{aligned}$$

where ϵ and ϵ_c are the complex dielectric constants of the core and the coating, respectively. Furthermore R and R_c are the radii of the core and of the core plus the coating. Substitution of these polarizabilities into the formulae given above is straightforward and will not be done explicitly. It is noted that one may also give these polarizabilities for multi-layered coatings.

The expressions, eqs.(5.80)–(5.83), for the polarizabilities of the spheres, modified by the proximity of the substrate, may be written in an alternative form, which will also be used in the rest of the book. This form is

$$\begin{aligned}
\alpha_z(0) &= \frac{\epsilon_a(\epsilon - \epsilon_a)V[\epsilon_a + L_z^1(\epsilon - \epsilon_a)]}{[\epsilon_a + L_z(\epsilon - \epsilon_a)][\epsilon_a + L_z^1(\epsilon - \epsilon_a)] + \Lambda_z^1(\epsilon - \epsilon_a)^2} \\
\alpha_z^{10}(0) &= \frac{\epsilon_a(\epsilon - \epsilon_a)^2\Lambda_z^{10}V\ R}{[\epsilon_a + L_z(\epsilon - \epsilon_a)][\epsilon_a + L_z^1(\epsilon - \epsilon_a)] + \Lambda_z^1(\epsilon - \epsilon_a)^2} \qquad (5.86)
\end{aligned}$$

and

$$\begin{aligned}\alpha_{\|}(0) &= \frac{\epsilon_a(\epsilon-\epsilon_a)V[\epsilon_a+L_{\|}^1(\epsilon-\epsilon_a)]}{[\epsilon_a+L_{\|}(\epsilon-\epsilon_a)][\epsilon_a+L_{\|}^1(\epsilon-\epsilon_a)]+\Lambda_{\|}^1(\epsilon-\epsilon_a)^2}\\ \alpha_{\|}^{10}(0) &= \frac{\epsilon_a(\epsilon-\epsilon_a)^2\Lambda_{\|}^{10}V\ R}{[\epsilon_a+L_{\|}(\epsilon-\epsilon_a)][\epsilon_a+L_{\|}^1(\epsilon-\epsilon_a)]+\Lambda_{\|}^1(\epsilon-\epsilon_a)^2}\end{aligned} \tag{5.87}$$

where $V \equiv \frac{4}{3}\pi\ R^3$ is the volume of the sphere. The coefficients L and Λ are found, after some simple calculations:

$$\begin{aligned}L_z &= \frac{1}{3}+\frac{1}{12}\left(\frac{\epsilon_a-\epsilon_s}{\epsilon_a+\epsilon_s}\right)\left(\frac{R}{d}\right)^3\\ L_z^1 &= \frac{2}{5}+\frac{3}{40}\left(\frac{\epsilon_a-\epsilon_s}{\epsilon_a+\epsilon_s}\right)\left(\frac{R}{d}\right)^5\\ \Lambda_z^1 &= -\frac{3}{640}\left(\frac{\epsilon_a-\epsilon_s}{\epsilon_a+\epsilon_s}\right)^2\left(\frac{R}{d}\right)^8\\ \Lambda_z^{10} &= -\frac{3}{80}\left(\frac{\epsilon_a-\epsilon_s}{\epsilon_a+\epsilon_s}\right)\left(\frac{R}{d}\right)^4\end{aligned} \tag{5.88}$$

and

$$\begin{aligned}L_{\|} &= \frac{1}{3}+\frac{1}{24}\left(\frac{\epsilon_a-\epsilon_s}{\epsilon_a+\epsilon_s}\right)\left(\frac{R}{d}\right)^3\\ L_{\|}^1 &= \frac{2}{5}+\frac{1}{20}\left(\frac{\epsilon_a-\epsilon_s}{\epsilon_a+\epsilon_s}\right)\left(\frac{R}{d}\right)^5\\ \Lambda_{\|}^1 &= -\frac{1}{640}\left(\frac{\epsilon_a-\epsilon_s}{\epsilon_a+\epsilon_s}\right)^2\left(\frac{R}{d}\right)^8\\ \Lambda_{\|}^{10} &= -\frac{1}{40}\left(\frac{\epsilon_a-\epsilon_s}{\epsilon_a+\epsilon_s}\right)\left(\frac{R}{d}\right)^4\end{aligned} \tag{5.89}$$

These coefficients will be called depolarization factors. This is because in the special case of the dipole model, where the polarizabilities become:

$$\begin{aligned}\alpha_z(0) &= \frac{\epsilon_a(\epsilon-\epsilon_a)V}{\epsilon_a+L_z(\epsilon-\epsilon_a)}\\ \alpha_{\|}(0) &= \frac{\epsilon_a(\epsilon-\epsilon_a)V}{\epsilon_a+L_{\|}(\epsilon-\epsilon_a)}\end{aligned} \tag{5.90}$$

the coefficients L_z and $L_{\|}$ are called depolarization factors. For a sphere in a homogeneous medium, which may be considered as the limiting case, where the substrate is infinitely far away ($d \to \infty$), $L_z = L_{\|} = \frac{1}{3}$, which is a well-known result. For the sum of the usual depolarization factors, the above expressions give

$$L_z + 2L_{\|} = 1 + \frac{1}{6}\left(\frac{\epsilon_a-\epsilon_s}{\epsilon_a+\epsilon_s}\right)\left(\frac{R}{d}\right)^3 \tag{5.91}$$

This result was first given by Yamaguchi, Yoshida and Kinbara [10].

The expressions for the interfacial susceptibilities may now also be written with the help of the above depolarization factors. From eqs.(5.64), (5.60), (5.86) and (5.87) one obtains:

$$\beta_e(d) = \rho\ \alpha_z(0)/\ \epsilon_a^2 = \frac{\rho\ \epsilon_a^{-1}(\epsilon - \epsilon_a)V[\epsilon_a + L_z^1(\epsilon - \epsilon_a)]}{[\epsilon_a + L_z(\epsilon - \epsilon_a)][\epsilon_a + L_z^1(\epsilon - \epsilon_a)] + \Lambda_z^1(\epsilon - \epsilon_a)^2} \tag{5.92}$$

$$\gamma_e(d) = \rho\ \alpha_\parallel(0) = \frac{\rho\ \epsilon_a(\epsilon - \epsilon_a)V[\epsilon_a + L_\parallel^1(\epsilon - \epsilon_a)]}{[\epsilon_a + L_\parallel(\epsilon - \epsilon_a)][\epsilon_a + L_\parallel^1(\epsilon - \epsilon_a)] + \Lambda_\parallel^1(\epsilon - \epsilon_a)^2} \tag{5.93}$$

$$\begin{aligned} \delta_e(d) &= -\rho[\alpha_z^{10}(0) + \alpha_\parallel^{10}(0) - d\ \alpha_z(0) - d\ \alpha_\parallel(0)]/\epsilon_a \\ &= \rho(\epsilon - \epsilon_a)V\ d\left\{\frac{\epsilon_a + (L_z^1 - \Lambda_z^{10}R/d)\,(\epsilon - \epsilon_a)}{[\epsilon_a + L_z(\epsilon - \epsilon_a)][\epsilon_a + L_z^1(\epsilon - \epsilon_a)] + \Lambda_z^1(\epsilon - \epsilon_a)^2}\right. \\ &\quad \left. + \frac{\epsilon_a + \left(L_\parallel^1 - \Lambda_\parallel^{10}R/d\right)(\epsilon - \epsilon_a)}{[\epsilon_a + L_\parallel(\epsilon - \epsilon_a)][\epsilon_a + L_\parallel^1(\epsilon - \epsilon_a)] + \Lambda_\parallel^1(\epsilon - \epsilon_a)^2}\right\} \end{aligned} \tag{5.94}$$

$$\begin{aligned} \tau(d) &= -\rho[\alpha_\parallel^{10}(0) - d\ \alpha_\parallel(0)] \\ &= \frac{\rho\ \epsilon_a(\epsilon - \epsilon_a)V\ d\left[\epsilon_a + \left(L_\parallel^1 - \Lambda_\parallel^{10}R/d\right)(\epsilon - \epsilon_a)\right]}{[\epsilon_a + L_\parallel(\epsilon - \epsilon_a)][\epsilon_a + L_\parallel^1(\epsilon - \epsilon_a)] + \Lambda_\parallel^1(\epsilon - \epsilon_a)^2} \end{aligned} \tag{5.95}$$

When the expressions, eqs.(5.88) and (5.89), are substituted into the above formulae, one of course finds back, after some simple calculations, the results eq.(5.84).

It is of interest to construct the solution to arbitrary order in the multipole expansion. Even though the dipole or the quadrupole model will in many cases be enough the general solution is convenient as a check on the limitations of the model description. Consider the response of the spherical particle in a constant electric field given in the ambient, which will be chosen to be transparent, by

$$\mathbf{E}_0 = E_0(\sin\theta_0\cos\phi_0, \sin\theta_0\sin\phi_0, \cos\theta_0) \tag{5.96}$$

where θ_0 is the angle of the field with the z-axis, and ϕ_0 the angle between its projection on the substrate and the x-axis. The corresponding incident potential is

$$\begin{aligned} \psi_0(\mathbf{r}) &= -\mathbf{E}_0.\mathbf{r} = -r\ E_0\sqrt{2\pi/3}[\cos\theta_0 Y_1^0(\theta,\phi)\sqrt{2} \\ &\quad + \sin\theta_0\{\exp(i\phi_0)Y_1^{-1}(\theta,\phi) - \exp(-i\phi_0)Y_1^1(\theta,\phi)\}] \end{aligned} \tag{5.97}$$

where the definition, eq.(5.26) with (5.27), of $Y_\ell^m(\theta,\phi)$ has been applied. Using spherical coordinates, the general solution in the ambient may be written as a sum

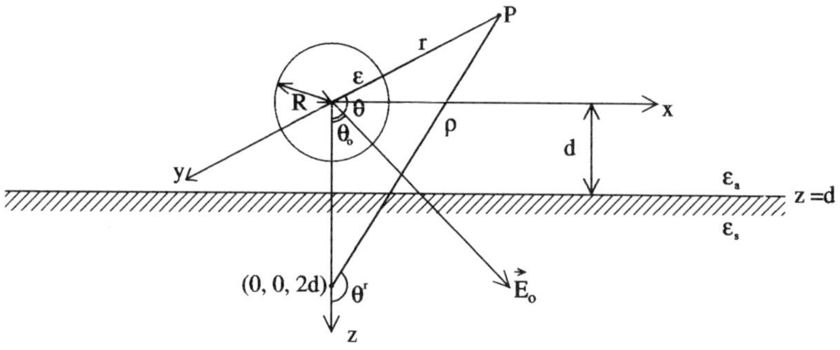

Figure 5.2 A sphere above a substrate

of the incident potential, the potential due to the induced charge distribution in the island and the image charge distribution in the substrate:

$$\begin{aligned}\psi_a(\mathbf{r}) &= -r\ E_0\sqrt{2\pi/3}[\cos\theta_0 Y_1^0(\theta,\phi)\sqrt{2} \\ &\quad +\sin\theta_0\{\exp(i\phi_0)Y_1^{-1}(\theta,\phi)-\exp(-i\phi_0)Y_1^1(\theta,\phi)\}] \\ &\quad +\sum_{\ell m}{}' A_{\ell m} r^{-\ell-1} Y_\ell^m(\theta,\phi)+\sum_{\ell m}{}' A_{\ell m}^r \rho^{-\ell-1} Y_\ell^m(\theta^r,\phi^r)\end{aligned} \tag{5.98}$$

The prime as index of the summation indicates that $\ell = 0$ is excluded. The center of the coordinate system has been chosen in the center of the sphere at a distance d from the substrate (see fig.5.2). The multipoles describing the direct field of the island are located at this point. The image multipoles are located in the substrate at (0,0,2d). Furthermore (r, θ, ϕ) are the usual spherical coordinates and (ρ, θ^r, ϕ^r) are the spherical coordinates with (0,0,2d) as the center of the coordinate system and the same z-axis. The surface of the substrate is again given by $z = d$.

For the potential in the substrate one may write similarly

$$\begin{aligned}\psi_s(\mathbf{r}) &= E_0\{d((\epsilon_a/\epsilon_s)-1)\cos\theta_0 - r\sqrt{2\pi/3}[(\epsilon_a/\epsilon_s)\cos\theta_0 Y_1^0(\theta,\phi)\sqrt{2} \\ &\quad +\sin\theta_0(\exp(i\phi_0)Y_1^{-1}(\theta,\phi)-\exp(-i\phi_0)Y_1^1(\theta,\phi))]\} \\ &\quad +\sum_{\ell m}{}' A_{\ell m}^t r^{-\ell-1} Y_\ell^m(\theta,\phi)\end{aligned} \tag{5.99}$$

As a first step to calculate the polarizabilities in this general case one must choose the amplitudes of the various multipoles such that the boundary conditions at the surface of the substrate are satisfied. For this purpose it is most convenient to use the fact that one finds for a point charge, that the image charge is equal to the original charge times $(\epsilon_a - \epsilon_s)/(\epsilon_a + \epsilon_s)$ and the charge, giving the transmitted field, the original charge times $2\epsilon_s/(\epsilon_a + \epsilon_s)$. The same relations may be used for a distribution of charges and one may then conclude that

$$A^r_{\ell m} = (-1)^{\ell+m}\frac{\epsilon_a - \epsilon_s}{\epsilon_a + \epsilon_s}A_{\ell m} \text{ and } A^t_{\ell m} = \frac{2\epsilon_a}{\epsilon_a + \epsilon_s}A_{\ell m} \tag{5.100}$$

One may, of course, verify directly that these relations make the potential and the dielectric constant times the normal derivative of the potential continuous at the surface of the substrate at $z = d$.

In order to relate the amplitudes $A_{\ell m}$ to the multipole polarizabilities for the particle in free space, one must use the potential for the local field acting on the particle, which is given as the sum of the potentials of the incident and the reflected fields

$$\begin{aligned}\psi_{loc}(\mathbf{r}) &= E_0\{-r\sqrt{2\pi/3}[\cos\theta_0 Y_1^0(\theta,\phi)\sqrt{2} + \sin\theta_0(\exp(i\phi_0)Y_1^{-1}(\theta,\phi) \\ &\quad - \exp(-i\phi_0)Y_1^1(\theta,\phi))]\} + \sum_{\ell m}{}' A^r_{\ell m}\rho^{-\ell-1}Y_\ell^m(\theta^r,\phi^r)\end{aligned} \tag{5.101}$$

It is now necessary to expand the potentials originating from (0,0,2d) in terms of a complete set of fields around the origin. In appendix B it is shown that

$$(2d/\rho)^{\ell+1}Y_\ell^m(\theta^r,\phi^r) = \sum_{\ell_1=|m|}^{\infty} M^m_{\ell_1,\ell}(r/2d)^{\ell_1}Y^m_{\ell_1}(\theta,\phi) \tag{5.102}$$

where

$$M^m_{\ell_1,\ell} \equiv (-1)^{\ell+m}\left[\frac{2\ell+1}{2\ell_1+1}\right]^{1/2}(\ell+\ell_1)![(\ell+m)!(\ell-m)!(\ell_1+m)!(\ell_1-m)!]^{-1/2} \tag{5.103}$$

Substituting eq.(5.102) into eq.(5.101), and interchanging ℓ with ℓ_1, one obtains

$$\begin{aligned}\psi_{loc}(\mathbf{r}) &= -r\ E_0\sqrt{2\pi/3}[\cos\theta_0 Y_1^0(\theta,\phi)\sqrt{2} + \sin\theta_0(\exp(i\phi_0)Y_1^{-1}(\theta,\phi) \\ &\quad - \exp(-i\phi_0)Y_1^1(\theta,\phi))] \\ &\quad + \sum_{\ell m}\sum_{\ell_1=|m|}^{\infty\prime} A^r_{\ell_1 m}M^m_{\ell,\ell_1}(2d)^{-\ell_1-1}(r/2d)^\ell Y_\ell^m(\theta,\phi)\end{aligned} \tag{5.104}$$

where the prime as index of the summation again indicates that $\ell_1 = 0$ should be excluded. Using eqs.(5.69) and (5.100) one obtains

$$\begin{aligned}A_{\ell m} &= -\alpha_\ell\{-E_0\sqrt{2\pi/3}\delta_{\ell 1}[\cos\theta_0\delta_{m0}\sqrt{2} + \sin\theta_0(\exp(i\phi_0)\delta_{m,-1} - \exp(-i\phi_0)\delta_{m1})] \\ &\quad + (\frac{\epsilon_a - \epsilon_s}{\epsilon_a + \epsilon_s})\sum_{\ell_1=|m|}^{\infty\prime}(-1)^{\ell_1+m}A_{\ell_1 m}M^m_{\ell,\ell_1}(2d)^{-\ell-\ell_1-1}\}\end{aligned} \tag{5.105}$$

It follows from this equation that different m's do not couple to each other. This property is due to the rotational symmetry of the system around the z-axis. As a consequence

$$A_{\ell m} = 0 \text{ for } m \neq -1, 0, 1 \tag{5.106}$$

The above set of linear equations for the amplitudes may thus be written as the sum of the separate and independent contributions for $m = -1$, 0 and 1. In this way one obtains

$$A_{\ell 0} + \alpha_\ell(\frac{\epsilon_a - \epsilon_s}{\epsilon_a + \epsilon_s})\sum_{\ell_1=1}^{\infty}(-1)^{\ell_1} A_{\ell_1 0} M^0_{\ell,\ell_1}(2d)^{-\ell-\ell_1-1} = 2\alpha_1\sqrt{\pi/3}\delta_{\ell 1} E_0 \cos\theta_0 \quad (5.107)$$

and

$$\begin{aligned} A_{\ell 1} - \alpha_\ell(\frac{\epsilon_a - \epsilon_s}{\epsilon_a + \epsilon_s})\sum_{\ell_1=1}^{\infty}(-1)^{\ell_1} A_{\ell_1 1} M^1_{\ell,\ell_1}(2d)^{-\ell-\ell_1-1} \\ = -\alpha_1\sqrt{2\pi/3}\delta_{\ell 1} E_0 \sin\theta_0 \exp(-i\phi_0) \end{aligned} \quad (5.108)$$

It follows, by comparing the $m = -1$ equation with the $m = 1$ equation above and using the fact that $M^1_{\ell_1,\ell} = M^{-1}_{\ell_1,\ell}$, that $A_{\ell 1}\exp(i\phi_0) = -A_{\ell,-1}\exp(-i\phi_0)$. The $m = -1$ equation is therefore superfluous. The above equations may be solved numerically by neglecting the amplitudes larger than a suitably chosen large order. One then solves the finite set of equations

$$\begin{aligned} A_{\ell 0} + \alpha_\ell(\frac{\epsilon_a - \epsilon_s}{\epsilon_a + \epsilon_s})\sum_{\ell_1=1}^{M}(-1)^{\ell_1} A_{\ell_1 0} M^0_{\ell,\ell_1}(2d)^{-\ell-\ell_1-1} &= 2\alpha_1\sqrt{\pi/3}\delta_{\ell 1} E_0 \cos\theta_0 \\ \text{for } \ell &= 1, 2, ..., M \end{aligned} \quad (5.109)$$

and

$$\begin{aligned} A_{\ell 1} - \alpha_\ell(\frac{\epsilon_a - \epsilon_s}{\epsilon_a + \epsilon_s})\sum_{\ell_1=1}^{M}(-1)^{\ell_1} A_{\ell_1 1} M^1_{\ell,\ell_1}(2d)^{-\ell-\ell_1-1} &= -\alpha_1\sqrt{2\pi/3}\delta_{\ell 1} E_0 \sin\theta_0 \exp(-i\phi_0) \\ \text{for } \ell &= 1, 2, ..., M \end{aligned} \quad (5.110)$$

where M is the number of multipoles, taken into account.

The dipole and quadrupole polarizabilities of the spheres, modified by the presence of the substrate, are given, in terms of the amplitudes A_{10} , A_{11} , A_{20} and A_{21} , by

$$\begin{aligned} \alpha_\parallel(0) &= -4\pi\epsilon_a A_{11}/(\sqrt{2\pi/3}E_0 \sin\theta_0 \exp(-i\phi_0)) \\ \alpha_z(0) &= 2\pi\epsilon_a A_{10}/(\sqrt{\pi/3}E_0 \cos\theta_0) \\ \alpha^{10}_\parallel(0) &= -4\pi\epsilon_a A_{21}/(\sqrt{6\pi/5}E_0 \sin\theta_0 \exp(-i\phi_0)) \\ \alpha^{10}_z(0) &= \pi\epsilon_a A_{20}/(\sqrt{\pi/5}E_0 \cos\theta_0) \end{aligned} \quad (5.111)$$

This follows by substituting the expressions, eq.(5.67), for the polarizabilities of particles, which have a symmetry axis parallel to the z-axis, into eq.(5.30), and subsequently expressing the amplitudes $B_{\ell m}$ into the parallel and normal component of the incident field $\mathbf{E}_0$, by using eqs.(5.25) and (5.97). The argument 0 of the above polarizabilities denotes again the position of the center of the sphere, where the multipoles are located, on the z-axis. It can be shown that, if one solves the equations

(5.109) and (5.110), neglecting amplitudes of higher than dipole or quadrupole order (i.e. for $M = 2$), the results reproduce the expressions, eqs.(5.55) and (5.59), given in those models.

The surface constitutive coefficients $\beta_e(d)$, $\gamma_e(d)$, $\delta_e(d)$ and $\tau(d)$, at $z = d$, follow from (cf. eqs.(5.92)–(5.95))

$$\begin{aligned} \gamma_e(d) &= \rho\, \alpha_{\parallel}(0) \\ \beta_e(d) &= \rho\, \alpha_z(0)/\, \epsilon_a^2 \\ \tau(d) &= -\rho[\alpha_{\parallel}^{10}(0) - d\, \alpha_{\parallel}(0)] \\ \delta_e(d) &= -\rho[\alpha_z^{10}(0) + \alpha_{\parallel}^{10}(0) - d\, \alpha_z(0) - d\, \alpha_{\parallel}(0)]/\epsilon_a \end{aligned} \tag{5.112}$$

Here $\alpha_z(0)$, $\alpha_{\parallel}(0)$, $\alpha_z^{10}(0)$ and $\alpha_{\parallel}^{10}(0)$ are given by eq.(5.111). The invariants can be found by substitution of the above formulae into eq.(5.66). In the case of a polydisperse distribution of the sphere sizes in the low coverage limit, one must replace the polarizabilities in the above equation by their averages over this distribution.

5.7 Application: Spherical gold islands on sapphire

In order to demonstrate how the equations, derived above, can be used in practice, the special case of gold islands on sapphire will be considered below. In particular the case of spheres touching the substrate, $d = R$, is considered, for which the influence of the substrate is most pronounced. In subsequent chapters the case of gold islands will repeatedly be used to further illustrate the influence of shape and higher coverage. The reason for this choice was an article by Craighead and Niklasson who made such gold on sapphire films with a square array of islands, [34], [35]. For this case the dielectric constant of the ambient is unity, $\epsilon_a = 1$. Sapphire has a small dispersion in the visible domain. An average value is used as dielectric constant of the substrate, $\epsilon_s = 3.13$. The complex dielectric constant of gold has a large dispersion in the visible domain. In table 5.1 a list of values is given, for the optical domain. These were obtained by interpolation from the refractive index, n, and the attenuation, k, given by Johnson and Christy [40], using $\epsilon = (n + ik)^2$.

λ/nm	Re ϵ	Im ϵ
390	-1.607	5.738
400	-1.651	5.740
410	-1.686	5.735
420	-1.715	5.696
430	-1.714	5.636
440	-1.714	5.515
450	-1.731	5.350
460	-1.704	5.127
470	-1.721	4.885
480	-1.784	4.520
490	-2.031	4.060
500	-2.509	3.627
510	-3.184	3.038
520	-3.871	2.626
530	-4.554	2.401
540	-5.280	2.254
550	-5.937	2.077
560	-6.619	1.950
570	-7.273	1.821

λ/nm	Re ϵ	Im ϵ
580	-7.961	1.703
590	-8.706	1.570
600	-9.398	1.476
610	-10.125	1.436
620	-10.850	1.320
630	-11.613	1.229
640	-12.310	1.159
650	-13.010	1.083
660	-13.687	1.074
670	-14.345	1.061
680	-15.056	1.048
690	-15.764	1.033
700	-16.487	1.056
710	-17.185	1.078
720	-17.918	1.101
730	-18.689	1.125
740	-19.431	1.147
750	-20.166	1.168
760	-20.914	1.190

Table 5.1: Real and imaginary part of the dielectric constant of gold

An extensive list of the frequency dependent dielectric constants of solids is given in the Handbook of Optical Constants of Solids [41]

Consider first the quadrupole model. Substituting $d = R$ and the values of ϵ_a and ϵ_s into the depolarization factors, given in eqs.(5.88) and (5.89), one obtains the values

$$\begin{array}{llll} L_z = 0.2903\,, & L_z^1 = 0.3613\,, & \Lambda_z^1 = -0.0012\,, & \Lambda_z^{10} = 0.0194 \\ L_\| = 0.3118\,, & L_\|^1 = 0.3742\,, & \Lambda_\|^1 = -0.0004\,, & \Lambda_\|^{10} = 0.0129 \end{array} \tag{5.113}$$

In this example the dipole and quadrupole polarizabilities, eqs.(5.86) and (5.87) with $\epsilon_a = 1$, therefore reduce to

$$\begin{aligned} \widehat{\alpha}_z &\equiv \frac{1}{V}\alpha_z(0) = \frac{(\epsilon-1)[1+L_z^1(\epsilon-1)]}{[1+L_z(\epsilon-1)][1+L_z^1(\epsilon-1)]+\Lambda_z^1(\epsilon-1)^2} \\ \widehat{\alpha}_z^{10} &\equiv \frac{1}{VR}\alpha_z^{10}(0) = \frac{(\epsilon-1)^2\Lambda_z^{10}}{[1+L_z(\epsilon-1)][1+L_z^1(\epsilon-1)]+\Lambda_z^1(\epsilon-1)^2} \end{aligned} \tag{5.114}$$

and

$$\begin{aligned} \widehat{\alpha}_\| &\equiv \frac{1}{V}\alpha_\|(0) = \frac{(\epsilon-1)[1+L_\|^1(\epsilon-1)]}{[1+L_\|(\epsilon-1)][1+L_\|^1(\epsilon-1)]+\Lambda_\|^1(\epsilon-1)^2} \\ \widehat{\alpha}_\|^{10} &\equiv \frac{1}{VR}\alpha_\|^{10}(0) = \frac{(\epsilon-1)^2\Lambda_\|^{10}}{[1+L_\|(\epsilon-1)][1+L_\|^1(\epsilon-1)]+\Lambda_\|^1(\epsilon-1)^2} \end{aligned} \tag{5.115}$$

where $V \equiv \frac{4}{3}\pi R^3$ is the volume of the sphere. It is not correct to neglect Λ_z^1 and $\Lambda_\parallel^1$. The size of ϵ in the neighbourhood of the resonance is large and makes the corresponding terms of comparable size as other contributions in the denominator. In all the sections on applications dimensionless quantities, like the $\widehat{\alpha}$'s, are introduced. This makes both the calculation and the results more transparent. The surface constitutive coefficients, eq.(5.112), similarly reduce to

$$\begin{aligned}
\widehat{\gamma}_e &\equiv \frac{1}{\rho V}\gamma_e(d) = \widehat{\alpha}_\parallel \\
\widehat{\beta}_e &\equiv \frac{1}{\rho V}\beta_e(d) = \widehat{\alpha}_z \\
\widehat{\tau} &\equiv \frac{1}{\rho V d}\tau(d) = \widehat{\alpha}_\parallel - \widehat{\alpha}_\parallel^{10} \\
\widehat{\delta}_e &\equiv \frac{1}{\rho V d}\delta_e(d) = \widehat{\alpha}_z + \widehat{\alpha}_\parallel - \widehat{\alpha}_z^{10} - \widehat{\alpha}_\parallel^{10}
\end{aligned} \tag{5.116}$$

for this case, where $d = R$ and $\epsilon_a = 1$. The invariants, eq.(5.66), become

$$\begin{array}{ll}
\widehat{I}_e \equiv I_e/\rho V = \widehat{\gamma}_e - \epsilon_s \widehat{\beta}_e & \widehat{I}_c \equiv (\epsilon_s - 1) I_c/\rho V = \mathrm{Im}\,\widehat{\gamma}_e \\
\widehat{I}_\tau \equiv I_\tau/\rho V d = \widehat{\tau} & \widehat{I}_{\delta,e} \equiv I_{\delta,e}/\rho V d = \widehat{\delta}_e
\end{array} \tag{5.117}$$

where it was used that ϵ_s is real.

It is interesting how the results, obtained above for the quadrupole/dipole model, deviate from the results obtained by solving eqs.(5.109)–(5.111) to a sufficiently large order in the multipole expansion. It is convenient to introduce dimensionless amplitudes by

$$\begin{aligned}
\widehat{A}_{\ell 0} &\equiv A_{\ell 0}/\left[V R^{\ell-1} 2\sqrt{\pi/3} E_0 \cos\theta_0\right] \\
\widehat{A}_{\ell 1} &\equiv -A_{\ell 1}/\left[V R^{\ell-1}\sqrt{2\pi/3} E_0 \sin\theta_0 \exp(-i\phi_0)\right]
\end{aligned} \tag{5.118}$$

Eqs.(5.109) and (5.110) then reduce to, using $\epsilon_a = 1$,

$$\begin{aligned}
\widehat{A}_{\ell 0}/\widehat{\alpha}_\ell + \frac{4\pi}{3}\left(\frac{1-\epsilon_s}{1+\epsilon_s}\right)\sum_{\ell_1=1}^{M}(-1)^{\ell_1}\widehat{A}_{\ell_1 0} M_{\ell,\ell_1}^0 (2\widehat{d})^{-\ell-\ell_1-1} &= \delta_{\ell 1} \\
\text{for } \ell &= 1, 2, \ldots\ldots, M
\end{aligned} \tag{5.119}$$

and

$$\begin{aligned}
\widehat{A}_{\ell 1}/\widehat{\alpha}_\ell - \frac{4\pi}{3}\left(\frac{1-\epsilon_s}{1+\epsilon_s}\right)\sum_{\ell_1=1}^{M}(-1)^{\ell_1}\widehat{A}_{\ell_1 1} M_{\ell,\ell_1}^1 (2\widehat{d})^{-\ell-\ell_1-1} &= \delta_{\ell 1} \\
\text{for } \ell &= 1, 2, \ldots\ldots, M
\end{aligned} \tag{5.120}$$

where

$$\widehat{\alpha}_\ell \equiv \alpha_\ell / V R^{2(\ell-1)} \quad \text{and} \quad \widehat{d} \equiv d/R \tag{5.121}$$

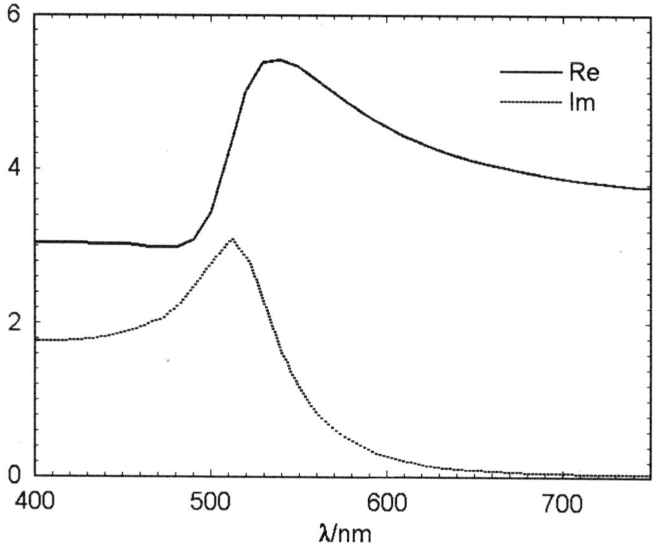

Figure 5.3 The real and imaginary parts of $\widehat{\gamma}_e = \widehat{\alpha}_\parallel$. Note that $\text{Im}\widehat{\gamma}_e = \widehat{I}_c$.

and M^1_{ℓ,ℓ_1} is given by eq.(5.103). The numerical calculation shows that for these films the value of the amplitudes has sufficiently converged for $M = 16$. Eq.(5.111) for the dipole and quadrupole polarizabilities of the spheres, modified by the presence of the substrate, in terms of the amplitudes becomes, using $\epsilon_a = 1$,

$$\begin{aligned}\widehat{\alpha}_\parallel &= 4\pi\widehat{A}_{11}\ , \quad \widehat{\alpha}_\parallel^{10} = \frac{4\pi}{3}\widehat{A}_{21}\sqrt{5} \\ \widehat{\alpha}_z &= 4\pi\widehat{A}_{10}\ , \quad \widehat{\alpha}_z^{10} = 2\pi\widehat{A}_{20}\sqrt{5/3}\end{aligned} \tag{5.122}$$

It is also found that the quadrupole approximation given above is accurate within a few procent, cf. [42] table 1 and 2, the biggest error being due to a small shift of the resonance frequency. Given the fact that this is less than the accuracy in the experimental values of the refractive index and the attenuation in this domain, these differences are not really significant for spheres. The figures 5.3-5.9 are all given in the quadrupole approximation. In fig.5.3 the real and imaginary parts of $\widehat{\gamma}_e = \widehat{\alpha}_\parallel$ are plotted as a function of the wavelength in the visual region. Note that $\text{Im}\widehat{\gamma}_e = \widehat{I}_c$. In fig.5.4 the same is done for $\widehat{\beta}_e = \widehat{\alpha}_z$. In fig.5.5 the real and imaginary parts of the invariant $\widehat{I}_e$ are plotted. The real and imaginary parts of the quadrupole polarizability $\widehat{\alpha}_\parallel^{10}$, are plotted in fig. 5.6. The real and imaginary parts of the quadrupole polarizability $\widehat{\alpha}_z^{10}$, are plotted in fig. 5.7. The constitutive coefficient and invariant $\widehat{\delta}_e = \widehat{I}_{\delta,e}$ is plotted in fig. 5.8. The constitutive coefficient and invariant $\widehat{\tau} = \widehat{I}_\tau$ is plotted in fig. 5.9. From the plots of the imaginary parts one sees that, in very good approximation, there is only a single resonance frequency

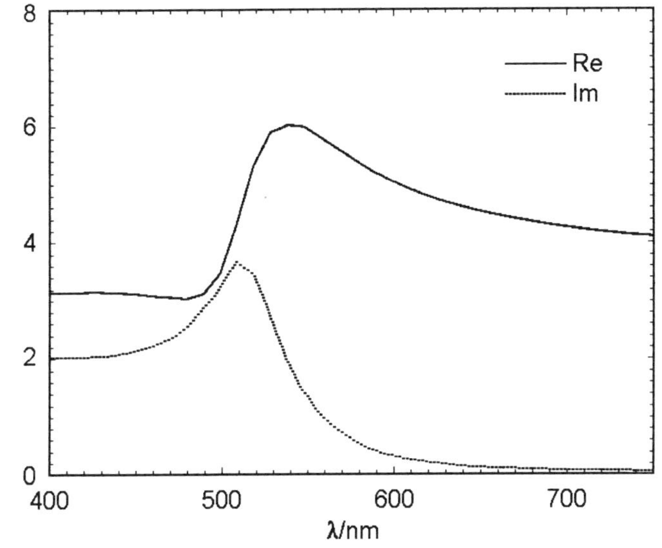

Figure 5.4 The real and imaginary parts of $\widehat{\beta}_e = \widehat{\alpha}_z$.

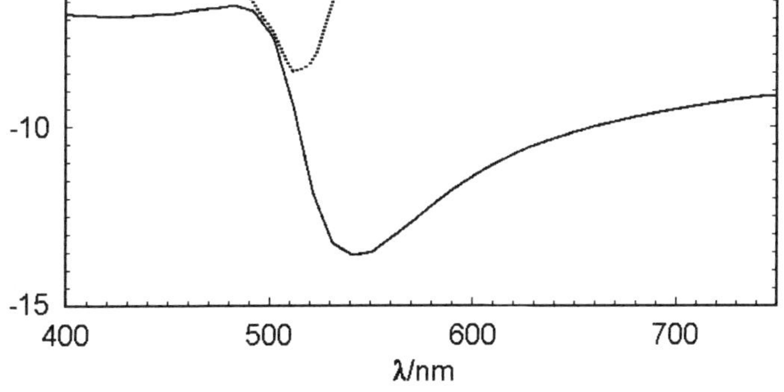

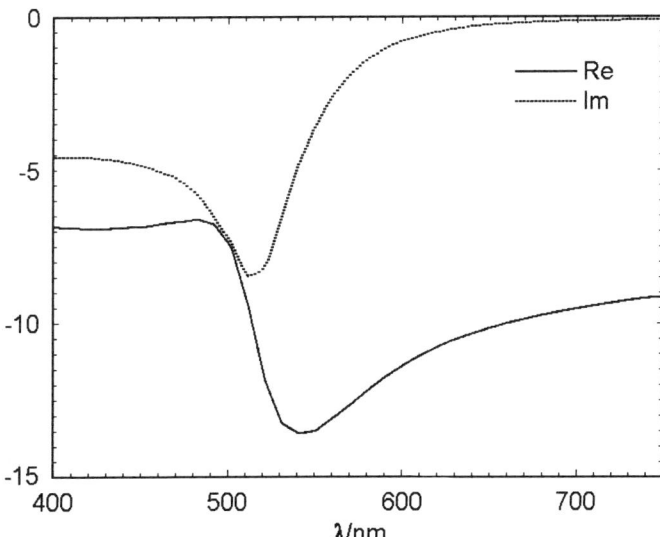

Figure 5.5 The real and imaginary parts of $\widehat{I}_e$.

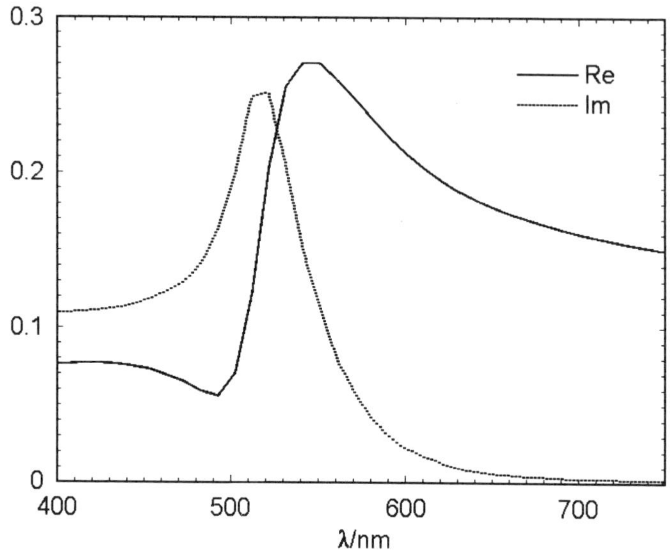

Figure 5.6 The real and imaginary parts of $\widehat{\alpha}_{\parallel}^{10}$.

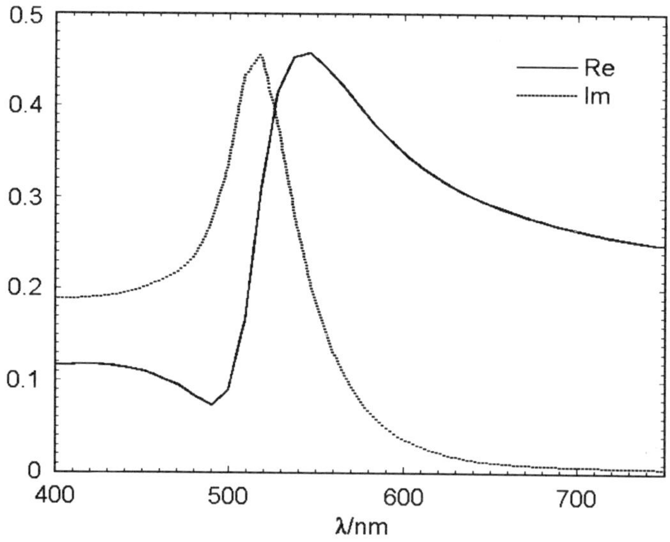

Figure 5.7 The real and imaginary part of $\widehat{\alpha}_{z}^{10}$.

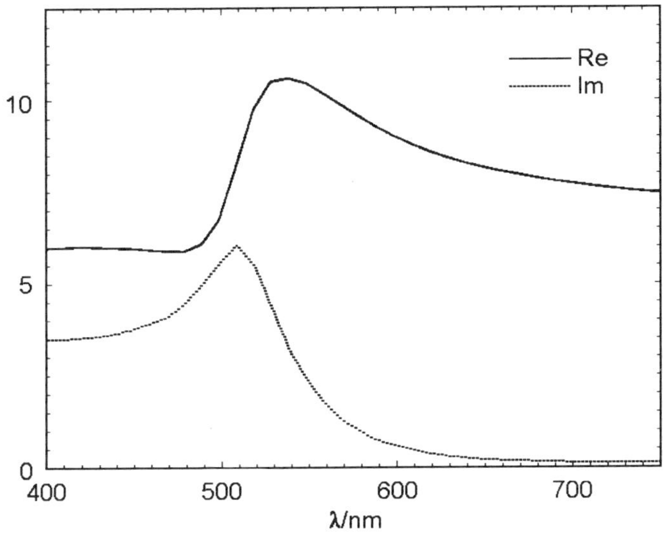

Figure 5.8 The real and imaginary parts of $\widehat{\delta}_e = \widehat{I}_{\delta,e}$.

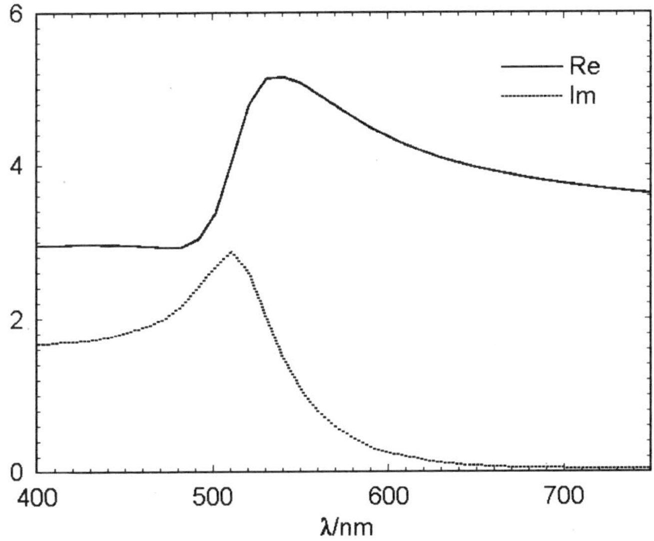

Figure 5.9 The real and imaginary parts of $\widehat{\tau} = \widehat{I}_\tau$.

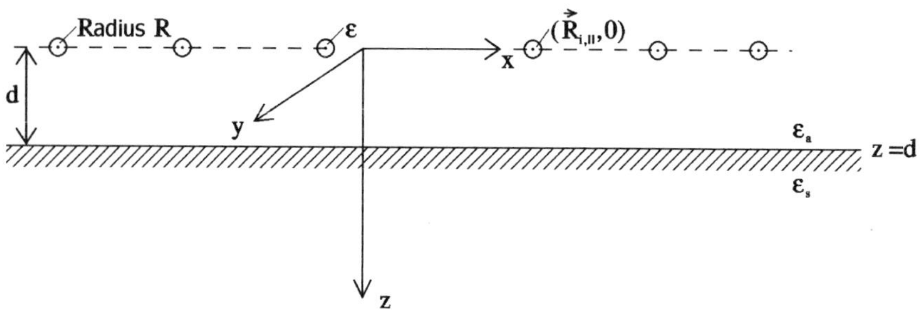

Figure 5.10 Model of a thin island film, consisting of identical spheres with a radius R much smaller than d.

with a wavelength of 515 nm, not only for the polarizabilities normal and parallel to the surface but also for the coefficients $\widehat{\delta}_e$ and $\widehat{\tau}$ which are of quadrupolar order. The influence of the substrate does lead to a slight increase in the size of the polarizability normal to the surface and a slight decrease of the polarizability parallel to the surface. The fact that the parallel and the normal polarizablities differ due to the interaction with the substrate was first given by Yamaguchi, Yoshida and Kinbara [10] to dipolar order. The contribution of the quadrupole polarizabilities $\widehat{\alpha}_z^{10}$ and $\widehat{\alpha}_\|^{10}$ to $\widehat{\delta}_e$ and $\widehat{\tau}$ is not more than 5 to 10 %. This shows that most of the contribution to the constitutive coefficients of quadrupolar order is due to moving the dipole from the center of the sphere to the surface of the substrate.

5.8 Appendix A

In this appendix the derivation of the surface constitutive coefficients $\gamma_e(d)$, eq.(5.39), and $\beta_e(d)$, eq.(5.40), will be given for an island film, consisting of identical particles, in the low density limit. In the polarizable dipole model, introduced in section 5.3, these particles were replaced by polarizable point dipoles. In the following, however, instead of point dipoles, small dielectric spheres will be considered, situated in the ambient (see fig.5.10). This is necessary, if one wants to apply the method of excess fields, introduced in section 2.4. The radius of these spheres is R and their centers are at a distance d above the surface of the substrate ($R << d$). These centers are assumed to be positioned in the XOY-plane of a Cartesian coordinate system, of which the positive z-axis is pointing into the substrate. The position of the i th particle is given by $(\mathbf{R}_{i,\|}, 0)$, the position of the surface of the substrate by $z = d$. The dielectric constants of ambient, substrate and particles are given by ϵ_a, ϵ_s and ϵ respectively (see fig.5.10).

The calculation of the coefficients γ_e and β_e for this system will be on the basis of the general method to introduce excess fields, as was given in section 2.4. It follows from the definition, eq.(2.32), that one then has to calculate both the total electric field $\mathbf{E}(\mathbf{r})$ and the extrapolated fields $\mathbf{E}^+(\mathbf{r})$ and $\mathbf{E}^-(\mathbf{r})$ in the $z > d$ and $z < d$ region. In the analysis in this appendix the dividing surface is chosen to coincide with the surface $z = d$ of the substrate. In the following first this total field will be evaluated in the ambient, the substrate and within the spheres. The external electric field applied to this system, is assumed to be homogeneous and will be denoted by $\mathbf{E}_0$ in the ambient. The total field in the ambient is the sum of this external field, the dipole fields of the spheres, with centers in $(\mathbf{R}_{i,\|}, 0)$, and of the image dipoles in $(\mathbf{R}_{i,\|}, 2d)$:

$$\mathbf{E}_a(\mathbf{r}) = \mathbf{E}_0 - \sum_i [4\pi\epsilon_a|(\mathbf{r}_\| - \mathbf{R}_{i,\|}, z)|^3]^{-1}\left[\mathbf{1} - 3\frac{(\mathbf{r}_\| - \mathbf{R}_{i,\|}\ , z)(\mathbf{r}_\| - \mathbf{R}_{i,\|}\ , z)}{|(\mathbf{r}_\| - \mathbf{R}_{i,\|},\ z)|^2}\right].\mathbf{p}$$
$$-\sum_i [4\pi\epsilon_a|(\mathbf{r}_\| - \mathbf{R}_{i,\|}, z - 2d)|^3]^{-1}$$
$$\times\left[\mathbf{1} - 3\frac{(\mathbf{r}_\| - \mathbf{R}_{i,\|}\ , z - 2d)(\mathbf{r}_\| - \mathbf{R}_{i,\|}\ , z - 2d)}{|(\mathbf{r}_\| - \mathbf{R}_{i,\|},\ z - 2d)|^2}\right].\mathbf{p}_r \tag{5.123}$$

(cf. eq.(5.31)). The dipole moments $\mathbf{p}$ of the identical spheres are the same, and given by

$$\mathbf{p} = [\alpha_\|(0)(\mathbf{1} - \hat{\mathbf{z}}\hat{\mathbf{z}}) + \alpha_z(0)\hat{\mathbf{z}}\hat{\mathbf{z}}].\mathbf{E}_0 \tag{5.124}$$

with the polarizabilities $\alpha_\|(0)$, eq.(5.36), and $\alpha_z(0)$, eq.(5.37), which contain the interaction with the substrate by means of the factor A, eq.(5.38). The polarizabilities $\alpha_\|$ and α_z at the right-hand sides of eqs.(5.36) and (5.37), are polarizabilities of free spheres in an infinite medium and given by eq.(5.72), together with eq.(5.70):

$$\alpha_z = \alpha_\| = 4\pi\epsilon_a R^3 \frac{\epsilon - \epsilon_a}{\epsilon + 2\epsilon_a} \equiv \alpha_d \tag{5.125}$$

and eqs.(5.36) and (5.37) become in this case

$$\alpha_\|(0) = [1 + A\alpha_d]^{-1}\alpha_d, \quad \alpha_z(0) = [1 + 2A\alpha_d]^{-1}\alpha_d \tag{5.126}$$

The dipole moment $\mathbf{p}_r$ of the image dipoles at $(\mathbf{R}_{i,\|}, 2d)$ is given by eq.(5.33). It will be clear that the interaction between the different spheres is neglected in the low density limit.

The total field in the substrate is given by (cf. eq.(5.32))

$$\mathbf{E}_s(\mathbf{r}) = (\mathbf{E}_{0,\|}, (\epsilon_a/\epsilon_s)E_{0,z}) - \sum_i [4\pi\epsilon_s|(\mathbf{r}_\| - \mathbf{R}_{i,\|}, z)|^3]^{-1}$$
$$\times\left[\mathbf{1} - 3\frac{(\mathbf{r}_\| - \mathbf{R}_{i,\|}\ , z)(\mathbf{r}_\| - \mathbf{R}_{i,\|}\ , z)}{|(\mathbf{r}_\| - \mathbf{R}_{i,\|},\ z)|^2}\right].\mathbf{p}_t \tag{5.127}$$

where $\mathbf{p}_t$ is given by eq.(5.33).

In order to calculate the electric field inside the spheres, one has to use the boundary conditions of this field on their surface. It is found that the electric field $\mathbf{E}_{inside}$, just inside a sphere, is related to the electric field $\mathbf{E}_{outside}$, just outside this sphere, by means of

$$\mathbf{E}_{inside} = [(\mathbf{1} - \mathbf{nn}) + (\epsilon_a/\epsilon)\mathbf{nn}].\mathbf{E}_{outside} = [\mathbf{1} - \epsilon^{-1}(\epsilon - \epsilon_a)\mathbf{nn}].\mathbf{E}_{outside} \tag{5.128}$$

where $\mathbf{n}$ is the normal to the surface of the sphere, which is a function of the position on this surface. $\mathbf{E}_{outside}$ follows from eqs.(5.123), (5.124) and (5.33):

$$\mathbf{E}_{outside} = -\frac{1}{4\pi\epsilon_a} R^{-3}(\mathbf{1} - 3\mathbf{nn}).[\alpha_{\|}(0)(\mathbf{1} - \hat{\mathbf{z}}\hat{\mathbf{z}}) + \alpha_z(0)\hat{\mathbf{z}}\hat{\mathbf{z}}].\mathbf{E}_0$$

$$+ \left[\mathbf{1} - \frac{1}{4\pi\epsilon_a}\left(\frac{1}{2d}\right)^3 \left(\frac{\epsilon_a - \epsilon_s}{\epsilon_a + \epsilon_s}\right)[\alpha_{\|}(0)(\mathbf{1} - \hat{\mathbf{z}}\hat{\mathbf{z}}) + 2\alpha_z(0)\hat{\mathbf{z}}\hat{\mathbf{z}}]\right].\mathbf{E}_0 \tag{5.129}$$

Here it has been used that $R << d$ and furthermore that in the low density limit one may neglect the fields of all other particles at the surface of the sphere under consideration. It now follows from the last two equations, that

$$\begin{aligned}\mathbf{E}_{inside} &= -\frac{1}{4\pi\epsilon_a} R^{-3}[\mathbf{1} - \epsilon^{-1}(\epsilon + 2\epsilon_a)\mathbf{nn}].[\alpha_{\|}(0)(\mathbf{1} - \hat{\mathbf{z}}\hat{\mathbf{z}}) + \alpha_z(0)\hat{\mathbf{z}}\hat{\mathbf{z}}].\mathbf{E}_0 \\ &\quad +[\mathbf{1} - \epsilon^{-1}(\epsilon - \epsilon_a)\mathbf{nn}].\{\mathbf{1} - \frac{1}{4\pi\epsilon_a}\left(\frac{1}{2d}\right)^3\left(\frac{\epsilon_a - \epsilon_s}{\epsilon_a + \epsilon_s}\right) \\ &\quad \times[\alpha_{\|}(0)(\mathbf{1} - \hat{\mathbf{z}}\hat{\mathbf{z}}) + 2\alpha_z(0)\hat{\mathbf{z}}\hat{\mathbf{z}}]\}.\mathbf{E}_0\end{aligned} \tag{5.130}$$

It is easily verified, by using eqs.(5.125), (5.126) and (5.38), that

$$\begin{aligned}&\frac{1}{4\pi\epsilon_a} R^{-3}\left(\frac{\epsilon + 2\epsilon_a}{\epsilon}\right)[\alpha_{\|}(0)(\mathbf{1} - \hat{\mathbf{z}}\hat{\mathbf{z}}) + \alpha_z(0)\hat{\mathbf{z}}\hat{\mathbf{z}}] = \left(\frac{\epsilon - \epsilon_a}{\epsilon}\right) \\ &\times\left[\mathbf{1} - \frac{1}{4\pi\epsilon_a}\left(\frac{1}{2d}\right)^3\left(\frac{\epsilon_a - \epsilon_s}{\epsilon_a + \epsilon_s}\right)[\alpha_{\|}(0)(\mathbf{1} - \hat{\mathbf{z}}\hat{\mathbf{z}}) + 2\alpha_z(0)\hat{\mathbf{z}}\hat{\mathbf{z}}]\right]\end{aligned} \tag{5.131}$$

The terms proportional to $\mathbf{nn}$ in eq.(5.130) therefore vanish, and one obtains

$$\begin{aligned}\mathbf{E}_{inside} &= -\frac{1}{4\pi\epsilon_a} R^{-3}[\alpha_{\|}(0)(\mathbf{1} - \hat{\mathbf{z}}\hat{\mathbf{z}}) + \alpha_z(0)\hat{\mathbf{z}}\hat{\mathbf{z}}].\mathbf{E}_0 \\ &\quad + \left[\mathbf{1} - A[\alpha_{\|}(0)(\mathbf{1} - \hat{\mathbf{z}}\hat{\mathbf{z}}) + 2\alpha_z(0)\hat{\mathbf{z}}\hat{\mathbf{z}}]\right].\mathbf{E}_0\end{aligned} \tag{5.132}$$

where the definition, eq.(5.38), for A has again been used. It will be clear that, since $\mathbf{E}_{inside}$ does not change along the surface of the sphere according to the above result, the electric field $\mathbf{E}_{sphere}$ at any point inside this sphere, will also be given by eq.(5.132). One therefore obtains for this field

$$\mathbf{E}_{sphere} = \mathbf{E}_0 - \left[\frac{1}{4\pi\epsilon_a} R^{-3}\mathbf{1} + A(\mathbf{1} + \hat{\mathbf{z}}\hat{\mathbf{z}})\right].\boldsymbol{\alpha}(0).\mathbf{E}_0 \tag{5.133}$$

with the polarizability tensor

$$\boldsymbol{\alpha}(0) \equiv [\alpha_{\|}(0)(\mathbf{1} - \hat{\mathbf{z}}\hat{\mathbf{z}}) + \alpha_z(0)\hat{\mathbf{z}}\hat{\mathbf{z}}] \tag{5.134}$$

Notice, that it follows with eqs.(5.125) and (5.126), that for $\epsilon_a = \epsilon_s$, and therefore $A = 0$, one obtains the well-known result

$$\mathbf{E}_{sphere} = \frac{3\epsilon_a}{\epsilon + 2\epsilon_a}\mathbf{E}_0 \tag{5.135}$$

for the electric field within a sphere with dielectric constant ϵ in an infinite medium with dielectric constant ϵ_a in a homogeneous external electric field $\mathbf{E}_0$.

Below it will appear, that the extrapolated fields $\mathbf{E}^+(\mathbf{r})$ and $\mathbf{E}^-(\mathbf{r})$, which are also needed for the calculation of the excess field

$$\mathbf{E}_{ex}(\mathbf{r}) \equiv \mathbf{E}(\mathbf{r}) - \mathbf{E}^+(\mathbf{r})\theta(z-d) - \mathbf{E}^-(\mathbf{r})\theta(d-z)$$

(cf. eq.(2.32)), may be approximated here by the external fields in the ambient and the substrate respectively:

$$\begin{aligned} \mathbf{E}^-(\mathbf{r}) &= \mathbf{E}_0 \text{ for } z < d \\ \mathbf{E}^+(\mathbf{r}) &= (\mathbf{E}_{0,\parallel}, (\epsilon_a/\epsilon_s)E_{0,z}) \text{ for } z > d \end{aligned} \tag{5.136}$$

From eqs.(5.123), (5.127), (5.133) and (5.136), together with eqs.(5.33), (5.124) and (5.134), one therefore obtains for this excess electric field:

$$\begin{aligned} \mathbf{E}_{ex}(\mathbf{r}) &= -\frac{1}{4\pi\epsilon_a}\theta(d-z)\sum_i \theta([(\mathbf{r}_\parallel - \mathbf{R}_{i,\parallel})^2 + z^2]^{1/2} - R) \\ &\times\{[(\mathbf{r}_\parallel - \mathbf{R}_{i,\parallel})^2 + z^2]^{-3/2}\left(\mathbf{1} - 3\frac{(\mathbf{r}_\parallel - \mathbf{R}_{i,\parallel}\ ,\ z)(\mathbf{r}_\parallel - \mathbf{R}_{i,\parallel}\ ,\ z)}{(\mathbf{r}_\parallel - \mathbf{R}_{i,\parallel})^2 + z^2}\right) \end{aligned}$$

$$\begin{aligned} &.[\alpha_\parallel(0)(\mathbf{1} - \hat{\mathbf{z}}\hat{\mathbf{z}}) + \alpha_z(0)\hat{\mathbf{z}}\hat{\mathbf{z}}].\mathbf{E}_0 + \left(\frac{\epsilon_a - \epsilon_s}{\epsilon_a + \epsilon_s}\right)[(\mathbf{r}_\parallel - \mathbf{R}_{i,\parallel})^2 + (z-2d)^2]^{-3/2} \\ &\times\left(\mathbf{1} - 3\frac{(\mathbf{r}_\parallel - \mathbf{R}_{i,\parallel}\ ,\ z-2d)(\mathbf{r}_\parallel - \mathbf{R}_{i,\parallel}\ ,\ z-2d)}{(\mathbf{r}_\parallel - \mathbf{R}_{i,\parallel})^2 + (z-2d)^2}\right).[\alpha_\parallel(0)(\mathbf{1} - \hat{\mathbf{z}}\hat{\mathbf{z}}) - \alpha_z(0)\hat{\mathbf{z}}\hat{\mathbf{z}}].\mathbf{E}_0\} \\ &-\sum_i \theta(R - [(\mathbf{r}_\parallel - \mathbf{R}_{i,\parallel})^2 + z^2]^{1/2})\left[\frac{1}{4\pi\epsilon_a}R^{-3}\mathbf{1} + A(\mathbf{1} + \hat{\mathbf{z}}\hat{\mathbf{z}})\right] \\ &.[\alpha_\parallel(0)(\mathbf{1} - \hat{\mathbf{z}}\hat{\mathbf{z}}) + \alpha_z(0)\hat{\mathbf{z}}\hat{\mathbf{z}}].\mathbf{E}_0 - \frac{1}{4\pi\epsilon_s}\theta(z-d)\sum_i\left(\frac{2\epsilon_s}{\epsilon_a + \epsilon_s}\right)[(\mathbf{r}_\parallel - \mathbf{R}_{i,\parallel})^2 + z^2]^{-3/2} \\ &\times\left(\mathbf{1} - 3\frac{(\mathbf{r}_\parallel - \mathbf{R}_{i,\parallel}\ ,\ z)(\mathbf{r}_\parallel - \mathbf{R}_{i,\parallel}\ ,\ z)}{(\mathbf{r}_\parallel - \mathbf{R}_{i,\parallel})^2 + z^2}\right).[\alpha_\parallel(0)(\mathbf{1} - \hat{\mathbf{z}}\hat{\mathbf{z}}) + \alpha_z(0)\hat{\mathbf{z}}\hat{\mathbf{z}}].\mathbf{E}_0 \end{aligned} \tag{5.137}$$

where the Heaviside functions in the first sum restrict the fields to the ambient around the spheres, in the second sum to the region within the spheres and in the last sum to the substrate (see fig.5.2).

The above field can now be evaluated explicitly, if one makes the assumption that the sums $\sum_i$ in eq.(5.137) may be replaced by integrals $\rho \int d\mathbf{R}_\parallel$, where ρ is the average number of spheres per unit of surface area of the substrate. This approximation means, in fact, that all correlations between the particles in the film are neglected. Only the z-component of this excess electric field is needed for the calculation of the z-component of the surface polarization density $\mathbf{P}^s$, according to eq.(2.44), which quantity on its turn is needed for the calculation of the coefficient

β_e, according to eq.(3.39). One obtains

$$\begin{aligned}
P_z^s &= -\int_{-\infty}^{\infty} E_{ex,z}(\mathbf{r})dz = \int_{-\infty}^{\infty} dz[\frac{1}{4\pi\epsilon_a}\theta(d-z)\rho \int d\mathbf{R}_\| \theta([(\mathbf{r}_\| - \mathbf{R}_\|)^2 + z^2]^{1/2} - R) \\
&\times\{[(\mathbf{r}_\| - \mathbf{R}_\|)^2 + z^2]^{-3/2}\left(1 - 3\frac{z^2}{(\mathbf{r}_\| - \mathbf{R}_\|)^2 + z^2}\right)\alpha_z(0)E_{0,z} - \left(\frac{\epsilon_a - \epsilon_s}{\epsilon_a + \epsilon_s}\right) \\
&\times[(\mathbf{r}_\| - \mathbf{R}_\|)^2 + (z-2d)^2]^{-3/2}\left(1 - 3\frac{(z-2d)^2}{(\mathbf{r}_\| - \mathbf{R}_\|)^2 + (z-2d)^2}\right)\alpha_z(0)E_{0,z}\} \\
&+\frac{1}{4\pi\epsilon_s}\theta(z-d)\rho\int d\mathbf{R}_\|\left(\frac{2\epsilon_s}{\epsilon_a + \epsilon_s}\right)[(\mathbf{r}_\| - \mathbf{R}_\|)^2 + z^2]^{-3/2} \\
&\times\left(1 - 3\frac{z^2}{(\mathbf{r}_\| - \mathbf{R}_\|)^2 + z^2}\right)\alpha_z(0)E_{0,z}] \\
&+\int_{-\infty}^{\infty} dz\ \rho\int d\mathbf{R}_\|\theta(R - [(\mathbf{r}_\| - \mathbf{R}_\|)^2 + z^2]^{1/2})\left(\frac{1}{4\pi\epsilon_a}R^{-3} + 2A\right)\alpha_z(0)E_{0,z}
\end{aligned} \tag{5.138}$$

where it has been used that terms, containing non-diagonal elements of matrices in eq.(5.137), vanish after integration over $\mathbf{R}_\|$ for reasons of symmetry. The last integral at the right-hand side of eq.(5.138) is easily performed, since

$$\int_{-\infty}^{\infty} dz \int d\mathbf{R}_\|\theta(R - [(\mathbf{r}_\| - \mathbf{R}_\|)^2 + z^2]^{1/2}) = \frac{4}{3}\pi\ R^3 \tag{5.139}$$

is the volume of a sphere. In order to calculate the other integrals at the right-hand side of eq.(5.138), first the new integration variables $\mathbf{R}_\|' \equiv \mathbf{R}_\| - \mathbf{r}_\|$ are introduced, and next planar polar coordinates $R_\|'$ and ϕ'. Integrating over ϕ' from 0 to 2π, one then obtains

$$\begin{aligned}
P_z^s/\rho\alpha_z(0)E_{0,z} &= \int_{-\infty}^{\infty} dz\{\frac{1}{2\epsilon_a}\theta(d-z)\int_0^{\infty} R_\|' dR_\|' \theta([R_\|'^2 + z^2]^{1/2} - R) \\
&\times[[R_\|'^2 + z^2]^{-3/2}\left(1 - 3\frac{z^2}{R_\|'^2 + z^2}\right) \\
&-\left(\frac{\epsilon_a - \epsilon_s}{\epsilon_a + \epsilon_s}\right)[R_\|'^2 + (z-2d)^2]^{-3/2}\left(1 - 3\frac{(z-2d)^2}{R_\|'^2 + (z-2d)^2}\right)] \\
&+\frac{1}{2\epsilon_s}\theta(z-d)\int_0^{\infty} R_\|' dR_\|'\left(\frac{2\epsilon_s}{\epsilon_a + \epsilon_s}\right)[R_\|'^2 + z^2]^{-3/2}\left(1 - 3\frac{z^2}{R_\|'^2 + z^2}\right)\} \\
&+\frac{4}{3}\pi\ R^3\left(\frac{1}{4\pi\epsilon_a}R^{-3} + 2A\right)
\end{aligned} \tag{5.140}$$

where both members have been divided by the common factor $\rho\alpha_z(0)E_{0,z}$. Introducing the new variable $t \equiv R_{\|}'^2$ and writing $\theta(d-z) = 1 - \theta(z-d)$, eq.(5.140) becomes

$$\begin{aligned} P_z^s/\rho\alpha_z(0)E_{0,z} &= \int_{-\infty}^{\infty} dz\{\frac{1}{4\epsilon_a}\int_0^{\infty} dt\ \theta([t+z^2]^{1/2} - R) \\ &\times[[t+z^2]^{-3/2}\left(1 - 3\frac{z^2}{t+z^2}\right) \\ &-\left(\frac{\epsilon_a - \epsilon_s}{\epsilon_a + \epsilon_s}\right)[t+(z-2d)^2]^{-3/2}\left(1 - 3\frac{(z-2d)^2}{t+(z-2d)^2}\right)] \\ &-\frac{1}{4\epsilon_a}\theta(z-d)\int_0^{\infty} dt\left(\frac{\epsilon_a - \epsilon_s}{\epsilon_a + \epsilon_s}\right)[t+z^2]^{-3/2}\left(1 - 3\frac{z^2}{t+z^2}\right)\} \\ &+\frac{1}{3\epsilon_a} + \frac{8}{3}\pi A\ R^3 \end{aligned} \tag{5.141}$$

where it has also been used that

$$\theta(z-d)\theta([t+z^2]^{1/2} - R) = \theta(z-d) \tag{5.142}$$

since $R < d$. Eq.(5.141) may alternatively be written as

$$\begin{aligned} &P_z^s/\rho\alpha_z(0)E_{0,z} \\ &= \frac{1}{4\epsilon_a}\int_{-\infty}^{\infty} dz\{\int_{Max[R^2-z^2,0]}^{\infty} dt[[t+z^2]^{-3/2}\left(1 - 3\frac{z^2}{t+z^2}\right) \\ &-\left(\frac{\epsilon_a - \epsilon_s}{\epsilon_a + \epsilon_s}\right)[t+(z-2d)^2]^{-3/2}\left(1 - 3\frac{(z-2d)^2}{t+(z-2d)^2}\right)] \\ &-\frac{1}{4\epsilon_a}\theta(z-d)\int_0^{\infty} dt\left(\frac{\epsilon_a - \epsilon_s}{\epsilon_a + \epsilon_s}\right)[t+z^2]^{-3/2}\left(1 - 3\frac{z^2}{t+z^2}\right)\} \\ &+\frac{1}{3\epsilon_a} + \frac{8}{3}\pi A\ R^3 \end{aligned} \tag{5.143}$$

Using the following result

$$\int_{t_0}^{\infty} dt[t+(z+D)^2]^{-3/2}\left(1 - 3\frac{(z+D)^2}{t+(z+D)^2}\right) = \frac{2t_0}{[t_0+(z+D)^2]^{3/2}} \tag{5.144}$$

one finds that the second integral at the right-hand side of eq.(5.143) vanishes (since $t_0 = 0$), whereas in the first integral the integration over z may be restricted to the interval $-R < z < R$, where $R^2 - z^2 > 0$. One therefore obtains

$$\begin{aligned} P_z^s/\rho\alpha_z(0)E_{0,z} &= \frac{1}{2\epsilon_a}\int_{-R}^{R} dz\left[\frac{R^2 - z^2}{R^3} - \left(\frac{\epsilon_a - \epsilon_s}{\epsilon_a + \epsilon_s}\right)\frac{R^2 - z^2}{[R^2 - z^2 + (z - 2\ d)^2]^{3/2}}\right] \\ &+\frac{1}{3\epsilon_a} + \frac{8}{3}\pi\ AR^3 \end{aligned} \tag{5.145}$$

The first integral is easily calculated:

$$\frac{1}{2\epsilon_a}\int_{-R}^{R} dz\frac{R^2 - z^2}{R^3} = \frac{2}{3\epsilon_a} \tag{5.146}$$

Since $R << d$, the second integral may be simplified and gives

$$\begin{aligned}
& \frac{1}{2\epsilon_a}\left(\frac{\epsilon_a-\epsilon_s}{\epsilon_a+\epsilon_s}\right)\int_{-R}^{R} dz \frac{R^2-z^2}{[R^2-z^2+(z-2d)^2]^{3/2}} \\
\simeq\ & \frac{1}{2\epsilon_a}\left(\frac{\epsilon_a-\epsilon_s}{\epsilon_a+\epsilon_s}\right)\int_{-R}^{R} dz \frac{R^2-z^2}{8d^3} \\
=\ & \frac{2}{3\epsilon_a}\left(\frac{\epsilon_a-\epsilon_s}{\epsilon_a+\epsilon_s}\right)\frac{R^3}{8d^3} = \frac{8}{3}\pi\, AR^3
\end{aligned} \tag{5.147}$$

where also the definition, eq.(5.38), of A has been used. It is then found from eqs.(5.145)–(5.147), that one obtains the simple result

$$P_z^s/\rho\alpha_z(0)E_{0,z} = \epsilon_a^{-1} \tag{5.148}$$

This can be written as

$$P_z^s = \rho\, \epsilon_a^{-1}\alpha_z(0)E_{0,z} = \rho\, \epsilon_a^{-2}\alpha_z(0)D_{0,z} \tag{5.149}$$

where $D_{0,z}$ is the z-component of the external electric displacement field $\mathbf{D}_0 \equiv \epsilon_a \mathbf{E}_0$ in the ambient.

It now follows from the analysis given above, in particular from the fact that the integration in eq.(5.145) could be restricted to the interval $-R < z < R$, that the z-component of the excess electric field vanishes in the region, where $z > R$ and where $z < -R$. This field therefore has indeed the correct limiting property of an excess field for $z \to \pm\infty$. It is also an indirect proof of the validity eq.(5.136) for the z-component of the electric field. From this equation one finds for the z-component of the extrapolated electric displacement fields $\mathbf{D}^+(\mathbf{r})$ and $\mathbf{D}^-(\mathbf{r})$:

$$D_z^-(\mathbf{r}) = D_z^+(\mathbf{r}) = \epsilon_a E_{0,z} = D_{0,z} \tag{5.150}$$

In the notation of eq.(3.34) this result may be written as

$$D_{z,+} = D_{0,z} \tag{5.151}$$

and eq.(5.149) therefore also becomes

$$P_z^s = \rho\, \epsilon_a^{-2}\alpha_z(0)D_{z,+} \tag{5.152}$$

from which one finds with eqs.(3.39) and (5.37), that the constitutive coefficient β_e is indeed given by eq.(5.40). Since the dividing surface was chosen at $z = d$ in the analysis above, one obtains $\beta_e(d)$. On the other hand it also follows, from the fact that the integration in eq.(5.145) is restricted to a narrow interval around $z = 0$ (width $2R << d$), that β_e may be interpreted as the constitutive coefficient of the surface $z = 0$, where the small spheres (dipoles) are located (see fig.5.2). In section 5.4 this coefficient is denoted by $\beta_{e,1}(0)$, and one therefore finds, that $\beta_{e,1}(0) = \beta_e(d)$.

It follows from eqs.(3.38) and (2.44), that for the evaluation of γ_e one must also calculate the excess electric displacement field for the present system. It follows from eq.(5.123) that in the ambient the total electric displacement field is given by

$$\begin{aligned}\mathbf{D}_a(\mathbf{r}) &= \epsilon_a \mathbf{E}_a(\mathbf{r}) = \mathbf{D}_0 \\ &- \sum_i [4\pi|(\mathbf{r}_{\|} - \mathbf{R}_{i,\|}, z)|^3]^{-1} \left[\mathbf{1} - 3\frac{(\mathbf{r}_{\|} - \mathbf{R}_{i,\|}\ ,\ z)(\mathbf{r}_{\|} - \mathbf{R}_{i,\|}\ ,\ z)}{|(\mathbf{r}_{\|} - \mathbf{R}_{i,\|},\ z)|^2}\right] \cdot \mathbf{p} \\ - \sum_i [4\pi|(\mathbf{r}_{\|} - \mathbf{R}_{i,\|}, z - 2d)|^3]^{-1} &\left[\mathbf{1} - 3\frac{(\mathbf{r}_{\|} - \mathbf{R}_{i,\|}\ ,\ z - 2d)(\mathbf{r}_{\|} - \mathbf{R}_{i,\|}\ ,\ z - 2d)}{|(\mathbf{r}_{\|} - \mathbf{R}_{i,\|},\ z - 2d)|^2}\right] \cdot \mathbf{p}_r\end{aligned} \tag{5.153}$$

where $\mathbf{p}$ and $\mathbf{p}_r$ are given by eqs.(5.124) and (5.33), and where $\mathbf{D}_0 = \epsilon_a \mathbf{E}_0$ is the external electric displacement field in the ambient. Furthermore it follows from eq.(5.127) that the total electric displacement field in the substrate is given by

$$\begin{aligned}\mathbf{D}_s(\mathbf{r}) &= \epsilon_s \mathbf{E}_s(r) = ((\epsilon_s/\epsilon_a)\mathbf{D}_{0,\|}, D_{0,z}) - \sum_i [4\pi|(\mathbf{r}_{\|} - \mathbf{R}_{i,\|}, z)|^3]^{-1} \\ &\times \left[\mathbf{1} - 3\frac{(\mathbf{r}_{\|} - \mathbf{R}_{i,\|}\ ,\ z)(\mathbf{r}_{\|} - \mathbf{R}_{i,\|}\ ,\ z)}{|(\mathbf{r}_{\|} - \mathbf{R}_{i,\|}}\ ,\ z)|^2\right] \cdot \mathbf{p}_t\end{aligned} \tag{5.154}$$

where $\mathbf{p}_t$ is given by eq.(5.33). Finally the electric displacement field in the spheres is obtained from eq.(5.133) and may be written, using also eqs.(5.125) and (5.126), as

$$\mathbf{D}_{sphere} = \epsilon \mathbf{E}_{sphere} = \mathbf{D}_0 + \left[\frac{1}{2\pi} R^{-3}\mathbf{1} - \epsilon_a A\left(\mathbf{1} + \hat{\mathbf{z}}\hat{\mathbf{z}}\right)\right] . \boldsymbol{\alpha}(0) . \mathbf{E}_0 \tag{5.155}$$

The excess electric displacement field is now given by (cf. eq.(5.137)

$$\begin{aligned}\mathbf{D}_{ex}(\mathbf{r}) &= -\frac{1}{4\pi}\theta(d - z)\sum_i \theta([(\mathbf{r}_{\|} - \mathbf{R}_{i,\|})^2 + z^2]^{1/2} - R) \\ &\times\{[(\mathbf{r}_{\|} - \mathbf{R}_{i,\|})^2 + z^2]^{-3/2}\left(\mathbf{1} - 3\frac{(\mathbf{r}_{\|} - \mathbf{R}_{i,\|}\ ,\ z)(\mathbf{r}_{\|} - \mathbf{R}_{i,\|}\ ,\ z)}{(\mathbf{r}_{\|} - \mathbf{R}_{i,\|})^2 + z^2}\right) \\ &.[\alpha_{\|}(0)(\mathbf{1} - \hat{\mathbf{z}}\hat{\mathbf{z}}) + \alpha_z(0)\hat{\mathbf{z}}\hat{\mathbf{z}}].\mathbf{E}_0 + \left(\frac{\epsilon_a - \epsilon_s}{\epsilon_a + \epsilon_s}\right)[(\mathbf{r}_{\|} - \mathbf{R}_{i,\|})^2 + (z - 2d)^2]^{-3/2} \\ &\times\left(\mathbf{1} - 3\frac{(\mathbf{r}_{\|} - \mathbf{R}_{i,\|}\ ,\ z - 2d)(\mathbf{r}_{\|} - \mathbf{R}_{i,\|}\ ,\ z - 2d)}{(\mathbf{r}_{\|} - \mathbf{R}_{i,\|})^2 + (z - 2d)^2}\right) \\ &.[\alpha_{\|}(0)(\mathbf{1} - \hat{\mathbf{z}}\hat{\mathbf{z}}) - \alpha_z(0)\hat{\mathbf{z}}\hat{\mathbf{z}}].\mathbf{E}_0\} + \sum_i \theta(R - [(\mathbf{r}_{\|} - \mathbf{R}_{i,\|})^2 + z^2]^{1/2}) \\ &\times\left[\frac{1}{2\pi}R^{-3}\mathbf{1} - \epsilon_a A\left(\mathbf{1} + \hat{\mathbf{z}}\hat{\mathbf{z}}\right)\right].[\alpha_{\|}(0)(\mathbf{1} - \hat{\mathbf{z}}\hat{\mathbf{z}}) + \alpha_z(0)\hat{\mathbf{z}}\hat{\mathbf{z}}].\mathbf{E}_0 \\ &-\frac{1}{4\pi}\theta(z - d)\sum_i\left(\frac{2\epsilon_s}{\epsilon_a + \epsilon_s}\right)[(\mathbf{r}_{\|} - \mathbf{R}_{i,\|})^2 + z)^2]^{-3/2}\times \\ &\left(\mathbf{1} - 3\frac{(\mathbf{r}_{\|} - \mathbf{R}_{i,\|}\ ,\ z)(\mathbf{r}_{\|} - \mathbf{R}_{i,\|}\ ,\ z)}{(\mathbf{r}_{\|} - \mathbf{R}_{i,\|})^2 + z^2}\right).[\alpha_{\|}(0)(\mathbf{1} - \hat{\mathbf{z}}\hat{\mathbf{z}}) + \alpha_z(0)\hat{\mathbf{z}}\hat{\mathbf{z}}].\mathbf{E}_0\end{aligned} \tag{5.156}$$

Using eq.(2.44), one then obtains for the component of the surface polarization density parallel to the interface, replacing $\sum_i$ by $\rho \int d\mathbf{R}_\parallel$,

$$\begin{aligned}
\mathbf{P}^s_\parallel &= \int_{-\infty}^{\infty} \mathbf{D}_{ex,\parallel}(\mathbf{r}) dz \\
&= \int_{-\infty}^{\infty} dz \{ -\frac{1}{4\pi}\theta(d-z)\rho \int d\mathbf{R}_\parallel \theta([(\mathbf{r}_\parallel - \mathbf{R}_\parallel)^2 + z^2]^{1/2} - R) \\
&\times [[(\mathbf{r}_\parallel - \mathbf{R}_\parallel)^2 + z^2]^{-3/2} \left(1 - \frac{3}{2}\frac{(\mathbf{r}_\parallel - \mathbf{R}_\parallel)^2}{(\mathbf{r}_\parallel - \mathbf{R}_\parallel)^2 + z^2}\right) \alpha_\parallel(0)\mathbf{E}_{0,\parallel} \\
&+ \left(\frac{\epsilon_a - \epsilon_s}{\epsilon_a + \epsilon_s}\right) [(\mathbf{r}_\parallel - \mathbf{R}_\parallel)^2 + (z-2d)^2]^{-3/2} \\
&\times \left(1 - \frac{3}{2}\frac{(\mathbf{r}_\parallel - \mathbf{R}_\parallel)^2}{(\mathbf{r}_\parallel - \mathbf{R}_\parallel)^2 + (z-2d)^2}\right) \alpha_\parallel(0)\mathbf{E}_{0,\parallel}] \\
&- \frac{1}{4\pi}\theta(z-d)\rho \int d\mathbf{R}_\parallel \left(\frac{2\epsilon_s}{\epsilon_a + \epsilon_s}\right) [(\mathbf{r}_\parallel - \mathbf{R}_\parallel)^2 + z^2]^{-3/2} \\
&\times \left(1 - \frac{3}{2}\frac{(\mathbf{r}_\parallel - \mathbf{R}_\parallel)^2}{(\mathbf{r}_\parallel - \mathbf{R}_\parallel)^2 + z^2}\right) \alpha_\parallel(0)\mathbf{E}_{0,\parallel}\} \\
&+ \int_{-\infty}^{\infty} dz\, \rho \int d\mathbf{R}_\parallel \theta(R - [(\mathbf{r}_\parallel - \mathbf{R}_\parallel)^2 + z^2]^{1/2}) \\
&\times \left(\frac{1}{2\pi}R^{-3} + \epsilon_a A\right) \alpha_\parallel(0)\mathbf{E}_{0,\parallel}
\end{aligned} \tag{5.157}$$

In the derivation of this equation, it has again been used that terms, containing non-diagonal elements of the matrices in eq.(5.156), vanish after integration over $\mathbf{R}_\parallel$ for reasons of symmetry. For the same reasons, the xx- and yy-elements of these matrices are equal, and these elements in eq.(5.156) have been replaced by half of their sum in eq.(5.157).

The integrals at the right-hand side of eq.(5.157) can now be evaluated in a completely analogous way as was done above for eq.(5.138), by introducing first the new integration variables $\mathbf{R}'_\parallel \equiv \mathbf{R}_\parallel - \mathbf{r}_\parallel$ and next the planar polar coordinates $R'_\parallel$ and ϕ'. After integration over ϕ' from 0 to 2π, one obtains an integral, which may again easily be calculated by introducing the new variable $t \equiv R'^2_\parallel$. The calculation, which runs along similar lines as in eq.(5.141)–(5.149), will not be given here. One now obtains the following simple result:

$$\mathbf{P}^s_\parallel = \rho\, \alpha_\parallel(0)\mathbf{E}_{0,\parallel} \tag{5.158}$$

Since $\mathbf{E}_{0,\parallel}$ may be replaced by $\mathbf{E}_{\parallel,+}$ for the same reason as $D_{0,z}$ was replaced by $D_{z,+}$ (see the discussion above eq.(5.150)), one finds with eqs.(3.38) and (5.36), that the constitutive coefficient $\gamma_e(d)$ is indeed given by eq.(5.39). This completes the proof of the relations eqs.(5.39) and (5.40). In a completely analogous way as for β_e , one finds that $\gamma_e(d) = \gamma_{e,1}(0)$, where $\gamma_{e,1}(0)$ is the constitutive coefficient of the surface $z = 0$, where the small spheres (dipoles) are located.

5.9 Appendix B

In this appendix the validity of the expansion, eq.(5.102), with eq.(5.103), will be proved. This is an expansion of the potentials of multipole fields, originating from the point (0,0,2d), in terms of a complete set of fields around the origin (0,0,0) (see fig.5.1). This expansion can be derived from the more general expansion, given below.

Consider two centers with positions $\mathbf{R}_1$ and $\mathbf{R}_2$. The coordinates of the field point $\mathbf{r}$, relative to both centers, will be denoted by $\mathbf{r}_1 \equiv \mathbf{r} - \mathbf{R}_1$ and $\mathbf{r}_2 \equiv \mathbf{r} - \mathbf{R}_2$. The spherical coordinates, with angles measured relative to the fixed Cartesian coordinate system, are $\mathbf{r}_1 = (r_1, \theta_1, \phi_1)$ and $\mathbf{r}_2 = (r_2, \theta_2, \phi_2)$. Similarly the vector $\mathbf{R} \equiv \mathbf{R}_2 - \mathbf{R}_1$, with spherical coordinates $\mathbf{R} = (R, \theta_R, \phi_R)$ is introduced. Then one can prove that for $r_1 < R$ it is possible to expand the potential of a multipole field, originating from $\mathbf{R}_2$, in terms of a complete set of fields around $\mathbf{R}_1$ in the following way [43]:

$$\frac{1}{r_2^{\ell_2+1}} Y_{\ell_2}^{m_2}(\theta_2, \phi_2) = \sum_{\ell_1 m_1} H(\ell_1, m_1|\ell_2, m_2) \frac{Y_\ell^m(\theta_R, \phi_R)}{R^{\ell+1}} r_1^{\ell_1} Y_{\ell_1}^{m_1}(\theta_1, \phi_1) \tag{5.159}$$

with $\ell \equiv \ell_1 + \ell_2$ and $m \equiv m_2 - m_1$. The coefficients $H(\ell_1, m_1|\ell_2, m_2)$ in this series development are given by

$$\begin{aligned} H(\ell_1, m_1|\ell_2, m_2) \equiv &\sqrt{4\pi}(-1)^{\ell_2+m_1} \left[\frac{2\ell_2+1}{(2\ell_1+1)(2\ell+1)}\right]^{1/2} \\ &\times \left[\frac{(\ell+m)!(\ell-m)!}{(\ell_1+m_1)!(\ell_1-m_1)!(\ell_2+m_2)!(\ell_2-m_2)!}\right]^{1/2} \end{aligned} \tag{5.160}$$

With the help of the above equations the validity of eq.(5.102), with eq.(5.103), can easily be proved. One has here the special case, that $\mathbf{r}_1 = (r, \theta, \phi)$, $\mathbf{r}_2 = (\rho, \theta^r, \phi^r)$ and $R = 2d$, $\theta_R = 0$ (see fig.5.2 and text around this figure). With the definition, eq.(5.26), of spherical harmonics, it first of all follows that

$$\begin{aligned} Y_\ell^m(\theta_R = 0, \phi_R) &= \left[\frac{2\ell+1}{4\pi}\frac{(\ell-m)!}{(\ell+m)!}\right]^{1/2} P_\ell^m(1)(-1)^m e^{im\phi} \\ &= \left(\frac{2\ell+1}{4\pi}\right)^{1/2} \delta_{m0} = \left(\frac{2\ell+1}{4\pi}\right)^{1/2} \delta_{m_1,m_2} \end{aligned} \tag{5.161}$$

since $P_\ell^m(1) = \delta_{m0}$. Next it follows, with eqs.(5.159)–(5.161), that

$$\begin{aligned} \frac{1}{\rho^{\ell_2+1}} Y_{\ell_2}^{m_2}(\theta^r, \phi^r) = &\sum_{\ell_1=|m_2|}^{\infty} \sqrt{4\pi}(-1)^{\ell_2+m_2} \left[\frac{2\ell_2+1}{(2\ell_1+1)(2\ell+1)}\right]^{1/2} \\ &\times \left[\frac{\{(\ell_1+\ell_2)!\}^2}{(\ell_1+m_2)!(\ell_1-m_2)!(\ell_2+m_2)!(\ell_2-m_2)!}\right]^{1/2} \\ &\times \left(\frac{2\ell+1}{4\pi}\right)^{1/2} \frac{r^{\ell_1}}{(2d)^{\ell_1+\ell_2+1}} Y_{\ell_1}^{m_2}(\theta, \phi) \end{aligned} \tag{5.162}$$

Replacing ℓ_2 and m_2 by ℓ and m, respectively, one obtains

$$\frac{1}{\rho^{\ell+1}} Y_\ell^m(\theta^r, \phi^r) = \sum_{\ell_1=|m|}^{\infty} (-1)^{\ell+m} \left[\frac{2\ell+1}{2\ell_1+1}\right]^{1/2} (\ell+\ell_1)!$$

$$\times [(\ell+m)!(\ell-m)!(\ell_1+m)!(\ell_1-m)!]^{-1/2} \frac{r^{\ell_1}}{(2d)^{\ell+\ell_1+1}} Y_{\ell_1}^m(\theta, \phi) \quad (5.163)$$

This is, of course, the same as eq.(5.102), with eq.(5.103).

Chapter 6
SPHEROIDAL ISLAND FILMS IN THE LOW COVERAGE LIMIT

6.1 Introduction

In the previous chapter the general theory of discrete small island films in the low coverage limit has been developed. The electromagnetic interaction between the islands is neglected in this limit. The interaction of the islands with their image charges in the substrate was first studied in dipole approximation, and next in quadrupole approximation. In the first case all higher order multipole moments of the islands and their images are neglected. In the second case one considers only the first two multipole moments. The dipole and quadrupole polarizabilities of the islands were calculated, and with the help of these the constitutive coefficients γ_e, β_e, δ_e and τ of the film.

When the islands are brought close to the substrate, one must also account for the modification of their dipole and quadrupole polarizabilities due to the interaction with higher order multipoles in the substrate. It is sufficient to calculate only the modification of these polarizabilities, as long as one is interested in the constitutive coefficients γ_e, β_e, δ_e and τ. As has been found in the analysis in the previous chapter, it is necessary to calculate first the multipole polarizabilities of arbitrary order of an island in the ambient medium, far from the substrate. The problem arises, that this is, in practice, only possible for islands with a simple shape. In section 5.6 this analysis was performed for spherical particles. Here the multipole polarizabilities in the ambient are known (cf. eq.(5.70)). The interaction with the image charges in the substrate can then be taken into account to arbitrary order in the multipole moments.

One may also calculate multipole polarizabilities for ellipsoids in the ambient in an analytic form. This led to the use of extensions of, for instance, the Maxwell Garnett theory, [44], [45], [30], [46], [4], for spheroids. For the derivation of expressions for the multipole polarizabilities of ellipsoids, cf. [47] page 207, one uses ellipsoidal coordinates. If one subsequently wants to calculate the interaction with the image charge distribution, the use of such ellipsoidal coordinates poses problems. For the interesting special case of spheroids with their axis of symmetry orthogonal to the surface of the substrate this may, however, be done. This is the subject of the present chapter. In sections 6.2 and 6.3 the use of the dipole and quadrupole models is discussed using the exact dipole and quadrupole polarizabilities of the spheroids far from the surface, for oblate and prolate spheroids, respectively. In sections 6.4 and 6.5 the complete solutions, found by Bobbert and Vlieger [42], are given for oblate and prolate spheroids, respectively, introducing spheroidal multipole fields. In section 6.6 these solutions are applied to the special case of gold islands on a sapphire substrate. The exact solution for, in particular the case of oblate spheroids, gives a very clear insight in the limitations of the dipole and the quadrupole model. On the basis of this analysis it is found, that the use of the dipole or quadrupole approximation is in general inadequate for islands which are flat, i.e. islands for which the distance of the center to the surface of the substrate is smaller than 30% of the linear diameter along the surface.

6.2 Oblate spheroids; the polarizable quadrupole model

In section 5.6 of the previous chapter the relatively simple spherical model for an island film was treated. It will be clear that this model only provides a reasonable description of the optical properties of such film, if the islands do not deviate too much in shape from spheres.

A more general model system, describing island films, in which the particles are flattened, but still (more or less) symmetric for rotation around an axis normal to the substrate, consists of oblate spheroids, touching this substrate. (The axis of rotation of an oblate spheroid is the short axis.) The attractiveness of this model is again, that the polarizabilities, given by eq.(5.50), can be calculated to any order of accuracy. The reason for this is the fact, that a complete set of eigenfunctions of the Laplace equation (5.1) can be found for this form of particles. This is also the case for prolate spheroids, of which the axis of rotation is the long axis. The model will therefore be extended, in this section, to include oblate and, in the next section, to include prolate spheroidal particles. The axis of rotation of the spheroids is normal to the surface of the (flat) substrate.

Consider a homogeneous oblate spheroid with a dielectric constant ϵ, embedded in a homogeneous medium with dielectric constant ϵ_a, far from the substrate. Both ϵ and ϵ_a may be frequency dependent. The origin of the Cartesian coordinate system is chosen at the center of the spheroid and the z-axis as the axis of revolution (see fig.6.1). The radius of the ring of foci of the oblate spheroid will be denoted by a. The two long axes of the spheroid have length $2a\ (\xi_0^2+1)^{1/2}$ and the short axis has length $2a\xi_0$, which defines a flattening parameter $\xi_0(0 \leq \xi_0 < \infty)$ (see fig.6.1). The limit $\xi_0 \to \infty$, $a \to 0$, with $a\xi_0 = R$ =constant, corresponds to a sphere with radius R, whereas the limit $\xi_0 \to 0$ corresponds to a thin circular disk with radius a.

In the previous chapter the multipole polarizabilities, eq.(5.70), of a sphere were given. These quantities can be calculated, using the multipole expansion eq.(5.25) for the potential $\psi(\mathbf{r})$, which is the general solution of the Laplace equation (5.1) for this potential in spherical coordinates. An analogous method will be employed here, but in order to adapt the method to the different geometry of the particle, it is more convenient to introduce (oblate) spheroidal multipoles. For that purpose first oblate spheroidal coordinates ξ, η and ϕ $(0 \leq \xi < \infty,\ -1 \leq \eta \leq 1,\ 0 \leq \phi < 2\pi)$ are

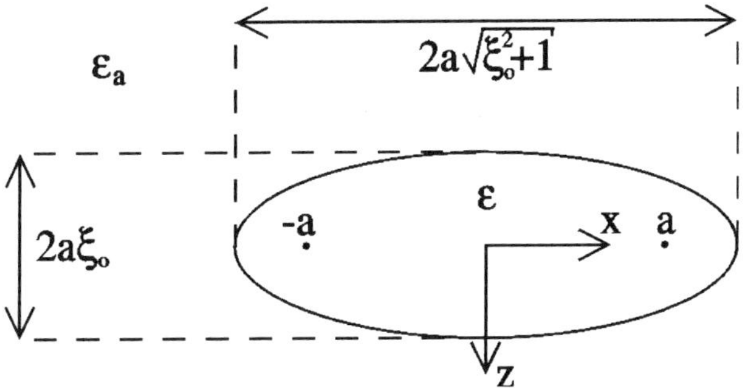

Figure 6.1 An oblate spheroid in an infinite surrounding medium. Cross-section through x-z plane.

introduced. These coordinates are defined as follows [37]:

$$\xi \equiv \left[\left(\frac{\rho_1+\rho_2}{2a}\right)^2 - 1\right]^{1/2}, \quad \eta \equiv \pm\left[1-\left(\frac{\rho_1-\rho_2}{2a}\right)^2\right]^{1/2}, \quad \phi \equiv \arctan\left(\frac{y}{x}\right) \tag{6.1}$$

with

$$\begin{aligned} \rho_1 &\equiv [z^2 + (x + a\ \cos\ \phi)^2 + (y + a\ \sin\ \phi)^2]^{1/2} \\ \rho_2 &\equiv [z^2 + (x - a \cos \phi)^2 + (y - a \sin\ \phi)^2]^{1/2} \end{aligned} \tag{6.2}$$

Here ρ_1 and ρ_2 are the distances of the point (x, y, z) to the points of intersection of the ring of foci with the plane through (x, y, z) and the z-axis, whereas ϕ is the angle of orientation of this plane with respect to the $x - z$ plane. The positive sign in eq.(6.1) should be used if $z > 0$ and the negative sign if $z < 0$. The inverse of the transformation, eqs.(6.1) and (6.2), is

$$\begin{aligned} x &= a[(\xi^2+1)(1-\eta^2)]^{1/2} \cos\ \phi \\ y &= a[(\xi^2+1)(1-\eta^2)]^{1/2} \sin\ \phi \\ z &= a\xi\eta \end{aligned} \tag{6.3}$$

In terms of these spheroidal coordinates the surface of the spheroid is given by $\xi = \xi_0$.

The Laplace equation (5.1) is separable in these coordinates and the solutions split up into two types: solutions that satisfy the Laplace equation everywhere, but are unequal to zero if $\xi \to \infty$ and solutions that are zero if $\xi \to \infty$, but satisfy the Laplace equation only for $\xi > 0$. The first type of solutions describe the "incoming" field due to sources outside the spheroid and can be expressed in terms of the functions [37]

$$P_\ell^m(i\xi)Y_\ell^m(\arccos\eta, \phi) \quad \text{with } \ell = 0, 1, 2, ... \text{ and } m = 0, \pm1, \pm2, ..., \pm\ell \tag{6.4}$$

where (cf. eq.(5.26))

$$Y_\ell^m(\arccos\eta, \phi) \equiv \left[\frac{2\ell+1}{4\pi}\frac{(\ell-m)!}{(\ell+m)!}\right]^{1/2} (-1)^m P_\ell^m(\eta)e^{im\phi} \tag{6.5}$$

Here P_ℓ^m are the associated Legendre functions of the first kind of degree ℓ and order m, defined for complex arguments and $m \geq 0$ by [37]

$$P_\ell^m(z) \equiv \frac{(1-z^2)^{m/2}}{2^\ell \ell!} \left(\frac{d}{dz}\right)^{\ell+m} (z^2-1)^\ell \tag{6.6}$$

Along the positive imaginary axis one has

$$P_\ell^m(i\xi) \equiv (-1)^m i^{\ell+m} \frac{(1+\xi^2)^{m/2}}{2^\ell \ell!} \left(\frac{d}{d\xi}\right)^{\ell+m} (1+\xi^2)^\ell \quad \text{for} \quad 0 \leq \xi < \infty \tag{6.7}$$

For $m < 0$ the definition is

$$P_\ell^m(z) \equiv (-1)^m \frac{(\ell+m)!}{(\ell-m)!} P_\ell^{-m}(z) \tag{6.8}$$

The second type of solution of the Laplace equation (5.1) describes the field outside the spheroid due to the charge distribution inside and can, after its separation in the coordinates ξ, η and ϕ, be expressed in terms of the functions

$$Q_\ell^m(i\xi) Y_\ell^m(\arccos\eta, \phi) \quad \text{with } \ell = 0, 1, 2, ... \text{ and } m = 0, \pm 1, \pm 2, ..., \pm \ell \tag{6.9}$$

where Q_ℓ^m are associated Legendre functions of the second kind, defined along the positive imaginary axis and for $m \geq 0$ by [37]

$$\begin{aligned} Q_\ell^m(i\xi) \; \equiv \; & (-1)^{m+1} i^{\ell+1} \frac{(1+\xi^2)^{m/2}}{2^\ell \ell!} \left(\frac{d}{d\xi}\right)^m \{2 \left(\frac{d}{d\xi}\right)^\ell [\arctan(1/\xi)(1+\xi^2)^\ell] \\ & - \arctan(1/\xi) \left(\frac{d}{d\xi}\right)^\ell (1+\xi^2)^\ell\} \end{aligned} \tag{6.10}$$

and for $m < 0$ by

$$Q_\ell^m(i\xi) \equiv \frac{(\ell+m)!}{(\ell-m)!} Q_\ell^{-m}(i\xi) \tag{6.11}$$

The fields given in eq.(6.9) will be referred to as the spheroidal multipole fields. The source of such a field, which is located in the center of the spheroid, will be called a spheroidal multipole. The general solution of the Laplace equation in oblate spheroidal coordinates ξ, η and ϕ in a region $\xi_1 < \xi < \xi_2$, where there are no sources, can now be written as

$$\psi(\mathbf{r}) = \sum_{\ell m} A_{\ell m} Z_\ell^m(\xi, a) Y_\ell^m(\arccos\eta, \phi) + \sum_{\ell m} B_{\ell m} X_\ell^m(\xi, a) Y_\ell^m(\arccos\eta, \phi) \tag{6.12}$$

where the summation is from $\ell = 0$ to ∞ and $m = -\ell$ through ℓ. For convenience the functions

$$X_\ell^m(\xi, a) \equiv i^{m-\ell} \frac{(\ell-m)!}{(2\ell-1)!!} a^\ell P_\ell^m(i\xi) \tag{6.13}$$

and

$$Z_\ell^m(\xi, a) \equiv i^{\ell+1} \frac{(2\ell+1)!!}{(\ell+m)!} a^{-\ell-1} Q_\ell^m(i\xi) \tag{6.14}$$

have been introduced. Here $(n)!! \equiv 1 \times 3 \times ... \times (n-2) \times n$ for n odd and $(n)!! \equiv 2 \times 4 \times ... \times (n-2) \times n$ for n is even; by definition $(-1)!! \equiv 1$. The asymptotic (large ξ) behavior of the associated Legendre functions along the positive imaginary axis is given by [37], [48]

$$P_\ell^m(i\xi) \simeq \frac{(2\ell-1)!!}{(\ell-m)!} i^{-m} (i\xi)^\ell \quad \text{for } \xi \to \infty \tag{6.15}$$

$$Q_\ell^m(i\xi) \simeq \frac{(\ell+m)!}{(2\ell+1)!!} \frac{1}{(i\xi)^{\ell+1}} \quad \text{for } \xi \to \infty \tag{6.16}$$

It then follows from the above equations that

$$X_\ell^m(\xi, a) \simeq (a\xi)^\ell \simeq r^\ell \quad \text{for } \xi \to \infty \tag{6.17}$$

$$Z_\ell^m(\xi, a) \simeq (a\xi)^{-\ell-1} \simeq r^{-\ell-1} \quad \text{for } \xi \to \infty \tag{6.18}$$

Here it was used that, cf. eq.(6.3),

$$r \equiv (x^2 + y^2 + z^2)^{1/2} = a(\xi^2 - \eta^2 + 1)^{1/2} \simeq a\xi \quad \text{for } \xi \to \infty \tag{6.19}$$

Using eq.(6.3) one furthermore finds the asymptotic relation

$$\cos\,\theta \equiv z/r = z(x^2 + y^2 + z^2)^{-1/2} = \xi\eta(\xi^2 - \eta^2 + 1)^{-1/2} \simeq \eta \quad \text{for } \xi \to \infty \tag{6.20}$$

If one now compares the oblate spheroidal multipole expansion eq.(6.12) of the potential $\psi(\mathbf{r})$ with the usual (spherical) multipole expansion eq.(5.25), one finds, using the results eqs.(6.17), (6.18) and (6.20), that asymptotically the spheroidal and the spherical multipole fields become identical. Of course, for spherical particles the two expansions eqs.(6.12) and (5.25) coincide for all distances. In general a spheroidal multipole field contains at shorter distances, in addition to the spherical multipole field of the same order, contributions due to higher order spherical multipoles. One might therefore wonder, whether formulae, derived in the previous chapter for the quadrupole model, using the spherical multipole expansion, need not be corrected, if one uses the spheroidal multipole expansion. It can be proved, however, from eqs.(6.12) and (5.25), together with eqs.(6.5), (6.14), (6.10), (6.19) and (6.20), that the spheroidal dipole field contains at shorter distances, in addition to the spherical dipole field, as first higher order contribution a spherical octupole field. In fact this is a consequence of the cylindrical symmetry around the normal on the substrate through the center of the particles. Since the latter contribution is neglected in the quadrupole model, it means that one can apply all formulae, derived in the previous chapter for this model, also in the present case of a spheroidal island.

One can now derive, using the (oblate) spheroidal multipole expansion eq.(6.12) for the potential together with the boundary conditions for this potential and its normal derivative on the surface of the spheroid, the following relation between induced and incident amplitudes $A_{\ell m}$ and $B_{\ell m}$:

$$A_{\ell m} = -\alpha_{\ell m} B_{\ell m} \tag{6.21}$$

where

$$\alpha_{\ell m} \equiv (\epsilon - \epsilon_a)[\epsilon \frac{Z_\ell^m(\xi_0, a)}{X_\ell^m(\xi_0, a)} - \epsilon_a \frac{\partial\ Z_\ell^m(\xi_0, a)/\partial \xi_0}{\partial\ X_\ell^m(\xi_0, a)/\partial \xi_0}]^{-1} \tag{6.22}$$

Note that the polarizability matrix $\alpha_{\ell m,\ell' m'}$ is again diagonal:

$$\alpha_{\ell m,\ell' m'} = \alpha_{\ell m} \delta_{\ell \ell'} \delta_{m m'} \tag{6.23}$$

However, in contrast to the spherical case (cf. eq.(5.68)), the diagonal elements now depend on both ℓ and m as a consequence of the fact that spheroids have a lower rotational symmetry than spheres. Again one does not have to consider the case $\ell = 0$ for uncharged spheroids. Using the definitions eqs.(6.13) and (6.14), one obtains for eq.(6.22):

$$\begin{aligned} \alpha_{\ell m} \equiv\ & (\epsilon - \epsilon_a) a^{2\ell+1} (-1)^{\ell+m} (i)^{-m-1} \left(\frac{(\ell - m)!\,(\ell + m)!}{(2\ell - 1)!!\,(2\ell + 1)!!} \right) \\ & \times [\epsilon \frac{Q_\ell^m(i\xi_0)}{P_\ell^m(i\xi_0)} - \epsilon_a \frac{dQ_\ell^m(i\xi_0)/d\xi_0}{dP_\ell^m(i\xi_0)/d\xi_0}]^{-1} \end{aligned} \tag{6.24}$$

It follows with eqs.(6.8) and (6.11) that

$$\alpha_{\ell m} = \alpha_{\ell,-m} \tag{6.25}$$

One should first of all note that, with eqs.(6.15) and (6.16), one finds in the limit that the oblate spheroid becomes spherical

$$\lim_{\xi_0 \to \infty} \alpha_{\ell m} = \frac{\ell(\epsilon - \epsilon_a) R^{2\ell+1}}{\ell\ \epsilon + (\ell + 1)\epsilon_a} = \alpha_\ell \tag{6.26}$$

The limit should be taken such that $a\xi_0 = R$ remains constant. This is the familiar result given also in eq.(5.70) for a sphere with radius R.

With eqs.(6.7), (6.8), (6.10) and (6.11) one can in principle calculate $\alpha_{\ell m}$ for any value of ℓ and m. In the polarizable dipole model one is only interested in $\alpha_{\ell m}$ for the values $\ell = 1$ and $m = 0, \pm 1$. In the polarizable quadrupole model also the values $\ell = 2$ and $m = 0, \pm 1, \pm 2$ are considered, whereas $\alpha_{\ell m}$ is put equal to zero for $\ell > 2$. One obtains

$$\begin{aligned} \alpha_{10} &= \frac{(1/3)(\epsilon - \epsilon_a) a^3 \xi_0 (1 + \xi_0^2)}{\epsilon_a + (\epsilon - \epsilon_a)(1 + \xi_0^2)[1 - \xi_0 \arctan(1/\xi_0)]} \\ \alpha_{1,\pm 1} &= \frac{(2/3)(\epsilon - \epsilon_a) a^3 \xi_0 (1 + \xi_0^2)}{2\epsilon_a + (\epsilon - \epsilon_a)(1 + \xi_0^2)[\xi_0 \arctan(1/\xi_0) - \xi_0^2 (1 + \xi_0^2)^{-1}]} \\ \alpha_{20} &= \frac{(4/15)(\epsilon - \epsilon_a) a^5 \xi_0 (1 + \xi_0^2)(1 + 3\xi_0^2)}{2\epsilon_a + 3(\epsilon - \epsilon_a)(1 + \xi_0^2)[\xi_0 (1 + 3\xi_0^2) \arctan(1/\xi_0) - 3\xi_0^2]} \\ \alpha_{2,\pm 1} &= \frac{(2/5)(\epsilon - \epsilon_a) a^5 \xi_0 (1 + \xi_0^2)(1 + 2\xi_0^2)}{2\epsilon_a + (\epsilon - \epsilon_a)(1 + 2\xi_0^2)[3\xi_0^2 + 2 - 3\xi_0 (1 + \xi_0^2) \arctan(1/\xi_0)]} \\ \alpha_{2,\pm 2} &= \frac{(8/5)(\epsilon - \epsilon_a) a^5 \xi_0 (1 + \xi_0^2)^2}{4\epsilon_a + (\epsilon - \epsilon_a)[3\xi_0 (1 + \xi_0^2)^2 \arctan(1/\xi_0) - \xi_0^2 (5 + 3\xi_0^2)]} \end{aligned} \tag{6.27}$$

It furthermore follows, from eqs.(5.67) and (6.23), that the polarizabilities in the above mentioned dipole and quadrupole models, are given for oblate spheroidal particles by

$$\begin{aligned}
\alpha_z &= 4\pi\ \epsilon_a\alpha_{10}, \quad \alpha_\| = 4\pi\ \epsilon_a\alpha_{11} = 4\pi\ \epsilon_a\alpha_{1,-1} \\
\alpha_z^{10} &= \alpha_\|^{10} = 0, \quad 2\alpha_z^{11} + 3\ \alpha^{11} = 2\pi\ \epsilon_a\alpha_{20} \\
\alpha_\|^{11} + 2\alpha^{11} &= \frac{4}{3}\pi\ \epsilon_a\alpha_{21} = \frac{4}{3}\pi\ \epsilon_a\alpha_{2,-1} \\
\alpha^{11} &= \frac{2}{3}\pi\ \epsilon_a\alpha_{22} = \frac{2}{3}\pi\ \epsilon_a\alpha_{2,-2}
\end{aligned} \tag{6.28}$$

Introducing the volume V of the oblate spheroid (see also fig.6.1):

$$V \equiv \frac{4}{3}\pi\ a^3\xi_0(1+\xi_0^2) \tag{6.29}$$

and using eq.(6.27), one can write these polarizabilities in the following form:

$$\begin{aligned}
\alpha_z &= \frac{\epsilon_a(\epsilon-\epsilon_a)V}{\epsilon_a + L_{a,z}(\xi_0)(\epsilon-\epsilon_a)} \\
\alpha_\| &= \frac{\epsilon_a(\epsilon-\epsilon_a)V}{\epsilon_a + L_{a,\|}(\xi_0)(\epsilon-\epsilon_a)} \\
\alpha_z^{10} &= \alpha_\|^{10} = 0 \\
2\alpha_z^{11} + 3\ \alpha^{11} &= \frac{\epsilon_a(\epsilon-\epsilon_a)(1+3\xi_0^2)a^2V}{5[\epsilon_a + (\epsilon-\epsilon_a)L_{a,z}^1(\xi_0)]} \\
\alpha_\|^{11} + 2\alpha^{11} &= \frac{\epsilon_a(\epsilon-\epsilon_a)(1+2\xi_0^2)a^2V}{5[\epsilon_a + (\epsilon-\epsilon_a)L_{a,\|}^1(\xi_0)]} \\
\alpha^{11} &= \frac{\epsilon_a(\epsilon-\epsilon_a)(1+\xi_0^2)a^2V}{5[\epsilon_a + (\epsilon-\epsilon_a)L_a^{11}(\xi_0)]}
\end{aligned} \tag{6.30}$$

Here the depolarization factors of the oblate spheroid in the ambient (infinitely) far from the substrate are given by

$$\begin{aligned}
L_{a,z}(\xi_0) &\equiv (1+\xi_0^2)[1-\xi_0\arctan(1/\xi_0)] \\
L_{a,\|}(\xi_0) &\equiv \frac{1}{2}(1+\xi_0^2)[\xi_0\arctan(1/\xi_0) - \xi_0^2(1+\xi_0^2)^{-1}] \\
L_{a,z}^1(\xi_0) &\equiv \frac{3}{2}(1+\xi_0^2)[(1+3\xi_0^2)\xi_0\arctan(1/\xi_0) - 3\xi_0^2] \\
L_{a,\|}^1(\xi_0) &\equiv \frac{1}{2}(1+2\xi_0^2)[2+3\xi_0^2 - 3(1+\xi_0^2)\xi_0\arctan(1/\xi_0)] \\
L_a^{11}(\xi_0) &\equiv \frac{1}{4}[3(1+\xi_0^2)^2\xi_0\arctan(1/\xi_0) - \xi_0^2(5+3\xi_0^2)]
\end{aligned} \tag{6.31}$$

In the context of the quadrupole model discussed in section 5.5 of the previous chapter one must now substitute $\alpha_z^{10} = 0$, cf. eq.(6.30), into eq.(5.55). Using also

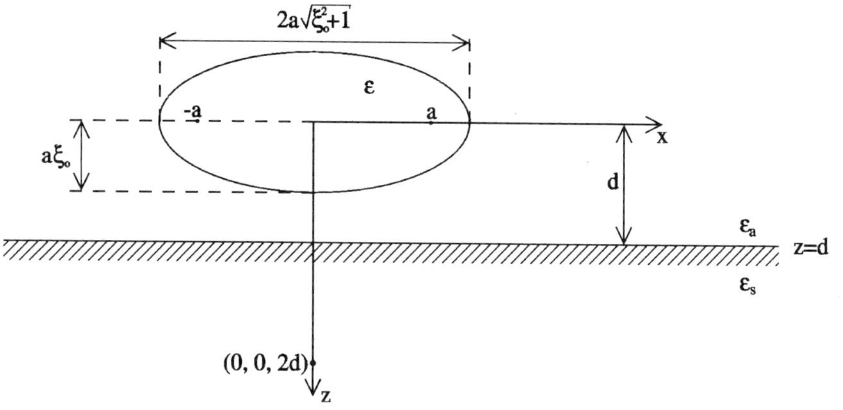

Figure 6.2 Oblate spheroid above a substrate. Cross-section through the $x - z$ plane.

eq.(5.53), one obtains for the polarizabilities, as modified by the proximity of the substrate, in a homogeneous electric field normal to the substrate (see fig.6.2)

$$\begin{aligned} \alpha_z(0) &= D_z^{-1}\alpha_z[1 + (3A/d^2)(2\ \alpha_z^{11} + 3\ \alpha^{11})] \\ \alpha_z^{10}(0) &= -\frac{3}{2}D_z^{-1}(A/d)\alpha_z(2\alpha_z^{11} + 3\ \alpha^{11}) \end{aligned} \tag{6.32}$$

where

$$D_z \equiv 1 + 2A\ \alpha_z + (3A/2d^2)(2 + A\ \alpha_z)(2\alpha_z^{11} + 3\alpha^{11}) \tag{6.33}$$

and (cf. eq.(5.54))

$$A \equiv (32\ \pi\ \epsilon_a d^3)^{-1}\frac{\epsilon_a - \epsilon_s}{\epsilon_a + \epsilon_s} \tag{6.34}$$

Here d is the distance from the center of the spheroid to the substrate ($d \geq a\xi_0$) and ϵ_s is the dielectric constant of the substrate (see fig. 6.2). The argument 0 of the polarizabilities, eq.(6.32), indicates the position $z = 0$ of this center. Just as for spheres, the quadrupole polarizability $\alpha_z^{10}(0)$, which is zero if the spheroid is far from the surface of the substrate, becomes finite when the spheroid approaches this surface.

If the electric field is directed parallel to the surface of the substrate one finds in a similar way, using eqs.(5.59) and (5.58), for the polarizabilities $\alpha_\parallel(0)$ and $\alpha_\parallel^{10}(0)$ of the spheroid above the substrate:

$$\begin{aligned} \alpha_\parallel(0) &= D_\parallel^{-1}\alpha_\parallel[1 + (3A/d^2)(\alpha_\parallel^{11} + 2\ \alpha^{11})] \\ \alpha_\parallel^{10}(0) &= -\frac{3}{2}D_\parallel^{-1}(A/d)\alpha_\parallel(\alpha_\parallel^{11} + 2\ \alpha^{11}) \end{aligned} \tag{6.35}$$

where

$$D_\parallel \equiv 1 + A\ \alpha_\parallel + (3A/4d^2)(4 + A\ \alpha_\parallel)(\alpha_\parallel^{11} + 2\alpha^{11}) \tag{6.36}$$

Again the quadrupole polarizability $\alpha_{\parallel}^{10}(0)$, which is zero if the spheroid is far away from the surface of the substrate, becomes finite, when the spheroid approaches the substrate.

If all spheroids have the same size, the resulting electric susceptibilities of the film are given by eq.(5.64), which reduces for this case to

$$
\begin{aligned}
\beta_e(d) &= \rho\, \epsilon_a^{-2} D_z^{-1}\alpha_z[1+(3A/d^2)(2\alpha_z^{11}+3\,\alpha^{11})] \\
\gamma_e(d) &= \rho\, D_{\parallel}^{-1}\alpha_{\parallel}[1+(3A/d^2)(\alpha_{\parallel}^{11}+2\alpha^{11})] \\
\delta_e(d) &= \rho\, \epsilon_a^{-1} d\{D_z^{-1}\alpha_z[1+\frac{9}{2}(A/d^2)(2\alpha_z^{11}+3\alpha^{11})] \\
&\quad +D_{\parallel}^{-1}\alpha_{\parallel}[1+\frac{9}{2}(A/d^2)(\alpha_{\parallel}^{11}+2\alpha^{11})]\} \\
\tau(d) &= \rho\, d\, D_{\parallel}^{-1}\alpha_{\parallel}[1+\frac{9}{2}(A/d^2)(\alpha_{\parallel}^{11}+2\alpha^{11})]
\end{aligned}
\tag{6.37}
$$

where A, D_z and $D_{\parallel}$ are given by eqs.(6.34), (6.33) and (6.36) respectively. The argument d of the above coefficients denotes the position $z = d$ of the dividing surface, which coincides with the surface of the substrate (see fig.6.3). If one considers a film in which the spheroidal particles have a certain dispersion both in size and shape (ξ_0), the expressions on the right-hand sides of eq.(6.37) have to be averaged. One then has to apply eq.(5.65), instead of eq.(5.64). The invariants can be found by substitution of the above formulae, eq.(6.37), into eq.(5.66). If one changes the direction of the z-axis one finds again that γ_e and β_e are independent of the direction of the z-axis while δ_e and τ change sign if the z-axis changes direction. If one uses the above formulae to calculate the reflection and transmission amplitudes, it is important to take the right sign. If, as in chapter 4, the light is incident through the ambient one may use the above expressions. If the light is incident through the substrate one must, in order to use the formulae given in chapter 4, interchange ϵ_a and ϵ_s in these formulae and change the sign of δ_e and τ in the expressions given above, before substituting them into the formulae in that chapter.

Substituting the explicit expressions of the various polarizabilities and depolarization factors in the ambient far from the substrate (cf. eqs.(6.30) and (6.31)) into eqs.(6.32) and (6.33) one obtains, using eqs.(6.34) and (6.29), for oblate spheroids at the distance $d \equiv a\xi_1$ on or above the substrate ($\xi_1 \geq \xi_0$) :

$$
\begin{aligned}
\alpha_z(0) &= \frac{\epsilon_a(\epsilon-\epsilon_a)V[\epsilon_a+L_z^1(\epsilon-\epsilon_a)]}{[\epsilon_a+L_z(\epsilon-\epsilon_a)][\epsilon_a+L_z^1(\epsilon-\epsilon_a)]+\Lambda_z^1(\epsilon-\epsilon_a)^2} \\
\alpha_z^{10}(0) &= \frac{\epsilon_a(\epsilon-\epsilon_a)^2\Lambda_z^{10}V\ a\xi_0}{[\epsilon_a+L_z(\epsilon-\epsilon_a)][\epsilon_a+L_z^1(\epsilon-\epsilon_a)]+\Lambda_z^1(\epsilon-\epsilon_a)^2}
\end{aligned}
\tag{6.38}
$$

The factors L and Λ, which will again be called depolarization factors, are the following functions of ξ_0, ξ_1, ϵ_a and ϵ_s for the oblate spheroid

$$
\begin{aligned}
L_z &\equiv (1+\xi_0^2)\left[1-\xi_0 \arctan\left(\frac{1}{\xi_0}\right)+\frac{1}{12}\left(\frac{\epsilon_a-\epsilon_s}{\epsilon_a+\epsilon_s}\right)\frac{\xi_0}{\xi_1^3}\right] \\
L_z^1 &\equiv (1+\xi_0^2)(1+3\xi_0^2)\left[\frac{3}{2}\xi_0 \arctan\left(\frac{1}{\xi_0}\right)-\frac{9}{2}\frac{\xi_0^2}{1+3\xi_0^2}+\frac{1}{40}\left(\frac{\epsilon_a-\epsilon_s}{\epsilon_a+\epsilon_s}\right)\frac{\xi_0}{\xi_1^5}\right] \\
\Lambda_z^1 &\equiv -\frac{1}{640}\left(\frac{\epsilon_a-\epsilon_s}{\epsilon_a+\epsilon_s}\right)^2(1+\xi_0^2)^2(1+3\xi_0^2)\frac{\xi_0^2}{\xi_1^8} \\
\Lambda_z^{10} &\equiv -\frac{1}{80}\left(\frac{\epsilon_a-\epsilon_s}{\epsilon_a+\epsilon_s}\right)(1+\xi_0^2)(1+3\xi_0^2)\frac{1}{\xi_1^4}
\end{aligned} \tag{6.39}
$$

In a completely analogous way one finds, by substituting the explicit expressions of the various polarizabilities and depolarization factors in the ambient far from the substrate (cf. eqs.(6.30) and (6.31)) into eqs.(6.35) and (6.36), using eqs.(6.34) and (6.29), for oblate spheroids at the distance $d \equiv a\xi_1(\xi_1 \geq \xi_0)$ on or above the substrate:

$$
\begin{aligned}
\alpha_{\parallel}(0) &= \frac{\epsilon_a(\epsilon-\epsilon_a)V[\epsilon_a+L_{\parallel}^1(\epsilon-\epsilon_a)]}{[\epsilon_a+L_{\parallel}(\epsilon-\epsilon_a)][\epsilon_a+L_{\parallel}^1(\epsilon-\epsilon_a)]+\Lambda_{\parallel}^1(\epsilon-\epsilon_a)^2} \\
\alpha_{\parallel}^{10}(0) &= \frac{\epsilon_a(\epsilon-\epsilon_a)^2\Lambda_{\parallel}^{10}V\ a\xi_0}{[\epsilon_a+L_{\parallel}(\epsilon-\epsilon_a)][\epsilon_a+L_{\parallel}^1(\epsilon-\epsilon_a)]+\Lambda_{\parallel}^1(\epsilon-\epsilon_a)^2}
\end{aligned} \tag{6.40}
$$

with

$$
\begin{aligned}
L_{\parallel} &\equiv \frac{1}{2}(1+\xi_0^2)\left[\xi_0 \arctan\left(\frac{1}{\xi_0}\right)-\frac{\xi_0^2}{1+\xi_0^2}+\frac{1}{12}\left(\frac{\epsilon_a-\epsilon_s}{\epsilon_a+\epsilon_s}\right)\frac{\xi_0}{\xi_1^3}\right] \\
L_{\parallel}^1 &\equiv \frac{1}{2}(1+\xi_0^2)(1+2\xi_0^2)\left[\frac{2+3\xi_0^2}{1+\xi_0^2}-3\ \xi_0 \arctan\left(\frac{1}{\xi_0}\right)+\frac{1}{20}\left(\frac{\epsilon_a-\epsilon_s}{\epsilon_a+\epsilon_s}\right)\frac{\xi_0}{\xi_1^5}\right] \\
\Lambda_{\parallel}^1 &\equiv -\frac{1}{1280}\left(\frac{\epsilon_a-\epsilon_s}{\epsilon_a+\epsilon_s}\right)^2(1+\xi_0^2)^2(1+2\xi_0^2)\frac{\xi_0^2}{\xi_1^8} \\
\Lambda_{\parallel}^{10} &\equiv -\frac{1}{80}\left(\frac{\epsilon_a-\epsilon_s}{\epsilon_a+\epsilon_s}\right)(1+\xi_0^2)(1+2\xi_0^2)\frac{1}{\xi_1^4}
\end{aligned} \tag{6.41}
$$

It can be shown, using the above expressions for the depolarization factors of oblate spheroids in the limit $\xi_0 \to \infty$, $\xi_1 \to \infty$ and $a \to 0$, with $a\xi_0 = R$ and $a\xi_1 = d$, that for spheres with radius R at distance d with $d \geq R$ above or on the substrate one obtains the results, eqs.(5.88) and (5.89). Notice that the expressions, eqs.(6.38) and (6.40), for the polarizabilities of spheroids are of the same form as those for spheres, i.e. eqs.(5.86) and (5.87), since $a\xi_0 = R$ and $V \to \frac{4}{3}\pi\ R^3$ in the above limit.

On the other hand one can also consider the limit $\xi_0 \to 0$ for an oblate spheroid, giving the results for a circular disk of radius a, parallel to the surface of the substrate and at the distance $d = a\xi_1$ above this substrate. From eqs.(6.39) and (6.41) one then obtains

$$
L_z = 1,\ \ L_z^1 = 0,\ \ \Lambda_z^1 = 0 \text{ and } \Lambda_z^{10} = -\frac{1}{80}\left(\frac{\epsilon_a-\epsilon_s}{\epsilon_a+\epsilon_s}\right)\frac{1}{\xi_1^4} \tag{6.42}
$$

$$L_{\|} = 0, \quad L_{\|}^{1} = 1, \quad \Lambda_{\|}^{1} = 0 \text{ and } \Lambda_{\|}^{10} = -\frac{1}{80}\left(\frac{\epsilon_a - \epsilon_s}{\epsilon_a + \epsilon_s}\right)\frac{1}{\xi_1^4} \tag{6.43}$$

Note that the quantities Λ^{10} given above diverge when $\xi_1 \to 0$, i.e. when the circular disk approaches the substrate. This is, in fact, the case with all depolarization factors L and Λ, if one considers oblate spheroids touching the substrate and takes the limit $\xi_1(= \xi_0) \to 0$. This follows from eqs.(6.39) and (6.41), putting $\xi_1 = \xi_0$. Then all terms, describing the interaction of the spheroid with the substrate, i.e. the terms containing $(\epsilon_a - \epsilon_s)/(\epsilon_a + \epsilon_s)$, are proportional to a negative power of ξ_0. As has also been discussed in sections 5.3 and 5.5 of the previous chapter, this is a consequence of the fact, that in the polarizable dipole and quadrupole model the electromagnetic interaction of the particle with the substrate is described as the interaction between a point dipole and quadrupole located at the center of this particle, with an image point dipole and quadrupole in the substrate located at the point (0,0,2d) in fig.6.2. When $d = a\xi_1 \to 0$ this interaction becomes infinite. As a result the polarizabilities and the constitutive coefficients become zero in this limit. In section 6.4 it will be shown, using the exact solution, that this result is unphysical. It will therefore be clear that the formulae, derived above for the depolarization factors of oblate spheroids and the polarizabilities calculated with these factors, are incorrect in the case that the spheroids are too flat. The value of the flattening parameter ξ_0, above which the description with the polarizable dipole and quadrupole model is still a reasonable approximation, depends, of course, on the values of the various dielectric constants (see also section 6.6).

Finally one can also derive expressions for the surface constitutive coefficients $\beta_e(d)$, $\gamma_e(d)$, $\delta_e(d)$ and $\tau(d)$ with the depolarization factors found above. Using eqs.(5.64) and (5.60), together with eqs.(6.38) and (6.40), one obtains:

$$\beta_e(d) = \frac{\rho\, \epsilon_a^{-1}(\epsilon - \epsilon_a)V[\epsilon_a + L_z^1(\epsilon - \epsilon_a)]}{[\epsilon_a + L_z(\epsilon - \epsilon_a)][\epsilon_a + L_z^1(\epsilon - \epsilon_a)] + \Lambda_z^1(\epsilon - \epsilon_a)^2} \tag{6.44}$$

$$\gamma_e(d) = \frac{\rho\, \epsilon_a(\epsilon - \epsilon_a)V[\epsilon_a + L_{\|}^1(\epsilon - \epsilon_a)]}{[\epsilon_a + L_{\|}(\epsilon - \epsilon_a)][\epsilon_a + L_{\|}^1(\epsilon - \epsilon_a)] + \Lambda_{\|}^1(\epsilon - \epsilon_a)^2} \tag{6.45}$$

$$\begin{aligned}\delta_e(d) = \rho(\epsilon - \epsilon_a)V\, d\{&\frac{\epsilon_a + (L_z^1 - \Lambda_z^{10}\xi_0/\xi_1)(\epsilon - \epsilon_a)}{[\epsilon_a + L_z(\epsilon - \epsilon_a)][\epsilon_a + L_z^1(\epsilon - \epsilon_a)] + \Lambda_z^1(\epsilon - \epsilon_a)^2} \\ &+ \frac{\epsilon_a + \left(L_{\|}^1 - \Lambda_{\|}^{10}\xi_0/\xi_1\right)(\epsilon - \epsilon_a)}{[\epsilon_a + L_{\|}(\epsilon - \epsilon_a)][\epsilon_a + L_{\|}^1(\epsilon - \epsilon_a)] + \Lambda_{\|}^1(\epsilon - \epsilon_a)^2}\}\end{aligned} \tag{6.46}$$

$$\tau(d) = \frac{\rho\, \epsilon_a(\epsilon - \epsilon_a)V\, d\left[\epsilon_a + \left(L_{\|}^1 - \Lambda_{\|}^{10}\xi_0/\xi_1\right)(\epsilon - \epsilon_a)\right]}{[\epsilon_a + L_{\|}(\epsilon - \epsilon_a)][\epsilon_a + L_{\|}^1(\epsilon - \epsilon_a)] + \Lambda_{\|}^1(\epsilon - \epsilon_a)^2} \tag{6.47}$$

assuming that all spheroids have the same size and shape. Usually the particles will be touching the substrate, in which case $\xi_0 = \xi_1$ and $d = a\xi_0$ in the above formulae. If there is size and shape dispersion of the particles (V and ξ_0 different), the right-hand sides of the above equations have to be averaged. Notice that the above expressions for the constitutive coefficients have the same form as the expressions, eqs.(5.92)–(5.95),

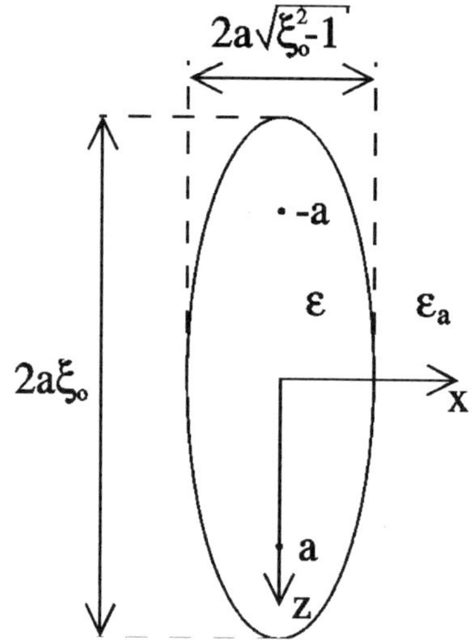

Figure 6.3 A prolate spheroid in an infinite surrounding medium. Cross-section through x-z plane.

for spheres, since $a\xi_0 = R$ (radius) in this case and $a\xi_1 = d$. This is the advantage of writing the expressions for the constitutive coefficients and for the polarizabilities in the form with depolarization factors. The invariants can be found by substitution of the above formulae, eqs.(6.44)–(6.47), into eq.(5.66). If one changes the direction of the z-axis one finds again that γ_e and β_e are independent of the direction of the z-axis while δ_e and τ change sign if the z-axis changes direction. If one uses the above formulae to calculate the reflection and transmission amplitudes, it is important to take the right sign. If, as in chapter 4, the light is incident through the ambient one may use the above expressions. If, however, the light is incident through the substrate one must, in order to use the formulae given in chapter 4, interchange ϵ_a and ϵ_s in these formulae and change the sign of δ_e and τ in the expressions given above, before substituting them into the formulae in that chapter.

6.3 Prolate spheroids; the polarizable quadrupole model

The calculation of the polarizabilities of prolate spheroids is analogous to that of oblate spheroids, given in the previous section. Consider a prolate spheroidal particle, see fig.6.3, with a dielectric constant ϵ surrounded by the ambient, with a dielectric constant ϵ_a. The z-axis is chosen as the axis of revolution, with $z = 0$ in the center of the spheroid. This is the long axis and has a length $2a\xi_0$. For the prolate spheroid the two foci lie on this axis at a distance a from its center. The two short axes have the lengths $2a(\xi_0^2 - 1)^{1/2}$ $(1 \leq \xi_0 < \infty)$, in terms of the elongation parameter ξ_0. The limit $\xi_0 \rightarrow \infty$, $a \rightarrow 0$, with $a\xi_0 = R$ again corresponds to a sphere with radius R. The limit $\xi_0 \rightarrow 1$ corresponds to a needle with length $2a$. The prolate spheroidal coordinates are defined by [37]

$$\xi \equiv \frac{\rho_1 + \rho_2}{2a}, \quad \eta \equiv \frac{\rho_1 - \rho_2}{2a}, \quad \phi \equiv \arctan\left(\frac{y}{x}\right) \tag{6.48}$$

with

$$\rho_1 \equiv [(z+a)^2 + x^2 + y^2]^{1/2} \ \text{ and } \ \rho_2 \equiv [(z-a)^2 + x^2 + y^2]^{1/2} \tag{6.49}$$

Here ρ_1 and ρ_2 are the distances of the point (x, y, z) to the two foci and ϕ describes the orientation of the plane through (x, y, z) and the z-axis with respect to the $x - z$ plane. The value range is $1 \le \xi < \infty, \ -1 \le \eta \le 1, \ 0 \le \phi < 2\pi$. The inverse of the above transformation is

$$x = a[(\xi^2 - 1)(1 - \eta^2)]^{1/2} \cos\phi, \quad y = a[(\xi^2 - 1)(1 - \eta^2)]^{1/2} \sin\phi, \quad z = a\xi\eta \tag{6.50}$$

At the surface of the particle $\xi = \xi_0$(see fig.6.3).

In these coordinates the Laplace equation (5.1) is again separable, and one can repeat the analysis, given in the previous section, now for prolate spheroids. Instead of the functions, eqs.(6.4) and (6.9), one should use here

$$P_\ell^m(\xi) Y_\ell^m(\arccos\eta, \phi) \ \text{ for } \ \ell = 0, 1, 2, ...; m = 0, \pm 1, \pm 2, ..., \pm \ell \tag{6.51}$$

and

$$Q_\ell^m(\xi) Y_\ell^m(\arccos\eta, \phi) \ \text{ for } \ \ell = 0, 1, 2, ...; m = 0, \pm 1, \pm 2, ..., \pm \ell \tag{6.52}$$

to describe the fields inside and outside the prolate spheroid. The spherical harmonics $Y_\ell^m(\arccos\eta, \phi)$ are again defined by eq.(6.5) and the associated Legendre functions $P_\ell^m(\xi)$ and $Q_\ell^m(\xi)$, for $m \ge 0$, by [37]

$$P_\ell^m(\xi) \equiv (-i)^m \frac{(\xi^2 - 1)^{m/2}}{2^\ell \ell!} \left(\frac{d}{d\xi}\right)^{\ell+m} (\xi^2 - 1)^\ell \ \text{ with } \ 1 \le \xi < \infty \tag{6.53}$$

and

$$\begin{aligned} Q_\ell^m(\xi) \ &\equiv \ (-1)^m \frac{(\xi^2 - 1)^{m/2}}{2^\ell \ell!} \left(\frac{d}{d\xi}\right)^m \{ \left(\frac{d}{d\xi}\right)^\ell \left[\ln\left(\frac{\xi+1}{\xi-1}\right)(\xi^2 - 1)^\ell\right] \\ &-\frac{1}{2} \ln\left(\frac{\xi+1}{\xi-1}\right) \left(\frac{d}{d\xi}\right)^\ell (\xi^2 - 1)^\ell \} \ \text{ for } \ 1 < \xi < \infty \end{aligned} \tag{6.54}$$

For $m < 0$, one has to use the definitions

$$P_\ell^m(\xi) \equiv (-1)^m \frac{(\ell+m)!}{(\ell-m)!} P_\ell^{-m}(\xi), \quad Q_\ell^m(\xi) \equiv \frac{(\ell+m)!}{(\ell-m)!} Q_\ell^{-m}(\xi) \tag{6.55}$$

cf. eqs.(6.8) and (6.11). Note that the functions $Q_\ell^m(\xi)$ are singular in $\xi = 1$. Notice furthermore, that the functions $P_\ell^m(\xi)$ and $Q_\ell^m(\xi)$ can also be obtained from eqs.(6.7) and (6.10), by replacing ξ by $-i\xi$ and using the relation

$$\arctan\left(\frac{1}{-i\xi}\right) = \arctan\left(\frac{i}{\xi}\right) = \frac{1}{2} i \ \ln\left(\frac{\xi+1}{\xi-1}\right) \tag{6.56}$$

The general solution of the Laplace equation in prolate spheroidal coordinates ξ, η and ϕ in a region $\xi_1 < \xi < \xi_2$, where there are no sources, can now be written as

$$\psi(\mathbf{r}) = \sum_{\ell m} A_{\ell m} \widetilde{Z}_\ell^m(\xi, a) Y_\ell^m(\arccos\eta, \phi) + \sum_{\ell m} B_{\ell m} \widetilde{X}_\ell^m(\xi, a) Y_\ell^m(\arccos\eta, \phi) \quad (6.57)$$

where the summation is from $\ell = 0$ to ∞ and $m = -\ell$ through ℓ. Here the functions

$$\widetilde{X}_\ell^m(\xi, a) \equiv i^m \frac{(\ell - m)!}{(2\ell - 1)!!} a^\ell P_\ell^m(\xi) \quad (6.58)$$

and

$$\widetilde{Z}_\ell^m(\xi, a) \equiv \frac{(2\ell + 1)!!}{(\ell + m)!} a^{-\ell-1} Q_\ell^m(\xi) \quad (6.59)$$

have been introduced. See below eq.(6.14) for the definition of the double factorial.

The asymptotic (large ξ) behavior of the associated Legendre functions along the positive real axis is given by [37], [48]

$$P_\ell^m(\xi) \simeq \frac{(2\ell - 1)!!}{(\ell - m)!} i^{-m} \xi^\ell \quad \text{for} \quad \xi \to \infty \quad (6.60)$$

$$Q_\ell^m(\xi) \simeq \frac{(\ell + m)!}{(2\ell + 1)!!} \xi^{-\ell-1} \quad \text{for} \quad \xi \to \infty \quad (6.61)$$

It then follows, that the functions $\widetilde{X}_\ell^m(\xi, a)$ and $\widetilde{Z}_\ell^m(\xi, a)$ have the same asymptotic behavior for $\xi \to \infty$ as the functions $X_\ell^m(\xi, a)$ and $Z_\ell^m(\xi, a)$, (cf. eqs.(6.17) and (6.18)):

$$\widetilde{X}_\ell^m(\xi, a) \simeq (a\xi)^\ell \simeq r^\ell \quad \text{for} \quad \xi \to \infty \quad (6.62)$$

$$\widetilde{Z}_\ell^m(\xi, a) \simeq (a\xi)^{-\ell-1} \simeq r^{-\ell-1} \quad \text{for} \quad \xi \to \infty \quad (6.63)$$

Here it was used that

$$r \equiv (x^2 + y^2 + z^2)^{1/2} = a(\xi^2 + \eta^2 - 1)^{1/2} \simeq a\xi \quad \text{for} \quad \xi \to \infty \quad (6.64)$$

which follows from eq.(6.50). Using this equation, one also finds the asymptotic relation

$$\cos\theta \equiv z/r = z(x^2 + y^2 + z^2)^{-1/2} = \xi\eta(\xi^2 + \eta^2 - 1)^{-1/2} \simeq \eta \quad \text{for} \quad \xi \to \infty \quad (6.65)$$

It finally follows from the definitions, eqs.(6.13), (6.14), (6.58) and (6.59), that the following relations hold between the prolate and oblate spheroidal functions X and Z:

$$\begin{aligned} \widetilde{X}_\ell^m(\xi, a) &= i^\ell X_\ell^m(-i\xi, a) \\ \widetilde{Z}_\ell^m(\xi, a) &= i^{-\ell-1} Z_\ell^m(-i\xi, a) \end{aligned} \quad (6.66)$$

As discussed in the previous section the spheroidal and the spherical multipole fields become identical at large distances. In general a spheroidal multipole field contains at shorter distances, in addition to the spherical multipole field of the same order, contributions due to higher order spherical multipoles. One might therefore wonder, whether formulae, derived in the previous chapter for the quadrupole model, using the spherical multipole expansion, need not be corrected, if one uses the spheroidal multipole expansion. As discussed in the previous section it can be proved that the spheroidal dipole field contains at shorter distances, in addition to the spherical dipole field, as first higher order contribution a spherical octupole field. In fact this is a consequence of the cylindrical symmetry around the normal on the substrate through the center of the particles. Since the latter contribution is neglected in the quadrupole model, it means, that one can apply all formulae, derived in the previous chapter for this model, also in the present case of a prolate spheroidal island.

One can now derive, using the (prolate) spheroidal multipole expansion eq.(6.57) for the potential together with the boundary conditions for this potential and its normal derivative on the surface of the spheroid, the following relation between induced and incident amplitudes $A_{\ell m}$ and $B_{\ell m}$:

$$A_{\ell m} = -\widetilde{\alpha}_{\ell m} B_{\ell m} \tag{6.67}$$

where

$$\widetilde{\alpha}_{\ell m} \equiv (\epsilon - \epsilon_a)[\epsilon \frac{\widetilde{Z}_\ell^m(\xi_0, a)}{\widetilde{X}_\ell^m(\xi_0, a)} - \epsilon_a \frac{\partial\, \widetilde{Z}_\ell^m(\xi_0, a)/\partial \xi_0}{\partial\, \widetilde{X}_\ell^m(\xi_0, a)/\partial \xi_0}]^{-1} \tag{6.68}$$

Note that the polarizability matrix $\widetilde{\alpha}_{\ell m, \ell' m'}$ is again diagonal:

$$\widetilde{\alpha}_{\ell m, \ell' m'} = \widetilde{\alpha}_{\ell m} \delta_{\ell \ell'} \delta_{m m'} \tag{6.69}$$

However, in contrast to the spherical case (cf. eq.(5.68)), the diagonal elements now depend on both ℓ and m as a consequence of the fact that spheroids have a lower rotational symmetry than spheres. Again one does not have to consider the case $\ell = 0$ for uncharged spheroids. Using the definitions eqs.(6.58) and (6.59), one obtains:

$$\begin{aligned} \widetilde{\alpha}_{\ell m} &\equiv (\epsilon - \epsilon_a) a^{2\ell+1} i^m \left(\frac{(\ell - m)!\,(\ell + m)!}{(2\ell - 1)!!\,(2\ell + 1)!!} \right) \\ &\times [\epsilon \frac{Q_\ell^m(\xi_0)}{P_\ell^m(\xi_0)} - \epsilon_a \frac{dQ_\ell^m(\xi_0)/d\xi_0}{dP_\ell^m(\xi_0)/d\xi_0}]^{-1} \end{aligned} \tag{6.70}$$

It follows with eq.(6.55) that

$$\widetilde{\alpha}_{\ell m} = \widetilde{\alpha}_{\ell, -m} \tag{6.71}$$

One should first of all note that one finds in the limit that the prolate spheroid becomes spherical

$$\lim_{\xi_0 \to \infty} \widetilde{\alpha}_{\ell m} = \frac{\ell(\epsilon - \epsilon_a) R^{2\ell+1}}{\ell\, \epsilon + (\ell + 1)\epsilon_a} = \alpha_\ell \tag{6.72}$$

The limit should be taken such that $a\xi_0 = R$ remains constant. This is the familiar result given also in eq.(5.70) for a sphere with radius R.

With eqs.(6.53)–(6.55), one can in principle calculate $\widetilde{\alpha}_{\ell m}$ for any value of ℓ and m. It is, however, also possible, and in fact more convenient, to calculate the polarizabilities for prolate spheroids from the expressions found in the previous section for oblate spheroids. For this purpose one must use the following relation

$$\widetilde{\alpha}_{\ell m}(\xi_0) = i(-1)^{\ell}\alpha_{\ell m}(-i\xi_0) \tag{6.73}$$

This may be derived from eqs.(6.68), (6.66) and (6.22). Using this relation and eq.(6.56) one obtains from eq.(6.27) the following polarizabilities for prolate spheroids:

$$\begin{aligned}
\widetilde{\alpha}_{10} &= \frac{(2/3)(\epsilon-\epsilon_a)a^3\xi_0(\xi_0^2-1)}{2\epsilon_a+(\epsilon-\epsilon_a)(1-\xi_0^2)[2-\xi_0\ln((\xi_0+1)/(\xi_0-1))]}\\
\widetilde{\alpha}_{1,\pm1} &= \frac{(4/3)(\epsilon-\epsilon_a)a^3\xi_0(\xi_0^2-1)}{4\epsilon_a+(\epsilon-\epsilon_a)(1-\xi_0^2)[2\xi_0^2(1-\xi_0^2)^{-1}+\xi_0\ln((\xi_0+1)/(\xi_0-1))]}\\
\widetilde{\alpha}_{20} &= \frac{(8/15)(\epsilon-\epsilon_a)a^5\xi_0(\xi_0^2-1)(3\xi_0^2-1)}{4\epsilon_a+3(\epsilon-\epsilon_a)(1-\xi_0^2)[\xi_0(1-3\xi_0^2)\ln((\xi_0+1)/(\xi_0-1))+6\xi_0^2]}\\
\widetilde{\alpha}_{2,\pm1} &= \frac{(4/5)(\epsilon-\epsilon_a)a^5\xi_0(\xi_0^2-1)(2\xi_0^2-1)}{4\epsilon_a+(\epsilon-\epsilon_a)(1-2\xi_0^2)[4-6\xi_0^2-3\xi_0(1-\xi_0^2)\ln((\xi_0+1)/(\xi_0-1))]}\\
\widetilde{\alpha}_{2,\pm2} &= \frac{(16/5)(\epsilon-\epsilon_a)a^5\xi_0(\xi_0^2-1)^2}{8\epsilon_a+(\epsilon-\epsilon_a)[3\xi_0(1-\xi_0^2)^2\ln((\xi_0+1)/(\xi_0-1))+2\xi_0^2(5-3\xi_0^2)]}
\end{aligned} \tag{6.74}$$

Now it follows, from eqs.(5.67) and (6.69), that the polarizabilities in the above mentioned dipole and quadrupole models, are given also for prolate spheroidal particles by

$$\begin{aligned}
\alpha_z &= 4\pi\ \epsilon_a\widetilde{\alpha}_{10}, \quad \alpha_{\parallel} = 4\pi\ \epsilon_a\widetilde{\alpha}_{11} = 4\pi\ \epsilon_a\widetilde{\alpha}_{1,-1}\\
\alpha_z^{10} &= \widetilde{\alpha}_{\parallel}^{10} = 0, \quad 2\alpha_z^{11} + 3\ \alpha^{11} = 2\pi\ \epsilon_a\widetilde{\alpha}_{20}\\
\alpha_{\parallel}^{11} + 2\alpha^{11} &= \frac{4}{3}\pi\ \epsilon_a\widetilde{\alpha}_{21} = \frac{4}{3}\pi\ \epsilon_a\widetilde{\alpha}_{2,-1}\\
\alpha^{11} &= \frac{2}{3}\pi\ \epsilon_a\widetilde{\alpha}_{22} = \frac{2}{3}\pi\ \epsilon_a\widetilde{\alpha}_{2,-2}
\end{aligned} \tag{6.75}$$

As there is no further danger of confusion with the formulae in the previous section, the tilde as a super index is further dropped. Introducing the volume V of the prolate spheroid (see also fig.6.3):

$$V \equiv \frac{4}{3}\pi\ a^3\xi_0(\xi_0^2-1) \tag{6.76}$$

and using eq.(6.74), one can write these polarizabilities in the following form:

$$\begin{aligned}
\alpha_z &= \frac{\epsilon_a(\epsilon-\epsilon_a)V}{\epsilon_a + L_{a,z}(\xi_0)(\epsilon-\epsilon_a)} \\
\alpha_{\|} &= \frac{\epsilon_a(\epsilon-\epsilon_a)V}{\epsilon_a + L_{a,\|}(\xi_0)(\epsilon-\epsilon_a)} \\
\alpha_z^{10} &= \alpha_{\|}^{10} = 0 \\
2\alpha_z^{11} + 3\,\alpha^{11} &= \frac{\epsilon_a(\epsilon-\epsilon_a)(3\xi_0^2-1)a^2V}{5[\epsilon_a + (\epsilon-\epsilon_a)L_{a,z}^1(\xi_0)]} \\
\alpha_{\|}^{11} + 2\alpha^{11} &= \frac{\epsilon_a(\epsilon-\epsilon_a)(2\xi_0^2-1)a^2V}{5[\epsilon_a + (\epsilon-\epsilon_a)L_{a,\|}^1(\xi_0)]} \\
\alpha^{11} &= \frac{\epsilon_a(\epsilon-\epsilon_a)(\xi_0^2-1)a^2V}{5[\epsilon_a + (\epsilon-\epsilon_a)L_a^{11}(\xi_0)]}
\end{aligned} \tag{6.77}$$

Here one then finds for prolate spheroids, in an infinite surrounding medium, the following depolarization factors

$$\begin{aligned}
L_{a,z}(\xi_0) &\equiv (1-\xi_0^2)\left[1 - \frac{1}{2}\xi_0 \ln\left(\frac{\xi_0+1}{\xi_0-1}\right)\right] \\
L_{a,\|}(\xi_0) &\equiv \frac{1}{2}(1-\xi_0^2)\left[\frac{\xi_0^2}{1-\xi_0^2} + \frac{1}{2}\xi_0 \ln\left(\frac{\xi_0+1}{\xi_0-1}\right)\right] \\
L_{a,z}^1(\xi_0) &\equiv \frac{3}{2}(1-\xi_0^2)\left[3\,\xi_0^2 + \frac{1}{2}(1-3\xi_0^2)\xi_0 \ln\left(\frac{\xi_0+1}{\xi_0-1}\right)\right] \\
L_{a,\|}^1(\xi_0) &\equiv \frac{1}{2}(1-2\xi_0^2)\left[2 - 3\xi_0^2 - \frac{3}{2}(1-\xi_0^2)\xi_0 \ln\left(\frac{\xi_0+1}{\xi_0-1}\right)\right] \\
L_a^{11}(\xi_0) &\equiv \frac{1}{4}\left[\xi_0^2(5-3\xi_0^2) + \frac{3}{2}(1-\xi_0^2)^2\xi_0 \ln\left(\frac{\xi_0+1}{\xi_0-1}\right)\right]
\end{aligned} \tag{6.78}$$

In the context of the quadrupole model discussed in section 5.5 of the previous chapter one must now substitute $\alpha_z^{10} = 0$. One obtains, cf. section 6.2, for the polarizabilities, as modified by the proximity of the substrate, in a homogeneous electric field normal to the substrate (see fig.6.4)

$$\begin{aligned}
\alpha_z(0) &= D_z^{-1}\alpha_z[1 + (3A/d^2)(2\,\alpha_z^{11} + 3\,\alpha^{11})] \\
\alpha_z^{10}(0) &= -\frac{3}{2}D_z^{-1}(A/d)\alpha_z(2\alpha_z^{11} + 3\,\alpha^{11})
\end{aligned} \tag{6.79}$$

where

$$D_z \equiv 1 + 2A\,\alpha_z + (3A/2d^2)(2 + A\,\alpha_z)(2\alpha_z^{11} + 3\alpha^{11}) \tag{6.80}$$

and

$$A \equiv (32\,\pi\,\epsilon_a d^3)^{-1}\frac{\epsilon_a - \epsilon_s}{\epsilon_a + \epsilon_s} \tag{6.81}$$

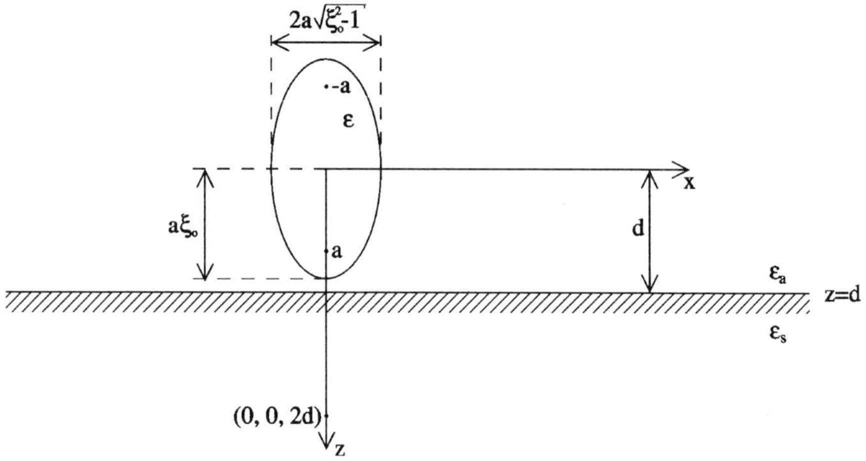

Figure 6.4 Prolate spheroid above substrate. Cross-section through the $x - z$ plane.

cf. eqs.(6.32)–(6.34). Here d is the distance from the center of the spheroid to the substrate ($d \geq a\xi_0$) and ϵ_s is the dielectric constant of the substrate (see fig.6.4). The argument 0 of the polarizabilities, eq.(6.79), indicates the position $z = 0$ of this center. Just as for spheres, the quadrupole polarizability $\alpha_z^{10}(0)$, which is zero if the spheroid is far from the surface of the substrate, becomes finite when the spheroid approaches this surface.

If the electric field is directed parallel to the surface of the substrate one finds, cf. eqs.(6.35) and (6.36), for the polarizabilities $\alpha_\parallel(0)$ and $\alpha_\parallel^{10}(0)$ of the prolate spheroid above the substrate:

$$\begin{aligned} \alpha_\parallel(0) &= D_\parallel^{-1}\alpha_\parallel[1 + (3A/d^2)(\alpha_\parallel^{11} + 2\,\alpha^{11})] \\ \alpha_\parallel^{10}(0) &= -\frac{3}{2}D_\parallel^{-1}(A/d)\alpha_\parallel(\alpha_\parallel^{11} + 2\,\alpha^{11}) \end{aligned} \tag{6.82}$$

where

$$D_\parallel \equiv 1 + A\,\alpha_\parallel + (3A/4d^2)(4 + A\,\alpha_\parallel)(\alpha_\parallel^{11} + 2\alpha^{11}) \tag{6.83}$$

Again the quadrupole polarizability $\alpha_\parallel^{10}(0)$, which is zero if the spheroid is far away from the surface of the substrate, becomes finite, when the spheroid approaches the substrate.

If all spheroids have the same size, the resulting electric susceptibilities of the film are given by eq.(5.64), which reduces for this case to, cf. eq.(6.37),

$$\begin{aligned}
\beta_e(d) &= \rho\, \epsilon_a^{-2} D_z^{-1}\alpha_z[1+(3A/d^2)(2\alpha_z^{11}+3\,\alpha^{11})] \\
\gamma_e(d) &= \rho\, D_\parallel^{-1}\alpha_\parallel[1+(3A/d^2)(\alpha_\parallel^{11}+2\alpha^{11})] \\
\delta_e(d) &= \rho\, \epsilon_a^{-1} d\{D_z^{-1}\alpha_z[1+\frac{9}{2}(A/d^2)(2\alpha_z^{11}+3\alpha^{11})] \\
&\quad +D_\parallel^{-1}\alpha_\parallel[1+\frac{9}{2}(A/d^2)(\alpha_\parallel^{11}+2\alpha^{11})]\} \\
\tau(d) &= \rho\, d\, D_\parallel^{-1}\alpha_\parallel[1+\frac{9}{2}(A/d^2)(\alpha_\parallel^{11}+2\alpha^{11})]
\end{aligned} \tag{6.84}$$

The argument d of the above coefficients denotes the position $z = d$ of the dividing surface, which coincides with the surface of the substrate (see fig.6.4). If one considers a film in which the spheroidal particles have a certain dispersion both in size and shape (ξ_0), the expressions on the right-hand sides of the above equation have to be averaged. One then has to apply eq.(5.65), instead of eq.(5.64). The invariants can be found using eq.(5.66). If one changes the direction of the z-axis one finds again that γ_e and β_e are independent of the direction of the z-axis while δ_e and τ change sign if the z-axis changes direction. If one uses the above formulae to calculate the reflection and transmission amplitudes, it is important to take the right sign. If, as in chapter 4, the light is incident through the ambient one may use the above expressions. If the light is incident through the substrate one must, in order to use the formulae given in chapter 4, interchange ϵ_a and ϵ_s in these formulae and change the sign of δ_e and τ in the expressions given above, before substituting them into the formulae in that chapter.

Substituting the explicit expressions of the various polarizabilities, eq.(6.77), and depolarization factors, eq.(6.78), in the ambient far from the substrate, into eqs.(6.79) and (6.80), one obtains, using also eqs.(6.81) and (6.76), for prolate spheroids at the distance $d \equiv a\xi_1$ on or above the substrate ($\xi_1 \geq \xi_0$) :

$$\begin{aligned}
\alpha_z(0) &= \frac{\epsilon_a(\epsilon-\epsilon_a)V[\epsilon_a+L_z^1(\epsilon-\epsilon_a)]}{[\epsilon_a+L_z(\epsilon-\epsilon_a)][\epsilon_a+L_z^1(\epsilon-\epsilon_a)]+\Lambda_z^1(\epsilon-\epsilon_a)^2} \\
\alpha_z^{10}(0) &= \frac{\epsilon_a(\epsilon-\epsilon_a)^2\Lambda_z^{10}V\ a\xi_0}{[\epsilon_a+L_z(\epsilon-\epsilon_a)][\epsilon_a+L_z^1(\epsilon-\epsilon_a)]+\Lambda_z^1(\epsilon-\epsilon_a)^2}
\end{aligned} \tag{6.85}$$

The depolarization factors for the prolate spheroid, are the following functions of ξ_0, ξ_1, ϵ_a and ϵ_s

$$\begin{aligned}
L_z &\equiv (1-\xi_0^2)\left[1-\frac{1}{2}\xi_0\ln\left(\frac{\xi_0+1}{\xi_0-1}\right)-\frac{1}{12}\left(\frac{\epsilon_a-\epsilon_s}{\epsilon_a+\epsilon_s}\right)\frac{\xi_0}{\xi_1^3}\right] \\
L_z^1 &\equiv (1-\xi_0^2)(1-3\xi_0^2)\left[\frac{3}{4}\xi_0\ln\left(\frac{\xi_0+1}{\xi_0-1}\right)+\frac{9}{2}\frac{\xi_0^2}{1-3\xi_0^2}+\frac{1}{40}\left(\frac{\epsilon_a-\epsilon_s}{\epsilon_a+\epsilon_s}\right)\frac{\xi_0}{\xi_1^5}\right] \\
\Lambda_z^1 &\equiv \frac{1}{640}\left(\frac{\epsilon_a-\epsilon_s}{\epsilon_a+\epsilon_s}\right)^2(1-\xi_0^2)^2(1-3\xi_0^2)\frac{\xi_0^2}{\xi_1^8} \\
\Lambda_z^{10} &\equiv -\frac{1}{80}\left(\frac{\epsilon_a-\epsilon_s}{\epsilon_a+\epsilon_s}\right)(1-\xi_0^2)(1-3\xi_0^2)\frac{1}{\xi_1^4}
\end{aligned} \tag{6.86}$$

In a completely analogous way as above one finds for the polarizabilities and depolarization factors along the surface:

$$\begin{aligned}\alpha_{\parallel}(0) &= \frac{\epsilon_a(\epsilon-\epsilon_a)V[\epsilon_a+L_{\parallel}^1(\epsilon-\epsilon_a)]}{[\epsilon_a+L_{\parallel}(\epsilon-\epsilon_a)][\epsilon_a+L_{\parallel}^1(\epsilon-\epsilon_a)]+\Lambda_{\parallel}^1(\epsilon-\epsilon_a)^2}\\ \alpha_{\parallel}^{10}(0) &= \frac{\epsilon_a(\epsilon-\epsilon_a)^2\Lambda_{\parallel}^{10}V\ a\xi_0}{[\epsilon_a+L_{\parallel}(\epsilon-\epsilon_a)][\epsilon_a+L_{\parallel}^1(\epsilon-\epsilon_a)]+\Lambda_{\parallel}^1(\epsilon-\epsilon_a)^2}\end{aligned} \tag{6.87}$$

with

$$\begin{aligned}L_{\parallel} &\equiv \frac{1}{2}(1-\xi_0^2)\left[\frac{\xi_0^2}{1-\xi_0^2}+\frac{1}{2}\xi_0\ln\left(\frac{\xi_0+1}{\xi_0-1}\right)-\frac{1}{12}\left(\frac{\epsilon_a-\epsilon_s}{\epsilon_a+\epsilon_s}\right)\frac{\xi_0}{\xi_1^3}\right]\\ L_{\parallel}^1 &\equiv \frac{1}{2}(1-\xi_0^2)(1-2\xi_0^2)\left[\frac{2-3\xi_0^2}{1-\xi_0^2}-\frac{3}{2}\xi_0\ln\left(\frac{\xi_0+1}{\xi_0-1}\right)+\frac{1}{20}\left(\frac{\epsilon_a-\epsilon_s}{\epsilon_a+\epsilon_s}\right)\frac{\xi_0}{\xi_1^5}\right]\\ \Lambda_{\parallel}^1 &\equiv \frac{1}{1280}\left(\frac{\epsilon_a-\epsilon_s}{\epsilon_a+\epsilon_s}\right)^2(1-\xi_0^2)^2(1-2\xi_0^2)\frac{\xi_0^2}{\xi_1^8}\\ \Lambda_{\parallel}^{10} &\equiv -\frac{1}{80}\left(\frac{\epsilon_a-\epsilon_s}{\epsilon_a+\epsilon_s}\right)(1-\xi_0^2)(1-2\xi_0^2)\frac{1}{\xi_1^4}\end{aligned} \tag{6.88}$$

It can first of all be shown, using the above expressions for the depolarization factors of prolate spheroids in the limit $\xi_0 \to \infty$, $\xi_1 \to \infty$ and $a \to 0$, with $a\xi_0 = R$ and $a\xi_1 = d$, that for spheres with radius R at distance d with $d \geq R$ above or on the substrate one obtains the results, eqs.(5.88) and (5.89). Notice that the expressions, eqs.(6.85) and (6.87), for the polarizabilities of spheroids are of the same form as those for spheres, i.e. eqs.(5.86) and (5.87), since $a\xi_0 = R$ and $V \to \frac{4}{3}\pi\ R^3$ in the above limit.

For prolate spheroids the depolarization factors, found above, do provide a good description for all values of ξ_0, as will be proved in section 6.5. This is in strong contrast to the case of oblate spheroids, cf. the discussion below eq.(6.43) in the previous section. In the case of prolate spheroids the center of the particle and the image point are at least at the finite distance $2a$ from each other. This is the case also for a "needle", i.e. a prolate spheroid, for which $\xi_0 \downarrow 1$. The values of the depolarization factors for a needle of length $2a$, normal to the surface of the substrate, are easily obtained from eqs.(6.86) and (6.88) in the limit $\xi_0 \downarrow 1$. One finds for all values of a and $d \geq a$:

$$L_z = L_z^1 = \Lambda_z^1 = \Lambda_z^{10} = 0 \tag{6.89}$$

and

$$L_{\parallel} = L_{\parallel}^1 = \frac{1}{2},\quad \Lambda_{\parallel}^1 = \Lambda_{\parallel}^{10} = 0 \tag{6.90}$$

For the polarizabilities normal and parallel to the surface of the substrate one therefore finds for the needle, using eqs.(6.85) and (6.87) (cf. also ref.[42]):

$$\alpha_z(0) = (\epsilon-\epsilon_a)V,\quad \alpha_z^{10}(0) = 0 \tag{6.91}$$

$$\alpha_{\parallel}(0) = 2\epsilon_a \left(\frac{\epsilon - \epsilon_a}{\epsilon + \epsilon_a} \right) V, \quad \alpha_{\parallel}^{10}(0) = 0 \tag{6.92}$$

where V is the "volume" of the needle: $V = \frac{4}{3}\pi\ a^3\xi_0(\xi_0^2 - 1)$ in the limit $\xi_0 \downarrow 1$. Note that the polarizability per unit of volume remains finite ($\neq 0$) in this limit. The influence of the substrate also vanishes in this limit.

Finally one can also derive expressions for the surface constitutive coefficients $\beta_e(d)$, $\gamma_e(d)$, $\delta_e(d)$ and $\tau(d)$ with the depolarization factors found above. Using eqs.(5.64) and (5.60), together with eqs.(6.85) and (6.87), one obtains:

$$\beta_e(d) = \frac{\rho\ \epsilon_a^{-1}(\epsilon - \epsilon_a)V[\epsilon_a + L_z^1(\epsilon - \epsilon_a)]}{[\epsilon_a + L_z(\epsilon - \epsilon_a)][\epsilon_a + L_z^1(\epsilon - \epsilon_a)] + \Lambda_z^1(\epsilon - \epsilon_a)^2} \tag{6.93}$$

$$\gamma_e(d) = \frac{\rho\ \epsilon_a(\epsilon - \epsilon_a)V[\epsilon_a + L_{\parallel}^1(\epsilon - \epsilon_a)]}{[\epsilon_a + L_{\parallel}(\epsilon - \epsilon_a)][\epsilon_a + L_{\parallel}^1(\epsilon - \epsilon_a)] + \Lambda_{\parallel}^1(\epsilon - \epsilon_a)^2} \tag{6.94}$$

$$\begin{aligned} \delta_e(d) = \rho(\epsilon - \epsilon_a)V\ d\{ & \frac{\epsilon_a + (L_z^1 - \Lambda_z^{10}\xi_0/\xi_1)(\epsilon - \epsilon_a)}{[\epsilon_a + L_z(\epsilon - \epsilon_a)][\epsilon_a + L_z^1(\epsilon - \epsilon_a)] + \Lambda_z^1(\epsilon - \epsilon_a)^2} \\ & + \frac{\epsilon_a + \left(L_{\parallel}^1 - \Lambda_{\parallel}^{10}\xi_0/\xi_1\right)(\epsilon - \epsilon_a)}{[\epsilon_a + L_{\parallel}(\epsilon - \epsilon_a)][\epsilon_a + L_{\parallel}^1(\epsilon - \epsilon_a)] + \Lambda_{\parallel}^1(\epsilon - \epsilon_a)^2}\} \end{aligned} \tag{6.95}$$

$$\tau(d) = \frac{\rho\ \epsilon_a(\epsilon - \epsilon_a)V\ d\left[\epsilon_a + \left(L_{\parallel}^1 - \Lambda_{\parallel}^{10}\xi_0/\xi_1\right)(\epsilon - \epsilon_a)\right]}{[\epsilon_a + L_{\parallel}(\epsilon - \epsilon_a)][\epsilon_a + L_{\parallel}^1(\epsilon - \epsilon_a)] + \Lambda_{\parallel}^1(\epsilon - \epsilon_a)^2} \tag{6.96}$$

assuming that all spheroids have the same size and shape. Usually the particles will be touching the substrate, in which case $\xi_0 = \xi_1$ and $d = a\xi_0$ in the above formulae. If there is size and shape dispersion of the particles (V and ξ_0 different), the right-hand sides of the above equations have to be averaged. Notice that the above expressions for the constitutive coefficients have, similar to those for oblate spheroids, the same form as the expressions, eqs.(5.92)–(5.95), for spheres, since $a\xi_0 = R$ (radius) in this case and $a\xi_1 = d$. This is the advantage of writing the expressions for the constitutive coefficients and for the polarizabilities in the form with depolarization factors. The invariants can be found by substitution of the above formulae into eq.(5.66). If one changes the direction of the z-axis one finds again that γ_e and β_e are independent of the direction of the z-axis while δ_e and τ change sign if the z-axis changes direction. If one uses the above formulae to calculate the reflection and transmission amplitudes, it is important to take the right sign. If, as in chapter 4, the light is incident through the ambient one may use the above expressions. If, however, the light is incident through the substrate one must, in order to use the formulae given in chapter 4, interchange ϵ_a and ϵ_s in these formulae and change the sign of δ_e and τ in the expressions given above, before substituting them into the formulae in that chapter.

6.4 Oblate spheroids; spheroidal multipole expansions

In the previous sections the polarizabilities of a spheroidal island, with axis of rotation normal to the surface of the substrate, were derived in the context of the polarizable quadrupole model, discussed in section 5.5 of the previous chapter. In this model

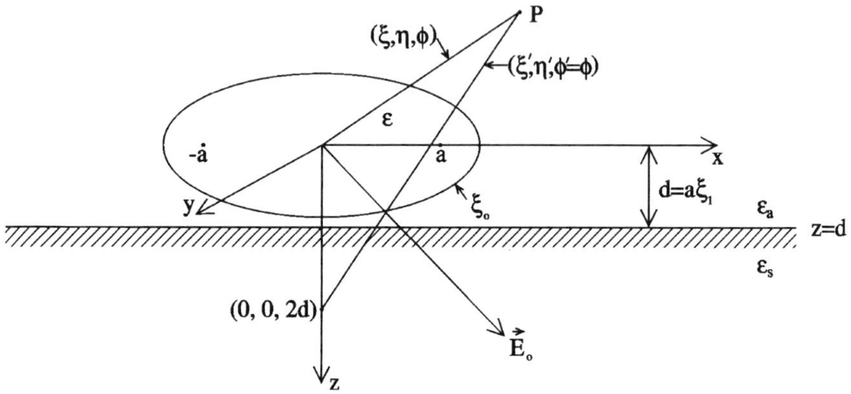

Figure 6.5 An oblate spheroid above a substrate.

the electromagnetic interaction of the particle with the substrate is described as the interaction between a point dipole and quadrupole, located at the center of this particle, and an image point dipole and quadrupole, located in the substrate at the distance $2d$ from this center (d is the distance between the center of the particle and the surface of the substrate: see figs.6.2 and 6.4 in the previous sections). First the dipole and quadrupole polarizabilities of a homogeneous spheroid in an infinite dielectric medium were calculated. The outgoing field, far away from this spheroid, is that of a point dipole and quadrupole, located at the center of this particle, with the same polarizabilities. These (free) polarizabilities are then used in order to calculate the polarizabilities of the spheroid, as modified by the proximity of the substrate, in homogeneous electric fields normal and parallel to the substrate. To this end general formulae, derived in section 5.5 for the polarizable quadrupole model, have been applied. It was found that the results obtained in that way are not correct for oblate spheroids, which are too flat and too near to the surface of the substrate. This is a consequence of the fact, that the interaction of these particles with the substrate was calculated in spherical (point) dipole and quadrupole approximation.

In the present section this restriction will be removed by describing the interaction of the spheroid with the substrate, using spheroidal multipoles, instead of spherical multipoles. A completely analogous method of solution, as was given at the end of section 5.6 for spherical particles, will be developed here for spheroidal particles. Instead of spherical multipole fields, one now has to use spheroidal multipole fields. In this section the solution of the problem will be given for oblate spheroids above the substrate (see fig.6.5) and in the next section for prolate spheroids. The analysis is similar to the one given by Bobbert and Vlieger [42]. The notation is somewhat different, however.

Consider the response of this spheroid in a constant electric field, given in the ambient by

$$\mathbf{E}_0 = (E_{0,x}, E_{0,y}, E_{0,z}) \tag{6.97}$$

The corresponding incident potential is

$$\psi_0(\mathbf{r}) = -\mathbf{E}_0.\mathbf{r} = (2\pi/3)^{1/2}[-2^{1/2}E_{0,z}X_1^0(\xi,a)Y_1^0(\arccos\eta,\phi)$$

$$+(E_{0,x} - i\ E_{0,y})X_1^1(\xi,a)Y_1^1(\arccos\eta,\phi) - (E_{0,x} + i\ E_{0,y})X_1^{-1}(\xi,a)Y_1^{-1}(\arccos\eta,\phi)] \tag{6.98}$$

where the definitions, eqs.(6.5)–(6.8) and (6.13), of the functions Y_ℓ^m and X_ℓ^m have been used and where oblate spheroidal coordinates ξ, η and ϕ, defined by eqs.(6.1) and (6.2) (or the inverse transformation eq.(6.3)) have been introduced. The general solution in the ambient may be written as a sum of the incident potential, the potential due to the charge distribution, induced in the island, and the image charge distribution in the substrate:

$$\psi_a(\mathbf{r}) = (2\pi/3)^{1/2}[-2^{1/2}E_{0,z}X_1^0(\xi,a)Y_1^0(\arccos\eta,\phi)$$

$$+(E_{0,x} - i\ E_{0,y})X_1^1(\xi,a)Y_1^1(\arccos\eta,\phi) - (E_{0,x} + i\ E_{0,y})X_1^{-1}(\xi,a)Y_1^{-1}(\arccos\eta,\phi)]$$

$$+\sum_{\ell m}{}' A_{\ell m}Z_\ell^m(\xi,a)Y_\ell^m(\arccos\eta,\phi) + \sum_{\ell m}{}' A_{\ell m}^r Z_\ell^m(\xi'(\xi,\eta),a)Y_\ell^m(\arccos\eta'(\xi,\eta),\phi) \tag{6.99}$$

The first sum at the right-hand side of this equation is of the same form as the first sum of eq.(6.12), which gives the general solution of the Laplace equation in oblate spheroidal coordinates. It is a superposition of spheroidal multipole fields. The multipoles describing this direct field are located in the centre O of the spheroid (see fig.6.5). The functions Z_ℓ^m and Y_ℓ^m were defined in section 6.2 by means of eqs.(6.14), (6.10), (6.11), (6.5), (6.6) and (6.8). The prime in the summation denotes again the exclusion of the term with $\ell = m = 0$, since the total charge of the particle is assumed to be zero. The second sum at the right-hand side of eq.(6.99) is a superposition of spheroidal multipole fields, whose origin is located at the point (0,0,2d) within the substrate (see fig.6.5); ξ' and η' are the first two oblate spheroidal coordinates of the point (x, y, z) in the shifted coordinate frame with origin at (0,0,2d). These fields may be attributed to virtual spheroidal multipoles, located at (0,0,2d).

For the potential in the substrate one may write similarly

$$\begin{aligned}\psi_s(\mathbf{r}) \quad &= \quad \psi_s' + c_1X_1^0(\xi,a)Y_1^0(\arccos\eta,\phi) + c_2X_1^1(\xi,a)Y_1^1(\arccos\eta,\phi) + \\ & \quad c_3X_1^{-1}(\xi,a)Y_1^{-1}(\arccos\eta,\phi) + \sum_{\ell m}{}' A_{\ell m}^t Z_\ell^m(\xi,a)Y_\ell^m(\arccos\eta,\phi)\end{aligned} \tag{6.100}$$

As a first step to calculate the polarizabilities of the oblate spheroid, the amplitudes $A_{\ell m}$, $A_{\ell m}^r$, $A_{\ell m}^t$, c_1, c_2 and c_3 of the various spheroidal multipoles, as well as the constant ψ_s', have to be chosen such that the boundary conditions of the potentials

ψ_a and ψ_s at the surface of the substrate are satisfied. This means that $\psi_a = \psi_s$ and $\epsilon_a \partial\, \psi_a/\partial z = \epsilon_s \partial \psi_s/\partial z$ for $z = d$, i.e. for

$$z = a\,\xi\,\eta = d = a\,\xi_1 \quad \text{or} \quad \eta = \frac{\xi_1}{\xi} = \frac{d}{a\xi} \quad (\xi_1 \leq \xi < \infty) \tag{6.101}$$

where eq.(6.3) has been applied. Using the relations

$$\xi'(\xi, \frac{\xi_1}{\xi}) = \xi, \quad \eta'(\xi, \frac{\xi_1}{\xi}) = -\frac{\xi_1}{\xi} \tag{6.102}$$

and

$$Y_\ell^m(\arccos(-\eta), \phi) = (-1)^{\ell+m} Y_\ell^m(\arccos\eta, \phi) \tag{6.103}$$

one then obtains the following results:

$$\begin{aligned} c_1 &= -\frac{\epsilon_a}{\epsilon_s}\left(\frac{4\pi}{3}\right)^{\frac{1}{2}} E_{0,z} \\ c_2 &= \left(\frac{2\pi}{3}\right)^{\frac{1}{2}} (E_{0,x} - i\,E_{0,y}), \quad c_3 = -\left(\frac{2\pi}{3}\right)^{\frac{1}{2}} (E_{0,x} + i\,E_{0,y}) \end{aligned} \tag{6.104}$$

$$A_{\ell m}^r = (-1)^{\ell+m}\left(\frac{\epsilon_a - \epsilon_s}{\epsilon_a + \epsilon_s}\right) A_{\ell m}, \quad A_{\ell m}^t = \left(\frac{2\epsilon_a}{\epsilon_a + \epsilon_s}\right) A_{\ell m} \tag{6.105}$$

and

$$\psi_s' = -\left(1 - \frac{\epsilon_a}{\epsilon_s}\right) d\, E_{0,z} \tag{6.106}$$

In order to relate the amplitudes $A_{\ell m}$ to the multipole polarizabilities for the oblate spheroid in free space, the potential of the so-called local field acting on this particle has to be introduced, which is given as the sum of the potentials of the incident and the reflected fields

$$\begin{aligned} \psi_{loc}(\mathbf{r}) &= (2\pi/3)^{1/2}[-2^{1/2} E_{0,z} X_1^0(\xi, a) Y_1^0(\arccos\eta, \phi) \\ &+ (E_{0,x} - i\,E_{0,y}) X_1^1(\xi, a) Y_1^1(\arccos\eta, \phi) - (E_{0,x} + i\,E_{0,y}) X_1^{-1}(\xi, a) Y_1^{-1}(\arccos\eta, \phi)] \\ &+ \sum_{\ell m}{}' A_{\ell m}^r Z_\ell^m(\xi'(\xi, \eta), a) Y_\ell^m(\arccos\eta'(\xi, \eta), \phi) \end{aligned} \tag{6.107}$$

It is now possible to expand the potentials originating from (0,0,2d) in terms of a complete set of potentials around the origin O (see fig.6.5):

$$\begin{aligned} &Z_\ell^m(\xi'(\xi, \eta), a) Y_\ell^m(\arccos\eta'(\xi, \eta), \phi) \\ &= \sum_{\ell_1 = |m|}^{\infty} K_{\ell_1, \ell}^m(\xi_1)(2d)^{-\ell-\ell_1-1} X_{\ell_1}^m(\xi, a) Y_{\ell_1}^m(\arccos\eta, \phi) \quad \text{for} \quad 0 \leq \xi < 2\xi_1 \end{aligned} \tag{6.108}$$

This is in fact a generalization of the expansion, eq.(5.102), which was valid for spherical particles. It may indeed be checked, using eqs.(6.17), (6.18) and (6.20), that eq.(5.102) follows from eq.(6.108), with

$$\lim_{\xi_1\to\infty} K^m_{\ell_1,\ell}(\xi_1) = M^m_{\ell_1,\ell} \tag{6.109}$$

In contrast to this limiting case of a sphere, however, it does not appear to be possible to give a straightforward generalization of the formula eq.(5.103) for $M^m_{\ell_1,\ell}$, which is valid in the case of an oblate spheroid. As is shown in Appendix A, it is still possible to obtain a complete set of relations and recurrence relations, by which in principle all coefficients $K^m_{\ell_1,\ell}(\xi_1)$ can be determined explicitly. These are the following:

$$K^0_{0,0}(\xi_1) = 2\xi_1 \arctan\left(\frac{1}{\xi_1}\right) - \xi_1^2 \ln\left(1+\frac{1}{\xi_1^2}\right) \tag{6.110}$$

$$\begin{aligned} K^0_{0,\ell}(\xi_1) &= \frac{(-1)^\ell(2\ell-1)!!}{\ell!}(2\ell+1)^{1/2}(2\xi_1)^{\ell+1}\,\mathrm{Re}\{i^{\ell+1}Q^0_{\ell+1}(1+2i\xi_1) \\ &\quad + i^{\ell-1}Q^0_{\ell-1}(1+2i\xi_1)\} \ \text{ for } \ \ell>0 \end{aligned} \tag{6.111}$$

$$\begin{aligned} K^0_{1,\ell}(\xi_1) &= \frac{(-1)^\ell(2\ell-1)!!}{\ell!}[3(2\ell+1)]^{1/2}(2\xi_1)^{\ell+2}\,\mathrm{Im}\{i^{\ell+1}Q^0_{\ell+1}(1+2i\xi_1) \\ &\quad + i^{\ell-1}Q^0_{\ell-1}(1+2i\xi_1)\} - \left(\frac{\ell+1}{2\ell+3}\right)\left(\frac{3}{(2\ell+1)(2\ell+3)}\right)^{1/2} K^0_{0,\ell+1}(\xi_1) \\ &\quad - \frac{4}{\ell}\left(\frac{3(2\ell+1)}{2\ell-1}\right)^{1/2} \xi_1^2 K^0_{0,\ell-1}(\xi_1) \text{ for } \ell>0 \end{aligned} \tag{6.112}$$

$$\begin{aligned} 0 &= \frac{\ell+1}{(2\ell+1)(2\ell+3)}\left(\frac{2\ell_1+1}{2\ell+3}\right)^{1/2} K^0_{\ell_1,\ell+1}(\xi_1) + \frac{4}{\ell}\xi_1^2\left(\frac{2\ell_1+1}{2\ell-1}\right)^{1/2} K^0_{\ell_1,\ell-1}(\xi_1) \\ &\quad + \frac{\ell_1+1}{(2\ell_1+1)(2\ell_1+3)}\left(\frac{2\ell_1+3}{2\ell+1}\right)^{1/2} K^0_{\ell_1+1,\ell}(\xi_1) \\ &\quad + \frac{4}{\ell_1}\xi_1^2\left(\frac{2\ell_1-1}{2\ell+1}\right)^{1/2} K^0_{\ell_1-1,\ell}(\xi_1) \ \text{ for } \ \ell>0, \ell_1>0 \end{aligned} \tag{6.113}$$

The above is sufficient to determine all the coefficients $K^0_{\ell_1,\ell}(\xi_1)$. The following recurrence relation for the coefficients is given for completeness

$$\begin{aligned} -4\,\xi_1^2\delta_{\ell_1,0}\delta_{\ell 0} &= \frac{(\ell+1)^2}{(2\ell+1)(2\ell+3)}\left(\frac{2\ell_1+1}{2\ell+3}\right)^{1/2} K^0_{\ell_1,\ell+1}(\xi_1) \\ &\quad - \frac{(\ell_1+1)^2}{(2\ell_1+1)(2\ell_1+3)}\left(\frac{2\ell_1+3}{2\ell+1}\right)^{1/2} K^0_{\ell_1+1,\ell}(\xi_1) \\ &\quad - 4\xi_1^2\left(\frac{2\ell_1+1}{2\ell+1}\right)^{1/2} K^0_{\ell_1,\ell}(\xi_1) + 4\,\xi_1^2\{\left(\frac{2\ell_1-1}{2\ell+1}\right)^{1/2} K^0_{\ell_1-1,\ell}(\xi_1) \\ &\quad - \left(\frac{2\ell_1+1}{2\ell-1}\right)^{1/2} K^0_{\ell_1,\ell-1}(\xi_1)\} \ \text{ for } \ \ell\geq 0, \ \ell_1\geq 0 \end{aligned} \tag{6.114}$$

In this equation $K_{-1,\ell}(\xi_1)$ and $K_{\ell_1,-1}(\xi_1)$ are assumed to vanish. For the evaluation of the coefficients $K^0_{0,\ell}(\xi_1)$ and $K^0_{1,\ell}(\xi_1)$ by means of eqs.(6.111) and (6.112) one needs the functions $Q^0_\ell(z)$ for complex z, which are defined by (see [37]):

$$Q^0_\ell(z) \equiv \frac{1}{2^\ell \ell!}\left\{\left(\frac{d}{dz}\right)^\ell\left[(z^2-1)^\ell \ln\left(\frac{z+1}{z-1}\right)\right] - \frac{1}{2}\left[\left(\frac{d}{dz}\right)^\ell (z^2-1)^\ell\right]\ln\left(\frac{z+1}{z-1}\right)\right\}$$

$$\text{for } \ell = 0, 1, 2, \ldots \tag{6.115}$$

One can furthermore make use of the symmetry relation (see appendix A)

$$K^0_{\ell_1,\ell}(\xi_1) = (-1)^{\ell+\ell_1}\left(\frac{2\ell+1}{2\ell_1+1}\right) K^0_{\ell,\ell_1}(\xi_1) \tag{6.116}$$

The recurrence relations, eqs.(6.113), or alternatively (6.114), can then be used to evaluate $K^0_{\ell_1,\ell}(\xi_1)$ for ℓ_1 and $\ell \geq 2$. In the following one also needs $K^m_{\ell_1,\ell}(\xi_1)$ for $m = \pm 1$. In appendix A it will be shown, that

$$K^m_{\ell_1,\ell}(\xi_1) = K^{-m}_{\ell_1,\ell}(\xi_1) \tag{6.117}$$

so that it is only necessary to give these coefficients for positive values of m. For $m = 1$ one has (see appendix A):

$$K^1_{\ell_1,\ell}(\xi_1) = -\left[\frac{\ell_1 \ell}{(\ell_1+1)(\ell+1)}\right]^{1/2} K^0_{\ell_1,\ell}(\xi_1) \quad \text{for} \quad \ell_1, \ell = 1, 2, \ldots. \tag{6.118}$$

It is therefore sufficient to concentrate on the evaluation of $K^0_{\ell_1,\ell}(\xi_1)$. Finally in the appendix A explicit expressions for these coefficients are given for ℓ, $\ell_1 = 0$, 1 and 2, which will be used below.

Substituting eq.(6.108) into eq.(6.107) and interchanging ℓ with ℓ_1, one obtains

$$\psi_{loc}(\mathbf{r}) = (2\pi/3)^{1/2}[-2^{1/2}E_{0,z}X^0_1(\xi,a)Y^0_1(\arccos\eta,\phi)$$

$$+(E_{0,x} - i\ E_{0,y})X^1_1(\xi,a)Y^1_1(\arccos\eta,\phi) - (E_{0,x} + i\ E_{0,y})X^{-1}_1(\xi,a)Y^{-1}_1(\arccos\eta,\phi)]$$
$$+\sum_{\ell m}\sum_{\ell_1=|m|}^{\infty\prime} A^r_{\ell_1,m}K^m_{\ell,\ell_1}(\xi_1)(2d)^{-\ell-\ell_1-1}X^m_\ell(\xi,a)Y^m_\ell(\arccos\eta,\phi) \tag{6.119}$$

where the prime as index of the summation indicates that $\ell_1 = 0$ should be excluded. Using eqs.(6.21) and (6.105) one obtains

$$A_{\ell m} = -\alpha_{\ell m}\{(2\pi/3)^{1/2}\delta_{\ell 1}[-2^{1/2}E_{0,z}\delta_{m0} + (E_{0,x} - i\ E_{0,y})\delta_{m1} - (E_{0,x} + i\ E_{0,y})\delta_{m,-1}]$$
$$+\left(\frac{\epsilon_a - \epsilon_s}{\epsilon_a + \epsilon_s}\right)\sum_{\ell_1=|m|}^{\infty\prime}(-1)^{\ell_1+m}K^m_{\ell,\ell_1}(\xi_1)A_{\ell_1,m}(2d)^{-\ell-\ell_1-1}\} \tag{6.120}$$

where $\alpha_{\ell m}$ is given by eq.(6.24). It follows from this equation that different m's do not couple to each other. Just as in the case of a sphere on (or above) a flat substrate in section 5.6 of the previous chapter, this property is, of course, due to the rotational symmetry of the system around the z-axis. As a consequence

$$A_{\ell m} = 0 \quad \text{for} \quad m \neq -1, 0, 1 \tag{6.121}$$

(cf. eq.(5.106)). The above set of linear equations for the amplitudes may thus be written as the sum of the separate and independent contributions for $m = -1$, 0 and 1. In this way one obtains

$$A_{\ell 0} + \alpha_{\ell 0}\left(\frac{\epsilon_a - \epsilon_s}{\epsilon_a + \epsilon_s}\right)\sum_{\ell_1=1}^{\infty}(-1)^{\ell_1}K^0_{\ell,\ell_1}(\xi_1)A_{\ell_1,0}(2d)^{-\ell-\ell_1-1} = 2(\pi/3)^{1/2}\alpha_{10}E_{0,z}\delta_{\ell 1} \quad \text{for } \ell = 1, 2, 3, ... \tag{6.122}$$

$$A_{\ell 1} - \alpha_{\ell 1}\left(\frac{\epsilon_a - \epsilon_s}{\epsilon_a + \epsilon_s}\right)\sum_{\ell_1=1}^{\infty}(-1)^{\ell_1}K^1_{\ell,\ell_1}(\xi_1)A_{\ell_1,1}(2d)^{-\ell-\ell_1-1}$$
$$= -(2\pi/3)^{1/2}\alpha_{11}(E_{0,x} - i\ E_{0,y})\delta_{\ell 1} \quad \text{for } \ell = 1, 2, 3, ... \tag{6.123}$$

and

$$A_{\ell,-1} - \alpha_{\ell,-1}\left(\frac{\epsilon_a - \epsilon_s}{\epsilon_a + \epsilon_s}\right)\sum_{\ell_1=1}^{\infty}(-1)^{\ell_1}K^{-1}_{\ell,\ell_1}(\xi_1)A_{\ell_1,-1}(2d)^{-\ell-\ell_1-1}$$
$$= (2\pi/3)^{1/2}\alpha_{1,-1}(E_{0,x} + i\ E_{0,y})\delta_{\ell 1} \quad \text{for } \ell = 1, 2, 3, ... \tag{6.124}$$

Using the symmetry relations, eqs.(6.25) and (6.117), the last equation may also be written as

$$A_{\ell,-1} - \alpha_{\ell 1}\left(\frac{\epsilon_a - \epsilon_s}{\epsilon_a + \epsilon_s}\right)\sum_{\ell_1=1}^{\infty}(-1)^{\ell_1}K^1_{\ell,\ell_1}(\xi_1)A_{\ell_1,-1}(2d)^{-\ell-\ell_1-1}$$
$$= (2\pi/3)^{1/2}\alpha_{11}(E_{0,x} + i\ E_{0,y})\delta_{\ell 1} \quad \text{for } \ell = 1, 2, 3, ... \tag{6.125}$$

By comparing this equation with eq.(6.123), it follows that

$$A_{\ell,-1}(E_{0,x} - i\ E_{0,y}) = -A_{\ell 1}(E_{0,x} + i\ E_{0,y}) \tag{6.126}$$

which means that eq.(6.124), or eq.(6.125), is superfluous, and it is sufficient to consider only eqs.(6.122) and (6.123).

If one finally introduces spherical coordinates for the external electric field $\mathbf{E}_0$

$$E_{0,x} = E_0 \sin\theta_0 \cos\phi_0, \quad E_{0,y} = E_0 \sin\theta_0 \sin\phi_0, \quad E_{0,z} = E_0 \cos\theta_0 \tag{6.127}$$

one can write eqs.(6.122) and (6.123) in the following form:

$$A_{\ell 0} + \alpha_{\ell 0}\left(\frac{\epsilon_a - \epsilon_s}{\epsilon_a + \epsilon_s}\right)\sum_{\ell_1=1}^{\infty}(-1)^{\ell_1}K^0_{\ell,\ell_1}(\xi_1)A_{\ell_1,0}(2d)^{-\ell-\ell_1-1}$$
$$= 2(\pi/3)^{1/2}\alpha_{10}E_0 \cos\theta_0 \delta_{\ell 1} \quad \text{for } \ell = 1, 2, 3, ... \tag{6.128}$$

and

$$\begin{aligned} A_{\ell 1} - \alpha_{\ell 1} \left(\frac{\epsilon_a - \epsilon_s}{\epsilon_a + \epsilon_s} \right) \sum_{\ell_1=1}^{\infty} (-1)^{\ell_1} K^1_{\ell,\ell_1}(\xi_1) A_{\ell_1,1} (2d)^{-\ell-\ell_1-1} \\ = -(2\pi/3)^{1/2} \alpha_{11} E_0 \sin\theta_0 e^{-i\phi_0} \delta_{\ell 1} \text{ for } \ell = 1,2,3,.... \end{aligned} \quad (6.129)$$

For spherical particles, one has to take the limit $\xi_1 \to \infty$. With eqs.(6.109) and (6.26) one then finds back the results eqs.(5.107) and (5.108).

Eqs.(6.128) and (6.129) may be solved numerically by neglecting the amplitudes larger than a suitable chosen order. One then solves the set of equations

$$\begin{aligned} A_{\ell 0} + \alpha_{\ell 0} \left(\frac{\epsilon_a - \epsilon_s}{\epsilon_a + \epsilon_s} \right) \sum_{\ell_1=1}^{M} (-1)^{\ell_1} K^0_{\ell,\ell_1}(\xi_1) A_{\ell_1,0} (2d)^{-\ell-\ell_1-1} \\ = 2(\pi/3)^{1/2} \alpha_{10} E_0 \cos\theta_0 \delta_{\ell 1} \quad \text{for } \ell = 1,2,...,M \end{aligned} \quad (6.130)$$

and

$$\begin{aligned} A_{\ell 1} - \alpha_{\ell 1} \left(\frac{\epsilon_a - \epsilon_s}{\epsilon_a + \epsilon_s} \right) \sum_{\ell_1=1}^{M} (-1)^{\ell_1} K^1_{\ell,\ell_1}(\xi_1) A_{\ell_1,1} (2d)^{-\ell-\ell_1-1} \\ = -(2\pi/3)^{1/2} \alpha_{11} E_0 \sin\theta_0 e^{-i\phi_0} \delta_{\ell 1} \quad \text{for } \ell = 1,2,...,M \end{aligned} \quad (6.131)$$

where M is the number of spheroidal multipoles, taken into account.

The polarizabilities $\alpha_{\|}(0)$, $\alpha_z(0)$, $\alpha_{\|}^{10}(0)$ and $\alpha_z^{10}(0)$ of the oblate spheroids, modified by the proximity of the substrate (see fig.6.5), can now be calculated again with the help of eq.(5.111), just as in the case of spheres. Notice that it is possible to apply this equation, which has been derived in the previous chapter, using a spherical multipole expansion of the field, to the present case, where one uses an oblate spheroidal multipole expansion, since one considers only dipole and quadrupole polarizabilities. As has been seen in section 6.2, these quantities are the same for the two expansions, as long as one neglects octupole polarizabilities. The constitutive coefficients $\beta_e(d)$, $\gamma_e(d)$, $\delta_e(d)$ and $\tau(d)$ then follow with eq.(5.112), and the invariants with eq.(5.66).

If one solves the above equations for the case $M = 2$, one obtains, using the expressions, eq.(6.27), for $\alpha_{\ell 0}$ and $\alpha_{\ell 1}$ with $\ell = 1$, 2 and the expressions, eqs.(6.243) – (6.245) and eq.(6.118), for $K^0_{\ell,\ell_1}(\xi_1)$ and $K^1_{\ell,\ell_1}(\xi_1)$ with ℓ, $\ell_1 = 1$, 2, together with eq.(5.111), the following results for the polarizabilities $\alpha_z(0)$, $\alpha_z^{10}(0)$, $\alpha_{\|}(0)$ and $\alpha_{\|}^{10}(0)$ of an oblate spheroid:

$$\begin{aligned} \alpha_z(0) &= \frac{\epsilon_a(\epsilon - \epsilon_a) V [\epsilon_a + L_z^1(\epsilon - \epsilon_a)]}{[\epsilon_a + L_z(\epsilon - \epsilon_a)][\epsilon_a + L_z^1(\epsilon - \epsilon_a)] + \Lambda_z^1(\epsilon - \epsilon_a)^2} \\ \alpha_z^{10}(0) &= \frac{\epsilon_a(\epsilon - \epsilon_a)^2 \Lambda_z^{10} V \ a\xi_0}{[\epsilon_a + L_z(\epsilon - \epsilon_a)][\epsilon_a + L_z^1(\epsilon - \epsilon_a)] + \Lambda_z^1(\epsilon - \epsilon_a)^2} \end{aligned} \quad (6.132)$$

and

$$
\begin{aligned}
\alpha_{\|}(0) &= \frac{\epsilon_a(\epsilon-\epsilon_a)V[\epsilon_a+L_{\|}^1(\epsilon-\epsilon_a)]}{[\epsilon_a+L_{\|}(\epsilon-\epsilon_a)][\epsilon_a+L_{\|}^1(\epsilon-\epsilon_a)]+\Lambda_{\|}^1(\epsilon-\epsilon_a)^2} \\
\alpha_{\|}^{10}(0) &= \frac{\epsilon_a(\epsilon-\epsilon_a)^2\Lambda_{\|}^{10}V\ a\xi_0}{[\epsilon_a+L_{\|}(\epsilon-\epsilon_a)][\epsilon_a+L_{\|}^1(\epsilon-\epsilon_a)]+\Lambda_{\|}^1(\epsilon-\epsilon_a)^2}
\end{aligned} \tag{6.133}
$$

The above expressions for these polarizabilities are of the same form as the expressions, eqs.(6.38) and (6.40), derived for oblate spheroids in the polarizable quadrupole model, (and, of course, also of the same form as the expressions, eqs.(5.86) and (5.87), derived for spheres in the same model).

The depolarization factors for the normal polarizabilities are now found to be:

$$
\begin{aligned}
L_z &\equiv (1+\xi_0^2)\left\{1-\xi_0\arctan\left(\frac{1}{\xi_0}\right)-\right. \\
&\left.\left(\frac{\epsilon_a-\epsilon_s}{\epsilon_a+\epsilon_s}\right)\frac{\xi_0}{\xi_1}[\left(\frac{3}{2}+\xi_1^2\right)\xi_1^2\ln\left(1+\frac{1}{\xi_1^2}\right)-\xi_1\arctan\left(\frac{1}{\xi_1}\right)-\xi_1^2]\right\}
\end{aligned} \tag{6.134}
$$

$$
\begin{aligned}
L_z^1 &\equiv (1+\xi_0^2)(1+3\xi_0^2)\left\{\frac{3}{2}\xi_0\arctan\left(\frac{1}{\xi_0}\right)-\frac{9}{2}\left(\frac{\xi_0^2}{1+3\xi_0^2}\right)\right. \\
&+\left(\frac{\epsilon_a-\epsilon_s}{\epsilon_a+\epsilon_s}\right)\frac{\xi_0}{\xi_1}\left[\frac{3}{2}\xi_1\arctan\left(\frac{1}{\xi_1}\right)-\frac{3}{4}(6\xi_1^4+10\xi_1^2+5)\xi_1^2\ln\left(1+\frac{1}{\xi_1^2}\right)\right. \\
&\left.\left.+\frac{3}{4}(6\xi_1^2+7)\xi_1^2\right]\right\}
\end{aligned} \tag{6.135}
$$

$$
\begin{aligned}
\Lambda_z^1 &\equiv -\frac{45}{2}\left(\frac{\epsilon_a-\epsilon_s}{\epsilon_a+\epsilon_s}\right)^2(1+\xi_0^2)^2(1+3\xi_0^2)\xi_0^2 \\
&[\xi_1\arctan\left(\frac{1}{\xi_1}\right)-\frac{1}{2}(\xi_1^2+2)\xi_1^2\ln\left(1+\frac{1}{\xi_1^2}\right)-\frac{1}{4}(1-2\xi_1^2)]^2
\end{aligned} \tag{6.136}
$$

$$
\begin{aligned}
\Lambda_z^{10} &\equiv \frac{3}{2}\left(\frac{\epsilon_a-\epsilon_s}{\epsilon_a+\epsilon_s}\right)(1+\xi_0^2)(1+3\xi_0^2)\left[\xi_1\arctan\left(\frac{1}{\xi_1}\right)-\frac{1}{2}(\xi_1^2+2)\xi_1^2\ln\left(1+\frac{1}{\xi_1^2}\right)\right. \\
&\left.-\frac{1}{4}(1-2\xi_1^2)\right]
\end{aligned} \tag{6.137}
$$

Similarly the depolarization factors for the parallel polarizabilities are given by

$$
\begin{aligned}
L_{\|} &\equiv \frac{1}{2}(1+\xi_0^2)\left\{\xi_0\arctan\left(\frac{1}{\xi_0}\right)-\frac{\xi_0^2}{1+\xi_0^2}+\left(\frac{\epsilon_a-\epsilon_s}{\epsilon_a+\epsilon_s}\right)\frac{\xi_0}{\xi_1}\left[\xi_1\arctan\left(\frac{1}{\xi_1}\right)\right.\right. \\
&\left.\left.-(\xi_1^2+\frac{3}{2})\xi_1^2\ln\left(1+\frac{1}{\xi_1^2}\right)+\xi_1^2\right]\right\}
\end{aligned} \tag{6.138}
$$

$$
\begin{aligned}
L_{\|}^1 &\equiv \frac{1}{2}(1+\xi_0^2)(1+2\xi_0^2)\left\{\frac{2+3\xi_0^2}{1+\xi_0^2}-3\ \xi_0\arctan\left(\frac{1}{\xi_0}\right)\right. \\
&+\left(\frac{\epsilon_a-\epsilon_s}{\epsilon_a+\epsilon_s}\right)\frac{\xi_0}{\xi_1}\left[3\xi_1\arctan\left(\frac{1}{\xi_1}\right)-\frac{3}{2}(6\xi_1^4+10\xi_1^2+5)\xi_1^2\ln\left(1+\frac{1}{\xi_1^2}\right)\right. \\
&\left.\left.+\frac{3}{2}(6\xi_1^2+7)\xi_1^2\right]\right\}
\end{aligned} \tag{6.139}
$$

$$\begin{aligned}\Lambda_{\parallel}^{1} &\equiv -\frac{45}{4}\left(\frac{\epsilon_a-\epsilon_s}{\epsilon_a+\epsilon_s}\right)^2(1+\xi_0^2)^2(1+2\xi_0^2)\xi_0^2 \\ &\quad [\xi_1 \arctan\left(\frac{1}{\xi_1}\right)-\frac{1}{2}(\xi_1^2+2)\xi_1^2\ln\left(1+\frac{1}{\xi_1^2}\right)-\frac{1}{4}(1-2\xi_1^2)]^2 \quad (6.140)\\ \Lambda_{\parallel}^{10} &\equiv \frac{3}{2}\left(\frac{\epsilon_a-\epsilon_s}{\epsilon_a+\epsilon_s}\right)(1+\xi_0^2)(1+2\xi_0^2) \\ &\quad [\xi_1 \arctan\left(\frac{1}{\xi_1}\right)-\frac{1}{2}(\xi_1^2+2)\xi_1^2\ln\left(1+\frac{1}{\xi_1^2}\right)-\frac{1}{4}(1-2\xi_1^2)] \quad (6.141)\end{aligned}$$

The volume V of the oblate spheroid is given by (cf. eq.(6.29))

$$V \equiv \frac{4}{3}\pi\, a^3\xi_0(1+\xi_0^2) \quad (6.142)$$

One may again verify that $L_z + 2L_{\parallel}$ differs from one due to the interaction with the substrate. It can furthermore be shown, using the expressions, eqs.(6.134) −(6.141), for the depolarization factors of oblate spheroids, in the limit $\xi_0 \to \infty$, $\xi_1 \to \infty$ and $a \to 0$, with $a\xi_0 = R$ and $a\xi_1 = d$, that for spheres with radius R at distance $d(d \geq R)$ above or on the substrate one obtains the expressions, eqs.(5.88) and (5.89). For spheres one therefore obtains the same results as in the polarizable quadrupole model.

If one now considers the limit $\xi_0 \to 0$ for an oblate spheroid, which gives the results for a circular disk of radius a, parallel to the surface of the substrate at the distance $d = a\xi_1$ above this substrate, one obtains from eqs.(6.134)−(6.141)

$$\begin{aligned}L_z &= 1, \quad L_z^1 = 0, \quad \Lambda_z^1 = 0,\\ \Lambda_z^{10} &= \frac{3}{2}\left(\frac{\epsilon_a-\epsilon_s}{\epsilon_a+\epsilon_s}\right)\times \\ &\quad [\xi_1 \arctan\left(\frac{1}{\xi_1}\right)-\frac{1}{2}(\xi_1^2+2)\xi_1^2\ln\left(1+\frac{1}{\xi_1^2}\right)-\frac{1}{4}(1-2\xi_1^2)] \quad (6.143)\end{aligned}$$

$$\begin{aligned}L_{\parallel} &= 0, \quad L_{\parallel}^1 = 1, \quad \Lambda_{\parallel}^1 = 0,\\ \Lambda_{\parallel}^{10} &= \frac{3}{2}\left(\frac{\epsilon_a-\epsilon_s}{\epsilon_a+\epsilon_s}\right)\times \\ &\quad [\xi_1 \arctan\left(\frac{1}{\xi_1}\right)-\frac{1}{2}(\xi_1^2+2)\xi_1^2\ln\left(1+\frac{1}{\xi_1^2}\right)-\frac{1}{4}(1-2\xi_1^2)] \quad (6.144)\end{aligned}$$

In contrast to the expressions, eqs.(6.42) and (6.43), obtained in the polarizable quadrupole model, the above depolarization factors all remain finite when $\xi_1 \to 0$, i.e. when the circular disk approaches the substrate. In this limit

$$\Lambda_z^{10} = \Lambda_{\parallel}^{10} = -\frac{3}{8}\left(\frac{\epsilon_a-\epsilon_s}{\epsilon_a+\epsilon_s}\right) \quad (6.145)$$

whereas

$$L_z = L_{\parallel}^1 = 1, \quad L_z^1 = L_{\parallel} = \Lambda_z^1 = \Lambda_{\parallel}^1 = 0 \quad (6.146)$$

Note that, if one considers oblate spheroids touching the substrate ($\xi_0 = \xi_1$) and takes the limit $\xi_0 \to 0$, one also obtains the above results from eqs.(6.134)–(6.141). The behavior of the depolarization factors, derived above for oblate spheroids, is therefore completely different from that found in section 6.2 with the polarizable quadrupole model: no divergencies are found anymore for the depolarization factors in the above described limit, as was the case in the polarizable dipole and quadrupole model. It is therefore to be expected and will be shown in section 6.6 in more detail, that the depolarization factors, derived above, provide a much better description of the optical behavior of oblate spheroidal particles than this is the case in the polarizable quadrupole model, in particular for flat particles.

Substituting the results eqs.(6.145) and (6.146) into eqs.(6.132) and (6.133), one finds for the polarizabilities of a circular disk on a substrate (cf. also ref. [42])

$$\begin{aligned}\alpha_z(0) &= \frac{\epsilon_a}{\epsilon}(\epsilon - \epsilon_a)V \\ \alpha_{\parallel}(0) &= (\epsilon - \epsilon_a)V \\ \alpha_z^{10}(0) &= \alpha_{\parallel}^{10}(0) = -\frac{3}{8}\frac{(\epsilon - \epsilon_a)^2}{\epsilon}\left(\frac{\epsilon_a - \epsilon_s}{\epsilon_a + \epsilon_s}\right) V a \xi_0 \end{aligned} \tag{6.147}$$

with $V \equiv \frac{4}{3}\pi\, a^3 \xi_0 (1+\xi_0^2)$ the "volume" of the disk which is, of course, zero for $\xi_0 \to 0$. It is therefore better to consider the quantities $\alpha_z(0)/\,V$, $\alpha_{\parallel}(0)/\,V$, $\alpha_z^{10}(0)/\,Va\xi_0$ and $\alpha_{\parallel}^{10}(0)/\,Va\xi_0$, which remain finite (and unequal to zero) in this limit.

Finally one can again write down expressions for the surface constitutive coefficients $\beta_e(d)$, $\gamma_e(d)$, $\delta_e(d)$ and $\tau(d)$ with the depolarization factors found above. It will be clear that, by using eqs.(5.64), (5.60), (6.132) and (6.133), one then obtains:

$$\beta_e(d) = \frac{\rho\, \epsilon_a^{-1}(\epsilon - \epsilon_a)V[\epsilon_a + L_z^1(\epsilon - \epsilon_a)]}{[\epsilon_a + L_z(\epsilon - \epsilon_a)][\epsilon_a + L_z^1(\epsilon - \epsilon_a)] + \Lambda_z^1(\epsilon - \epsilon_a)^2} \tag{6.148}$$

$$\gamma_e(d) = \frac{\rho\, \epsilon_a(\epsilon - \epsilon_a)V[\epsilon_a + L_{\parallel}^1(\epsilon - \epsilon_a)]}{[\epsilon_a + L_{\parallel}(\epsilon - \epsilon_a)][\epsilon_a + L_{\parallel}^1(\epsilon - \epsilon_a)] + \Lambda_{\parallel}^1(\epsilon - \epsilon_a)^2} \tag{6.149}$$

$$\begin{aligned}\delta_e(d) = \rho(\epsilon - \epsilon_a)Va\xi_1 \{ & \frac{\epsilon_a + (L_z^1 - \Lambda_z^{10}\xi_0/\xi_1)\,(\epsilon - \epsilon_a)}{[\epsilon_a + L_z(\epsilon - \epsilon_a)][\epsilon_a + L_z^1(\epsilon - \epsilon_a)] + \Lambda_z^1(\epsilon - \epsilon_a)^2} \\ & + \frac{\epsilon_a + \left(L_{\parallel}^1 - \Lambda_{\parallel}^{10}\xi_0/\xi_1\right)(\epsilon - \epsilon_a)}{[\epsilon_a + L_{\parallel}(\epsilon - \epsilon_a)][\epsilon_a + L_{\parallel}^1(\epsilon - \epsilon_a)] + \Lambda_{\parallel}^1(\epsilon - \epsilon_a)^2}\} \end{aligned} \tag{6.150}$$

$$\tau(d) = \frac{\rho\, \epsilon_a(\epsilon - \epsilon_a)V\, a\, \xi_1 \left[\epsilon_a + \left(L_{\parallel}^1 - \Lambda_{\parallel}^{10}\xi_0/\xi_1\right)(\epsilon - \epsilon_a)\right]}{[\epsilon_a + L_{\parallel}(\epsilon - \epsilon_a)][\epsilon_a + L_{\parallel}^1(\epsilon - \epsilon_a)] + \Lambda_{\parallel}^1(\epsilon - \epsilon_a)^2} \tag{6.151}$$

which are, of course, of the same form as the eqs.(6.44)–(6.47). Here it is assumed that all spheroids have the same size and shape. If the particles are touching the substrate, as will usually be the case, $\xi_0 = \xi_1$ in the above formulae. When there is size and shape dispersion of the particles (V and ξ_0 different), the right-hand sides of the above equations have to be averaged. The invariants can be found by substitution of the above formulae into eq.(5.66). If one changes the direction of the z-axis one finds again that γ_e and β_e are independent of the direction of the z-axis, while δ_e and τ change sign if the z-axis changes direction. If one uses the above formulae to calculate the reflection and transmission amplitudes, it is important to take the right sign. If, as in chapter 4, the light is incident through the ambient one may use the above expressions. If the light is incident through the substrate one must, in order to use the formulae given in chapter 4, interchange ϵ_a and ϵ_s in these formulae and change the sign of δ_e and τ in the expressions given above, before substituting them into the formulae in that chapter.

In the limit that the height of the island approaches zero, one obtains, using the above equations

$$\begin{aligned}
\beta_e(d) &= \frac{\epsilon-\epsilon_a}{\epsilon\epsilon_a}\rho V \\
\gamma_e(d) &= (\epsilon-\epsilon_a)\rho V \\
\delta_e(d) &= \frac{\epsilon-\epsilon_a}{\epsilon\epsilon_a}\left[\epsilon+\epsilon_a+\frac{3}{4}(\epsilon-\epsilon_a)\left(\frac{\epsilon_a-\epsilon_s}{\epsilon_a+\epsilon_s}\right)\right]\rho V a\xi_0 \\
\tau(d) &= (\epsilon-\epsilon_a)\left[1+\frac{3}{8}\frac{\epsilon-\epsilon_a}{\epsilon}\left(\frac{\epsilon_a-\epsilon_s}{\epsilon_a+\epsilon_s}\right)\right]\rho V a\xi_0 \qquad (6.152)
\end{aligned}$$

Due to the fact that the volume is linear in the height, $2a\xi_0$, the constitutive coefficients β_e and γ_e are linear, while δ_e and τ are quadratic, in the height.

6.5 Prolate spheroids; spheroidal multipole expansions

The solution of the problem for a prolate spheroid, see fig.6.6, runs along the same lines as the one given in the previous section for an oblate spheroid. Consider the response in a constant electric field $\mathbf{E}_0 = (E_{0,x}, E_{0,y}, E_{0,z})$. The corresponding incident potential is now written as

$$\psi_0(\mathbf{r}) = -\mathbf{E}_0.\mathbf{r} = (2\pi/3)^{1/2}[-2^{1/2}E_{0,z}\widetilde{X}_1^0(\xi,a)Y_1^0(\arccos\eta,\phi)$$

$$+(E_{0,x}-i\ E_{0,y})\widetilde{X}_1^1(\xi,a)Y_1^1(\arccos\eta,\phi)-(E_{0,x}+i\ E_{0,y})\widetilde{X}_1^{-1}(\xi,a)Y_1^{-1}(\arccos\eta,\phi)] \qquad (6.153)$$

where the definitions, eqs.(6.5), (6.6), (6.8), (6.58), (6.53) and (6.55), of the functions Y_ℓ^m and $\widetilde{X}_\ell^m$ have been used and where prolate spheroidal coordinates ξ, η and ϕ, defined by eqs.(6.48) and (6.49) (or the inverse transformation eq.(6.50)) have been introduced. The general solution in the ambient may be written, in complete analogy with the corresponding expression for an oblate spheroid, as a sum of the incident potential, the potential due to the charge distribution, induced in the island, and the image charge distribution in the substrate:

$$\psi_a(\mathbf{r}) = (2\pi/3)^{1/2}[-2^{1/2}E_{0,z}\widetilde{X}_1^0(\xi,a)Y_1^0(\arccos\eta,\phi)$$

$$+(E_{0,x}-i\ E_{0,y})\widetilde{X}_1^1(\xi,a)Y_1^1(\arccos\eta,\phi)-(E_{0,x}+i\ E_{0,y})\widetilde{X}_1^{-1}(\xi,a)Y_1^{-1}(\arccos\eta,\phi)]$$

$$+\sum_{\ell m}{}' A_{\ell m}\widetilde{Z}_\ell^m(\xi,a)Y_\ell^m(\arccos\eta,\phi)+\sum_{\ell m}{}' A_{\ell m}^r\widetilde{Z}_\ell^m(\xi'(\xi,\eta),a)Y_\ell^m(\arccos\eta'(\xi,\eta),\phi) \qquad (6.154)$$

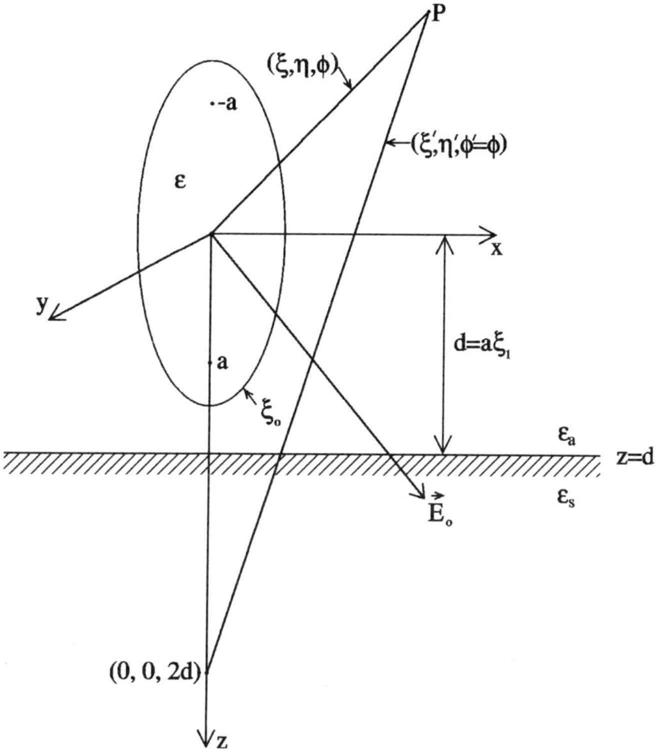

Figure 6.6 A prolate spheroid above a substrate.

Here ξ' and η' are the first two prolate spheroidal coordinates of the point (x, y, z) in the shifted coordinate frame with origin at (0,0,2d), see fig.6.6. The functions $\widetilde{Z}_\ell^m$ are defined by eqs.(6.59), (6.54) and (6.55). For the potential in the substrate one may write similarly

$$\begin{aligned}\psi_s(\mathbf{r}) &= \psi_s' + c_1\widetilde{X}_1^0(\xi,a)Y_1^0(\arccos\eta,\phi) + c_2\widetilde{X}_1^1(\xi,a)Y_1^1(\arccos\eta,\phi) + \\ &\quad c_3\widetilde{X}_1^{-1}(\xi,a)Y_1^{-1}(\arccos\eta,\phi) + \sum_{\ell m}{}' A_{\ell m}^t\widetilde{Z}_\ell^m(\xi,a)Y_\ell^m(\arccos\eta,\phi)\end{aligned} \tag{6.155}$$

It can be verified that eqs.(6.101)–(6.106), given in the previous section, remain valid also for prolate spheroids.

The potential of the so-called local field acting on the prolate spheroid is given as the sum of the potentials of the incident and the reflected fields

$$\psi_{loc}(\mathbf{r}) = (2\pi/3)^{1/2}[-2^{1/2}E_{0,z}\widetilde{X}_1^0(\xi,a)Y_1^0(\arccos\eta,\phi)$$

$$+(E_{0,x} - i\,E_{0,y})\widetilde{X}_1^1(\xi,a)Y_1^1(\arccos\eta,\phi) - (E_{0,x} + i\,E_{0,y})\widetilde{X}_1^{-1}(\xi,a)Y_1^{-1}(\arccos\eta,\phi)]$$
$$+\sum_{\ell m}{}' A_{\ell m}^r\widetilde{Z}_\ell^m(\xi'(\xi,\eta),a)Y_\ell^m(\arccos\eta'(\xi,\eta),\phi) \tag{6.156}$$

It is now possible to expand the potentials originating from (0,0,2d) in terms of a complete set of potentials around the origin O (see fig.6.6):

$$\begin{aligned}&\widetilde{Z}_\ell^m(\xi'(\xi,\eta),a)Y_\ell^m(\arccos\eta'(\xi,\eta),\phi) \\ &= \sum_{\ell_1=|m|}^{\infty}\widetilde{K}_{\ell_1,\ell}^m(\xi_1)(2d)^{-\ell-\ell_1-1}\widetilde{X}_{\ell_1}^m(\xi,a)Y_{\ell_1}^m(\arccos\eta,\phi) \quad \text{for} \quad 1 \le \xi < 2\xi_1 - 1\end{aligned} \tag{6.157}$$

Contrary to the oblate spheroid case the coordinate ξ should now be restricted to the interval $1 \le \xi < 2\xi_1 - 1$, where the functions used in the above expansion are finite. It may be checked, using eqs.(6.62), (6.63) and (6.65), that eq.(5.102) for the spherical case follows from eq.(6.157), with

$$\lim_{\xi_1\to\infty}\widetilde{K}_{\ell_1,\ell}^m(\xi_1) = M_{\ell_1,\ell}^m \tag{6.158}$$

Just as in the case of oblate spheroids it is not possible to give a straightforward generalization of the formula eq.(5.103) for $M_{\ell_1,\ell}^m$, which is valid for a prolate spheroid. A complete set of relations and recurrence relations, similar to the set of eqs.(6.110)–(6.114), may be derived for the coefficients $\widetilde{K}_{\ell_1,\ell}^m(\xi_1)$, following the rather laborious route sketched in appendix A for oblate spheroids. It is more convenient, however, to use the identity

$$\widetilde{K}_{\ell_1,\ell}^m(\xi_1) = K_{\ell_1,\ell}^m(-i\xi_1) \tag{6.159}$$

The validity of this identity follows using eqs.(6.108), (6.157) and (6.66). Using furthermore eq.(6.56) one finds for a prolate spheroid:

$$\widetilde{K}^0_{0,0}(\xi_1) = \xi_1 \ln\left(\frac{\xi_1+1}{\xi_1-1}\right) + \xi_1^2 \ln\left(1-\frac{1}{\xi_1^2}\right) \tag{6.160}$$

$$\begin{aligned}\widetilde{K}^0_{0,\ell}(\xi_1) &= \frac{(-1)^\ell(2\ell-1)!!}{2\,(\ell!)}(2\ell+1)^{1/2}(2\xi_1)^{\ell+1}\{Q^0_{\ell+1}(2\xi_1+1) - Q^0_{\ell+1}(2\xi_1-1) \\ &\quad -Q^0_{\ell-1}(2\xi_1+1) + Q^0_{\ell-1}(2\xi_1-1)\} \quad \text{for } \ell > 0\end{aligned} \tag{6.161}$$

$$\begin{aligned}\widetilde{K}^0_{1,\ell}(\xi_1) &= -\frac{(-1)^\ell(2\ell-1)!!}{2\,(\ell!)}[3(2\ell+1)]^{1/2}(2\xi_1)^{\ell+2}\{Q^0_{\ell+1}(2\xi_1+1) \\ &\quad +Q^0_{\ell+1}(2\xi_1-1) - Q^0_{\ell-1}(2\xi_1+1) - Q^0_{\ell-1}(2\xi_1-1)\} \\ &\quad -\left(\frac{\ell+1}{2\ell+3}\right)\left(\frac{3}{(2\ell+1)(2\ell+3)}\right)^{1/2}\widetilde{K}^0_{0,\ell+1}(\xi_1) \\ &\quad +\frac{4}{\ell}\left(\frac{3(2\ell+1)}{2\ell-1}\right)^{1/2}\xi_1^2\widetilde{K}^0_{0,\ell-1}(\xi_1) \quad \text{for } \ell > 0\end{aligned} \tag{6.162}$$

$$\begin{aligned}0 &= \frac{\ell+1}{(2\ell+1)(2\ell+3)}\left(\frac{2\ell_1+1}{2\ell+3}\right)^{1/2}\widetilde{K}^0_{\ell_1,\ell+1}(\xi_1) - \frac{4}{\ell}\xi_1^2\left(\frac{2\ell_1+1}{2\ell-1}\right)^{1/2}\widetilde{K}^0_{\ell_1,\ell-1}(\xi_1) \\ &\quad +\frac{\ell_1+1}{(2\ell_1+1)(2\ell_1+3)}\left(\frac{2\ell_1+3}{2\ell+1}\right)^{1/2}\widetilde{K}^0_{\ell_1+1,\ell}(\xi_1) \\ &\quad -\frac{4}{\ell_1}\xi_1^2\left(\frac{2\ell_1-1}{2\ell+1}\right)^{1/2}\widetilde{K}^0_{\ell_1-1,\ell}(\xi_1) \quad \text{for } \ell > 0, \ell_1 > 0\end{aligned} \tag{6.163}$$

The above is sufficient to determine all the coefficients $\widetilde{K}^0_{\ell_1,\ell}(\xi_1)$. The following identity for the coefficients is given for completeness

$$\begin{aligned}4\,\xi_1^2\delta_{\ell_1,0}\delta_{\ell 0} &= \frac{(\ell+1)^2}{(2\ell+1)(2\ell+3)}\left(\frac{2\ell_1+1}{2\ell+3}\right)^{1/2}\widetilde{K}^0_{\ell_1,\ell+1}(\xi_1) \\ &\quad -\frac{(\ell_1+1)^2}{(2\ell_1+1)(2\ell_1+3)}\left(\frac{2\ell_1+3}{2\ell+1}\right)^{1/2}\widetilde{K}^0_{\ell_1+1,\ell}(\xi_1) \\ &\quad +4\xi_1^2\left(\frac{2\ell_1+1}{2\ell+1}\right)^{1/2}\widetilde{K}^0_{\ell_1,\ell}(\xi_1) - 4\,\xi_1^2\{\left(\frac{2\ell_1-1}{2\ell+1}\right)^{1/2}\widetilde{K}^0_{\ell_1-1,\ell}(\xi_1) \\ &\quad -\left(\frac{2\ell_1+1}{2\ell-1}\right)^{1/2}\widetilde{K}^0_{\ell_1,\ell-1}(\xi_1)\} \quad \text{for } \ell \geq 0,\ \ell_1 \geq 0\end{aligned} \tag{6.164}$$

In this equation $\widetilde{K}_{-1,\ell}(\xi_1)$ and $\widetilde{K}_{\ell_1,-1}(\xi_1)$ are assumed to vanish. For the evaluation of the coefficients $\widetilde{K}^0_{0,\ell}(\xi_1)$ and $\widetilde{K}^0_{1,\ell}(\xi_1)$ by means of eqs.(6.161) and (6.162) one needs the functions $Q^0_\ell(z)$, which are defined in eq.(6.115). It is furthermore found, using eq.(6.159), that the symmetry relations (6.116) and (6.117) also hold:

$$\widetilde{K}^0_{\ell_1,\ell}(\xi_1) = (-1)^{\ell+\ell_1}\left(\frac{2\ell+1}{2\ell_1+1}\right)\widetilde{K}^0_{\ell,\ell_1}(\xi_1) \tag{6.165}$$

and

$$\widetilde{K}^{m}_{\ell_1,\ell}(\xi_1) = \widetilde{K}^{-m}_{\ell_1,\ell}(\xi_1) \qquad (6.166)$$

It is therefore only necessary to give these coefficients for positive values of m. For $m = 1$ one now has:

$$\widetilde{K}^{1}_{\ell_1,\ell}(\xi_1) = -\left[\frac{\ell_1 \ell}{(\ell_1+1)(\ell+1)}\right]^{1/2} \widetilde{K}^{0}_{\ell_1,\ell}(\xi_1) \text{ for } \ell_1, \ell = 1, 2, ... \qquad (6.167)$$

It is therefore sufficient to concentrate on the evaluation of $\widetilde{K}^{0}_{\ell_1,\ell}(\xi_1)$.

Substituting eq.(6.157) into eq.(6.156) and interchanging ℓ with ℓ_1, one obtains

$$\psi_{loc}(\mathbf{r}) = (2\pi/3)^{1/2}[-2^{1/2}E_{0,z}\widetilde{X}^0_1(\xi, a)Y^0_1(\arccos\eta, \phi)$$

$$+(E_{0,x} - i\ E_{0,y})\widetilde{X}^1_1(\xi, a)Y^1_1(\arccos\eta, \phi) - (E_{0,x} + i\ E_{0,y})\widetilde{X}^{-1}_1(\xi, a)Y^{-1}_1(\arccos\eta, \phi)]$$
$$+\sum_{\ell m}\sum_{\ell_1=|m|}^{\infty\prime} A^r_{\ell_1,m}\widetilde{K}^m_{\ell,\ell_1}(\xi_1)(2d)^{-\ell-\ell_1-1}\widetilde{X}^m_\ell(\xi, a)Y^m_\ell(\arccos\eta, \phi) \qquad (6.168)$$

where the prime as index of the summation indicates that $\ell_1 = 0$ should be excluded. Using eqs.(6.67) and (6.105) one obtains

$$\begin{aligned} A_{\ell m} &= -\widetilde{\alpha}_{\ell m}\{(2\pi/3)^{1/2}\delta_{\ell 1}[-2^{1/2}E_{0,z}\delta_{m0} + (E_{0,x} - i\ E_{0,y})\delta_{m1} - (E_{0,x} + i\ E_{0,y})\delta_{m,-1}] \\ &+ \left(\frac{\epsilon_a - \epsilon_s}{\epsilon_a + \epsilon_s}\right)\sum_{\ell_1=|m|}^{\infty\prime}(-1)^{\ell_1+m}\widetilde{K}^m_{\ell,\ell_1}(\xi_1)A_{\ell_1,m}(2d)^{-\ell-\ell_1-1}\} \end{aligned} \qquad (6.169)$$

where $\widetilde{\alpha}_{\ell m}$ is given by eq.(6.70). It follows that eqs.(6.121) and (6.126) are also valid for a prolate spheroid.

If one finally introduces spherical coordinates for the external electric field $\mathbf{E}_0$

$$E_{0,x} = E_0 \sin\theta_0 \cos\phi_0, \quad E_{0,y} = E_0 \sin\theta_0 \sin\phi_0, \quad E_{0,z} = E_0 \cos\theta_0 \qquad (6.170)$$

one finds the following set of inhomogeneous linear equations for the amplitudes:

$$\begin{aligned} & A_{\ell 0} + \widetilde{\alpha}_{\ell 0}\left(\frac{\epsilon_a - \epsilon_s}{\epsilon_a + \epsilon_s}\right)\sum_{\ell_1=1}^{\infty}(-1)^{\ell_1}\widetilde{K}^0_{\ell,\ell_1}(\xi_1)A_{\ell_1,0}(2d)^{-\ell-\ell_1-1} \\ &= 2(\pi/3)^{1/2}\widetilde{\alpha}_{10}E_0\cos\theta_0\delta_{\ell 1} \quad \text{for} \quad \ell = 1, 2, 3, ... \end{aligned} \qquad (6.171)$$

and

$$\begin{aligned} & A_{\ell 1} - \widetilde{\alpha}_{\ell 1}\left(\frac{\epsilon_a - \epsilon_s}{\epsilon_a + \epsilon_s}\right)\sum_{\ell_1=1}^{\infty}(-1)^{\ell_1}\widetilde{K}^1_{\ell,\ell_1}(\xi_1)A_{\ell_1,1}(2d)^{-\ell-\ell_1-1} \\ &= -(2\pi/3)^{1/2}\widetilde{\alpha}_{11}E_0\sin\theta_0 e^{-i\phi_0}\delta_{\ell 1} \quad \text{for} \quad \ell = 1, 2, 3, ... \end{aligned} \qquad (6.172)$$

For spherical particles, one has to take the limit $\xi_1 \to \infty$. With eqs.(6.158) and (6.72) one then finds back the results eqs.(5.107) and (5.108).

Eqs.(6.171) and (6.172) may be solved numerically by neglecting the amplitudes larger than a suitable chosen order. One then solves the set of equations

$$\begin{aligned} & A_{\ell 0} + \widetilde{\alpha}_{\ell 0} \left(\frac{\epsilon_a - \epsilon_s}{\epsilon_a + \epsilon_s} \right) \sum_{\ell_1=1}^{M} (-1)^{\ell_1} \widetilde{K}^0_{\ell,\ell_1}(\xi_1) A_{\ell_1,0} (2d)^{-\ell-\ell_1-1} \\ = & \ 2(\pi/3)^{1/2} \widetilde{\alpha}_{10} E_0 \cos\theta_0 \delta_{\ell 1} \ \text{ for } \ \ell = 1, 2, ..., M \end{aligned} \tag{6.173}$$

and

$$\begin{aligned} & A_{\ell 1} - \widetilde{\alpha}_{\ell 1} \left(\frac{\epsilon_a - \epsilon_s}{\epsilon_a + \epsilon_s} \right) \sum_{\ell_1=1}^{M} (-1)^{\ell_1} \widetilde{K}^1_{\ell,\ell_1}(\xi_1) A_{\ell_1,1} (2d)^{-\ell-\ell_1-1} \\ = & \ -(2\pi/3)^{1/2} \widetilde{\alpha}_{11} E_0 \sin\theta_0 e^{-i\phi_0} \delta_{\ell 1} \ \text{ for } \ \ell = 1, 2, ..., M \end{aligned} \tag{6.174}$$

where M is the number of spheroidal multipoles, taken into account.

The polarizabilities $\alpha_{\parallel}(0)$, $\alpha_z(0)$, $\alpha_{\parallel}^{10}(0)$ and $\alpha_z^{10}(0)$ of the prolate spheroids, modified by the proximity of the substrate (see fig.6.6), can now be calculated again with the help of eq.(5.111), just as in the case of spheres. Notice that it is again possible to apply this equation, derived in the previous chapter using a spherical multipole expansion, in the present case, where one uses a prolate spheroidal multipole expansion, since one considers only dipole and quadrupole polarizabilities. As has been seen in section 6.2, these quantities are the same for the two expansions, as long as one neglects octupole polarizabilities. The constitutive coefficients $\beta_e(d)$, $\gamma_e(d)$, $\delta_e(d)$ and $\tau(d)$ then follow with eq.(5.112), and the invariants with eq.(5.66).

If one solves the above equations for the case $M = 2$, one obtains the following results for the polarizabilities $\alpha_z(0)$, $\alpha_z^{10}(0)$, $\alpha_{\parallel}(0)$ and $\alpha_{\parallel}^{10}(0)$ of a prolate spheroid:

$$\begin{aligned} \alpha_z(0) &= \frac{\epsilon_a(\epsilon - \epsilon_a)V[\epsilon_a + L_z^1(\epsilon - \epsilon_a)]}{[\epsilon_a + L_z(\epsilon - \epsilon_a)][\epsilon_a + L_z^1(\epsilon - \epsilon_a)] + \Lambda_z^1(\epsilon - \epsilon_a)^2} \\ \alpha_z^{10}(0) &= \frac{\epsilon_a(\epsilon - \epsilon_a)^2 \Lambda_z^{10} V \ a\xi_0}{[\epsilon_a + L_z(\epsilon - \epsilon_a)][\epsilon_a + L_z^1(\epsilon - \epsilon_a)] + \Lambda_z^1(\epsilon - \epsilon_a)^2} \end{aligned} \tag{6.175}$$

and

$$\begin{aligned} \alpha_{\parallel}(0) &= \frac{\epsilon_a(\epsilon - \epsilon_a)V[\epsilon_a + L_{\parallel}^1(\epsilon - \epsilon_a)]}{[\epsilon_a + L_{\parallel}(\epsilon - \epsilon_a)][\epsilon_a + L_{\parallel}^1(\epsilon - \epsilon_a)] + \Lambda_{\parallel}^1(\epsilon - \epsilon_a)^2} \\ \alpha_{\parallel}^{10}(0) &= \frac{\epsilon_a(\epsilon - \epsilon_a)^2 \Lambda_{\parallel}^{10} V \ a\xi_0}{[\epsilon_a + L_{\parallel}(\epsilon - \epsilon_a)][\epsilon_a + L_{\parallel}^1(\epsilon - \epsilon_a)] + \Lambda_{\parallel}^1(\epsilon - \epsilon_a)^2} \end{aligned} \tag{6.176}$$

In the derivation one uses the expressions, eq.(6.74), for $\widetilde{\alpha}_{\ell 0}$ and $\widetilde{\alpha}_{\ell 1}$ with $\ell = 1, 2$, the expressions for $\widetilde{K}^0_{\ell,\ell_1}(\xi_1)$ and $\widetilde{K}^1_{\ell,\ell_1}(\xi_1)$, which follow from eqs.(6.243) – (6.245), using eqs.(6.159), (6.56) and (6.165), with ℓ, $\ell_1 = 1, 2$, together with eq.(5.111). The above expressions are similar to those found for the polarizable dipole and quadrupole models. The difference is that the depolarization factors given below are now modified and improved.

The depolarization factors in the normal direction are:

$$
\begin{aligned}
L_z &\equiv (1-\xi_0^2)\left\{1-\frac{1}{2}\xi_0\ln\left(\frac{\xi_0+1}{\xi_0-1}\right)+\right. \\
&\left.\left(\frac{\epsilon_a-\epsilon_s}{\epsilon_a+\epsilon_s}\right)\frac{\xi_0}{\xi_1}[(\frac{3}{2}-\xi_1^2)\xi_1^2\ln\left(1-\frac{1}{\xi_1^2}\right)+\frac{1}{2}\xi_1\ln\left(\frac{\xi_1+1}{\xi_1-1}\right)-\xi_1^2]\right\} \qquad (6.177) \\
L_z^1 &\equiv (1-\xi_0^2)(1-3\xi_0^2)\left\{\frac{3}{4}\xi_0\ln\left(\frac{\xi_0+1}{\xi_0-1}\right)+\frac{9}{2}\left(\frac{\xi_0^2}{1-3\xi_0^2}\right)\right. \\
&+\left(\frac{\epsilon_a-\epsilon_s}{\epsilon_a+\epsilon_s}\right)\frac{\xi_0}{\xi_1}\left[\frac{3}{4}\xi_1\ln\left(\frac{\xi_1+1}{\xi_1-1}\right)+\frac{3}{4}(6\xi_1^4-10\xi_1^2+5)\xi_1^2\ln\left(1-\frac{1}{\xi_1^2}\right)\right. \\
&\left.\left.-\frac{3}{4}(7-6\xi_1^2)\xi_1^2\right]\right\} \qquad (6.178)
\end{aligned}
$$

$$
\begin{aligned}
\Lambda_z^1 &\equiv -\frac{45}{2}\left(\frac{\epsilon_a-\epsilon_s}{\epsilon_a+\epsilon_s}\right)^2(1-\xi_0^2)^2(3\xi_0^2-1)\xi_0^2\left[\frac{1}{2}\xi_1\ln\left(\frac{\xi_1+1}{\xi_1-1}\right)\right. \\
&\left.+\frac{1}{2}(2-\xi_1^2)\xi_1^2\ln\left(1-\frac{1}{\xi_1^2}\right)-\frac{1}{4}(1+2\xi_1^2)\right]^2 \qquad (6.179) \\
\Lambda_z^{10} &\equiv \frac{3}{2}\left(\frac{\epsilon_a-\epsilon_s}{\epsilon_a+\epsilon_s}\right)(\xi_0^2-1)(3\xi_0^2-1)\left[\frac{1}{2}\xi_1\ln\left(\frac{\xi_1+1}{\xi_1-1}\right)\right. \\
&\left.+\frac{1}{2}(2-\xi_1^2)\xi_1^2\ln\left(1-\frac{1}{\xi_1^2}\right)-\frac{1}{4}(1+2\xi_1^2)\right] \qquad (6.180)
\end{aligned}
$$

The depolarization factors in the parallel direction are

$$
\begin{aligned}
L_\parallel &\equiv \frac{1}{2}(1-\xi_0^2)\left\{\frac{1}{2}\xi_0\ln\left(\frac{\xi_0+1}{\xi_0-1}\right)+\frac{\xi_0^2}{1-\xi_0^2}\right. \\
&\left.+\left(\frac{\epsilon_a-\epsilon_s}{\epsilon_a+\epsilon_s}\right)\frac{\xi_0}{\xi_1}[(\frac{3}{2}-\xi_1^2)\xi_1^2\ln\left(1-\frac{1}{\xi_1^2}\right)+\frac{1}{2}\xi_1\ln\left(\frac{\xi_1+1}{\xi_1-1}\right)-\xi_1^2]\right\} \qquad (6.181) \\
L_\parallel^1 &\equiv \frac{1}{2}(1-\xi_0^2)(1-2\xi_0^2)\left\{\frac{2-3\xi_0^2}{1-\xi_0^2}-\frac{3}{2}\xi_0\ln\left(\frac{\xi_0+1}{\xi_0-1}\right)\right. \\
&+\left(\frac{\epsilon_a-\epsilon_s}{\epsilon_a+\epsilon_s}\right)\frac{\xi_0}{\xi_1}\left[\frac{3}{2}\xi_1\ln\left(\frac{\xi_1+1}{\xi_1-1}\right)+\frac{3}{2}(6\xi_1^4-10\xi_1^2+5)\xi_1^2\ln\left(1-\frac{1}{\xi_1^2}\right)\right. \\
&\left.\left.-\frac{3}{2}(7-6\xi_1^2)\xi_1^2\right]\right\} \qquad (6.182)
\end{aligned}
$$

$$\Lambda_{\parallel}^{1} \equiv -\frac{45}{4}\left(\frac{\epsilon_a-\epsilon_s}{\epsilon_a+\epsilon_s}\right)^2(1-\xi_0^2)^2(2\xi_0^2-1)\xi_0^2\left[\frac{1}{2}\xi_1\ln\left(\frac{\xi_1+1}{\xi_1-1}\right)\right.$$
$$\left.+\frac{1}{2}(2-\xi_1^2)\xi_1^2\ln\left(1-\frac{1}{\xi_1^2}\right)-\frac{1}{4}(1+2\xi_1^2)\right]^2 \qquad (6.183)$$

$$\Lambda_{\parallel}^{10} \equiv \frac{3}{2}\left(\frac{\epsilon_a-\epsilon_s}{\epsilon_a+\epsilon_s}\right)(\xi_0^2-1)(2\xi_0^2-1)\left[\frac{1}{2}\xi_1\ln\left(\frac{\xi_1+1}{\xi_1-1}\right)\right.$$
$$\left.+\frac{1}{2}(2-\xi_1^2)\xi_1^2\ln\left(1-\frac{1}{\xi_1^2}\right)-\frac{1}{4}(1+2\xi_1^2)\right] \qquad (6.184)$$

The volume of the prolate spheroid is (cf. eq.(6.76))

$$V \equiv \frac{4}{3}\pi\, a^3\xi_0(\xi_0^2-1) \qquad (6.185)$$

One may again verify that $L_z + 2L_{\parallel}$ differs from one due to the interaction with the substrate. It can be shown, using the above depolarization factors for prolate spheroids in the limit $\xi_0 \to \infty$, $\xi_1 \to \infty$ and $a \to 0$, with $a\xi_0 = R$ and $a\xi_1 = d$, that for spheres with radius R at distance $d(d \geq R)$ above or on the substrate one finds eqs.(5.88) and (5.89). For spheres one therefore obtains the same results as in the polarizable quadrupole model.

For prolate spheroids the above obtained depolarization factors and those found with the polarizable quadrupole model, i.e. eqs.(6.86) and (6.88), will in general not be very different. This can be checked by explicit calculation (cf. also ref. [42] and the next section). This is in sharp contrast to the case of oblate spheroids. In particular for "needles" normal to the substrate ($\xi_0 \downarrow 1$), above ($\xi_1 > 1$) or touching ($\xi_1 \downarrow 1$) the substrate, one obtains the same result(cf. eqs.(6.89) and (6.90)):

$$L_z = L_z^1 = \Lambda_z^1 = \Lambda_z^{10} = 0 \qquad (6.186)$$

and

$$L_{\parallel} = L_{\parallel}^1 = \frac{1}{2}, \quad \Lambda_{\parallel}^1 = \Lambda_{\parallel}^{10} = 0 \qquad (6.187)$$

For the polarizabilities normal and parallel to the surface of the substrate one therefore obtains the same results, eqs.(6.91) and (6.92), as with the quadrupole model.

Finally one can again write down expressions for the surface constitutive coefficients $\beta_e(d)$, $\gamma_e(d)$, $\delta_e(d)$ and $\tau(d)$ with the depolarization factors found above. It will be clear that, by using eqs.(5.64), (5.60), (6.175) and (6.176), one then obtains:

$$\beta_e(d) = \frac{\rho\, \epsilon_a^{-1}(\epsilon-\epsilon_a)V[\epsilon_a + L_z^1(\epsilon-\epsilon_a)]}{[\epsilon_a + L_z(\epsilon-\epsilon_a)][\epsilon_a + L_z^1(\epsilon-\epsilon_a)] + \Lambda_z^1(\epsilon-\epsilon_a)^2} \tag{6.188}$$

$$\gamma_e(d) = \frac{\rho\, \epsilon_a(\epsilon-\epsilon_a)V[\epsilon_a + L_\parallel^1(\epsilon-\epsilon_a)]}{[\epsilon_a + L_\parallel(\epsilon-\epsilon_a)][\epsilon_a + L_\parallel^1(\epsilon-\epsilon_a)] + \Lambda_\parallel^1(\epsilon-\epsilon_a)^2} \tag{6.189}$$

$$\delta_e(d) = \rho(\epsilon-\epsilon_a)Va\xi_1 \left\{ \frac{\epsilon_a + \left(L_z^1 - \Lambda_z^{10}\xi_0/\xi_1\right)(\epsilon-\epsilon_a)}{[\epsilon_a + L_z(\epsilon-\epsilon_a)][\epsilon_a + L_z^1(\epsilon-\epsilon_a)] + \Lambda_z^1(\epsilon-\epsilon_a)^2} + \frac{\epsilon_a + \left(L_\parallel^1 - \Lambda_\parallel^{10}\xi_0/\xi_1\right)(\epsilon-\epsilon_a)}{[\epsilon_a + L_\parallel(\epsilon-\epsilon_a)][\epsilon_a + L_\parallel^1(\epsilon-\epsilon_a)] + \Lambda_\parallel^1(\epsilon-\epsilon_a)^2} \right\} \tag{6.190}$$

$$\tau(d) = \frac{\rho\, \epsilon_a(\epsilon-\epsilon_a)V\; a\; \xi_1 \left[\epsilon_a + \left(L_\parallel^1 - \Lambda_\parallel^{10}\xi_0/\xi_1\right)(\epsilon-\epsilon_a)\right]}{[\epsilon_a + L_\parallel(\epsilon-\epsilon_a)][\epsilon_a + L_\parallel^1(\epsilon-\epsilon_a)] + \Lambda_\parallel^1(\epsilon-\epsilon_a)^2} \tag{6.191}$$

which are, of course, of the same form as the eqs.(6.93)–(6.96). Here it is assumed that all spheroids have the same size and shape. If the particles are touching the substrate, as will usually be the case, $\xi_0 = \xi_1$ in the above formulae. When there is size and shape dispersion of the particles (V and ξ_0 different), the right-hand sides of the above equations have to be averaged. The invariants can be found by substitution of the above formulae into eq.(5.66). If one changes the direction of the z-axis one finds again that γ_e and β_e are independent of the direction of the z-axis, while δ_e and τ change sign if the z-axis changes direction. If one uses the above formulae to calculate the reflection and transmission amplitudes, it is important to take the right sign. If, as in chapter 4, the light is incident through the ambient one may use the above expressions. If the light is incident through the substrate one must, in order to use the formulae given in chapter 4, interchange ϵ_a and ϵ_s in these formulae and change the sign of δ_e and τ in the expressions given above, before substituting them into the formulae in that chapter.

6.6 Application: Spheroidal gold islands on sapphire

In the present section the theory, developed above, will be applied to the special case of spheroidal gold islands on a flat sapphire substrate. The spheroids are touching the substrate and their axis of revolution is normal to the surface of this substrate. The axial ratio, AR, is defined for the spheroid as the ratio of its diameter along the substrate divided by its diameter along the symmetry axis, normal to the substrate. An axial ratio of 1 thus corresponds to a sphere, axial ratios larger than 1 correspond to oblate spheroids and axial ratios smaller than 1 to prolate spheroids. In section 5.7 the special case of the sphere, $AR = 1$, was discussed. In particular, the dependence of the polarizabilities, the constitutive coefficients and the invariants on the wavelength was calculated in the visible domain, where gold has a plasmon resonance. In this section the dependence of these quantities on the shape, AR, will be calculated. For the value of the dielectric constant of the particle the complex number $\epsilon = -5.28 + 2.25i$ is chosen, which is the dielectric constant of gold at the wavelength

$\lambda = 540$ nm, cf. table 5.1. This wavelength is chosen because these gold films have a minimum in the transmission roughly at this wavelength. The shape dependence is expected to be most pronounced in that case. The dielectric constant of sapphire is again chosen to be equal to ϵ_s =3.13. The ambient is again vacuum, $\epsilon_a = 1$.

The relation between the axial ratio and the parameter ξ_0 is, cf. figs.6.2 and 6.4,

$$AR = \sqrt{1 + \xi_0^{-2}} \text{ for an oblate spheroid, where } 0 \leq \xi_0 \leq \infty \tag{6.192}$$

and

$$AR = \sqrt{1 - \xi_0^{-2}} \text{ for a prolate spheroid, where } 1 \leq \xi_0 \leq \infty \tag{6.193}$$

In the plots it is convenient to use the alternative variable

$$s \equiv 1 \,/(1 + AR) \text{ with the range } 0 \leq s \leq 1 \tag{6.194}$$

This is the diameter in the direction normal to the surface, along the symmetry axis of the spheroid, divided by the sum of the diameters in the directions normal and parallel to the surface. For a disk this quantity is 0 ($AR = \infty$), for a needle it is 1 ($AR = 0$), whereas it is 0.5 for a sphere ($AR = 1$).

First the depolarization factors L_z, L_z^1, Λ_z^1, Λ_z^{10}, $L_\|$, $L_\|^1$, $\Lambda_\|^1$ and $\Lambda_\|^{10}$, given by eqs.(6.134)–(6.141) for an oblate spheroid and by eqs.(6.177)–(6.184) for a prolate spheroid, are calculated [42]. Since the spheroids are assumed to touch the substrate, ξ_0 is equal to ξ_1 in the formulae for the depolarization factors. The results are plotted as solid curves in figs.6.7–6.10 as a function of s. In order to see the influence of the substrate, also the depolarization factors for the spheroids in the ambient, infinitely far from the substrate, are plotted as dashed curves . These values were obtained from the same equations, substituting $\epsilon_s = \epsilon_a = 1$ and $\xi_0 = \xi_1$. Notice that the depolarization factors Λ_z^1, Λ_z^{10}, $\Lambda_\|^1$ and $\Lambda_\|^{10}$ vanish in this case. A remarkable feature is that the influence of the substrate on L_z, $L_\|$, L_z^1 and $L_\|^1$, though substantial, is at most in the order of 0.1. In limits where one of these depolarization factors approaches zero this is most important. A completely different behavior is found, if one uses eqs.(6.39), (6.41), (6.86) and (6.88), derived for the polarizable quadrupole model, to calculate the same depolarization factors. The results are plotted as dotted curves. As is clear from the figures this approximation is completely incorrect for the oblate spheroid, $0 \leq s < 0.5$. The physical reason of this is that the distance to the image spheroid is too small in this case. For the sphere, $s = 0.5$, the same values are obtained for these factors, since the spherical and spheroidal multipole expansions are identical in this special case. For a prolate spheroid, $0.5 < s \leq 1$, the results obtained using the polarizable quadrupole model, are an improvement compared to the values obtained with the spheroid far from the surface. For the oblate spheroids one should not use the spherical dipole or quadrupole model. The values for the spheroid far from the surface are then much better.

The further results, i.e. for the polarizabilities and the surface constitutive coefficients, will all be given, using the results from the spheroidal multipole expansion. To quadrupolar order one then obtains eqs.(6.132) and (6.133) or eqs.(6.175)

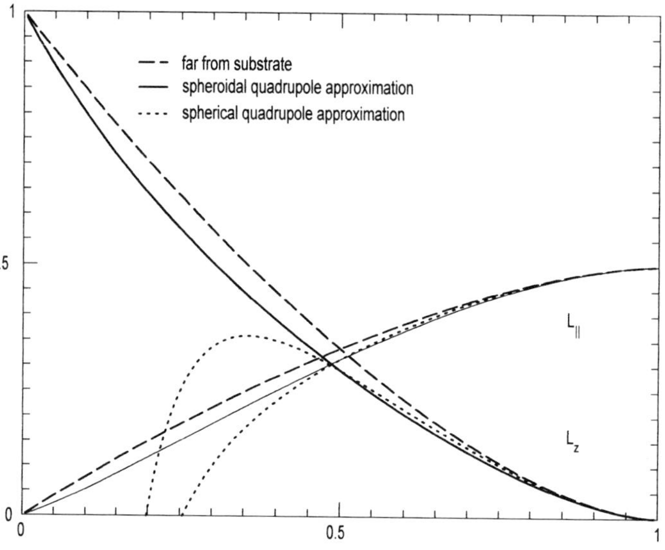

Figure 6.7 Depolarization factors L_z and $L_{\|}$ as function of $1/(1+AR)$.

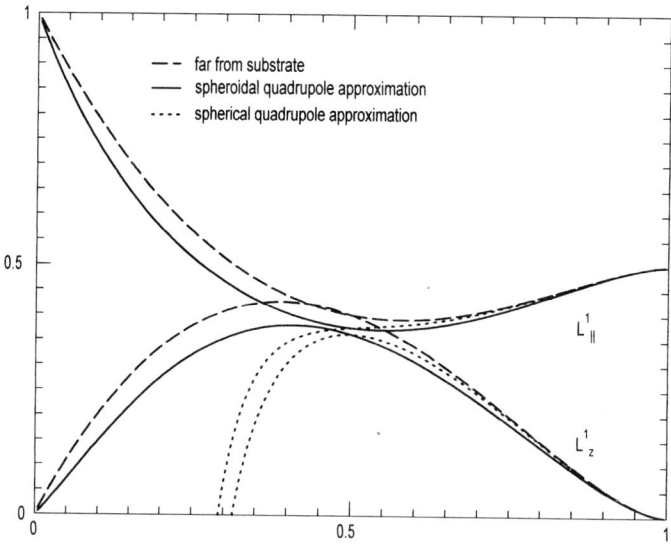

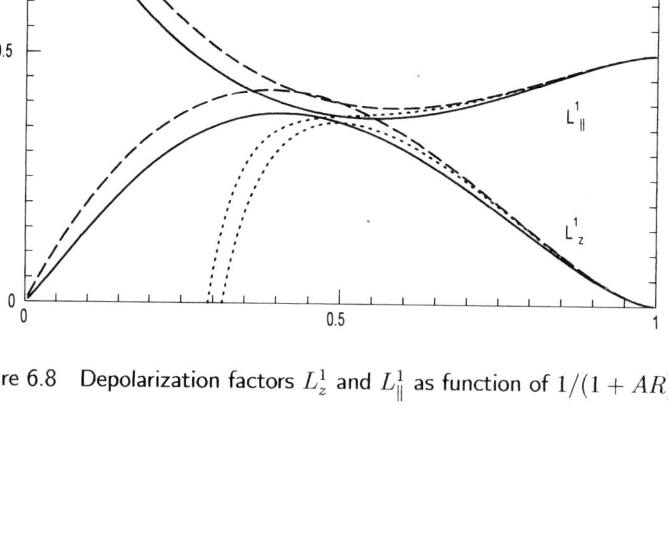

Figure 6.8 Depolarization factors L_z^1 and $L_{\|}^1$ as function of $1/(1+AR)$.

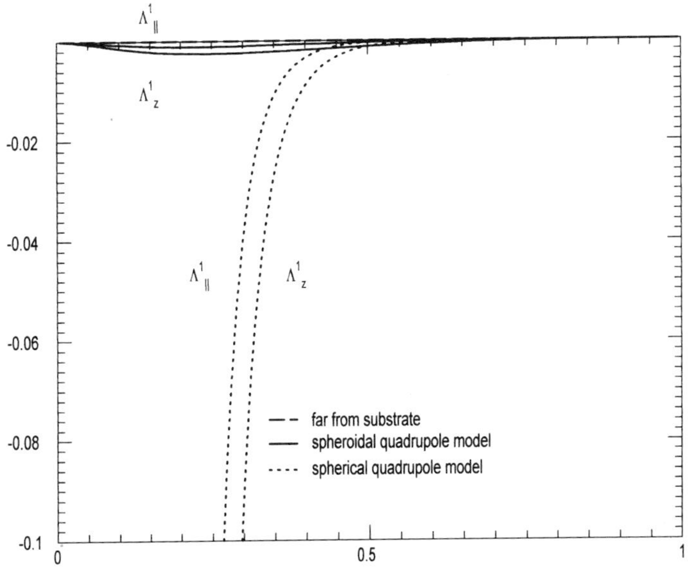

Figure 6.9 Depolarization factors Λ_z^1 and $\Lambda_{\parallel}^1$ as function of $1/(1+AR)$.

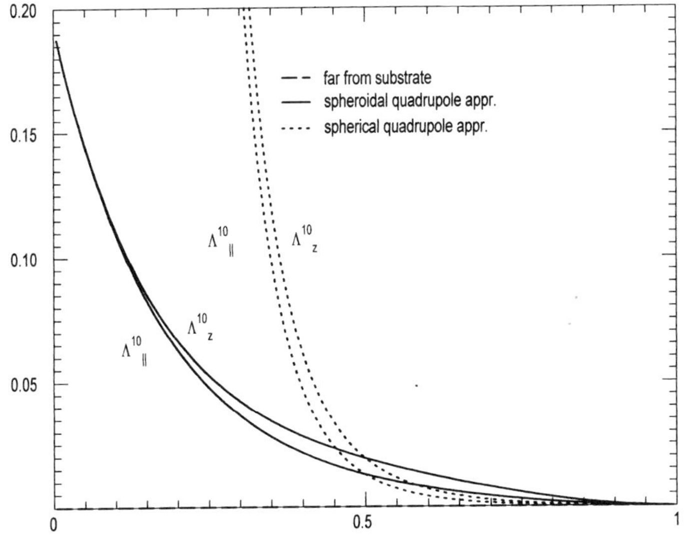

Figure 6.10 Depolarization factors Λ_z^{10} and $\Lambda_{\parallel}^{10}$ as function of $1/(1+AR)$.

and (6.176). Replacing ϵ_a by unity, the dipole and quadrupole polarizabilities are given by

$$\begin{aligned}\widehat{\alpha}_z &\equiv \frac{1}{V}\alpha_z(0) = \frac{(\epsilon-1)[1+L_z^1(\epsilon-1)]}{[1+L_z(\epsilon-1)][1+L_z^1(\epsilon-1)]+\Lambda_z^1(\epsilon-1)^2} \\ \widehat{\alpha}_z^{10} &\equiv \frac{1}{Va\xi_0}\alpha_z^{10}(0) = \frac{(\epsilon-1)^2\Lambda_z^{10}}{[1+L_z(\epsilon-1)][1+L_z^1(\epsilon-1)]+\Lambda_z^1(\epsilon-1)^2} \end{aligned} \tag{6.195}$$

and

$$\begin{aligned}\widehat{\alpha}_\| &\equiv \frac{1}{V}\alpha_\|(0) = \frac{(\epsilon-1)[1+L_\|^1(\epsilon-1)]}{[1+L_\|(\epsilon-1)][1+L_\|^1(\epsilon-1)]+\Lambda_\|^1(\epsilon-1)^2} \\ \widehat{\alpha}_\|^{10} &\equiv \frac{1}{Va\xi_0}\alpha_\|^{10}(0) = \frac{(\epsilon-1)^2\Lambda_\|^{10}}{[1+L_\|(\epsilon-1)][1+L_\|^1(\epsilon-1)]+\Lambda_\|^1(\epsilon-1)^2} \end{aligned} \tag{6.196}$$

where V is the volume of the spheroid. These expressions are similar to those given for the sphere in the previous chapter, eqs.(5.114) and (5.115). For the sphere $a\xi_0$ is equal to R, so that the expressions become identical. It is again not correct to neglect Λ_z^1 and $\Lambda_\|^1$. The size of ϵ in the neighborhood of the resonance is large and makes the corresponding terms of comparable size as other contributions in the denominator. The surface constitutive coefficients, eqs.(6.148)–(6.151) for oblate spheroids and eqs.(6.188)–(6.191) for prolate spheroids, similarly reduce to

$$\begin{aligned}\widehat{\gamma}_e &\equiv \frac{1}{\rho V}\gamma_e(d) = \widehat{\alpha}_\| \\ \widehat{\beta}_e &\equiv \frac{1}{\rho V}\beta_e(d) = \widehat{\alpha}_z \\ \widehat{\tau} &\equiv \frac{1}{\rho V d}\tau(d) = \widehat{\alpha}_\| - \widehat{\alpha}_\|^{10} \\ \widehat{\delta}_e &\equiv \frac{1}{\rho V d}\delta_e(d) = \widehat{\alpha}_z + \widehat{\alpha}_\| - \widehat{\alpha}_z^{10} - \widehat{\alpha}_\|^{10} \end{aligned} \tag{6.197}$$

for this case, where $d = a\xi_0$ and $\epsilon_a = 1$. The invariants, eq.(5.66), become, for ϵ_s real,

$$\begin{aligned}\widehat{I}_e \equiv I_e/\rho V = \widehat{\gamma}_e - \epsilon_s\widehat{\beta}_e \quad & \widehat{I}_c \equiv (\epsilon_s - 1)I_c/\rho V = \operatorname{Im}\widehat{\gamma}_e \\ \widehat{I}_\tau \equiv I_\tau/\rho V d = \widehat{\tau} \quad & \widehat{I}_{\delta,e} \equiv I_{\delta,e}/\rho V d = \widehat{\delta}_e \end{aligned} \tag{6.198}$$

In order to obtain the solutions of eqs.(6.130) and (6.131) for oblate spheroids and eqs.(6.173) and (6.174) for prolate spheroids, to higher than quadrupole order, it is convenient to introduce dimensionless amplitudes by

$$\begin{aligned}\widehat{A}_{\ell 0} &\equiv A_{\ell 0}/V(a\xi_0)^{\ell-1}2\sqrt{\pi/3}E_0\cos\theta_0 \\ \widehat{A}_{\ell 1} &\equiv -A_{\ell 1}/V(a\xi_0)^{\ell-1}\sqrt{2\pi/3}E_0\sin\theta_0\exp(-i\phi_0) \end{aligned} \tag{6.199}$$

Eqs.(6.130) and (6.131) for oblate spheroids, then reduce to, using $\epsilon_a = 1$,

$$\widehat{A}_{\ell 0}/\widehat{\alpha}_{\ell 0} + \frac{4\pi}{3}\left(\frac{1-\epsilon_s}{1+\epsilon_s}\right)\sum_{\ell_1=1}^{M}(-1)^{\ell_1}\widehat{A}_{\ell_1,0}K^0_{\ell,\ell_1}(\xi_1)(2\widehat{d})^{-\ell-\ell_1-1} = \delta_{\ell 1}$$
$$\text{for } \ell = 1,2,...,M \tag{6.200}$$

and

$$\widehat{A}_{\ell 1}/\widehat{\alpha}_{\ell 1} - \frac{4\pi}{3}\left(\frac{1-\epsilon_s}{1+\epsilon_s}\right)\sum_{\ell_1=1}^{M}(-1)^{\ell_1}\widehat{A}_{\ell_1,1}K^1_{\ell,\ell_1}(\xi_1)(2\widehat{d})^{-\ell-\ell_1-1} = \delta_{\ell 1}$$
$$\text{for } \ell = 1,2,...,M \qquad (6.201)$$

where

$$\widehat{\alpha}_{\ell m} \equiv \alpha_{\ell m}/V(a\xi_0)^{2(\ell-1)} \text{ and } \widehat{d} \equiv d/(a\xi_0) = \xi_1/\xi_0 \qquad (6.202)$$

Here V is the volume of the oblate spheroid, cf. eq.(6.29).

Similarly eqs.(6.173) and (6.174) for prolate spheroids reduce to, using $\epsilon_a = 1$,

$$\widehat{A}_{\ell 0}/\widehat{\widetilde{\alpha}}_{\ell 0} + \frac{4\pi}{3}\left(\frac{1-\epsilon_s}{1+\epsilon_s}\right)\sum_{\ell_1=1}^{M}(-1)^{\ell_1}\widehat{A}_{\ell_1,0}\widetilde{K}^0_{\ell,\ell_1}(\xi_1)(2\widehat{d})^{-\ell-\ell_1-1} = \delta_{\ell 1}$$
$$\text{for } \ell = 1,2,...,M \qquad (6.203)$$

and

$$\widehat{A}_{\ell 1}/\widehat{\widetilde{\alpha}}_{\ell 1} - \frac{4\pi}{3}\left(\frac{1-\epsilon_s}{1+\epsilon_s}\right)\sum_{\ell_1=1}^{M}(-1)^{\ell_1}\widehat{A}_{\ell_1,1}\widetilde{K}^1_{\ell,\ell_1}(\xi_1)(2\widehat{d})^{-\ell-\ell_1-1} = \delta_{\ell 1}$$
$$\text{for } \ell = 1,2,...,M \qquad (6.204)$$

where

$$\widehat{\widetilde{\alpha}}_{\ell m} \equiv \widetilde{\alpha}_{\ell m}/V(a\xi_0)^{2(\ell-1)} \text{ and } \widehat{d} \equiv d/(a\xi_0) = \xi_1/\xi_0 \qquad (6.205)$$

Here V is the volume of the prolate spheroid, cf. eq.(6.76). The dipole and quadrupole polarizabilities, modified by the presence of the substrate, become, for the spheroids, in terms of the amplitudes:

$$\begin{aligned} \widehat{\alpha}_{\parallel} &= 4\pi\widehat{A}_{11}\ , \quad \widehat{\alpha}^{10}_{\parallel} = \frac{4\pi}{3}\widehat{A}_{21}\sqrt{5} \\ \widehat{\alpha}_z &= 4\pi\widehat{A}_{10}\ , \quad \widehat{\alpha}^{10}_z = 2\pi\widehat{A}_{20}\sqrt{5/3} \end{aligned} \qquad (6.206)$$

using $\epsilon_a = 1$. These are the same equations (5.111) as found for the sphere.

The numerical calculations below are done with the spheroid touching the substrate, so that $\xi_1 = \xi_0$. It is found for these films that the value of the amplitudes has sufficiently converged for $M = 16$. The difference with the quadrupole approximation is only a few percent, cf. [42] table 1 and 2, so that one may in practice use the analytical expressions given above for that case. In the figures 6.11 through 6.17 the quadrupole approximation, $M = 2$, has been used.

In fig.6.11 the real and imaginary parts of $\widehat{\gamma}_e = \widehat{\alpha}_{\parallel}$ are plotted as a function of the shape variable $s = 1/(1+AR)$. The invariant $\widehat{I}_c$ is equal to the imaginary part of $\widehat{\gamma}_e$, plotted in fig.6.11. In fig.6.12 the same is done for $\widehat{\beta}_e = \widehat{\alpha}_z$. In fig.6.13 the real and imaginary part of the invariant $\widehat{I}_e$ are plotted. This invariant is a combination of the

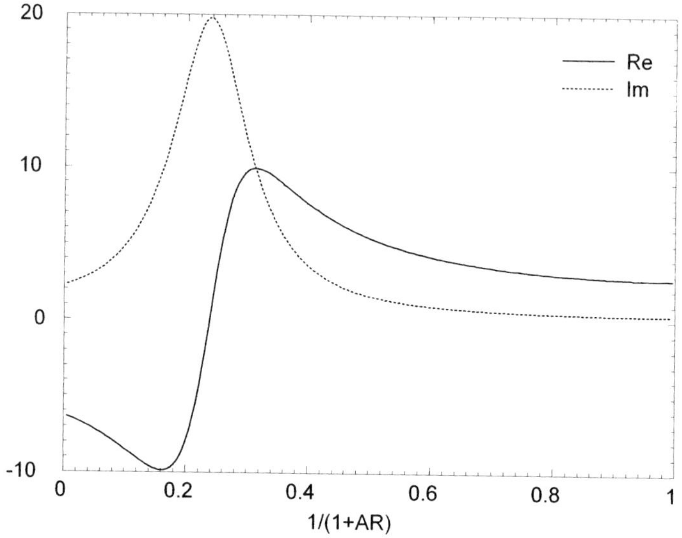

Figure 6.11 Real and imaginary parts of $\widehat{\alpha}_{\|} = \widehat{\gamma}_e$, as function of $1/(1+AR)$. Notice that Im$\widehat{\gamma}_e$=$\widehat{I}_c$.

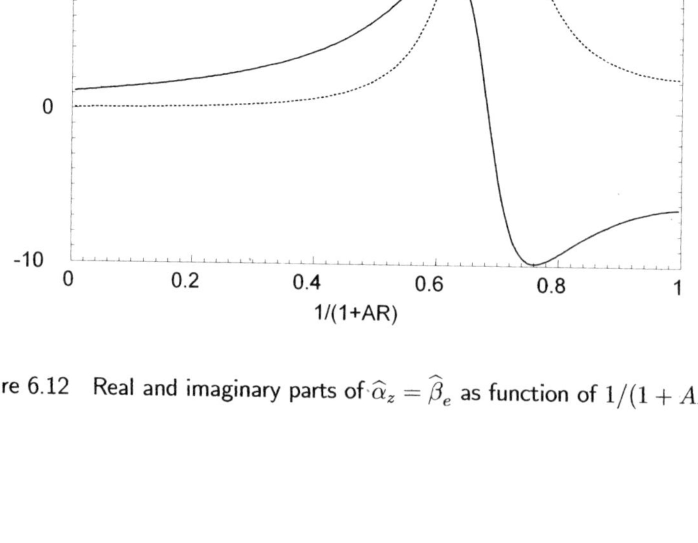

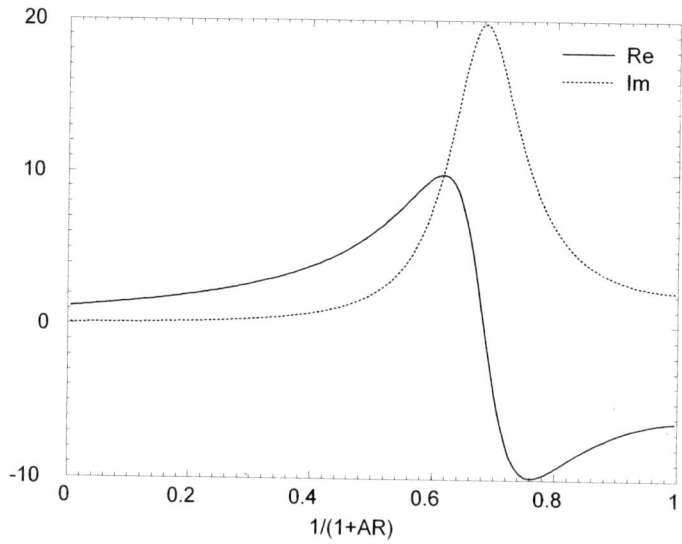

Figure 6.12 Real and imaginary parts of $\widehat{\alpha}_z = \widehat{\beta}_e$ as function of $1/(1+AR)$.

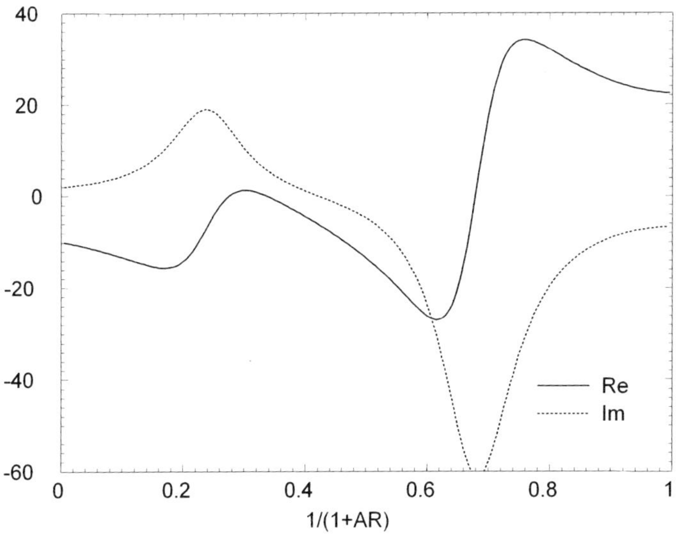

Figure 6.13 Real and imaginary parts of $\widehat{I_e}$ as function of $1/(1+AR)$.

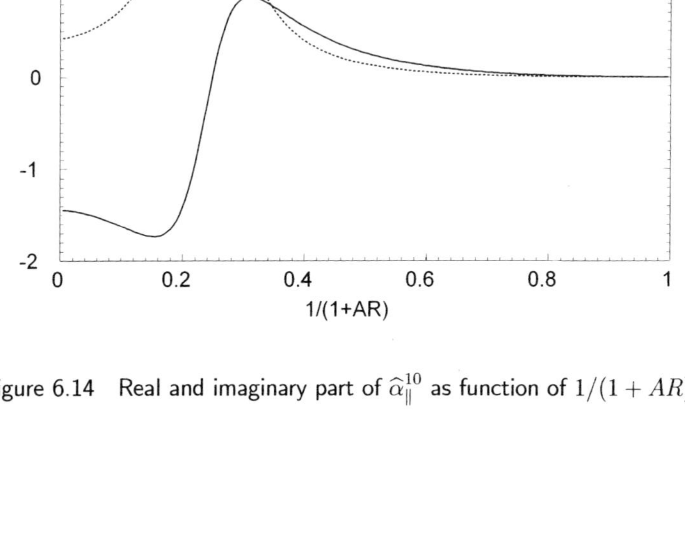

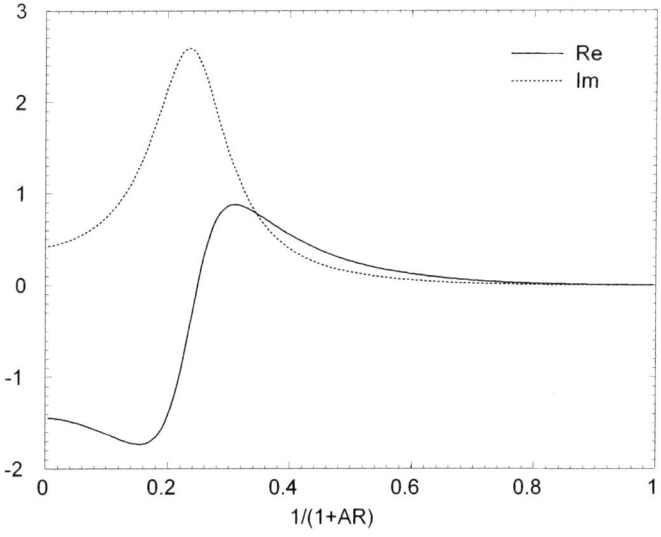

Figure 6.14 Real and imaginary part of $\widehat{\alpha}_{\parallel}^{10}$ as function of $1/(1+AR)$.

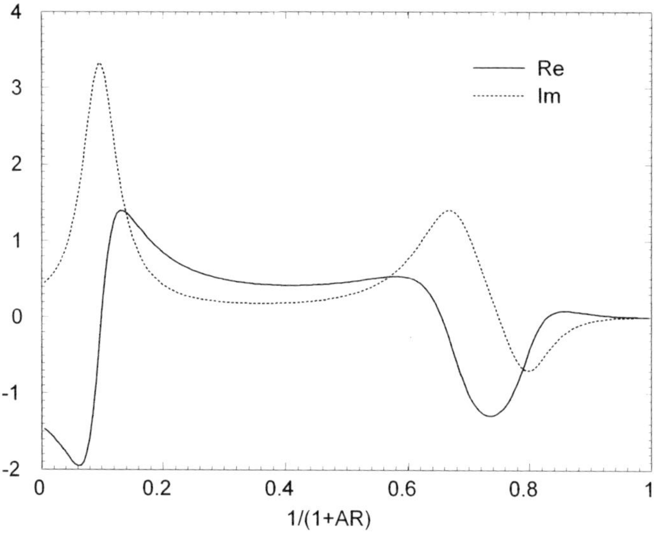

Figure 6.15 Real and imaginary parts of $\widehat{\alpha}_z^{10}$ as function of $1/(1+AR)$.

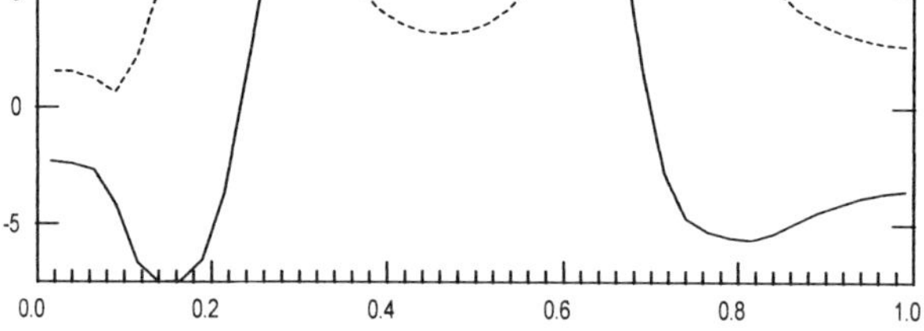

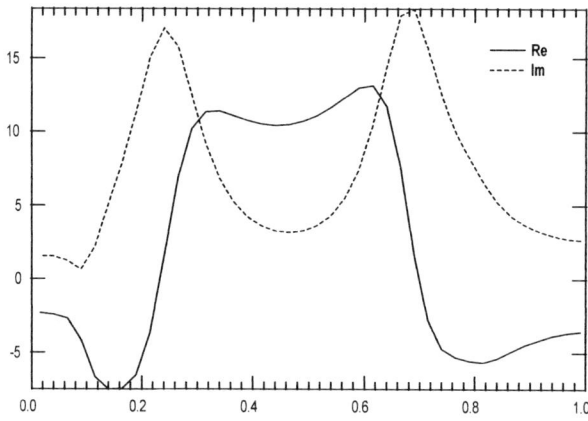

Figure 6.16 Real and imaginary parts of $\widehat{\delta}_e = \widehat{I}_{\delta,e}$ as function of $1/(1+AR)$

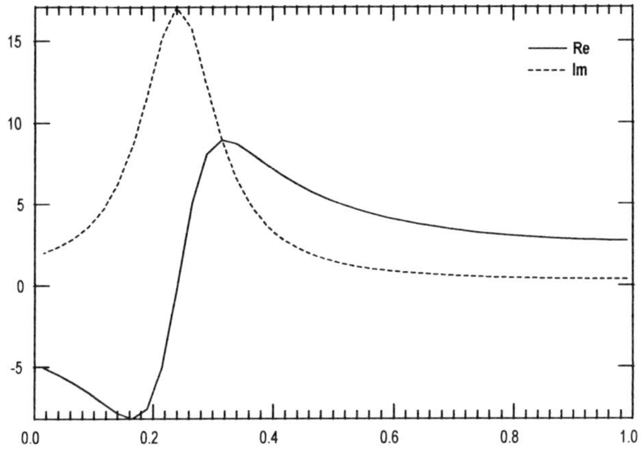

Figure 6.17 Real and imaginary parts of $\widehat{\tau} = \widehat{I}_\tau$ as function of $1/(1+AR)$

constitutive coefficients $\widehat{\beta}_e$ and $\widehat{\gamma}_e$. The real and imaginary parts of the quadrupole polarizabilities, $\widehat{\alpha}_{\|}^{10}$ and $\widehat{\alpha}_z^{10}$, are plotted in figs.6.14 and 6.15. The real and imaginary parts of the resulting constitutive coefficients and invariants, $\widehat{\delta}_e = \widehat{I}_{\delta,e}$ and $\widehat{\tau} = \widehat{I}_\tau$, are plotted in figs.6.16 and 6.17, all as a function of the shape variable s.

The location of the resonance is different for the coefficients characterizing the parallel and the normal directions. It depends strongly on the shape parameter $s = 1/(1+AR)$. The reason for this is the dependence of the depolarization factors, given in figs.6.7 through 6.10, on this parameter. In view of the relatively small change of the depolarization factors due to the interaction with the substrate, this location is found not to be substantially affected by the presence of the substrate. The figures above show the strong dependence of the various quantities on the shape parameter. The wavelength is kept constant at 540 nm. The maxima of the imaginary part now are found at the shape parameter for which the resonance wavelength is equal to 540 nm.

6.7 Appendix A

In this appendix recurrence relations for the coefficients $K^m_{\ell_1,\ell}(\xi_1)$, defined for oblate spheroids by eq.(6.108), and further useful formulae for these coefficients will be derived, cf. ref. [42] where a slightly different notation is used.

It may, first of all, be remarked that eq.(6.108) is in particular valid on the negative z-axis ($\eta = -1$) with $|z| < 2a\xi_1$ (see fig.6.18). It is easily seen that on that axis

$$\xi'(\xi, -1) = \xi + 2\xi_1, \quad \eta'(\xi, -1) = -1 \tag{6.207}$$

where $\xi'(\xi, \eta)$ and $\eta'(\xi, \eta)$ are the first two oblate spheroidal coordinates of a point in

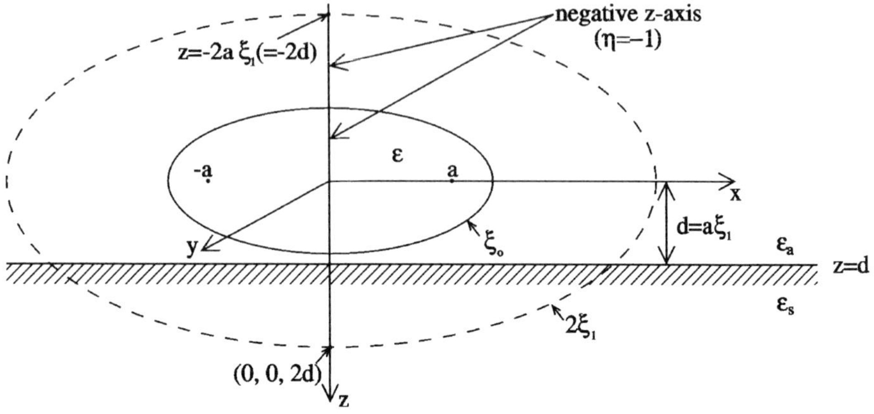

Figure 6.18 Region of validity of expansion eq. (6.108), within dashed oblate spheroid.

the shifted coordinate frame with origin at (0,0,2d).

It furthermore follows with eqs.(6.14) and (6.13) that

$$Z_\ell^m(\xi'(\xi,-1),a) \equiv i^{\ell+1}\frac{(2\ell+1)!!}{(\ell+m)!}a^{-\ell-1}Q_\ell^m(i\xi+2i\xi_1) \tag{6.208}$$

and

$$X_{\ell_1}^m(\xi,a) \equiv i^{m-\ell_1}\frac{(\ell_1-m)!}{(2\ell_1-1)!!}a^{\ell_1}P_{\ell_1}^m(i\xi) \tag{6.209}$$

Using eqs.(6.207) and (6.103) one obtains

$$\begin{aligned} Y_\ell^m(\arccos\eta'(\xi,-1),\phi) &= Y_\ell^m(\arccos(-1),\phi) = (-1)^{\ell+m}Y_\ell^m(\arccos 1,\phi) \\ &= (-1)^{\ell+m}Y_\ell^m(0,\phi) \end{aligned} \tag{6.210}$$

From eq.(6.5) one has

$$Y_\ell^m(0,\phi) = \left[\frac{2\ell+1}{4\pi}\frac{(\ell-m)!}{(\ell+m)!}\right]^{1/2}(-1)^m P_\ell^m(1)e^{im\phi} \tag{6.211}$$

Since $P_\ell^m(1)=\delta_{m0}$, one finally finds from the last two equations

$$Y_\ell^m(\arccos\eta'(\xi,-1),\phi) = Y_\ell^m(\arccos(-1),\phi) = \left[\frac{2\ell+1}{4\pi}\right]^{1/2}(-1)^\ell\delta_{m0} \tag{6.212}$$

Substituting eqs.(6.208), (6.209) and (6.212) into eq.(6.108) for $\eta=-1$, one obtains, using $d=a\xi_1$, for $m=0$ the formula

$$\begin{aligned} &Q_\ell^0(i\xi+2i\xi_1) \\ &= \sum_{\ell_1=0}^{\infty}\frac{\ell!\ell_1!(-1)^{\ell+\ell_1}}{(2\ell+1)!!(2\ell_1-1)!!}\left(\frac{2\ell_1+1}{2\ell+1}\right)^{1/2}(2i\xi_1)^{-\ell-\ell_1-1}P_{\ell_1}^0(i\xi)K_{\ell_1,\ell}^0(\xi_1) \\ &\quad \text{for } \ 0\le\xi<2\xi_1 \ \text{ and } \ \ell=0,1,2,..... \end{aligned} \tag{6.213}$$

For $m = 1$ one has to divide eq.(6.108) by $(1-\eta^2)^{1/2}$ and then take the limit $\eta \to -1$. One obtains

$$Z_\ell^1(\xi'(\xi,-1),a) \lim_{\eta\to-1} \frac{Y_\ell^1(\arccos\ \eta'(\xi,\eta),\phi)}{\{1-[\eta'(\xi,\eta)]^2\}^{1/2}} \lim_{\eta\to-1} \frac{\{1-[\eta'(\xi,\eta)]^2\}^{1/2}}{(1-\eta^2)^{1/2}}$$
$$= \sum_{\ell_1=1}^{\infty} K_{\ell_1,\ell}^1(\xi_1)(2d)^{-\ell-\ell_1-1} X_{\ell_1}^1(\xi,a) \lim_{\eta\to-1} \frac{Y_{\ell_1}^1(\arccos\eta,\phi)}{(1-\eta^2)^{1/2}} \tag{6.214}$$

With eqs.(6.103) and (6.5) it follows that

$$\lim_{\eta\to-1} \frac{Y_{\ell_1}^1(\arccos\eta,\phi)}{(1-\eta^2)^{1/2}} = \lim_{\eta\to1} \frac{Y_{\ell_1}^1(\arccos(-\eta),\phi)}{(1-\eta^2)^{1/2}} = \lim_{\eta\to1}(-1)^{\ell_1+1}\frac{Y_{\ell_1}^1(\arccos\eta,\phi)}{(1-\eta^2)^{1/2}}$$
$$= (-1)^{\ell_1}\left[\frac{2\ell_1+1}{4\pi}\frac{(\ell_1-1)!}{(\ell_1+1)!}\right]^{1/2} \lim_{\eta\to1}\frac{P_{\ell_1}^1(\eta)}{(1-\eta^2)^{1/2}} e^{i\phi} \tag{6.215}$$

The limit in the last member of this equation is found to be

$$\lim_{\eta\to1}\frac{P_{\ell_1}^1(\eta)}{(1-\eta^2)^{1/2}} = \frac{1}{2}\ell_1(\ell_1+1) \tag{6.216}$$

so that

$$\lim_{\eta\to-1} \frac{Y_{\ell_1}^1(\arccos\eta,\phi)}{(1-\eta^2)^{1/2}} = \frac{1}{2}\ell_1(\ell_1+1)(-1)^{\ell_1}\left[\frac{2\ell_1+1}{4\pi}\frac{(\ell_1-1)!}{(\ell_1+1)!}\right]^{1/2} e^{i\phi} \tag{6.217}$$

It furthermore follows, with eqs.(6.207) and (6.217), that

$$\lim_{\eta\to-1} \frac{Y_\ell^1\ (\arccos\ \eta'(\xi,\eta),\phi)}{\{1-[\eta'(\xi,\eta)]^2\}^{1/2}} = \lim_{\eta'\to-1} \frac{Y_\ell^1(\arccos\eta',\phi)}{(1-\eta'^2)^{1/2}}$$
$$= \frac{1}{2}\ell(\ell+1)(-1)^{\ell}\left[\frac{2\ell+1}{4\pi}\frac{(\ell-1)!}{(\ell+1)!}\right]^{1/2} e^{i\phi} \tag{6.218}$$

Finally one obtains, using the definitions of η and $\eta'(\xi,\eta)$, given in sections 6.2 and 6.4, the following relation

$$\lim_{\eta\to-1} \frac{\{1-[\eta'(\xi,\eta)]^2\}^{1/2}}{(1-\eta^2)^{1/2}} = \frac{(1+\xi^2)^{1/2}}{[1+(\xi+2\xi_1)^2]^{1/2}} \tag{6.219}$$

Using the above limits in eq.(6.214), and also eqs.(6.208) and (6.209) and $d = a\xi_1$, one obtains

$$\frac{Q_\ell^1(i\xi+2i\xi_1)}{[1+(\xi+2\xi_1)^2]^{1/2}}$$
$$= \sum_{\ell_1=1}^{\infty} \frac{(\ell-1)!(\ell_1+1)!(-1)^{\ell+\ell_1}}{(2\ell+1)!!(2\ell_1-1)!!}\left(\frac{2\ell_1+1}{2\ell+1}\frac{\ell(\ell+1)}{\ell_1(\ell_1+1)}\right)^{1/2} i(2i\xi_1)^{-\ell-\ell_1-1}$$
$$\times\frac{P_{\ell_1}^1(i\xi)}{(1+\xi^2)^{1/2}}K_{\ell_1,\ell}^1(\xi_1) \quad \text{for} \quad 0 \le \xi < 2\xi_1 \quad \text{and} \quad \ell = 1,2,..... \tag{6.220}$$

Now from eqs.(6.7) and (6.10) it can be derived that

$$P_\ell^1(i\xi) = -i(1+\xi^2)^{1/2}\frac{d}{d\xi}P_\ell^0(i\xi) \text{ for } \ell = 1, 2, ... \tag{6.221}$$

and

$$Q_\ell^1(i\xi) = -(1+\xi^2)^{1/2}\frac{d}{d\xi}Q_\ell^0(i\xi) \text{ for } \ell = 1, 2, ... \tag{6.222}$$

So differentiating eq.(6.213) with respect to ξ and comparing the results with eq.(6.220), one can make the following identification:

$$K_{\ell_1,\ell}^1(\xi_1) = -\left[\frac{\ell_1\ell}{(\ell_1+1)(\ell+1)}\right]^{1/2} K_{\ell_1,\ell}^0(\xi_1) \text{ for } \ell_1, \ell = 1, 2, ... \tag{6.223}$$

This is the relation, eq.(6.118), by means of which one can evaluate $K_{\ell_1,\ell}^m(\xi_1)$ for $m = 1$ in a simple way from $K_{\ell_1,\ell}^0(\xi_1)$.

For $m = -1$ one obtains the same result for $K_{\ell_1,\ell}^m(\xi_1)$ as for $m = 1$, as a consequence of eq.(6.117). This equation can be derived in the following way. From eq.(6.108) one finds

$$\begin{aligned} &Z_\ell^{-m}(\xi'(\xi,\eta), a)Y_\ell^{-m}(\arccos\eta'(\xi,\eta), \phi) \\ &= \sum_{\ell_1=|m|}^{\infty} K_{\ell_1,\ell}^{-m}(\xi_1)(2d)^{-\ell-\ell_1-1}X_{\ell_1}^{-m}(\xi, a)Y_{\ell_1}^{-m}(\arccos\eta, \phi) \end{aligned} \tag{6.224}$$

From eqs.(6.14) and (6.11) it follows that

$$Z_\ell^{-m}(\xi', a) = Z_\ell^m(\xi', a) \tag{6.225}$$

and from eqs. (6.13) and (6.8) that

$$X_\ell^{-m}(\xi, a) = X_\ell^m(\xi, a) \tag{6.226}$$

From eqs.(6.5) and (6.8) one furthermore obtains

$$Y_\ell^{-m}(\arccos\eta', \phi) = (-1)^m Y_\ell^m(\arccos\eta', \phi)e^{-2im\phi} \tag{6.227}$$

and an analogous relation for $Y_{\ell_1}^{-m}(\arccos\eta, \phi)$. Substituting eqs.(6.225)–(6.227) into eq.(6.224) and comparing the resulting equation with the original eq.(6.108), one obtains the relation, eq.(6.117).

In order to derive the formulae and recurrence relations, eqs.(6.110) through (6.114), by means of which the coefficients $K_{\ell_1,\ell}^0(\xi_1)$ may be evaluated, it should first of all be noticed, that the functions $P_{\ell_1}^0(i\xi)$ and $Q_\ell^0(i\xi + 2i\xi_1)$ in eq.(6.213) can be analytically continued to complex $z = i\xi$ in a region around the origin. For $P_{\ell_1}^0(i\xi)$ one can do this by means of eq.(6.6), and for $Q_\ell^0(i\xi+2i\xi_1)$ by means of eq.(6.115). One can then verify that the function $Q_\ell^0(z + 2i\xi_1)$ is analytic within and on a suitably chosen ellipse C in the complex z-plane with foci at $z = \pm 1$. Choosing for z in particular real values in the closed interval [−1,1] and denoting this variable by x, one obtains from eq.(6.213) the following relation:

$$\begin{aligned} & Q_\ell^0(x+2i\xi_1) \\ = & \sum_{\ell_1=0}^{\infty} \frac{\ell!\ell_1!(-1)^{\ell+\ell_1}}{(2\ell+1)!!(2\ell_1-1)!!} \left(\frac{2\ell_1+1}{2\ell+1}\right)^{1/2} (2i\xi_1)^{-\ell-\ell_1-1} P_{\ell_1}^0(x) K_{\ell_1,\ell}^0(\xi_1) \\ & \text{for } -1 \le x \le 1, \quad \xi_1 > 0, \quad \ell = 0,1,2,\ldots.. \end{aligned} \tag{6.228}$$

Using the orthogonality relation

$$\int_{-1}^{1} dx \; P_\ell^0(x) P_{\ell_1}^0(x) = \frac{2}{2\ell+1}\delta_{\ell\ell_1} \tag{6.229}$$

one then finds from eq.(6.228) the following formula

$$\begin{aligned} K_{\ell_1,\ell}^0(\xi_1) &= \left(\frac{2\ell+1}{2\ell_1+1}\right)^{1/2} \frac{(2\ell+1)!(2\ell_1+1)!}{(\ell!)^2(\ell_1!)^2}(-1)^{\ell+\ell_1}(i\xi_1)^{\ell+\ell_1+1} \\ & \times \int_{-1}^{1} dx \; Q_\ell^0(x+2i\xi_1) P_{\ell_1}^0(x) \end{aligned} \tag{6.230}$$

The recurrence relation eq.(6.114) can now be derived by applying alternatively the following recurrence relations to eq.(6.230):

$$(2\ell+1)z \; P_\ell^0(z) = (\ell+1)P_{\ell+1}^0(z) + \ell \; P_{\ell-1}^0(z) \tag{6.231}$$

$$(2\ell+1)z \; Q_\ell^0(z) = (\ell+1)Q_{\ell+1}^0(z) + \ell \; Q_{\ell-1}^0(z) + \delta_{\ell 0} \tag{6.232}$$

where it is assumed that P_ℓ^0 and Q_ℓ^0 vanish for negative ℓ.

The recurrence relation, eq.(6.113), can be derived in an analogous way, now by applying alternatively the recurrence relations

$$(2\ell+1)P_\ell^0(z) = \frac{d}{dz}P_{\ell+1}^0(z) - \frac{d}{dz}P_{\ell-1}^0(z) \tag{6.233}$$

$$(2\ell+1)Q_\ell^0(z) = \frac{d}{dz}Q_{\ell+1}^0(z) - \frac{d}{dz}Q_{\ell-1}^0(z) + \delta_{\ell 0}\frac{z}{z^2-1} \tag{6.234}$$

to eq.(6.230) and using partial integration. (Again P_ℓ^0 and Q_ℓ^0 are assumed to vanish for negative ℓ.) The boundary terms resulting from this partial integration cancel each other, since $P_\ell^0(1) = 1$ and $P_\ell^0(-1) = (-1)^\ell$.

The relation, eq.(6.111), and the recurrence relation, eq.(6.112), can be derived in an analogous way as the recurrence relation, eq.(6.113). However, in the right-hand sides of eqs.(6.111) and (6.112) terms depending on $Q_\ell^0(1+2i\xi_1)$ appear, resulting from the fact that boundary terms of partial integration do not cancel each other anymore, as was the case in the derivation of eq.(6.113). To evaluate these boundary terms the following property was used:

$$Q_\ell^0(-z^*) = (-1)^{\ell+1}[Q_\ell^0(z)]^* \tag{6.235}$$

where the asterisk denotes complex conjugation. (This property may be proved with eq.(6.115).)

The explicit form, eq.(6.110), of $K^0_{0,0}(\xi_1)$ can be derived also from eq.(6.230), by replacing $P^0_0(x)$ by $dP^0_1(x)/dx$ and integrating partially, which gives:

$$K^0_{0,0}(\xi_1) = -i\xi_1 \int_{-1}^{1} x\, dx\, dQ^0_0(x + 2i\xi_1)/dx + i\xi_1[Q^0_0(1 + 2i\xi_1) + Q^0_0(-1 + 2i\xi_1)] \tag{6.236}$$

Using the definition, eq.(6.115), of $Q^0_\ell(z)$, one then obtains the result, eq.(6.110).

The symmetry relation, eq.(6.116), can be derived from eqs.(6.113) and (6.114) by observing that, if one considers these equations for the coefficients

$$k^0_{\ell_1,\ell}(\xi_1) \equiv \left(\frac{2\ell_1 + 1}{2\ell + 1}\right)^{1/2} K^0_{\ell_1,\ell}(\xi_1) \tag{6.237}$$

they may be written as follows:

$$\begin{aligned} & \frac{\ell + 1}{(2\ell + 1)(2\ell + 3)} k^0_{\ell_1,\ell+1}(\xi_1) + \frac{\ell_1 + 1}{(2\ell_1 + 1)(2\ell_1 + 3)} k^0_{\ell_1+1,\ell}(\xi_1) \\ + \quad & \frac{4}{\ell}\xi_1^2 k^0_{\ell_1,\ell-1}(\xi_1) + \frac{4}{\ell_1}\xi_1^2 k^0_{\ell_1-1,\ell}(\xi_1) = 0 \quad \text{for } \ell,\ \ell_1 > 0 \end{aligned} \tag{6.238}$$

$$\begin{aligned} & \frac{(\ell + 1)^2}{(2\ell + 1)(2\ell + 3)} k^0_{\ell_1,\ell+1}(\xi_1) - \frac{(\ell_1 + 1)^2}{(2\ell_1 + 1)(2\ell_1 + 3)} k^0_{\ell_1+1,\ell}(\xi_1) - 4\ \xi_1^2 k^0_{\ell_1,\ell}(\xi_1) \\ + \quad & 4\ \xi_1^2[k^0_{\ell_1-1,\ell}(\xi_1) - k^0_{\ell_1,\ell-1}(\xi_1)] = -\xi_1^2 \delta_{\ell_1,0}\delta_{\ell 0} \quad \text{for } \ell,\ \ell_1 \geq 0 \end{aligned} \tag{6.239}$$

From the symmetry of these equations in ℓ and ℓ_1 it is found that the solutions must possess the following symmetry:

$$k^0_{\ell_1,\ell}(\xi_1) = (-1)^{\ell_1+\ell} k^0_{\ell,\ell_1}(\xi_1) \tag{6.240}$$

With eq.(6.237) one then proves the validity of eq.(6.116).

In this appendix finally explicit expressions for the coefficients $K^0_{\ell_1,\ell}(\xi_1)$ for $\ell,\ \ell_1 = 0, 1$ and 2 will be given. The derivation from eqs.(6.110)–(6.113), together with eqs.(6.115)–(6.117), is straightforward and will not be given here in detail. One obtains:

$$K^0_{0,1}(\xi_1) = -3\ K^0_{1,0}(\xi_1) = 6\sqrt{3}\left[2\xi_1^3 \arctan\left(\frac{1}{\xi_1}\right) - \xi_1^4 \ln\left(1 + \frac{1}{\xi_1^2}\right) - \xi_1^2\right] \tag{6.241}$$

$$K^0_{0,2}(\xi_1) = 5\ K^0_{2,0}(\xi_1) = 30\sqrt{5}\left[6\xi_1^5 \arctan\left(\frac{1}{\xi_1}\right) + (1 - 2\xi_1^2)\xi_1^4 \ln\left(1 + \frac{1}{\xi_1^2}\right) - 4\xi_1^4\right] \tag{6.242}$$

$$K^0_{1,1}(\xi_1) = 24\left[-\xi_1^3 \arctan\left(\frac{1}{\xi_1}\right) + (\frac{3}{2} + \xi_1^2)\xi_1^4 \ln\left(1 + \frac{1}{\xi_1^2}\right) - \xi_1^4\right] \tag{6.243}$$

$$K^0_{1,2}(\xi_1) = -\frac{5}{3}K^0_{2,1}(\xi_1) = 90\sqrt{5/3}$$
$$\times \left[-4\xi_1^5 \arctan\left(\frac{1}{\xi_1}\right) + 2(\xi_1^2 + 2)\xi_1^6 \ln\left(1 + \frac{1}{\xi_1^2}\right) + (1 - 2\xi_1^2)\xi_1^4\right] \tag{6.244}$$

$$K^0_{2,2}(\xi_1) = 180\left[2\xi_1^5 \arctan\left(\frac{1}{\xi_1}\right) - (6\xi_1^4 + 10\xi_1^2 + 5)\xi_1^6 \ln\left(1 + \frac{1}{\xi_1^2}\right) + (6\xi_1^2 + 7)\xi_1^6\right] \tag{6.245}$$

whereas $K^0_{0,0}(\xi_1)$ is given by eq.(6.110).

Chapter 7
ISLANDS FILMS FOR A FINITE COVERAGE

7.1 Introduction

In this chapter the analysis, given in the previous two chapters for a low coverage, will be extended to a higher coverage following Haarmans and Bedeaux [49], [50], [23]. In this case the electromagnetic interaction between different small particles lying on the substrate has to be taken into account. As in the two previous chapters the electric field is described using the Laplace equation for the potential. Retardation effects are neglected. On the one hand, because the size of the particles is small compared to the wavelength. On the other hand it is used, that the interaction between particles affects the polarization substantially only, if the particles are close together compared to the wavelength. For the low coverage limit considered in the two previous chapters the interaction may be neglected. If the coverage is increased it is at first sufficient to account only for the dipole interactions between the particles, because the range of the dipole interaction is longest. For high coverages one must take multipole interactions into account up to a rather high order to obtain accurate results. In order to make an explicit analysis possible the particles will all be assumed to be identical.

In the analysis in sections 7.2–7.6 of this chapter the islands are assumed to be spherically symmetric. A complete analysis to arbitrary order in the multipole expansion is given in sections 7.2–7.4. If the coverage is not too high it is usually sufficient to use the dipole or the quadrupole approximation to calculate the interaction between the different islands. This is discussed in sections 7.5 and 7.6. In the literature the interaction between islands along the substrate is always taken into account to dipolar order [10].

Regarding the distribution of the islands on the substrate, one may have on the one hand a regular array of these particles and on the other hand a random array. For the case of a regular array a solution of the problem is constructed along somewhat familiar lines using lattice sums. For the random array an assumption is made, expressing the higher order distribution functions in terms of the pair correlation function, using a product approximation. With this assumption the analysis may then further be carried out with the same rigor as for the regular lattice. Even though the product approximation may be questionable it is not likely that, given the relatively low value of the coverage for random close packing in two dimensions, this will lead to sizeable errors. It is found by explicit calculation, that the optical properties for the regular and the random arrays at the same coverage are, up to a coverage of about 50%, almost (within a few percent) identical, [49], [50] and [23]. If extensive clustering of the islands occurs, as for instance in fractal structures, the optical properties are expected to become different. This last case will not be considered in this book.

In section 7.7 the theory is extended to two-dimensional arrays of identical spheroids (both oblate and prolate). The interaction between the spheroids and their own images in the substrate is described, just as in sections 6.4 and 6.5, with spheroidal multipoles. This gives the exact polarizabilities of the islands as modified by the interaction with the substrate. The interaction between different islands, as well as that between the islands and images of different islands, however, is taken into account using a spherical multipole expansion. Mathematical techniques to treat the interaction between different spheroids along the substrate analytically, using spheroidal multipole potentials, do not seem to be available. The use of the spherical multipole potentials is correct, as long as the particles are not too densely packed in the two-dimensional array. The approximation will be better for oblate spheroids. In view of the fact that the spheroids are not be too densely packed, the analysis is restricted to the dipole and the quadrupole approximation for the interaction between the spheroids, cf. sections 7.8 and 7.9. In section 7.10 the theory is applied to gold islands on sapphire.

7.2 Two-dimensional arrays of spheres

Consider a two-dimensional array of identical spherical islands with a radius R at a distance d above a flat substrate. This array may either be regular or random. The positions of the spheres are given by $\mathbf{R}_i = (\mathbf{R}_{i,\|}, 0)$ and the positions of the images by $\mathbf{R}_i^r = (\mathbf{R}_{i,\|}, 2d)$, (see fig.7.1). In the static approximation the incident field is constant and given in the ambient by

$$\mathbf{E}_0 = E_0(\sin\theta_0\cos\phi_0, \sin\theta_0\sin\phi_0, \cos\theta_0) \tag{7.1}$$

where θ_0 is the angle of the field with the z-axis, and ϕ_0 the angle between its projection on the substrate and the x-axis. The corresponding incident potential is (cf. eq.(5.97))

$$\begin{aligned}\psi_0(\mathbf{r}) &= -\mathbf{E}_0.\mathbf{r} = -r\ E_0\sqrt{2\pi/3}[\cos\theta_0 Y_1^0(\theta,\phi)\sqrt{2} + \sin\theta_0\{\exp(i\phi_0)Y_1^{-1}(\theta,\phi) \\ &\quad - \exp(-i\phi_0)Y_1^1(\theta,\phi)\}]\end{aligned} \tag{7.2}$$

The general solution in the ambient may be written as a sum of the incident potential, the potential due to the induced charge distribution in the islands and the image charge distribution in the substrate:

$$\begin{aligned}\psi_a(\mathbf{r}) &= -r\ E_0\sqrt{2\pi/3}[\cos\theta_0 Y_1^0(\theta,\phi)\sqrt{2} \\ &\quad + \sin\theta_0\{\exp(i\phi_0)Y_1^{-1}(\theta,\phi) - \exp(-i\phi_0)Y_1^1(\theta,\phi)\}] \\ &\quad + \sum_{\ell m}{}' \sum_i A_{\ell m,i} r_i^{-\ell-1} Y_\ell^m(\theta_i,\phi_i) + \sum_{\ell m}{}' \sum_i A_{\ell m,i}^r \rho_i^{-\ell-1} Y_\ell^m(\theta_i^r,\phi_i^r)\end{aligned} \tag{7.3}$$

The first term is the incident field. The second term gives the contributions of multipoles of arbitrary order originating from the spheres with positions $\mathbf{R}_i$ and the third

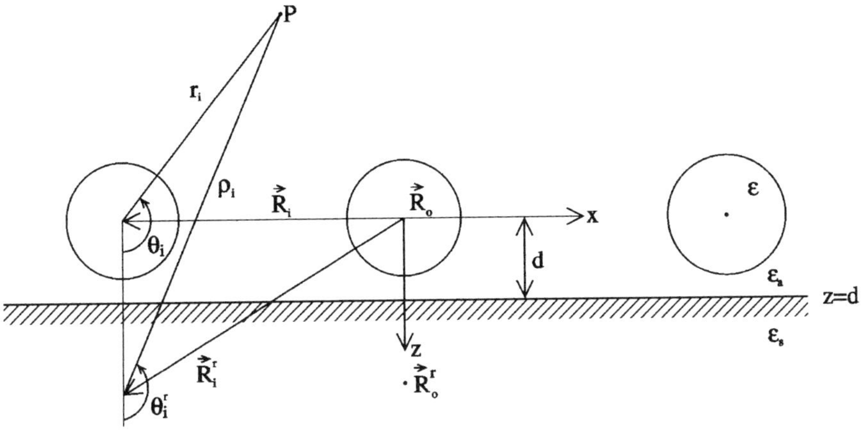

Figure 7.1 Array of spheres above the substrate.

term gives the contributions from the image multipoles, which are located in the positions $\mathbf{R}_i^r$. Furthermore (r_i, θ_i, ϕ_i) and $(\rho_i, \theta_i^r, \phi_i^r)$ are the spherical coordinates with $\mathbf{R}_i$ and $\mathbf{R}_i^r$ as center of the coordinate system, respectively (see fig.7.1). As in the two previous chapters, the prime in the summation indicates that the $\ell = 0$ term is excluded in view of the fact, that there are no free charges in the system. For the potential in the substrate one may write similarly

$$\begin{aligned}\psi_s(\mathbf{r}) &= E_0\{d((\epsilon_a/\epsilon_s) - 1)\cos\theta_0 - r\sqrt{2\pi/3}[(\epsilon_a/\epsilon_s)\cos\theta_0 Y_1^0(\theta,\phi)\sqrt{2} \\ &\quad + \sin\theta_0(\exp(i\phi_0)Y_1^{-1}(\theta,\phi) - \exp(-i\phi_0)Y_1^1(\theta,\phi))]\} \\ &\quad + {\sum_{\ell m}}' \sum_i A_{\ell m,i}^t r_i^{-\ell-1} Y_\ell^m(\theta_i,\phi_i)\end{aligned} \tag{7.4}$$

where ϵ_a and ϵ_s are the (frequency dependent) dielectric constants of ambient and substrate, respectively.

As a first step to calculate the polarizabilities in this general case, one must choose the amplitudes of the various multipoles such, that the boundary conditions at the surface of the substrate are satisfied. For this purpose it is most convenient to use the fact that one finds for a point charge, that the image charge is equal to the original charge times $(\epsilon_a - \epsilon_s)/(\epsilon_a + \epsilon_s)$, and the charge giving the transmitted field is equal to the original charge times $2\epsilon_s/(\epsilon_a + \epsilon_s)$, [26]. The same relations may be used for a distribution of charges and one may then conclude that

$$A_{\ell m,i}^r = (-1)^{\ell+m}\frac{\epsilon_a - \epsilon_s}{\epsilon_a + \epsilon_s}A_{\ell m,i} \text{ and } A_{\ell m,i}^t = \frac{2\epsilon_a}{\epsilon_a + \epsilon_s}A_{\ell m,i} \tag{7.5}$$

One may, of course, also verify directly that these relations make the potential and the dielectric constant times the normal derivative of the potential continuous at the surface $z = d$ of the substrate (see fig.7.1).

In order to relate the amplitudes to the multipole polarizabilities in free space, one must use the potential for the so-called local field acting on particle j and given by the sum of the incident field, the direct field from the other particles and the reflected fields of all particles:

$$\begin{aligned}\psi_{j,loc}(\mathbf{r}) &= -\mathbf{E}_0.\mathbf{R}_j - r_j E_0\sqrt{2\pi/3}[\cos\theta_0 Y_1^0(\theta_j,\phi_j)\sqrt{2} \\ &+\sin\theta_0\{\exp(i\phi_0)Y_1^{-1}(\theta_j,\phi_j) - \exp(-i\phi_0)Y_1^1(\theta_j,\phi_j)\}] \\ &+\sum_{\ell m}{}^{'}\sum_{i\neq j} A_{\ell m,i} r_i^{-\ell-1} Y_\ell^m(\theta_i,\phi_i) + \sum_{\ell m}{}^{'}\sum_{i} A_{\ell m,i}^r \rho_i^{-\ell-1} Y_\ell^m(\theta_i^r,\phi_i^r)\end{aligned} \quad (7.6)$$

where it is used that $\mathbf{E}_0.\mathbf{r} = \mathbf{E}_0.\mathbf{R}_j + \mathbf{E}_0.(\mathbf{r}-\mathbf{R}_j)$. The second term is then written in terms of spherical coordinates around $\mathbf{R}_j$. It is now necessary to expand the fields from the other particles and from the images in terms of a complete set of fields around the center of particle j. It follows from eq.(5.159), with eq.(5.160) (see appendix B of chapter 5) that

$$\begin{aligned}\psi_{j,loc}(\mathbf{r}) &= -\mathbf{E}_0.\mathbf{R}_j - r_j E_0\sqrt{2\pi/3}[\cos\theta_0 Y_1^0(\theta_j,\phi_j)\sqrt{2} \\ &+\sin\theta_0\{\exp(i\phi_0)Y_1^{-1}(\theta_j,\phi_j) - \exp(-i\phi_0)Y_1^1(\theta_j,\phi_j)\}] \\ &+\sum_{\ell m}{}^{'}\sum_{\ell_2 m_2}{}^{'} H(\ell,m|\ell_2,m_2)[\sum_{i\neq j} A_{\ell_2 m_2,i} T_{\ell_1 m_1}(\mathbf{R}_i - \mathbf{R}_j) \\ &+\sum_i A_{\ell_2 m_2,i}^r T_{\ell_1 m_1}(\mathbf{R}_i^r - \mathbf{R}_j)] r_j^\ell Y_\ell^m(\theta_j,\phi_j)\end{aligned} \quad (7.7)$$

where $\ell_1 \equiv \ell + \ell_2$ and $m_1 \equiv m_2 - m$, while furthermore

$$T_{\ell_1 m_1}(\mathbf{R}) \equiv [r^{-\ell_1-1} Y_{\ell_1}^{m_1}(\theta,\phi)]_{\mathbf{r}=\mathbf{R}} \quad (7.8)$$

and

$$\begin{aligned}H(\ell,m|\ell_2,m_2) &\equiv (-1)^{\ell_2+m}\sqrt{4\pi}\left[\frac{2\ell_2+1}{(2\ell+1)(2\ell_1+1)}\right]^{1/2} \\ &\times\left[\frac{(\ell_1+m_1)!(\ell_1-m_1)!}{(\ell+m)!(\ell-m)!(\ell_2+m_2)!(\ell_2-m_2)!}\right]^{1/2}\end{aligned} \quad (7.9)$$

Using eqs.(5.69) and (7.7) one then obtains for the amplitudes

$$\begin{aligned}A_{\ell m,j} &= -\alpha_\ell\{-E_0\sqrt{2\pi/3}\delta_{\ell 1}[\cos\theta_0\delta_{m0}\sqrt{2} + \sin\theta_0(\exp(i\phi_0)\delta_{m,-1} - \exp(-i\phi_0)\delta_{m1})] \\ &+\sum_{\ell_2 m_2}{}^{'} H(\ell,m|\ell_2,m_2)[\sum_{i\neq j} A_{\ell_2 m_2,i} T_{\ell_1 m_1}(\mathbf{R}_i - \mathbf{R}_j) \\ &+(-1)^{\ell_2+m_2}(\frac{\epsilon_a-\epsilon_s}{\epsilon_a+\epsilon_s})\sum_i A_{\ell_2 m_2,i} T_{\ell_1 m_1}(\mathbf{R}_i^r - \mathbf{R}_j)]\}\end{aligned} \quad (7.10)$$

Notice the fact, that it was used, that all the spherical islands are identical and that the multipole polarizabilities of the islands in the ambient far from the substrate α_ℓ

are independent of m. Furthermore the reflected amplitudes were eliminated using eq.(7.5).

The above expression should be solved to obtain all the amplitudes $A_{\ell m,j}$. As this is an infinite number of coefficients one needs some further assumptions about, for instance, the positions of the particles. Before entering into this matter it is interesting to compare the equation with the corresponding equation in chapter 5. For that purpose one must remove the contribution due to the image of sphere j out of the sum on the right-hand side of the equation. This then results in

$$\begin{aligned} A_{\ell m,j} &= -\alpha_\ell\{-E_0\sqrt{2\pi/3}\delta_{\ell 1}[\cos\theta_0\delta_{m0}\sqrt{2}+\sin\theta_0(\exp(i\phi_0)\delta_{m,-1} \\ &\quad -\exp(-i\phi_0)\delta_{m1})] + (\frac{\epsilon_a-\epsilon_s}{\epsilon_a+\epsilon_s})\sum_{\ell'=|m|}^{\infty\prime}(-1)^{\ell'+m}A_{\ell' m,j}M_{\ell,\ell'}^{m}(2d)^{-\ell-\ell'-1} \\ &\quad +\sum_{\ell_2 m_2}{}' H(\ell,m|\ell_2,m_2)\sum_{i\neq j}A_{\ell_2 m_2,i}[T_{\ell_1 m_1}(\mathbf{R}_i-\mathbf{R}_j) \\ &\quad +(-1)^{\ell_2+m_2}(\frac{\epsilon_a-\epsilon_s}{\epsilon_a+\epsilon_s})T_{\ell_1 m_1}(\mathbf{R}_i^r-\mathbf{R}_j)]\} \end{aligned} \tag{7.11}$$

where eqs.(7.8), (7.9), (5.102) and (5.103) have been used. In the approximation of the two previous chapters the distance between the particles was assumed to be so large compared to the radius, that the last term in this expression, describing the coupling between the particles, may be neglected. The equations for the amplitudes of the different particles then decouple and the above equation reduces to eq.(5.105) for each sphere. In this chapter the interest is, in particular, in the consequences of the coupling between the particles and the further analysis will be based on eq.(7.10). In the following section the special case of a regular array of spheres will first be discussed, as the solution for this case is most straightforward.

7.3 Regular arrays of spheres

The islands are assumed to be located on a regular array with only one sphere per unit cell. The x-axis is chosen along one of the lattice vectors. It is not difficult to generalize the analysis to the case of more than one sphere per unit cell. Using the invariance of the problem for translation over a lattice vector it follows that the multipole amplitudes of all the spheres are identical

$$A_{\ell m,j} = A_{\ell m} \tag{7.12}$$

Substitution of this relation into eq.(7.10) then gives for these amplitudes

$$\begin{aligned} A_{\ell m} &= -\alpha_\ell\{-E_0\sqrt{2\pi/3}\delta_{\ell 1}[\cos\theta_0\delta_{m0}\sqrt{2}+\sin\theta_0(\exp(i\phi_0)\delta_{m,-1} \\ &\quad -\exp(-i\phi_0)\delta_{m1})] + \sum_{\ell_2 m_2}{}' H(\ell,m|\ell_2,m_2)A_{\ell_2 m_2}[\sum_{i\neq j}T_{\ell_1 m_1}(\mathbf{R}_i-\mathbf{R}_j) \\ &\quad +(-1)^{\ell_2+m_2}(\frac{\epsilon_a-\epsilon_s}{\epsilon_a+\epsilon_s})\sum_i T_{\ell_1 m_1}(\mathbf{R}_i^r-\mathbf{R}_j)]\} \end{aligned} \tag{7.13}$$

where $\ell_1 = \ell + \ell_2$ and $m_1 = m_2 - m$. To simplify this equation it is now convenient to introduce the following lattice sums over a two-dimensional array of spheres

$$S_{\ell m} \equiv L^{\ell+1} \sum_{i \neq j} T_{\ell m}(\mathbf{R}_i - \mathbf{R}_j) = L^{\ell+1} \sum_{i \neq 0} T_{\ell m}(\mathbf{R}_i) = \sum_{i \neq 0} \left(\frac{L}{r}\right)^{\ell+1} Y_\ell^m(\theta, \phi)|_{\mathbf{r}=\mathbf{R}_i} \tag{7.14}$$

and over the images of the array

$$S^r_{\ell m} \equiv L^{\ell+1} \sum_{i} T_{\ell m}(\mathbf{R}^r_i - \mathbf{R}_j) = L^{\ell+1} \sum_{i} T_{\ell m}(\mathbf{R}^r_i) = \sum_{i} \left(\frac{L}{r}\right)^{\ell+1} Y_\ell^m(\theta, \phi)|_{\mathbf{r}=\mathbf{R}^r_i} \tag{7.15}$$

A dimensionless definition is chosen, where L is the lattice constant. For square and triangular lattices, there is only one lattice constant. For other lattices one should use the square root of the product of the lattice constants. An important simplification is that for lattices with a 2-, 4- or 6- fold symmetry $S_{\ell m} = S^r_{\ell m} = 0$ if m is unequal to a multiple of 2, 4 or 6 respectively. Substitution of these lattice sums into eq.(7.13) results in

$$\begin{aligned} & A_{\ell m} + \alpha_\ell \sum_{\ell_2 m_2}{}' H(\ell, m|\ell_2, m_2) A_{\ell_2 m_2} L^{-\ell_1 - 1}[S_{\ell_1 m_1} + (-1)^{\ell_2 + m_2}(\frac{\epsilon_a - \epsilon_s}{\epsilon_a + \epsilon_s}) S^r_{\ell_1 m_1}] \\ = & -\alpha_\ell\{-E_0\sqrt{2\pi/3}\delta_{\ell 1}[\cos\theta_0 \delta_{m0}\sqrt{2} + \sin\theta_0(\exp(i\phi_0)\delta_{m,-1} - \exp(-i\phi_0)\delta_{m1})]\} \end{aligned} \tag{7.16}$$

The solution of this equation is somewhat more involved than the solution of eq.(5.105) for the independent particle case in chapter 5. In the first place one must now calculate the lattice sums; the procedure to do this is explained in appendix A. In the second place there is no longer a symmetry for rotation around the normal on the substrate through the center of the sphere. This implies that different values of m are now coupled. For a lattice with a 4-fold symmetry, $A_{\ell 0}$ couples to $A_{\ell_2, 4\nu}$, $A_{\ell 1}$ to $A_{\ell_2, 4\nu+1}$ and $A_{\ell,-1}$ to $A_{\ell_2, 4\nu-1}$,where ν is a natural number. Replace the 4 in this expression by a 6 for a 6-fold symmetry. This means that one needs to solve three independent sets of equations to obtain $A_{\ell 0}$, $A_{\ell 1}$ and $A_{\ell,-1}$, respectively. One may again show that $A_{\ell 1}\exp(i\phi_0) = -A_{\ell,-1}\exp(-i\phi_0)$. It is crucial in this context that the x-axis has been chosen along one of the lattice vectors. As a consequence two sets of equations remain to be solved. One may furthermore use the identities $A_{\ell,4\nu} = A_{\ell,-4\nu}$ and $A_{\ell,1+4\nu}\exp(i\phi_0) = -A_{\ell,-1-4\nu}\exp(-i\phi_0)$. This reduces the number of independent amplitudes considerably. In section 10 this solution is discussed in more detail. For 2-fold symmetry the situation is more involved because then $A_{\ell 1}$ and $A_{\ell,-1}$ couple to each other. In this book 2-fold symmetry will not be further analyzed. The resulting equations for the amplitudes may, similarly to the solution of eq.(5.105), or eqs.(5.107) and (5.108), be solved numerically. For this purpose one takes only a finite number M of multipoles unequal to zero and solves the equations (cf. eqs.(5.109) and (5.110)). Subsequently one increases the number of multipoles and solves the equations again, until the values of the dipole and quadrupole amplitudes A_{10} , A_{11} , A_{20} and A_{21} no longer change appreciably. After numerically solving the above equations one may, for 4- and 6-fold symmetry, identify the relevant polarizabilities using eq.(5.111). The reason that this identification remains valid, for these symmetries, is that the same elements of the polarizability tensor, cf. eq.(5.30), needed to prove the relation, are zero. The numerical analysis will be restricted to square and triangular arrays, which have 4- and 6-fold symmetry respectively.

7.4 Random arrays of spheres

In this case one must start from eq.(7.10):

$$\begin{aligned} A_{\ell m,j} &= -\alpha_\ell\{-E_0\sqrt{2\pi/3}\delta_{\ell 1}[\cos\theta_0\delta_{m0}\sqrt{2}+\sin\theta_0(\exp(i\phi_0)\delta_{m,-1} \\ &\quad -\exp(-i\phi_0)\delta_{m1})] + \sum_{\ell_2 m_2}{}' H(\ell,m|\ell_2,m_2)[\sum_{i\neq j} A_{\ell_2 m_2,i}T_{\ell_1 m_1}(\mathbf{R}_i-\mathbf{R}_j) \\ &\quad +(-1)^{\ell_2+m_2}(\frac{\epsilon_a-\epsilon_s}{\epsilon_a+\epsilon_s})\sum_i A_{\ell_2 m_2,i}T_{\ell_1 m_1}(\mathbf{R}_i^r-\mathbf{R}_j)]\} \end{aligned} \quad (7.17)$$

where $\ell_1 = \ell+\ell_2$ and $m_1 = m_2 - m$. A general way to solve this equation is by iteration. This leads to an expression, which may then be averaged over the distribution functions of arbitrary order, giving an expression for the average multipole polarizability of the spheres. This procedure is most appropriate, if one wants to obtain results in a (virial) expansion in the coverage. As the influence of the coverage is most pronounced when the coverage is high, the virial expansion as a tool to study these effects is not very useful. In view of this fact a different procedure is followed, which reduces to the proper low coverage limit including the first virial correction and which becomes exact for all volume fractions, if one substitutes the pair distribution for a regular array. In the iterated form this procedure corresponds to a product approximation for the higher order correlation functions. It is more convenient to introduce the approximation directly in this equation. Averaging then gives

$$\begin{aligned} <A_{\ell m}> &= -\alpha_\ell\{-E_0\sqrt{2\pi/3}\delta_{\ell 1}[\cos\theta_0\delta_{m0}\sqrt{2}+\sin\theta_0(\exp(i\phi_0)\delta_{m,-1} \\ &\quad -\exp(-i\phi_0)\delta_{m1})] + \sum_{\ell_2 m_2}{}' H(\ell,m|\ell_2,m_2)[\sum_{i\neq j} < A_{\ell_2 m_2,i}T_{\ell_1 m_1}(\mathbf{R}_i-\mathbf{R}_j) > \\ &\quad +(-1)^{\ell_2+m_2}(\frac{\epsilon_a-\epsilon_s}{\epsilon_a+\epsilon_s})\sum_i < A_{\ell_2 m_2,i}T_{\ell_1 m_1}(\mathbf{R}_i^r-\mathbf{R}_j) >]\} \end{aligned} \quad (7.18)$$

The average is indicated by $< ... >$. Note furthermore that the average amplitude, in view of the assumed homogeneity of the film, is independent of j. The assumption is now equivalent to replacing $A_{\ell_2 m_2,i}$ in the averages on the right-hand side of the above equation by $< A_{\ell_2 m_2} >$. In this way one obtains for an isotropic film

$$\begin{aligned} \sum_{i\neq j} < A_{\ell_2 m_2,i}T_{\ell_1 m_1}(\mathbf{R}_i-\mathbf{R}_j) > &= < A_{\ell_2 m_2} > \rho\int d\mathbf{r}_\parallel g(r_\parallel)T_{\ell_1 m_1}(\mathbf{r}_\parallel,0) \\ &= < A_{\ell_2 m_2} > \delta_{m_1 0}\,\rho\int d\mathbf{r}_\parallel g(r_\parallel)T_{\ell_1 0}(\mathbf{r}_\parallel,0) \end{aligned} \quad (7.19)$$

and

$$\begin{aligned}\sum_i &< A_{\ell_2 m_2,i} T_{\ell_1 m_1}(\mathbf{R}_i^r - \mathbf{R}_j) > \\ &= < A_{\ell_2 m_2} > [T_{\ell_1 m_1}(0,2d) + \rho \int d\mathbf{r}_{\|} g(r_{\|}) T_{\ell_1 m_1}(\mathbf{r}_{\|}, 2d)] \\ &= < A_{\ell_2 m_2} > \delta_{m_1 0}[T_{\ell_1 0}(0,2d) + \rho \int d\mathbf{r}_{\|} g(r_{\|}) T_{\ell_1 0}(\mathbf{r}_{\|}, 2d)] \end{aligned} \quad (7.20)$$

where the factor $\delta_{m_1 0}$ results from the isotropy and ρ is the number of islands per unit of surface area. Furthermore $g(r_{\|})$ is the pair correlation function defined by

$$g(|\mathbf{r}_{\|} - \mathbf{r}'_{\|}|) = g(\mathbf{r}_{\|}, \mathbf{r}'_{\|}) \equiv \rho^{-2} < \sum_{i \neq j} \delta(\mathbf{r}_{\|} - \mathbf{R}_{i,\|}) \delta(\mathbf{r}'_{\|} - \mathbf{R}_{j,\|}) > \quad (7.21)$$

The pair correlation function depends only on $|\mathbf{r}_{\|} - \mathbf{r}'_{\|}|$ due to the isotropy of the film. It is now convenient to introduce the following dimensionless "distribution integrals"

$$I_\ell \equiv \rho^{-(\ell-1)/2} \int d\mathbf{r}_{\|} g(r_{\|}) T_{\ell 0}(\mathbf{r}_{\|}, 0) = 2\pi \rho^{-(\ell-1)/2} \int_0^\infty dr_{\|} g(r_{\|}) r_{\|}^{-\ell} P_\ell^0(0) \sqrt{(2\ell+1)/4\pi} \quad (7.22)$$

and

$$\begin{aligned} I_\ell^r &\equiv \rho^{-(\ell+1)/2} \left[T_{\ell 0}(0,2d) + \rho \int d\mathbf{r}_{\|} g(r_{\|}) T_{\ell 0}(\mathbf{r}_{\|}, 2d) \right] = \rho^{-(\ell+1)/2} [(2d)^{-\ell-1} P_\ell^0(1) \\ &+ 2\pi\rho \int_0^\infty dr_{\|} g(r_{\|}) r_{\|} (r_{\|}^2 + 4d^2)^{-(\ell+1)/2} P_\ell^0(2d/(r_{\|}^2 + 4d^2)^{1/2})] \sqrt{(2\ell+1)/4\pi} \end{aligned} \quad (7.23)$$

where eqs.(7.8) and (5.26) have been used. These distribution integrals play the same role as the lattice sums introduced in the previous section. The analogy in the definition is most clearly seen for the square array, for which $\rho = L^{-2}$. Due to the isotropic nature of the film the distribution integrals are independent of m, whereas the lattice sums do depend on m. Substitution of eqs.(7.19) through (7.23) into eq.(7.18) reduces this equation to

$$\begin{aligned} < A_{\ell m} > &= -\alpha_\ell \{ -E_0 \sqrt{2\pi/3} \delta_{\ell 1} [\cos\theta_0 \delta_{m0} \sqrt{2} + \sin\theta_0 (\exp(i\phi_0) \delta_{m,-1} - \exp(-i\phi_0) \delta_{m1})] \\ &+ \sum_{\ell_2 = |m|}^{\infty\prime} H(\ell, m | \ell_2, m) < A_{\ell_2 m} > \rho^{(\ell_1+1)/2} [I_{\ell_1} + (-1)^{\ell_2+m} (\frac{\epsilon_a - \epsilon_s}{\epsilon_a + \epsilon_s}) I_{\ell_1}^r] \} \end{aligned} \quad (7.24)$$

It follows from this equation that the different m's do not couple. This a consequence, on the one hand, of the approximation in which the multipole amplitudes are replaced by their average and, on the other hand, by the isotropy of the film. As a consequence

$$< A_{\ell m} > = < A_{\ell m,j} > = 0 \quad \text{for} \quad m \neq -1, 0, 1 \quad (7.25)$$

The above set of linear equations for the average amplitudes may thus be written as the sum of separate contributions for $m = -1$, 0 and 1. In this way one obtains for $m = 0$

$$\begin{aligned} < \ & A_{\ell 0} > +\alpha_\ell \sum_{\ell_2=1}^{\infty} H(\ell, 0|\ell_2, 0) < A_{\ell_2 0} > \rho^{(\ell_1+1)/2}[I_{\ell_1} + (-1)^{\ell_2}(\frac{\epsilon_a - \epsilon_s}{\epsilon_a + \epsilon_s})I^r_{\ell_1}] \\ = \ & \alpha_\ell E_0 \sqrt{4\pi/3}\delta_{\ell 1} \cos\theta_0 \qquad \text{for } \ell = 1, 2, 3, ... \end{aligned} \tag{7.26}$$

and for $m = 1$

$$\begin{aligned} < \ & A_{\ell 1} > +\alpha_\ell \sum_{\ell_2=1}^{\infty} H(\ell, 1|\ell_2, 1) < A_{\ell_2 1} > \rho^{(\ell_1+1)/2}[I_{\ell_1} - (-1)^{\ell_2}(\frac{\epsilon_a - \epsilon_s}{\epsilon_a + \epsilon_s})I^r_{\ell_1}] \\ = \ & -\alpha_\ell E_0 \sqrt{2\pi/3}\delta_{\ell 1} \sin\theta_0 \exp(-i\phi_0) \qquad \text{for } \ell = 1, 2, 3, ... \end{aligned} \tag{7.27}$$

As $< A_{\ell,-1} > \exp(-i\phi_0) = - < A_{\ell 1} > \exp(i\phi_0)$ the $m = -1$ equation is superfluous. After numerically solving the above equations in the usual way, one may identify the relevant polarizabilities using eq.(5.111).

7.5 Dipole approximation for spheres

In the dipole approximation all multipole moments of order $\ell \geq 2$ are neglected, so that eq. (7.16) becomes a finite set of linear equations for the amplitudes A_{10}, A_{11} and $A_{1,-1}$. For the regular array case one furthermore has for lattices with 4- or 6-fold symmetry, that

$$S_{2,-1} = S_{21} = S^r_{2,-1} = S^r_{21} = S_{2,-2} = S_{22} = S^r_{2,-2} = S^r_{22} = 0 \tag{7.28}$$

Substitution of all these zeros into eq.(7.16) and using eq.(7.9) leads to an expression, in which $A_{\ell,-1}$, $A_{\ell 0}$ and $A_{\ell 1}(\ell = 1)$ are again decoupled and one finds

$$A_{10} = \alpha_1\{E_0 \cos\theta_0 \sqrt{4\pi/3} + 2\ A_{10}L^{-3}[S_{20} - (\frac{\epsilon_a - \epsilon_s}{\epsilon_a + \epsilon_s})S^r_{20}]\sqrt{4\pi/5}\} \tag{7.29}$$

and

$$A_{11} = -\alpha_1\{E_0 \sin\theta_0 \exp(-i\phi_0)\sqrt{2\pi/3} + A_{11}L^{-3}[S_{20} + (\frac{\epsilon_a - \epsilon_s}{\epsilon_a + \epsilon_s})S^r_{20}]\sqrt{4\pi/5}\} \tag{7.30}$$

where $A_{1,-1}\exp(-i\phi_0) = -A_{11}\exp(i\phi_0)$, cf. section 7.3. Notice that for lattices with 4- or 6-fold symmetry different m's do not couple. Solving these equations is straightforward and using eq.(5.111) one finds for the polarizabilities

$$\alpha_z(0) = 4\pi\epsilon_a\alpha_1\{1 - 2\alpha_1 L^{-3}[S_{20} - (\frac{\epsilon_a - \epsilon_s}{\epsilon_a + \epsilon_s})S^r_{20}]\sqrt{4\pi/5}\}^{-1} \tag{7.31}$$

$$\alpha_{\|}(0) = 4\pi\epsilon_a\alpha_1\{1 + \alpha_1 L^{-3}[S_{20} + (\frac{\epsilon_a - \epsilon_s}{\epsilon_a + \epsilon_s})S^r_{20}]\sqrt{4\pi/5}\}^{-1} \tag{7.32}$$

As indicated by the argument 0 these polarizabilities are for a dipole located in the center of the spheres.

For the special case of a homogeneous spherical particle one has, cf. eq.(5.70),

$$\alpha_1 = \frac{\epsilon - \epsilon_a}{\epsilon + 2\epsilon_a} R^3 \tag{7.33}$$

where R is the radius of the sphere and ϵ its (frequency dependent) dielectric constant. If this expression is substituted into eqs.(7.31) and (7.32) one finds

$$\alpha_z(0) = \epsilon_a V [\frac{\epsilon - \epsilon_a}{\epsilon_a + L_z(\epsilon - \epsilon_a)}] \tag{7.34}$$

$$\alpha_\parallel(0) = \epsilon_a V [\frac{\epsilon - \epsilon_a}{\epsilon_a + L_\parallel(\epsilon - \epsilon_a)}] \tag{7.35}$$

where $V \equiv (4\pi/3)R^3$ is the volume of the sphere. Furthermore the depolarization factors are found to be given by

$$L_z = \frac{1}{3}\{1 - 4\left(\frac{R}{L}\right)^3 [S_{20} - (\frac{\epsilon_a - \epsilon_s}{\epsilon_a + \epsilon_s}) S_{20}^r] \sqrt{\pi/5}\} \tag{7.36}$$

$$L_\parallel = \frac{1}{3}\{1 + 2\left(\frac{R}{L}\right)^3 [S_{20} + (\frac{\epsilon_a - \epsilon_s}{\epsilon_a + \epsilon_s}) S_{20}^r] \sqrt{\pi/5}\} \tag{7.37}$$

In the low coverage limit, where S_{20} reduces to zero and $2S_{20}^r\sqrt{\pi/5}$ to $(L/2d)^3$, these depolarization factors become equal to L_z and $L_\parallel$, as given by eqs.(5.88) and (5.89) in chapter 5. Notice also the fact that, due to the presence of the substrate, the sum of the three depolarization factors is no longer equal to one:

$$L_z + 2L_\parallel = 1 + \frac{8}{3}\left(\frac{R}{L}\right)^3 (\frac{\epsilon_a - \epsilon_s}{\epsilon_a + \epsilon_s}) S_{20}^r \sqrt{\pi/5} \tag{7.38}$$

A similar feature also holds in the low coverage limit.

The above formulae were all given for the regular array case with 4- or 6-fold symmetry. From these expressions one may obtain the analogous expressions for the random array by replacing S_{20} and S_{20}^r by I_2 and I_2^r respectively.

In chapter 5 expressions were derived, giving the interfacial susceptibilities in terms of the polarizabilities. In that case these polarizabilities were calculated, neglecting the interaction with the other islands along the substrate. In the analysis in this chapter the modifications, due to this interaction, have been calculated. The polarization of the film is again given in terms of the modified polarizabilities times the number of islands per unit of area. Thus one may again use eqs.(5.39), (5.40) and (5.43), which give for the interfacial susceptibilities

$$\begin{aligned}
\beta_e(d) &= \rho\, \epsilon_a^{-2}\alpha_z(0) = \epsilon_a^{-1}\rho V[\frac{\epsilon-\epsilon_a}{\epsilon_a+L_z(\epsilon-\epsilon_a)}] \\
\gamma_e(d) &= \rho\, \alpha_{\|}(0) = \epsilon_a\rho V[\frac{\epsilon-\epsilon_a}{\epsilon_a+L_{\|}(\epsilon-\epsilon_a)}] \\
\delta_e(d) &= \rho\, d\, \epsilon_a^{-1}[\alpha_{\|}(0)+\alpha_z(0)] = \rho V d[\frac{\epsilon-\epsilon_a}{\epsilon_a+L_{\|}(\epsilon-\epsilon_a)} + \frac{\epsilon-\epsilon_a}{\epsilon_a+L_z(\epsilon-\epsilon_a)}] \\
\tau(d) &= \rho\, d\, \alpha_{\|}(0) = \epsilon_a\rho V d[\frac{\epsilon-\epsilon_a}{\epsilon_a+L_{\|}(\epsilon-\epsilon_a)}] \qquad (7.39)
\end{aligned}$$

The argument d denotes the position of the dividing surface at $z = d$. All these expressions for the susceptibilities reduce to those given in chapter 5 for spheres, if one neglects the terms due to the interactions between the islands. In that case S_{20} reduces to zero and $2S_{20}^{\tau}\sqrt{\pi/5}$ to $(L/2d)^3$, so that the depolarization factors become equal to those given in eqs.(5.88) and (5.89). Just as in the previous chapters one finds that, if one changes the direction of the z-axis, γ_e and β_e remain unchanged, whereas δ_e and τ change sign. If one uses the above formulae to calculate the reflection and transmission amplitudes, it is important to take the right sign. If, as in chapter 4, the light is incident through the ambient, one may use the above expressions for the constitutive coefficients. If, however, the light is incident through the substrate one must, in order to apply the formulae of chapter 4, interchange ϵ_a and ϵ_s in these formulae and change the sign of δ_e and τ in the expressions (7.39), before substituting them into the formulae of chapter 4.

7.6 Quadrupole approximation for spheres

In the quadrupole approximation all multipole moments of order $\ell > 2$ are neglected, so that eq.(7.16) becomes a finite set of linear equations for $A_{1,-1}$, A_{10}, A_{11}, $A_{2,-2}$, $A_{2,-1}$, A_{20}, A_{21} and A_{22}. For the regular array case with a 4-fold or 6-fold symmetry (square and triangular lattices respectively)

$$S_{\ell m} = S_{\ell m}^{\tau} = 0 \;\text{ for }\; m \neq \text{multiple of } \;4 \text{ or } 6\; \text{ respectively} \qquad (7.40)$$

Substitution of all these zeros into the above mentioned finite set of equations leads, with the definition, eq. (7.9), of $H(\ell, m|\ell_2, m_2)$ to

$$\begin{aligned}
A_{10} &= \alpha_1(2E_0\cos\theta_0\sqrt{\pi/3} + 4S_{20}^{-}A_{10}L^{-3}\sqrt{\pi/5} - 2S_{30}^{+}A_{20}L^{-4}\sqrt{15\pi/7}) \\
A_{20} &= \alpha_2(6S_{30}^{-}A_{10}L^{-4}\sqrt{3\pi/35} - 4S_{40}^{+}A_{20}L^{-5}\sqrt{\pi}) \\
A_{11} &= \alpha_1(-E_0\sin\theta_0 e^{-i\phi_0}\sqrt{2\pi/3} - 2S_{20}^{+}A_{11}L^{-3}\sqrt{\pi/5} + 2S_{30}^{-}A_{21}L^{-4}\sqrt{5\pi/7}) \\
A_{21} &= \alpha_2(-6S_{30}^{+}A_{11}L^{-4}\sqrt{\pi/35} + \frac{8}{3}S_{30}^{-}A_{21}L^{-4}\sqrt{\pi}) \qquad (7.41)
\end{aligned}$$

Here the abbreviations

$$\begin{aligned}
S_{\ell m}^{+} &\equiv S_{\ell m} + \left(\frac{\epsilon_a-\epsilon_s}{\epsilon_a+\epsilon_s}\right)S_{\ell m}^{\tau} \\
S_{\ell m}^{-} &\equiv S_{\ell m} - \left(\frac{\epsilon_a-\epsilon_s}{\epsilon_a+\epsilon_s}\right)S_{\ell m}^{\tau} \qquad (7.42)
\end{aligned}$$

have been introduced. Notice that for lattices with 4- or 6-fold symmetry different m's do not couple. The equations for $A_{1,-1}$ and $A_{2,-1}$ are superfluous, since $A_{\ell,-1}\exp(-i\phi_0) = -A_{\ell 1}\exp(i\phi_0)$, ($\ell = 1, 2$). It is furthermore found that in quadrupole approximation

$$A_{22} = A_{2,-2} = 0 \tag{7.43}$$

Solving eq.(7.41) and using eq.(5.111), one finds for the polarizabilities normal to the substrate

$$\begin{aligned}
\alpha_z(0) &= \frac{4\,\pi\,\epsilon_a\alpha_1[1 + 4\alpha_2 L^{-5}S^+_{40}\sqrt{\pi}]}{[1 - 4\alpha_1 L^{-3}S^-_{20}\sqrt{\pi/5}][1 + 4\alpha_2 L^{-5}S^+_{40}\sqrt{\pi}] + (36/7)\pi\alpha_1\alpha_2 L^{-8}S^+_{30}S^-_{30}} \\
\alpha_z^{10}(0) &= \frac{12\,\pi\,\epsilon_a\alpha_1\alpha_2 L^{-4}S^-_{30}\sqrt{\pi/7}}{[1 - 4\alpha_1 L^{-3}S^-_{20}\sqrt{\pi/5}][1 + 4\alpha_2 L^{-5}S^+_{40}\sqrt{\pi}] + (36/7)\pi\alpha_1\alpha_2 L^{-8}S^+_{30}S^-_{30}}
\end{aligned} \tag{7.44}$$

and for the polarizabilities parallel to the substrate

$$\begin{aligned}
\alpha_\parallel(0) &= \frac{4\,\pi\,\epsilon_a\alpha_1[1 - \frac{8}{3}\alpha_2 L^{-5}S^-_{40}\sqrt{\pi}]}{[1 + 2\alpha_1 L^{-3}S^+_{20}\sqrt{\pi/5}][1 - \frac{8}{3}\alpha_2 L^{-5}S^-_{40}\sqrt{\pi}] + (12/7)\pi\alpha_1\alpha_2 L^{-8}S^-_{30}S^+_{30}} \\
\alpha_\parallel^{10}(0) &= \frac{-8\pi\,\epsilon_a\alpha_1\alpha_2 L^{-4}S^+_{30}\sqrt{\pi/7}}{[1 + 2\alpha_1 L^{-3}S^+_{20}\sqrt{\pi/5}][1 - \frac{8}{3}\alpha_2 L^{-5}S^-_{40}\sqrt{\pi}] + (12/7)\pi\alpha_1\alpha_2 L^{-8}S^-_{30}S^+_{30}}
\end{aligned} \tag{7.45}$$

As indicated by the argument 0, these polarizabilities are for a dipole and a quadrupole located in the center of the spheres.

For the special case of a homogeneous spherical particle one has, cf. eq.(5.70),

$$\alpha_1 = \frac{\epsilon - \epsilon_a}{\epsilon + 2\epsilon_a}R^3 \text{ and } \alpha_2 = \frac{2(\epsilon - \epsilon_a)}{2\epsilon + 3\epsilon_a}R^5 \tag{7.46}$$

If these expressions are substituted into eqs. (7.44) and (7.45), one finds

$$\begin{aligned}
\alpha_z(0) &= \frac{\epsilon_a(\epsilon - \epsilon_a)V[\epsilon_a + L_z^1(\epsilon - \epsilon_a)]}{[\epsilon_a + L_z(\epsilon - \epsilon_a)][\epsilon_a + L_z^1(\epsilon - \epsilon_a)] + \Lambda_z^1(\epsilon - \epsilon_a)^2} \\
\alpha_z^{10}(0) &= \frac{\epsilon_a(\epsilon - \epsilon_a)^2\Lambda_z^{10}V\,R}{[\epsilon_a + L_z(\epsilon - \epsilon_a)][\epsilon_a + L_z^1(\epsilon - \epsilon_a)] + \Lambda_z^1(\epsilon - \epsilon_a)^2}
\end{aligned} \tag{7.47}$$

and

$$\begin{aligned}
\alpha_\parallel(0) &= \frac{\epsilon_a(\epsilon - \epsilon_a)V[\epsilon_a + L_\parallel^1(\epsilon - \epsilon_a)]}{[\epsilon_a + L_\parallel(\epsilon - \epsilon_a)][\epsilon_a + L_\parallel^1(\epsilon - \epsilon_a)] + \Lambda_\parallel^1(\epsilon - \epsilon_a)^2} \\
\alpha_\parallel^{10}(0) &= \frac{\epsilon_a(\epsilon - \epsilon_a)^2\Lambda_\parallel^{10}V\,R}{[\epsilon_a + L_\parallel(\epsilon - \epsilon_a)][\epsilon_a + L_\parallel^1(\epsilon - \epsilon_a)] + \Lambda_\parallel^1(\epsilon - \epsilon_a)^2}
\end{aligned} \tag{7.48}$$

where $V \equiv (4\pi/3)R^3$ is the volume of the sphere. The depolarization factors in these equations are found to be given by

$$\begin{aligned}
L_z &= \frac{1}{3}\{1 - 4\left(\frac{R}{L}\right)^3 [S_{20} - (\frac{\epsilon_a - \epsilon_s}{\epsilon_a + \epsilon_s})S_{20}^r]\sqrt{\pi/5}\} \\
L_z^1 &= \frac{2}{5}\{1 + 4\left(\frac{R}{L}\right)^5 [S_{40} + (\frac{\epsilon_a - \epsilon_s}{\epsilon_a + \epsilon_s})S_{40}^r]\sqrt{\pi}\} \\
\Lambda_z^1 &= \frac{24}{35}\pi\left(\frac{R}{L}\right)^8 [(S_{30})^2 - (\frac{\epsilon_a - \epsilon_s}{\epsilon_a + \epsilon_s})^2 (S_{30}^r)^2] \\
\Lambda_z^{10} &= \frac{6}{5}\left(\frac{R}{L}\right)^4 [S_{30} - (\frac{\epsilon_a - \epsilon_s}{\epsilon_a + \epsilon_s})S_{30}^r]\sqrt{\pi/7}
\end{aligned} \tag{7.49}$$

and

$$\begin{aligned}
L_\parallel &= \frac{1}{3}\{1 + 2\left(\frac{R}{L}\right)^3 [S_{20} + (\frac{\epsilon_a - \epsilon_s}{\epsilon_a + \epsilon_s})S_{20}^r]\sqrt{\pi/5}\} \\
L_\parallel^1 &= \frac{2}{5}\{1 - \frac{8}{3}\left(\frac{R}{L}\right)^5 [S_{40} - (\frac{\epsilon_a - \epsilon_s}{\epsilon_a + \epsilon_s})S_{40}^r]\sqrt{\pi}\} \\
\Lambda_\parallel^1 &= \frac{8}{35}\pi\left(\frac{R}{L}\right)^8 [(S_{30})^2 - (\frac{\epsilon_a - \epsilon_s}{\epsilon_a + \epsilon_s})^2 (S_{30}^r)^2] \\
\Lambda_\parallel^{10} &= -\frac{4}{5}\left(\frac{R}{L}\right)^4 [S_{30} + (\frac{\epsilon_a - \epsilon_s}{\epsilon_a + \epsilon_s})S_{30}^r]\sqrt{\pi/7}
\end{aligned} \tag{7.50}$$

where the definitions (7.42) have been used. The depolarization factors L_z and $L_\parallel$ are the same as those found in the dipole model. If one takes the other depolarization factors zero the results of the quadrupole model reduces to those of the dipole model.

The above formulae, eqs.(7.49) and (7.50), were all given for the regular array case, with 4- and 6-fold symmetry (square and triangular lattices). From these expressions one may obtain again the analogous expressions for the random array case, by replacing $S_{\ell 0}$ and $S_{\ell 0}^r$ by I_ℓ and I_ℓ^r, respectively ($\ell = 2, 3, 4$). It can furthermore be checked that, if one neglects the interaction between the islands, in which case $S_{\ell 0}$ reduces to zero and $S_{\ell 0}^r$ to $\frac{1}{2}(L/2d)^{\ell+1}\sqrt{(2\ell+1)/\pi}$, ($\ell = 2, 3, 4$), the above depolarization factors become equal to those given by eqs.(5.88) and (5.89).

In a completely analogous way as in the previous section the interfacial susceptibilities may be derived, using the polarizabilities found above. From eqs.(5.64), (5.60), (7.47) and (7.48) one then obtains the following result:

$$\begin{aligned}
\beta_e(d) &= \rho\,\epsilon_a^{-2}\alpha_z(0) = \frac{\rho\,\epsilon_a^{-1}(\epsilon - \epsilon_a)V[\epsilon_a + L_z^1(\epsilon - \epsilon_a)]}{[\epsilon_a + L_z(\epsilon - \epsilon_a)][\epsilon_a + L_z^1(\epsilon - \epsilon_a)] + \Lambda_z^1(\epsilon - \epsilon_a)^2} \\
\gamma_e(d) &= \rho\,\alpha_\parallel(0) = \frac{\rho\,\epsilon_a(\epsilon - \epsilon_a)V[\epsilon_a + L_\parallel^1(\epsilon - \epsilon_a)]}{[\epsilon_a + L_\parallel(\epsilon - \epsilon_a)][\epsilon_a + L_\parallel^1(\epsilon - \epsilon_a)] + \Lambda_\parallel^1(\epsilon - \epsilon_a)^2}
\end{aligned}$$

$$
\begin{aligned}
\delta_e(d) &= -\rho\, \epsilon_a^{-1}[\alpha_z^{10}(0) + \alpha_{\|}^{10}(0) - d\alpha_z(0) - d\alpha_{\|}(0)] \\
&= \rho(\epsilon - \epsilon_a)V\; d\{\frac{\epsilon_a + (L_z^1 - \Lambda_z^{10}R/d)\,(\epsilon - \epsilon_a)}{[\epsilon_a + L_z(\epsilon - \epsilon_a)][\epsilon_a + L_z^1(\epsilon - \epsilon_a)] + \Lambda_z^1(\epsilon - \epsilon_a)^2} \\
&\quad + \frac{\epsilon_a + \left(L_{\|}^1 - \Lambda_{\|}^{10}R/d\right)(\epsilon - \epsilon_a)}{[\epsilon_a + L_{\|}(\epsilon - \epsilon_a)][\epsilon_a + L_{\|}^1(\epsilon - \epsilon_a)] + \Lambda_{\|}^1(\epsilon - \epsilon_a)^2}\} \\
\tau(d) &= -\rho[\alpha_{\|}^{10}(0) - d\alpha_{\|}(0)] \\
&= \frac{\rho\, \epsilon_a(\epsilon - \epsilon_a)V\; d\left[\epsilon_a + \left(L_{\|}^1 - \Lambda_{\|}^{10}R/d\right)(\epsilon - \epsilon_a)\right]}{[\epsilon_a + L_{\|}(\epsilon - \epsilon_a)][\epsilon_a + L_{\|}^1(\epsilon - \epsilon_a)] + \Lambda_{\|}^1(\epsilon - \epsilon_a)^2}
\end{aligned}
\tag{7.51}
$$

These expressions are of the same form as obtained in chapter 5: see eqs.(5.92)–(5.95). The argument d denotes the position of the dividing surface at $z = d$. Notice again that, if one changes the direction of the z-axis, γ_e and β_e remain unchanged, whereas δ_e and τ change sign. If one uses the above formulae to calculate the reflection and transmission amplitudes, it is important to take the right sign. If, as in chapter 4, the light is incident through the ambient, one may use the expressions, eq.(7.51). If, however, the light is incident through the substrate, one must, in order to apply the formulae of chapter 4, interchange ϵ_a and ϵ_s in these formulae and change the sign of δ_e and τ in the above expressions, before substituting them into the formulae of chapter 4.

7.7 Two-dimensional arrays of spheroids

The calculation of the polarizabilities in arrays of identical spherical islands will now be extended to arrays of identical spheroidal islands. For this case it is not a good idea, in particular for oblate spheroids, to use the spherical multipole expansion to describe the interaction with the image of the island in the substrate, cf. the discussion about the limitation of the polarizable dipole and quadrupole model after eq.(6.43). The theory will therefore use as basic element the polarizability of the single particle, interacting with its image, as calculated in sections 6.4 and 6.5, using the spheroidal multipole expansion.

It would be most rigorous if the interaction between these elements could be treated also using the spheroidal multipole expansion. The mathematical techniques to do this analytically, regretfully enough, do not seem to be available. If the islands are not too densely packed, it is sufficient to take this interaction along in a spherical multipole expansion. This will be done below.

First a 2-dimensional array of identical oblate spheroids, with axis of revolution normal to the substrate, and centers at a distance d from the substrate, will be considered, cf. fig. 7.2. Using eq.(6.107), one finds for the potential of the local field

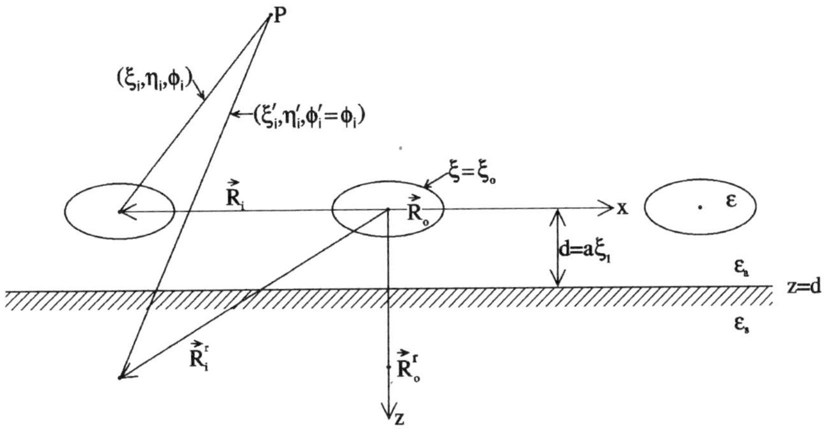

Figure 7.2 Array of oblate spheroids above the substrate.

acting on the oblate spheroid j:

$$\begin{aligned}\psi_{j,loc}(\mathbf{r}) &= -\mathbf{E}_0.\mathbf{R}_j + (2\pi/3)^{1/2}[-2^{1/2}E_{0,z}X_1^0(\xi_j,a)Y_1^0(\arccos\eta_j,\phi_j)\\ &\quad +(E_{0,x}-i\ E_{0,y})X_1^1(\xi_j,a)Y_1^1(\arccos\eta_j,\phi_j)\\ &\quad -(E_{0,x}+i\ E_{0,y})X_1^{-1}(\xi_j,a)Y_1^{-1}(\arccos\eta_j,\phi_j)]\\ &\quad +\sum_{\ell m}{}' A^r_{\ell m,j}Z_\ell^m(\xi'(\xi_j,\eta_j),a)Y_\ell^m(\arccos\eta'(\xi_j,\eta_j),\phi_j)\\ &\quad +\sum_{\ell m}{}'\sum_{i\neq j} A_{\ell m,i}Z_\ell^m(\xi_i,a)Y_\ell^m(\arccos\eta_i,\phi_i)\\ &\quad +\sum_{\ell m}{}'\sum_{i\neq j} A^r_{\ell m,i}Z_\ell^m(\xi'(\xi_i,\eta_i),a)Y_\ell^m(\arccos\eta'(\xi_i,\eta_i),\phi_i) \qquad (7.52)\end{aligned}$$

For the meaning of the various symbols and functions the reader is referred to sections 6.2 and 6.4. The subindexes of the oblate spheroidal coordinates indicate the particle around whose center these coordinates are taken, cf. fig. 7.2. The first three lines, of the above equation, give the incident and reflected electric fields, in the absence of the island. The fourth line gives the field due to the image of particle j. The prime as summation index indicates, as usual, that $\ell = 0$ is excluded. The fifth line gives the field due to the other islands along the substrate and the last line gives the field due to their images. In the last two contributions, it is then assumed that the distance between the islands along the substrate is long enough so that one may use eqs.(6.18)–(6.20). This gives

$$\begin{aligned}\psi_{j,loc}(\mathbf{r}) &= -\mathbf{E}_0.\mathbf{R}_j + (2\pi/3)^{1/2}[-2^{1/2}E_{0,z}X_1^0(\xi_j,a)Y_1^0(\arccos\eta_j,\phi_j) \\ &+(E_{0,x} - i\ E_{0,y})X_1^1(\xi_j,a)Y_1^1(\arccos\eta_j,\phi_j) \\ &-(E_{0,x} + i\ E_{0,y})X_1^{-1}(\xi_j,a)Y_1^{-1}(\arccos\eta_j,\phi)] \\ &+\sum_{\ell m}{}' A^r_{\ell m,j} Z_\ell^m(\xi'(\xi_j,\eta_j),a)Y_\ell^m(\arccos\eta'(\xi_j,\eta_j),\phi_j) \\ &+\sum_{\ell m}{}'\sum_{i\neq j} A_{\ell m,i} r_i^{-\ell-1} Y_\ell^m(\theta_i,\phi_i) + \sum_{\ell m}{}'\sum_{i\neq j} A^r_{\ell m,i}\rho_i^{-\ell-1}Y_\ell^m(\theta_i^r,\phi_i^r)\end{aligned} \tag{7.53}$$

Following the same method, as used to obtain eq.(7.11) from eq.(7.6), one now obtains, using eqs.(6.21) and (6.108),

$$\begin{aligned}A_{\ell m,j} &= -\alpha_{\ell,m}\{(2\pi/3)^{1/2}\delta_{l1}[-2^{1/2}E_{0,z}\delta_{m0} + (E_{0,x} - i\ E_{0,y})\delta_{m1} \\ &-(E_{0,x} + i\ E_{0,y})\delta_{m,-1}] \\ &+\left(\frac{\epsilon_a - \epsilon_s}{\epsilon_a + \epsilon_s}\right)\sum_{\ell'=|m|}^{\infty\prime}(-1)^{\ell'+m}K^m_{\ell\ell'}(\xi_1)\,A_{\ell' m,j}(2d)^{-\ell-\ell'-1} \\ &+\sum_{\ell_2 m_2}{}' H(\ell,m|\ell_2,m_2)\sum_{i\neq j} A_{\ell_2 m_2,i}[T_{\ell_1 m_1}(\mathbf{R}_i - \mathbf{R}_j) \\ &+(-1)^{\ell_2+m_2}\left(\frac{\epsilon_a - \epsilon_s}{\epsilon_a + \epsilon_s}\right)T_{\ell_1 m_1}(\mathbf{R}_i^r - \mathbf{R}_j)]\}\end{aligned} \tag{7.54}$$

where $\ell_1 \equiv \ell + \ell_2$ and $m_1 \equiv m_2 - m$. Furthermore $\alpha_{\ell,m}$ are the multipole polarizabilities of the oblate spheroids in the ambient (with dielectric constant ϵ_a), far away from the substrate (with dielectric constant ϵ_s), cf. eq.(6.24). The matrices $H(\ell,m|\ell_2,m_2)$ and $T_{\ell_1 m_1}(\mathbf{R})$ are defined by eqs.(7.9) and (7.8) respectively. In section 6.4 it is discussed how the matrix elements $K^m_{\ell\ell_1}$ can be found using recursion relations. In the approximation of the previous chapter the distance between the particles was assumed to be so large compared to the size, that the last term in the above expression, describing the coupling between the particles, may be neglected. The equations for the amplitudes of the different particles then decouple and the above equation reduces to eq.(6.120) for each oblate spheroid. In this chapter the interest is, in particular, in the consequences of the coupling between the particles and the further analysis will be based on the above equation.

In order to obtain a more convenient form of the above equation it is first written as

$$\sum_{\ell'=|m|}^{\infty\prime}[\delta_{\ell\ell'}+\left(\frac{\epsilon_a-\epsilon_s}{\epsilon_a+\epsilon_s}\right)\alpha_{\ell m}(-1)^{\ell'+m}K^m_{\ell\ell'}(\xi_1)(2d)^{-\ell-\ell'-1}]A_{\ell' m,j}=$$
$$-\alpha_{\ell,m}\{(2\pi/3)^{1/2}\delta_{l1}[-2^{1/2}E_{0,z}\delta_{m0}+(E_{0,x}-i\,E_{0,y})\delta_{m1}-(E_{0,x}+i\,E_{0,y})\delta_{m,-1}]$$
$$+\sum_{\ell_2 m_2}^{\prime}H(\ell,m|\ell_2,m_2)\sum_{i\neq j}A_{\ell_2 m_2,i}[T_{\ell_1 m_1}(\mathbf{R}_i-\mathbf{R}_j)$$
$$+(-1)^{\ell_2+m_2}\left(\frac{\epsilon_a-\epsilon_s}{\epsilon_a+\epsilon_s}\right)T_{\ell_1 m_1}(\mathbf{R}_i^r-\mathbf{R}_j)]\} \tag{7.55}$$

Introducing the inverse matrix $(\Omega^{-1})^m_{\ell\ell_1}$ of

$$\Omega^m_{\ell\ell_1}\equiv\delta_{\ell\ell_1}+\left(\frac{\epsilon_a-\epsilon_s}{\epsilon_a+\epsilon_s}\right)\alpha_{\ell m}(-1)^{\ell_1+m}K^m_{\ell\ell_1}(2d)^{-\ell-\ell_1-1} \tag{7.56}$$

by means of

$$\sum_{\ell'=|m|}^{\infty\prime}(\Omega^{-1})^m_{\ell\ell'}\Omega^m_{\ell'\ell_1}=\delta_{\ell\ell_1}\quad\text{for}\quad \ell,\ell_1=|m|,|m|+1,\ldots.. \tag{7.57}$$

the above equation for the amplitudes can be transformed into

$$A_{\ell m,j} \;=\; -\sum_{\ell'=|m|}^{\infty\prime}(\Omega^{-1})^m_{\ell\ell'}\alpha_{\ell' m}\{\sqrt{2\pi/3}\delta_{\ell' 1}$$
$$\times[-2^{1/2}E_{0,z}\delta_{m0}+(E_{0,x}-i\,E_{0,y})\delta_{m1}-(E_{0,x}+i\,E_{0,y})\delta_{m,-1}]$$
$$+\sum_{\ell_2 m_2}^{\prime}H(\ell',m|\ell_2,m_2)\sum_{i\neq j}A_{\ell_2 m_2,i}[T_{\ell_1 m_1}(\mathbf{R}_i-\mathbf{R}_j)$$
$$+(-1)^{\ell_2+m_2}\left(\frac{\epsilon_a-\epsilon_s}{\epsilon_a+\epsilon_s}\right)T_{\ell_1 m_1}(\mathbf{R}_i^r-\mathbf{R}_j)]\} \tag{7.58}$$

where $\ell_1\equiv\ell'+\ell_2$ and $m_1\equiv m_2-m$. Introducing the multipole polarizabilities in the low coverage limit of an oblate spheroidal island, interacting with its image in the substrate:

$$\alpha_{\ell m,\ell' m}\equiv(\Omega^{-1})^m_{\ell\ell'}\alpha_{\ell' m}\quad\text{for}\quad \ell,\ell'=|m|,|m|+1,\ldots.. \tag{7.59}$$

and writing the external field in polar coordinates, cf. eq.(7.1), one obtains

$$A_{\ell m,j} \;=\; -\sum_{\ell'=|m|}^{\infty\prime}\alpha_{\ell m,\ell' m}\{-E_0\sqrt{2\pi/3}\delta_{\ell' 1}[\cos\theta_0\delta_{m0}\sqrt{2}$$
$$+\sin\theta_0(\exp(i\phi_0)\delta_{m,-1}-\exp(-i\phi_0)\delta_{m1})]$$
$$+\sum_{\ell_2 m_2}^{\prime}H(\ell',m|\ell_2,m_2)\sum_{i\neq j}A_{\ell_2 m_2,i}[T_{\ell_1 m_1}(\mathbf{R}_i-\mathbf{R}_j)$$
$$+(-1)^{\ell_2+m_2}\left(\frac{\epsilon_a-\epsilon_s}{\epsilon_a+\epsilon_s}\right)T_{\ell_1 m_1}(\mathbf{R}_i^r-\mathbf{R}_j)]\} \tag{7.60}$$

The lower multipole polarizabilities for oblate spheroids in the low coverage limit, $\alpha_{\ell m,\ell' m}$, appearing in this equation can be calculated using the analysis outlined in section 6.4. Note that $\alpha_{\ell m,\ell' m'} = 0$ for $m \neq m'$ due to symmetry for rotation around the short axis. Higher multipole polarizabilities can be calculated by extending the theory developed in section 6.4 to non-homogeneous external fields. This is done in appendix B up to the order needed. For arrays, in which the particles are not packed too densely, the dipole or the quadrupole approximation, discussed in the following sections, for the interaction along the substrate are sufficient.

For arrays of prolate spheroids, with the axis of revolution normal to the substrate, the derivation, sketched above for oblate spheroids, follows the same lines. The resulting equation (7.60) is in fact the same. The needed multipole polarizabilities should now be calculated using section 6.5 and appendix B.

As has been discussed at the end of section 7.2, the solution of eq.(7.60), with an infinite number of coefficients $A_{\ell m,j}$, is only possible if one makes further assumptions about the positions of the particles. If the islands are located on a regular array with only one spheroid per unit cell

$$A_{\ell m,j} = A_{\ell m} \tag{7.61}$$

(cf. eq.(7.12)). Though this will not be done in this book, the analysis can also be given for the case of more particles per unit cell. Substituting the above relation into eq.(7.60) one obtains the following infinite set of linear inhomogeneous equations for the amplitudes:

$$\begin{aligned} A_{\ell m} &= -\sum_{\ell'=|m|}^{\infty\prime} \alpha_{\ell m,\ell' m}\{-E_0\sqrt{2\pi/3}\delta_{\ell' 1}[\cos\theta_0\delta_{m0}\sqrt{2} \\ &+ \sin\theta_0(\exp(i\phi_0)\delta_{m,-1} - \exp(-i\phi_0)\delta_{m1})] \\ &+ \sum_{\ell_2 m_2}{}^{\prime} H(\ell', m|\ell_2, m_2) A_{\ell_2 m_2} L^{-\ell_1 - 1}[S_{\ell_1 m_1} + (-1)^{\ell_2+m_2}\left(\frac{\epsilon_a - \epsilon_s}{\epsilon_a + \epsilon_s}\right)\widetilde{S}^r_{\ell_1 m_1}]\} \end{aligned} \tag{7.62}$$

Here the lattice sums $S_{\ell m}$ has been defined by eq.(7.14), while the other lattice sum is given by

$$\begin{aligned} \widetilde{S}^r_{\ell m} &\equiv L^{\ell+1}\sum_{i\neq j} T_{\ell m}(\mathbf{R}^r_i - \mathbf{R}_j) = L^{\ell+1}\sum_{i\neq 0} T_{\ell m}(\mathbf{R}^r_i) = \sum_{i\neq 0}(L/r)^{\ell+1} Y^m_\ell(\theta,\phi)|_{\mathbf{r}=\mathbf{R}_i} \\ &= S^r_{\ell m} - \frac{1}{2}\delta_{m0}\left(\frac{L}{2d}\right)^{\ell+1}\sqrt{\frac{2\ell+1}{\pi}} \end{aligned} \tag{7.63}$$

cf. eq.(7.15) for the last identity. The solution of equation (7.62) for the amplitudes, may be performed in an analogous way as that of eq.(7.16), described at the end of section 7.3. First one must calculate the lattice sums. For the procedure to do this, the reader is referred to appendix A. Notice again that an important simplification is the fact, that for lattices with 2-, 4- and 6-fold symmetry $S_{\ell m} = \widetilde{S}^r_{\ell m} = 0$, if m is unequal to a multiple of 2, 4 or 6 respectively. As a consequence of the interaction with the other islands, there is no longer a symmetry for rotation around the normal on the substrate through the center of a spheroid. This implies that the interaction along the surface, the last term in eq.(7.62), leads to coupling of different values of m.

For spheroids the solution of eq.(7.62) to arbitrary order in the multipole expansion is complicated by the need to first calculate the polarizabilities $\alpha_{\ell m,\ell' m}$ to the appropriate order. This is very complicated, and will therefore be done only to quadrupolar order in the interaction between the islands in the following two sections. For spheres the needed polarizabilities are known to arbitrary order. For that case the influence of higher order multipoles on the solution may be more easily assessed. See section 7.10 for a further discussion of this matter.

For a random array of identical spheroids, distributed homogeneously and isotropically on a flat substrate one obtains, using the same statistical approximations as in section 7.4, from eq.(7.60) the following set of equations for the average amplitudes

$$< A_{\ell m} > = < A_{\ell m,j} > = \sum_{\ell'=|m|}^{\infty\prime} \alpha_{\ell m,\ell' m} \{ \sqrt{2\pi/3} \delta_{\ell' 1} E_0 [\cos\theta_0 \delta_{m0} \sqrt{2}$$
$$+ \sin\theta_0 (\exp(i\phi_0)\delta_{m,-1} - \exp(-i\phi_0)\delta_{m1})]$$

$$- \sum_{\ell_2=|m|}^{\infty\prime} H(\ell', m|\ell_2, m) < A_{\ell_2 m} > \rho^{(\ell_1+1)/2} [I_{\ell_1} + (-1)^{\ell_2+m} \left(\frac{\epsilon_a - \epsilon_s}{\epsilon_a + \epsilon_s}\right) \widetilde{I}^r_{\ell_1}]\} \qquad (7.64)$$

cf. eq.(7.24). Here the "distribution integral" I_ℓ is given by eq.(7.22), and the distribution integral $\widetilde{I}^r_\ell$ by

$$\begin{aligned}
\widetilde{I}^r_\ell &\equiv \rho^{-(\ell-1)/2} \int d\mathbf{r}_\| g(\mathbf{r}_\|) T_{\ell 0}(\mathbf{r}_\|, 2d) \\
&= [2\pi\rho^{-(\ell-1)/2} \int_0^\infty dr_\| g(r_\|) r_\| (r_\|^2 + 4d^2)^{-(\ell+1)/2} \\
&\quad \times P_\ell^0 (2d/(r_\|^2 + 4d^2)^{1/2})] \sqrt{(2\ell+1)/4\pi} = \\
&= I_\ell^r - \rho^{-(\ell+1)/2} \left[(2d)^{-\ell-1} P_\ell^0(1)\right] \qquad (7.65)
\end{aligned}$$

where the last identity was obtained using eq.(7.23). The pair correlation function $g(r_\|)$ of the islands is defined by eq.(7.21) and the Legendre function $P_\ell^0(x)$ by eq.(5.27). It follows from eq.(7.64) that amplitudes with a different m do not couple. As has been remarked in section 4, this is a consequence of the isotropy of the film and the statistical approximations made in the derivation of eq.(7.64). As a consequence

$$< A_{\ell m} > = < A_{\ell m,j} > = 0 \ \text{ for } \ m \neq 0, \pm 1 \qquad (7.66)$$

(cf. eq.(7.25)). Eq.(7.64) may therefore be written as the sum of separate contributions for $m = 0, \pm 1$. In this way one obtains for $m = 0$

$$< A_{\ell 0} >= \sum_{\ell'=1}^{\infty} \alpha_{\ell 0,\ell' 0}\{\sqrt{4\pi/3}E_0 \cos\theta_0 \delta_{\ell' 1}$$

$$- \sum_{\ell_2=1}^{\infty} H(\ell', 0|\ell_2, 0) < A_{\ell_2 0} > \rho^{(\ell_1+1)/2}[I_{\ell_1} + (-1)^{\ell_2} \left(\frac{\epsilon_a - \epsilon_s}{\epsilon_a + \epsilon_s}\right) \widetilde{I}^r_{\ell_1}]\} \quad (7.67)$$

and for $m = 1$

$$< A_{\ell 1} >= -\sum_{\ell'=1}^{\infty} \alpha_{\ell 1,\ell' 1}\{\sqrt{2\pi/3}E_0 \sin\theta_0 \exp(-i\phi_0)\delta_{\ell' 1}$$

$$+ \sum_{\ell_2=1}^{\infty} H(\ell', 1|\ell_2, 1) < A_{\ell_2 1} > \rho^{(\ell_1+1)/2}[I_{\ell_1} - (-1)^{\ell_2} \left(\frac{\epsilon_a - \epsilon_s}{\epsilon_a + \epsilon_s}\right) \widetilde{I}^r_{\ell_1}]\} \quad (7.68)$$

where $\ell_1 \equiv \ell' + \ell_2$. The $m = -1$ equation is superfluous, as

$$< A_{\ell,-1} > \exp(-i\phi_0) = - < A_{\ell 1} > \exp(i\phi_0) \quad (7.69)$$

After solving the above equations in the way, explained above for eq.(7.62), one can find the relevant polarizabilities, using eq.(5.111).

In following sections eq.(7.62) for regular arrays and eqs.(7.67) and (7.68) for random arrays of spheroids will be solved in dipole and quadrupole approximation. This is sufficient, if the particles are not too densely packed on the substrate.

7.8 Dipole approximation for spheroids

First the case of a regular array of identical spheroids will be considered. In the dipole approximation all multipole moments of order $\ell \geq 2$ are neglected, so that eq.(7.62) becomes a finite set of linear equations for the amplitudes A_{10} , A_{11} and $A_{1,-1}$. For lattices with 4- or 6-fold symmetry (square or triangular lattices respectively) one has

$$S_{2,-1} = S_{21} = \widetilde{S}^r_{2,-1} = \widetilde{S}^r_{21} = S_{2,-2} = S_{22} = \widetilde{S}^r_{2,-2} = \widetilde{S}^r_{22} = 0 \quad (7.70)$$

Substitution of all these zeros into eq.(7.62) and using eq.(7.9) leads to an expression in which $A_{\ell 0}$, $A_{\ell 1}$ and $A_{\ell,-1} (\ell = 1)$ are again decoupled and one finds:

$$A_{10} = \alpha_{10,10} \left\{ E_0\sqrt{4\pi/3}\cos\theta_0 + 2A_{10}L^{-3}\left[S_{20} - \left(\frac{\epsilon_a - \epsilon_s}{\epsilon_a + \epsilon_s}\right)\widetilde{S}^r_{20}\right]\sqrt{4\pi/5}\right\}$$

$$A_{11} = -\alpha_{11,11} \left\{ E_0\sqrt{2\pi/3}\sin\theta_0 \exp(-i\phi_0) + A_{11}L^{-3}\left[S_{20} + \left(\frac{\epsilon_a - \epsilon_s}{\epsilon_a + \epsilon_s}\right)\widetilde{S}^r_{20}\right]\sqrt{4\pi/5}\right\}$$
$$(7.71)$$

where again $A_{1,-1}\exp(-i\phi_0) = -A_{11}\exp(i\phi_0)$. Notice that for lattices with 4- or 6-fold symmetry different m's do not couple. Solving these equations is straightforward and using eq.(5.111), one finds for the dipole polarizabilities of the spheroids on the substrate, as changed by the interaction with the other particles:

$$\begin{aligned}\alpha_z(0) &= 4\pi\epsilon_a\alpha_{10,10}\left\{1 - 2\alpha_{10,10}L^{-3}\left[S_{20} - \left(\frac{\epsilon_a - \epsilon_s}{\epsilon_a + \epsilon_s}\right)\widetilde{S}^{\tau}_{20}\right]\sqrt{4\pi/5}\right\}^{-1} \\ \alpha_{\|}(0) &= 4\pi\epsilon_a\alpha_{11,11}\left\{1 + \alpha_{11,11}L^{-3}\left[S_{20} + \left(\frac{\epsilon_a - \epsilon_s}{\epsilon_a + \epsilon_s}\right)\widetilde{S}^{\tau}_{20}\right]\sqrt{4\pi/5}\right\}^{-1} \qquad (7.72)\end{aligned}$$

As indicated by the argument 0, these polarizabilities are for a dipole located in the center of the spheroid. The above results and eqs.(7.31) and (7.32), valid for spheres, have an analogous form. In two aspects there are, however, differences: first the lattice sum $\widetilde{S}^{\tau}_{20}$ differs from S^{τ}_{20}. Furthermore the polarizabilities $\alpha_{10,10}$ and $\alpha_{11,11}$ on the right-hand side in eq.(7.72) are polarizabilities of an island, interacting with the substrate, whereas α_1 in eqs.(7.31) and (7.32) are free polarizabilities. The methods to obtain the polarizabilities $\alpha_{10,10}$ and $\alpha_{11,11}$ to arbitrary order in the multipole expansion have been discussed in sections 6.4 and 6.5. The procedure to calculate the lattice sums is treated in appendix A.

The interfacial susceptibilities $\beta_e(d)$, $\gamma_e(d)$, $\delta_e(d)$ and $\tau(d)$ of a film of interacting identical spheroidal islands can now be obtained from the expressions for the polarizabilities $\alpha_z(0)$ and $\alpha_{\|}(0)$, obtained above, using the relations

$$\begin{aligned}\beta_e(d) &= \rho\,\epsilon_a^{-2}\alpha_z(0) \\ \gamma_e(d) &= \rho\,\alpha_{\|}(0) \\ \delta_e(d) &= \rho\, d\,\epsilon_a^{-1}[\alpha_{\|}(0) + \alpha_z(0)] \\ \tau(d) &= \rho\, d\alpha_{\|}(0) \qquad (7.73)\end{aligned}$$

This has been proved at the end of section 7.5. The argument d denotes the position of the dividing surface at $z = d = a\xi_1$.

If one simplifies the expressions for the polarizabilities of the spheroids, by accounting for the interaction with their image only to dipolar order, one obtains the usual form

$$\begin{aligned}\alpha_z(0) &= \epsilon_a V\left[\frac{\epsilon - \epsilon_a}{\epsilon_a + L_z(\epsilon - \epsilon_a)}\right] \\ \alpha_{\|}(0) &= \epsilon_a V\left[\frac{\epsilon - \epsilon_a}{\epsilon_a + L_{\|}(\epsilon - \epsilon_a)}\right] \qquad (7.74)\end{aligned}$$

where V is the volume of the spheroid. The depolarization factors are given by

$$\begin{aligned}L_z &= (1+\xi_0^2)\left\{1 - \xi_0 \arctan\left(\frac{1}{\xi_0}\right)\right. \\ &\quad - \left(\frac{\epsilon_a - \epsilon_s}{\epsilon_a + \epsilon_s}\right)\frac{\xi_0}{\xi_1}\left[\left(\frac{3}{2} + \xi_1^2\right)\xi_1^2 \ln\left(1 + \frac{1}{\xi_1^2}\right) - \xi_1 \arctan\left(\frac{1}{\xi_1}\right) - \xi_1^2\right] \\ &\quad \left. - \frac{4}{3}\left(\frac{a}{L}\right)^3 \xi_0\left[S_{20} - \left(\frac{\epsilon_a - \epsilon_s}{\epsilon_a + \epsilon_s}\right)\widetilde{S}^{\tau}_{20}\right]\sqrt{\pi/5}\right\}\end{aligned}$$

$$\begin{aligned} L_{\|} &= \frac{1}{2}(1+\xi_0^2)\left\{\xi_0 \arctan\left(\frac{1}{\xi_0}\right) - \frac{\xi_0^2}{1+\xi_0^2}\right. \\ &+ \left(\frac{\epsilon_a - \epsilon_s}{\epsilon_a + \epsilon_s}\right)\frac{\xi_0}{\xi_1}[\xi_1 \arctan\left(\frac{1}{\xi_1}\right) - \left(\frac{3}{2} + \xi_1^2\right)\xi_1^2 \ln\left(1 + \frac{1}{\xi_1^2}\right) + \xi_1^2] \\ &\left. + \frac{4}{3}\left(\frac{a}{L}\right)^3 \xi_0 [S_{20} + \left(\frac{\epsilon_a - \epsilon_s}{\epsilon_a + \epsilon_s}\right)\widetilde{S}_{20}^r]\sqrt{\pi/5}\right\} \end{aligned} \tag{7.75}$$

for oblate spheroids and by

$$\begin{aligned} L_z &= (1-\xi_0^2)\left\{1 - \frac{1}{2}\xi_0 \ln\left(\frac{\xi_0+1}{\xi_0-1}\right)\right. \\ &+ \left(\frac{\epsilon_a - \epsilon_s}{\epsilon_a + \epsilon_s}\right)\frac{\xi_0}{\xi_1}\left[\left(\frac{3}{2} - \xi_1^2\right)\xi_1^2 \ln\left(1 - \frac{1}{\xi_1^2}\right) + \frac{1}{2}\xi_1 \ln\left(\frac{\xi_1+1}{\xi_1-1}\right) - \xi_1^2\right] \\ &\left. + \frac{4}{3}\left(\frac{a}{L}\right)^3 \xi_0 [S_{20} - \left(\frac{\epsilon_a - \epsilon_s}{\epsilon_a + \epsilon_s}\right)\widetilde{S}_{20}^r]\sqrt{\pi/5}\right\} \end{aligned}$$

$$\begin{aligned} L_{\|} &= \frac{1}{2}(1-\xi_0^2)\left\{\frac{1}{2}\xi_0 \ln\left(\frac{\xi_0+1}{\xi_0-1}\right) + \frac{\xi_0^2}{1-\xi_0^2}\right. \\ &+ \left(\frac{\epsilon_a - \epsilon_s}{\epsilon_a + \epsilon_s}\right)\frac{\xi_0}{\xi_1}\left[\frac{1}{2}\xi_1 \ln\left(\frac{\xi_1+1}{\xi_1-1}\right) + \left(\frac{3}{2} - \xi_1^2\right)\xi_1^2 \ln\left(1 - \frac{1}{\xi_1^2}\right) - \xi_1^2\right] \\ &\left. - \frac{4}{3}\left(\frac{a}{L}\right)^3 \xi_0 [S_{20} + \left(\frac{\epsilon_a - \epsilon_s}{\epsilon_a + \epsilon_s}\right)\widetilde{S}_{20}^r]\sqrt{\pi/5}\right\} \end{aligned} \tag{7.76}$$

for prolate spheroids. For the definitions of ξ_0 and ξ_1 see figures 6.5 and 6.6. The above results can be found by solution of eqs.(6.130) and (6.131), for oblate spheroids, and of eqs.(6.173) and (6.174), for prolate spheroids, with $M = 1$. Substitution of the results into eq.(5.111) and using eq.(5.67) gives $\alpha_{10,10}$ and $\alpha_{11,11}$. These must then be used in eq.(7.72). The depolarization factors now contain contributions due to the interaction along the substrate. If these are neglected, by taking the low density limit $(a/L \to 0)$, one finds the depolarization factors given in eqs.(6.134) and (6.138) for oblate speroids, and in eqs.(6.177) and (6.181) for prolate spheroids.

For random arrays one finds in a similar way as above, in dipolar approximation from eqs.(7.67) and (7.68) expressions analogous to eq.(7.72), with S_{20} and $\widetilde{S}_{20}^r$ replaced by I_2 and $\widetilde{I}_2^r$ respectively, and L^{-3} replaced by $\rho^{3/2}$. The same should be done in the depolarization factors.

If one substitutes eq.(7.74) into eq.(7.73) one obtains

$$\begin{aligned} \beta_e(d) &= \epsilon_a^{-1}\rho V\left[\frac{\epsilon - \epsilon_a}{\epsilon_a + L_z(\epsilon - \epsilon_a)}\right] \\ \gamma_e(d) &= \epsilon_a \rho V\left[\frac{\epsilon - \epsilon_a}{\epsilon_a + L_{\|}(\epsilon - \epsilon_a)}\right] \\ \delta_e(d) &= \rho V d\left[\frac{\epsilon - \epsilon_a}{\epsilon_a + L_{\|}(\epsilon - \epsilon_a)} + \frac{\epsilon - \epsilon_a}{\epsilon_a + L_z(\epsilon - \epsilon_a)}\right] \\ \tau(d) &= \epsilon_a \rho V d\left[\frac{\epsilon - \epsilon_a}{\epsilon_a + L_{\|}(\epsilon - \epsilon_a)}\right] \end{aligned} \tag{7.77}$$

As usual one finds that, if one changes the direction of the z-axis, γ_e and β_e remain unchanged, whereas δ_e and τ change sign. If one uses the above formulae to calculate the reflection and transmission amplitudes, it is important to take the right sign. If, as in chapter 4, the light is incident through the ambient, one may use the above expressions for the constitutive coefficients. If, however, the light is incident through the substrate one must, in order to apply the formulae of chapter 4, interchange ϵ_a and ϵ_s in these formulae and change the sign of δ_e and τ, before substituting them into the formulae of chapter 4.

7.9 Quadrupole approximation for spheroids

As in the previous section the regular array case is considered first. In the quadrupole approximation for the interaction along the substrate, all multipole moments of order $\ell > 2$ are neglected, so that eq.(7.62) becomes a finite set of linear equations for $A_{1,-1}$, A_{10} , A_{11} , $A_{2,-2}$, $A_{2,-1}$, A_{20} , A_{21} and A_{22}. For lattices with 4- or 6-fold symmetry (square and triangular lattices respectively)

$$S_{\ell m} = S_{\ell m}^{r'} = 0 \quad \text{for m} \neq \text{multiple of } 4 \text{ or } 6 \text{ respectively} \tag{7.78}$$

Substitution of these zeros into the above mentioned finite set of equations leads, with the definition, eq.(7.9), of $H(\ell, m|\ell_2, m_2)$ to

$$\begin{aligned} A_{10} &= \alpha_{10,10}\{2E_0\cos\theta_0\sqrt{\pi/3} + 4\widetilde{S}_{20}^{-}A_{10}L^{-3}\sqrt{\pi/5} - 2\widetilde{S}_{30}^{+}A_{20}L^{-4}\sqrt{15\pi/7}\} \\ &\quad +\alpha_{10,20}\{6\widetilde{S}_{30}^{-}A_{10}L^{-4}\sqrt{3\pi/35} - 4\widetilde{S}_{40}^{+}A_{20}L^{-5}\sqrt{\pi}\} \\ A_{20} &= \alpha_{20,10}\{2E_0\cos\theta_0\sqrt{\pi/3} + 4\widetilde{S}_{20}^{-}A_{10}L^{-3}\sqrt{\pi/5} - 2\widetilde{S}_{30}^{+}A_{20}L^{-4}\sqrt{15\pi/7}\} \\ &\quad +\alpha_{20,20}\{6\widetilde{S}_{30}^{-}A_{10}L^{-4}\sqrt{3\pi/35} - 4\widetilde{S}_{40}^{+}A_{20}L^{-5}\sqrt{\pi}\} \end{aligned}$$

$$\begin{aligned} A_{11} &= \alpha_{11,11}\{-E_0\sin\theta_0 e^{-i\phi_0}\sqrt{2\pi/3} - 2\widetilde{S}_{20}^{+}A_{11}L^{-3}\sqrt{\pi/5} + 2\widetilde{S}_{30}^{-}A_{21}L^{-4}\sqrt{5\pi/7}\} \\ &\quad +\alpha_{11,21}\{-6\widetilde{S}_{30}^{+}A_{11}L^{-4}\sqrt{\pi/35} + \frac{8}{3}\widetilde{S}_{40}^{-}A_{21}L^{-5}\sqrt{\pi}\} \\ A_{21} &= \alpha_{21,11}\{-E_0\sin\theta_0 e^{-i\phi_0}\sqrt{2\pi/3} - 2\widetilde{S}_{20}^{+}A_{11}L^{-3}\sqrt{\pi/5} + 2\widetilde{S}_{30}^{-}A_{21}L^{-4}\sqrt{5\pi/7}\} \\ &\quad +\alpha_{21,21}\{-6\widetilde{S}_{30}^{+}A_{11}L^{-4}\sqrt{\pi/35} + \frac{8}{3}\widetilde{S}_{40}^{-}A_{21}L^{-5}\sqrt{\pi}\} \end{aligned} \tag{7.79}$$

Here the abbreviations

$$\widetilde{S}_{\ell m}^{\pm} \equiv S_{\ell m} \pm \left(\frac{\epsilon_a - \epsilon_s}{\epsilon_a + \epsilon_s}\right)\widetilde{S}_{\ell m}^{r} \tag{7.80}$$

analogous to eq.(7.42)), have been used. For lattices with 4- or 6-fold symmetry different m's do not couple. The x-axis has been chosen along one of the lattice vectors. The equations for $A_{1,-1}$ and $A_{2,-1}$ are superfluous, since one can prove again, that $A_{\ell,-1}\exp(-i\phi_0) = -A_{\ell 1}\exp(i\phi_0)$, $(\ell = 1, 2)$. It is furthermore found that in quadrupole approximation $A_{22} = A_{2,-2} = 0$ (cf. eq.(7.43)).

Solving eq.(7.79) and using eq.(5.111), one finds for the polarizabilities normal to the surface of the substrate

$$\begin{aligned}\alpha_z(0) &= 4\pi\epsilon_a D_z^{-1}\{[1+4\alpha_{20,20}\widetilde{S}_{40}^{+}L^{-5}\sqrt{\pi}]\alpha_{10,10}\\&\quad -4\alpha_{10,20}\alpha_{20,10}\widetilde{S}_{40}^{+}L^{-5}\sqrt{\pi}\}\\ \alpha_z^{10}(0) &= 2\pi\epsilon_a\sqrt{5/3}D_z^{-1}\{[1-6\alpha_{10,20}\widetilde{S}_{30}^{-}L^{-4}\sqrt{3\pi/35}]\alpha_{20,10}\\&\quad +6\alpha_{20,20}\alpha_{10,10}\widetilde{S}_{30}^{-}L^{-4}\sqrt{3\pi/35}\}\end{aligned} \tag{7.81}$$

where

$$\begin{aligned}D_z &\equiv [1-4\alpha_{10,10}\widetilde{S}_{20}^{-}L^{-3}\sqrt{\pi/5}-6\alpha_{10,20}\widetilde{S}_{30}^{-}L^{-4}\sqrt{3\pi/35}]\\&\quad\times[1+2\alpha_{20,10}\widetilde{S}_{30}^{+}L^{-4}\sqrt{15\pi/7}+4\alpha_{20,20}\widetilde{S}_{40}^{+}L^{-5}\sqrt{\pi}]\\&\quad+[4\alpha_{20,10}\widetilde{S}_{20}^{-}L^{-3}\sqrt{\pi/5}+6\alpha_{20,20}\widetilde{S}_{30}^{-}L^{-4}\sqrt{3\pi/35}]\\&\quad\times[2\alpha_{10,10}\widetilde{S}_{30}^{+}L^{-4}\sqrt{15\pi/7}+4\alpha_{10,20}\widetilde{S}_{40}^{+}L^{-5}\sqrt{\pi}]\end{aligned} \tag{7.82}$$

and for the polarizabilities parallel to the surface of the substrate

$$\begin{aligned}\alpha_\parallel(0) &= 4\pi\epsilon_a D_\parallel^{-1}\{[1-\frac{8}{3}\alpha_{21,21}\widetilde{S}_{40}^{-}L^{-5}\sqrt{\pi}]\alpha_{11,11}\\&\quad+\frac{8}{3}\alpha_{11,21}\alpha_{21,11}\widetilde{S}_{40}^{-}L^{-5}\sqrt{\pi}\}\\ \alpha_\parallel^{10}(0) &= \frac{4}{3}\pi\ \epsilon_a\sqrt{5}D_\parallel^{-1}\{[1+6\alpha_{11,21}\widetilde{S}_{30}^{+}L^{-4}\sqrt{\pi/35}]\alpha_{21,11}\\&\quad-6\alpha_{21,21}\alpha_{11,11}\widetilde{S}_{30}^{+}L^{-4}\sqrt{\pi/35}\}\end{aligned} \tag{7.83}$$

where

$$\begin{aligned}D_\parallel &\equiv [1+2\alpha_{11,11}\widetilde{S}_{20}^{+}L^{-3}\sqrt{\pi/5}+6\alpha_{11,21}\widetilde{S}_{30}^{+}L^{-4}\sqrt{\pi/35}]\\&\quad\times[1-2\alpha_{21,11}\widetilde{S}_{30}^{-}L^{-4}\sqrt{5\pi/7}-\frac{8}{3}\alpha_{21,21}\widetilde{S}_{40}^{-}L^{-5}\sqrt{\pi}]\\&\quad+[2\alpha_{21,11}\widetilde{S}_{20}^{+}L^{-3}\sqrt{\pi/5}+6\alpha_{21,21}\widetilde{S}_{30}^{+}L^{-4}\sqrt{\pi/35}]\\&\quad\times[2\alpha_{11,11}\widetilde{S}_{30}^{-}L^{-4}\sqrt{5\pi/7}+\frac{8}{3}\alpha_{11,21}\widetilde{S}_{40}^{-}L^{-5}\sqrt{\pi}]\end{aligned} \tag{7.84}$$

As indicated by the argument 0, these polarizabilities are for a dipole and a quadrupole located in the center of the spheroids.

The above results and eqs.(7.44) and (7.45), valid for spheres, have an analogous form. In two aspects there are, however, differences. First the lattice sums $\widetilde{S}_{\ell m}^{\pm}$ differ from $S_{\ell m}^{\pm}$. Furthermore the polarizabilities $\alpha_{\ell m,\ell' m}$ on the right-hand side in the above equations are polarizabilities of a spheroid, interacting with the substrate, whereas α_ℓ in eqs.(7.44) and (7.45) are polarizabilities of spheres far from the substrate.

In order to calculate the above polarizabilities $\alpha_z(0)$, $\alpha_z^{10}(0)$, $\alpha_\parallel(0)$ and $\alpha_\parallel^{10}(0)$ for regular arrays (with 4- or 6-fold symmetry), one first has to evaluate the lattice sums $\widetilde{S}_{\ell 0}^{\pm}(\ell$=2, 3 and 4). For the procedure to do this, the reader is referred to appendix A of this chapter. Next the multipole polarizabilities $\alpha_{\ell m,\ell' m}$ for $\ell, \ell' = 1, 2$ and $m = 0, 1$ have to be calculated for oblate and prolate spheroids, interacting with the substrate. With the method, developed in the previous chapters, in which a homogeneous external field is used, one can only calculate $\alpha_{10,10}$, $\alpha_{11,11}$, $\alpha_{10,20}$, $\alpha_{20,10}$, $\alpha_{11,21}$ and $\alpha_{21,11}$. This follows from eqs.(6.130) and (6.131) for oblate and eqs.(6.173) and (6.174) for prolate spheroids, using also eqs.(5.111) and (5.67). In order to be able to evaluate also the polarizabilities $\alpha_{20,20}$ and $\alpha_{21,21}$, the method of the two previous chapters has to be extended to the case of a non-homogeneous external electric field, which may be taken as a linear function in space. This problem is treated in appendix B of this chapter.

The calculation of the polarizabilities of the spheroids in interaction with their image, can also be restricted to quadrupolar order. This has been explained in detail in sections 6.4, 6.5 and in appendix B of this chapter. Using these results in the formulae above one finds, after a lengthy but straightforward calculation, that $\alpha_z(0)$, $\alpha_z^{10}(0)$, $\alpha_\parallel(0)$ and $\alpha_\parallel^{10}(0)$, given by eqs.(7.81)–(7.84), can again be written in the form given in eqs.(6.132) and (6.133)

$$\begin{aligned}
\alpha_z(0) &= \frac{\epsilon_a(\epsilon-\epsilon_a)V[\epsilon_a + L_z^1(\epsilon-\epsilon_a)]}{[\epsilon_a + L_z(\epsilon-\epsilon_a)][\epsilon_a + L_z^1(\epsilon-\epsilon_a)] + \Lambda_z^1(\epsilon-\epsilon_a)^2} \\
\alpha_z^{10}(0) &= \frac{\epsilon_a(\epsilon-\epsilon_a)^2\Lambda_z^{10} V a\xi_0}{[\epsilon_a + L_z(\epsilon-\epsilon_a)][\epsilon_a + L_z^1(\epsilon-\epsilon_a)] + \Lambda_z^1(\epsilon-\epsilon_a)^2}
\end{aligned} \tag{7.85}$$

and

$$\begin{aligned}
\alpha_\parallel(0) &= \frac{\epsilon_a(\epsilon-\epsilon_a)V[\epsilon_a + L_\parallel^1(\epsilon-\epsilon_a)]}{[\epsilon_a + L_\parallel(\epsilon-\epsilon_a)][\epsilon_a + L_\parallel^1(\epsilon-\epsilon_a)] + \Lambda_\parallel^1(\epsilon-\epsilon_a)^2} \\
\alpha_\parallel^{10}(0) &= \frac{\epsilon_a(\epsilon-\epsilon_a)^2\Lambda_\parallel^{10} V a\xi_0}{[\epsilon_a + L_\parallel(\epsilon-\epsilon_a)][\epsilon_a + L_\parallel^1(\epsilon-\epsilon_a)] + \Lambda_\parallel^1(\epsilon-\epsilon_a)^2}
\end{aligned} \tag{7.86}$$

where V is the volume of the spheroid, given for oblate speroids by eq.(6.29) and for prolate spheroids by eq.(6.76).

The depolarization factors are found to be given for oblate spheroids by

$$\begin{aligned}
L_z &= (1+\xi_0^2)\left\{1 - \xi_0 \arctan\left(\frac{1}{\xi_0}\right)\right. \\
&\quad - \left(\frac{\epsilon_a-\epsilon_s}{\epsilon_a+\epsilon_s}\right)\frac{\xi_0}{\xi_1}[\left(\frac{3}{2}+\xi_1^2\right)\xi_1^2 \ln\left(1+\frac{1}{\xi_1^2}\right) - \xi_1 \arctan\left(\frac{1}{\xi_1}\right) - \xi_1^2] \\
&\quad \left. -\frac{4}{3}\left(\frac{a}{L}\right)^3 \xi_0 \left[S_{20} - \left(\frac{\epsilon_a-\epsilon_s}{\epsilon_a+\epsilon_s}\right)\widetilde{S}_{20}^r\right]\sqrt{\pi/5}\right\}
\end{aligned} \tag{7.87}$$

$$\begin{aligned}
L_z^1 &= (1+\xi_0^2)(1+3\xi_0^2)\left\{\frac{3}{2}\xi_0 \arctan\left(\frac{1}{\xi_0}\right) - \frac{9}{2}\left(\frac{\xi_0^2}{1+3\xi_0^2}\right)\right. \\
&\quad + \left(\frac{\epsilon_a-\epsilon_s}{\epsilon_a+\epsilon_s}\right)\frac{\xi_0}{\xi_1}\left[\frac{3}{2}\xi_1 \arctan\left(\frac{1}{\xi_1}\right) - \frac{3}{4}(6\,\xi_1^4 + 10\,\xi_1^2 + 5)\xi_1^2 \ln\left(1+\frac{1}{\xi_1^2}\right)\right. \\
&\quad \left.\left. +\frac{3}{4}(6\xi_1^2+7)\xi_1^2\right] + \frac{8}{15}\left(\frac{a}{L}\right)^5 \xi_0[S_{40} + \left(\frac{\epsilon_a-\epsilon_s}{\epsilon_a+\epsilon_s}\right)\widetilde{S}_{40}^r]\sqrt{\pi}\right\}
\end{aligned} \tag{7.88}$$

$$\Lambda_z^1 = \frac{45}{2}\xi_0^2(1+\xi_0^2)^2(1+3\xi_0^2)\left\{\left(\frac{4}{15}\left(\frac{a}{L}\right)^4 S_{30}\sqrt{\pi/7}\right)^2 - \left(\frac{\epsilon_a-\epsilon_s}{\epsilon_a+\epsilon_s}\right)^2\left[\xi_1\arctan\left(\frac{1}{\xi_1}\right) - \frac{1}{2}(\xi_1^2+2)\xi_1^2\ln\left(1+\frac{1}{\xi_1^2}\right) - \frac{1}{4}(1-2\,\xi_1^2) - \frac{4}{15}\left(\frac{a}{L}\right)^4\widetilde{S}_{30}^r\sqrt{\pi/7}\right]^2\right\} \quad (7.89)$$

$$\Lambda_z^{10} = \frac{3}{2}(1+\xi_0^2)(1+3\,\xi_0^2)\left\{\left(\frac{\epsilon_a-\epsilon_s}{\epsilon_a+\epsilon_s}\right)\left[\xi_1\arctan\left(\frac{1}{\xi_1}\right) - \frac{1}{2}(\xi_1^2+2)\xi_1^2\ln\left(1+\frac{1}{\xi_1^2}\right) - \frac{1}{4}(1-2\,\xi_1^2)\right] + \frac{4}{15}\left(\frac{a}{L}\right)^4[S_{30} - \left(\frac{\epsilon_a-\epsilon_s}{\epsilon_a+\epsilon_s}\right)\widetilde{S}_{30}^r]\sqrt{\pi/7}\right\} \quad (7.90)$$

in the direction normal to the substrate, and by

$$L_\parallel = \frac{1}{2}(1+\xi_0^2)\left\{\xi_0\arctan\left(\frac{1}{\xi_0}\right) - \frac{\xi_0^2}{1+\xi_0^2} + \left(\frac{\epsilon_a-\epsilon_s}{\epsilon_a+\epsilon_s}\right)\frac{\xi_0}{\xi_1}[\xi_1\arctan\left(\frac{1}{\xi_1}\right) - \left(\frac{3}{2}+\xi_1^2\right)\xi_1^2\ln\left(1+\frac{1}{\xi_1^2}\right) + \xi_1^2] + \frac{4}{3}\left(\frac{a}{L}\right)^3\xi_0\left[S_{20} + \left(\frac{\epsilon_a-\epsilon_s}{\epsilon_a+\epsilon_s}\right)\widetilde{S}_{20}^r\right]\sqrt{\pi/5}\right\} \quad (7.91)$$

$$L_\parallel^1 = \frac{1}{2}(1+\xi_0^2)(1+2\xi_0^2)\left\{\frac{2+3\xi_0^2}{1+\xi_0^2} - 3\,\xi_0\arctan\left(\frac{1}{\xi_0}\right) + \left(\frac{\epsilon_a-\epsilon_s}{\epsilon_a+\epsilon_s}\right)\frac{\xi_0}{\xi_1}\left[3\xi_1\arctan\left(\frac{1}{\xi_1}\right) - \frac{3}{2}(6\,\xi_1^4+10\,\xi_1^2+5)\xi_1^2\ln\left(1+\frac{1}{\xi_1^2}\right) + \frac{3}{2}(6\xi_1^2+7)\xi_1^2\right] - \frac{16}{15}\left(\frac{a}{L}\right)^5\xi_0\left[S_{40} - \left(\frac{\epsilon_a-\epsilon_s}{\epsilon_a+\epsilon_s}\right)\widetilde{S}_{40}^r\right]\sqrt{\pi}\right\} \quad (7.92)$$

$$\Lambda_\parallel^1 = \frac{45}{4}\xi_0^2(1+\xi_0^2)^2(1+2\xi_0^2)\left\{\left(\frac{4}{15}\left(\frac{a}{L}\right)^4 S_{30}\sqrt{\pi/7}\right)^2 - \left(\frac{\epsilon_a-\epsilon_s}{\epsilon_a+\epsilon_s}\right)^2\left[\xi_1\arctan\left(\frac{1}{\xi_1}\right) - \frac{1}{2}(\xi_1^2+2)\xi_1^2\ln\left(1+\frac{1}{\xi_1^2}\right) - \frac{1}{4}(1-2\,\xi_1^2) - \frac{4}{15}\left(\frac{a}{L}\right)^4\widetilde{S}_{30}^r\sqrt{\pi/7}\right]^2\right\} \quad (7.93)$$

$$\Lambda_\parallel^{10} = \frac{3}{2}(1+\xi_0^2)(1+2\,\xi_0^2)\left\{\left(\frac{\epsilon_a-\epsilon_s}{\epsilon_a+\epsilon_s}\right)\left[\xi_1\arctan\left(\frac{1}{\xi_1}\right) - \frac{1}{2}(\xi_1^2+2)\xi_1^2\ln\left(1+\frac{1}{\xi_1^2}\right) - \frac{1}{4}(1-2\,\xi_1^2)\right] - \frac{4}{15}\left(\frac{a}{L}\right)^4\left[S_{30} + \left(\frac{\epsilon_a-\epsilon_s}{\epsilon_a+\epsilon_s}\right)\widetilde{S}_{30}^r\right]\sqrt{\pi/7}\right\} \quad (7.94)$$

parallel to the surface of the substrate. The depolarization factors L_z and $L_\|$ in the quadrupole approximation, given above, are identical to those found in the previous section in the dipole approximation. Notice that, if one neglects the interaction between the particles, $a/L \to 0$, one obtains eqs.(6.134)–(6.141), which were derived for that case.

For prolate spheroids the expressions, eqs.(7.85) and (7.86), for the polarizabilities are again valid, now with V the volume of the prolate spheroid. The depolarization factors can be derived in an analogous way as above. In the direction normal to the surface of the substrate one finds:

$$\begin{aligned} L_z &= (1-\xi_0^2)\left\{1-\frac{1}{2}\xi_0 \ln\left(\frac{\xi_0+1}{\xi_0-1}\right)\right. \\ &+\left(\frac{\epsilon_a-\epsilon_s}{\epsilon_a+\epsilon_s}\right)\frac{\xi_0}{\xi_1}\left[\left(\frac{3}{2}-\xi_1^2\right)\xi_1^2\ln\left(1-\frac{1}{\xi_1^2}\right)+\frac{1}{2}\xi_1\ln\left(\frac{\xi_1+1}{\xi_1-1}\right)-\xi_1^2\right] \\ &\left.+\frac{4}{3}\left(\frac{a}{L}\right)^3\xi_0\left[S_{20}-\left(\frac{\epsilon_a-\epsilon_s}{\epsilon_a+\epsilon_s}\right)\widetilde{S}^r_{20}\right]\sqrt{\pi/5}\right\} \end{aligned} \tag{7.95}$$

$$\begin{aligned} L_z^1 &= (1-\xi_0^2)(1-3\xi_0^2)\left\{\frac{3}{4}\xi_0\ln\left(\frac{\xi_0+1}{\xi_0-1}\right)+\frac{9}{2}\left(\frac{\xi_0^2}{1-3\xi_0^2}\right)\right. \\ &+\left(\frac{\epsilon_a-\epsilon_s}{\epsilon_a+\epsilon_s}\right)\frac{\xi_0}{\xi_1}\left[\frac{3}{4}\xi_1\ln\left(\frac{\xi_1+1}{\xi_1-1}\right)+\frac{3}{4}(6\,\xi_1^4-10\,\xi_1^2+5)\xi_1^2\ln\left(1-\frac{1}{\xi_1^2}\right)\right. \\ &\left.\left.-\frac{3}{4}(7-6\xi_1^2)\xi_1^2\right]+\frac{8}{15}\left(\frac{a}{L}\right)^5\xi_0\left[S_{40}+\left(\frac{\epsilon_a-\epsilon_s}{\epsilon_a+\epsilon_s}\right)\widetilde{S}^r_{40}\right]\sqrt{\pi}\right\} \end{aligned} \tag{7.96}$$

$$\begin{aligned} \Lambda_z^1 &= \frac{45}{2}\xi_0^2(1-\xi_0^2)^2(3\xi_0^2-1)\left\{\left(\frac{4}{15}\left(\frac{a}{L}\right)^4 S_{30}\sqrt{\pi/7}\right)^2\right. \\ &-\left(\frac{\epsilon_a-\epsilon_s}{\epsilon_a+\epsilon_s}\right)^2\left[\frac{1}{2}\xi_1\ln\left(\frac{\xi_1+1}{\xi_1-1}\right)+\frac{1}{2}(2-\xi_1^2)\xi_1^2\ln\left(1-\frac{1}{\xi_1^2}\right)\right. \\ &\left.\left.-\frac{1}{4}(1+2\,\xi_1^2)+\frac{4}{15}\left(\frac{a}{L}\right)^4\widetilde{S}^r_{30}\sqrt{\pi/7}\right]^2\right\} \end{aligned} \tag{7.97}$$

$$\begin{aligned} \Lambda_z^{10} &= \frac{3}{2}(1-\xi_0^2)(1-3\,\xi_0^2)\left\{\left(\frac{\epsilon_a-\epsilon_s}{\epsilon_a+\epsilon_s}\right)\left[\frac{1}{2}\xi_1\ln\left(\frac{\xi_1+1}{\xi_1-1}\right)+\frac{1}{2}(2-\xi_1^2)\xi_1^2\ln\left(1-\frac{1}{\xi_1^2}\right)\right.\right. \\ &\left.\left.-\frac{1}{4}(1+2\,\xi_1^2)\right]+\frac{4}{15}\left(\frac{a}{L}\right)^4\left[S_{30}-\left(\frac{\epsilon_a-\epsilon_s}{\epsilon_a+\epsilon_s}\right)\widetilde{S}^r_{30}\right]\sqrt{\pi/7}\right\} \end{aligned} \tag{7.98}$$

and parallel to the surface:

$$\begin{aligned} L_\| &= \frac{1}{2}(1-\xi_0^2)\left\{\frac{1}{2}\xi_0\ln\left(\frac{\xi_0+1}{\xi_0-1}\right)+\frac{\xi_0^2}{1-\xi_0^2}\right. \\ &+\left(\frac{\epsilon_a-\epsilon_s}{\epsilon_a+\epsilon_s}\right)\frac{\xi_0}{\xi_1}\left[\frac{1}{2}\xi_1\ln\left(\frac{\xi_1+1}{\xi_1-1}\right)+\left(\frac{3}{2}-\xi_1^2\right)\xi_1^2\ln\left(1-\frac{1}{\xi_1^2}\right)-\xi_1^2\right] \\ &\left.-\frac{4}{3}\left(\frac{a}{L}\right)^3\xi_0\left[S_{20}+\left(\frac{\epsilon_a-\epsilon_s}{\epsilon_a+\epsilon_s}\right)\widetilde{S}^r_{20}\right]\sqrt{\pi/5}\right\} \end{aligned} \tag{7.99}$$

$$
\begin{aligned}
L_{\|}^{1} &= \frac{1}{2}(1-\xi_0^2)(1-2\xi_0^2)\left\{\frac{2-3\xi_0^2}{1-\xi_0^2}-\frac{3}{2}\xi_0\ln\left(\frac{\xi_0+1}{\xi_0-1}\right)\right. \\
&+\left(\frac{\epsilon_a-\epsilon_s}{\epsilon_a+\epsilon_s}\right)\frac{\xi_0}{\xi_1}\left[\frac{3}{2}\xi_1\ln\left(\frac{\xi_1+1}{\xi_1-1}\right)+\frac{3}{2}(6\,\xi_1^4-10\,\xi_1^2+5)\xi_1^2\ln\left(1-\frac{1}{\xi_1^2}\right)\right. \\
&\left.\left.-\frac{3}{2}(7-6\xi_1^2)\xi_1^2\right]-\frac{16}{15}\left(\frac{a}{L}\right)^5\xi_0\left[S_{40}-\left(\frac{\epsilon_a-\epsilon_s}{\epsilon_a+\epsilon_s}\right)\widetilde{S}_{40}^r\right]\sqrt{\pi}\right\} \qquad (7.100)
\end{aligned}
$$

$$
\begin{aligned}
\Lambda_{\|}^{1} &= \frac{45}{4}\xi_0^2(1-\xi_0^2)^2(2\xi_0^2-1)\left\{\left(\frac{4}{15}\left(\frac{a}{L}\right)^4 S_{30}\sqrt{\pi/7}\right)^2\right. \\
&-\left(\frac{\epsilon_a-\epsilon_s}{\epsilon_a+\epsilon_s}\right)^2\left[\frac{1}{2}\xi_1\ln\left(\frac{\xi_1+1}{\xi_1-1}\right)+\frac{1}{2}(2-\xi_1^2)\xi_1^2\ln\left(1-\frac{1}{\xi_1^2}\right)\right. \\
&\left.\left.-\frac{1}{4}(1-2\,\xi_1^2)+\frac{4}{15}\left(\frac{a}{L}\right)^4\widetilde{S}_{30}^r\sqrt{\pi/7}\right]^2\right\} \qquad (7.101)
\end{aligned}
$$

$$
\begin{aligned}
\Lambda_{\|}^{10} &= \frac{3}{2}(1-\xi_0^2)(1-2\,\xi_0^2)\left\{\left(\frac{\epsilon_a-\epsilon_s}{\epsilon_a+\epsilon_s}\right)\left[\frac{1}{2}\xi_1\ln\left(\frac{\xi_1+1}{\xi_1-1}\right)+\frac{1}{2}(2-\xi_1^2)\xi_1^2\ln\left(1-\frac{1}{\xi_1^2}\right)\right.\right. \\
&\left.\left.-\frac{1}{4}(1+2\,\xi_1^2)\right]-\frac{4}{15}\left(\frac{a}{L}\right)^4\left[S_{30}+\left(\frac{\epsilon_a-\epsilon_s}{\epsilon_a+\epsilon_s}\right)\widetilde{S}_{30}^r\right]\sqrt{\pi/7}\right\} \qquad (7.102)
\end{aligned}
$$

It is, of course, assumed that the prolate spheroids are not packed too densely in the two-dimensional array. The depolarization factors L_z and $L_{\|}$ in the quadrupole approximation, given above, are identical to those found in the previous section in the dipole approximation. Notice that, if one neglects the interaction between the particles, $a/L \to 0$, one obtains eqs.(6.177) –(6.184), which were derived for that case.

For random arrays one finds in a similar way as above, in quadrupolar approximation from eqs.(7.67) and (7.68) expressions analogous to those above, with $S_{\ell 0}$ and $\widetilde{S}_{\ell 0}^r$ replaced by I_ℓ and $\widetilde{I}_\ell^r$ respectively, and L^{-1} replaced by $\rho^{1/2}$.

In a completely analogous way as in the previous sections the interfacial susceptibilities can be derived, using the polarizabilities found above. One obtains, cf. eq.(7.51),

$$
\begin{aligned}
\beta_e(d) &= \rho\,\epsilon_a^{-2}\alpha_z(0) = \frac{\rho\,\epsilon_a^{-1}(\epsilon-\epsilon_a)V[\epsilon_a+L_z^1(\epsilon-\epsilon_a)]}{[\epsilon_a+L_z(\epsilon-\epsilon_a)][\epsilon_a+L_z^1(\epsilon-\epsilon_a)]+\Lambda_z^1(\epsilon-\epsilon_a)^2} \\
\gamma_e(d) &= \rho\,\alpha_{\|}(0) = \frac{\rho\,\epsilon_a(\epsilon-\epsilon_a)V[\epsilon_a+L_{\|}^1(\epsilon-\epsilon_a)]}{[\epsilon_a+L_{\|}(\epsilon-\epsilon_a)][\epsilon_a+L_{\|}^1(\epsilon-\epsilon_a)]+\Lambda_{\|}^1(\epsilon-\epsilon_a)^2}
\end{aligned}
$$

$$
\begin{aligned}
\delta_e(d) &= -\rho\,\epsilon_a^{-1}[\alpha_z^{10}(0) + \alpha_\parallel^{10}(0) - d\alpha_z(0) - d\alpha_\parallel(0)] \\
&= \rho(\epsilon-\epsilon_a)V\,a\xi_1 \left\{ \frac{\epsilon_a + (L_z^1 - \Lambda_z^{10}\xi_0/\xi_1)(\epsilon-\epsilon_a)}{[\epsilon_a + L_z(\epsilon-\epsilon_a)][\epsilon_a + L_z^1(\epsilon-\epsilon_a)] + \Lambda_z^1(\epsilon-\epsilon_a)^2} \right. \\
&\quad \left. + \frac{\epsilon_a + \left(L_\parallel^1 - \Lambda_\parallel^{10}\xi_0/\xi_1\right)(\epsilon-\epsilon_a)}{[\epsilon_a + L_\parallel(\epsilon-\epsilon_a)][\epsilon_a + L_\parallel^1(\epsilon-\epsilon_a)] + \Lambda_\parallel^1(\epsilon-\epsilon_a)^2} \right\} \\
\tau(d) &= -\rho[\alpha_\parallel^{10}(0) - d\alpha_\parallel(0)] \\
&= \frac{\rho\,\epsilon_a(\epsilon-\epsilon_a)V\,a\xi_1\left[\epsilon_a + \left(L_\parallel^1 - \Lambda_\parallel^{10}\xi_0/\xi_1\right)(\epsilon-\epsilon_a)\right]}{[\epsilon_a + L_\parallel(\epsilon-\epsilon_a)][\epsilon_a + L_\parallel^1(\epsilon-\epsilon_a)] + \Lambda_\parallel^1(\epsilon-\epsilon_a)^2}
\end{aligned}
\tag{7.103}
$$

These expressions have the same form as those given by eqs.(6.44)–(6.47) in the case of non-interacting spheroids. Only the depolarization coefficients are different, due to the interaction. Notice that, if one changes the direction of the z-axis, γ_e and β_e remain unchanged, while δ_e and τ change sign. If one uses the above formulae to calculate the reflection and transmission amplitudes, it is important to take the right sign. If, as in chapter 4, the light is incident through the ambient, one may also use the above expressions. If, however, the light is incident through the substrate, one must, in order to apply the formulae of chapter 4, interchange ϵ_a and ϵ_s in these formulae and change the sign of δ_e and τ in the above expressions for the constitutive coefficients, before substituting them into the formulae of chapter 4.

7.10 Application: Gold islands on sapphire

In order to demonstrate how the equations, derived above, can be used in practice, the special case of gold islands on sapphire will again be considered. In particular the case of spheres and spheroids touching the substrate, for which $d = R$ and $d = a\xi_1 = a\xi_0$ respectively, is considered. When the islands touch the surface the influence of the substrate is most pronounced. For this case the dielectric constant of the ambient is unity, $\epsilon_a = 1$. Sapphire has a small dispersion in the visible domain. An average value is used as dielectric constant of the substrate, $\epsilon_s = 3.13$. For the complex dielectric constant of gold a list of values due to Johnson and Christy [40] is given in table 5.1, for the optical domain.

Consider first the model in which all interactions are taken into account until quadrupolar order. In that case the dipole and quadrupole polarizabilities, eqs.(7.47), (7.48) and eqs. (7.85), (7.86), with $\epsilon_a = 1$, reduce, in their dimensionless form, to

$$
\begin{aligned}
\widehat{\alpha}_z &\equiv \frac{1}{V}\alpha_z(0) = \frac{(\epsilon-1)[1 + L_z^1(\epsilon-1)]}{[1 + L_z(\epsilon-1)][1 + L_z^1(\epsilon-1)] + \Lambda_z^1(\epsilon-1)^2} \\
\widehat{\alpha}_z^{10} &\equiv \frac{1}{Vd}\alpha_z^{10}(0) = \frac{(\epsilon-1)^2\Lambda_z^{10}}{[1 + L_z(\epsilon-1)][1 + L_z^1(\epsilon-1)] + \Lambda_z^1(\epsilon-1)^2}
\end{aligned}
\tag{7.104}
$$

and

$$\begin{aligned}\widehat{\alpha}_{\|} &\equiv \frac{1}{V}\alpha_{\|}(0) = \frac{(\epsilon-1)[1+L^1_{\|}(\epsilon-1)]}{[1+L_{\|}(\epsilon-1)][1+L^1_{\|}(\epsilon-1)]+\Lambda^1_{\|}(\epsilon-1)^2} \\ \widehat{\alpha}^{10}_{\|} &\equiv \frac{1}{Vd}\alpha^{10}_{\|}(0) = \frac{(\epsilon-1)^2\Lambda^{10}_{\|}}{[1+L_{\|}(\epsilon-1)][1+L^1_{\|}(\epsilon-1)]+\Lambda^1_{\|}(\epsilon-1)^2}\end{aligned} \qquad (7.105)$$

where V is the volume of the sphere or spheroid. In all the application dimensionless quantities, like the $\widehat{\alpha}$'s, are introduced. This makes both the calculation and the results more transparent. The surface constitutive coefficients, eq.(5.112), similarly reduce to

$$\begin{aligned}\widehat{\gamma}_e &\equiv \frac{1}{\rho V}\gamma_e(d) = \widehat{\alpha}_{\|} \\ \widehat{\beta}_e &\equiv \frac{1}{\rho V}\beta_e(d) = \widehat{\alpha}_z \\ \widehat{\tau} &\equiv \frac{1}{\rho V d}\tau(d) = \widehat{\alpha}_{\|} - \widehat{\alpha}^{10}_{\|} \\ \widehat{\delta}_e &\equiv \frac{1}{\rho V d}\delta_e(d) = \widehat{\alpha}_z + \widehat{\alpha}_{\|} - \widehat{\alpha}^{10}_z - \widehat{\alpha}^{10}_{\|}\end{aligned} \qquad (7.106)$$

for this case, where $\epsilon_a = 1$. The invariants, eq.(5.66), become

$$\begin{aligned}\widehat{I}_e &\equiv \frac{1}{\rho V}I_e = \widehat{\gamma}_e - \epsilon_s\widehat{\beta}_e \ , \quad \widehat{I}_c \equiv \frac{1}{\rho V}(\epsilon_s - 1)I_c = \mathrm{Im}\,\widehat{\gamma}_e \\ \widehat{I}_\tau &\equiv \frac{1}{\rho V d}I_\tau = \widehat{\tau} \ , \quad\quad \widehat{I}_{\delta,e} \equiv \frac{1}{\rho V d}I_{\delta,e} = \widehat{\delta}_e\end{aligned} \qquad (7.107)$$

where ϵ_s was taken real. All the above expressions for the quadrupole and the dipole model, including interaction, are in fact identical to those given in the previous chapters without interaction along the substrate. The differences are in the expressions for the depolarization factors. These contain the lattice sums which were defined to be dimensionless. The resulting depolarization factors are given in eqs.(7.49) and (7.50) for spheres, in eqs.(7.87)–(7.94) for oblate spheroids, and in eqs.(7.95)–(7.102) for prolate spheroids, in the quadrupole approximation.

In order to investigate how the higher order multipoles contribute, consider eq.(7.16) for a regular array of spheres. To obtain the solution, it is again convenient to introduce dimensionless amplitudes. If the lattice has a 4-fold symmetry one introduces

$$\begin{aligned}\widehat{A}_{\ell,4\nu} &\equiv A_{\ell,4\nu}/VR^{\ell-1}2\sqrt{\pi/3}E_0\cos\theta_0 \\ \widehat{A}_{\ell,4\nu+1} &\equiv -A_{\ell,4\nu+1}/VR^{\ell-1}\sqrt{2\pi/3}E_0\sin\theta_0\exp(-i\phi_0)\end{aligned} \qquad (7.108)$$

For 6-fold symmetry replace the 4ν in the lower index by 6ν. For 4-fold symmetry eq.(7.16) then reduces to, using $\epsilon_a = 1$,

$$\widehat{A}_{\ell,4\nu}/\widehat{\alpha}_\ell + \frac{4\pi}{3}\sum_{\ell_2\nu_2}{}' H(\ell,4\nu|\ell_2,4\nu_2)\widehat{A}_{\ell_2,4\nu_2}\widehat{L}^{-\ell_1-1} \times$$
$$\left[S_{\ell_1,4\nu_1} + (-1)^{\ell_2}\left(\frac{1-\epsilon_s}{1+\epsilon_s}\right)S^r_{\ell_1,4\nu_1}\right] = \delta_{\ell 1}\delta_{\nu 0} \qquad (7.109)$$

where $|4\nu| \leq \ell$ and $|4\nu_2| \leq \ell_2$, and

$$\widehat{A}_{\ell,4\nu+1}/\widehat{\alpha}_\ell + \frac{4\pi}{3}\sum_{\ell_2\nu_2}{}' H(\ell,4\nu+1|\ell_2,4\nu_2+1)\widehat{A}_{\ell_2,4\nu_2+1}\widehat{L}^{-\ell_1-1} \times$$
$$\left[S_{\ell_1,4\nu_1} - (-1)^{\ell_2}(\frac{1-\epsilon_s}{1+\epsilon_s})S^r_{\ell_1,4\nu_1}\right] = \delta_{\ell 1}\delta_{\nu 0} \qquad (7.110)$$

where $|4\nu+1| \leq \ell$ and $|4\nu_2+1| \leq \ell_2$. Furthermore $\widehat{L} \equiv L/R$, $\widehat{\alpha}_\ell \equiv \alpha_\ell/VR^{2(\ell-1)}$, $\ell_1 \equiv \ell+\ell_2$ and $\nu_1 \equiv \nu_2 - \nu$. H is given by eq.(7.9). Eq.(7.109) makes it possible to calculate $\widehat{A}_{1,0}$ and $\widehat{A}_{2,0}$, and eq.(7.110) makes it possible to calculate $\widehat{A}_{1,1}$ and $\widehat{A}_{2,1}$, in the usual manner. For 6-fold symmetry all the 4's before the ν's should be replaced by 6. Eq.(5.111) for the dipole and quadrupole polarizabilities of the spheres, modified by the presence of the substrate and the interaction along the substrate, in terms of the amplitudes, remains valid and becomes, using $\epsilon_a = 1$,

$$\begin{aligned}\widehat{\alpha}_\parallel &= 4\pi\widehat{A}_{11}\ , \quad \widehat{\alpha}_\parallel^{10} = \frac{4\pi}{3}\widehat{A}_{21}\sqrt{5}\\ \widehat{\alpha}_z &= 4\pi\widehat{A}_{10}\ , \quad \widehat{\alpha}_z^{10} = 2\pi\widehat{A}_{20}\sqrt{5/3}\end{aligned} \qquad (7.111)$$

in terms of the dimensionless quantities, cf. eq.(5.122).

In order to investigate, how the higher order multipoles contribute for a random array of spheres, consider eqs.(7.26) and (7.27). It is again convenient to introduce dimensionless amplitudes

$$\begin{aligned}\widehat{A}_{\ell,0} &\equiv \ < A_{\ell,0} > /VR^{\ell-1}2E_0\sqrt{\pi/3}\cos\theta_0\\ \widehat{A}_{\ell,1} &\equiv \ - < A_{\ell,1} > /VR^{\ell-1}E_0\sqrt{2\pi/3}\sin\theta_0\exp(-i\phi_0)\end{aligned} \qquad (7.112)$$

Eqs.(7.26) and (7.27) then reduce to, using $\epsilon_a = 1$ and neglecting amplitudes for $\ell > M$,

$$\widehat{A}_{\ell,0}/\widehat{\alpha}_\ell + \frac{4\pi}{3}\sum_{\ell_2=1}^{M} H(\ell,0|\ell_2,0)\widehat{A}_{\ell_2,0}\,(R\sqrt{\rho})^{\ell_1+1} \times$$
$$[I_{\ell_1} + (-1)^{\ell_2}(\frac{1-\epsilon_s}{1+\epsilon_s})I^r_{\ell_1}] = \delta_{\ell 1} \qquad (7.113)$$

and

$$\widehat{A}_{\ell,1}/\widehat{\alpha}_\ell + \frac{4\pi}{3}\sum_{\ell_2=1}^{M} H(\ell,1|\ell_2,1)\widehat{A}_{\ell_2,0}\,(R\sqrt{\rho})^{\ell_1+1} \times$$
$$[I_{\ell_1} - (-1)^{\ell_2}(\frac{1-\epsilon_s}{1+\epsilon_s})I^r_{\ell_1}] = \delta_{\ell 1} \qquad (7.114)$$

for $\ell = 1, 2, ...$, and where $\ell_1 = \ell + \ell_2$. Eq.(7.113) makes it possible to calculate $\widehat{A}_{1,0}$ and $\widehat{A}_{2,0}$, and eq.(7.114) makes it possible to calculate $\widehat{A}_{1,1}$ and $\widehat{A}_{2,1}$, in the usual manner. Eq.(7.111), expressing the polarizabilities in the amplitudes, remains valid.

For spheres touching the substrate the results are given in figs. 7.3 through 7.10 for a square array. The interaction along the substrate was taken into account to quadrupolar order. The interaction with the image charge distribution was taken to the order $M = 16$. Eqs.(7.109) and (7.110) then, using eq.(7.63), reduce to

$$\widehat{A}_{\ell,4\nu}/\widehat{\alpha}_\ell + \frac{2\pi}{3}(\frac{1-\epsilon_s}{1+\epsilon_s})\sum_{\ell_2=1}^{16} H(\ell,4\nu|\ell_2,4\nu)\widehat{A}_{\ell_2,4\nu} \times$$
$$\left[(-1)^{\ell_2}\left(\frac{1}{2}\right)^{\ell_1+1}\sqrt{\frac{2\ell_1+1}{\pi}}\right] + \frac{4\pi}{3}\sum_{\ell_2=1}^{2} H(\ell,4\nu|\ell_2,0)\widehat{A}_{\ell_2,0}\widehat{L}^{-\ell_1-1} \times$$
$$\left[S_{\ell_1,-4\nu} + (-1)^{\ell_2}(\frac{1-\epsilon_s}{1+\epsilon_s})\widetilde{S}^{r}_{\ell_1,-4\nu}\right] = \delta_{\ell 1}\delta_{\nu 0} \tag{7.115}$$

where $|4\nu| \leq \ell$ and

$$\widehat{A}_{\ell,4\nu+1}/\widehat{\alpha}_\ell + \frac{2\pi}{3}(\frac{1-\epsilon_s}{1+\epsilon_s})\sum_{\ell_2=1}^{16} H(\ell,4\nu+1|\ell_2,4\nu+1)\widehat{A}_{\ell_2,4\nu+1} \times$$
$$\left[(-1)^{\ell_2}\left(\frac{1}{2}\right)^{\ell_1+1}\sqrt{\frac{2\ell_1+1}{\pi}}\right] + \frac{4\pi}{3}\sum_{\ell_2=1}^{2} H(\ell,4\nu+1|\ell_2,1)\widehat{A}_{\ell_2,1}\widehat{L}^{-\ell_1-1} \times$$
$$\left[S_{\ell_1,-4\nu} - (-1)^{\ell_2}(\frac{1-\epsilon_s}{1+\epsilon_s})\widetilde{S}^{r}_{\ell_1,-4\nu}\right] = \delta_{\ell 1}\delta_{\nu 0} \tag{7.116}$$

where $|4\nu+1| \leq \ell$ and $\ell_1 \equiv \ell + \ell_2$. For 6-fold symmetry replace the 4ν in the lower index by 6ν. The calculations were done for scaled densities $\widehat{\rho} \equiv \rho(2R)^2 = (2R/L)^2 = (2\widehat{R})^2$ of 0 (no interaction), 0.2, 0.4 and 0.8, L is the lattice constant and $\widehat{R} \equiv R/L$. The reason to go to a higher order in the multipole expansion in the interaction with the substrate is that, close to touching the substrate, the higher order multipoles start to give corrections of about 5 to 10 %. As is analyzed in great detail by Haarmans and Bedeaux [49], higher order multipole corrections for the interaction along the substrate become important for $\widehat{\rho} > 0.9$. They use as density parameter the coverage, which is equal to $(\pi/4)\widehat{\rho}$ for the square array. They also find that the results for a triangular lattice and for the random array are essentially the same as a function of the coverage for scaled densities up to $\widehat{\rho} = 0.8$. As already mentioned the use of the quadrupole approximation for the interactions along the substrate as well as with the images, cf eqs.(7.104) and (7.105), gives very similar results. These equations have the advantage of a simple analytic form and are correct within 5 to 10%. The interaction along the substrate has a substantial effect on the size of the polarizabilities and the resulting constitutive coefficients. The polarizability along the substrate approximately doubles from $\widehat{\rho} = 0$ to 0.8, while the polarizability normal to the surface decreases by a factor of 2. The resonance frequency shifts a little bit down for the polarizability parallel and up for the polarizability normal to the surface. The contributions due to the quadrupole polarizabilities to the coefficients $\widehat{\delta}_e$ and $\widehat{\tau}$ are small compared to the contributions due to the dipole polarizabilities.

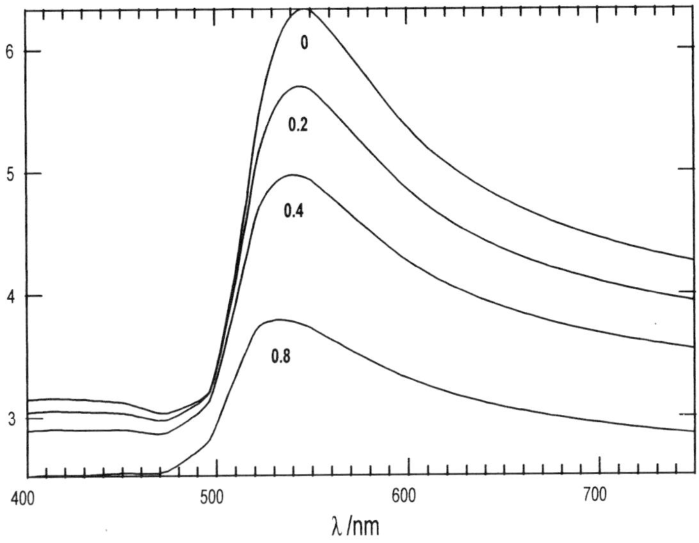

Figure 7.3 Real part of $\widehat{\alpha}_z = \widehat{\beta}_e$ as a function of the wavelength for coverages $\widehat{\rho} = 0; 0.2; 0.4$ and 0.8.

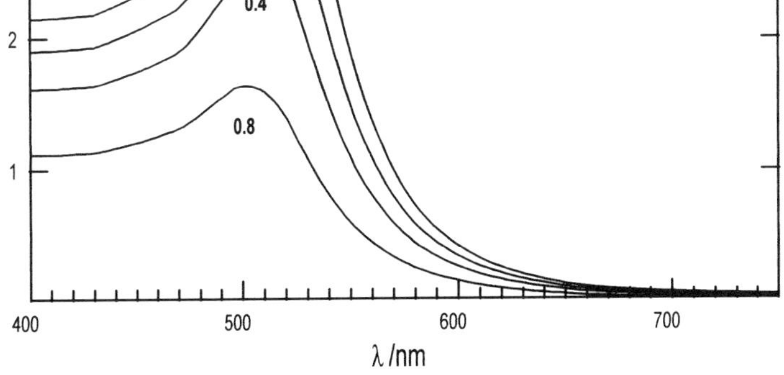

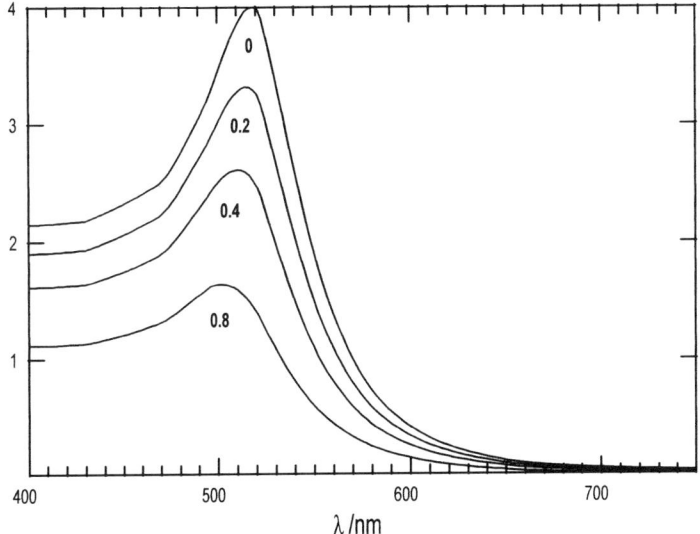

Figure 7.4 Imaginary part of $\widehat{\alpha}_z = \widehat{\beta}_e$ as a function of the wavelength for coverages $\widehat{\rho} = 0; 0.2; 0.4$ and 0.8.

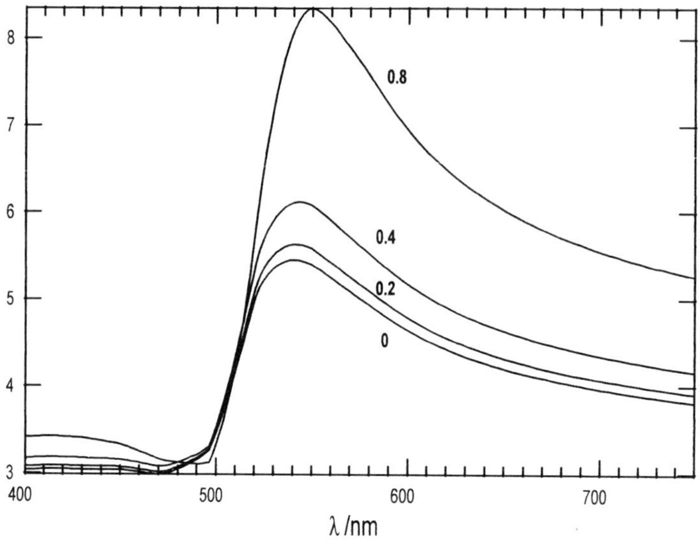

Figure 7.5 Real part of $\widehat{\alpha}_{\|} = \widehat{\gamma}_e$ as a function of the wavelength for coverages $\widehat{\rho} = 0; 0.2; 0.4$ and 0.8.

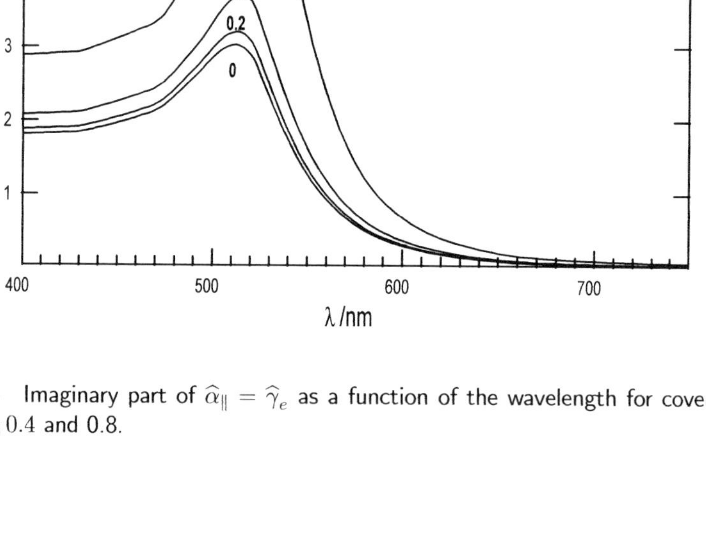

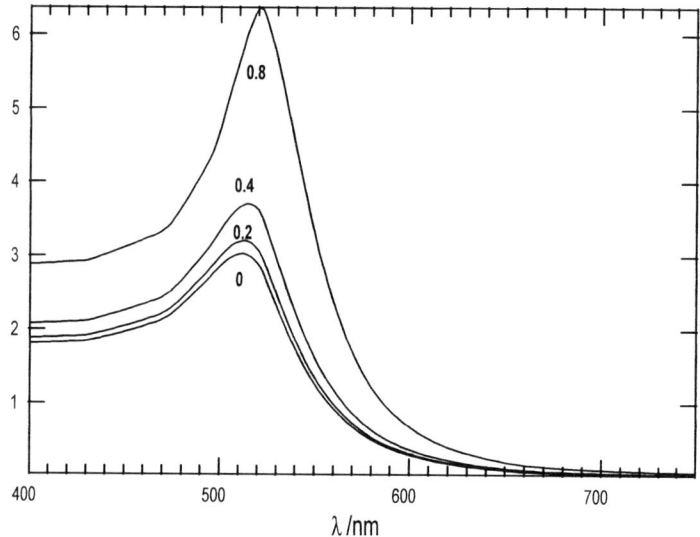

Figure 7.6 Imaginary part of $\widehat{\alpha}_{\|} = \widehat{\gamma}_e$ as a function of the wavelength for coverages $\widehat{\rho} = 0; 0.2; 0.4$ and 0.8.

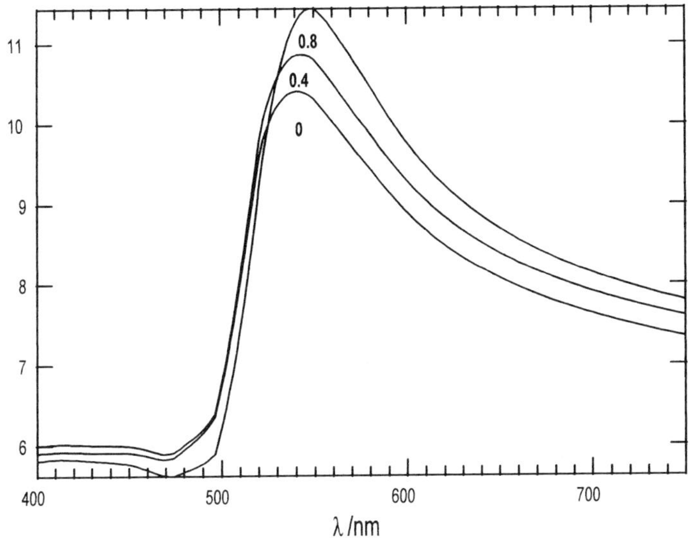

Figure 7.7 Real part of $\widehat{\delta}_e = \widehat{I}_{\delta,e}$ as a function of the wavelength for coverages $\widehat{\rho} = 0; 0.4$ and 0.8.

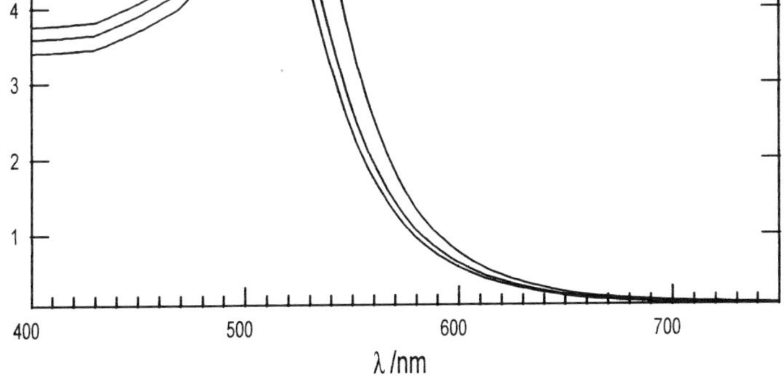

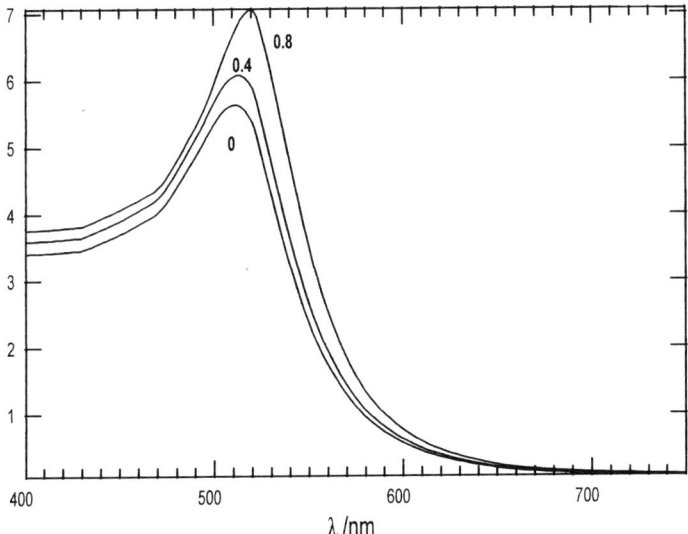

Figure 7.8 Imaginary part of $\widehat{\delta}_e = \widehat{I}_{\delta,e}$ as a function of the wavelength for coverages $\widehat{\rho} = 0; 0.4$ and 0.8.

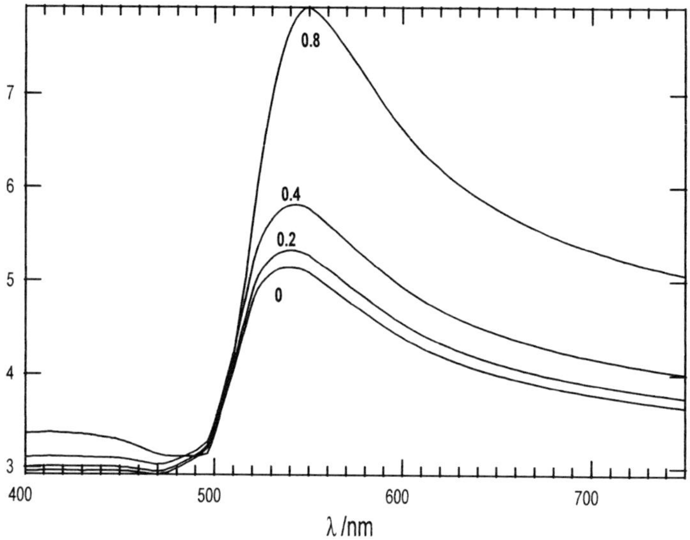

Figure 7.9 Real part of $\widehat{\tau} = \widehat{I}_\tau$ as a function of the wavelength for coverages $\widehat{\rho} = 0; 0.2; 0.4$ and 0.8.

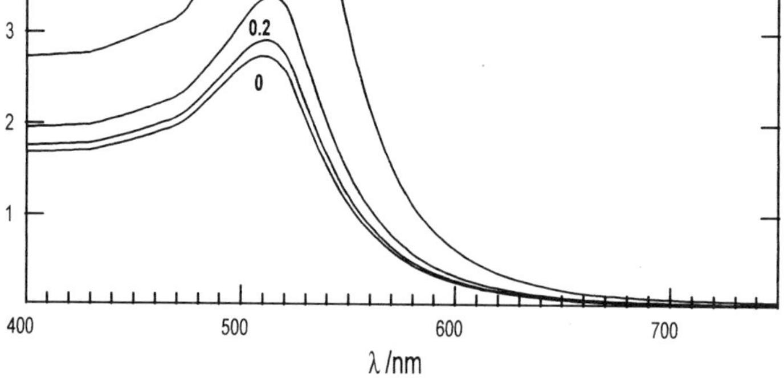

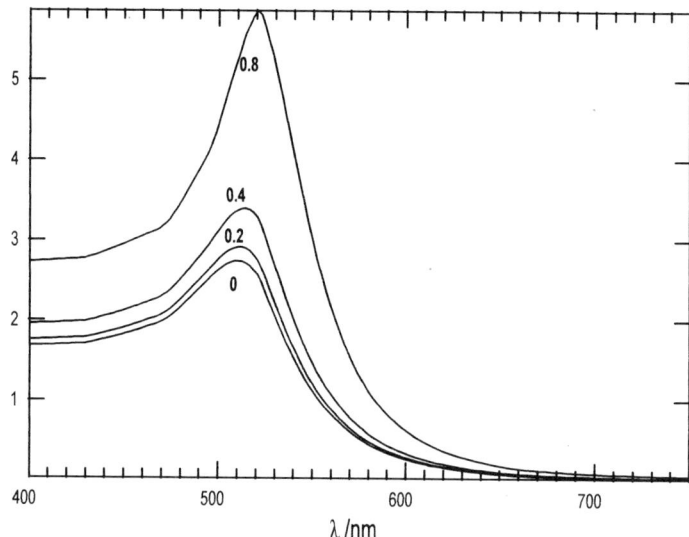

Figure 7.10 Imaginary part of $\widehat{\tau} = \widehat{I}_\tau$ as a function of the wavelength for coverages $\widehat{\rho} = 0; 0.2; 0.4$ and 0.8.

These coefficients are therefore mainly due to the shift of the induced dipole to the surface of the substrate.

For spheroids touching the substrate the results are given in figs. 7.11 through 7.18 for a square array. The interaction along the substrate is now taken into account to dipolar order. In that case eqs.(7.81) until (7.84) in scaled form reduce to

$$\begin{aligned}\widehat{\alpha}_z &\equiv \frac{\alpha_z(0)}{V} = 4\pi D_z^{-1}\widehat{\alpha}_{10,10} \\ \widehat{\alpha}_z^{10} &\equiv \frac{\alpha_z^{10}(0)}{Vd} = 2\pi\sqrt{\frac{5}{3}}D_z^{-1}\widehat{\alpha}_{20,10}\end{aligned} \tag{7.117}$$

with

$$D_z \equiv 1 - 4\widehat{V}\widehat{\alpha}_{10,10}\widetilde{S}_{20}^{-}\sqrt{\pi/5} \tag{7.118}$$

and

$$\begin{aligned}\widehat{\alpha}_{\parallel} &\equiv \frac{\alpha_{\parallel}(0)}{V} = 4\pi D_{\parallel}^{-1}\widehat{\alpha}_{11,11} \\ \widehat{\alpha}_{\parallel}^{10} &\equiv \frac{\alpha_{\parallel}^{10}(0)}{VR} = \frac{4}{3}\pi\sqrt{5}D_{\parallel}^{-1}\widehat{\alpha}_{21,11}\end{aligned} \tag{7.119}$$

with

$$D_{\parallel} \equiv 1 + 2\widehat{V}\widehat{\alpha}_{11,11}\widetilde{S}_{20}^{+}\sqrt{\pi/5} \tag{7.120}$$

Here $\widehat{V} \equiv V/L^3$, where V is the volume of the spheroid, and $\widehat{R} \equiv d/L = a\xi_0/L$. Furthermore the abbreviations $\widetilde{S}_{\ell m}^{\pm}$, eq.(7.80), have been used, with $\epsilon_a = 1$. The following new dimensionless quadrupolar polarizabilities are defined

$$\begin{aligned}\widehat{\alpha}_{10,10} &\equiv \alpha_{10,10}/V, \quad \widehat{\alpha}_{11,11} \equiv \alpha_{11,11}/V \\ \widehat{\alpha}_{10,20} &\equiv \alpha_{10,20}/Va\xi_0, \quad \widehat{\alpha}_{20,10} \equiv \alpha_{20,10}/Va\xi_0 \\ \widehat{\alpha}_{11,21} &\equiv \alpha_{11,21}/Va\xi_0, \quad \widehat{\alpha}_{21,11} \equiv \alpha_{21,11}/Va\xi_0\end{aligned} \tag{7.121}$$

The calculations were done for scaled densities $\widehat{\rho} \equiv \rho\,[2a\xi_0(AR)]^2 = [2a\xi_0\,(AR)\,/L]^2$ of 0 (no interaction), 0.2, 0.4 and 0.8, L is the lattice constant. Here $2a\xi_0\,(AR)$ is the diameter of the spheroid parallel to the substrate. The interaction with the image charge distribution was taken to the order $M = 16$. The procedure to do this is to use eqs.(6.200) through (6.206) to the order $M = 16$ together with eq.(5.67) to obtain the scaled polarizabilities in eq.(7.121). These should subsequently be substituted into eqs.(7.117) through (7.120). The reason to go to a higher order in the multipole expansion in the interaction with the substrate is that close to touching the substrate the higher order multipoles start to give corrections of about 5 to 10 %, see table 1 and 2 in ref.[42]. Due to the use of spheroidal multipoles this interaction can be analyzed accurately also for touching spheroids. This is not the case for the interaction along the substrate, for which spherical multipoles had to be used. When the distance along the substrate becomes small, the convergence becomes unsatisfactory. When the

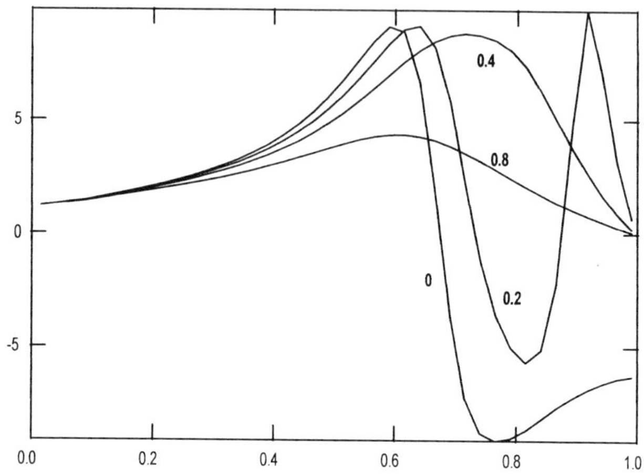

Figure 7.11 Real part of $\widehat{\alpha}_z = \widehat{\beta}_e$ as a function of $1/(1{+}AR)$ for coverages $\widehat{\rho} = 0; 0.2; 0.4$ and 0.8.

dipole approximation breaks down, one would need many more spherical multipoles to describe this interaction. If one could use spheroidal multipoles for the interaction along the substrate this would be much better. The mathematical tools for such an analysis are, however, not yet available. For the value of the dielectric constant of the particle the complex number $\epsilon = -5.28 + 2.25i$ is chosen, which is the dielectric constant of gold at the wavelength $\lambda = 540$ nm, cf. table 5.1. This wavelength is chosen because these gold films have a minimum in the transmission roughly at this wavelength. The shape dependence is expected to be most pronounced in that case. The dielectric constant of sapphire is again chosen to be equal to ϵ_s =3.13. The ambient is again vacuum, $\epsilon_a = 1$. The figures give the polarizabilities and the constitutive coefficients as a function of $1/(1 + AR)$. In this way it is possible to see how these quantities depend on the axial ratio. One sees in the figures that the polarizabilities and constitutive coefficients of oblate spheroids, 0<1/(1+AR)<0.5, do not depend very much on the coverage. Only the axial ratio for which 540 nm is the resonance wavelength changes a little bit. The reason for this is that the centers of the oblate spheroids remain relatively far apart, even for high densities. For prolate spheroids the dependence on the coverage becomes dramatic. The reason for this is that the prolate spheroids approach each other for a higher coverage to distances smaller than the long axis. It is to be expected that these results are unreliable for the combination of a high density and a small axial ratio. If one would set as a criterium that the distance between the particles should be larger than half the long axis, one finds that for the densities 0.0, 0.2, 0.4 and 0.8 the maximum values of 1/(1+AR) are 1.00, 0.69, 0.61 and 0.53 respectively. It follows that all the new maxima appearing for larger values of $\widehat{\rho}$ and 1/(1+AR) are outside the range of

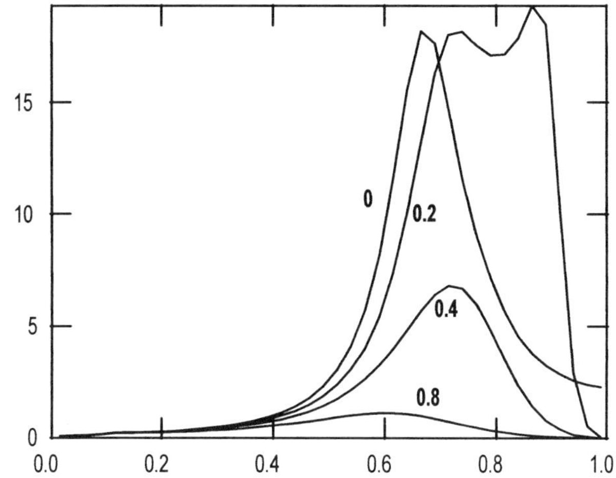

Figure 7.12 Imaginary part of $\widehat{\alpha}_z = \widehat{\beta}_e$ as a function of $1/(1 + AR)$ for coverages $\widehat{\rho} = 0; 0.2; 0.4$ and 0.8.

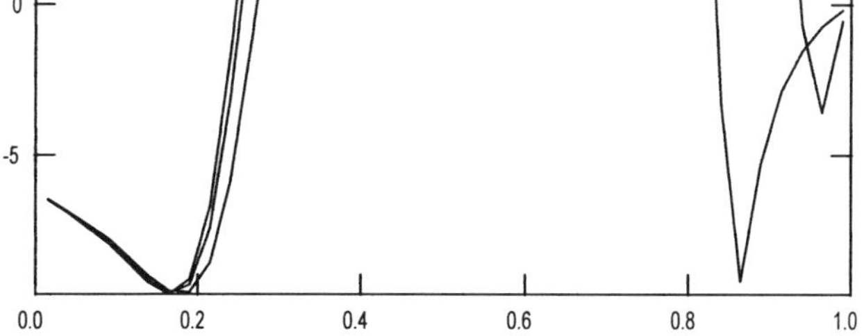

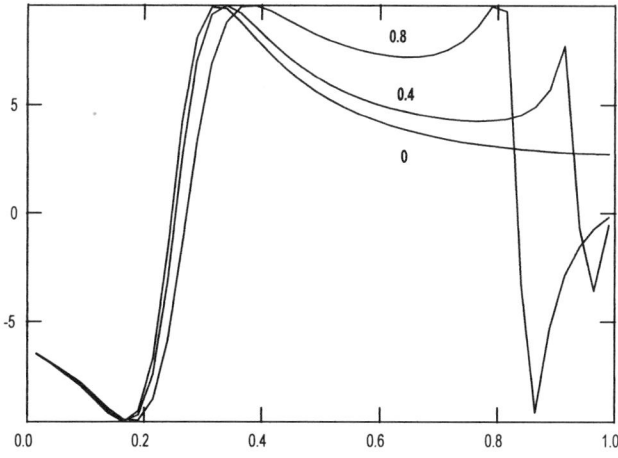

Figure 7.13 Real part of $\widehat{\alpha}_{\|} = \widehat{\gamma}_e$ as a function of $1/(1 + AR)$ for coverages $\widehat{\rho} = 0; 0.4$ and 0.8.

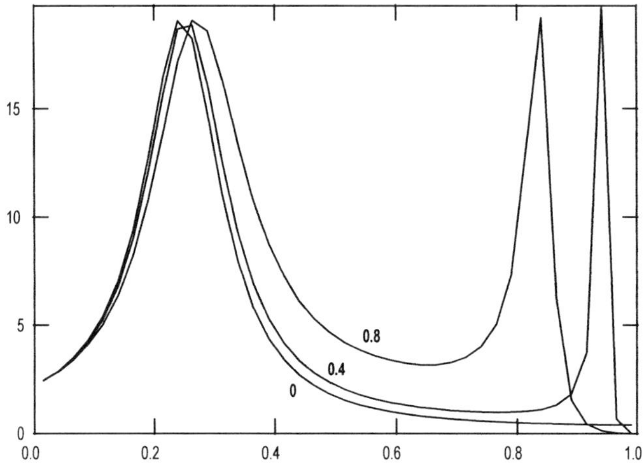

Figure 7.14 Imaginary part of $\widehat{\alpha}_{\parallel} = \widehat{\gamma}_e$ as a function of $1/(1 + AR)$ for coverages $\widehat{\rho} = 0; 0.4$ and 0.8.

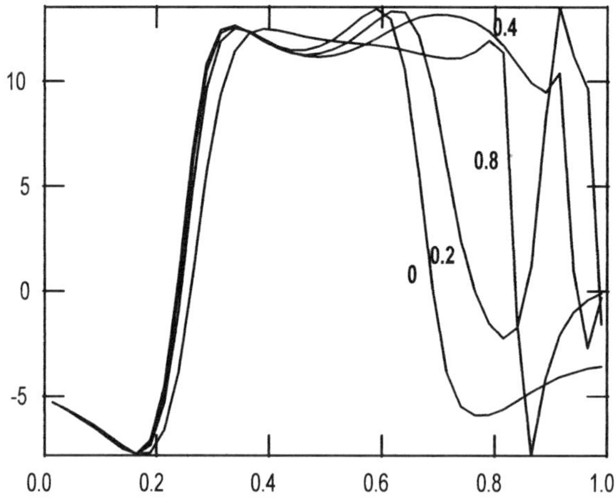

Figure 7.15 Real part of $\widehat{\delta}_e = \widehat{I}_{\delta,e}$ as a function of $1/(1 + AR)$ for coverages $\widehat{\rho} = 0; 0.2; 0.4$ and 0.8.

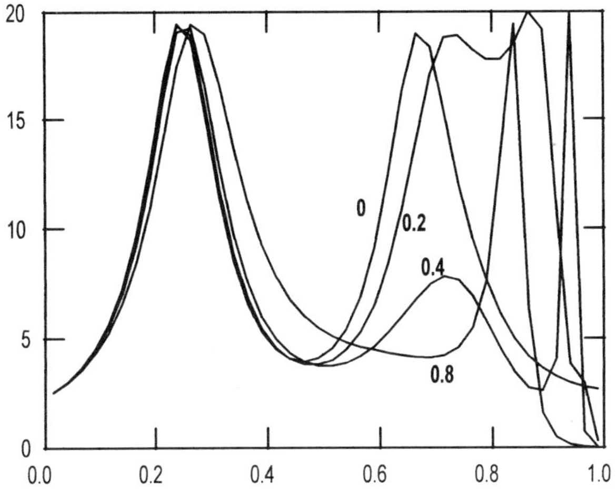

Figure 7.16 Imaginary part of $\widehat{\delta}_e = \widehat{I}_{\delta,e}$ as a function of $1/(1+AR)$ for coverages $\widehat{\rho} = 0; 0.2; 0.4$ and 0.8.

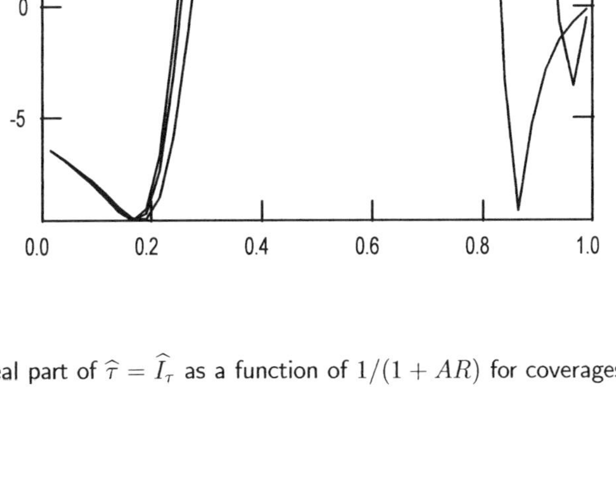

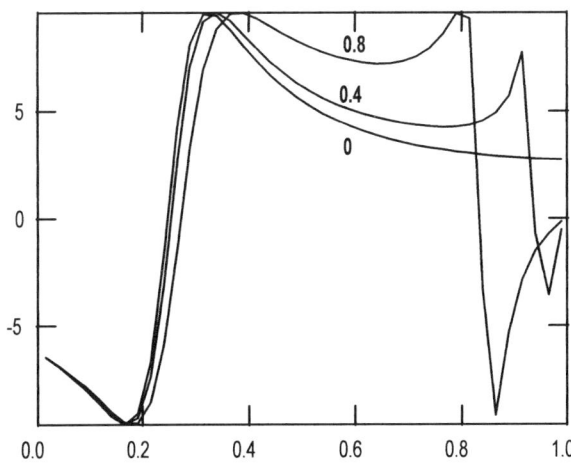

Figure 7.17 Real part of $\widehat{\tau} = \widehat{I}_\tau$ as a function of $1/(1+AR)$ for coverages $\widehat{\rho} = 0; 0.4$ and 0.8.

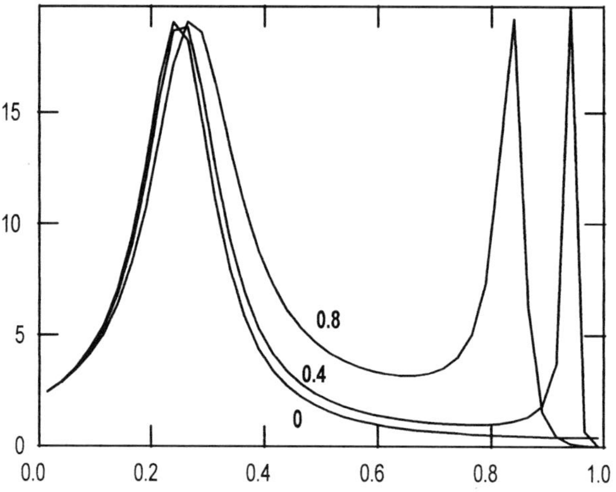

Figure 7.18 Imaginary part of $\widehat{\tau} = \widehat{I}_{\tau}$ as a function of $1/(1+AR)$ for coverages $\widehat{\rho} = 0; 0.4$ and 0.8.

reliability of the dipole approximation for the interaction along the substrate. Since the shape of the particles in island films is usually oblate rather than prolate, the dipole approximation to describe the interaction between the particles will in practice suffice.

7.11 Appendix A: Lattice sums

The lattice sums are defined by:

$$\begin{aligned} S_{\ell m} &= \sum_{i \neq 0} \left(\frac{L}{r}\right)^{\ell+1} Y_{\ell}^{m}(\theta,\phi)|_{\mathbf{r}=\mathbf{R}_i} \\ S_{\ell m}^{\tau} &= \sum_{i} \left(\frac{L}{r}\right)^{\ell+1} Y_{\ell}^{m}(\theta,\phi)|_{\mathbf{r}=\mathbf{R}_i^{\tau}} \end{aligned} \tag{7.122}$$

Using eq.(5.28) it follows that

$$S_{\ell m} = (-1)^m S_{\ell,-m}^{*} \quad \text{and} \quad S_{\ell m}^{\tau} = (-1)^m S_{\ell,-m}^{\tau *} \tag{7.123}$$

It is therefore sufficient to consider only positive values of m in this appendix. For $\ell > 2$ the sums converge reasonably well. For $\ell = 2$ and $m = 0$ the convergence is poor. It helps, in order to improve the convergence, to add lattice points, starting from the central point, layer by layer in the two-dimensional Bravais lattice, see fig.7.19,

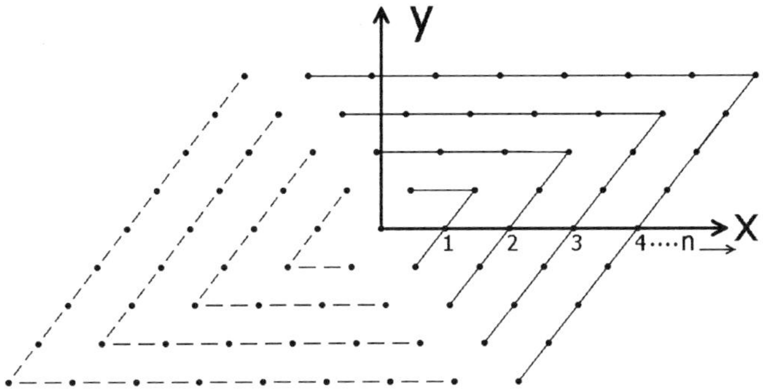

Figure 7.19 Illustration of the layer by layer summation method in a two-dimensional Bravais lattice.

$$
\begin{aligned}
S_{\ell m} &= \sum_{n=1}^{\infty} \sum_{par\ n} \left(\frac{L}{r}\right)^{\ell+1} Y_\ell^m(\theta,\phi)|_{\mathbf{r}=\mathbf{R}_i} = \sum_{n=1}^{\infty} \frac{C_{\ell m}^n}{n^\ell} \\
S_{\ell m}^r &= \sum_{n=0}^{\infty} \sum_{par\ n} \left(\frac{L}{r}\right)^{\ell+1} Y_\ell^m(\theta,\phi)|_{\mathbf{r}=\mathbf{R}_i^r} = \frac{1}{2}\delta_{m0}\left(\frac{L}{2d}\right)^{\ell+1}\sqrt{\frac{2\ell+1}{\pi}} + \sum_{n=1}^{\infty} \frac{C_{\ell m}^{n,r}}{n^\ell}
\end{aligned}
\tag{7.124}
$$

The subindex *par n* indicates that the sum is over the n-th parallelogram. For a two-dimensional Bravais lattice two sides of the parallelogram are sufficient, the other two sides give the same contribution to $C_{\ell m}^n$ and $C_{\ell m}^{n,r}$, the sums over the parallelogram, see 7.19. For a square lattice a quarter of the points is enough, and for a triangular lattice one sixth of the points. Using the fact that $C_{\ell m}^n$ and $C_{\ell m}^{n,r}$ approach constants for large n one may use the following convergence improving scheme:

$$
\begin{aligned}
S_{\ell m} &= \sum_{n=1}^{j} \frac{C_{\ell m}^n}{n^\ell} + C_{\ell m}^j \left(c_\ell - \sum_{n=1}^{j} \frac{1}{n^\ell}\right) \\
S_{\ell m}^r &= \frac{1}{2}\delta_{m0}\left(\frac{L}{2d}\right)^{\ell+1}\sqrt{\frac{2\ell+1}{\pi}} + \sum_{n=1} \frac{C_{\ell m}^{n,r}}{n^\ell} + C_{\ell m}^{j,r}\left(c_\ell - \sum_{n=1}^{j} \frac{1}{n^\ell}\right)
\end{aligned}
\tag{7.125}
$$

Here the constants are defined as

$$
c_\ell = \sum_{n=1}^{\infty} \frac{1}{n^\ell} \tag{7.126}
$$

This convergence improving trick is in particular needed to calculate S_{20}, which was not conveniently possible otherwise. For this case one has $c_2 = \pi^2/6$.

For the $S_{\ell m}^r$'s one may calculate the lattice sum as a sum over the reciprocal lattice. For small ℓ this converges much faster than the procedure above. In order to obtain a sum over the reciprocal lattice one may use the identity

$$
\sum_i f(\mathbf{R}_i^r) = O^{-1} \sum_\lambda \int d\mathbf{R}_\parallel \; e^{-i\mathbf{k}_\lambda \cdot \mathbf{R}_\parallel} f(\mathbf{R}^r) \tag{7.127}
$$

which expresses the sum over the two-dimensional direct lattice of a function f in terms of a sum of its Fourier transform over the two-dimensional reciprocal lattice. Here $\mathbf{k}_\lambda$ are the reciprocal lattice vectors. Furthermore O is the surface area of the Bravais lattice cell. In view of eq.(7.123) it is sufficient to give this derivation only for m positive. The image position is defined as $\mathbf{R}^r = (\mathbf{R}_\|, 2d)$. The Fourier transform needed is

$$O^{-1}\int d\mathbf{R}_\| \; e^{-i\mathbf{k}_\lambda.\mathbf{R}_\|}\left(\frac{L}{r}\right)^{\ell+1} Y_\ell^m(\theta,\phi)|_{\mathbf{r}=\mathbf{R}^r} = (-1)^m O^{-1}\sqrt{\frac{(2\ell+1)\,(\ell-m)!}{4\pi\,(\ell+m)!}}$$
$$\times\int_0^\infty\int_0^{2\pi}\left[\left(\frac{L^2}{r^2+4d^2}\right)^{(\ell+1)/2} P_\ell^m\left(2d/\sqrt{r^2+4d^2}\right)e^{im\phi}e^{-ik_\lambda r\cos(\phi-\phi_\lambda)}\right] rdrd\phi$$

Here ϕ_λ and ϕ are the angles of k_λ and $\mathbf{R}_\|$ with the x-axis. Introducing the angle between k_λ and $\mathbf{R}_\|$, given by $\varphi = \phi - \phi_\lambda$, one has

$$\int_0^{2\pi} e^{im\phi}e^{-ik_\lambda r\cos(\phi-\phi_\lambda)}d\phi = e^{im\phi_\lambda}\int_0^{2\pi} e^{im\varphi}e^{-ik_\lambda r\cos\varphi}d\varphi = 2\pi e^{im\phi_\lambda}\,(-i)^m\, J_m\,(k_\lambda r)$$

using one of the definitions of the Bessel function. Substitution in the previous equation gives

$$O^{-1}\int d\mathbf{R}_\| \; e^{-i\mathbf{k}_\lambda.\mathbf{R}_\|}\left(\frac{L}{r}\right)^{\ell+1} Y_\ell^m(\theta,\phi)|_{\mathbf{r}=\mathbf{R}^r} = i^m e^{im\phi_\lambda}\sqrt{\frac{\pi\,(2\ell+1)\,(\ell-m)!}{(\ell+m)!}}$$
$$\times O^{-1}\int_0^\infty\left[\left(\frac{L^2}{r^2+4d^2}\right)^{(\ell+1)/2} P_\ell^m\left(2d/\sqrt{r^2+4d^2}\right)J_m\,(k_\lambda r)\right] rdr$$
$$= (-i)^m\, e^{im\phi_\lambda}\,(Lk_\lambda)^{\ell-1}\, e^{-2dk_\lambda}\sqrt{\frac{\pi\,(2\ell+1)}{(\ell+m)!\,(\ell-m)!}} \qquad (7.128)$$

The last identity is a known Hankel transform, in ref.[48] this identity is only given for positive values of m . In view of eq.(7.123) this is sufficient. For the lattice sum one obtains in this way, using eq.(7.127),

$$S_{\ell m}^r = (-i)^m\sqrt{\frac{\pi\,(2\ell+1)}{(\ell+m)!\,(\ell-m)!}}\sum_\lambda e^{im\phi_\lambda}\,(Lk_\lambda)^{\ell-1}\, e^{-2dk_\lambda} \qquad (7.129)$$

It is clear that this relation satisfies eq.(7.123) and can therefore be used also for negative values of m. From the above expression one sees that for finite values of d the sum over the inverse lattice has an exponential convergence factor. The analogous expression for $S_{\ell m}$ does not have this convergence factor and is therefore useless.

For three-dimensional lattice sums it is convenient to use the method due to Nijboer en de Wette[51]. For two-dimensional lattice sums it seems impossible to give a convenient help function like they do. Since the two-dimensional sums converge better anyway, it is not worthwhile to pursue this point any further.

The lattice sums used in the calculations for a square and a triangular lattice are given as a function of $2d/L$ in the table below

$2d/L$	$\tilde{S}_{20}$	$\tilde{S}_{30}$	$\tilde{S}_{40}$
0.00	-2.849	0	1.615
0.05	-2.831	-0.282	1.572
0.10	-2.778	-0.550	1.447
0.15	-2.693	-0.789	1.253
0.20	-2.580	-0.989	1.012
0.25	-2.444	-1.144	0.746
0.30	-2.292	-1.252	0.477
0.35	-2.129	-1.313	0.226
0.40	-1.961	-1.333	0.005
0.45	-1.792	-1.317	-0.177
0.50	-1.628	-1.273	-0.317
0.55	-1.470	-1.208	-0.416
0.60	-1.322	-1.129	-0.479
0.65	-1.184	-1.041	-0.511
0.70	-1.058	-0.950	-0.519
0.75	-0.944	-0.859	-0.508
0.80	-0.840	-0.772	-0.484
0.85	-0.748	-0.689	-0.452
0.90	-0.665	-0.612	-0.416
0.95	-0.592	-0.542	-0.378
1.00	-0.528	-0.479	-0.340

Table 7.1: Quadratic lattice

$2d/L$	$\tilde{S}_{20}$	$\tilde{S}_{30}$	$\tilde{S}_{40}$
0.00	-3.480	0	2.146
0.05	-3.456	-0.375	2.085
0.10	-3.386	-0.729	1.910
0.15	-3.273	-1.043	1.640
0.20	-3.124	-1.303	1.304
0.25	-2.945	-1.501	0.937
0.30	-2.746	-1.634	0.569
0.35	-2.534	-1.703	0.227
0.40	-2.316	-1.717	-0.068
0.45	-2.101	-1.683	-0.307
0.50	-1.891	-1.612	-0.486
0.55	-1.693	-1.515	-0.607
0.60	-1.508	-1.400	-0.679
0.65	-1.338	-1.278	-0.707
0.70	-1.184	-1.153	-0.704
0.75	-1.046	-1.031	-0.678
0.80	-0.922	-0.914	-0.637
0.85	-0.813	-0.807	-0.585
0.90	-0.718	-0.708	-0.530
0.95	-0.633	-0.620	-0.475
1.00	-0.560	-0.542	-0.417

Triangular lattice

7.12 Appendix B

In this appendix it will be discussed, how the method, for the calculation of the polarizabilities of spheroids on or above a flat substrate, developed in the previous chapter, sections 6.4 and 6.5, must be changed, in order to be able to calculate the multipole polarizabilities $\alpha_{20,20}$, $\alpha_{21,21}$ and $\alpha_{22,22}$ for these particles. First the case of a oblate spheroid, with an axis of rotation normal to the surface of the substrate, will be considered, cf. fig.6.5. Instead of a constant external field $\mathbf{E}_0$, an external field will be taken, which is a linear function of space and therefore the gradient of a potential of the form

$$\psi_0(\mathbf{r}) = -\frac{1}{2}\boldsymbol{\phi} : \mathbf{rr} \equiv -\frac{1}{2}\sum_{i,j=1}^{3} \phi_{ij} r_i r_j \tag{7.130}$$

Here $r_1 \equiv x$, $r_2 \equiv y, r_3 \equiv z$ and ϕ_{ij} is a symmetric traceless matrix . The indices i and j denote the Cartesian components x, y and z. Introducing oblate spheroidal coordinates ξ, η and ϕ, defined by eqs.(6.1) and (6.2), and eliminating $\phi_{zz} = -\left(\phi_{xx} + \phi_{yy}\right)$, this potential can be written in the form:

$$\begin{aligned}\psi_0(\mathbf{r}) &= \sqrt{2\pi/15}[\frac{1}{2}(\phi_{xx}+\phi_{yy})X_2^0(\xi,a)Y_2^0(\arccos\eta,\phi)\sqrt{6}-(\phi_{xz}+i\ \phi_{yz})\\ &\times X_2^{-1}(\xi,a)Y_2^{-1}(\arccos\eta,\phi)+(\phi_{xz}-i\ \phi_{yz})X_2^1(\xi,a)Y_2^1(\arccos\eta,\phi)\\ &-\left(\frac{1}{2}\phi_{xx}-\frac{1}{2}\phi_{yy}+i\ \phi_{xy}\right)X_2^{-2}(\xi,a)Y_2^{-2}(\arccos\eta,\phi)\\ &-\left(\frac{1}{2}\phi_{xx}-\frac{1}{2}\phi_{yy}-i\ \phi_{xy}\right)X_2^2(\xi,a)Y_2^2(\arccos\eta,\phi)]\end{aligned}\tag{7.131}$$

where an additive constant has been omitted. For the definition of the spherical harmonics Y_ℓ^m and the functions X_ℓ^m, the reader is referred to section 6.2.

For the general solution in the ambient one now obtains, instead of eq.(6.99),

$$\begin{aligned}\psi_a(\mathbf{r}) &= \sqrt{2\pi/15}[\frac{1}{2}(\phi_{xx}+\phi_{yy})X_2^0(\xi,a)Y_2^0(\arccos\eta,\phi)\sqrt{6}-(\phi_{xz}+i\ \phi_{yz})\\ &\times X_2^{-1}(\xi,a)Y_2^{-1}(\arccos\eta,\phi)+(\phi_{xz}-i\ \phi_{yz})X_2^1(\xi,a)Y_2^1(\arccos\eta,\phi)\\ &-\left(\frac{1}{2}\phi_{xx}-\frac{1}{2}\phi_{yy}+i\ \phi_{xy}\right)X_2^{-2}(\xi,a)Y_2^{-2}(\arccos\eta,\phi)\\ &-\left(\frac{1}{2}\phi_{xx}-\frac{1}{2}\phi_{yy}-i\ \phi_{xy}\right)X_2^2(\xi,a)Y_2^2(\arccos\eta,\phi)]\\ &+\sum_{\ell m}{}' A_{\ell m}Z_\ell^m(\xi,a)Y_\ell^m(\arccos\eta,\phi)\\ &+\sum_{\ell m}{}' A_{\ell m}^r Z_\ell^m(\xi'(\xi,\eta),a)Y_\ell^m(\arccos\eta'(\xi,\eta),\phi)\end{aligned}\tag{7.132}$$

For the origin of the various contributions the reader is referred to the discussion below eq.(6.99).

The potential in the substrate, which was given by eqs.(6.100) with eqs.(6.104) and (6.105), now becomes

$$\begin{aligned}\psi_s(\mathbf{r}) &= \left(1-\frac{\epsilon_a}{\epsilon_s}\right)(\phi_{xx}+\phi_{yy})d^2-\sqrt{2\pi/3}\left(1-\frac{\epsilon_a}{\epsilon_s}\right)d[(\phi_{xx}+\phi_{yy})X_1^0(\xi,a)\\ &\times Y_1^0(\arccos\eta,\phi)\sqrt{2}+(\phi_{xz}+i\phi_{yz})X_1^{-1}(\xi,a)Y_1^{-1}(\arccos\eta,\phi)\\ &-(\phi_{xz}-i\phi_{yz})X_1^1(\xi,a)Y_1^1(\arccos\eta,\phi)]+\sqrt{2\pi/15}[\frac{1}{2}(\phi_{xx}+\phi_{yy})\\ &\times X_2^0(\xi,a)Y_2^0(\arccos\eta,\phi)\sqrt{6}-\frac{\epsilon_a}{\epsilon_s}(\phi_{xz}+i\ \phi_{yz})X_2^{-1}(\xi,a)Y_2^{-1}(\arccos\eta,\phi)\\ &+\frac{\epsilon_a}{\epsilon_s}(\phi_{xz}-i\ \phi_{yz})X_2^1(\xi,a)Y_2^1(\arccos\eta,\phi)-\left(\frac{1}{2}\phi_{xx}-\frac{1}{2}\phi_{yy}+i\ \phi_{xy}\right)\\ &\times X_2^{-2}(\xi,a)Y_2^{-2}(\arccos\eta,\phi)-\left(\frac{1}{2}\phi_{xx}-\frac{1}{2}\phi_{yy}-i\ \phi_{xy}\right)\\ &\times X_2^2(\xi,a)Y_2^2(\arccos\eta,\phi)]+\sum_{\ell m}{}' A_{\ell m}^t Z_\ell^m(\xi,a)Y_\ell^m(\arccos\eta,\phi)\end{aligned}\tag{7.133}$$

It can be checked, using eqs.(6.5)–(6.8), (6.13), (6.101)–(6.103), that the external potential, appearing in the right-hand sides of the above equations, satisfies the boundary conditions at the surface of the substrate, see text below eq.(6.100). Furthermore it follows from these boundary conditions that eq.(6.105) holds for the multipole coefficients in eqs.(7.132) and (7.133).

In order to relate the amplitudes $A_{\ell m}$ to the multipole polarizabilities for the oblate spheroids in free space, the potential of the so-called local field, acting on the particles, is again introduced. Instead of eq.(6.107), one now has

$$\begin{aligned}\psi_{loc}(\mathbf{r}) &= \sqrt{2\pi/15}[\frac{1}{2}(\phi_{xx}+\phi_{yy})X_2^0(\xi,a)Y_2^0(\arccos\eta,\phi)\sqrt{6}-(\phi_{xz}+i\ \phi_{yz}) \\ &\times X_2^{-1}(\xi,a)Y_2^{-1}(\arccos\eta,\phi)+(\phi_{xz}-i\ \phi_{yz})X_2^1(\xi,a)Y_2^1(\arccos\eta,\phi) \\ &-\left(\frac{1}{2}\phi_{xx}-\frac{1}{2}\phi_{yy}+i\ \phi_{xy}\right)X_2^{-2}(\xi,a)Y_2^{-2}(\arccos\eta,\phi) \\ &-\left(\frac{1}{2}\phi_{xx}-\frac{1}{2}\phi_{yy}-i\ \phi_{xy}\right)X_2^{2}(\xi,a)Y_2^{2}(\arccos\eta,\phi)] \\ &+\sum_{\ell m}{}' A_{\ell m}^r Z_\ell^m(\xi'(\xi,\eta),a)Y_\ell^m(\arccos\eta'(\xi,\eta),\phi)\end{aligned} \tag{7.134}$$

Substitution of the series expansion (6.108), into this equation and interchanging ℓ with ℓ_1, one obtains, instead of eq.(6.119),

$$\begin{aligned}\psi_{loc}(\mathbf{r}) &= \sqrt{2\pi/15}[\frac{1}{2}(\phi_{xx}+\phi_{yy})X_2^0(\xi,a)Y_2^0(\arccos\eta,\phi)\sqrt{6}-(\phi_{xz}+i\ \phi_{yz}) \\ &\times X_2^{-1}(\xi,a)Y_2^{-1}(\arccos\eta,\phi)+(\phi_{xz}-i\ \phi_{yz})X_2^1(\xi,a)Y_2^1(\arccos\eta,\phi) \\ &-\left(\frac{1}{2}\phi_{xx}-\frac{1}{2}\phi_{yy}+i\ \phi_{xy}\right)X_2^{-2}(\xi,a)Y_2^{-2}(\arccos\eta,\phi) \\ &-\left(\frac{1}{2}\phi_{xx}-\frac{1}{2}\phi_{yy}-i\ \phi_{xy}\right)X_2^{2}(\xi,a)Y_2^{2}(\arccos\eta,\phi)] \\ &+\sum_{\ell m}\sum_{\ell_1=|m|}^{\infty\prime} A_{\ell_1,m}^r K_{\ell,\ell_1}^m(\xi_1)(2d)^{-\ell-\ell_1-1}X_\ell^m(\xi,a)Y_\ell^m(\arccos\eta,\phi)\end{aligned} \tag{7.135}$$

where the prime as index of summation indicates that $\ell_1 = 0$ should be excluded. Using eqs.(6.21) and (6.105), one now obtains

$$\begin{aligned}A_{\ell m} &= -\alpha_{\ell m}\Big\{\delta_{\ell 2}\sqrt{2\pi/15}[\frac{1}{2}(\phi_{xx}+\phi_{yy})\delta_{m0}\sqrt{6}-(\phi_{xz}+i\ \phi_{yz})\delta_{m,-1} \\ &+(\phi_{xz}-i\ \phi_{yz})\delta_{m1}-\left(\frac{1}{2}\phi_{xx}-\frac{1}{2}\phi_{yy}+i\ \phi_{xy}\right)\delta_{m,-2} \\ &-\left(\frac{1}{2}\phi_{xx}-\frac{1}{2}\phi_{yy}-i\ \phi_{xy}\right)\delta_{m2}] \\ &+\left(\frac{\epsilon_a-\epsilon_s}{\epsilon_a+\epsilon_s}\right)\sum_{\ell_1=|m|}^{\infty\prime}(-1)^{\ell_1+m}K_{\ell,\ell_1}^m(\xi_1)A_{\ell_1,m}(2d)^{-\ell-\ell_1-1}\Big\}\end{aligned} \tag{7.136}$$

where $\alpha_{\ell m}$ is given by eq.(6.24).

It follows that different m's do not couple to each other. This is due to the rotational symmetry of the system around the z-axis. As a consequence

$$A_{\ell m} = 0 \quad \text{for} \quad m \neq -2, -1, 0, 1, 2 \tag{7.137}$$

The above set of linear equations for the amplitudes may thus be written as five sets of independent linear equations:

$$\begin{aligned} & A_{\ell 0} + \alpha_{\ell 0} \left(\frac{\epsilon_a - \epsilon_s}{\epsilon_a + \epsilon_s} \right) \sum_{\ell_1=1}^{\infty} (-1)^{\ell_1} K^0_{\ell,\ell_1}(\xi_1) A_{\ell_1,0} (2d)^{-\ell-\ell_1-1} \\ = \; & -\alpha_{20} \delta_{\ell 2} (\phi_{xx} + \phi_{yy}) \sqrt{\pi/5} \quad \text{for} \quad \ell = 1, 2, ... \end{aligned} \tag{7.138}$$

$$\begin{aligned} & A_{\ell 1} - \alpha_{\ell 1} \left(\frac{\epsilon_a - \epsilon_s}{\epsilon_a + \epsilon_s} \right) \sum_{\ell_1=1}^{\infty} (-1)^{\ell_1} K^1_{\ell,\ell_1}(\xi_1) A_{\ell_1,1} (2d)^{-\ell-\ell_1-1} \\ = \; & -\alpha_{21} \delta_{\ell 2} (\phi_{xz} - i\phi_{yz}) \sqrt{2\pi/15} \quad \text{for} \quad \ell = 1, 2, ... \end{aligned} \tag{7.139}$$

$$\begin{aligned} & A_{\ell,-1} - \alpha_{\ell,-1} \left(\frac{\epsilon_a - \epsilon_s}{\epsilon_a + \epsilon_s} \right) \sum_{\ell_1=1}^{\infty} (-1)^{\ell_1} K^{-1}_{\ell,\ell_1}(\xi_1) A_{\ell_1,-1} (2d)^{-\ell-\ell_1-1} \\ = \; & \alpha_{2,-1} \delta_{\ell 2} (\phi_{xz} + i\phi_{yz}) \sqrt{2\pi/15} \quad \text{for} \quad \ell = 1, 2, ... \end{aligned} \tag{7.140}$$

$$\begin{aligned} & A_{\ell 2} + \alpha_{\ell 2} \left(\frac{\epsilon_a - \epsilon_s}{\epsilon_a + \epsilon_s} \right) \sum_{\ell_1=2}^{\infty} (-1)^{\ell_1} K^2_{\ell,\ell_1}(\xi_1) A_{\ell_1,2} (2d)^{-\ell-\ell_1-1} \\ = \; & \alpha_{22} \delta_{\ell 2} \left(\frac{1}{2}\phi_{xx} - \frac{1}{2}\phi_{yy} - i\ \phi_{xy} \right) \sqrt{2\pi/15} \quad \text{for} \quad \ell = 2, 3, ... \end{aligned} \tag{7.141}$$

$$\begin{aligned} & A_{\ell,-2} + \alpha_{\ell,-2} \left(\frac{\epsilon_a - \epsilon_s}{\epsilon_a + \epsilon_s} \right) \sum_{\ell_1=2}^{\infty} (-1)^{\ell_1} K^{-2}_{\ell,\ell_1}(\xi_1) A_{\ell_1,-2} (2d)^{-\ell-\ell_1-1} \\ = \; & \alpha_{2,-2} \delta_{\ell 2} \left(\frac{1}{2}\phi_{xx} - \frac{1}{2}\phi_{yy} + i\ \phi_{xy} \right) \sqrt{2\pi/15} \quad \text{for} \quad \ell = 2, 3, ... \end{aligned} \tag{7.142}$$

In an analogous way as in section 6.4, one can now prove that eq.(7.140) for $m = -1$ and (7.142) for $m = -2$ are redundant. Using the symmetry relations, eqs.(6.25) and (6.117), in eqs.(7.140) and (7.142), and comparing the results with eqs.(7.139) and (7.141), respectively, one finds that

$$(\phi_{xz} + i\ \phi_{yz}) A_{\ell 1} = -(\phi_{xz} - i\ \phi_{yz}) A_{\ell,-1} \tag{7.143}$$

and

$$\left(\frac{1}{2}\phi_{xx} - \frac{1}{2}\phi_{yy} + i\ \phi_{xy} \right) A_{\ell 2} = \left(\frac{1}{2}\phi_{xx} - \frac{1}{2}\phi_{yy} - i\ \phi_{xy} \right) A_{\ell,-2} \tag{7.144}$$

It is therefore sufficient to consider only the $m = 0, m = 1$ and the $m = 2$ equations.

These equations may now be solved numerically, by neglecting the amplitudes for $\ell > M$. One then solves the finite sets of inhomogeneous equations:

$$\begin{aligned} A_{\ell 0} + \alpha_{\ell 0} \left(\frac{\epsilon_a - \epsilon_s}{\epsilon_a + \epsilon_s} \right) \sum_{\ell_1=1}^{M} (-1)^{\ell_1} K^0_{\ell,\ell_1}(\xi_1) A_{\ell_1,0} (2d)^{-\ell-\ell_1-1} \\ = -\alpha_{20}\delta_{\ell 2}(\phi_{xx} + \phi_{yy})\sqrt{\pi/5} \quad \text{for} \quad \ell = 1, 2, ..., M \end{aligned} \tag{7.145}$$

$$\begin{aligned} A_{\ell 1} - \alpha_{\ell 1} \left(\frac{\epsilon_a - \epsilon_s}{\epsilon_a + \epsilon_s} \right) \sum_{\ell_1=1}^{M} (-1)^{\ell_1} K^1_{\ell,\ell_1}(\xi_1) A_{\ell_1,1} (2d)^{-\ell-\ell_1-1} \\ = -\alpha_{21}\delta_{\ell 2}(\phi_{xz} - i\phi_{yz})\sqrt{2\pi/15} \quad \text{for} \quad \ell = 1, 2, ..., M \end{aligned} \tag{7.146}$$

$$\begin{aligned} A_{\ell 2} + \alpha_{\ell 2} \left(\frac{\epsilon_a - \epsilon_s}{\epsilon_a + \epsilon_s} \right) \sum_{\ell_1=2}^{M} (-1)^{\ell_1} K^2_{\ell,\ell_1}(\xi_1) A_{\ell_1,2} (2d)^{-\ell-\ell_1-1} \\ = \alpha_{22}\delta_{\ell 2} \left(\frac{1}{2}\phi_{xx} - \frac{1}{2}\phi_{yy} - i\,\phi_{xy} \right) \sqrt{2\pi/15} \quad \text{for} \quad \ell = 2, 3, ..., M \end{aligned} \tag{7.147}$$

In this way the amplitudes A_{10}, A_{20}, A_{11}, A_{21} and A_{22} may be calculated to any desired degree of accuracy.

It follows from the above equations, that A_{10} and A_{20} are proportional to $(\phi_{xx} + \phi_{yy})$, A_{11} and A_{21} to $(\phi_{xz} - i\phi_{yz})$ and A_{22} to $(\frac{1}{2}\phi_{xx} - \frac{1}{2}\phi_{yy} - i\,\phi_{xy})$, appearing in the corresponding terms in the potential of the external field, eq.(7.131). If one writes this potential as

$$\psi_0(\mathbf{r}) = \sum_{m=-2}^{2} B^{(0)}_{2m} X^m_2(\xi, a) Y^m_2(\arccos \eta, \phi) \tag{7.148}$$

it is found that

$$\begin{aligned} B^{(0)}_{20} &= (\phi_{xx} + \phi_{yy})\sqrt{\pi/5}, \quad B^{(0)}_{2,\pm1} = \pm(\phi_{xz} \mp i\,\phi_{yz})\sqrt{2\pi/15} \\ B^{(0)}_{2,\pm2} &= -\left(\frac{1}{2}\phi_{xx} - \frac{1}{2}\phi_{yy} \mp i\,\phi_{xy} \right) \sqrt{2\pi/15} \end{aligned} \tag{7.149}$$

Therefore A_{10} and A_{20} are proportional to $B^{(0)}_{20}$, A_{11} and A_{21} to $B^{(0)}_{21}$ and A_{22} to $B^{(0)}_{22}$. The proportionality constants are, according to eq.(5.30), the multipole polarizabilities $\alpha_{\ell m,\ell' m'}$:

$$\begin{aligned} A_{10} &= -\alpha_{10,20} B^{(0)}_{20}, \quad A_{20} = -\alpha_{20,20} B^{(0)}_{20}, \quad A_{11} = -\alpha_{11,21} B^{(0)}_{21} \\ A_{21} &= -\alpha_{21,21} B^{(0)}_{21}, \quad A_{22} = -\alpha_{22,22} B^{(0)}_{22} \end{aligned} \tag{7.150}$$

These are the multipole polarizabilities of the oblate spheroid, interacting with its image in the substrate. The multipole polarizabilities $\alpha_{10,20}$ and $\alpha_{11,21}$ could also be found using the method developed in the previous chapter, in which a homogeneous external field is used. New are the multipole polarizabilities $\alpha_{20,20}$, $\alpha_{21,21}$ and $\alpha_{22,22}$. These quantities are therefore found by dividing the amplitudes A_{20}, A_{21} and A_{22} by $B_{20}^{(0)}$, $B_{21}^{(0)}$ and $B_{22}^{(0)}$ respectively or, more explicitly, from

$$\begin{aligned}\alpha_{20,20} &= -A_{20}/(\phi_{xx}+\phi_{yy})\sqrt{\pi/5}, \quad \alpha_{21,21} = -A_{21}/(\phi_{xz} - i\ \phi_{yz})\sqrt{2\pi/15} \\ \alpha_{22,22} &= A_{22}/\left(\frac{1}{2}\phi_{xx} - \frac{1}{2}\phi_{yy} - i\ \phi_{xy}\right)\sqrt{2\pi/15} \end{aligned} \tag{7.151}$$

With eq.(5.67), one can then also calculate the quadrupole polarizabilities:

$$\begin{aligned}\alpha_z^{11}(0) &= -\frac{\pi\ \epsilon_a A_{20}}{(\phi_{xx}+\phi_{yy})\sqrt{\pi/5}} - \frac{6\pi\ \epsilon_a A_{22}}{(\phi_{xx}-\phi_{yy}-2i\phi_{xy})\sqrt{6\pi/5}} \\ \alpha_{\|}^{11}(0) &= -\frac{4\pi\ \epsilon_a A_{21}}{(\phi_{xz}-i\phi_{yz})\sqrt{6\pi/5}} - \frac{8\pi\ \epsilon_a A_{22}}{(\phi_{xx}-\phi_{yy}-2i\phi_{xy})\sqrt{6\pi/5}} \\ \alpha^{11}(0) &= \frac{4\pi\ \epsilon_a A_{22}}{(\phi_{xx}-\phi_{yy}-2i\phi_{xy})\sqrt{6\pi/5}} \end{aligned} \tag{7.152}$$

of an oblate spheroid interacting with the substrate.

In solving eqs.(7.145)–(7.147), one must use eq.(6.24) for the free polarizabilities $\alpha_{\ell 0}$, $\alpha_{\ell 1}$ and $\alpha_{\ell 2}$ of the oblate spheroid. Furthermore one needs eqs.(6.110)–(6.113) to find $K^0_{\ell,\ell_1}(\xi_1)$ and eq.(6.118) for $K^1_{\ell,\ell_1}(\xi_1)$. An analogous formula may be derived for $K^2_{\ell,\ell_1}(\xi_1)$. This quantity is needed if one wants to calculate $\alpha_{22,22}$. Since the latter quantity does not contribute to the interfacial polarizabilities needed in the analysis in this book, cf. eqs.(7.81)–(7.84), this formula will not be given here.

Solving eqs.(7.145) and (7.146) for $M = 2$, using eqs.(6.27), for $\alpha_{\ell 0}$ and $\alpha_{\ell 1}$ with $\ell = 1, 2$, (6.243)–(6.245) and (6.118), for $K^0_{\ell,\ell_1}(\xi_1)$ and $K^1_{\ell,\ell_1}(\xi_1)$ with $\ell, \ell_1 = 1, 2$, one obtains a solution which accounts only for the spheroidal dipolar and quadrupolar interaction between the oblate spheroid and its image in the substrate. Substitution of the resulting amplitudes into eq.(7.151) gives

$$\begin{aligned}\alpha_{20,20} &= \frac{2(\epsilon-\epsilon_a)a^5\xi_0(1+\xi_0^2)(1+3\xi_0^2)[\epsilon_a + L_z(\epsilon-\epsilon_a)]}{15\{[\epsilon_a+L_z(\epsilon-\epsilon_a)][\epsilon_a+L_z^1(\epsilon-\epsilon_a)]+\Lambda_z^1(\epsilon-\epsilon_a)^2\}} \\ \alpha_{21,21} &= \frac{(\epsilon-\epsilon_a)a^5\xi_0(1+\xi_0^2)(1+2\xi_0^2)[\epsilon_a + L_{\|}(\epsilon-\epsilon_a)]}{5\{[\epsilon_a+L_{\|}(\epsilon-\epsilon_a)][\epsilon_a+L_{\|}^1(\epsilon-\epsilon_a)]+\Lambda_{\|}^1(\epsilon-\epsilon_a)^2\}} \end{aligned} \tag{7.153}$$

Here the depolarization factors are again given by eqs.(6.134)–(6.141).

Next the case of a prolate spheroid, with the axis of rotation normal to the surface of the substrate, will be considered, cf. fig. 6.6. One now has to use the prolate spheroidal coordinates ξ, η and ϕ defined in eqs.(6.48) and (6.49). In terms of these coordinates, the external potential $\psi_0(\mathbf{r})$ has exactly the same form as for oblate spheroidal coordinates, eq.(7.131), the only difference being that one should replace $X_\ell^m(\xi, a)$ by $\widetilde{X}_\ell^m(\xi, a)$, defined by eq.(6.58). The expressions for $\psi_s(\mathbf{r})$ and $\psi_{loc}(\mathbf{r})$ can similarly be obtained from the eqs. (7.133) and (7.134), respectively, by replacing $X_\ell^m(\xi, a)$ by $\widetilde{X}_\ell^m(\xi, a)$ and $Z_\ell^m(\xi, a)$ by $\widetilde{Z}_\ell^m(\xi, a)$, defined by eqs.(6.58) and (6.59). In the expression for $\psi_{loc}(\mathbf{r})$ one then substitutes expansion (6.157) and obtains eq.(7.135) with $X_\ell^m(\xi, a)$ replaced by $\widetilde{X}_\ell^m(\xi, a)$ and K_{ℓ,ℓ_1}^m by $\widetilde{K}_{\ell,\ell_1}^m$, cf. section 6.5. This results in equations (7.136), (7.138)– (7.142) for the amplitudes with $\alpha_{\ell m}$ replaced by $\widetilde{\alpha}_{\ell m}$, defined by eq.(6.70) and K_{ℓ,ℓ_1}^m by $\widetilde{K}_{\ell,\ell_1}^m$. Eqs.(7.145)– (7.147) then become

$$\begin{aligned} & A_{\ell 0} + \widetilde{\alpha}_{\ell 0}\left(\frac{\epsilon_a - \epsilon_s}{\epsilon_a + \epsilon_s}\right)\sum_{\ell_1=1}^{M}(-1)^{\ell_1}\widetilde{K}_{\ell,\ell_1}^0(\xi_1)A_{\ell_1,0}(2d)^{-\ell-\ell_1-1} \\ = & -\widetilde{\alpha}_{20}\delta_{\ell 2}(\phi_{xx} + \phi_{yy})\sqrt{\pi/5} \quad \text{for} \quad \ell = 1, 2, ..., M \end{aligned} \tag{7.154}$$

$$\begin{aligned} & A_{\ell 1} - \widetilde{\alpha}_{\ell 1}\left(\frac{\epsilon_a - \epsilon_s}{\epsilon_a + \epsilon_s}\right)\sum_{\ell_1=1}^{M}(-1)^{\ell_1}\widetilde{K}_{\ell,\ell_1}^1(\xi_1)A_{\ell_1,1}(2d)^{-\ell-\ell_1-1} \\ = & -\widetilde{\alpha}_{21}\delta_{\ell 2}(\phi_{xz} - i\phi_{yz})\sqrt{2\pi/15} \quad \text{for} \quad \ell = 1, 2, ..., M \end{aligned} \tag{7.155}$$

$$\begin{aligned} & A_{\ell 2} + \widetilde{\alpha}_{\ell 2}\left(\frac{\epsilon_a - \epsilon_s}{\epsilon_a + \epsilon_s}\right)\sum_{\ell_1=2}^{M}(-1)^{\ell_1}\widetilde{K}_{\ell,\ell_1}^2(\xi_1)A_{\ell_1,2}(2d)^{-\ell-\ell_1-1} \\ = & \ \widetilde{\alpha}_{22}\delta_{\ell 2}\left(\frac{1}{2}\phi_{xx} - \frac{1}{2}\phi_{yy} - i\ \phi_{xy}\right)\sqrt{2\pi/15} \quad \text{for} \quad \ell = 2, 3, ..., M \end{aligned} \tag{7.156}$$

In this way the amplitudes A_{10}, A_{20}, A_{11}, A_{21} and A_{22} may be calculated to any desired degree of accuracy. The polarizabilities for prolate spheroids are subsequently obtained using eqs.(7.151) and (7.152).

If one solves eqs.(7.154) and (7.155) for $M = 2$, one obtains the amplitudes to quadrupolar order in the interaction with their image. One then uses eq.(6.74) for $\widetilde{\alpha}_{\ell 0}$ and $\widetilde{\alpha}_{\ell 1}$ with $\ell = 1$, 2 and the expressions for $\widetilde{K}_{\ell,\ell_1}^0(\xi_1)$ and $\widetilde{K}_{\ell,\ell_1}^1(\xi_1)$, which follow from eqs.(6.243) – (6.245) using eqs.(6.159), (6.56) and (6.167), with ℓ, $\ell_1 = 1$, 2. Together with eq.(7.151), one now obtains

$$\begin{aligned} \alpha_{20,20} &= \frac{2(\epsilon - \epsilon_a)a^5\xi_0(\xi_0^2 - 1)(3\xi_0^2 - 1)[\epsilon_a + L_z(\epsilon - \epsilon_a)]}{15\{[\epsilon_a + L_z(\epsilon - \epsilon_a)][\epsilon_a + L_z^1(\epsilon - \epsilon_a)] + \Lambda_z^1(\epsilon - \epsilon_a)^2\}} \\ \alpha_{21,21} &= \frac{(\epsilon - \epsilon_a)a^5\xi_0(\xi_0^2 - 1)(2\xi_0^2 - 1)[\epsilon_a + L_\parallel(\epsilon - \epsilon_a)]}{5\{[\epsilon_a + L_\parallel(\epsilon - \epsilon_a)][\epsilon_a + L_\parallel^1(\epsilon - \epsilon_a)] + \Lambda_\parallel^1(\epsilon - \epsilon_a)^2\}} \end{aligned} \tag{7.157}$$

The depolarization factors are now given by eqs.(6.177)– (6.184) for prolate spheroids.

Chapter 8
FILMS OF TRUNCATED SPHERES FOR A LOW COVERAGE

8.1 Introduction

In the spherical and spheroidal models for island films, the contact between island and substrate takes place in at most one single point. In many cases, in particular for liquid droplets, the region of contact is finite. In such cases, the bottom of the island is planar. There is a finite contact angle between the sides of the island and the substrate. This affects the optical response of the island film. The models will be extended in the present chapter to truncated spheres, following Wind et al. [52], [53], and in the next chapter to truncated spheroids. The first is the simplest model for islands with a planar contact region with the substrate. As in the previous chapters, the size of the islands is assumed to be small compared to the wavelength. As a consequence the electric potential satisfies the Laplace equation

$$\Delta\, \psi(\mathbf{r}) = 0 \quad \text{with} \quad \mathbf{E}(\mathbf{r}) = -\nabla\psi(\mathbf{r}) \tag{8.1}$$

As in the spherical model, treated in chapter 5, one might start by considering a free island (i.e. no substrate) and, after solving this model, include the effects of the substrate on the dipole and quadrupole polarizabilities of this island using the so-called dipole or quadrupole models. Unfortunately, for the truncated particle shape no complete set of eigenfunctions of the above Laplace equation can be found. Therefore there is little point in calculating first the response of a free truncated sphere, and next to incorporate the effects of the substrate by means of the dipole or quadrupole model. The approach in the present chapter will therefore be to include the substrate directly in the calculations. Because part of the island surface is spherical, it is most convenient to expand the potential ψ into the same complete set of spherical harmonic functions within and outside the islands, as was done in chapter 5. Using the boundary conditions for the potential and its normal derivative on the surface of the island and of the substrate, an infinite set of inhomogeneous linear equations for the multipole coefficients in this expansion is derived in the next section. The matrix elements in these equations are given in the form of integrals, involving Legendre functions. Approximate solutions can be obtained by truncating the set and solving the now finite number of linear equations. In section 8.2 the case will be studied of a truncated sphere with its center above the substrate. The case of a truncated sphere with its center in the substrate, which will be referred to as the spherical cap, will be treated in section 8.3.

In section 8.4 two special cases will be considered, where the above mentioned integrals can easily be evaluated analytically. First the case of a sphere, touching the substrate, is treated. It is shown that the inhomogeneous linear equations, derived in section 8.2, reduce to the set of equations (5.109) and (5.110). Furthermore the case of a hemisphere on a substrate is considered in section 8.4. For an island with dielectric constant equal to that of the substrate this is identical to the system considered by Berreman[54]. The more general case of a hemispherical particle on a substrate was also treated by Chavaux and Meessen [55].

In the general case of a truncated sphere on a substrate the evaluation of the above mentioned matrix elements is more complicated. Nevertheless it is possible to express the integrals, in which these matrix elements can be expressed, as polynomials in the truncation parameter (the distance of the center of the truncated sphere to the substrate divided by its radius). This has been proved by Wind, Bobbert, Vlieger and Bedeaux [53], who furthermore derived a complete set of recurrence relations for these integrals, by means of which all matrix elements can be evaluated explicitly. Since for accurate calculations of the polarizabilities of truncated spheres a sufficiently large number of matrix elements is needed, which have to be calculated on a computer, it is as simple to evaluate these matrix elements directly by numerical integration. One then uses well-known recurrence relations to obtain the Legendre functions appearing in the integrands. Both parallel and normal components of the dipole and quadrupole polarizability of such particles are calculated in the optical region, as well as the surface constitutive coefficients β_e, γ_e, δ_e and τ for a film of such particles in the low coverage limit.

8.2 Truncated spheres on a substrate

Consider a truncated sphere of radius R with a frequency dependent dielectric constant $\epsilon \equiv \epsilon_3(\omega)$ on a substrate with a dielectric constant $\epsilon_s \equiv \epsilon_2(\omega)$ and surrounded by an ambient with dielectric constant $\epsilon_a \equiv \epsilon_1(\omega)$. (The advantage of the introduction of the new notation for the dielectric constants ϵ, ϵ_s and ϵ_a will become clear below). The surface of the substrate is planar. See also figures 8.1 and 8.2 below for a cross section of the system normal to the substrate and through the center of the sphere. This center is chosen as the origin of a Cartesian coordinate system with positive z-axis normal to the surface of the substrate and pointing downward into or in the substrate. The surface of the substrate is given by $z = \pm d$, where $0 \leq d \leq R$.

In fig.8.1 a situation has been drawn, where the center of the sphere lies above the surface of the substrate ($z = d$) and in fig.8.2 the case, that this center lies in the substrate below that surface ($z = -d$). The latter case will be referred to as a spherical cap, whereas from now on the name truncated sphere is only used for the case that the center of the sphere lies above the substrate. In the analysis below it will be convenient to replace the substrate in region 4 by a different material with a dielectric constant $\epsilon_4(\omega)$. It is then clear that the two situations, drawn in figs.8.1 and 8.2, may be obtained from each other, if one interchanges ϵ_1 and ϵ_2 (of the ambient and the substrate) and also ϵ_3 and ϵ_4 (of the particle and the material within the part of the sphere in the substrate). In the present section only the case of the truncated

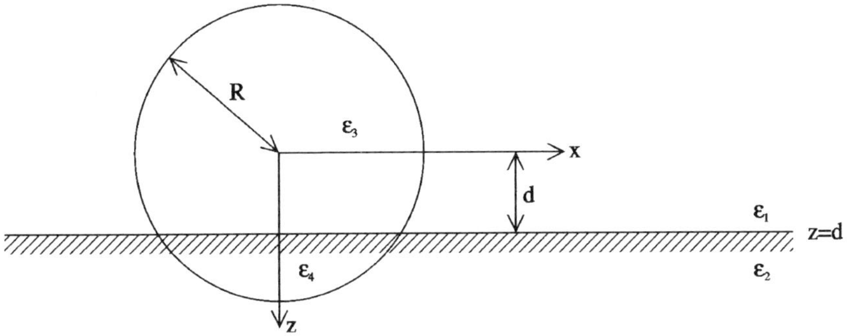

Figure 8.1 Cross section of a truncated sphere on a substrate, with center above the substrate.

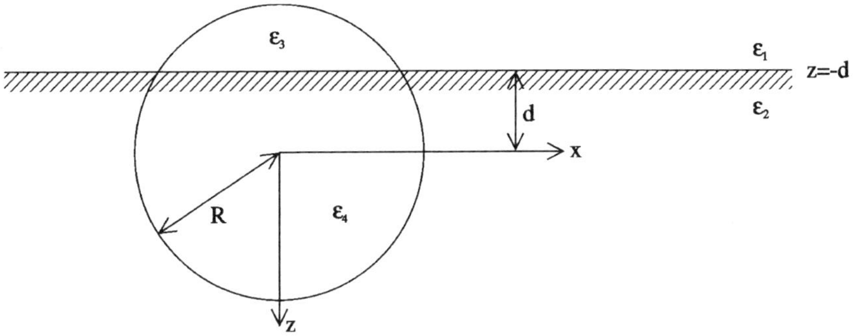

Figure 8.2 Cross section of a truncated sphere on a substrate, with center in the substrate (to be referred to as spherical cap).

sphere will be studied. The case of a spherical cap will be treated in the next section.

For islands small compared to the minimum wavelength of the incident radiation in the media involved, retardation effects become unimportant in and around the particle. One may therefore consider the incident field to be homogeneous and use the Laplace equation eq.(8.1) for the potential $\psi(\mathbf{r})$ in the particle, the ambient and the substrate. At the surface of the island and the substrate, the potential and the dielectric constant times the normal derivative of the potential must be continuous.

Just as in the previous chapters image multipoles in (0,0,2d) will be introduced to describe the field due to the charge distribution in the substrate underneath the island (see fig.8.3). One now has to distinguish the four different regions denoted by 1, 2, 3 and 4, indicated in fig.8.3 by their dielectric constants, ϵ_1, ϵ_2, ϵ_3 and ϵ_4. The

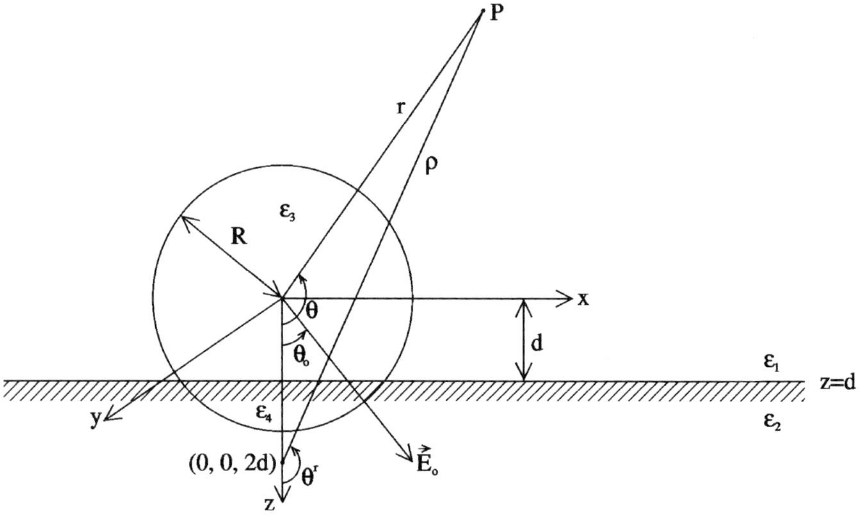

Figure 8.3 Truncated sphere on a substrate.

potential in region 1 (ambient) can be written in spherical coordinates as

$$
\begin{aligned}
\psi_1(\mathbf{r}) &= -rE_0\sqrt{2\pi/3}[\cos\theta_0 Y_1^0(\theta,\phi)\sqrt{2} \\
&+ \sin\theta_0\{\exp(i\phi_0)Y_1^{-1}(\theta,\phi) - \exp(-i\phi_0)Y_1^1(\theta,\phi)\}] \\
&+ \sum_{\ell m}{}' A_{\ell m} r^{-\ell-1} Y_\ell^m(\theta,\phi) + \sum_{\ell m}{}' A_{\ell m}^r \rho^{-\ell-1} Y_\ell^m(\theta^r,\phi^r) \qquad (8.2)
\end{aligned}
$$

(cf. eq.(5.98)). The first term on the right-hand side gives the potential of the constant electric field:

$$
\mathbf{E}_0 = E_0(\sin\theta_0\cos\phi_0, \sin\theta_0\sin\phi_0, \cos\theta_0) \qquad (8.3)
$$

Here θ_0 is the angle of this field with the z-axis, and ϕ_0 the angle between its projection on the substrate and the x-axis (see fig.8.3). The ambient is chosen to be transparent. The second term (first sum) at the right-hand side of eq.(8.2) is the potential due to the induced charge distribution in the island and the third term (second sum) is that of the image charge distribution in the substrate. The prime as super index in these sums indicates that $\ell \neq 0$. The direct field is described by multipoles, located at O, and the field of the image charge distribution by image multipoles in (0,0,2d), (see fig.8.3). Furthermore (r,θ,ϕ) are the usual spherical coordinates, describing the position of a point P with Cartesian coordinates $\mathbf{r}$ and (ρ,θ^r,ϕ^r) are the spherical coordinates with (0,0,2d) as the center of the coordinate system and the same z-axis (see fig.8.3). The spherical harmonics $Y_\ell^m(\theta,\phi)$ are again defined by eqs.(5.26) and (5.27).

For the potential in region 2 (substrate) one may write similarly

$$\psi_2(\mathbf{r}) = E_0[d\{(\epsilon_1/\epsilon_2) - 1\}\cos\theta_0 - r\sqrt{2\pi/3}[(\epsilon_1/\epsilon_2)\cos\theta_0 Y_1^0(\theta,\phi)\sqrt{2}$$

$$+ \sin\theta_0\{\exp(i\phi_0)Y_1^{-1}(\theta,\phi) - \exp(-i\phi_0)Y_1^1(\theta,\phi)\}]] + \sum_{\ell m}{}' A_{\ell m}^t r^{-\ell-1} Y_\ell^m(\theta,\phi) \quad (8.4)$$

(cf. eq.(5.99)). Furthermore one can write the potential in regions 3 (island) and 4 (substrate) as follows

$$\psi_3(\mathbf{r}) = \psi_0 + \sum_{\ell m}{}' B_{\ell m} r^\ell Y_\ell^m(\theta,\phi) + \sum_{\ell m}{}' B_{\ell m}^r \rho^\ell Y_\ell^m(\theta^r,\phi^r) \quad (8.5)$$

$$\psi_4(\mathbf{r}) = \psi_0 + \sum_{\ell m}{}' B_{\ell m}^t r^\ell Y_\ell^m(\theta,\phi) \quad (8.6)$$

where ψ_0 is a constant, which is the same in 3 and 4 as a consequence of the continuity of the potential.

In a completely analogous way as in section 5.6, one can derive from the boundary conditions at the surface of the substrate (i.e. continuity of ψ at the interface between region 1 and 2, as well as that of the normal derivative of ψ times the dielectric constant) the following relations between the multipole coefficients $A_{\ell m}$, $A_{\ell m}^t$ and $A_{\ell m}^r$:

$$A_{\ell m}^r = (-1)^{\ell+m}\frac{\epsilon_1 - \epsilon_2}{\epsilon_1 + \epsilon_2}A_{\ell m} \quad \text{and} \quad A_{\ell m}^t = \frac{2\epsilon_1}{\epsilon_1 + \epsilon_2}A_{\ell m} \quad (8.7)$$

cf. eq.(5.100). Moreover, using similar boundary conditions at the interface between regions 3 and 4, one obtains analogous relations between the multipole coefficients $B_{\ell m}$, $B_{\ell m}^t$ and $B_{\ell m}^r$:

$$B_{\ell m}^r = (-1)^{\ell+m}\frac{\epsilon_3 - \epsilon_4}{\epsilon_3 + \epsilon_4}B_{\ell m} \quad \text{and} \quad B_{\ell m}^t = \frac{2\epsilon_3}{\epsilon_3 + \epsilon_4}B_{\ell m} \quad (8.8)$$

At the surface of the sphere $r = R$, one can use the fact that the spherical harmonics $Y_\ell^m(\theta,\phi)$ form a complete orthonormal set of functions on the sphere (cf. eq.(5.29)). Multiplication of the boundary conditions on the surface of this sphere between regions 1 and 3 and between 2 and 4 with complex conjugated spherical harmonics and integration over all directions gives:

$$\int_{-1}^{t_r} d(\cos\theta) \int_0^{2\pi} d\phi(\psi_1 - \psi_3)_{r=R}[Y_{\ell'}^{m'}(\theta,\phi)]^*$$

$$+ \int_{t_r}^{1} d(\cos\theta) \int_0^{2\pi} d\phi(\psi_2 - \psi_4)_{r=R}[Y_{\ell'}^{m'}(\theta,\phi)]^* = 0 \quad (8.9)$$

and

$$\int_{-1}^{t_r} d(\cos\theta) \int_0^{2\pi} d\phi \left[\frac{\partial}{\partial r}(\epsilon_1\psi_1 - \epsilon_3\psi_3)_{r=R}\right] [Y_{\ell'}^{m'}(\theta,\phi)]^*$$

$$+ \int_{t_r}^{1} d(\cos\theta) \int_0^{2\pi} d\phi \left[\frac{\partial}{\partial r}(\epsilon_2\psi_2 - \epsilon_4\psi_4)_{r=R}\right] [Y_{\ell'}^{m'}(\theta,\phi)]^* = 0 \qquad (8.10)$$

Here the truncation parameter

$$t_r \equiv d/R \qquad (8.11)$$

has been introduced (e.g. $t_r = 1$ for a sphere and $t_r = 0$ for a hemisphere). It follows from $0 \leq d \leq R$ that $0 \leq t_r \leq 1$.

Upon substitution of the potentials, eqs.(8.2), (8.4)–(8.6), into eqs.(8.9) and (8.10), using also eqs.(8.7) and (8.8), one obtains the following infinite set of inhomogeneous linear equations for the multipole coefficients $A_{\ell m}$ and $B_{\ell m}$:

$$\sum_{\ell_1=|m|}^{\infty\prime} C_{\ell\ell_1}^m R^{-\ell_1-2} A_{\ell_1 m} + \sum_{\ell_1=|m|}^{\infty\prime} D_{\ell\ell_1}^m R^{\ell_1-1} B_{\ell_1 m} = H_\ell^m$$

$$\sum_{\ell_1=|m|}^{\infty\prime} F_{\ell\ell_1}^m R^{-\ell_1-2} A_{\ell_1 m} + \sum_{\ell_1=|m|}^{\infty\prime} G_{\ell\ell_1}^m R^{\ell_1-1} B_{\ell_1 m} = J_\ell^m \qquad (8.12)$$

where $\ell = 0, 1, 2, 3, ...$ and $m = 0, \pm1, \pm2, \pm3, ..., \pm\ell$. R is the radius of the (truncated) sphere. The prime as super index of the summation symbols denotes again that the term with $\ell_1 = 0$ should be omitted. Using the definition, eqs.(5.26) and (5.27), of spherical harmonics $Y_\ell^m(\theta,\phi)$ and the orthogonality relations, eqs.(5.29), of these functions, one finds:

$$C_{\ell\ell_1}^m \equiv \left(\frac{2\epsilon_1}{\epsilon_1+\epsilon_2}\right)\delta_{\ell\ell_1} - \left(\frac{\epsilon_1-\epsilon_2}{\epsilon_1+\epsilon_2}\right)\zeta_{\ell\ell_1}^m [Q_{\ell\ell_1}^m(t_r) - (-1)^{\ell_1+m} S_{\ell\ell_1}^m(t_r)]$$

$$D_{\ell\ell_1}^m \equiv -\left(\frac{2\epsilon_3}{\epsilon_3+\epsilon_4}\right)\delta_{\ell\ell_1} + \left(\frac{\epsilon_3-\epsilon_4}{\epsilon_3+\epsilon_4}\right)\zeta_{\ell\ell_1}^m [Q_{\ell\ell_1}^m(t_r) - (-1)^{\ell_1+m} T_{\ell\ell_1}^m(t_r)]$$

$$\begin{aligned} F_{\ell\ell_1}^m \equiv & -\left(\frac{2\epsilon_1\epsilon_2}{\epsilon_1+\epsilon_2}\right)(\ell_1+1)\delta_{\ell\ell_1} \\ & -\epsilon_1\left(\frac{\epsilon_1-\epsilon_2}{\epsilon_1+\epsilon_2}\right)\zeta_{\ell\ell_1}^m \left\{(\ell_1+1)Q_{\ell\ell_1}^m(t_r) - (-1)^{\ell_1+m}\left[\frac{\partial}{\partial t}S_{\ell\ell_1}^m(t,t_r)\right]_{t=1}\right\} \end{aligned}$$

$$\begin{aligned} G_{\ell\ell_1}^m \equiv & -\left(\frac{2\epsilon_3\epsilon_4}{\epsilon_3+\epsilon_4}\right)\ell_1\delta_{\ell\ell_1} \\ & -\epsilon_3\left(\frac{\epsilon_3-\epsilon_4}{\epsilon_3+\epsilon_4}\right)\zeta_{\ell\ell_1}^m \left\{\ell_1 Q_{\ell\ell_1}^m(t_r) + (-1)^{\ell_1+m}\left[\frac{\partial}{\partial t}T_{\ell\ell_1}^m(t,t_r)\right]_{t=1}\right\} \end{aligned} \qquad (8.13)$$

where

$$\zeta^m_{\ell\ell_1} \equiv \frac{1}{2}\left[\frac{(2\ell+1)(2\ell_1+1)(\ell-m)!(\ell_1-m)!}{(\ell+m)!(\ell_1+m)!}\right]^{\frac{1}{2}} \tag{8.14}$$

and

$$Q^m_{\ell\ell_1}(t_r) \equiv \int_{-1}^{t_r} P^m_\ell(x)P^m_{\ell_1}(x)dx \tag{8.15}$$

with the associated Legendre functions, eq.(5.27). Here the new variable

$$x \equiv \cos\theta \tag{8.16}$$

has been introduced. The matrices $S^m_{\ell\ell_1}(t_r)$ and $T^m_{\ell\ell_1}(t_r)$ are defined by

$$S^m_{\ell\ell_1}(t_r) \equiv [S^m_{\ell\ell_1}(t,t_r)]_{t=1} \text{ and } T^m_{\ell\ell_1}(t_r) \equiv [T^m_{\ell\ell_1}(t,t_r)]_{t=1} \tag{8.17}$$

where

$$\begin{aligned} S^m_{\ell\ell_1}(t,t_r) \equiv & \int_{-1}^{t_r} P^m_\ell(x)P^m_{\ell_1}([tx-2t_r][t^2-4t_rtx+4t_r^2]^{-1/2}) \\ & \times(t^2-4t_rtx+4t_r^2)^{-(\ell_1+1)/2}dx \end{aligned} \tag{8.18}$$

and

$$\begin{aligned} T^m_{\ell\ell_1}(t,t_r) \equiv & \int_{-1}^{t_r} P^m_\ell(x)P^m_{\ell_1}([tx-2t_r][t^2-4t_rtx+4t_r^2]^{-1/2}) \\ & \times(t^2-4t_rtx+4t_r^2)^{\ell_1/2}dx \end{aligned} \tag{8.19}$$

with

$$t \equiv \frac{r}{R} \tag{8.20}$$

Notice that also the partial derivatives of these matrices with respect to t for $t = 1$ appear in the expressions, eq.(8.13), for $F^m_{\ell\ell_1}$ and $G^m_{\ell\ell_1}$. In the derivation of the above integrals use has also been made of the transformation formulae from the spherical coordinates (r, θ, ϕ) into (ρ, θ^r, ϕ^r), which are given by

$$\begin{aligned} \rho &= \sqrt{r^2-4rd\,\cos\theta+4d^2} \\ \cos\theta^r &= (r\,\cos\theta-2d)/\sqrt{r^2-4rd\,\cos\theta+4d^2} \\ \phi^r &= \phi \end{aligned} \tag{8.21}$$

The terms H^m_ℓ and J^m_ℓ at the right-hand sides of eq.(8.12) are found to be only unequal to zero, if $m = 0$, 1, or -1. Since terms with different m do not couple in this equation, as a consequence of the rotational symmetry of the system around the z-axis, one only has to consider the case $m = 0$, 1 or -1, (see also chapter 5). H^0_0 is found to contain the constant ψ_0, from eqs.(8.5) and (8.6). Eq.(8.12a), for $\ell = 0$, can therefore be used to determine this unknown quantity ψ_0 in terms of the multipole coefficients. This results in

$$\begin{aligned}\psi_0 &= R\left(\frac{\epsilon_1}{\epsilon_2}-1\right)\left\{\frac{1}{\sqrt{3}}\zeta^0_{01}Q^0_{01}(t_r)+t_r[1-\zeta^0_{00}Q^0_{00}(t_r)]\right\}E_0\cos\theta_0\\ &-\frac{1}{2\sqrt{\pi}}\sum_{\ell=1}^{\infty}A_{\ell 0}R^{-\ell-1}\left(\frac{\epsilon_1-\epsilon_2}{\epsilon_1+\epsilon_2}\right)\zeta^0_{0\ell}[Q^0_{0\ell}(t_r)-(-1)^\ell S^0_{0\ell}(t_r)]\\ &+\frac{1}{2\sqrt{\pi}}\sum_{\ell=1}^{\infty}B_{\ell 0}R^{\ell}\left(\frac{\epsilon_3-\epsilon_4}{\epsilon_3+\epsilon_4}\right)\zeta^0_{0\ell}[Q^0_{0\ell}(t_r)-(-1)^\ell T^0_{0\ell}(t_r)] \qquad (8.22)\end{aligned}$$

For the calculation of the multipole coefficients, one can further simply discard eq.(8.12), a as well as b, for $\ell = 0$. If free charge were present in the system this would not be the case.

Eq.(8.12) can now be written as

$$\sum_{\ell_1=1}^{\infty}C^m_{\ell\ell_1}R^{-\ell_1-2}A_{\ell_1 m}+\sum_{\ell_1=1}^{\infty}D^m_{\ell\ell_1}R^{\ell_1-1}B_{\ell_1 m}=H^m_\ell$$

$$\sum_{\ell_1=1}^{\infty}F^m_{\ell\ell_1}R^{-\ell_1-2}A_{\ell_1 m}+\sum_{\ell_1=1}^{\infty}G^m_{\ell\ell_1}R^{\ell_1-1}B_{\ell_1 m}=J^m_\ell$$

$$\text{for}\quad \ell=1,2,3,.....\ \text{and}\ m=0,\pm 1 \qquad (8.23)$$

and it is found that

$$\begin{aligned}H^m_\ell &\equiv \sqrt{4\pi/3}E_0\cos\theta_0\delta_{m0}\left\{\frac{\epsilon_1}{\epsilon_2}\delta_{\ell 1}+\left(\frac{\epsilon_1-\epsilon_2}{\epsilon_2}\right)[\sqrt{3}t_r\zeta^0_{\ell 0}Q^0_{\ell 0}(t_r)-\zeta^0_{\ell 1}Q^0_{\ell 1}(t_r)]\right\}\\ &+\sqrt{2\pi/3}E_0\sin\theta_0\delta_{\ell 1}[\exp(i\phi_0)\delta_{m,-1}-\exp(-i\phi_0)\delta_{m1}] \qquad (8.24)\end{aligned}$$

$$\begin{aligned}J^m_\ell &\equiv \sqrt{4\pi/3}E_0\cos\theta_0\epsilon_1\delta_{\ell 1}\delta_{m0}\\ &+\sqrt{2\pi/3}E_0\sin\theta_0\{\epsilon_2\delta_{\ell 1}[\exp(i\phi_0)\delta_{m,-1}-\exp(-i\phi_0)\delta_{m1}]\\ &+(\epsilon_1-\epsilon_2)[\exp(i\phi_0)\zeta^{-1}_{\ell 1}Q^{-1}_{\ell 1}(t_r)\delta_{m,-1}-\exp(-i\phi_0)\zeta^1_{\ell 1}Q^1_{\ell 1}(t_r)\delta_{m1}]\} \qquad (8.25)\end{aligned}$$

Just as in chapter 5 (cf. text below eqs.(5.107) and (5.108)), it will be shown that eq.(8.23) for $m = -1$ is redundant. It follows from the definitions, eqs.(8.13)–(8.20) and (6.8), that

$$C^m_{\ell\ell_1}=C^{-m}_{\ell\ell_1},\quad D^m_{\ell\ell_1}=D^{-m}_{\ell\ell_1},\quad F^m_{\ell\ell_1}=F^{-m}_{\ell\ell_1},\quad G^m_{\ell\ell_1}=G^{-m}_{\ell\ell_1} \qquad (8.26)$$

Furthermore one proves from the definitions, eqs.(8.24) and (8.25), together with eqs.(8.14), (8.15) and (6.8), that

$$H^1_\ell\exp(i\phi_0)=-H^{-1}_\ell\exp(-i\phi_0)\ \text{ and }\ J^1_\ell\exp(i\phi_0)=-J^{-1}_\ell\exp(-i\phi_0) \qquad (8.27)$$

From eqs.(8.23), (8.26) and (8.27) one can then prove that

$$A_{\ell 1}\exp(i\phi_0) = -A_{\ell,-1}\exp(-i\phi_0) \text{ and } B_{\ell 1}\exp(i\phi_0) = -B_{\ell,-1}\exp(-i\phi_0) \quad (8.28)$$

The $m = -1$ equation (8.23) is therefore superfluous.

Eq.(8.23) may now be solved numerically by neglecting the amplitudes $A_{\ell m}$ and $B_{\ell m}$ larger than a suitably chosen order. One then solves the finite set of inhomogeneous linear equations

$$\begin{aligned}
\sum_{\ell_1=1}^{M} C^m_{\ell\ell_1} R^{-\ell_1-2} A_{\ell_1 m} + \sum_{\ell_1=1}^{M} D^m_{\ell\ell_1} R^{\ell_1-1} B_{\ell_1 m} &= H^m_\ell \\
\sum_{\ell_1=1}^{M} F^m_{\ell\ell_1} R^{-\ell_1-2} A_{\ell_1 m} + \sum_{\ell_1=1}^{M} G^m_{\ell\ell_1} R^{\ell_1-1} B_{\ell_1 m} &= J^m_\ell \\
\text{for } \ell &= 1,2,3,...,M \text{ and } m = 0,1 \quad (8.29)
\end{aligned}$$

where

$$\begin{aligned}
H^0_\ell &\equiv \sqrt{4\pi/3}E_0\cos\theta_0\left\{\frac{\epsilon_1}{\epsilon_2}\delta_{\ell 1} + \left(\frac{\epsilon_1-\epsilon_2}{\epsilon_2}\right)[\sqrt{3}t_r\zeta^0_{\ell 0}Q^0_{\ell 0}(t_r) - \zeta^0_{\ell 1}Q^0_{\ell 1}(t_r)]\right\} \\
J^0_\ell &\equiv \sqrt{4\pi/3}E_0\cos\theta_0\epsilon_1\delta_{\ell 1} \\
H^1_\ell &\equiv -\sqrt{2\pi/3}E_0\sin\theta_0\exp(-i\phi_0)\delta_{\ell 1} \\
J^1_\ell &\equiv -\sqrt{2\pi/3}E_0\sin\theta_0\exp(-i\phi_0)[\epsilon_2\delta_{\ell 1} + (\epsilon_1-\epsilon_2)\zeta^1_{\ell 1}Q^1_{\ell 1}(t_r)] \quad (8.30)
\end{aligned}$$

and where the matrix elements $C^m_{\ell\ell_1}$, $D^m_{\ell\ell_1}$, $F^m_{\ell\ell_1}$ and $G^m_{\ell\ell_1}$ $(\ell, \ell_1 = 1,2,3,.....,M,\ m = 0,1)$ are again given by eq.(8.13).

The dipole and quadrupole polarizabilities of the truncated sphere on the substrate are now given in terms of the amplitudes A_{10}, A_{11}, A_{20} and A_{21} by the same formulae as for the sphere in chapter 5, i.e. by eq.(5.111):

$$\begin{aligned}
\alpha_z(0) &= 2\pi\epsilon_1 A_{10}/(\sqrt{\pi/3}E_0\cos\theta_0) \\
\alpha_\parallel(0) &= -4\pi\epsilon_1 A_{11}/[\sqrt{2\pi/3}E_0\sin\theta_0\exp(-i\phi_0)] \\
\alpha^{10}_z(0) &= \pi\epsilon_1 A_{20}/(\sqrt{\pi/5}E_0\cos\theta_0) \\
\alpha^{10}_\parallel(0) &= -4\pi\epsilon_1 A_{21}/[\sqrt{6\pi/5}E_0\sin\theta_0\exp(-i\phi_0)] \quad (8.31)
\end{aligned}$$

The argument 0 of these polarizabilities denotes again the position of the dipole and the quadrupole, which are located on the z-axis at the center of the truncated sphere. Since this center lies above the substrate, just as this was the case for the sphere, treated in section 5.6, the validity of the above equation may be proved in a completely analogous way, as that of eq.(5.111), (see text below that equation).

Using eqs.(5.64) and (5.60) one finds for the surface constitutive coefficients $\beta_e(d)$, $\gamma_e(d)$, $\delta_e(d)$ and $\tau(d)$ of an island film, consisting of identical truncated spheres, in the low coverage limit (cf. eqs.(5.92)–(5.95)):

$$\begin{aligned}
\beta_e(d) &= \rho\alpha_z(0)/\epsilon_1^2 \\
\gamma_e(d) &= \rho\alpha_\parallel(0) \\
\delta_e(d) &= -\rho[\alpha^{10}_z(0) + \alpha^{10}_\parallel(0) - d\alpha_z(0) - d\alpha_\parallel(0)]/\epsilon_1 \\
\tau(d) &= -\rho[\alpha^{10}_\parallel(0) - d\alpha_\parallel(0)] \quad (8.32)
\end{aligned}$$

where ρ is the number of particles per unit of surface area. The argument d of these coefficients denotes again the position $z = d$ of the dividing surface, which is chosen to coincide with the surface of the substrate. For truncated spherical particles of different sizes and shapes the polarizabilities in the above formulae have to be replaced by averages, i.e. the sum of these polarizabilities divided by their number, (cf. eq.(5.65)). Finally the invariants I_e, $I_{\delta,e}$, I_τ and I_c can be found by substitution of the above formulae, eq.(8.32), into eq.(5.66).

The main problem is therefore the solution of eq.(8.29) for such value of M, that the multipole coefficients A_{10}, A_{11}, A_{20} and A_{21}, necessary for the calculation of the polarizabilities, are obtained sufficiently accurate. One simply increases M until the values of these coefficients become stable within the desired accuracy. It appears that the value $M = 16$ is usually sufficient, depending, of course, on the dielectric constants ϵ_1, ϵ_2, ϵ_3 and ϵ_4, as well as on the value of the truncation parameter t_r. It will be clear, that this solution can be given only on a computer. Since the multipole coefficients $B_{\ell m}$ can not be eliminated in general (i.e. for arbitrary values of the truncation parameter t_r) from eq.(8.29), the number of equations to be solved is twice the number, that has to be solved e.g. in the special case of a sphere on a substrate (see chapter 5, section 5.6). Section 8.6 deals with this problem, in particular with the way, in which the integrals, eqs.(8.15), (8.18) and (8.19), appearing in the matrix elements, eq.(8.13), and in the expressions eqs.(8.30), can be evaluated. In sections 8.4 and 8.5 a number of special cases will be considered, where these integrals can be evaluated more easily.

8.3 Spherical caps on a substrate

In the previous section the case has been studied, that the center of the truncated sphere lies above the substrate (see figs.8.1 and 8.3 in that section). Now the case will be considered, that this center lies in the substrate (spherical cap). In fig.8.2 in the previous section this situation has been drawn. As has been mentioned in the text below that figure, the two situations, drawn in figs.8.1 and 8.2, could be obtained from each other by interchanging the dielectric constants ϵ_1 and ϵ_2, and also ϵ_3 and ϵ_4. The disadvantage of this trick is that it gives the final result without the explicit expressions for the potentials in the various regions. In order to be able to give these potentials this trick is not used.

The formulae derived in the previous section have been used with a negative truncation parameter to calculate the polarizability of caps. This leads to convergence problems. In order to improve the convergence an alternative method was developed in which the point around which the multipole expansion is performed is an arbitrary point chosen inside the cap along the symmetry axis [56]. This leads to a certain improvement of the convergence but in particular around resonances the method still has problems. The method in the present section is different and uses a complete new expansion of the fields. This method has also been used and converges better (Remi Lazzari, private communication). This numerical calculation for the caps is more time consuming than for the truncated islands. The reason is that the ratio of the smallest and the largest eigenvalues of the to be inverted matrix becomes larger

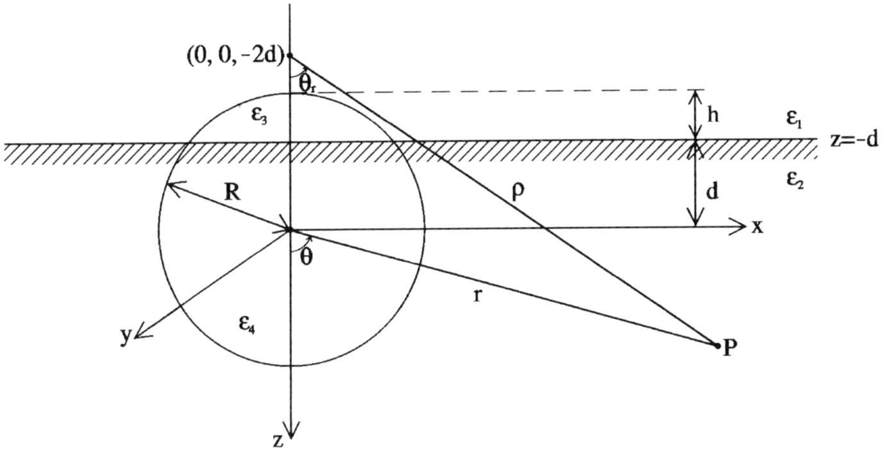

Figure 8.4 Spherical cap on substrate.

and larger when the order of the multipole expansion is increased.

Consider a spherical cap of radius R with a frequency dependent dielectric constant $\epsilon \equiv \epsilon_3(\omega)$ on a substrate with a dielectric constant $\epsilon_s \equiv \epsilon_2(\omega)$ and surrounded by an ambient with dielectric constant $\epsilon_a \equiv \epsilon_1(\omega)$. The surface of the substrate is planar and given by $z = -d$, where $0 \leq d \leq R$. See fig. 8.4. The center is chosen as the origin of a Cartesian coordinate system with positive z-axis normal to the surface of the substrate and pointing downward in the substrate.

Just as in the previous section image multipoles, which are now positioned in (0,0,$-2d$), will be introduced to describe the fields. One distinguishes again four different regions denoted by 1, 2, 3 and 4, indicated in the figure by their dielectric constants ϵ_1, ϵ_2, ϵ_3 and ϵ_4. The potential in region 1 (ambient) can be written in spherical coordinates as

$$\begin{aligned}\psi_1(\mathbf{r}) &= -rE_0\sqrt{2\pi/3}[\cos\theta_0 Y_1^0(\theta,\phi)\sqrt{2} \\ &\quad + \sin\theta_0\{\exp(i\phi_0)Y_1^{-1}(\theta,\phi) - \exp(-i\phi_0)Y_1^1(\theta,\phi)\}] \\ &\quad + \sum_{\ell m}{}' A_{\ell m}^t r^{-\ell-1} Y_\ell^m(\theta,\phi) \qquad (8.33)\end{aligned}$$

The first term on the right-hand side gives the potential of the constant electric field $\mathbf{E}_0$, cf. eq.(8.3). The sum is the potential due to the induced charge distribution in the island. The superindex t indicates that the corresponding multipoles are located in (0,0,0), which is in the substrate. The potential is then "transmitted" to the ambient from this position. The prime as super index in these sums indicates that $\ell \neq 0$. Furthermore (r, θ, ϕ) are the usual spherical coordinates. The spherical harmonics $Y_\ell^m(\theta, \phi)$ are again defined by eqs.(5.26) and (5.27).

For the potential in the region 2 (substrate) one may write

$$\begin{aligned}\psi_2(\mathbf{r}) &= E_0[-d\{(\epsilon_1/\epsilon_2)-1\}\cos\theta_0 - r\sqrt{2\pi/3}[(\epsilon_1/\epsilon_2)\cos\theta_0 Y_1^0(\theta,\phi)\sqrt{2} \\ &\quad + \sin\theta_0\{\exp(i\phi_0)Y_1^{-1}(\theta,\phi) - \exp(-i\phi_0)Y_1^1(\theta;\phi)\}]] \\ &\quad + \sum_{\ell m}{}' A_{\ell m} r^{-\ell-1} Y_\ell^m(\theta,\phi) + \sum_{\ell m}{}' A_{\ell m}^r \rho^{-\ell-1} Y_\ell^m(\theta^r,\phi^r)\end{aligned} \tag{8.34}$$

The first sum on the right-hand side originates from the center of the sphere, (0,0,0), whereas the second sum originates from the image, which is now located above the substrate in the point (0,0,$-2d$). Furthermore one can write the potential in the regions 3 and 4 as follows

$$\psi_3(\mathbf{r}) = \psi_0 + \sum_{\ell m}{}' B_{\ell m}^t r^\ell Y_\ell^m(\theta,\phi) \tag{8.35}$$

$$\psi_4(\mathbf{r}) = \psi_0 + \sum_{\ell m}{}' B_{\ell m} r^\ell Y_\ell^m(\theta,\phi) + \sum_{\ell m}{}' B_{\ell m}^r \rho^\ell Y_\ell^m(\theta^r,\phi^r) \tag{8.36}$$

where ψ_0 is a constant, which is the same in 3 and 4 as a consequence of the continuity of the potential.

In a completely analogous way as in the previous section, one can derive from the boundary conditions at the surface of the substrate (i.e. continuity of ψ at the interface between region 1 and 2, as well as that of the normal derivative of ψ times the dielectric constant) the following relations between the multipole coefficients $A_{\ell m}$, $A_{\ell m}^t$ and $A_{\ell m}^r$:

$$A_{\ell m}^r = (-1)^{\ell+m}\frac{\epsilon_2-\epsilon_1}{\epsilon_2+\epsilon_1}A_{\ell m} \text{ and } A_{\ell m}^t = \frac{2\epsilon_2}{\epsilon_2+\epsilon_1}A_{\ell m} \tag{8.37}$$

Moreover, using similar boundary conditions at the interface between regions 3 and 4, one obtains analogous relations between the multipole coefficients $B_{\ell m}$, $B_{\ell m}^t$ and $B_{\ell m}^r$:

$$B_{\ell m}^r = (-1)^{\ell+m}\frac{\epsilon_4-\epsilon_3}{\epsilon_4+\epsilon_3}B_{\ell m} \text{ and } B_{\ell m}^t = \frac{2\epsilon_4}{\epsilon_4+\epsilon_3}B_{\ell m} \tag{8.38}$$

At the surface of the sphere $r = R$, one can use the fact that the spherical harmonics $Y_\ell^m(\theta,\phi)$ form a complete orthonormal set of functions on the sphere (cf. eq.(5.29)). Multiplication of the boundary conditions on the surface of this sphere between regions 1 and 3 and between 2 and 4 with complex conjugated spherical harmonics and integration over all directions gives:

$$\begin{aligned}&\int_{-1}^{-t_r} d(\cos\theta)\int_0^{2\pi} d\phi(\psi_1-\psi_3)_{r=R}[Y_{\ell'}^{m'}(\theta,\phi)]^* \\ &+ \int_{-t_r}^{1} d(\cos\theta)\int_0^{2\pi} d\phi(\psi_2-\psi_4)_{r=R}[Y_{\ell'}^{m'}(\theta,\phi)]^* = 0\end{aligned} \tag{8.39}$$

and

$$\int_{-1}^{-t_r} d(\cos\theta) \int_0^{2\pi} d\phi \left[\frac{\partial}{\partial r}(\epsilon_1\psi_1 - \epsilon_3\psi_3)_{r=R}\right] [Y_{\ell'}^{m'}(\theta,\phi)]^*$$
$$+ \int_{-t_r}^{1} d(\cos\theta) \int_0^{2\pi} d\phi \left[\frac{\partial}{\partial r}(\epsilon_2\psi_2 - \epsilon_4\psi_4)_{r=R}\right] [Y_{\ell'}^{m'}(\theta,\phi)]^* = 0 \qquad (8.40)$$

Here the truncation parameter

$$t_r \equiv d/R \qquad (8.41)$$

has been introduced. One has $t_r = 1$ in the limit of an infinitely thin spherical cap and $t_r = 0$ for a hemisphere. It follows from $0 \le d \le R$ that $0 \le t_r \le 1$. It is noted that the truncation parameter t_r has been defined such that it is positive both for the truncated sphere and the spherical cap. In some papers, cf. [56], one uses a negative t_r in the case of a cap. In the equations in this section one should then use the absolute value of this parameter.

Upon substitution of the potentials, eqs.(8.33)–(8.36), into eqs.(8.39) and (8.40), using also eqs.(8.37) and (8.38), one obtains the following infinite set of inhomogeneous linear equations for the multipole coefficients $A_{\ell m}$ and $B_{\ell m}$:

$$\begin{aligned}
\sum_{\ell_1=|m|}^{\infty\prime} C_{\ell\ell_1}^m R^{-\ell_1-2} A_{\ell_1 m} + \sum_{\ell_1=|m|}^{\infty\prime} D_{\ell\ell_1}^m R^{\ell_1-1} B_{\ell_1 m} &= H_\ell^m \\
\sum_{\ell_1=|m|}^{\infty\prime} F_{\ell\ell_1}^m R^{-\ell_1-2} A_{\ell_1 m} + \sum_{\ell_1=|m|}^{\infty\prime} G_{\ell\ell_1}^m R^{\ell_1-1} B_{\ell_1 m} &= J_\ell^m
\end{aligned} \qquad (8.42)$$

where $\ell = 0, 1, 2, 3, ...$ and $m = 0, \pm1, \pm2, \pm3, ..., \pm\ell$. R is the radius of the spherical cap. The prime as super index of the summation symbols denotes again that the term with $\ell_1 = 0$ should be omitted. Using the definition, eqs.(5.26) and (5.27), of spherical harmonics $Y_\ell^m(\theta,\phi)$ and the orthogonality relations, eqs.(5.29), of these functions, one finds:

$$C_{\ell\ell_1}^m \equiv \left(\frac{2\epsilon_2}{\epsilon_2+\epsilon_1}\right)\delta_{\ell\ell_1} - \left(\frac{\epsilon_2-\epsilon_1}{\epsilon_2+\epsilon_1}\right)\zeta_{\ell\ell_1}^m(-1)^{\ell+\ell_1}[Q_{\ell\ell_1}^m(t_r) - (-1)^{\ell_1+m}S_{\ell\ell_1}^m(t_r)]$$

$$D_{\ell\ell_1}^m \equiv -\left(\frac{2\epsilon_4}{\epsilon_4+\epsilon_3}\right)\delta_{\ell\ell_1} + \left(\frac{\epsilon_4-\epsilon_3}{\epsilon_4+\epsilon_3}\right)\zeta_{\ell\ell_1}^m(-1)^{\ell+\ell_1}[Q_{\ell\ell_1}^m(t_r) - (-1)^{\ell_1+m}T_{\ell\ell_1}^m(t_r)]$$

$$\begin{aligned}
F_{\ell\ell_1}^m \equiv& -\left(\frac{2\epsilon_2\epsilon_1}{\epsilon_2+\epsilon_1}\right)(\ell_1+1)\delta_{\ell\ell_1} \\
&-\epsilon_2\left(\frac{\epsilon_2-\epsilon_1}{\epsilon_2+\epsilon_1}\right)\zeta_{\ell\ell_1}^m(-1)^{\ell+\ell_1}\left\{(\ell_1+1)Q_{\ell\ell_1}^m(t_r) - (-1)^{\ell_1+m}\left[\frac{\partial}{\partial t}S_{\ell\ell_1}^m(t,t_r)\right]_{t=1}\right\}
\end{aligned}$$

$$
\begin{aligned}
G^m_{\ell\ell_1} \;\equiv\; & -\left(\frac{2\epsilon_4\epsilon_3}{\epsilon_4+\epsilon_3}\right)\ell_1\delta_{\ell\ell_1} - \epsilon_4\left(\frac{\epsilon_4-\epsilon_3}{\epsilon_4+\epsilon_3}\right)\zeta^m_{\ell\ell_1}(-1)^{\ell+\ell_1} \\
& \times\left\{\ell_1 Q^m_{\ell\ell_1}(t_r) + (-1)^{\ell_1+m}\left[\frac{\partial}{\partial t}T^m_{\ell\ell_1}(t,t_r)\right]_{t=1}\right\} \qquad (8.43)
\end{aligned}
$$

where the various matrices on the right-hand sides are given by eqs.(8.14)–(8.20). Notice that, in the derivation of these formulae, one should now use, instead of eq.(8.21) the following transformation formulae:

$$
\begin{aligned}
\rho &= \sqrt{r^2+4rd\;\cos\theta+4d^2} \\
\cos\theta^r &= (r\;\cos\theta+2d)/\sqrt{r^2+4rd\;\cos\theta+4d^2} \\
\phi^r &= \phi \qquad (8.44)
\end{aligned}
$$

The terms H^m_ℓ and J^m_ℓ at the right-hand sides of eq.(8.42) are found to be only unequal to zero, if $m = 0$, 1, or -1. Since terms with ψ''_3 different m do not couple in this equation, as a consequence of the rotational symmetry of the system around the z-axis, one only has to consider the case $m = 0$, 1 or -1, (see also the previous section). H^0_0 is found to contain the constant ψ_0, from eqs.(8.35) and (8.36). Eq.(8.42a), for $\ell = 0$, can therefore be used to determine this unknown quantity ψ_0 in terms of the multipole coefficients. This results in

$$
\begin{aligned}
\psi_0 \;=\; & R\left(\frac{\epsilon_1}{\epsilon_2}-1\right)\left[\frac{1}{\sqrt{3}}\zeta^0_{01}Q^0_{01}(t_r) - t_r\zeta^0_{00}Q^0_{00}(t_r)\right]E_0\cos\theta_0 \\
& -\frac{1}{2\sqrt{\pi}}\sum_{\ell=1}^{\infty}A_{\ell 0}R^{-\ell-1}\left(\frac{\epsilon_2-\epsilon_1}{\epsilon_2+\epsilon_1}\right)\zeta^0_{0\ell}[(-1)^\ell Q^0_{0\ell}(t_r) - S^0_{0\ell}(t_r)] \\
& +\frac{1}{2\sqrt{\pi}}\sum_{\ell=1}^{\infty}B_{\ell 0}R^{\ell}\left(\frac{\epsilon_4-\epsilon_3}{\epsilon_4+\epsilon_3}\right)\zeta^0_{0\ell}[(-1)^\ell Q^0_{0\ell}(t_r) - T^0_{0\ell}(t_r)] \qquad (8.45)
\end{aligned}
$$

For the calculation of the multipole coefficients, one can simply discard eq.(8.12), a as well as b, for $\ell = 0$. If free charge were present in the system this would not be the case.

Eq.(8.42) can now be written as

$$
\begin{aligned}
\sum_{\ell_1=1}^{\infty}C^m_{\ell\ell_1}R^{-\ell_1-2}A_{\ell_1 m} + \sum_{\ell_1=1}^{\infty}D^m_{\ell\ell_1}R^{\ell_1-1}B_{\ell_1 m} &= H^m_\ell \\
\sum_{\ell_1=1}^{\infty}F^m_{\ell\ell_1}R^{-\ell_1-2}A_{\ell_1 m} + \sum_{\ell_1=1}^{\infty}G^m_{\ell\ell_1}R^{\ell_1-1}B_{\ell_1 m} &= J^m_\ell
\end{aligned}
$$

$$
\text{for}\quad \ell = 1,2,3,\ldots\ldots \text{ and } m = 0,\pm 1 \qquad (8.46)
$$

and it is found that

$$
\begin{aligned}
H^m_\ell \;\equiv\; & \sqrt{4\pi/3}E_0\cos\theta_0\delta_{m0}\left\{\delta_{\ell 1} + \left(\frac{\epsilon_2-\epsilon_1}{\epsilon_2}\right)(-1)^{\ell+1}[\sqrt{3}t_r\zeta^0_{\ell 0}Q^0_{\ell 0}(t_r) - \zeta^0_{\ell 1}Q^0_{\ell 1}(t_r)]\right\} \\
& +\sqrt{2\pi/3}E_0\sin\theta_0\delta_{\ell 1}[\exp(i\phi_0)\delta_{m,-1} - \exp(-i\phi_0)\delta_{m1}] \qquad (8.47)
\end{aligned}
$$

$$
\begin{aligned}
J_\ell^m \equiv\ & \sqrt{4\pi/3}E_0\cos\theta_0\epsilon_1\delta_{\ell 1}\delta_{m0} \\
& +\sqrt{2\pi/3}E_0\sin\theta_0\{\epsilon_1\delta_{\ell 1}[\exp(i\phi_0)\delta_{m,-1}-\exp(-i\phi_0)\delta_{m1}]+(\epsilon_2-\epsilon_1)(-1)^{\ell+1} \\
& \times[\exp(i\phi_0)\zeta_{\ell 1}^1 Q_{\ell 1}^1(t_r)\delta_{m,-1}-\exp(-i\phi_0)\zeta_{\ell 1}^{-1}Q_{\ell 1}^{-1}(t_r)\delta_{m1}]\}
\end{aligned}
\tag{8.48}
$$

In a completely analogous way as in the previous section for truncated spheres, one can prove here that the $m = -1$ equation (8.46) is superfluous. This equation may be solved again numerically by neglecting the amplitudes $A_{\ell m}$ and $B_{\ell m}$ larger than a suitably chosen order M. One then has to solve the finite set of linear equations

$$
\begin{aligned}
\sum_{\ell_1=1}^{M} C_{\ell\ell_1}^m R^{-\ell_1-2}A_{\ell_1 m}+\sum_{\ell_1=1}^{M} D_{\ell\ell_1}^m R^{\ell_1-1}B_{\ell_1 m} &= H_\ell^m \\
\sum_{\ell_1=1}^{M} F_{\ell\ell_1}^m R^{-\ell_1-2}A_{\ell_1 m}+\sum_{\ell_1=1}^{M} G_{\ell\ell_1}^m R^{\ell_1-1}B_{\ell_1 m} &= J_\ell^m
\end{aligned}
$$

$$\text{for } \ell = 1,2,3,...,M \text{ and } m=0,1 \tag{8.49}$$

where

$$
\begin{aligned}
H_\ell^0 &\equiv \sqrt{4\pi/3}E_0\cos\theta_0\left\{\delta_{\ell 1}+\left(\frac{\epsilon_2-\epsilon_1}{\epsilon_2}\right)(-1)^{\ell+1}[\sqrt{3}t_r\zeta_{\ell 0}^0 Q_{\ell 0}^0(t_r)-\zeta_{\ell 1}^0 Q_{\ell 1}^0(t_r)]\right\} \\
J_\ell^0 &\equiv \sqrt{4\pi/3}E_0\cos\theta_0\epsilon_1\delta_{\ell 1} \\
H_\ell^1 &\equiv -\sqrt{2\pi/3}E_0\sin\theta_0\exp(-i\phi_0)\delta_{\ell 1} \\
J_\ell^1 &\equiv -\sqrt{2\pi/3}E_0\sin\theta_0\exp(-i\phi_0)[\epsilon_1\delta_{\ell 1}+(\epsilon_2-\epsilon_1)(-1)^{\ell+1}\zeta_{\ell 1}^1 Q_{\ell 1}^1(t_r)]
\end{aligned}
\tag{8.50}
$$

Here it has also been used that $\zeta_{\ell\ell_1}^{-m}Q_{\ell\ell_1}^{-m}(t_r) = \zeta_{\ell\ell_1}^m Q_{\ell\ell_1}^m(t_r)$, which property may be derived from the definitions, eqs.(8.14) and (8.15), of these quantities, together with eq.(6.8). The matrix elements $C_{\ell\ell_1}^m$, $D_{\ell\ell_1}^m$, $F_{\ell\ell_1}^m$ and $G_{\ell\ell_1}^m (\ell, \ell_1 = 1, 2, 3, ..., M$ and $m = 0, 1)$ are again given by eq.(8.43).

The multipole coefficients $A_{\ell m}$ and $B_{\ell m}$, obtained in this way for the spherical cap, refer to fields in the substrate, in contrast with the analogous coefficients for the truncated sphere in the previous section. This is, of course, a consequence of the fact, that the center of the spherical cap now lies in the substrate. The dipole and quadrupole polarizabilities $\alpha_z(0)$, $\alpha_\|(0)$, $\alpha_z^{10}(0)$ and $\alpha_\|^{10}(0)$ will therefore be given by expressions, different from eq.(8.31). It is not difficult to find these expressions for the spherical cap, if one realizes, that the dielectric constant ϵ_1 in eq.(8.31) should be replaced by the dielectric constant ϵ_2 of the substrate in the case of the spherical cap. Furthermore the external field in the ambient at the right-hand side of eq.(8.31), has to be replaced by this field in the substrate, which means that the normal (z) component $E_0\cos\theta_0$ has to be replaced by $(\epsilon_1/\epsilon_2)E_0\cos\theta_0$, whereas the parallel component remains unchanged. In the case of the spherical cap one therefore has, instead of eq.(8.31),

$$\begin{aligned}
\alpha_z(0) &= 2\pi\epsilon_2 A_{10}/[(\epsilon_1/\epsilon_2)\sqrt{\pi/3}E_0\cos\theta_0] \\
\alpha_\parallel(0) &= -4\pi\epsilon_2 A_{11}/[\sqrt{2\pi/3}E_0\sin\theta_0\exp(-i\phi_0)] \\
\alpha_z^{10}(0) &= \pi\epsilon_2 A_{20}/[(\epsilon_1/\epsilon_2)\sqrt{\pi/5}E_0\cos\theta_0] \\
\alpha_\parallel^{10}(0) &= -4\pi\epsilon_2 A_{21}/[\sqrt{6\pi/5}E_0\sin\theta_0\exp(-i\phi_0)]
\end{aligned} \tag{8.51}$$

In order to give the expressions for the surface constitutive coefficients β_e , γ_e , δ_e and τ for the spherical cap, the position of the dipole and quadrupole has to be shifted from the center of the sphere (at $z = 0$) to the dividing surface, which will again be chosen to coincide with the surface $z = -d$ of the substrate (see also fig.8.4). Just as in the case of a truncated sphere, this dividing surface will be chosen in the ambient, at an infinitesimal distance above the surface of the substrate (i.e. at $z = -d - 0$). Therefore shifting the dipole and quadrupole from $z = 0$ to $z = -d - 0$ means in the case of the spherical cap, that one intersects the surface of the substrate, where the dielectric constant changes from ϵ_2 to ϵ_1.

Shifting the dipole and quadrupole first in the substrate from the position $z = 0$ to the position $z = -d + 0$, so that they stay in the substrate, one can apply eq.(5.60), and one obtains:

$$\begin{array}{ll}
\alpha_z(-d+0) = \alpha_z(0) & \alpha_\parallel(-d+0) = \alpha_\parallel(0) \\
\alpha_z^{10}(-d+0) = \alpha_z^{10}(0) + d\,\alpha_z(0) & \alpha_\parallel^{10}(-d+0) = \alpha_\parallel^{10}(0) + d\,\alpha_\parallel(0)
\end{array} \tag{8.52}$$

Next the dipole and quadrupole have to be shifted from $z = -d + 0$ to $z = -d - 0$, intersecting the surface of the substrate from the substrate, with dielectric constant ϵ_2, to the ambient, with dielectric constant ϵ_1. In appendix A it is proved, that the dipole and quadrupole polarizabilities then change in the following way:

$$\begin{array}{ll}
\alpha_z(-d-0) = (\epsilon_1/\epsilon_2)^2\,\alpha_z(-d+0) & \alpha_\parallel(-d-0) = \alpha_\parallel(-d+0) \\
\alpha_z^{10}(-d-0) = (\epsilon_1/\epsilon_2)\,\alpha_z^{10}(-d+0) & \alpha_\parallel^{10}(-d-0) = (\epsilon_1/\epsilon_2)\,\alpha_\parallel^{10}(-d+0)
\end{array} \tag{8.53}$$

Combining the above equations, one finds

$$\begin{array}{ll}
\alpha_z(-d-0) = (\epsilon_1/\epsilon_2)^2\,\alpha_z(0) & \alpha_\parallel(-d-0) = \alpha_\parallel(0) \\
\alpha_z^{10}(-d-0) = (\epsilon_1/\epsilon_2)\,[\alpha_z^{10}(0) + d\,\alpha_z(0)] & \alpha_\parallel^{10}(-d-0) = (\epsilon_1/\epsilon_2)\,[\alpha_\parallel^{10}(0) + d\,\alpha_\parallel(0)]
\end{array} \tag{8.54}$$

The surface constitutive coefficients β_e , γ_e , δ_e and τ for the spherical cap can now be found, using eq.(5.64), which in the present case becomes

$$\begin{aligned}
\beta_e(-d-0) &= \rho\,\alpha_z(-d-0)/\,\epsilon_1^2 \\
\gamma_e(-d-0) &= \rho\,\alpha_\parallel(-d-0) \\
\delta_e(-d-0) &= -\rho[\alpha_\parallel^{10}(-d-0) + \alpha_z^{10}(-d-0)]/\,\epsilon_1 \\
\tau(-d-0) &= -\rho\,\alpha_\parallel^{10}(-d-0)
\end{aligned} \tag{8.55}$$

where ρ is the number of identical spherical caps per unit of surface area. From now on the notation of these surface coefficients will be changed into $\beta_e(-d)$, $\gamma_e(-d)$,

$\delta_e(-d)$ and $\tau(-d)$, which will not lead to confusion any more. One then obtains, from the last two the following expressions for the surface constitutive coefficients of a thin film of identical spherical caps in the low density limit:

$$\begin{aligned}
\beta_e(-d) &= \rho\,\alpha_z(0)/\,\epsilon_2^2 \\
\gamma_e(-d) &= \rho\,\alpha_{\|}(0) \\
\delta_e(-d) &= -\rho[\alpha_{\|}^{10}(0)+\alpha_z^{10}(0)+d\,\alpha_{\|}(0)+d\,\alpha_z(0)]/\,\epsilon_2 \\
\tau(-d) &= -\rho\frac{\epsilon_1}{\epsilon_2}[\alpha_{\|}^{10}(0)+d\,\alpha_{\|}(0)]
\end{aligned} \tag{8.56}$$

For spherical caps of different sizes and shapes, the polarizabilities in the above formulae have to be replaced by averages, i.e. the sum of these polarizabilities, divided by their number (cf. eq.(5.65)). Finally the invariants I_e , $I_{\delta,e}$, I_τ and I_c can be found by substitution of the above formulae into eq.(5.66) with d replaced by $-d$.

The main problem is therefore here the solution of eq.(8.49) for such values of M, that the multipole coefficients A_{10} , A_{11} , A_{20} and A_{21} , necessary for the calculation of the polarizabilities, eq.(8.51), for the spherical cap, are obtained sufficiently accurate. The surface constitutive coefficients are then found, using eq.(8.56). For a further discussion the reader is referred to the last paragraph of the preceding section.

8.4 Spheres and hemispheres on a substrate

In this section two special cases will be considered. The first case is the sphere. Of course this case has already been considered in chapter 5. It is instructive to see how the more general solution in this chapter reduces to the one given for the sphere. The other special case is the hemisphere. In this case the solution also simplifies considerably. It is therefore interesting to give also this case some special attention.

In the case of a sphere the truncation parameter $t_r = 1$. Since in this case the region 4 in fig. 8.3 of section 8.2 disappears, the multipole coefficients are independent of the value of ϵ_4. One can therefore conveniently put $\epsilon_4 = \epsilon_3$. The matrix elements $C_{\ell\ell_1}^m$, $D_{\ell\ell_1}^m$, $F_{\ell\ell_1}^m$ and $G_{\ell\ell_1}^m$, eq.(8.13), then become

$$\begin{aligned}
C_{\ell\ell_1}^m &= \left(\frac{2\epsilon_1}{\epsilon_1+\epsilon_2}\right)\delta_{\ell\ell_1} - \left(\frac{\epsilon_1-\epsilon_2}{\epsilon_1+\epsilon_2}\right)\zeta_{\ell\ell_1}^m[Q_{\ell\ell_1}^m(1)-(-1)^{\ell_1+m}S_{\ell\ell_1}^m(1)] \\
D_{\ell\ell_1}^m &= -\delta_{\ell\ell_1}
\end{aligned}$$

$$\begin{aligned}
F_{\ell\ell_1}^m &= -\left(\frac{2\epsilon_1\epsilon_2}{\epsilon_1+\epsilon_2}\right)(\ell+1)\delta_{\ell\ell_1} \\
&\quad -\epsilon_1\left(\frac{\epsilon_1-\epsilon_2}{\epsilon_1+\epsilon_2}\right)\zeta_{\ell\ell_1}^m\left\{(\ell_1+1)Q_{\ell\ell_1}^m(1)-(-1)^{\ell_1+m}\left[\frac{\partial}{\partial t}S_{\ell\ell_1}^m(t,1)\right]_{t=1}\right\} \\
G_{\ell\ell_1}^m &= -\epsilon_3\ell_1\delta_{\ell\ell_1}
\end{aligned} \tag{8.57}$$

where $\zeta_{\ell\ell_1}^m$ is given by eq.(8.14) and $Q_{\ell\ell_1}^m(1)$, $S_{\ell\ell_1}^m(1)$ and $S_{\ell\ell_1}^m(t,1)$ by eqs.(8.15), (8.17) and (8.18), respectively, with $t_r = 1$. From the normalization and orthogonality condition, eq.(5.29), for spherical harmonics, and their definition, eq.(5.26), in terms of Legendre functions $P_\ell^m(\cos\theta)$, one obtains, after integration over the angle ϕ , and using the definition, eq.(8.14), of $\zeta_{\ell\ell_1}^m$:

$$\zeta_{\ell\ell_1}^m Q_{\ell\ell_1}^m(1) = \delta_{\ell\ell_1} \tag{8.58}$$

Multiplication of the expansion, eq.(5.102), on both sides with the complex conjugate of a spherical harmonic, and integration over all directions in space, gives with eq.(5.29)

$$\int_0^{2\pi} d\phi \int_0^{\pi} d\theta \, \sin\theta \left(\frac{2d}{\rho}\right)^{\ell+1} [Y_{\ell'}^{m'}(\theta,\phi)]^* Y_\ell^m(\theta^r,\phi^r) = M_{\ell',\ell}^m \left(\frac{r}{2d}\right)^{\ell'} \delta_{mm'} \tag{8.59}$$

where the matrix elements $M_{\ell',\ell}^m$ are defined by eq.(5.103). Expressing the spherical harmonics in this equation by means of eq.(5.26) in terms of Legendre functions and using eqs.(8.21), (8.11), (8.14), (8.16) and (8.20), one obtains the following result:

$$\begin{aligned}\left(\frac{t}{2t_r}\right)^{\ell'} M_{\ell',\ell}^m &= (2t_r)^{\ell+1} \zeta_{\ell',\ell}^m \int_{-1}^{1} P_{\ell'}^m(x) P_\ell^m([tx - 2t_r][t^2 - 4tt_r x + 4t_r^2]^{-1/2}) \\ &\times (t^2 - 4tt_r x + 4t_r^2)^{-(\ell+1)/2} dx \end{aligned} \tag{8.60}$$

Introducing for ℓ' and ℓ the new notation ℓ and ℓ_1, respectively, and using the definitions, eqs.(8.18) and (8.17), one finds for the value $t_r = 1$:

$$\zeta_{\ell\ell_1}^m S_{\ell\ell_1}^m(t,1) = \frac{t^\ell}{2^{\ell+\ell_1+1}} M_{\ell\ell_1}^m \tag{8.61}$$

and

$$\zeta_{\ell\ell_1}^m S_{\ell\ell_1}^m(1) = 2^{-\ell-\ell_1-1} M_{\ell\ell_1}^m \tag{8.62}$$

where (cf. eq.(5.103))

$$M_{\ell\ell_1}^m \equiv (-1)^{\ell_1+m} \left(\frac{2\ell_1+1}{2\ell+1}\right)^{1/2} (\ell+\ell_1)![(\ell+m)!(\ell-m)!(\ell_1+m)!(\ell_1-m)!]^{-1/2} \tag{8.63}$$

If the above formulae, eqs.(8.58), (8.61) and (8.62), are substituted into eq.(8.57), one obtains the following results for the matrix elements $C_{\ell\ell_1}^m$, $D_{\ell\ell_1}^m$, $F_{\ell\ell_1}^m$ and $G_{\ell\ell_1}^m$ in the case of a sphere on a substrate:

$$\begin{aligned} C_{\ell\ell_1}^m &= \delta_{\ell\ell_1} + \left(\frac{\epsilon_1-\epsilon_2}{\epsilon_1+\epsilon_2}\right) \frac{(-1)^{\ell_1+m}}{2^{\ell+\ell_1+1}} M_{\ell\ell_1}^m \\ D_{\ell\ell_1}^m &= -\delta_{\ell\ell_1} \\ F_{\ell\ell_1}^m &= -\epsilon_1(\ell+1)\delta_{\ell\ell_1} + \epsilon_1 \left(\frac{\epsilon_1-\epsilon_2}{\epsilon_1+\epsilon_2}\right) \ell \frac{(-1)^{\ell_1+m}}{2^{\ell+\ell_1+1}} M_{\ell\ell_1}^m \\ G_{\ell\ell_1}^m &= -\epsilon_3 \ell \, \delta_{\ell\ell_1} \text{ for } \ell,\ell_1 = 1,2,3,\ldots \text{ and } m = 0,1 \end{aligned} \tag{8.64}$$

Furthermore it follows from eqs.(8.30) and (8.58), using also the fact that $\ell \neq 0$, that

$$\begin{aligned} H_\ell^0 &= \sqrt{4\pi/3}E_0 \cos\theta_0 \delta_{\ell 1} \\ J_\ell^0 &= \sqrt{4\pi/3}E_0 \cos\theta_0 \epsilon_1 \delta_{\ell 1} \\ H_\ell^1 &= -\sqrt{2\pi/3}E_0 \sin\theta_0 \exp(-i\phi_0)\delta_{\ell 1} \\ J_\ell^1 &= -\sqrt{2\pi/3}E_0 \sin\theta_0 \exp(-i\phi_0)\epsilon_1 \delta_{\ell 1} \quad \text{for } \ell = 1,2,3,..... \end{aligned} \tag{8.65}$$

Using the above results, eqs.(8.64) and (8.65), in eq.(8.29), one can easily eliminate the multipole coefficients $B_{\ell m}$, by multiplying the first equation (8.29) with $\ell \epsilon_3$ and subsequently subtracting the second equation from the first. The result can then be written in the following form for $m = 0$:

$$\begin{aligned} & A_{\ell 0} + \left(\frac{\epsilon_1 - \epsilon_2}{\epsilon_1 + \epsilon_2}\right) \frac{\ell(\epsilon_3 - \epsilon_1)}{\ell\ \epsilon_3 + (\ell+1)\epsilon_1} R^{2\ell+1} \sum_{\ell_1=1}^{M} (-1)^{\ell_1} M_{\ell\ell_1}^0 (2R)^{-\ell-\ell_1-1} A_{\ell_1,0} \\ = & \sqrt{4\pi/3}E_0 \cos\theta_0 \left(\frac{\epsilon_3 - \epsilon_1}{\epsilon_3 + 2\epsilon_1}\right) R^3 \delta_{\ell 1} \quad \text{for } \ell = 1,2,3,....., M \end{aligned} \tag{8.66}$$

and for $m = 1$:

$$\begin{aligned} & A_{\ell 1} - \left(\frac{\epsilon_1 - \epsilon_2}{\epsilon_1 + \epsilon_2}\right) \frac{\ell(\epsilon_3 - \epsilon_1)}{\ell\ \epsilon_3 + (\ell+1)\epsilon_1} R^{2\ell+1} \sum_{\ell_1=1}^{M} (-1)^{\ell_1} M_{\ell\ell_1}^1 (2R)^{-\ell-\ell_1-1} A_{\ell_1,1} \\ = & -\sqrt{2\pi/3}E_0 \sin\theta_0 \exp(-i\phi_0) \left(\frac{\epsilon_3 - \epsilon_1}{\epsilon_3 + 2\epsilon_1}\right) R^3 \delta_{\ell 1} \quad \text{for } \ell = 1,2,3,..., M \end{aligned} \tag{8.67}$$

where M is any natural number. Since ϵ_1, ϵ_2 and ϵ_3 were denoted by ϵ_a, ϵ_s and ϵ, respectively, in chapter 5, one finds, using eq.(5.70) for the multipole polarizabilities α_ℓ of the sphere in the infinite ambient, that eqs.(8.66) and (8.67) are identical with eqs.(5.109) and (5.110). Notice that the latter equations have been derived for the more general case, that the spheres are at a distance $d(\geq R)$ on or above the substrate. Eqs.(8.66) and (8.67) were derived for spheres touching the substrate, so that $R = d$.

Next the case of a hemispherical island, touching with its flat side the substrate, will be considered. This case can be obtained by putting $t_r = 0$ and $\epsilon_4 = \epsilon_2$ in the general formulae for the matrix elements $C_{\ell\ell_1}^m$, $D_{\ell\ell_1}^m$, $F_{\ell\ell_1}^m$ and $G_{\ell\ell_1}^m$, and for H_ℓ^m and J_ℓ^m in section 8.2. It follows from eqs.(8.18) and (8.15) that

$$S_{\ell\ell_1}^m(t,0) = t^{-\ell_1-1} Q_{\ell\ell_1}^m(0) \tag{8.68}$$

and with eq.(8.17), that

$$S_{\ell\ell_1}^m(0) = Q_{\ell\ell_1}^m(0) \tag{8.69}$$

Similarly one obtains from eqs.(8.19), (8.15) and (8.17):

$$T_{\ell\ell_1}^m(t,0) = t^{\ell_1} Q_{\ell\ell_1}^m(0) \tag{8.70}$$

and

$$T^m_{\ell\ell_1}(0) = Q^m_{\ell\ell_1}(0) \tag{8.71}$$

Using these results in eq.(8.13), one finds, with $\epsilon_4 = \epsilon_2$,

$$\begin{aligned}
C^m_{\ell\ell_1} &= \left(\frac{2\epsilon_1}{\epsilon_1+\epsilon_2}\right)\delta_{\ell\ell_1} - \left(\frac{\epsilon_1-\epsilon_2}{\epsilon_1+\epsilon_2}\right)[1-(-1)^{\ell_1+m}]\zeta^m_{\ell\ell_1}Q^m_{\ell\ell_1}(0)\\
D^m_{\ell\ell_1} &= -\left(\frac{2\epsilon_3}{\epsilon_2+\epsilon_3}\right)\delta_{\ell\ell_1} - \left(\frac{\epsilon_2-\epsilon_3}{\epsilon_2+\epsilon_3}\right)[1-(-1)^{\ell_1+m}]\zeta^m_{\ell\ell_1}Q^m_{\ell\ell_1}(0)\\
F^m_{\ell\ell_1} &= -\left(\frac{2\epsilon_1\epsilon_2}{\epsilon_1+\epsilon_2}\right)(\ell_1+1)\delta_{\ell\ell_1} - \epsilon_1\left(\frac{\epsilon_1-\epsilon_2}{\epsilon_1+\epsilon_2}\right)(\ell_1+1)[1+(-1)^{\ell_1+m}]\zeta^m_{\ell\ell_1}Q^m_{\ell\ell_1}(0)\\
G^m_{\ell\ell_1} &= -\left(\frac{2\epsilon_2\epsilon_3}{\epsilon_2+\epsilon_3}\right)\ell_1\delta_{\ell\ell_1} + \epsilon_3\left(\frac{\epsilon_2-\epsilon_3}{\epsilon_2+\epsilon_3}\right)\ell_1[1+(-1)^{\ell_1+m}]\zeta^m_{\ell\ell_1}Q^m_{\ell\ell_1}(0)
\end{aligned} \tag{8.72}$$

Furthermore it follows from eq.(8.30), that for hemispheres

$$\begin{aligned}
H^0_\ell &= \sqrt{4\pi/3}E_0\cos\theta_0\left[\frac{\epsilon_1}{\epsilon_2}\delta_{\ell 1} - \left(\frac{\epsilon_1-\epsilon_2}{\epsilon_2}\right)\zeta^0_{\ell 1}Q^0_{\ell 1}(0)\right]\\
J^0_\ell &= \sqrt{4\pi/3}E_0\cos\theta_0\epsilon_1\delta_{\ell 1}\\
H^1_\ell &= -\sqrt{2\pi/3}E_0\sin\theta_0\exp(-i\phi_0)\delta_{\ell 1}\\
J^1_\ell &= -\sqrt{2\pi/3}E_0\sin\theta_0\exp(-i\phi_0)[\epsilon_2\delta_{\ell 1} + (\epsilon_1-\epsilon_2)\zeta^1_{\ell 1}Q^1_{\ell 1}(0)]
\end{aligned} \tag{8.73}$$

In the above equations $\zeta^m_{\ell\ell_1}$ is again given by eq.(8.14). The integrals (cf. eq.(8.15))

$$Q^m_{\ell\ell_1}(0) \equiv \int_{-1}^{0} P^m_\ell(x)P^m_{\ell_1}(x)dx \tag{8.74}$$

can be evaluated explicitly. For $\ell + \ell_1$ even, one easily finds [48]

$$Q^m_{\ell\ell_1}(0) = \frac{(\ell+m)!}{(2\ell+1)(\ell-m)!}\delta_{\ell\ell_1} \tag{8.75}$$

whereas for $\ell =$ even and $\ell_1 =$ odd:

$$Q^m_{\ell\ell_1}(0) = \frac{(-1)^{(\ell+\ell_1-1)/2}(\ell+1)!(\ell_1)!}{2^{\ell+\ell_1-2m-1}(\ell-m+1)(\ell-\ell_1)(\ell+\ell_1+1)}\left[\left(\frac{\ell-2m}{2}\right)!\left(\frac{\ell_1-1}{2}\right)!\right]^{-2} \quad \text{for} \quad m = 0, 1 \tag{8.76}$$

For $\ell =$ odd and $\ell_1 =$ even, one simply has to interchange ℓ and ℓ_1 in the latter formula, since $Q^m_{\ell\ell_1}$ is symmetric in the indices ℓ and ℓ_1.

In appendix B it will be proved, that one can eliminate the multipole coefficients $B_{\ell m}$ by using the above results, eqs.(8.72), (8.73), (8.75) and (8.14), in eq.(8.29). One then obtains for $m = 0$:

$$\frac{(\ell+1)\epsilon_1(\epsilon_2+\epsilon_3) + \ell\,\epsilon_3(\epsilon_1+\epsilon_2)}{\epsilon_1-\epsilon_3}R^{-\ell-2}A_{\ell 0}$$

$$+2\epsilon_1 \sum_{\ell_1\ even}^{M} (\ell_1+1)\zeta^0_{\ell\ell_1} Q^0_{\ell\ell_1}(0) R^{-\ell_1-2} A_{\ell_1,0} = -\sqrt{\pi/3}(\epsilon_1+\epsilon_2)\delta_{\ell 1} E_0 \cos\theta_0$$

$$\text{for } \ell = 1,3,5,...... \qquad (8.77)$$

where the sum is over $\ell_1 = 2, 4, ...$ until the largest even number smaller than or equal to M, and furthermore

$$\epsilon_1 \frac{(\ell+1)(\epsilon_1+\epsilon_2)+\ell(\epsilon_2+\epsilon_3)}{\ell(\epsilon_3-\epsilon_1)} R^{-\ell-2} A_{\ell 0}$$

$$+2\epsilon_2 \sum_{\ell_1 odd}^{M} \zeta^0_{\ell\ell_1} Q^0_{\ell\ell_1}(0) R^{-\ell_1-2} A_{\ell_1,0} = 2\sqrt{\pi/3}(\epsilon_1+\epsilon_2)\zeta^0_{\ell 1} Q^0_{\ell 1}(0) E_0 \cos\theta_0$$

$$\text{for } \ell = 2,4,6,...... \qquad (8.78)$$

where the sum is over $\ell_1 = 1, 3, ...$ until the largest odd number smaller than or equal to M. For $m = 1$ one obtains the following set of equations for the multipole coefficients $A_{\ell 1}$:

$$\epsilon_1 \frac{(\ell+1)(\epsilon_1+\epsilon_2)+\ell(\epsilon_2+\epsilon_3)}{\ell(\epsilon_1-\epsilon_{3})} R^{-\ell-2} A_{\ell 1}$$

$$-2\epsilon_2 \sum_{\ell_1 even}^{M} \zeta^1_{\ell\ell_1} Q^1_{\ell\ell_1}(0) R^{-\ell_1-2} A_{\ell_1,1} = \frac{1}{2}\sqrt{2\pi/3}(\epsilon_1+\epsilon_2)\delta_{\ell 1} E_0 \sin\theta_0 \exp(-i\phi_0)$$

$$\text{for } \ell = 1,3,5,...... \qquad (8.79)$$

and

$$\frac{(\ell+1)\epsilon_1(\epsilon_2+\epsilon_3)+\ell\ \epsilon_3(\epsilon_1+\epsilon_2)}{\epsilon_1-\epsilon_3} R^{-\ell-2} A_{\ell 1}$$

$$+2\epsilon_1 \sum_{\ell_1 odd}^{M} (\ell_1+1)\zeta^1_{\ell\ell_1} Q^1_{\ell\ell_1}(0) R^{-\ell_1-2} A_{\ell_1,1}$$

$$= \sqrt{2\pi/3}(\epsilon_1+\epsilon_2)\zeta^1_{\ell 1} Q^1_{\ell 1}(0) E_0 \sin\theta_0 \exp(-i\phi_0)$$

$$\text{for } \ell = 2,4,6,...... \qquad (8.80)$$

(see also appendix B for proof)

In the above equations $Q^m_{\ell\ell_1}(0)$ for $m = 0, 1$ is given by eq.(8.76) and $\zeta^m_{\ell\ell_1}$ by eq.(8.14). They can be solved on a computer (see the discussion at the end of section 8.2). With the values for A_{10}, A_{20}, A_{11} and A_{20}, obtained in this way, one can then calculate the polarizabilities, eq.(8.31), of hemispheres on a substrate and the surface constitutive coefficients, eq.(8.32), of an island film of identical hemispheres on a substrate in the limit of low coverage. Finally the invariants I_e, $I_{\delta,e}$, I_τ and I_c are found, using eq.(5.66).

It is, of course, also possible to calculate the polarizabilities of hemispheres using the equations in section 8.3 for spherical caps in the limit $t_r \to 0$. Starting from eqs.(8.49), (8.43) and (8.50), with $t_r = 0$ and $\epsilon_4 = \epsilon_2$, and using again eqs.(8.68)–(8.71) and (8.74)–(8.76), one can show, in an analogous way as in appendix B, that equations equivalent to eqs.(8.77)–(8.80) are valid. Replacing $\epsilon_2 A_{\ell m}$ by $\epsilon_1 A_{\ell m}$, for $\ell + m$ even in these equations one obtains eqs.(8.77)–(8.80). This is a consequence of the fact that the center of the hemisphere in the one case is above the substrate and in the other case below the substrate. Both sets of equations may alternatively be used if one combines them with the formulae for the polarizabilities and constitutive coefficients for the corresponding case. The resulting values for the constitutive coefficients are, of course, identical.

Berreman [54] was the first, who considered the problem of the electromagnetic response of a hemispherical particle on a substrate with the same dielectric constant ($\epsilon_2 = \epsilon_3$). This served as a model for surface roughness. The more general case of a hemisphere on a substrate with $\epsilon_2 \neq \epsilon_3$ was treated by Chauvaux and Meessen [55]. It can be shown, that the sets of equations, derived by these authors, are equivalent with the above equations (8.77) through (8.80), which have been obtained here as a special case of the more general equations (8.29), together with eqs.(8.30) and (8.13), for arbitrary truncated spheres on a substrate.

8.5 Thin spherical caps

A case of special interest is the case of thin spherical caps. For these thin caps the distance of an arbitrary point in the cap is small compared to the radius of the sphere. It is then convenient to introduce the relative height of the cap,

$$h_r \equiv \frac{h}{R} \equiv 1 - t_r \tag{8.81}$$

as a small parameter, cf. fig.8.4. For this case it is appropriate to transform the equations, found in section 3, for the multipole amplitudes, so that the small nature of this parameter becomes useful.

It follows from eqs.(8.14), (8.15) and (8.58) that

$$\zeta_{\ell\ell_1}^m Q_{\ell\ell_1}^m(t_r) = \delta_{\ell\ell_1} - \zeta_{\ell\ell_1}^m q_{\ell\ell_1}^m(h_r) \tag{8.82}$$

where

$$q_{\ell\ell_1}^m(h_r) \equiv \int_{t_r}^1 P_\ell^m(x) P_{\ell_1}^m(x) dx \tag{8.83}$$

Notice that $q_{\ell\ell_1}^m(h_r)$ is of order h_r or smaller. Furthermore it follows from eqs.(8.18) and (8.60) that

$$\zeta_{\ell\ell_1}^m S_{\ell\ell_1}^m(t, t_r) = \left[t^\ell / (2t_r)^{\ell+\ell_1+1}\right] M_{\ell\ell_1}^m - \zeta_{\ell\ell_1}^m s_{\ell\ell_1}^m(t, h_r) \tag{8.84}$$

where

$$s^m_{\ell\ell_1}(t,h_r) \equiv \int_{t_r}^{1} P^m_\ell(x)P^m_{\ell_1}([tx-2t_r][t^2-4t_rtx+4t_r^2]^{-1/2}) \times(t^2-4t_rtx+4t_r^2)^{-(\ell_1+1)/2}dx \quad (8.85)$$

Also $s^m_{\ell\ell_1}(t,h_r)$ is of order h_r or smaller. The matrix $M^m_{\ell\ell_1}$ is defined by eq.(5.103). Eq.(8.84) is valid for $t < 2t_r$. As has been seen in the previous sections, one only needs the above matrices and their partial derivatives with respect to t for $t = 1$. One finds

$$\zeta^m_{\ell\ell_1}S^m_{\ell\ell_1}(t_r) = (2t_r)^{-\ell-\ell_1-1}M^m_{\ell\ell_1} - \zeta^m_{\ell\ell_1}s^m_{\ell\ell_1}(h_r) \quad (8.86)$$

where $S^m_{\ell\ell_1}(t_r) \equiv S^m_{\ell\ell_1}(t=1,t_r)$ and $s^m_{\ell\ell_1}(h_r) \equiv s^m_{\ell\ell_1}(t=1,h_r)$ have been used. Furthermore

$$\zeta^m_{\ell\ell_1}\left[\frac{\partial}{\partial t}S^m_{\ell\ell_1}(t,t_r)\right]_{t=1} = \ell\,(2t_r)^{-\ell-\ell_1-1}M^m_{\ell\ell_1} - \zeta^m_{\ell\ell_1}\left[\frac{\partial}{\partial t}s^m_{\ell\ell_1}(t,h_r)\right]_{t=1} \quad (8.87)$$

The above equations are valid for $1/2 < t_r < 1$ or, in terms of the relative height, for $0 < h_r < 1/2$.

To transform the matrix $T^m_{\ell\ell_1}(t,t_r)$, defined in eq.(8.19), analogously one must use the following identity

$$P^m_{\ell_1}([tx-2t_r][t^2-4t_rtx+4t_r^2]^{-1/2})(t^2-4t_rtx+4t_r^2)^{\ell_1/2} = \sum_{\ell=|m|}^{\ell_1}(-1)^{\ell+\ell_1}\frac{(\ell_1+m)!}{(\ell+m)!\,(\ell_1-\ell)!}(2t_r)^{\ell_1-\ell}t^\ell P^m_\ell(x) \quad (8.88)$$

It then follows, with eqs.(8.19), (8.14), (8.15) and (8.58), that

$$\zeta^m_{\ell\ell_1}T^m_{\ell\ell_1}(t,t_r) = t^\ell\,(2t_r)^{\ell_1-\ell}N^m_{\ell\ell_1} - \zeta^m_{\ell\ell_1}t^m_{\ell\ell_1}(t,h_r) \quad (8.89)$$

where

$$t^m_{\ell\ell_1}(t,h_r) \equiv \int_{t_r}^{1} P^m_\ell(x)P^m_{\ell_1}([tx-2t_r][t^2-4t_rtx+4t_r^2]^{-1/2}) \times(t^2-4t_rtx+4t_r^2)^{\ell_1/2}dx \quad (8.90)$$

Again $t^m_{\ell\ell_1}(t,h_r)$ is of order h_r or smaller. The matrix $N^m_{\ell\ell_1}$ is defined by

$$N^m_{\ell\ell_1} \equiv (-1)^{\ell+\ell_1}\frac{1}{(\ell_1-\ell)!}\left[\frac{(2\ell_1+1)\,(\ell_1+m)!\,(\ell_1-m)!}{(2\ell+1)\,(\ell+m)!\,(\ell-m)!}\right]^{1/2} \quad \text{for} \quad \ell \le \ell_1$$
$$\equiv 0 \quad \text{for } \ell > \ell_1 \quad (8.91)$$

As was the case for $S^m_{\ell\ell_1}(t,t_r)$, one only needs $T^m_{\ell\ell_1}(t,t_r)$ and its partial derivative with respect to t for $t = 1$. It then follows that

$$\zeta^m_{\ell\ell_1}T^m_{\ell\ell_1}(t_r) = (2t_r)^{\ell_1-\ell}N^m_{\ell\ell_1} - \zeta^m_{\ell\ell_1}t^m_{\ell\ell_1}(h_r) \quad (8.92)$$

where $T^m_{\ell\ell_1}(t_r) \equiv T^m_{\ell\ell_1}(t=1,t_r)$ and $t^m_{\ell\ell_1}(h_r) \equiv t^m_{\ell\ell_1}(t=1,h_r)$ have been used. Furthermore

$$\zeta^m_{\ell\ell_1}\left[\frac{\partial}{\partial t}T^m_{\ell\ell_1}(t,t_r)\right]_{t=1} = \ell\,(2t_r)^{\ell_1-\ell}\,N^m_{\ell\ell_1} - \zeta^m_{\ell\ell_1}\left[\frac{\partial}{\partial t}t^m_{\ell\ell_1}(t,h_r)\right]_{t=1} \tag{8.93}$$

If one substitutes the above equations (8.82), (8.86), (8.87), (8.92) and (8.93) into eq.(8.43), one obtains for the matrices appearing in eq.(8.49)

$$\begin{aligned}
C^m_{\ell\ell_1} &= \delta_{\ell\ell_1} - \left(\frac{\epsilon_1-\epsilon_2}{\epsilon_1+\epsilon_2}\right)\{(-1)^{\ell+m}(2t_r)^{-\ell-\ell_1-1}M^m_{\ell\ell_1} \\
&\quad +(-1)^{\ell+\ell_1}\zeta^m_{\ell\ell_1}[q^m_{\ell\ell_1}(h_r)-(-1)^{\ell_1+m}s^m_{\ell\ell_1}(h_r)]\} \\
D^m_{\ell\ell_1} &= -\delta_{\ell\ell_1} + \left(\frac{\epsilon_3-\epsilon_4}{\epsilon_3+\epsilon_4}\right)\{(-1)^{\ell+m}(2t_r)^{\ell_1-\ell}N^m_{\ell\ell_1} \\
&\quad +(-1)^{\ell+\ell_1}\zeta^m_{\ell\ell_1}[q^m_{\ell\ell_1}(h_r)-(-1)^{\ell_1+m}t^m_{\ell\ell_1}(h_r)]\}
\end{aligned}$$

$$\begin{aligned}
F^m_{\ell\ell_1} &= -\epsilon_2(\ell+1)\delta_{\ell\ell_1} - \epsilon_2\left(\frac{\epsilon_1-\epsilon_2}{\epsilon_1+\epsilon_2}\right)\Bigg(\ell(-1)^{\ell+m}(2t_r)^{-\ell-\ell_1-1}M^m_{\ell\ell_1} \\
&\quad +(-1)^{\ell+\ell_1}\zeta^m_{\ell\ell_1}\left\{(\ell_1+1)q^m_{\ell\ell_1}(h_r)-(-1)^{\ell_1+m}\left[\frac{\partial}{\partial t}s^m_{\ell\ell_1}(t,h_r)\right]_{t=1}\right\}\Bigg) \\
G^m_{\ell\ell_1} &= -\epsilon_4\ell\delta_{\ell\ell_1} + \epsilon_4\left(\frac{\epsilon_3-\epsilon_4}{\epsilon_3+\epsilon_4}\right)\Bigg(\ell(-1)^{\ell+m}(2t_r)^{\ell_1-\ell}N^m_{\ell\ell_1} \\
&\quad -(-1)^{\ell+\ell_1}\zeta^m_{\ell\ell_1}\left\{\ell_1 q^m_{\ell\ell_1}(h_r)+(-1)^{\ell_1+m}\left[\frac{\partial}{\partial t}t^m_{\ell\ell_1}(t,h_r)\right]_{t=1}\right\}\Bigg)
\end{aligned} \tag{8.94}$$

For the terms at the right-hand side of eq.(8.49) one obtains, from eqs.(8.50), (8.82) and (8.83),

$$\begin{aligned}
H^0_\ell &= \sqrt{\frac{4\pi}{3}}E_0\cos\theta_0\left\{\left(\frac{\epsilon_1}{\epsilon_2}\right)\delta_{\ell 1} + \left(\frac{\epsilon_1-\epsilon_2}{\epsilon_2}\right)(-1)^\ell\zeta^0_{\ell 1}\int_{1-h_r}^1 P^0_\ell(x)(x-1+h_r)dx\right\} \\
J^0_\ell &= \sqrt{4\pi/3}E_0\cos\theta_0\epsilon_1\delta_{\ell 1} \\
H^1_\ell &= -\sqrt{2\pi/3}E_0\sin\theta_0\exp(-i\phi_0)\delta_{\ell 1} \\
J^1_\ell &= -\sqrt{2\pi/3}E_0\sin\theta_0\exp(-i\phi_0)[\epsilon_2\delta_{\ell 1}-(\epsilon_1-\epsilon_2)(-1)^\ell\zeta^1_{\ell 1}q^1_{\ell 1}(h_r)]
\end{aligned} \tag{8.95}$$

In the further calculation below ϵ_4 will be set equal to ϵ_2. In the limit $h_r \to 0$ the cap disappears and the only non-zero multipole coefficients are B_{10} and B_{11}. Solving eq.(8.49) for $h_r = 0$, one finds

$$\begin{aligned}
B_{10} &= -\sqrt{\frac{\pi}{3}}E_0\cos\theta_0\left(\frac{\epsilon_1(\epsilon_2+\epsilon_3)}{\epsilon_2\epsilon_3}\right) \\
B_{11} &= \sqrt{\frac{\pi}{6}}E_0\sin\theta_0\exp(-i\phi_0)\left(\frac{\epsilon_2+\epsilon_3}{\epsilon_2}\right)
\end{aligned} \tag{8.96}$$

These amplitudes give the external field inside the sphere, which has now completely disappeared in the substrate. If one calculates this field using these amplitudes the factor containing ϵ_3 cancels, and one gets the same field as one has outside the sphere in the substrate. Because of this it is convenient to introduce new amplitudes

$$\begin{aligned}\widetilde{B}_{\ell 0} &\equiv B_{\ell 0} + \delta_{\ell 1}\sqrt{\frac{\pi}{3}}E_0\cos\theta_0\left(\frac{\epsilon_1(\epsilon_2+\epsilon_3)}{\epsilon_2\epsilon_3}\right)\\ \widetilde{B}_{\ell 1} &\equiv B_{\ell 1} - \delta_{\ell 1}\sqrt{\frac{\pi}{6}}E_0\sin\theta_0\exp(-i\phi_0)\left(\frac{\epsilon_2+\epsilon_3}{\epsilon_2}\right)\end{aligned} \tag{8.97}$$

Eq.(8.49) then becomes

$$\begin{aligned}\sum_{\ell_1=1}^{M} C^m_{\ell\ell_1}R^{-\ell_1-2}A_{\ell_1 m} + \sum_{\ell_1=1}^{M} D^m_{\ell\ell_1}R^{\ell_1-1}\widetilde{B}_{\ell_1 m} &= \widetilde{H}^m_\ell\\ \sum_{\ell_1=1}^{M} F^m_{\ell\ell_1}R^{-\ell_1-2}A_{\ell_1 m} + \sum_{\ell_1=1}^{M} G^m_{\ell\ell_1}R^{\ell_1-1}\widetilde{B}_{\ell_1 m} &= \widetilde{J}^m_\ell\end{aligned}$$

$$\text{for } \ell = 1,2,3,...,M \text{ and } m = 0,1 \tag{8.98}$$

where

$$\begin{aligned}\widetilde{H}^0_\ell &\equiv H^0_\ell + \sqrt{\frac{\pi}{3}}E_0\cos\theta_0\left(\frac{\epsilon_1(\epsilon_2+\epsilon_3)}{\epsilon_2\epsilon_3}\right)D^0_{\ell 1}\\ \widetilde{J}^0_\ell &\equiv J^0_\ell + \sqrt{\frac{\pi}{3}}E_0\cos\theta_0\left(\frac{\epsilon_1(\epsilon_2+\epsilon_3)}{\epsilon_2}\right)G^0_{\ell 1}\\ \widetilde{H}^1_\ell &\equiv H^1_\ell - \sqrt{\frac{\pi}{6}}E_0\sin\theta_0\exp(-i\phi_0)\left(\frac{\epsilon_2+\epsilon_3}{\epsilon_2\epsilon_3}\right)D^1_{\ell 1}\\ \widetilde{J}^1_\ell &\equiv J^1_\ell - \sqrt{\frac{\pi}{6}}E_0\sin\theta_0\exp(-i\phi_0)\left(\frac{\epsilon_2+\epsilon_3}{\epsilon_2}\right)G^1_{\ell 1}\end{aligned} \tag{8.99}$$

It follows from eqs.(8.94), with $\epsilon_2 = \epsilon_4$, and (8.95) that

$$\begin{aligned}\widetilde{H}^0_\ell &= \sqrt{4\pi/3}E_0\cos\theta_0\left(\frac{\epsilon_1-\epsilon_3}{\epsilon_3}\right)(-1)^\ell\zeta^0_{\ell 1}\int_{1-h_r}^{1}P^0_\ell(x)(x-1+h_r)dx\\ \widetilde{J}^0_\ell &= \widetilde{H}^1_\ell = 0\\ \widetilde{J}^1_\ell &= \sqrt{2\pi/3}E_0\sin\theta_0\exp(-i\phi_0)(\epsilon_1-\epsilon_3)(-1)^\ell\zeta^1_{\ell 1}q^1_{\ell 1}(h_r)\end{aligned} \tag{8.100}$$

where $q^1_{\ell 1}(h_r)$ is defined by eq.(8.83).

It follows from the above equation that $\widetilde{H}^0_\ell$ and $\widetilde{J}^1_\ell$ are of second order in h_r. Since the matrix elements $C^m_{\ell\ell_1}$,$D^m_{\ell\ell_1}$,$F^m_{\ell\ell_1}$ and $G^m_{\ell\ell_1}$ are of the zeroth order, it follows from eq.(8.98) that the amplitudes $A_{\ell m}$ and $\widetilde{B}_{\ell m}$ are also of second order in h_r. Now all integrals at the right-hand sides of eq.(8.94) are at least of first order in h_r. These can therefore be neglected in lowest order. One can then eliminate the coefficients

$\widetilde{B}_{\ell m}$ in eq.(8.98). One then obtains up to second order in h_r

$$\begin{aligned} A_{\ell 0} &= \left(\frac{\ell}{2\ell+1}\right) R^{\ell+2}\sqrt{\frac{4\pi}{3}} E_0 \cos\theta_0 \left(\frac{\epsilon_1-\epsilon_3}{\epsilon_3}\right)(-1)^\ell \zeta^0_{\ell 1} \int_{1-h_r}^{1} P^0_\ell(x)(x-1+h_r)dx \\ A_{\ell 1} &= \left(\frac{1}{2\ell+1}\right) R^{\ell+2}\sqrt{2\pi/3} E_0 \sin\theta_0 \exp(-i\phi_0) \left(\frac{\epsilon_1-\epsilon_3}{\epsilon_2}\right)(-1)^{\ell+1}\zeta^1_{\ell 1} q^1_{\ell 1}(h_r) \end{aligned} \tag{8.101}$$

For the dipole and quadrupole coefficients one finds up to second order in h_r

$$\begin{aligned} A_{10} &= \frac{1}{2}R^3 (h_r)^2 \left(\frac{\epsilon_3-\epsilon_1}{\epsilon_3}\right)\sqrt{\frac{\pi}{3}} E_0 \cos\theta_0 \\ A_{20} &= -R^4 (h_r)^2 \left(\frac{\epsilon_3-\epsilon_1}{\epsilon_3}\right)\sqrt{\frac{\pi}{5}} E_0 \cos\theta_0 \\ A_{11} &= -\frac{1}{4}R^3 (h_r)^2 \left(\frac{\epsilon_3-\epsilon_1}{\epsilon_2}\right)\sqrt{\frac{2\pi}{3}} E_0 \sin\theta_0 \exp(-i\phi_0) \\ A_{21} &= \frac{1}{4}R^4 (h_r)^2 \left(\frac{\epsilon_3-\epsilon_1}{\epsilon_2}\right)\sqrt{\frac{6\pi}{5}} E_0 \sin\theta_0 \exp(-i\phi_0) \end{aligned} \tag{8.102}$$

The volume of the spherical cap is given (exactly) by

$$V = \pi R^3 (h_r)^2 \left(1 - \frac{1}{3}h_r\right) \tag{8.103}$$

and is therefore of second order in the height. The dipole and quadrupole polarizabilities of the cap can then be calculated using eq.(8.51). This results, to second order in h_r, in

$$\begin{aligned} \alpha_z(0) &= V\epsilon_2^2 \left(\frac{\epsilon_3-\epsilon_1}{\epsilon_1\epsilon_3}\right) \\ \alpha_z^{10}(0) &= -VR\epsilon_2^2 \left(\frac{\epsilon_3-\epsilon_1}{\epsilon_1\epsilon_3}\right) \\ \alpha_\parallel(0) &= V(\epsilon_3-\epsilon_1) \\ \alpha_\parallel^{10}(0) &= -VR(\epsilon_3-\epsilon_1) \end{aligned} \tag{8.104}$$

The interfacial constitutive coefficients for the caps are, using eq.(8.56) with $d = R$,

$$\begin{aligned} \beta_e(-d) &= \rho V \left(\frac{\epsilon_3-\epsilon_1}{\epsilon_1\epsilon_3}\right) \\ \gamma_e(-d) &= \rho V(\epsilon_3-\epsilon_1) \\ \delta_e(-d) &= \tau(-d) = 0 \end{aligned} \tag{8.105}$$

Notice that both $\beta_e(-d)$ and $\gamma_e(-d)$ are of second order in the height due to the fact that the volume is of second order. The coefficients $\delta_e(-d)$ and $\tau(-d)$ are of third order in the height, volume times height, and therefore vanish in this approximation.

Comparing the above results with those obtained in section 6.4 for the thin oblate spheroid, which was called the circular disk, one sees that the results are the same for β_e and γ_e, cf. eq.(6.152). In that case the coefficients δ_e and τ are unequal to zero, however. The reason for this is that the volume of the circular disk is proportional to the height $2a\xi_0$, cf. eq.(6.142), whereas the volume of the spherical cap is proportional to the height squared. Notice that $\epsilon_1 = \epsilon_a$, $\epsilon_2 = \epsilon_s$ and $\epsilon_3 = \epsilon$. In both cases the constitutive coefficients are calculated with respect to the surface of the substrate.

In the last chapter, on rough surfaces, a very different method is developed, which may alternatively be used for thin spherical caps. The results for these thin spherical caps are asymptotically the same. The leading corrections to this behavior are very different, however, and point to the fact that the various quantities are not analytic in $h_r = 0$. This will be further discussed in section 8 of chapter 14.

8.6 Application to truncated gold spheres and caps on sapphire

In this section the theory, developed above, will be applied to the special case of gold islands on a flat sapphire surface. In particular the dependence of the polarizabilities, the surface constitutive coefficients and the invariants on the shape, i.e. on the truncation parameter, t_r, will be considered. The dielectric constant of the island is a complex function of the wavelength, cf table 5.1. The dielectric constant of sapphire is $\epsilon_2 = 3.13$ and of the ambient, vacuum, is $\epsilon_1 = 1$. As in the previous chapters, the following dimensionless amplitudes are introduced, cf. eq.(5.118)

$$\begin{aligned}
\widehat{A}_{\ell 0} &\equiv A_{\ell 0}/\left[VR^{\ell-1}\sqrt{4\pi/3}E_0\cos\theta_0\right] \\
\widehat{B}_{\ell 0} &\equiv B_{\ell 0}R^{\ell+2}/\left[V\sqrt{4\pi/3}E_0\cos\theta_0\right] \\
\widehat{A}_{\ell 1} &\equiv -A_{\ell 1}/\left[VR^{\ell-1}\sqrt{2\pi/3}E_0\sin\theta_0\exp(-i\phi_0)\right] \\
\widehat{B}_{\ell 1} &\equiv -B_{\ell 1}R^{\ell+2}/\left[V\sqrt{2\pi/3}E_0\sin\theta_0\exp(-i\phi_0)\right]
\end{aligned} \tag{8.106}$$

where the volume of the truncated sphere may be written as

$$V = \frac{1}{3}\pi R^3\left(2 + 3t_r - t_r^3\right) \tag{8.107}$$

Eq.(8.29) can then be written as

$$\begin{aligned}
\sum_{\ell_1=1}^{M} C_{\ell\ell_1}^m \widehat{A}_{\ell_1 m} + \sum_{\ell_1=1}^{M} D_{\ell\ell_1}^m \widehat{B}_{\ell_1 m} &= \widehat{H}_\ell^m \\
\sum_{\ell_1=1}^{M} F_{\ell\ell_1}^m \widehat{A}_{\ell_1 m} + \sum_{\ell_1=1}^{M} G_{\ell\ell_1}^m \widehat{B}_{\ell_1 m} &= \widehat{J}_\ell^m \\
\text{for } \ell &= 1, 2, 3, ..., M \text{ and } m = 0, 1
\end{aligned} \tag{8.108}$$

where

$$\begin{aligned}
\widehat{H}_\ell^0 &\equiv \frac{3}{\pi}\left(2+3t_r-t_r^3\right)^{-1}\left\{\frac{\epsilon_1}{\epsilon_2}\delta_{\ell 1}+\left(\frac{\epsilon_1-\epsilon_2}{\epsilon_2}\right)\left[\sqrt{3}t_r\zeta_{\ell 0}^0 Q_{\ell 0}^0(t_r)-\zeta_{\ell 1}^0 Q_{\ell 1}^0(t_r)\right]\right\} \\
\widehat{J}_\ell^0 &\equiv \frac{3}{\pi}\left(2+3t_r-t_r^3\right)^{-1}\epsilon_1\delta_{\ell 1} \\
\widehat{H}_\ell^1 &\equiv \frac{3}{\pi}\left(2+3t_r-t_r^3\right)^{-1}\delta_{\ell 1} \\
\widehat{J}_\ell^1 &\equiv \frac{3}{\pi}\left(2+3t_r-t_r^3\right)^{-1}\left[\epsilon_2\delta_{\ell 1}+(\epsilon_1-\epsilon_2)\zeta_{\ell 1}^1 Q_{\ell 1}^1(t_r)\right]
\end{aligned} \tag{8.109}$$

as follows using eq.(8.30). The matrix elements $C_{\ell\ell_1}^m$, $D_{\ell\ell_1}^m$, $F_{\ell\ell_1}^m$ and $G_{\ell\ell_1}^m(\ell,\ell_1 = 1,2,3,.....,M,\ m=0,1)$ are given by eq.(8.13), with $\epsilon_4=\epsilon_2$.

After solving the linear set of eqs.(8.108), for a sufficiently large value of M, the dipole and quadrupole polarizabilities, in dimensionless form, are

$$\begin{aligned}
\widehat{\alpha}_z &\equiv \alpha_z(0)/V = 4\pi\widehat{A}_{10} \\
\widehat{\alpha}_\| &\equiv \alpha_\|(0)/V = 4\pi\widehat{A}_{11} \\
\widehat{\alpha}_z^{10} &\equiv \alpha_z^{10}(0)/VR = 2\pi\widehat{A}_{20}\sqrt{5/3} \\
\widehat{\alpha}_\|^{10} &\equiv \alpha_\|^{10}(0)/VR = \frac{4\pi}{3}\widehat{A}_{21}\sqrt{5}
\end{aligned} \tag{8.110}$$

This equation follows from eqs.(8.31) and (8.106). The surface constitutive coefficients in dimensionless form become

$$\begin{aligned}
\widehat{\gamma}_e &\equiv \frac{1}{\rho V}\gamma_e(d) = \widehat{\alpha}_\| \\
\widehat{\beta}_e &\equiv \frac{1}{\rho V}\beta_e(d) = \widehat{\alpha}_z \\
\widehat{\tau} &\equiv \frac{1}{\rho VR}\tau(d) = t_r\widehat{\alpha}_\| - \widehat{\alpha}_\|^{10} \\
\widehat{\delta}_e &\equiv \frac{1}{\rho VR}\delta_e(d) = t_r\widehat{\alpha}_z + t_r\widehat{\alpha}_\| - \widehat{\alpha}_z^{10} - \widehat{\alpha}_\|^{10}
\end{aligned} \tag{8.111}$$

This follows using eqs.(8.32), (8.11) and (8.110). The invariants, eq.(5.66), become

$$\begin{aligned}
\widehat{I}_e \equiv I_e/\rho V = \widehat{\gamma}_e - \epsilon_2\widehat{\beta}_e \quad & \widehat{I}_c \equiv (\epsilon_2-1)I_c/\rho V = \operatorname{Im}\widehat{\gamma}_e \\
\widehat{I}_\tau \equiv I_\tau/\rho VR = \widehat{\tau} \quad & \widehat{I}_{\delta,e} \equiv I_{\delta,e}/\rho VR = \widehat{\delta}_e
\end{aligned} \tag{8.112}$$

Here it has been used that ϵ_2 is real.

In the case of a spherical cap eq.(8.106) can again be used, now with the volume

$$V = \frac{1}{3}\pi R^3\left(2-3t_r+t_r^3\right) \tag{8.113}$$

The linear eq.(8.49) can again be written in the form (8.108), where

$$\begin{aligned}
\widehat{H}_\ell^0 &\equiv \frac{3}{\pi}\left(2-3t_r+t_r^3\right)^{-1}\left\{\delta_{\ell 1}+\left(\frac{\epsilon_2-\epsilon_1}{\epsilon_2}\right)(-1)^{\ell+1}\left[\sqrt{3}t_r\zeta_{\ell 0}^{0}Q_{\ell 0}^{0}(t_r)-\zeta_{\ell 1}^{0}Q_{\ell 1}^{0}(t_r)\right]\right\} \\
\widehat{J}_\ell^0 &\equiv \frac{3}{\pi}\left(2-3t_r+t_r^3\right)^{-1}\epsilon_1\delta_{\ell 1} \\
\widehat{H}_\ell^1 &\equiv \frac{3}{\pi}\left(2-3t_r+t_r^3\right)^{-1}\delta_{\ell 1} \\
\widehat{J}_\ell^1 &\equiv \frac{3}{\pi}\left(2-3t_r+t_r^3\right)^{-1}\left[\epsilon_1\delta_{\ell 1}+(\epsilon_2-\epsilon_1)(-1)^{\ell+1}\zeta_{\ell 1}^{1}Q_{\ell 1}^{1}(t_r)\right]
\end{aligned} \tag{8.114}$$

as follows using eq.(8.50). The matrix elements $C_{\ell\ell_1}^m$, $D_{\ell\ell_1}^m$, $F_{\ell\ell_1}^m$ and $G_{\ell\ell_1}^m(\ell,\ell_1 = 1,2,3,.....,M,\ m=0,1)$ are given by eq.(8.43), with $\epsilon_4=\epsilon_2$.

After solving the linear set of eqs.(8.108), for a sufficiently large value of M, the dipole and quadrupole polarizabilities, in dimensionless form, are

$$\begin{aligned}
\widehat{\alpha}_z &\equiv \alpha_z(0)/V = 4\pi\epsilon_2^2\widehat{A}_{10} \\
\widehat{\alpha}_\| &\equiv \alpha_\|(0)/V = 4\pi\epsilon_2\widehat{A}_{11} \\
\widehat{\alpha}_z^{10} &\equiv \alpha_z^{10}(0)/VR = 2\pi\epsilon_2^2\widehat{A}_{20}\sqrt{5/3} \\
\widehat{\alpha}_\|^{10} &\equiv \alpha_\|^{10}(0)/VR = \frac{4\pi}{3}\epsilon_2\widehat{A}_{21}\sqrt{5}
\end{aligned} \tag{8.115}$$

This equation follows from eqs.(8.51) and (8.106). The surface constitutive coefficients in dimensionless form become

$$\begin{aligned}
\widehat{\gamma}_e &\equiv \frac{1}{\rho V}\gamma_e(-d) = \widehat{\alpha}_\| \\
\widehat{\beta}_e &\equiv \frac{1}{\rho V}\beta_e(-d) = \widehat{\alpha}_z/\epsilon_2^2 \\
\widehat{\tau} &\equiv \frac{1}{\rho VR}\tau(-d) = -\left(t_r\widehat{\alpha}_\| + \widehat{\alpha}_\|^{10}\right)/\epsilon_2 \\
\widehat{\delta}_e &\equiv \frac{1}{\rho VR}\delta_e(-d) = -\left(t_r\widehat{\alpha}_z + t_r\widehat{\alpha}_\| + \widehat{\alpha}_z^{10} + \widehat{\alpha}_\|^{10}\right)/\epsilon_2
\end{aligned} \tag{8.116}$$

This follows using eqs.(8.56), (8.41) and (8.115). The invariants are again given by eq.(8.112).

In figures 8.5 through 8.12 the scaled polarizabilities and constitutive coefficients are given as a function of the wavelength in the optical region for a series of truncation parameters. For caps a minus sign was added to distinguish this case from the truncated sphere case with the same truncation parameter. In all calculations the absolute value is used. The interaction with the image has been taken into account to order $M = 16$ which was sufficient to obtain a converged result. The normal polarizability increases from $t_r = 1$ (sphere) to $t_r = 0.8$ and subsequently decreases. For the hemisphere, $t_r = 0$, a second resonance develops. The original resonance remains for all values of the truncation parameter. It has a small shift to lower wavelength from the sphere to the cap and decreases in size. The second resonance has a substantial shift to the infrared for caps and increases in size, when t_r decreases. A third resonance starts to develop for $t_r = -0.4$ which also shifts to the red for $t_r = -0.6$.

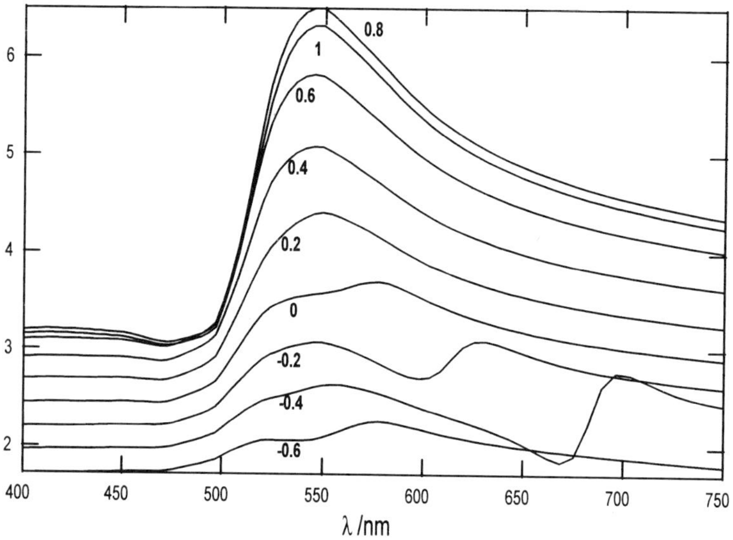

Figure 8.5 Real part of $\widehat{\alpha}_z = \widehat{\beta}_e$ as a function of the wavelength for truncation parameters from -0.6 through 1.0.

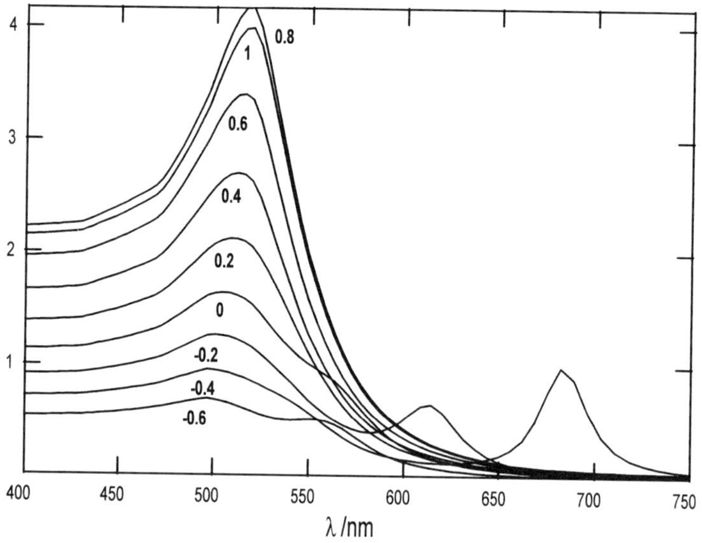

Figure 8.6 Imaginary part of $\widehat{\alpha}_z = \widehat{\beta}_e$ as a function of the wavelength for truncation parameters from -0.6 through 1.0.

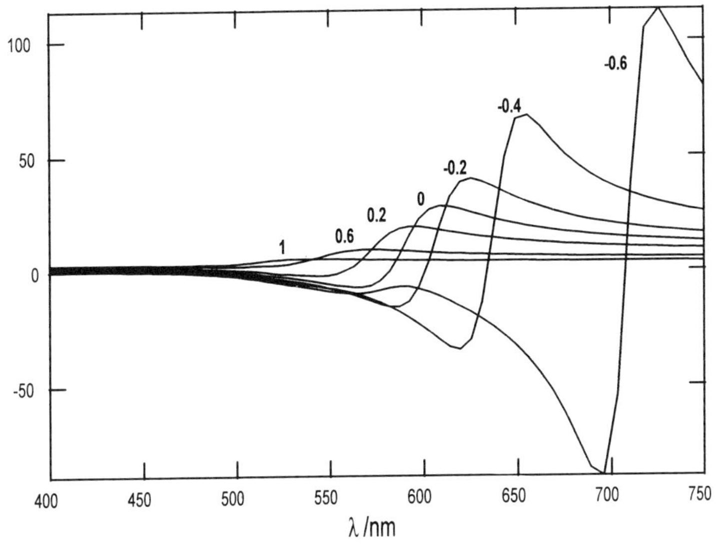

Figure 8.7 Real part of $\widehat{\alpha}_{\parallel} = \widehat{\gamma}_e$ as a function of the wavelength for truncation parameters from -0.6 through 1.0.

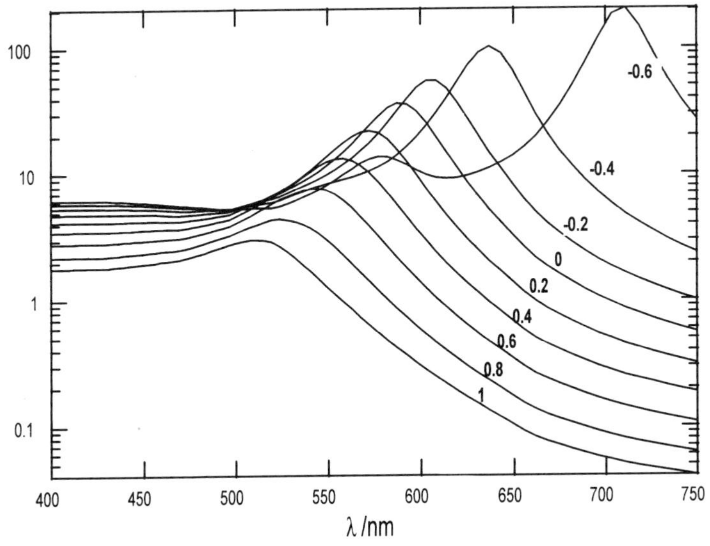

Figure 8.8 Imaginary part of $\widehat{\alpha}_{\parallel} = \widehat{\gamma}_e$ as a function of the wavelength for truncation parameters from -0.6 through 1.0.

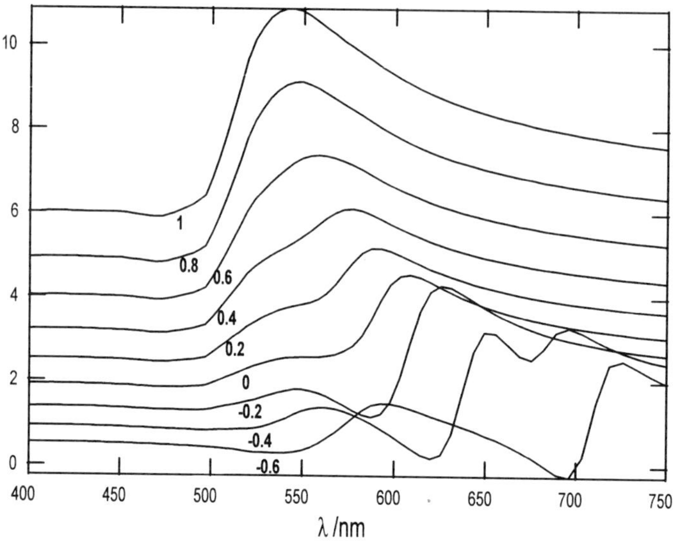

Figure 8.9 Real part of $\widehat{\delta}_e = \widehat{I}_{\delta,e}$ as a function of the wavelength for truncation parameters from -0.6 through 1.0.

The parallel polarizability increases orders of magnitude from $t_r = 1$ (sphere) to the thin cap limit. It becomes orders of magnitude larger than the normal polarizability. Given the change of the island to a flat disk like shape this is not unexpected. The resonance shifts to the infrared. A second resonance has developed for $t_r = -0.6$. The enormous gain in the parallel polarizability is in the optical domain most important around the hemisphere. For thin caps the gain shifts out of the optical region to the infrared, while the response in the optical region becomes the same as for a thin plane parallel layer of gold with the same weight thickness, cf. eq.(6.152). The results for truncated spheres are identical to those found by Wind et al [53].

The figures for $\widehat{\delta}_e$ and $\widehat{\tau}$ are very similar. In fact $\widehat{\delta}_e$ is about twice the size of $\widehat{\tau}$. This shows that $t_r\widehat{\alpha}_z - \widehat{\alpha}_z^{10} \simeq t_r\widehat{\alpha}_\| - \widehat{\alpha}_\|^{10}$, where a negative t_r is used for caps. The value of $\widehat{\tau}$ is for truncated spheres and caps, much smaller than $\widehat{\gamma}_e$. This shows that $t_r\widehat{\alpha}_\| \simeq \widehat{\alpha}_\|^{10}$. It is clear that the truncation of the spheres further decreases the importance of the quadrupolar constitutive coefficients $\widehat{\delta}_e$ and $\widehat{\tau}$ compared to the dipolar constitutive coefficients.

The analysis above evaluated all the multipole amplitudes up to $M = 16$. Using the formulae given in sections 8.2 and 8.3, it is then possible to plot equipotential surfaces for homogeneous external fields. This procedure was developed by Simonsen, Lazzari, Jupille and Roux [56]. This gives insight in for instance the induced charge distribution due to the field in the island. In figures 8.13 through 8.20 equipotential lines are plotted for cross sections through the symmetry axis of the islands. The

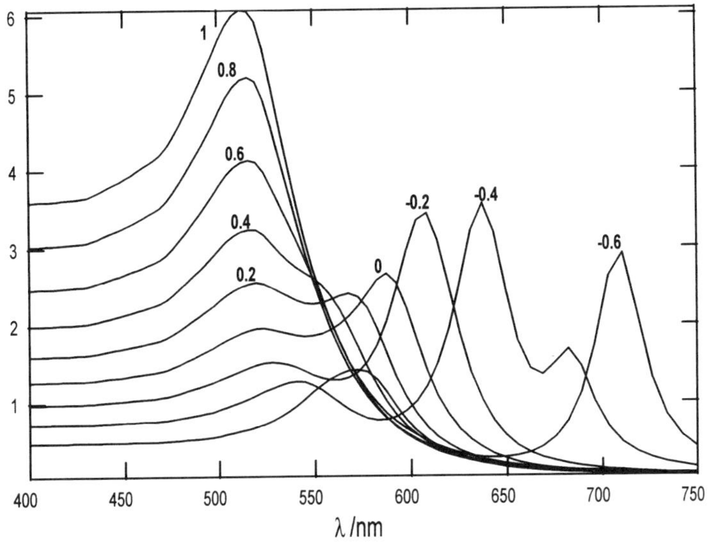

Figure 8.10 Imaginary part of $\widehat{\delta}_e = \widehat{I}_{\delta,e}$ as a function of the wavelength for truncation parameters from -0.6 through 1.0.

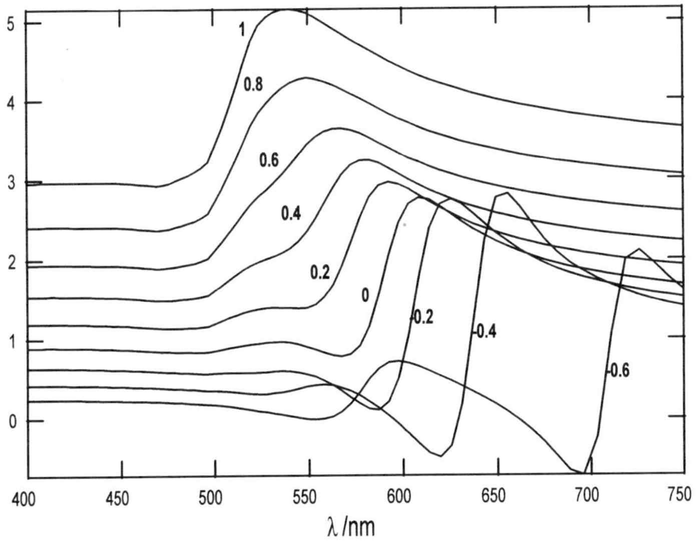

Figure 8.11 Real part of $\widehat{\tau} = \widehat{I}_\tau$ as a function of the wavelength for truncation parameters from -0.6 through 1.0.

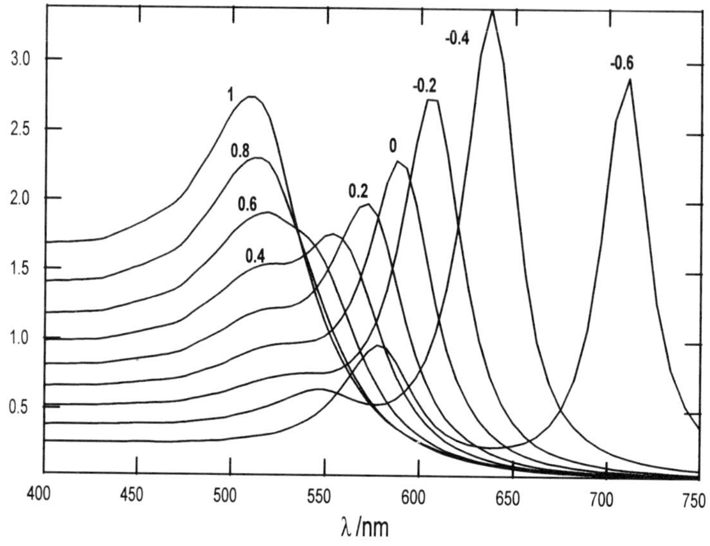

Figure 8.12 Imaginary part of $\widehat{\tau} = \widehat{I}_{\tau}$ as a function of the wavelength for truncation parameters from -0.6 through 1.0.

potential is in arbitrary units. The unit along the axes of the figures is the sphere radius. For a sphere in a field normal to the surface one sees that the equipotential planes inside the sphere near the top are more or less planar. This is similar to the behavior in the absence of the substrate and indicates a charge distribution which varies like a cosine around the top. In the lower region close to the substrate the equipotential planes are bent. This shows that the charge distribution has contracted towards the point where the sphere touches the substrate. For a field normal to the surface of the substrate the equipotential planes are rotationally symmetric around the symmetry axis. For a truncated sphere in a normal field the behavior near the top has not changed much. Along the plane of contact with the substrate one sees that the charge has moved away from the symmetry axis and is now closer to the three phase line. In this figure, as well as in the following, one can see that the convergence obtained to the 16th order in the multipole expansion is not everywhere equally good. If one follows the -0.3 line there is a small irregularity, where it crosses the boundary of the continuation of the spherical surface in the substrate. This irregularity, and similar irregularities in the subsequent figures, are not real. They will disappear if one goes to higher order in the multipole expansion. For the hemisphere the charge distribution around the top has not changed much. The charge along the substrate is still located away from the center but not as much as for the truncated sphere. For the cap the charge distribution around the top is still more or less the same. The charge along the substrate has returned to be centered around the symmetry axis.

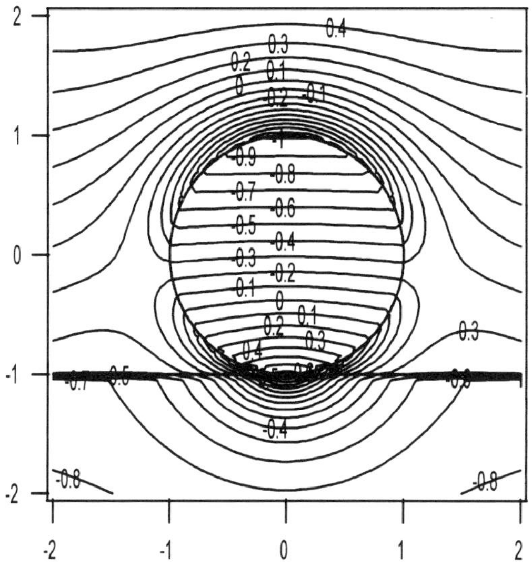

Figure 8.13 Potential distribution of a sphere, $t_r = 1.0$, in a normal electric field.

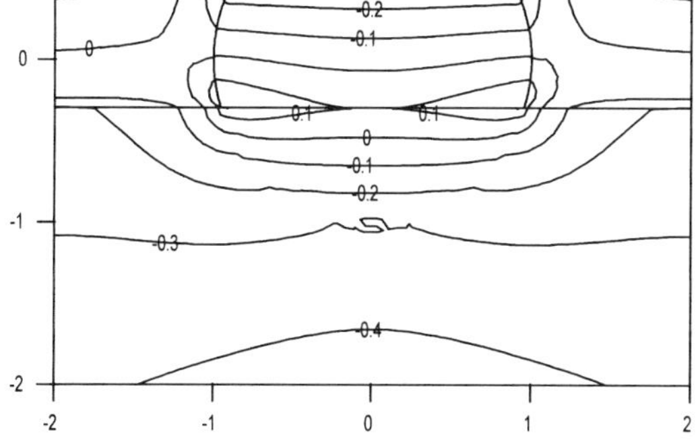

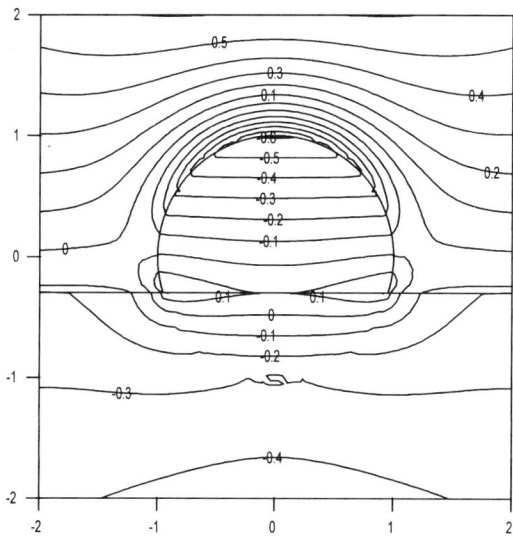

Figure 8.14 Potential distribution of a truncated sphere, $t_r = 0.3$, in a normal electric field.

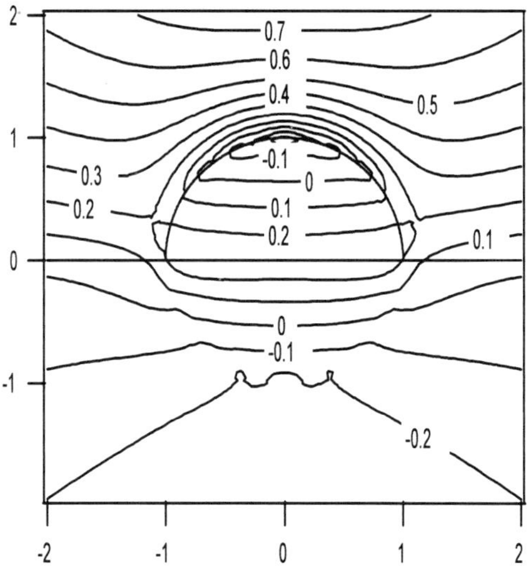

Figure 8.15 Potential distribution of a hemisphere, $t_r = 0.0$, in a normal electric field.

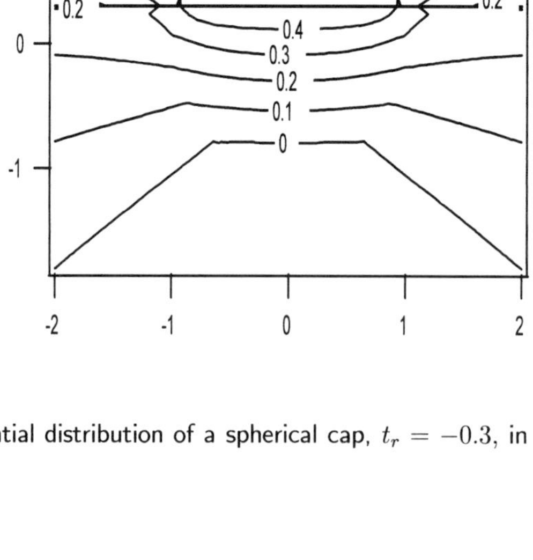

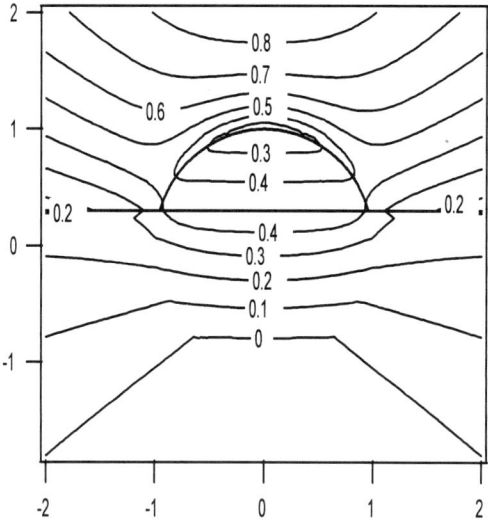

Figure 8.16 Potential distribution of a spherical cap, $t_r = -0.3$, in a normal electric field.

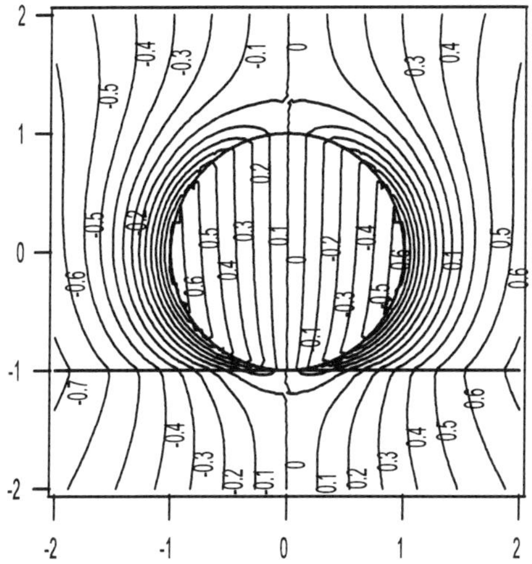

Figure 8.17 Potential distribution of a sphere, $t_r = 1.0$, in a parallel electric field.

An important conclusion is that for a normal field there is no line charge along the three phase line.

When the electric field is parallel to the surface the equipotential planes are no longer symmetric around the symmetry axis. The figures give equipotential lines in a cross section through the symmetry axis in the direction of the electric field. In this plane the potential can be set equal to zero along the z-axis. The value of the potential is then antisymmetric in the direction of the field. The equipotential lines for the sphere touching the substrate have not changed much from the ones for the case without a substrate. The charge is still centered around the two ends of the sphere in the direction of the field like a cosine. There is a shift in the direction of the substrate but not much. For the truncated sphere the charge distribution has changed dramatically. It is now located along the three phase line. This remains the same for the hemisphere and for the cap.

8.7 Appendix A

In this appendix equation (8.53) will be proved. Consider two dipoles, one located in the substrate at $z = -d + 0$, just below the surface $z = -d$ of this substrate, and the other in the ambient at $z = -d - 0$, just above this surface. The dipoles are assumed to produce the same field. It then follows from eqs.(5.31) and (5.32) that the following relations hold for their dipole moments:

$$\begin{aligned} \mathbf{p}(-d-0) + \mathbf{p}_r(-d-0) &= \mathbf{p}_t(-d+0) \\ \mathbf{p}_t(-d-0) &= \mathbf{p}(-d+0) + \mathbf{p}_r(-d+0) \end{aligned} \qquad (8.117)$$

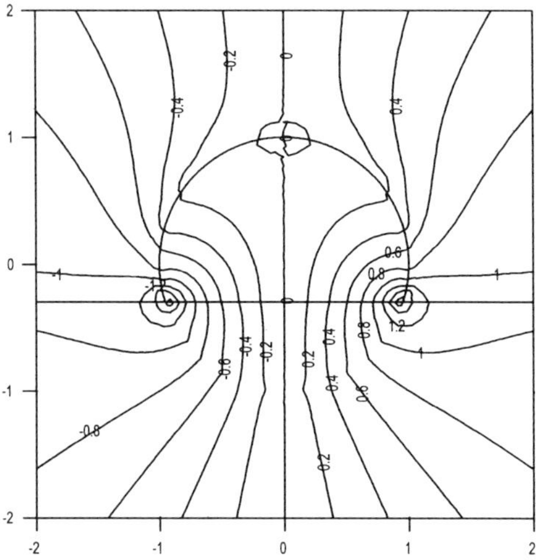

Figure 8.18 Potential distribution of a truncated sphere, $t_r = 0.3$, in a parallel electric field.

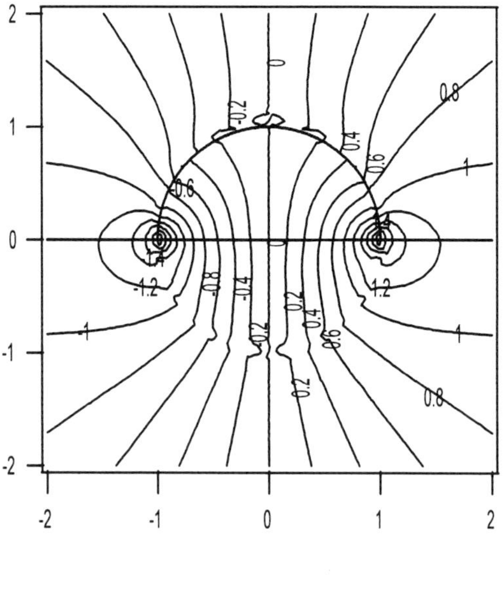

Figure 8.19 Potential distribution of a hemisphere, $t_r = 0.0$, in a parallel electric field.

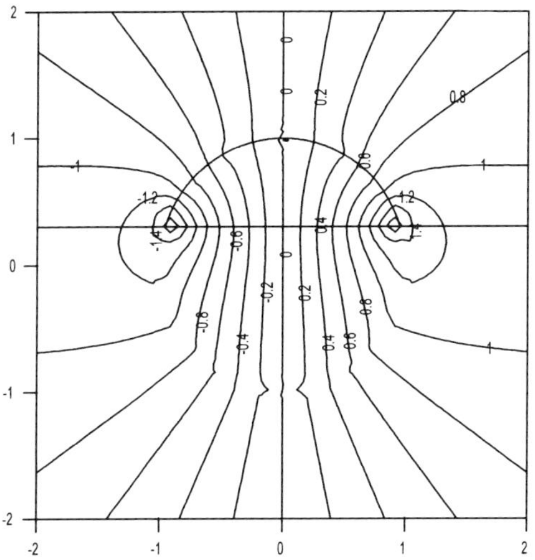

Figure 8.20 Potential distribution of a spherical cap, $t_r = -0.3$, in a parallel electric field.

If one now applies eq.(5.33), which holds for a dipole in the ambient (for a dipole in the substrate the dielectric constants of ambient and substrate have to be interchanged), one finds from eq.(8.117):

$$\begin{aligned}
\mathbf{p}_{\|}(-d-0) + \frac{\epsilon_1 - \epsilon_2}{\epsilon_1 + \epsilon_2}\mathbf{p}_{\|}(-d-0) &= \frac{2\epsilon_1}{\epsilon_1 + \epsilon_2}\mathbf{p}_{\|}(-d+0) \\
p_z(-d-0) - \frac{\epsilon_1 - \epsilon_2}{\epsilon_1 + \epsilon_2}p_z(-d-0) &= \frac{2\epsilon_1}{\epsilon_1 + \epsilon_2}p_z(-d+0) \\
\frac{2\epsilon_2}{\epsilon_1 + \epsilon_2}\mathbf{p}_{\|}(-d-0) &= \mathbf{p}_{\|}(-d+0) + \frac{\epsilon_2 - \epsilon_1}{\epsilon_1 + \epsilon_2}\mathbf{p}_{\|}(-d+0) \\
\frac{2\epsilon_2}{\epsilon_1 + \epsilon_2}p_z(-d-0) &= p_z(-d+0) - \frac{\epsilon_2 - \epsilon_1}{\epsilon_1 + \epsilon_2}p_z(-d+0) \quad (8.118)
\end{aligned}$$

From these relations, it follows that

$$\mathbf{p}_{\|}(-d-0) = \mathbf{p}_{\|}(-d+0), \qquad p_z(-d-0) = \frac{\epsilon_1}{\epsilon_2}p_z(-d+0) \tag{8.119}$$

With eq.(5.35) one now finds

$$\begin{aligned}
\mathbf{p}_{\|}(-d-0) &= \alpha_{\|}(-d-0)\mathbf{E}_{0,\|}, \quad \mathbf{p}_{\|}(-d+0) = \alpha_{\|}(-d+0)\mathbf{E}_{0,\|} \\
p_z(-d-0) &= \alpha_z(-d-0)E_{0,z}, \quad p_z(-d+0) = \alpha_z(-d+0)\frac{\epsilon_1}{\epsilon_2}E_{0,z} \quad (8.120)
\end{aligned}$$

since the z-component of the external field in the substrate is $(\epsilon_1/\ \epsilon_2)$ times the component $E_{0,z}$ in the ambient. From eqs.(8.119) and (8.120), one then derives the first equations (8.53).

The other two equations (8.53) may be derived in a completely analogous way. From eqs.(5.45) and (5.46), and the analogous equations for the transmitted field and quadrupole moment, one obtains (cf. eq.(8.117)):

$$\begin{aligned}\mathbf{Q}(-d-0)+\mathbf{Q}_r(-d-0) &= \mathbf{Q}_t(-d+0)\\ \mathbf{Q}_t(-d-0) &= \mathbf{Q}(-d+0)+\mathbf{Q}_r(-d+0)\end{aligned} \tag{8.121}$$

For Q_z , defined in eq.(5.51), and Q_{zx} , defined in eq.(5.56), one then finds the following relations (cf. eq.(8.119)):

$$Q_z(-d-0)=Q_z(-d+0),\quad Q_{zx}(-d-0)=\frac{\epsilon_1}{\epsilon_2}Q_{zx}(-d+0) \tag{8.122}$$

Using eqs.(5.52), (5.55), (5.57) and (5.59), one has furthermore:

$$\begin{aligned}Q_z(-d-0) &= \alpha_z^{10}(-d-0)E_{0,z},\quad Q_z(-d+0)=\alpha_z^{10}(-d+0)\frac{\epsilon_1}{\epsilon_2}E_{0,z}\\ Q_{zx}(-d-0) &= \alpha_\parallel^{10}(-d-0)E_{0,x},\quad Q_{zx}(-d+0)=\alpha_\parallel^{10}(-d+0)E_{0,x}\end{aligned} \tag{8.123}$$

From eqs.(8.122) and (8.123) one then also obtains the last two equations (8.53).

8.8 Appendix B

In this appendix eqs.(8.77) – (8.80) for the multipole coefficients $A_{\ell 0}$ and $A_{\ell 1}$ in the case of hemispheres on a flat substrate will be derived. Using the results, eqs.(8.72) and (8.73), valid for hemispheres, in the general equation (8.29), one obtains the following set of equations for $m=0$:

$$\begin{aligned}&\left(\frac{2\epsilon_1}{\epsilon_1+\epsilon_2}\right)R^{-\ell-2}A_{\ell 0}-2\left(\frac{\epsilon_1-\epsilon_2}{\epsilon_1+\epsilon_2}\right)\sum_{\ell_1\ odd}^{M}\zeta^0_{\ell\ell_1}Q^0_{\ell\ell_1}(0)R^{-\ell_1-2}A_{\ell_1,0}\\ &-\left(\frac{2\epsilon_3}{\epsilon_2+\epsilon_3}\right)R^{\ell-1}B_{\ell 0}-2\left(\frac{\epsilon_2-\epsilon_3}{\epsilon_2+\epsilon_3}\right)\sum_{\ell_1\ odd}^{M}\zeta^0_{\ell\ell_1}Q^0_{\ell\ell_1}(0)R^{\ell_1-1}B_{\ell_1,0}\\ =\ &\sqrt{4\pi/3}\left[\frac{\epsilon_1}{\epsilon_2}\delta_{\ell 1}-\left(\frac{\epsilon_1-\epsilon_2}{\epsilon_2}\right)\zeta^0_{\ell 1}Q^0_{\ell 1}(0)\right]E_0\cos\theta_0\end{aligned} \tag{8.124}$$

The sums are over $\ell_1=1,3,...$until the largest odd number smaller than or equal to M. Furthermore

$$\begin{aligned}&-\left(\frac{2\epsilon_1\epsilon_2}{\epsilon_1+\epsilon_2}\right)(\ell+1)R^{-\ell-2}A_{\ell 0}\\ &-2\epsilon_1\left(\frac{\epsilon_1-\epsilon_2}{\epsilon_1+\epsilon_2}\right)\sum_{\ell_1\ even}^{M}(\ell_1+1)\zeta^0_{\ell\ell_1}Q^0_{\ell\ell_1}(0)R^{-\ell_1-2}A_{\ell_1,0}\\ &-\left(\frac{2\epsilon_2\epsilon_3}{\epsilon_2+\epsilon_3}\right)\ell\ R^{\ell-1}B_{\ell 0}+2\epsilon_3\left(\frac{\epsilon_2-\epsilon_3}{\epsilon_2+\epsilon_3}\right)\sum_{\ell_1\ even}^{M}\ell_1\zeta^0_{\ell\ell_1}Q^0_{\ell\ell_1}(0)R^{\ell_1-1}B_{\ell_1,0}\\ =\ &\sqrt{4\pi/3}\epsilon_1\delta_{\ell 1}E_0\cos\theta_0\end{aligned} \tag{8.125}$$

The sums are over $\ell_1 = 2, 4, ...$until the largest even number smaller than or equal to M. It can now easily be derived, by applying eqs.(8.75) and (8.14) to eq.(8.124) for odd values of ℓ, and to eq.(8.125) for even values of ℓ, that the following relations hold

$$\begin{aligned} R^{\ell-1}B_{\ell 0} &= R^{-\ell-2}A_{\ell 0} - \sqrt{4\pi/3}\left(\frac{\epsilon_1+\epsilon_2}{2\epsilon_2}\right)\delta_{\ell 1}E_0\cos\theta_0 \\ \text{for } \ell &= 1,3,5,... \qquad (8.126) \\ R^{\ell-1}B_{\ell 0} &= -\left(\frac{\ell+1}{\ell}\right)\frac{\epsilon_1}{\epsilon_3}R^{-\ell-2}A_{\ell 0} \text{ for } \ell = 2,4,6,..... \qquad (8.127) \end{aligned}$$

With the help of these relations, one can eliminate the multipole coefficients $B_{\ell 0}$ from eqs.(8.124) and (8.125). Applying eqs.(8.126) and (8.127) to eq.(8.125) for odd values of ℓ, one then obtains eq.(8.77). Application of eqs.(8.126) and (8.127) to eq.(8.124) for even values of ℓ gives eq.(8.78).

For $m = 1$ one obtains, by using eqs.(8.72) and (8.73) in eq.(8.29), the following set of equations:

$$\begin{aligned} &\left(\frac{2\epsilon_1}{\epsilon_1+\epsilon_2}\right)R^{-\ell-2}A_{\ell 1} - 2\left(\frac{\epsilon_1-\epsilon_2}{\epsilon_1+\epsilon_2}\right)\sum_{\ell_1\, even}^{M}\zeta^1_{\ell\ell_1}Q^1_{\ell\ell_1}(0)R^{-\ell_1-2}A_{\ell_1,1} \\ &-\left(\frac{2\epsilon_3}{\epsilon_2+\epsilon_3}\right)R^{\ell-1}B_{\ell 1} - 2\left(\frac{\epsilon_2-\epsilon_3}{\epsilon_2+\epsilon_3}\right)\sum_{\ell_1\, even}^{M}\zeta^1_{\ell\ell_1}Q^1_{\ell\ell_1}(0)R^{\ell_1-1}B_{\ell_1,1} \\ = &-\sqrt{2\pi/3}\delta_{\ell 1}E_0\sin\theta_0\exp(-i\phi_0) \qquad (8.128) \end{aligned}$$

and

$$\begin{aligned} &-\left(\frac{2\epsilon_1\epsilon_2}{\epsilon_1+\epsilon_2}\right)(\ell+1)R^{-\ell-2}A_{\ell 1} \\ &-2\epsilon_1\left(\frac{\epsilon_1-\epsilon_2}{\epsilon_1+\epsilon_2}\right)\sum_{\ell_1\, odd}^{M}(\ell_1+1)\zeta^1_{\ell\ell_1}Q^1_{\ell\ell_1}(0)R^{-\ell_1-2}A_{\ell_1,1} \\ &-\left(\frac{2\epsilon_2\epsilon_3}{\epsilon_2+\epsilon_3}\right)\ell\, R^{\ell-1}B_{\ell 1} + 2\epsilon_3\left(\frac{\epsilon_2-\epsilon_3}{\epsilon_2+\epsilon_3}\right)\sum_{\ell_1\, odd}^{M}\ell_1\zeta^1_{\ell\ell_1}Q^1_{\ell\ell_1}(0)R^{\ell_1-1}B_{\ell_1,1} \\ = &-\sqrt{2\pi/3}[\epsilon_2\delta_{\ell 1} + (\epsilon_1-\epsilon_2)\zeta^1_{\ell 1}Q^1_{\ell 1}(0)]E_0\sin\theta_0\exp(-i\phi_0) \qquad (8.129) \end{aligned}$$

By applying eqs.(8.75) and (8.14) to eq.(8.129) for odd values of ℓ, and to eq.(8.128) for even values of ℓ, one finds the relations

$$\begin{aligned} R^{\ell-1}B_{\ell 1} &= -\left(\frac{\ell+1}{\ell}\right)\frac{\epsilon_1}{\epsilon_3}R^{-\ell-2}A_{\ell 1} + \frac{1}{2}\sqrt{2\pi/3}\left(\frac{\epsilon_1+\epsilon_2}{\epsilon_3}\right)\delta_{\ell 1}E_0\sin\theta_0\exp(-i\phi_0) \\ &\text{for } \ell = 1,3,5,..... \qquad (8.130) \\ R^{\ell-1}B_{\ell 1} &= R^{-\ell-2}A_{\ell 1} \text{ for } \ell = 2,4,6,..... \qquad (8.131) \end{aligned}$$

Using these relations, one can now eliminate the multipole coefficients $B_{\ell 1}$ from eqs.(8.128) and (8.129). Applying eqs.(8.130) and (8.131) to eq.(8.128) for odd values of ℓ, one then obtains eq.(8.79), and applying these equations to eq.(8.129) for even values of ℓ, one finds eq.(8.80).

Chapter 9

FILMS OF TRUNCATED SPHEROIDS IN THE LOW COVERAGE LIMIT

9.1 Introduction

In the previous chapter the optical properties were studied of thin island films, consisting of identical truncated spherical particles on a flat substrate. The size of the islands was assumed to small compared to the wave length of the light, so that the Laplace equation (8.1) for the electric potential could be used. This potential was expanded in terms of spherical harmonics within and outside the islands. Using the boundary conditions for the electric potential and its normal derivative on the surface of the sphere and of the substrate, an infinite set of inhomogeneous linear equations for the multipole coefficients in this expansion was derived. Approximate solutions could be obtained by truncating this set and solving the now finite set of linear equations for a finite number of multipole coefficients.

By comparing the results, obtained in section 8.6 for truncated spherical gold particles on a sapphire substrate with those, obtained in section 6.6 for spheroidal gold particles on the same substrate, it is observed that e.g. the absorption of light for particles of the same volume and the same ratio of the diameter of the particle as viewed from above and its height is much larger for truncated spheres than for oblate spheroids. As a consequence of the finite angle between the side of the truncated particle and its flat bottom part, the electric field inside this particle is affected more strongly than in the corresponding oblate spheroid. Since the results obtained with the two models are rather different, in the present chapter a new model will be introduced, which contains the two models as special cases. It is the model of a truncated spheroidal particle on a substrate. Because part of the island is spheroidal, it is now most convenient to use a spheroidal multipole expansion of the potential, as this was also used in sections 6.4 and 6.5. Analogously to the truncated sphere case, cf. section 8.2, an infinite set of inhomogeneous linear equations will be derived, for the coefficients in the spheroidal multipole expansion. The matrix elements in these equations are integrals over products of Legendre functions of the first and second kind, which can be calculated numerically. By truncating this infinite set, it can be solved with sufficient accuracy, if enough equations and matrix elements are used. In sections 9.2 and 9.3 truncated oblate and prolate spheroids are considered. In sections 9.4 and 9.5 oblate and prolate spheroidal caps will be studied.

The truncated spheroid model combines the effects of the truncation with the effects of the spheroidal deformation of the islands. In the interpretation of experimental result the hope is that one may, on the basis of the observed resonance behaviour, be able to decide to what extent the behaviour is a consequence of the truncation and to what extent it is a consequence of the spheroidal deformation. The lack of such a model made it difficult to come to a conclusion regarding these matters for the results of Craighead and Niklasson, [34], [35], [53], [42].

In section 9.6 two special cases will be considered, where the above mentioned integrals can be evaluated analytically. First it is shown, that the set of inhomogeneous linear equations, derived in sections 9.2 and 9.3, reduce, for the case of a spheroid touching the substrate, to the set of equations derived in sections 6.4 and 6.5, respectively. Furthermore the case of a hemispheroid on a substrate is considered in this section.

In the general case of truncated spheroids or spheroidal caps on a substrate the evaluation of the above mentioned matrix elements can only be done by numerical integration, using well known recurrence relations to calculate the Legendre functions of the first and second kind, appearing in the integrals in these matrix elements. In section 9.7, the theory is applied again to the special case of a gold island on a sapphire substrate.

9.2 Truncated oblate spheroids on a substrate

The truncated spheroidal model consists of a spheroid, with its axis of revolution normal to the surface of the substrate, and truncated by a plane, parallel to this surface (see figs.9.1 and 9.2 below). In this model the truncation parameter, as well as the eccentricity of the spheroids may be changed independently. Particles with the same ratio of the diameter, as viewed from above, over the height, may still have different values of the truncation parameter and the eccentricity. The model contains two independent parameters, which characterize the optical properties of the islands. Both the truncated spherical model of islands, treated in the previous chapter, and the spheroidal model of chapter 6 are special (limiting) cases of the truncated spheroidal model. The theory will be given for truncated oblate spheroids in this section. Truncated prolate spheroids will be treated in the next section. As will be seen below, the mathematics is a combination of that used in sections 6.4 and 8.2. Because the island is a part of a spheroid (see figs.9.1 and 9.2), it will be most convenient to use a spheroidal multipole expansion of the potential ψ in the Laplace equation (8.1), as was also done in section 6.4.

Consider a truncated oblate spheroid with a frequency dependent dielectric constant $\epsilon_3(\omega)$ on a substrate with a dielectric constant $\epsilon_2(\omega)$ and surrounded by an ambient with dielectric constant $\epsilon_1(\omega)$. The same notation will be used as in the previous chapter. The axis of rotation of the oblate spheroid is supposed to be normal to the surface of the substrate which is, as usual, planar. A Cartesian coordinate system is chosen with origin O at the center of the oblate spheroid and with positive z-axis along the axis of rotation, pointing downward into or in the substrate (see figs.9.1 and 9.2 below). The surface of the substrate is given by $z = \pm d$, with $0 \leq d \leq R_s$; R_s is the length of half the short axis of the spheroid. In fig.9.1 a situation has been drawn, where the center of the spheroid lies above the surface

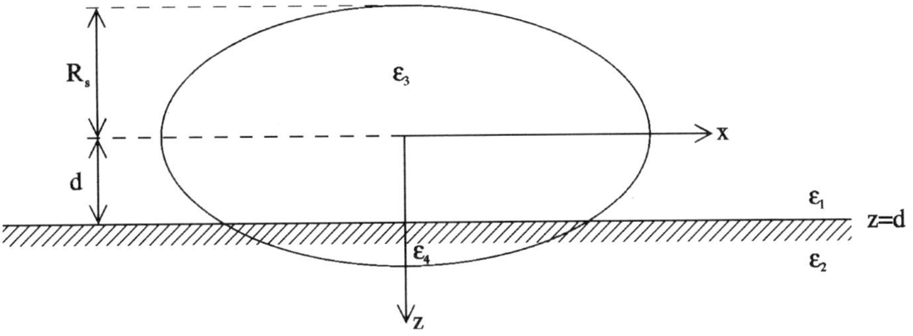

Figure 9.1 Cross section of a truncated oblate spheroid, with center above the substrate.

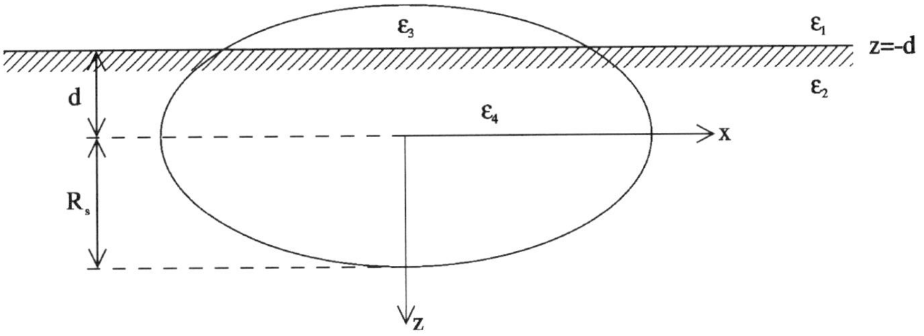

Figure 9.2 Cross section of a truncated oblate spheroid, with center in the substrate (to be referred to as spheroidal cap).

$(z = d)$ of the substrate, and in fig.9.2 the case that this center lies in the substrate below this surface $(z = -d)$. The latter case will be referred to as a spheroidal cap, whereas from now on the name truncated spheroid will only be used for the case that the center of the spheroid lies above the substrate. Just as in the previous chapter, it will be convenient in the following analysis to replace the substrate in region 4 by a different material with dielectric constant $\epsilon_4(\omega)$. It will then again be clear, that the two situations drawn in fig.9.1 and fig.9.2 may be obtained from each other, if one interchanges the dielectric constants ϵ_1 and ϵ_2 (of the ambient and the substrate) and also ϵ_3 and ϵ_4 (of the particle and of the material of the spheroid within the substrate). In the present section only the case of the truncated oblate spheroid will be studied. The case of an oblate spheroidal cap will be treated in the section 9.4. Truncated prolate spheroids and spheroidal caps will be treated in sections 9.3 and 9.5, respectively.

Since the islands are assumed to be small as compared to the minimum wave length of the incident radiation in the media involved, one may consider the incident field to be again homogeneous and use the Laplace equation, eq.(8.1), for the potential $\psi(\mathbf{r})$ in the particle, the ambient and the substrate. At the surface of the island and

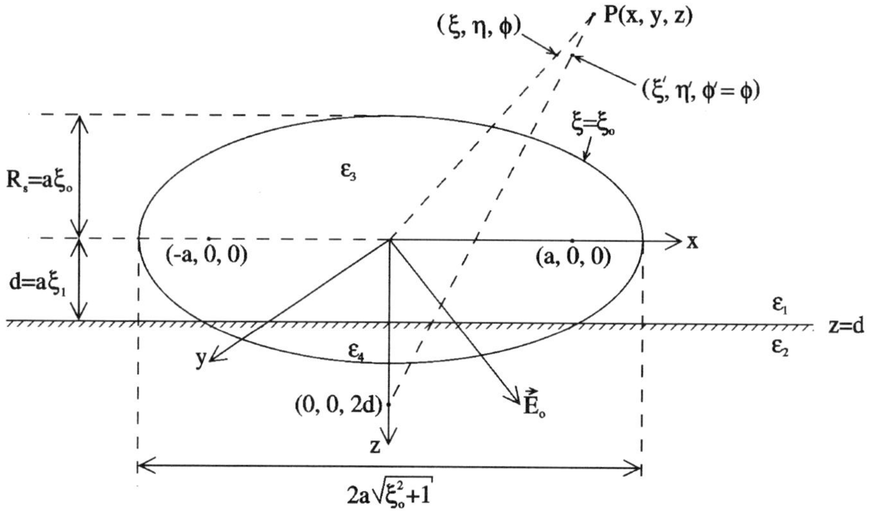

Figure 9.3 Truncated oblate spheroid on substrate.

the substrate, the potential and the dielectric constant times the normal derivative of this potential must be continuous. In order to solve this boundary value problem in the present case, oblate spheroidal coordinates ξ, η and ϕ $(0 \leq \xi < \infty,\ -1 \leq \eta \leq 1,$ $0 \leq \phi < 2\pi)$ will be introduced. These coordinates are defined by eqs.(6.1) and (6.2) or by the inverse transformation, eq.(6.3), (see section 6.2 for further explanation). The interface between the island (region 3, with dielectric constant ϵ_3) and the ambient (region 1, with dielectric constant ϵ_1), as well as between the regions 2, with dielectric constant ϵ_2, and 4, with dielectric constant ϵ_4, in the substrate, is then given by $\xi = \xi_0$, (see fig. 9.3). The flat surface of the substrate is given by $z = a\xi\eta = d = a\xi_1$(cf. eq.(6.101)).

More in particular one can distinguish the following four interfaces:

$$\begin{aligned}
&\text{Interface between regions 1 and 3} &:\quad & \xi = \xi_0, \quad -1 \leq \eta \leq \frac{\xi_1}{\xi_0} \\
&\text{Interface between regions 2 and 4} &:\quad & \xi = \xi_0, \quad \frac{\xi_1}{\xi_0} \leq \eta \leq 1 \\
&\text{Interface between regions 3 and 4} &:\quad & \eta = \frac{\xi_1}{\xi}, \quad \xi_1 \leq \xi \leq \xi_0 \\
&\text{Interface between regions 1 and 2} &:\quad & \eta = \frac{\xi_1}{\xi}, \quad \xi_0 \leq \xi < \infty
\end{aligned} \tag{9.1}$$

Notice, that it follows from the inequality $0 \leq d \leq R_s$, that $0 \leq \xi_1 \leq \xi_0$ (see fig.9.3). The quantity

$$t_r \equiv \frac{d}{a\xi_0} = \frac{d}{R_s} = \frac{\xi_1}{\xi_0} \quad \text{with} \quad 0 \leq t_r \leq 1 \tag{9.2}$$

is again introduced as a truncation parameter ($t_r = 0$ for oblate hemispheroids and $t_r = 1$ for spheroids touching the substrate). Since truncated spheres are the limiting case of truncated oblate spheroids, $a\xi_0 = R_s \to R$, the above definition of t_r reduces to the one given in eq.(8.11).

Just as in the previous chapters, image multipoles will now be introduced in (0,0,2d) to describe the fields due to the charge distribution in the substrate underneath the island (see fig.9.3). In the present case, these will be oblate spheroidal multipoles (cf. section 6.4). The potential in region 1 (ambient) can be written in oblate spheroidal coordinates as

$$\begin{aligned}
&\psi_1(\mathbf{r}) = (2\pi/3)^{1/2}[-2^{1/2}E_{0,z}X_1^0(\xi,a)Y_1^0(\arccos\eta,\phi) \\
&+(E_{0,x}-iE_{0,y})X_1^1(\xi,a)Y_1^1(\arccos\eta,\phi)-(E_{0,x}+iE_{0,y})X_1^{-1}(\xi,a)Y_1^{-1}(\arccos\eta,\phi)] \\
&+\sum_{\ell m}{}' A_{\ell m}Z_\ell^m(\xi,a)Y_\ell^m(\arccos\eta,\phi)+\sum_{\ell m}{}' A_{\ell m}^r Z_\ell^m(\xi'(\xi,\eta),a)Y_\ell^m(\arccos\eta'(\xi,\eta),\phi) \\
&\quad \text{for } \xi_0 \le \xi < \infty \text{ and } \eta \le \xi_1/\xi
\end{aligned} \tag{9.3}$$

(cf. eq.(8.2)). The first term at the right-hand side is the potential of the constant electric field $\mathbf{E}_0 = (E_{0,x}, E_{0,y}, E_{0,z})$ in the ambient (cf. eqs.(6.97) and (6.98)). The second term (first sum) at the right-hand side is the potential due to the induced charge distribution in the island and the third term (second sum) is that of the image charge distribution in the substrate. The prime as super index in these sums again indicates that $\ell \neq 0$. The direct field is described here by spheroidal multipoles, located at O, and the field of the image charge distribution by spheroidal image multipoles in (0,0,2d), (see fig.9.3). The functions X_ℓ^m, Z_ℓ^m and Y_ℓ^m are defined by eqs.(6.13), (6.14) and (6.5), together with eqs.(6.6)–(6.8), (6.10) and (6.11). Finally ξ' and η' are the first two oblate spheroidal coordinates of the point (x, y, z) in the shifted coordinate frame with origin at (0,0,2d), (see fig. 9.3). The functions $\xi'(\xi,\eta)$ and $\eta'(\xi,\eta)$ are found from eq.(6.3), by noting, that $x'^2+y'^2 = x^2+y^2$ and $z' = z-2d$ for this coordinate transformation. One obtains

$$\begin{aligned}
\xi'(\xi,\eta) &= \frac{1}{2}\sqrt{2}\xi\left\{1+4\frac{\xi_1^2}{\xi^2}-4\frac{\xi_1}{\xi}\eta-\frac{\eta^2}{\xi^2}\right. \\
&\quad \left.+\left[\left(1+4\frac{\xi_1^2}{\xi^2}-4\frac{\xi_1}{\xi}\eta-\frac{\eta^2}{\xi^2}\right)^2+4\frac{1}{\xi^2}\left(2\frac{\xi_1}{\xi}-\eta\right)^2\right]^{1/2}\right\}^{1/2}
\end{aligned} \tag{9.4}$$

$$\begin{aligned}
\eta'(\xi,\eta) &= \sqrt{2}\left(\eta-2\frac{\xi_1}{\xi}\right)\left\{1+4\frac{\xi_1^2}{\xi^2}-4\frac{\xi_1}{\xi}\eta-\frac{\eta^2}{\xi^2}\right. \\
&\quad \left.+\left[\left(1+4\frac{\xi_1^2}{\xi^2}-4\frac{\xi_1}{\xi}\eta-\frac{\eta^2}{\xi^2}\right)^2+4\frac{1}{\xi^2}\left(2\frac{\xi_1}{\xi}-\eta\right)^2\right]^{1/2}\right\}^{-1/2}
\end{aligned} \tag{9.5}$$

The transformation

$$\xi' = \xi'(\xi,\eta),\ \ \eta' = \eta'(\xi,\eta),\ \ \phi' = \phi \tag{9.6}$$

is the generalization for oblate spheroidal coordinates of the transformation, eq.(8.21), valid for spherical coordinates. In fact, it is easily proved by using the asymptotic relations, eqs.(6.19) and (6.20), that eq.(9.6) with eq.(9.4) and (9.5) becomes equal to eq.(8.21) in the limit as $\xi \to \infty$, $\xi_1 \to \infty$, $a \to 0$, with $a\xi_1 = d$.

For the potential in the region 2 (substrate) one may write in oblate spheroidal coordinates

$$\begin{aligned} \psi_2(\mathbf{r}) &= \psi_2' + c_1 X_1^0(\xi,a)Y_1^0(\arccos\eta,\phi) \\ &+ c_2 X_1^1(\xi,a)Y_1^1(\arccos\eta,\phi) + c_3 X_1^{-1}(\xi,a)Y_1^{-1}(\arccos\eta,\phi) \\ &+ \sum_{\ell m}' A_{\ell m}^t Z_\ell^m(\xi,a)Y_\ell^m(\arccos\eta,\phi) \text{ for } \xi_0 \le \xi < \infty,\ \eta \ge \xi_1/\xi \end{aligned} \tag{9.7}$$

(cf. eq.(6.100). Here ψ_2', c_1, c_2 and c_3 are constants, which will be determined below. Furthermore one can write the potential in the region 3 (island) and 4 (substrate) as follows:

$$\begin{aligned} \psi_3(\mathbf{r}) &= \psi_3' + \sum_{\ell m}' B_{\ell m} X_\ell^m(\xi,a)Y_\ell^m(\arccos\eta,\phi) \\ &+ \sum_{\ell m}' B_{\ell m}^r X_\ell^m(\xi'(\xi,\eta),a)Y_\ell^m(\arccos\eta'(\xi,\eta),\phi) \\ &\text{for } 0 \le \xi \le \xi_0,\ \eta \le \xi_1/\xi \end{aligned} \tag{9.8}$$

$$\psi_4(\mathbf{r}) = \psi_4' + \sum_{\ell m}' B_{\ell m}^t X_\ell^m(\xi,a)Y_\ell^m(\arccos\eta,\phi) \text{ for } 0 \le \xi \le \xi_0,\ \eta \ge \xi_1/\xi \tag{9.9}$$

where ψ_3' and ψ_4' are constants, which will be determined below.

The constants in the above potentials, eqs.(9.3) and (9.7), are found by applying the boundary conditions $\psi_1 = \psi_2$ and $\epsilon_1\partial\psi_1/\partial z = \epsilon_2\partial\psi_2/\partial z$ at the interface between the regions 1 and 2, which is given by $\eta = \xi_1/\xi$ and $\xi_0 \le \xi < \infty$ (cf. eq.(9.1)). Using eqs.(6.102) and (6.103) one then finds from eqs.(9.3) and (9.7):

$$\begin{aligned} c_1 &= -\frac{\epsilon_1}{\epsilon_2}\left(\frac{4\pi}{3}\right)^{1/2} E_{0,z}, \quad c_2 = \left(\frac{2\pi}{3}\right)^{1/2}(E_{0,x} - iE_{0,y}) \\ c_3 &= -\left(\frac{2\pi}{3}\right)^{1/2}(E_{0,x} + iE_{0,y}) \end{aligned} \tag{9.10}$$

and

$$\psi_2' = -\left(1 - \frac{\epsilon_1}{\epsilon_2}\right) d\, E_{0,z} \tag{9.11}$$

(cf. eqs.(6.104) and (6.106). Furthermore the following relations between the multipole coefficients $A_{\ell m}$, $A_{\ell m}^r$ and $A_{\ell m}^t$ in these potentials are found:

$$A_{\ell m}^r = (-1)^{\ell+m}\left(\frac{\epsilon_1 - \epsilon_2}{\epsilon_1 + \epsilon_2}\right) A_{\ell m}, \quad A_{\ell m}^t = \frac{2\epsilon_1}{\epsilon_1 + \epsilon_2} A_{\ell m} \tag{9.12}$$

(cf. eqs.(6.105) and (8.7)).

Applying the boundary conditions $\psi_3 = \psi_4$ and $\epsilon_3 \partial\psi_3/\partial z = \epsilon_4 \partial\psi_4/\partial z$ to the potentials, eqs.(9.8) and (9.9), at the interface between the regions 3 and 4, which is given by $\eta = \xi_1/\xi$ and $\xi_1 \leq \xi \leq \xi_0$ (cf. eq.(9.1)), and using again eqs.(6.102) and (6.103), one gets

$$\psi_3' = \psi_4' \equiv \psi_0 \tag{9.13}$$

(cf. eqs.(8.5) and (8.6)). Furthermore the relations

$$B^r_{\ell m} = (-1)^{\ell+m}\left(\frac{\epsilon_3 - \epsilon_4}{\epsilon_3 + \epsilon_4}\right) B_{\ell m}, \quad B^t_{\ell m} = \frac{2\epsilon_3}{\epsilon_3 + \epsilon_4} B_{\ell m} \tag{9.14}$$

are obtained, between the multipole coefficients $B_{\ell m}$, $B^r_{\ell m}$ and $B^t_{\ell m}$ in these potentials (cf. eq.(8.8)).

At the surface of the oblate spheroid $\xi = \xi_0$, one can use that the oblate spherical harmonics $Y_\ell^m(\arccos\eta, \phi)$ form a complete orthonormal set of functions of η and ϕ. Multiplication of the boundary conditions on the surface of the oblate spheroid, between regions 1 and 3 and between 2 and 4, with complex conjugated harmonics and integration over η and ϕ, gives

$$\begin{aligned} &\int_{-1}^{t_r} d\eta \int_0^{2\pi} d\phi (\psi_1 - \psi_3)_{\xi=\xi_0} [Y_{\ell'}^{m'}(\arccos\eta, \phi)]^* \\ &+ \int_{t_r}^{1} d\eta \int_0^{2\pi} d\phi (\psi_2 - \psi_4)_{\xi=\xi_0} [Y_{\ell'}^{m'}(\arccos\eta, \phi)]^* = 0 \end{aligned} \tag{9.15}$$

and

$$\begin{aligned} &\int_{-1}^{t_r} d\eta \int_0^{2\pi} d\phi \left[\frac{\partial}{\partial\xi}(\epsilon_1\psi_1 - \epsilon_3\psi_3)_{\xi=\xi_0}\right] [Y_{\ell'}^{m'}(\arccos\eta, \phi)]^* \\ &+ \int_{t_r}^{1} d\eta \int_0^{2\pi} d\phi \left[\frac{\partial}{\partial\xi}(\epsilon_2\psi_2 - \epsilon_4\psi_4)_{\xi=\xi_0}\right] [Y_{\ell'}^{m'}(\arccos\eta, \phi)]^* = 0 \end{aligned} \tag{9.16}$$

(cf. eqs.(8.9) and (8.10)).

Upon substitution of the potentials, eqs.(9.3), (9.7)–(9.9), into eqs.(9.15) and (9.16), using also eqs.(9.10)–(9.14), one obtains the following set of inhomogeneous linear equations for the multipole coefficients $A_{\ell m}$ and $B_{\ell m}$ (cf. eq.(8.12)):

$$\begin{aligned} \sum_{\ell_1=|m|}^{\infty\prime} C^m_{\ell\ell_1} R_s^{-\ell_1-2} A_{\ell_1 m} + \sum_{\ell_1=|m|}^{\infty\prime} D^m_{\ell\ell_1} R_s^{\ell_1-1} B_{\ell_1 m} &= H^m_\ell \\ \sum_{\ell_1=|m|}^{\infty\prime} F^m_{\ell\ell_1} R_s^{-\ell_1-2} A_{\ell_1 m} + \sum_{\ell_1=|m|}^{\infty\prime} G^m_{\ell\ell_1} R_s^{\ell_1-1} B_{\ell_1 m} &= J^m_\ell \end{aligned} \tag{9.17}$$

where $\ell = 0, 1, 2, 3, ...$ and $m = 0, \pm1, \pm2, \pm3, ..., \pm\ell$; $R_s = a\xi_0$ is half the length of the short axis of the oblate spheroid (see fig.9.3). The prime as superindex of the summation symbols denotes again, that the term with $\ell_1 = 0$ should be omitted.

Using the definition, eq.(6.5), of the spherical harmonics and their normalization and orthogonality relations (cf. eq.(5.29)):

$$\int_{-1}^{1} d\eta \int_{0}^{2\pi} d\phi [Y_\ell^m(\arccos\eta, \phi)]^* Y_{\ell'}^{m'}(\arccos\eta, \phi) = \delta_{\ell\ell'}\delta_{mm'} \tag{9.18}$$

one finds

$$\begin{aligned}
C_{\ell\ell_1}^m &\equiv \left(\frac{2\epsilon_1}{\epsilon_1+\epsilon_2}\right)\xi_0^{\ell_1+1} Z_{\ell_1}^m(\xi_0)\delta_{\ell\ell_1} \\
&\quad - \left(\frac{\epsilon_1-\epsilon_2}{\epsilon_1+\epsilon_2}\right)\zeta_{\ell\ell_1}^m \xi_0^{\ell_1+1}[Z_{\ell_1}^m(\xi_0)Q_{\ell\ell_1}^m(t_r) - (-1)^{\ell_1+m}V_{\ell\ell_1}^m(\xi_0,t_r)] \\
D_{\ell\ell_1}^m &\equiv -\left(\frac{2\epsilon_3}{\epsilon_3+\epsilon_4}\right)\xi_0^{-\ell_1} X_{\ell_1}^m(\xi_0)\delta_{\ell\ell_1} \\
&\quad + \left(\frac{\epsilon_3-\epsilon_4}{\epsilon_3+\epsilon_4}\right)\zeta_{\ell\ell_1}^m \xi_0^{-\ell_1}[X_{\ell_1}^m(\xi_0)Q_{\ell\ell_1}^m(t_r) - (-1)^{\ell_1+m}W_{\ell\ell_1}^m(\xi_0,t_r)] \\
F_{\ell\ell_1}^m &\equiv \left(\frac{2\epsilon_1\epsilon_2}{\epsilon_1+\epsilon_2}\right)\xi_0^{\ell_1+2}\frac{dZ_{\ell_1}^m(\xi_0)}{d\xi_0}\delta_{\ell\ell_1} \\
&\quad + \epsilon_1\left(\frac{\epsilon_1-\epsilon_2}{\epsilon_1+\epsilon_2}\right)\zeta_{\ell\ell_1}^m \xi_0^{\ell_1+2}\left[\frac{dZ_{\ell_1}^m(\xi_0)}{d\xi_0}Q_{\ell\ell_1}^m(t_r) + (-1)^{\ell_1+m}\frac{\partial V_{\ell\ell_1}^m(\xi_0,t_r)}{\partial\xi_0}\right] \\
G_{\ell\ell_1}^m &\equiv -\left(\frac{2\epsilon_3\epsilon_4}{\epsilon_3+\epsilon_4}\right)\xi_0^{-\ell_1+1}\frac{dX_{\ell_1}^m(\xi_0)}{d\xi_0}\delta_{\ell\ell_1} - \epsilon_3\left(\frac{\epsilon_3-\epsilon_4}{\epsilon_3+\epsilon_4}\right)\zeta_{\ell\ell_1}^m\xi_0^{-\ell_1+1} \\
&\quad \times \left[\frac{dX_{\ell_1}^m(\xi_0)}{d\xi_0}Q_{\ell\ell_1}^m(t_r) + (-1)^{\ell_1+m}\frac{\partial W_{\ell\ell_1}^m(\xi_0,t_r)}{\partial\xi_0}\right]
\end{aligned} \tag{9.19}$$

Here

$$X_\ell^m(\xi) \equiv X_\ell^m(\xi,a)a^{-\ell} = i^{m-\ell}\frac{(\ell-m)!}{(2\ell-1)!!}P_\ell^m(i\xi) \tag{9.20}$$

and

$$Z_\ell^m(\xi) \equiv Z_\ell^m(\xi,a)a^{\ell+1} = i^{\ell+1}\frac{(2\ell+1)!!}{(\ell+m)!}Q_\ell^m(i\xi) \tag{9.21}$$

where the definitions, eqs.(6.13) and (6.14), have been used. Here $(n)!! \equiv 1\times3\times\ldots\times(n-2)\times n$ for n odd and $2\times4\times\ldots\times(n-2)\times n$ for n even; by definition $(-1)!! \equiv 1$. Furthermore $Q_{\ell\ell_1}^m(t_r)$ is defined by (cf. eq.(8.15))

$$Q_{\ell\ell_1}^m(t_r) \equiv \int_{-1}^{t_r} P_\ell^m(\eta)P_{\ell_1}^m(\eta)d\eta \tag{9.22}$$

and $\zeta_{\ell\ell_1}^m$ by eq.(8.14). The matrices $V_{\ell\ell_1}^m(\xi,t_r)$ and $W_{\ell\ell_1}^m(\xi,t_r)$ are defined by the following equations:

$$V_{\ell\ell_1}^m(\xi,t_r) \equiv \int_{-1}^{t_r} P_\ell^m(\eta)P_{\ell_1}^m(\eta'(\xi,\eta))Z_{\ell_1}^m(\xi'(\xi,\eta))d\eta \tag{9.23}$$

and

$$W^m_{\ell\ell_1}(\xi, t_r) \equiv \int_{-1}^{t_r} P^m_\ell(\eta) P^m_{\ell_1}(\eta'(\xi,\eta)) X^m_{\ell_1}(\xi'(\xi,\eta)) d\eta \qquad (9.24)$$

These matrices are the generalization for truncated oblate spheroids of the matrices $S^m_{\ell\ell_1}(t,t_r)$ and $T^m_{\ell\ell_1}(t,t_r)$, eqs.(8.18) and (8.19), for truncated spheres. In a similar way the matrices $C^m_{\ell\ell_1}$, $D^m_{\ell\ell_1}$, $F^m_{\ell\ell_1}$ and $G^m_{\ell\ell_1}$, defined by eq.(9.19), are the generalization for truncated oblate spheroids, of the corresponding matrices for truncated spheres, defined by eq.(8.13). In appendix A it is shown, in which way the matrices $S^m_{\ell\ell_1}(t,t_r)$ and $T^m_{\ell\ell_1}(t,t_r)$ may be obtained from $V^m_{\ell\ell_1}(\xi,t_r)$ and $W^m_{\ell\ell_1}(\xi,t_r)$, respectively, in the limit as $\xi \to \infty$. It is furthermore proved in this appendix, that the expressions, eq.(9.19), for the matrices $C^m_{\ell\ell_1}$, $D^m_{\ell\ell_1}$, $F^m_{\ell\ell_1}$ and $G^m_{\ell\ell_1}$ become equal to the expressions, given by eq.(8.13), in the limit as $\xi_0 \to \infty$.

Just as for truncated spheres (see section 8.2), the terms H^m_ℓ and J^m_ℓ at the right-hand sides of eq.(9.17) are found to be only unequal to zero, if $m = 0, 1$ or -1. Since terms with different m do not couple in this equation, as a consequence of the rotational symmetry of the system around the z-axis, only the case $m = 0, 1$ or -1 has to be considered. H^0_0 is found to contain the constant ψ_0, from eqs.(9.13), appearing in the potentials, eqs.(9.8) and (9.9). Eq.(9.17a), for $\ell = 0$, can therefore be used to determine this unknown quantity ψ_0 in terms of the multipole coefficients. This results in

$$\begin{aligned}
\psi_0 &= R_s\left(\frac{\epsilon_1}{\epsilon_2}-1\right)\left\{\frac{1}{\sqrt{3}}\zeta^0_{01}Q^0_{01}(t_r) + t_r[1-\zeta^0_{00}Q^0_{00}(t_r)]\right\} E_0\cos\theta_0 \\
&\quad -\frac{1}{2\sqrt{\pi}}\sum_{\ell=1}^{\infty} A_{\ell 0}R_s^{-\ell-1}\left(\frac{\epsilon_1-\epsilon_2}{\epsilon_1+\epsilon_2}\right)\zeta^0_{0\ell}\xi_0^{\ell+1}[Z^0_\ell(\xi_0)Q^0_{0\ell}(t_r) - (-1)^\ell V^0_{0\ell}(\xi_0,t_r)] \\
&\quad +\frac{1}{2\sqrt{\pi}}\sum_{\ell=1}^{\infty} B_{\ell 0}R_s^{\ell}\left(\frac{\epsilon_3-\epsilon_4}{\epsilon_3+\epsilon_4}\right)\zeta^0_{0\ell}\xi_0^{-\ell}[X^0_\ell(\xi_0)Q^0_{0\ell}(t_r) - (-1)^\ell W^0_{0\ell}(\xi_0,t_r)]
\end{aligned} \qquad (9.25)$$

Using the limiting formulae, for $\xi_0 \to \infty$, eqs.(9.129) and (9.130) of appendix A, it follows that this expression gives eq.(8.22) for truncated spheres. Note that $R_s \to R$ in this limit.

For the calculation of the multipole coefficients, one can further simply discard eq.(9.17), a as well as b, for $\ell = 0$. If free charge were present in the system this would not be the case. Eq.(9.17) can now be written as

$$\begin{aligned}
\sum_{\ell_1=1}^{\infty} C^m_{\ell\ell_1}R_s^{-\ell_1-2}A_{\ell_1 m} + \sum_{\ell_1=1}^{\infty} D^m_{\ell\ell_1}R_s^{\ell_1-1}B_{\ell_1 m} &= H^m_\ell \\
\sum_{\ell_1=1}^{\infty} F^m_{\ell\ell_1}R_s^{-\ell_1-2}A_{\ell_1 m} + \sum_{\ell_1=1}^{\infty} G^m_{\ell\ell_1}R_s^{\ell_1-1}B_{\ell_1 m} &= J^m_\ell \\
\text{for } \ell &= 1,2,3,.....; m = 0, \pm 1
\end{aligned} \qquad (9.26)$$

and it is found that

$$\begin{aligned}
H_\ell^m &\equiv \sqrt{4\pi/3}E_0\cos\theta_0\delta_{m0}\left\{\frac{\epsilon_1}{\epsilon_2}\xi_0^{-1}X_1^0(\xi_0)\delta_{\ell 1}\right. \\
&\left.+\left(\frac{\epsilon_1-\epsilon_2}{\epsilon_2}\right)[\sqrt{3}t_r\zeta_{\ell 0}^0 Q_{\ell 0}^0(t_r)-\xi_0^{-1}X_1^0(\xi_0)\zeta_{\ell 1}^0 Q_{\ell 1}^0(t_r)]\right\} \\
&+\sqrt{2\pi/3}E_0\sin\theta_0\delta_{\ell 1}\left[\exp(i\phi_0)\xi_0^{-1}X_1^{-1}(\xi_0)\delta_{m,-1}\right. \\
&\left.-\exp(-i\phi_0)\xi_0^{-1}X_1^1(\xi_0)\delta_{m1}\right]
\end{aligned} \tag{9.27}$$

$$\begin{aligned}
J_\ell^m &\equiv \sqrt{4\pi/3}E_0\cos\theta_0\epsilon_1\frac{dX_1^0(\xi_0)}{d\xi_0}\delta_{\ell 1}\delta_{m0}+\sqrt{2\pi/3}E_0\sin\theta_0 \\
&\times\left\{\epsilon_2\delta_{\ell 1}\left[\exp(i\phi_0)\frac{dX_1^{-1}(\xi_0)}{d\xi_0}\delta_{m,-1}-\exp(-i\phi_0)\frac{dX_1^1(\xi_0)}{d\xi_0}\delta_{m1}\right]\right. \\
&+(\epsilon_1-\epsilon_2)\left[\exp(i\phi_0)\frac{dX_1^{-1}(\xi_0)}{d\xi_0}\zeta_{\ell 1}^{-1}Q_{\ell 1}^{-1}(t_r)\delta_{m,-1}\right. \\
&\left.\left.-\exp(-i\phi_0)\frac{dX_1^1(\xi_0)}{d\xi_0}\zeta_{\ell 1}^1 Q_{\ell 1}^1(t_r)\delta_{m1}\right]\right\}
\end{aligned} \tag{9.28}$$

In appendix A it is shown that the above expressions tend to those given by eqs.(8.24) and (8.25) for truncated spheres in the limit as $\xi_0 \to \infty$. Since $R_s \to R$ in this limit, it follows that eq.(9.26), with eqs.(9.19), (9.27) and (9.28), becomes equal to eq.(8.23), with eqs.(8.13), (8.24) and (8.25), in the limit as $\xi_0 \to \infty$.

Just as in section 8.2, it can be shown, that the symmetry relations, eqs.(8.26) and (8.27), also hold in the present case. To this end one now has to apply eqs.(6.8) and (6.11) to eqs.(9.19), (9.27) and (9.28), using furthermore eqs.(9.20)–(9.24) and (8.14). With eqs.(9.26), (8.26) and (8.27) one can then again prove the validity of eq.(8.28) for truncated oblate spheroids. The $m = -1$ equation (9.26) is therefore again superfluous.

Equation (9.26) may now be solved numerically by neglecting the amplitudes $A_{\ell m}$ and $B_{\ell m}$ larger than a suitably chosen order. One then solves the finite set of inhomogeneous linear equations (cf. eq.(8.29))

$$\begin{aligned}
\sum_{\ell_1=1}^{M} C_{\ell\ell_1}^m R_s^{-\ell_1-2}A_{\ell_1 m}+\sum_{\ell_1=1}^{M} D_{\ell\ell_1}^m R_s^{\ell_1-1}B_{\ell_1 m} &= H_\ell^m \\
\sum_{\ell_1=1}^{M} F_{\ell\ell_1}^m R_s^{-\ell_1-2}A_{\ell_1 m}+\sum_{\ell_1=1}^{M} G_{\ell\ell_1}^m R_s^{\ell_1-1}B_{\ell_1 m} &= J_\ell^m \\
\text{for } \ell &= 1,2,3,.....M;\ m=0,1
\end{aligned} \tag{9.29}$$

where the matrix elements $C_{\ell\ell_1}^m$, $D_{\ell\ell_1}^m$, $F_{\ell\ell_1}^m$ and $G_{\ell\ell_1}^m (\ell, \ell_1 = 1,2,3,....., M;\ m = 0,1)$

are again given by eq.(9.19), whereas it follows from eqs.(9.27) and (9.28) that

$$
\begin{aligned}
H_\ell^0 &\equiv \sqrt{4\pi/3}E_0\cos\theta_0\left\{\frac{\epsilon_1}{\epsilon_2}\xi_0^{-1}X_1^0(\xi_0)\delta_{\ell 1}\right. \\
&\left. + \left(\frac{\epsilon_1-\epsilon_2}{\epsilon_2}\right)[\sqrt{3}t_r\zeta_{\ell 0}^0 Q_{\ell 0}^0(t_r) - \xi_0^{-1}X_1^0(\xi_0)\zeta_{\ell 1}^0 Q_{\ell 1}^0(t_r)]\right\} \\
J_\ell^0 &\equiv \sqrt{4\pi/3}E_0\cos\theta_0\epsilon_1\frac{dX_1^0(\xi_0)}{d\xi_0}\delta_{\ell 1} \\
H_\ell^1 &\equiv -\sqrt{2\pi/3}E_0\sin\theta_0\exp(-i\phi_0)\xi_0^{-1}X_1^1(\xi_0)\delta_{\ell 1} \\
J_\ell^1 &\equiv -\sqrt{2\pi/3}E_0\sin\theta_0\exp(-i\phi_0)\frac{dX_1^1(\xi_0)}{d\xi_0}[\epsilon_2\delta_{\ell 1} + (\epsilon_1-\epsilon_2)\zeta_{\ell 1}^1 Q_{\ell 1}^1(t_r)] \quad (9.30)
\end{aligned}
$$

The dipole and quadrupole polarizabilities of the truncated oblate spheroid on the substrate are given, in terms of the amplitudes A_{10}, A_{11}, A_{20} and A_{21}, by (cf. e.g. eq.(8.31))

$$
\begin{aligned}
\alpha_z(0) &= 2\pi\epsilon_1 A_{10}/(\sqrt{\pi/3}E_0\cos\theta_0) \\
\alpha_\parallel(0) &= -4\pi\epsilon_1 A_{11}/[\sqrt{2\pi/3}E_0\sin\theta_0\exp(-i\phi_0)] \\
\alpha_z^{10}(0) &= \pi\epsilon_1 A_{20}/(\sqrt{\pi/5}E_0\cos\theta_0) \\
\alpha_\parallel^{10}(0) &= -4\pi\epsilon_1 A_{21}/[\sqrt{6\pi/5}E_0\sin\theta_0\exp(-i\phi_0)] \qquad (9.31)
\end{aligned}
$$

The argument 0 of these polarizabilities denotes the position of the dipole and the quadrupole along the z-axis, which are located at the center of the truncated oblate spheroid. Since this center lies above the substrate, just as in the case of the truncated sphere, treated in section 8.2, an equation of the same form as eq.(8.31) may be used here. The only difference between equations (9.31) and (8.31) is that the amplitudes A_{10} , A_{11} , A_{20} and A_{21} in eq.(9.31) refer to an oblate spheroidal multipole expansion, whereas in eq.(8.31) these amplitudes refer to a spherical multipole expansion. As has been discussed in chapter 6, it is possible to use here the same equations, since these two expansions coincide, as long as one neglects octupole and higher order moments (see e.g. text below eq.(6.20) and eq.(6.131) in chapter 6).

Using eqs.(5.64) and (5.60), one finds for the surface constitutive coefficients $\beta_e(d)$, $\gamma_e(d)$, $\delta_e(d)$ and $\tau(d)$ of an island film, consisting of identical truncated oblate spheroids, in the low coverage limit again the result, eq.(8.32):

$$
\begin{aligned}
\beta_e(d) &= \rho\alpha_z(0)/\epsilon_1^2 \\
\gamma_e(d) &= \rho\alpha_\parallel(0) \\
\delta_e(d) &= -\rho[\alpha_z^{10}(0) + \alpha_\parallel^{10}(0) - d\,\alpha_z(0) - d\,\alpha_\parallel(0)]/\epsilon_1 \\
\tau(d) &= -\rho[\alpha_\parallel^{10}(0) - d\,\alpha_\parallel(0)] \qquad (9.32)
\end{aligned}
$$

where ρ is the number of particles per unit of surface area. The argument d of these coefficients denotes again the position $z = d$ of the dividing surface, which is chosen to coincide with the substrate. For truncated oblate spheroidal islands of different sizes and shapes, the polarizabilities in the above formulae have to be replaced by their averages over the surface,(cf. eq.(5.65)). The invariants I_e, I_c, $I_{\delta,e}$ and I_τ are found by using eq.(5.66).

The main problem is now the solution of eq.(9.29) for such values of M, that the polarizabilities, eq.(9.31), are obtained sufficiently accurate. One simply increases M until the values of these coefficients become stable within the desired accuracy. It appears, that the value $M = 16$ is usually sufficient, depending, of course, on the values of the dielectric constants ϵ_1, ϵ_2, ϵ_3 and ϵ_4, as well as on the value of the truncation parameter t_r. As will be clear, this solution can only be given on a computer. Since the multipole coefficients $B_{\ell m}$ can not be eliminated in general (i.e. for arbitrary values of the truncation parameter t_r) from eq.(9.29), the number of equations to be solved is twice the number, that has to be solved in the special case of an oblate spheroid touching the substrate (see section 6.4). Section 9.7 deals with this problem, in particular with the way, in which the functions $X_\ell^m(\xi_0)$, $Z_\ell^m(\xi_0)$ and the integrals, eqs.(9.22)–(9.24), appearing in the matrix elements and right-hand sides of eq.(9.29) can be evaluated. In section 9.6 a number of special cases will be treated, where these integrals can be evaluated more easily.

9.3 Truncated prolate spheroids on a substrate

Consider a truncated prolate spheroidal particle on a substrate, with axis of revolution normal to the substrate (see fig.9.4). The center, which is located in the origin of the Cartesian coordinate frame, lies a distance d above the substrate. This distance is smaller than, or equal to, half the long axis. The case of the prolate spheroidal cap, which has its center in the substrate, will be treated in section 9.5. For the prolate spheroid the two foci (0,0,$-a$) and (0,0,a) lie on the symmetry axis, which is the long axis with length $2R_l = 2a\xi_0$. As in the previous section four regions are distinguished with different frequency dependent dielectric constants ϵ_1, ϵ_2, ϵ_3 and ϵ_4, cf. fig.9.4. The surface of the substrate is located at $z = d$. The two short axes have length $2a(\xi_0^2 - 1)^{1/2}$ $(1 \leq \xi_0 < \infty)$, in terms of the elongation parameter ξ_0. This is the value of the first of a set of prolate spheroidal coordinates ξ, η and ϕ $(1 \leq \xi < \infty,\ -1 \leq \eta \leq 1,\ 0 \leq \phi < 2\pi)$ at the surface of the particle (see fig.9.4). These coordinates are defined by eqs.(6.48)–(6.50).

For the truncated prolate spheroidal particle, drawn in fig.9.4, one can distinguish the following four interfaces:

$$\begin{aligned}
&\text{Interface between regions 1 and 3} &:\quad & \xi = \xi_0,\quad -1 \leq \eta \leq \frac{d}{a\xi_0} \\
&\text{Interface between regions 2 and 4} &:\quad & \xi = \xi_0,\quad \frac{d}{a\xi_0} \leq \eta \leq 1 \\
&\text{Interface between regions 3 and 4} &:\quad & \eta = \frac{d}{a\xi},\quad 1 \leq \xi \leq \xi_0 \\
&\text{Interface between regions 1 and 2} &:\quad & \eta = \frac{d}{a\xi},\quad \xi_0 \leq \xi < \infty
\end{aligned} \tag{9.33}$$

where $\xi_0 \geq 1$. Regions 1,2,3 and 4 are the ambient, the substrate, the part of the spheroid in the ambient and the part in the substrate, respectively. Notice that this

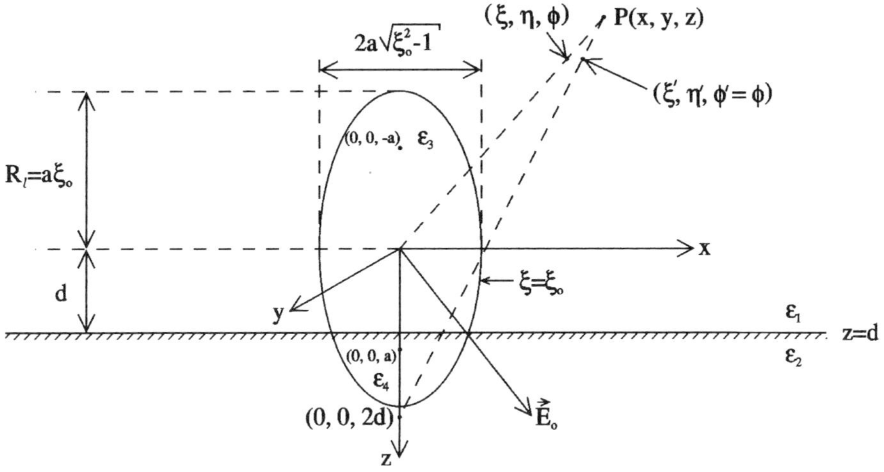

Figure 9.4 Truncated prolate spheroid on substrate.

is, in fact, the same subdivision as in eq.(9.1). As truncation parameter one must introduce here the quantity

$$t_r \equiv \frac{d}{a\xi_0} = \frac{d}{R_l} \quad \text{with} \quad 0 \leq t_r \leq 1 \tag{9.34}$$

cf. eq.(9.2). Notice that $0 \leq t_r \leq 1$, since $0 \leq d \leq R_l$. The case $t_r = 0$ now corresponds to a prolate hemispheroid and $t_r = 1$ to a prolate spheroid touching the substrate. Truncated spheres may again be obtained as the limiting case of truncated prolate spheroids, where $\xi_0 \to \infty$, $a \to 0$, with $a\xi_0 = R$ constant. In this limit eq.(9.34) becomes again equal to eq.(8.11).

As in the previous section, image multipoles are now introduced at $(0,0,2d)$ to describe the fields due to the charge distribution in the substrate. In the present case these will be prolate spheroidal multipoles, cf. section 6.5. The potentials in the regions 1,2,3 and 4 now become

$$\psi_1(\mathbf{r}) = (2\pi/3)^{1/2}[-2^{1/2}E_{0,z}\widetilde{X}_1^0(\xi,a)Y_1^0(\arccos\eta,\phi)$$

$$+(E_{0,x} - iE_{0,y})\widetilde{X}_1^1(\xi,a)Y_1^1(\arccos\eta,\phi) - (E_{0,x} + iE_{0,y})\widetilde{X}_1^{-1}(\xi,a)Y_1^{-1}(\arccos\eta,\phi)]$$

$$+\sum_{\ell m}{}' A_{\ell m}\widetilde{Z}_\ell^m(\xi,a)Y_\ell^m(\arccos\eta,\phi) + \sum_{\ell m}{}' A^r_{\ell m}\widetilde{Z}_\ell^m(\xi'(\xi,\eta),a)Y_\ell^m(\arccos\eta'(\xi,\eta),\phi)$$

for $\xi_0 \leq \xi < \infty$, $\eta \leq d/a\xi$,

$$\begin{aligned}\psi_2(\mathbf{r}) &= \psi_2' + c_1\widetilde{X}_1^0(\xi,a)Y_1^0(\arccos\eta,\phi) + c_2\widetilde{X}_1^1(\xi,a)Y_1^1(\arccos\eta,\phi)\\ &\quad +c_3\widetilde{X}_1^{-1}(\xi,a)Y_1^{-1}(\arccos\eta,\phi) + \sum_{\ell m}{}' A^t_{\ell m}\widetilde{Z}_\ell^m(\xi,a)Y_\ell^m(\arccos\eta,\phi)\end{aligned}$$

for $\xi_0 \leq \xi < \infty$, $\eta \geq d/a\xi$,

$$\begin{aligned}\psi_3(\mathbf{r}) &= \psi_3' + \sum_{\ell m}{}' B_{\ell m} \widetilde{X}_\ell^m(\xi, a) Y_\ell^m(\arccos\eta, \phi) \\ &\quad + \sum_{\ell m}{}' B^r_{\ell m} \widetilde{X}_\ell^m(\xi'(\xi,\eta), a) Y_\ell^m(\arccos\eta'(\xi,\eta), \phi) \\ &\quad \text{for } 1 \le \xi \le \xi_0,\ \eta \le d/a\xi\end{aligned}$$

$$\begin{aligned}\psi_4(\mathbf{r}) &= \psi_4' + \sum_{\ell m}{}' B^t_{\ell m} \widetilde{X}_\ell^m(\xi, a) Y_\ell^m(\arccos\eta, \phi) \\ &\quad \text{for } 1 \le \xi \le \xi_0,\ \eta \ge d/a\xi\end{aligned} \tag{9.35}$$

The functions $\widetilde{X}_\ell^m$, $\widetilde{Z}_\ell^m$ and Y_ℓ^m are defined by eqs.(6.58), (6.59) and (6.5). Furthermore ξ' and η' are the first two prolate spheroidal coordinates of the point (x, y, z) in the shifted coordinate frame with origin at (0,0,2d) (see fig.9.4). The functions $\xi'(\xi,\eta)$ and $\eta'(\xi,\eta)$ are found from eq.(6.50), by noting that $x'^2 + y'^2 = x^2 + y^2$ and $z' = z - 2d$ for the coordinate transformation. One then obtains

$$\begin{aligned}\xi'(\xi,\eta) &= \frac{1}{2}\sqrt{2}\xi\left\{1 + 4\frac{d^2}{a^2\xi^2} - 4\frac{d}{a\xi}\eta + \frac{\eta^2}{\xi^2}\right. \\ &\quad \left. + \left[\left(1 + 4\frac{d^2}{a^2\xi^2} - 4\frac{d}{a\xi}\eta + \frac{\eta^2}{\xi^2}\right)^2 - 4\frac{1}{\xi^2}\left(2\frac{d}{a\xi} - \eta\right)^2\right]^{1/2}\right\}^{1/2}\end{aligned}$$

$$\begin{aligned}\eta'(\xi,\eta) &= \sqrt{2}\left(\eta - 2\frac{d}{a\xi}\right)\left\{1 + 4\frac{d^2}{a^2\xi^2} - 4\frac{d}{a\xi}\eta + \frac{\eta^2}{\xi^2}\right. \\ &\quad \left. + \left[\left(1 + 4\frac{d^2}{a^2\xi^2} - 4\frac{d}{a\xi}\eta + \frac{\eta^2}{\xi^2}\right)^2 - 4\frac{1}{\xi^2}\left(2\frac{d}{a\xi} - \eta\right)^2\right]^{1/2}\right\}^{-1/2}\end{aligned} \tag{9.36}$$

If the boundary conditions $\psi_1 = \psi_2$ and $\epsilon_1\partial\psi_1/\partial z = \epsilon_2\partial\psi_2/\partial z$ are applied at the interface between the region 1 and 2, one finds, with eqs.(9.35), (9.36) and (6.103), that eqs.(9.10)–(9.12) are also valid for prolate spheroids. In the same way one finds, by applying the boundary conditions $\psi_3 = \psi_4$ and $\epsilon_3\partial\psi_3/\partial z = \epsilon_4\partial\psi_4/\partial z$ at the interface between regions 3 and 4, that eqs.(9.13) and (9.14) are valid too. Next the boundary conditions, eqs.(9.15) and (9.16), have to be applied at the surface $\xi = \xi_0$ of the prolate spheroid. Upon substitution of the potentials, eq.(9.35), into these boundary conditions, using also eqs.(9.10)–(9.14), one obtains

$$\begin{aligned}\sum_{\ell_1=|m|}^{\infty\prime} C^m_{\ell\ell_1} R_l^{-\ell_1-2} A_{\ell_1 m} + \sum_{\ell_1=|m|}^{\infty\prime} D^m_{\ell\ell_1} R_l^{\ell_1-1} B_{\ell_1 m} &= H^m_\ell \\ \sum_{\ell_1=|m|}^{\infty\prime} F^m_{\ell\ell_1} R_l^{-\ell_1-2} A_{\ell_1 m} + \sum_{\ell_1=|m|}^{\infty\prime} G^m_{\ell\ell_1} R_l^{\ell_1-1} B_{\ell_1 m} &= J^m_\ell\end{aligned} \tag{9.37}$$

where $\ell = 0, 1, 2, 3, ...$ and $m = 0, \pm1, \pm2, \pm3, ..., \pm\ell$; $R_l \equiv a\xi_0$ is the length of half the long axis of the prolate spheroid (see fig.9.4 and cf. eq.(9.17)). The prime as superindex of the summation symbol means again, that the term $\ell_1 = 0$ should be omitted. Using the definition of the spherical harmonics and their normalization and orthogonality relations, eqs.(6.5) and (9.18), one finds that the matrix elements $C^m_{\ell\ell_1}$, $D^m_{\ell\ell_1}$, $F^m_{\ell\ell_1}$ and $G^m_{\ell\ell_1}$ are now given by

$$\begin{aligned} C^m_{\ell\ell_1} &\equiv \left(\frac{2\epsilon_1}{\epsilon_1+\epsilon_2}\right)\xi_0^{\ell_1+1}\widetilde{Z}^m_{\ell_1}(\xi_0)\delta_{\ell\ell_1} \\ &- \left(\frac{\epsilon_1-\epsilon_2}{\epsilon_1+\epsilon_2}\right)\zeta^m_{\ell\ell_1}\xi_0^{\ell_1+1}[\widetilde{Z}^m_{\ell_1}(\xi_0)Q^m_{\ell\ell_1}(t_r) - (-1)^{\ell_1+m}\widetilde{V}^m_{\ell\ell_1}(\xi_0,t_r)] \end{aligned}$$

$$\begin{aligned} D^m_{\ell\ell_1} &\equiv -\left(\frac{2\epsilon_3}{\epsilon_3+\epsilon_4}\right)\xi_0^{-\ell_1}\widetilde{X}^m_{\ell_1}(\xi_0)\delta_{\ell\ell_1} \\ &+ \left(\frac{\epsilon_3-\epsilon_4}{\epsilon_3+\epsilon_4}\right)\zeta^m_{\ell\ell_1}\xi_0^{-\ell_1}[\widetilde{X}^m_{\ell_1}(\xi_0)Q^m_{\ell\ell_1}(t_r) - (-1)^{\ell_1+m}\widetilde{W}^m_{\ell\ell_1}(\xi_0,t_r)] \end{aligned}$$

$$\begin{aligned} F^m_{\ell\ell_1} &\equiv \left(\frac{2\epsilon_1\epsilon_2}{\epsilon_1+\epsilon_2}\right)\xi_0^{\ell_1+2}\frac{d\widetilde{Z}^m_{\ell_1}(\xi_0)}{d\xi_0}\delta_{\ell\ell_1} \\ &+\epsilon_1\left(\frac{\epsilon_1-\epsilon_2}{\epsilon_1+\epsilon_2}\right)\zeta^m_{\ell\ell_1}\xi_0^{\ell_1+2}\left[\frac{d\widetilde{Z}^m_{\ell_1}(\xi_0)}{d\xi_0}Q^m_{\ell\ell_1}(t_r) + (-1)^{\ell_1+m}\frac{\partial\widetilde{V}^m_{\ell\ell_1}(\xi_0,t_r)}{\partial\xi_0}\right] \end{aligned}$$

$$\begin{aligned} G^m_{\ell\ell_1} &\equiv -\left(\frac{2\epsilon_3\epsilon_4}{\epsilon_3+\epsilon_4}\right)\xi_0^{-\ell_1+1}\frac{d\widetilde{X}^m_{\ell_1}(\xi_0)}{d\xi_0}\delta_{\ell\ell_1} - \epsilon_3\left(\frac{\epsilon_3-\epsilon_4}{\epsilon_3+\epsilon_4}\right)\zeta^m_{\ell\ell_1}\xi_0^{-\ell_1+1} \\ &\times\left[\frac{d\widetilde{X}^m_{\ell_1}(\xi_0)}{d\xi_0}Q^m_{\ell\ell_1}(t_r) + (-1)^{\ell_1+m}\frac{\partial\widetilde{W}^m_{\ell\ell_1}(\xi_0,t_r)}{\partial\xi_0}\right] \end{aligned} \tag{9.38}$$

It will immediately be clear that these expressions are of the same form as those given by eq.(9.19). Only the functions $X^m_\ell(\xi)$ and $Z^m_\ell(\xi)$ have to be replaced by

$$\widetilde{X}^m_\ell(\xi) \equiv \widetilde{X}^m_\ell(\xi,a)a^{-\ell} \equiv i^m\frac{(\ell-m)!}{(2\ell-1)!!}P^m_\ell(\xi) = i^\ell X^m_\ell(-i\xi) \tag{9.39}$$

and

$$\widetilde{Z}^m_\ell(\xi) \equiv \widetilde{Z}^m_\ell(\xi,a)a^{\ell+1} \equiv \frac{(2\ell+1)!!}{(\ell+m)!}Q^m_\ell(\xi) = i^{-\ell-1}Z^m_\ell(-i\xi) \tag{9.40}$$

and the matrix elements $V^m_{\ell\ell_1}(\xi,t_r)$ and $W^m_{\ell\ell_1}(\xi,t_r)$ by

$$\widetilde{V}^m_{\ell\ell_1}(\xi,t_r) \equiv \int_{-1}^{t_r} P^m_\ell(\eta)P^m_{\ell_1}(\eta'(\xi,\eta))\widetilde{Z}^m_{\ell_1}(\xi'(\xi,\eta))d\eta \tag{9.41}$$

and

$$\widetilde{W}^m_{\ell\ell_1}(\xi, t_r) \equiv \int_{-1}^{t_r} P^m_\ell(\eta) P^m_{\ell_1}(\eta'(\xi,\eta)) \widetilde{X}^m_{\ell_1}(\xi'(\xi,\eta)) d\eta \tag{9.42}$$

Here eqs.(6.58), (6.59), (6.66), (9.20) and (9.21) have been used. The matrices $Q^m_{\ell\ell_1}(t_r)$ and $\zeta^m_{\ell\ell_1}$ are again defined by eqs.(9.22) and (8.14). The matrices $\widetilde{V}^m_{\ell\ell_1}(\xi, t_r)$ and $\widetilde{W}^m_{\ell\ell_1}(\xi, t_r)$ are, of course, the generalization, for truncated prolate spheroids, of the matrices $S^m_{\ell\ell_1}(t, t_r)$ and $T^m_{\ell\ell_1}(t, t_r)$, eqs.(8.18) and (8.19), for truncated spheres. This is shown in appendix A, where it is also proved that the matrices, defined by eq.(9.38) tend to the expressions, eq.(8.13), for truncated spheres in the limit as $\xi_0 \to \infty$.

Just as for truncated oblate spheroids (previous section), the terms H^m_ℓ and J^m_ℓ at the right-hand side of eq.(9.37) are found to be only unequal to zero, if $m = -1, 0, 1$. H^0_0 is found to contain the constant ψ_0, from eqs.(9.13), appearing in the potentials ψ_3 and ψ_4, eq.(9.35). Eq.(9.37), for $\ell = 0$, can therefore be used to determine this unknown quantity ψ_0 in terms of the multipole coefficients. This results in

$$\begin{aligned}
\psi_0 &= R_l\left(\frac{\epsilon_1}{\epsilon_2} - 1\right)\left\{\frac{1}{\sqrt{3}}\zeta^0_{01}Q^0_{01}(t_r) + t_r[1 - \zeta^0_{00}Q^0_{00}(t_r)]\right\} E_0 \cos\theta_0 \\
&\quad - \frac{1}{2\sqrt{\pi}} \sum_{\ell=1}^{\infty} A_{\ell 0} R_l^{-\ell-1}\left(\frac{\epsilon_1 - \epsilon_2}{\epsilon_1 + \epsilon_2}\right) \zeta^0_{0\ell}\xi_0^{\ell+1}[\widetilde{Z}^0_\ell(\xi_0)Q^0_{0\ell}(t_r) - (-1)^\ell \widetilde{V}^0_{0\ell}(\xi_0, t_r)] \\
&\quad + \frac{1}{2\sqrt{\pi}} \sum_{\ell=1}^{\infty} B_{\ell 0} R_l^{\ell}\left(\frac{\epsilon_3 - \epsilon_4}{\epsilon_3 + \epsilon_4}\right) \zeta^0_{0\ell}\xi_0^{-\ell}[\widetilde{X}^0_\ell(\xi_0)Q^0_{0\ell}(t_r) - (-1)^\ell \widetilde{W}^0_{0\ell}(\xi_0, t_r)]
\end{aligned} \tag{9.43}$$

Using the limiting formulae, for $\xi_0 \to \infty$, eqs.(9.134) and (9.135) of appendix A, it follows that this expression gives eq.(8.22) for truncated spheres. Note that $R_l \to R$ in this limit.

For the calculation of the multipole coefficients, one can further discard eq.(9.37), a as well as b, for $\ell = 0$. If free charge were present in the system this would not be the case. Eq.(9.37) can now be written as

$$\begin{aligned}
\sum_{\ell_1=1}^{\infty} C^m_{\ell\ell_1} R_l^{-\ell_1-2} A_{\ell_1 m} + \sum_{\ell_1=1}^{\infty} D^m_{\ell\ell_1} R_l^{\ell_1-1} B_{\ell_1 m} &= H^m_\ell \\
\sum_{\ell_1=1}^{\infty} F^m_{\ell\ell_1} R_l^{-\ell_1-2} A_{\ell_1 m} + \sum_{\ell_1=1}^{\infty} G^m_{\ell\ell_1} R_l^{\ell_1-1} B_{\ell_1 m} &= J^m_\ell \\
\text{for } \ell &= 1, 2, 3, ...; m = 0, \pm 1
\end{aligned} \tag{9.44}$$

For truncated prolate spheroids H^m_ℓ and J^m_ℓ are given by the expressions, eqs.(9.27) and (9.28), with $X_\ell^m(\xi_0)$ and $Z_\ell^m(\xi_0)$ replaced by $\widetilde{X}_\ell^m(\xi_0)$ and $\widetilde{Z}_\ell^m(\xi_0)$:

$$\begin{aligned} H_\ell^m &\equiv \sqrt{4\pi/3}E_0\cos\theta_0\delta_{m0}\left\{\frac{\epsilon_1}{\epsilon_2}\xi_0^{-1}\widetilde{X}_1^0(\xi_0)\delta_{\ell 1}\right. \\ &\left.+\left(\frac{\epsilon_1-\epsilon_2}{\epsilon_2}\right)\left[\sqrt{3}t_r\zeta_{\ell 0}^0 Q_{\ell 0}^0(t_r)-\xi_0^{-1}\widetilde{X}_1^0(\xi_0)\zeta_{\ell 1}^0 Q_{\ell 1}^0(t_r)\right]\right\} \\ &+\sqrt{2\pi/3}E_0\sin\theta_0\delta_{\ell 1}\left[\exp(i\phi_0)\xi_0^{-1}\widetilde{X}_1^{-1}(\xi_0)\delta_{m,-1}\right. \\ &\left.-\exp(-i\phi_0)\xi_0^{-1}\widetilde{X}_1^1(\xi_0)\delta_{m1}\right] \end{aligned} \tag{9.45}$$

$$\begin{aligned} J_\ell^m &\equiv \sqrt{4\pi/3}E_0\cos\theta_0\epsilon_1\frac{d\widetilde{X}_1^0(\xi_0)}{d\xi_0}\delta_{\ell 1}\delta_{m0}+\sqrt{2\pi/3}E_0\sin\theta_0 \\ &\times\left\{\epsilon_2\delta_{\ell 1}\left[\exp(i\phi_0)\frac{d\widetilde{X}_1^{-1}(\xi_0)}{d\xi_0}\delta_{m,-1}-\exp(-i\phi_0)\frac{d\widetilde{X}_1^1(\xi_0)}{d\xi_0}\delta_{m1}\right]\right. \\ &+(\epsilon_1-\epsilon_2)\left[\exp(i\phi_0)\frac{d\widetilde{X}_1^{-1}(\xi_0)}{d\xi_0}\zeta_{\ell 1}^{-1}Q_{\ell 1}^{-1}(t_r)\delta_{m,-1}\right. \\ &\left.\left.-\exp(-i\phi_0)\frac{d\widetilde{X}_1^1(\xi_0)}{d\xi_0}\zeta_{\ell 1}^1 Q_{\ell 1}^1(t_r)\delta_{m1}\right]\right\} \end{aligned} \tag{9.46}$$

In appendix A it is finally shown that the above expressions tend to the expressions, eqs.(8.24) and(8.25), for truncated spheres, in the limit as $\xi_0 \to \infty$. Since $R_l \to R$ in this limit, it follows that eq.(9.44), with eqs.(9.38), (9.45) and (9.46), becomes again equal to eq.(8.23), with eqs.(8.13), (8.24) and (8.25), in the limit as $\xi_0 \to \infty$.

Just as for truncated oblate spheroids, it can be shown for truncated prolate spheroids, that the symmetry relations, eqs.(8.26) and (8.27), hold again. Eq.(8.28) is therefore also valid for truncated prolate spheroids and the $m = -1$ equation (9.44) is again redundant.

Equation (9.44) may be solved numerically, by neglecting the amplitudes $A_{\ell m}$ and $B_{\ell m}$ larger than a suitably chosen order. One then solves the finite set of inhomogeneous linear equations:

$$\begin{aligned} \sum_{\ell_1=1}^{M} C_{\ell\ell_1}^m R_l^{-\ell_1-2}A_{\ell_1 m}+\sum_{\ell_1=1}^{M} D_{\ell\ell_1}^m R_l^{\ell_1-1}B_{\ell_1 m} &= H_\ell^m \\ \sum_{\ell_1=1}^{M} F_{\ell\ell_1}^m R_l^{-\ell_1-2}A_{\ell_1 m}+\sum_{\ell_1=1}^{M} G_{\ell\ell_1}^m R_l^{\ell_1-1}B_{\ell_1 m} &= J_\ell^m \\ \text{for } \ell &= 1,2,3,.....,M;\ m=0,1 \end{aligned} \tag{9.47}$$

where the matrix elements $C_{\ell\ell_1}^m$, $D_{\ell\ell_1}^m$, $F_{\ell\ell_1}^m$ and $G_{\ell\ell_1}^m$ $(\ell,\ell_1 = 1,2,3,.....,M;\ m=0,1)$

are given by eq.(9.38), whereas it follows from eqs.(9.45) and (9.46) that

$$\begin{aligned}
H_\ell^0 &= \sqrt{4\pi/3}E_0\cos\theta_0\{\frac{\epsilon_1}{\epsilon_2}\xi_0^{-1}\widetilde{X}_1^0(\xi_0)\delta_{\ell 1} \\
&+ \left(\frac{\epsilon_1-\epsilon_2}{\epsilon_2}\right)[\sqrt{3}t_r\zeta_{\ell 0}^0 Q_{\ell 0}^0(t_r) - \xi_0^{-1}\widetilde{X}_1^0(\xi_0)\zeta_{\ell 1}^0 Q_{\ell 1}^0(t_r)]\} \\
J_\ell^0 &= \sqrt{4\pi/3}E_0\cos\theta_0\epsilon_1\frac{d\widetilde{X}_1^0(\xi_0)}{d\xi_0}\delta_{\ell 1} \\
H_\ell^1 &= -\sqrt{2\pi/3}E_0\sin\theta_0\exp(-i\phi_0)\xi_0^{-1}\widetilde{X}_1^1(\xi_0)\delta_{\ell 1} \\
J_\ell^1 &= -\sqrt{2\pi/3}E_0\sin\theta_0\exp(-i\phi_0)\frac{d\widetilde{X}_1^1(\xi_0)}{d\xi_0}[\epsilon_2\delta_{\ell 1} + (\epsilon_1-\epsilon_2)\zeta_{\ell 1}^1 Q_{\ell 1}^1(t_r)] \quad (9.48)
\end{aligned}$$

The dipole and quadrupole polarizabilities $\alpha_z(0)$, $\alpha_\parallel(0)$, $\alpha_z^{10}(0)$ and $\alpha_\parallel^{10}(0)$ of the truncated prolate spheroid on a substrate are again given, in terms of the amplitudes A_{10} , A_{11} , A_{20} and A_{21}, by eq.(9.31). The surface constitutive coefficients $\beta_e(d)$, $\gamma_e(d)$, $\delta_e(d)$ and $\tau(d)$ of an island film, consisting of identical truncated prolate spheroids, in the low coverage limit, by eq.(9.32). For truncated prolate spheroidal islands of different sizes and shapes, the polarizabilities in eq.(9.32) have to be replaced by their averages over the surface, (cf. eq.(5.65)). The invariants I_e, I_c, $I_{\delta,e}$ and I_τ are found again by using eq.(5.66).

The main problem is now the solution of eq.(9.47) for such values of M, that the polarizabilities, eq.(9.31), are obtained sufficiently accurate. One simply increases M until the values of these coefficients become stable within the desired accuracy. It appears, that the value $M = 16$ is usually sufficient, depending, of course, on the values of the dielectric constants ϵ_1, ϵ_2, ϵ_3 and ϵ_4, as well as on the value of the truncation parameter t_r. As will be clear, this solution can only be found numerically. Since the multipole coefficients $B_{\ell m}$ can not be eliminated in general (i.e. for arbitrary values of the truncation parameter t_r) from eq.(9.47), the number of equations to be solved is twice the number, that has to be solved e.g. in the special case of a prolate spheroid on a substrate (see section 6.5). Section 9.7 deals with this problem, in particular with the way, in which the functions $\widetilde{X}_\ell^m(\xi_0)$, $\widetilde{Z}_\ell^m(\xi_0)$ and the integrals, eqs.(9.41) and (9.42), appearing in the matrix elements and the right-hand sides of eq.(9.47), can be evaluated. In section 9.6 a number of special cases will be treated, where these integrals can be evaluated more easily.

9.4 Oblate spheroidal caps on a substrate

In the previous sections the case has been studied, that the center of the truncated spheroid lies above the substrate, see figs.9.3 and 9.4. Now the case will be considered, that this center lies in the substrate (spheroidal cap). In fig.9.2 this situation has been drawn. As has been mentioned in the text below that figure, the two situations, drawn in figs.9.1 and 9.2, could be obtained from each other by interchanging the dielectric constants ϵ_1 and ϵ_2, and also ϵ_3 and ϵ_4. The disadvantage of this trick is that it gives the final result without the explicit expressions for the potentials in the various regions. In order to be able to give these potentials explicitly, this trick is not used.

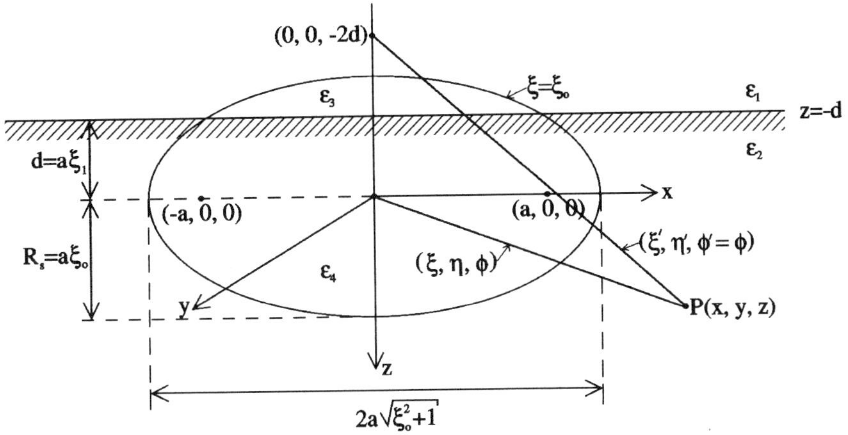

Figure 9.5 Oblate spheroidal cap on a substrate

In the present section oblate spheroidal caps will be treated, and in the next section prolate spheroidal caps.

Consider an oblate spheroidal cap with a frequency dependent dielectric constant $\epsilon_3(\omega)$ on a substrate with a dielectric constant $\epsilon_2(\omega)$ and surrounded by an ambient with dielectric constant $\epsilon_1(\omega)$, in regions 3, 2 and 1 respectively; see fig. 9.5, As in the previous sections the dielectric constant in region 4 is replaced by $\epsilon_4(\omega)$. The axis of rotation of the spheroid is assumed to be normal to the surface of the planar substrate. The origin of the Cartesian coordinate system is chosen at the center of the spheroid. The z-axis is along the axis of rotation and the positive z-axis points into the substrate. The surface of the substrate is given by $z = -d$, where $0 \leq d \leq R_s$. Here R_s is half the length of the short axis of the spheroid.

Oblate spheroidal coordinates ξ, η and ϕ $(0 \leq \xi < \infty,\ -1 \leq \eta \leq 1,\ 0 \leq \phi < 2\pi)$ are again introduced, defined by eqs.(6.1)–(6.3). One can distinguish the following four interfaces, cf. eq.(9.1):

$$\begin{aligned}
&\text{Interface between regions 1 and 3} \ : \quad \xi = \xi_0, \quad -1 \leq \eta \leq -\frac{\xi_1}{\xi_0} \\
&\text{Interface between regions 2 and 4} \ : \quad \xi = \xi_0, \quad -\frac{\xi_1}{\xi_0} \leq \eta \leq 1 \\
&\text{Interface between regions 3 and 4} \ : \quad \eta = -\frac{\xi_1}{\xi}, \quad \xi_1 \leq \xi \leq \xi_0 \\
&\text{Interface between regions 1 and 2} \ : \quad \eta = -\frac{\xi_1}{\xi}, \quad \xi_0 \leq \xi < \infty
\end{aligned} \tag{9.49}$$

A positive truncation parameter is defined by

$$t_r \equiv \frac{d}{a\xi_0} = \frac{d}{R_s} = \frac{\xi_1}{\xi_0} \quad \text{with} \quad 0 \leq t_r \leq 1 \tag{9.50}$$

Now $t_r = 1$ for an infinitely thin cap and $t_r = 0$ for a hemispheroid.

Just as in section 9.2, oblate spheroidal image multipoles will be introduced, now in the point (0,0,$-2d$), to describe the fields (see fig.9.5). The potential in region 1 (ambient) can be written in oblate spheroidal coordinates as

$$\psi_1(\mathbf{r}) = (2\pi/3)^{1/2}[-2^{1/2}E_{0,z}X_1^0(\xi,a)Y_1^0(\arccos\eta,\phi)$$
$$+(E_{0,x}-iE_{0,y})X_1^1(\xi,a)Y_1^1(\arccos\eta,\phi)-(E_{0,x}+iE_{0,y})X_1^{-1}(\xi,a)Y_1^{-1}(\arccos\eta,\phi)]$$
$$+\sum_{\ell m}{}' A_{\ell m}^t Z_\ell^m(\xi,a)Y_\ell^m(\arccos\eta,\phi) \text{ for } \xi_0\le\xi<\infty \text{ and } \eta\le-\xi_1/\xi \qquad (9.51)$$

The first term at the right-hand side is the potential of the constant electric field $\mathbf{E}_0 = (E_{0,x}, E_{0,y}, E_{0,z})$ in the ambient. The sum is the potential due to the induced charge distribution in the island. The super index t indicates that the corresponding multipoles are situated in the point (0,0,0), which is located in the substrate. The potential is then "transmitted" to the ambient from this position. The prime as super index in the sum again indicates that $\ell \neq 0$. The functions X_ℓ^m, Z_ℓ^m and Y_ℓ^m are defined by eqs.(6.13), (6.14) and (6.5), together with eqs.(6.6)–(6.8), (6.10) and (6.11).

For the potential in region 2, substrate, one may now write

$$\psi_2(\mathbf{r}) = \psi_2'' + c_1X_1^0(\xi,a)Y_1^0(\arccos\eta,\phi)$$
$$+c_2X_1^1(\xi,a)Y_1^1(\arccos\eta,\phi)+c_3X_1^{-1}(\xi,a)Y_1^{-1}(\arccos\eta,\phi)]$$
$$+\sum_{\ell m}{}' A_{\ell m}Z_\ell^m(\xi,a)Y_\ell^m(\arccos\eta,\phi)+\sum_{\ell m}{}' A_{\ell m}^r Z_\ell^m(\xi'(\xi,\eta),a)Y_\ell^m(\arccos\eta'(\xi,\eta),\phi)$$
$$\text{for } \xi_0\le\xi<\infty \text{ and } -\xi_1/\xi\le\eta \qquad (9.52)$$

Here ψ_2'', c_1, c_2 and c_3 are constants, which will be determined below. The first sum at the right-hand side originates from the center, (0,0,0), of the spheroid, whereas the second sum originates from the image, which is now located above the substrate in the point (0,0,$-2d$), see fig.9.5. Finally ξ' and η' are the first two oblate spheroidal coordinates of the point (x, y, z) in the shifted coordinate frame with as origin (0,0,$-2d$). The functions $\xi'(\xi,\eta)$ and $\eta'(\xi,\eta)$ are found from eq.(6.3), by noting, that $x'^2 + y'^2 = x^2 + y^2$ and $z' = z + 2d$ for this coordinate transformation. One now obtains

$$\xi'(\xi,\eta) = \frac{1}{2}\sqrt{2}\xi\left\{1+4\frac{\xi_1^2}{\xi^2}+4\frac{\xi_1}{\xi}\eta-\frac{\eta^2}{\xi^2}\right.$$
$$\left.+\left[\left(1+4\frac{\xi_1^2}{\xi^2}+4\frac{\xi_1}{\xi}\eta-\frac{\eta^2}{\xi^2}\right)^2+4\frac{1}{\xi^2}\left(2\frac{\xi_1}{\xi}+\eta\right)^2\right]^{1/2}\right\}^{1/2} \qquad (9.53)$$

$$\eta'(\xi,\eta) = \sqrt{2}\left(\eta+2\frac{\xi_1}{\xi}\right)\left\{1+4\frac{\xi_1^2}{\xi^2}+4\frac{\xi_1}{\xi}\eta-\frac{\eta^2}{\xi^2}\right.$$
$$\left.+\left[\left(1+4\frac{\xi_1^2}{\xi^2}+4\frac{\xi_1}{\xi}\eta-\frac{\eta^2}{\xi^2}\right)^2+4\frac{1}{\xi^2}\left(2\frac{\xi_1}{\xi}+\eta\right)^2\right]^{1/2}\right\}^{-1/2} \qquad (9.54)$$

The transformation

$$\xi' = \xi'(\xi,\eta), \quad \eta' = \eta'(\xi,\eta), \quad \phi' = \phi \tag{9.55}$$

is the generalization for oblate spheroidal coordinates of the transformation, eq.(8.44), valid for spherical coordinates. In fact, it follows, using the asymptotic relations, eqs.(6.19) and (6.20), that eqs.(9.53)–(9.55) becomes equal to eq.(8.44) in the limit as $\xi \to \infty$, $\xi_1 \to \infty$, $a \to 0$, with $a\xi_1 = d$.

The potentials in the regions 3 and 4 are given by

$$\psi_3(\mathbf{r}) = \psi_3'' + \sum_{\ell m}{}' B_{\ell m}^t X_\ell^m(\xi, a) Y_\ell^m(\arccos\eta, \phi) \quad \text{for} \quad 0 \leq \xi \leq \xi_0 \quad \text{and} \quad \eta \leq -\xi_1/\xi \tag{9.56}$$

and

$$\begin{aligned} \psi_4(\mathbf{r}) &= \psi_4'' + \sum_{\ell m}{}' B_{\ell m} X_\ell^m(\xi, a) Y_\ell^m(\arccos\eta, \phi) + \\ & \sum_{\ell m}{}' B_{\ell m}^r X_\ell^m(\xi'(\xi,\eta), a) Y_\ell^m(\arccos\eta'(\xi,\eta), \phi) \\ & \text{for } 0 \leq \xi \leq \xi_0 \text{ and } -\xi_1/\xi \leq \eta \end{aligned} \tag{9.57}$$

where ψ_3'' and ψ_4'' are constants, which will be determined below.

In a completely analogous way as in section 9.2, one can derive, from the boundary conditions at the surface of the substrate (i.e. continuity of ψ at the interface between region 1 and 2, as well as that of the normal derivative of ψ times the dielectric constant), the following results. First of all it is found that the constants c_1, c_2 and c_3 in eq.(9.52) are again given by eq.(9.10), whereas the constant ψ_2'' is now given by

$$\psi_2'' = \left(1 - \frac{\epsilon_1}{\epsilon_2}\right) d\, E_{0,z} \tag{9.58}$$

Furthermore the following relations hold between the multipole coefficients $A_{\ell m}$, $A_{\ell m}^t$ and $A_{\ell m}^r$:

$$A_{\ell m}^r = (-1)^{\ell+m} \frac{\epsilon_2 - \epsilon_1}{\epsilon_2 + \epsilon_1} A_{\ell m} \quad \text{and} \quad A_{\ell m}^t = \frac{2\epsilon_2}{\epsilon_2 + \epsilon_1} A_{\ell m} \tag{9.59}$$

cf. eq.(8.37). Moreover, using similar boundary conditions at the interface between regions 3 and 4, one finds that the constant contributions in these regions are equal

$$\psi_3'' = \psi_4'' \equiv \psi_0 \tag{9.60}$$

The value of ψ_0 will be determined below. Finally one obtains the following relations between the multipole coefficients $B_{\ell m}$, $B_{\ell m}^t$ and $B_{\ell m}^r$:

$$B_{\ell m}^r = (-1)^{\ell+m} \frac{\epsilon_4 - \epsilon_3}{\epsilon_4 + \epsilon_3} B_{\ell m} \quad \text{and} \quad B_{\ell m}^t = \frac{2\epsilon_4}{\epsilon_4 + \epsilon_3} B_{\ell m} \tag{9.61}$$

cf. eq.(8.38).

At the surface of the oblate spheroid $\xi = \xi_0$, one can use that the oblate spherical harmonics $Y_\ell^m(\arccos\eta, \phi)$ form a complete orthonormal set of functions of η and ϕ. Multiplication of the boundary conditions on the surface of the oblate spheroid, between regions 1 and 3 and between 2 and 4, with complex conjugated harmonics and integration over η and ϕ, gives

$$\int_{-1}^{-t_r} d\eta \int_0^{2\pi} d\phi (\psi_1 - \psi_3)_{\xi=\xi_0} [Y_{\ell'}^{m'}(\arccos\eta, \phi)]^* + \int_{-t_r}^{1} d\eta \int_0^{2\pi} d\phi (\psi_2 - \psi_4)_{\xi=\xi_0} [Y_{\ell'}^{m'}(\arccos\eta, \phi)]^* = 0 \tag{9.62}$$

and

$$\int_{-1}^{-t_r} d\eta \int_0^{2\pi} d\phi \left[\frac{\partial}{\partial\xi}(\epsilon_1\psi_1 - \epsilon_3\psi_3)_{\xi=\xi_0}\right] [Y_{\ell'}^{m'}(\arccos\eta, \phi)]^* + \int_{-t_r}^{1} d\eta \int_0^{2\pi} d\phi \left[\frac{\partial}{\partial\xi}(\epsilon_2\psi_2 - \epsilon_4\psi_4)_{\xi=\xi_0}\right] [Y_{\ell'}^{m'}(\arccos\eta, \phi)]^* = 0 \tag{9.63}$$

(cf. eqs.(8.39) and (8.40)). The truncation parameter is again given by eq.(9.2). It is important to realize that this definition is such that it is positive both for truncated spheroids and for spheroidal caps. In the work by, for instance, Simonsen, Lazzari, Jupille and Roux, [56], one uses negative truncation parameters for the caps. The truncation parameter used in this book is always equal to the absolute value of the truncation parameter used in those papers.

Upon substitution of the above potentials, ψ_1, ψ_2, ψ_3 and ψ_4, into the boundary conditions, eqs.(9.62) and (9.63), using also eqs.(9.58) and (9.61), one obtains the following infinite set of inhomogeneous linear equations for the multipole coefficients $A_{\ell m}$ and $B_{\ell m}$:

$$\begin{aligned} \sum_{\ell_1=|m|}^{\infty\prime} C_{\ell\ell_1}^m R_s^{-\ell_1-2} A_{\ell_1 m} + \sum_{\ell_1=|m|}^{\infty\prime} D_{\ell\ell_1}^m R_s^{\ell_1-1} B_{\ell_1 m} &= H_\ell^m \\ \sum_{\ell_1=|m|}^{\infty\prime} F_{\ell\ell_1}^m R_s^{-\ell_1-2} A_{\ell_1 m} + \sum_{\ell_1=|m|}^{\infty\prime} G_{\ell\ell_1}^m R_s^{\ell_1-1} B_{\ell_1 m} &= J_\ell^m \end{aligned} \tag{9.64}$$

where $\ell = 0, 1, 2, 3, ...$ and $m = 0, \pm1, \pm2, \pm3, ..., \pm\ell$; $R_s = a\xi_0$ is half the length of the short axis of the oblate spheroid (see fig.9.5). The prime as superindex of the summation symbols denotes again that the term with $\ell_1 = 0$ should be omitted. Using the definition, eq.(6.5), of the spherical harmonics and their normalization and orthogonality relations, eq.(9.18), one finds:

$$\begin{aligned} C_{\ell\ell_1}^m &\equiv \left(\frac{2\epsilon_2}{\epsilon_2+\epsilon_1}\right) \xi_0^{\ell_1+1} Z_{\ell_1}^m(\xi_0)\delta_{\ell\ell_1} \\ &- \left(\frac{\epsilon_2-\epsilon_1}{\epsilon_2+\epsilon_1}\right) \zeta_{\ell\ell_1}^m \xi_0^{\ell_1+1}(-1)^{\ell+\ell_1}[Z_{\ell_1}^m(\xi_0)Q_{\ell\ell_1}^m(t_r) - (-1)^{\ell_1+m}V_{\ell\ell_1}^m(\xi_0, t_r)] \end{aligned}$$

$$D^m_{\ell\ell_1} \equiv -\left(\frac{2\epsilon_4}{\epsilon_4+\epsilon_3}\right)\xi_0^{-\ell_1}X^m_{\ell_1}(\xi_0)\delta_{\ell\ell_1}$$
$$+\left(\frac{\epsilon_4-\epsilon_3}{\epsilon_4+\epsilon_3}\right)\zeta^m_{\ell\ell_1}\xi_0^{-\ell_1}(-1)^{\ell+\ell_1}[X^m_{\ell_1}(\xi_0)Q^m_{\ell\ell_1}(t_r)-(-1)^{\ell_1+m}W^m_{\ell\ell_1}(\xi_0,t_r)]$$

$$F^m_{\ell\ell_1} \equiv \left(\frac{2\epsilon_2\epsilon_1}{\epsilon_2+\epsilon_1}\right)\xi_0^{\ell_1+2}\frac{dZ^m_{\ell_1}(\xi_0)}{d\xi_0}\delta_{\ell\ell_1}$$
$$+\epsilon_2\left(\frac{\epsilon_2-\epsilon_1}{\epsilon_2+\epsilon_1}\right)\zeta^m_{\ell\ell_1}\xi_0^{\ell_1+2}(-1)^{\ell+\ell_1}\left[\frac{dZ^m_{\ell_1}(\xi_0)}{d\xi_0}Q^m_{\ell\ell_1}(t_r)+(-1)^{\ell_1+m}\frac{\partial V^m_{\ell\ell_1}(\xi_0,t_r)}{\partial\xi_0}\right]$$

$$G^m_{\ell\ell_1} \equiv -\left(\frac{2\epsilon_4\epsilon_3}{\epsilon_4+\epsilon_3}\right)\xi_0^{-\ell_1+1}\frac{dX^m_{\ell_1}(\xi_0)}{d\xi_0}\delta_{\ell\ell_1}-\epsilon_4\left(\frac{\epsilon_4-\epsilon_3}{\epsilon_4+\epsilon_3}\right)\zeta^m_{\ell\ell_1}\xi_0^{-\ell_1+1}(-1)^{\ell+\ell_1}$$
$$\times\left[\frac{dX^m_{\ell_1}(\xi_0)}{d\xi_0}Q^m_{\ell\ell_1}(t_r)+(-1)^{\ell_1+m}\frac{\partial W^m_{\ell\ell_1}(\xi_0,t_r)}{\partial\xi_0}\right] \tag{9.65}$$

where the various functions and matrices at the right-hand sides are given by eqs.(9.20) –(9.24), and (8.14).

Just as for truncated oblate spheroids (see section 9.2), the terms H^m_ℓ and J^m_ℓ at the right-hand sides of eq.(9.64), are found to be only unequal to zero, if $m = 0, 1$ or -1. Since terms with different m do not couple in this equation, as a consequence of the rotational symmetry of the system around the z-axis, only the case $m = 0, 1$ or -1 has to be considered. H^0_0 is again found to contain the constant ψ_0, from eqs.(9.60). Eq.(9.64a), for $\ell = 0$, can therefore be used to determine this unknown quantity ψ_0 in terms of the multipole coefficients. This results in

$$\psi_0 = R_s\left(\frac{\epsilon_1}{\epsilon_2}-1\right)\left\{\frac{1}{\sqrt{3}}\zeta^0_{01}Q^0_{01}(t_r)-t_r\zeta^0_{00}Q^0_{00}(t_r)\right\}E_0\cos\theta_0$$
$$-\frac{1}{2\sqrt{\pi}}\sum_{\ell=1}^{\infty}A_{\ell 0}R_s^{-\ell-1}\left(\frac{\epsilon_2-\epsilon_1}{\epsilon_2+\epsilon_1}\right)\zeta^0_{0\ell}\xi_0^{\ell+1}[(-1)^\ell Z^0_\ell(\xi_0)Q^0_{0\ell}(t_r)-V^0_{0\ell}(\xi_0,t_r)]$$
$$+\frac{1}{2\sqrt{\pi}}\sum_{\ell=1}^{\infty}B_{\ell 0}R_s^{\ell}\left(\frac{\epsilon_4-\epsilon_3}{\epsilon_4+\epsilon_3}\right)\zeta^0_{0\ell}\xi_0^{-\ell}[(-1)^\ell X^0_\ell(\xi_0)Q^0_{0\ell}(t_r)-W^0_{0\ell}(\xi_0,t_r)] \tag{9.66}$$

Using the limiting formulae, for $\xi_0 \to \infty$, eqs.(9.129) and (9.130) of appendix A, it follows that this expression gives eq.(8.45) for spherical caps. Note that $R_s \to R$ in this limit.

For the calculation of the multipole coefficients, one can simply discard eq.(9.64), a as well as b, for $\ell = 0$. If free charge were present in the system this would not be the case. Eq.(9.64) can now be written as

$$\sum_{\ell_1=1}^{\infty}C^m_{\ell\ell_1}R^{-\ell_1-2}A_{\ell_1 m}+\sum_{\ell_1=1}^{\infty}D^m_{\ell\ell_1}R^{\ell_1-1}B_{\ell_1 m} = H^m_\ell$$
$$\sum_{\ell_1=1}^{\infty}F^m_{\ell\ell_1}R^{-\ell_1-2}A_{\ell_1 m}+\sum_{\ell_1=1}^{\infty}G^m_{\ell\ell_1}R^{\ell_1-1}B_{\ell_1 m} = J^m_\ell$$

$$\text{for} \quad \ell = 1, 2, 3, ... \text{ and } \quad m = 0, \pm 1 \tag{9.67}$$

and it is found that

$$\begin{aligned} H_\ell^m &\equiv \sqrt{4\pi/3}E_0 \cos\theta_0 \delta_{m0} \{\xi_0^{-1} X_1^0(\xi_0)\delta_{\ell 1} + \left(\frac{\epsilon_2 - \epsilon_1}{\epsilon_2}\right)(-1)^{\ell+1} \\ &\times [\sqrt{3} t_r \zeta_{\ell 0}^0 Q_{\ell 0}^0(t_r) - \xi_0^{-1} X_1^0(\xi_0)\zeta_{\ell 1}^0 Q_{\ell 1}^0(t_r)]\} + \sqrt{2\pi/3}E_0 \sin\theta_0 \delta_{\ell 1} \\ &\times [\exp(i\phi_0)\xi_0^{-1} X_1^1(\xi_0)\delta_{m,-1} - \exp(-i\phi_0)\xi_0^{-1} X_1^{-1}(\xi_0)\delta_{m1}] \end{aligned} \tag{9.68}$$

$$\begin{aligned} J_\ell^m &\equiv \sqrt{4\pi/3}E_0 \cos\theta_0 \epsilon_1 \frac{dX_1^0(\xi_0)}{d\xi_0}\delta_{\ell 1}\delta_{m0} \\ &+ \sqrt{2\pi/3}E_0 \sin\theta_0 \left\{ \epsilon_1 \delta_{\ell 1} \left[\exp(i\phi_0)\frac{dX_1^1(\xi_0)}{d\xi_0}\delta_{m,-1} - \exp(-i\phi_0)\frac{dX_1^{-1}(\xi_0)}{d\xi_0}\delta_{m1} \right] \right. \\ &+ (\epsilon_2 - \epsilon_1)(-1)^{\ell+1} \left[\exp(i\phi_0)\frac{dX_1^1(\xi_0)}{d\xi_0}\zeta_{\ell 1}^1 Q_{\ell 1}^1(t_r)\delta_{m,-1} \right. \\ &\left.\left. - \exp(-i\phi_0)\frac{dX_1^{-1}(\xi_0)}{d\xi_0}\zeta_{\ell 1}^{-1} Q_{\ell 1}^{-1}(t_r)\delta_{m1} \right]\right\} \end{aligned} \tag{9.69}$$

If one applies the limiting formulae, (9.128)–(9.132) of appendix A, to the above expressions, (9.65), (9.68) and (9.69), one finds eqs.(8.43), (8.47) and (8.48), valid for the spherical cap, as should be the case in the limit $\xi_0 \to \infty$. The homogeneous linear equations (9.67) become equal to eq.(8.46) in this limit, since then $R_s \to R$.

In a completely analogous way as for truncated spheroids, one can prove here that the $m = -1$ equation (9.67) is superfluous. This equation may now be solved again numerically by neglecting amplitudes $A_{\ell m}$ and $B_{\ell m}$ larger than a suitably chosen order M. One then has to solve here the finite set of linear equations

$$\begin{aligned} \sum_{\ell_1=1}^{M} C_{\ell\ell_1}^m R_s^{-\ell_1-2} A_{\ell_1 m} + \sum_{\ell_1=1}^{M} D_{\ell\ell_1}^m R_s^{\ell_1-1} B_{\ell_1 m} &= H_\ell^m \\ \sum_{\ell_1=1}^{M} F_{\ell\ell_1}^m R_s^{-\ell_1-2} A_{\ell_1 m} + \sum_{\ell_1=1}^{M} G_{\ell\ell_1}^m R_s^{\ell_1-1} B_{\ell_1 m} &= J_\ell^m \\ \text{for } \ell &= 1, 2, 3, ...M; m = 0, 1 \end{aligned} \tag{9.70}$$

Here the matrix elements $C_{\ell\ell_1}^m$, $D_{\ell\ell_1}^m$, $F_{\ell\ell_1}^m$ and $G_{\ell\ell_1}^m$ $(\ell, \ell_1 = 1, 2, 3, ..., M;\ m = 0, 1)$

are again given by eq.(9.65), whereas it follows from eqs.(9.68) and (9.69) that

$$\begin{aligned}
H_\ell^0 &\equiv \sqrt{4\pi/3}E_0\cos\theta_0\{\xi_0^{-1}X_1^0(\xi_0)\delta_{\ell 1} \\
&+\left(\frac{\epsilon_2-\epsilon_1}{\epsilon_2}\right)(-1)^{\ell+1}[\sqrt{3}t_r\zeta_{\ell 0}^0 Q_{\ell 0}^0(t_r)-\xi_0^{-1}X_1^0(\xi_0)\zeta_{\ell 1}^0 Q_{\ell 1}^0(t_r)]\} \\
J_\ell^0 &\equiv \sqrt{4\pi/3}E_0\cos\theta_0\epsilon_1\frac{dX_1^0(\xi_0)}{d\xi_0}\delta_{\ell 1} \\
H_\ell^1 &\equiv -\sqrt{2\pi/3}E_0\sin\theta_0\exp(-i\phi_0)\xi_0^{-1}X_1^1(\xi_0)\delta_{\ell 1} \\
J_\ell^1 &\equiv -\sqrt{2\pi/3}E_0\sin\theta_0\exp(-i\phi_0)\frac{dX_1^1(\xi_0)}{d\xi_0} \\
&\times[\epsilon_1\delta_{\ell 1}+(\epsilon_2-\epsilon_1)(-1)^{\ell+1}\zeta_{\ell 1}^1 Q_{\ell 1}^1(t_r)]
\end{aligned} \tag{9.71}$$

Here it has been used, that $\zeta_{\ell\ell_1}^{-m}Q_{\ell\ell_1}^{-m}(t_r)=\zeta_{\ell\ell_1}^{m}Q_{\ell\ell_1}^{m}(t_r)$ and $X_\ell^{-m}(\xi_0)=X_\ell^m(\xi_0)$, which properties follow from their definitions, eqs.(8.14), (9.22) and (9.20), together with eq.(6.8). The matrix elements $C_{\ell\ell_1}^m$, $D_{\ell\ell_1}^m$, $F_{\ell\ell_1}^m$ and $G_{\ell\ell_1}^m(\ell,\ell_1=1,2,3,\ .....\ ,M;$ $m=0,1)$ are again given by eq.(9.65).

The multipole coefficients $A_{\ell m}$ and $B_{\ell m}$, found in this way for the oblate spheroidal cap, refer to fields in the substrate, just as this was the case for the spherical cap, treated in section 8.3. Moreover, since spheroidal and spherical multipole fields coincide, as long as octupole and higher order fields are neglected, the dipole and quadrupole polarizabilities $\alpha_z(0)$, $\alpha_\|(0)$, $\alpha_z^{10}(0)$ and $\alpha_\|^{10}(0)$ of the spheroidal cap on a substrate may be calculated with the same formulae as for the spherical cap, i.e. with eq.(8.51), (see also text in section 8.3 above this equation). Therefore

$$\begin{aligned}
\alpha_z(0) &= 2\pi\epsilon_2 A_{10}/[(\epsilon_1/\epsilon_2)\sqrt{\pi/3}E_0\cos\theta_0] \\
\alpha_\|(0) &= -4\pi\epsilon_2 A_{11}/[\sqrt{2\pi/3}E_0\sin\theta_0\exp(-i\phi_0)] \\
\alpha_z^{10}(0) &= \pi\epsilon_2 A_{20}/[(\epsilon_1/\epsilon_2)\sqrt{\pi/5}E_0\cos\theta_0] \\
\alpha_\|^{10}(0) &= -4\pi\epsilon_2 A_{21}/[\sqrt{6\pi/5}E_0\sin\theta_0\exp(-i\phi_0)]
\end{aligned} \tag{9.72}$$

Here the argument 0 refers again to the z-coordinate of the center of the cap in the substrate, where the dipole and quadrupole are located.

In a completely analogous way as for spherical caps in section 8.3, one derives that the surface constitutive coefficients $\beta_e(-d)$, $\gamma_e(-d)$, $\delta_e(-d)$ and $\tau(-d)$ for a thin film of identical spheroidal caps in the low density limit is given again by eq.(8.56), (see text in section 8.3 above that equation). One therefore finds:

$$\begin{aligned}
\beta_e(-d) &= \rho\alpha_z(0)/\ \epsilon_2^2 \\
\gamma_e(-d) &= \rho\alpha_\|(0) \\
\delta_e(-d) &= -\rho[\alpha_z^{10}(0)+\alpha_\|^{10}(0)+d\ \alpha_z(0)+d\ \alpha_\|(0)]/\epsilon_2 \\
\tau(-d) &= -\rho\frac{\epsilon_1}{\epsilon_2}[\alpha_\|^{10}(0)+d\ \alpha_\|(0)]
\end{aligned} \tag{9.73}$$

where ρ is the number of particles per unit of surface area. The argument $-d$ refers to the z-coordinate of the dividing surface at $z=-d$ (see fig.9.5). For oblate spheroidal

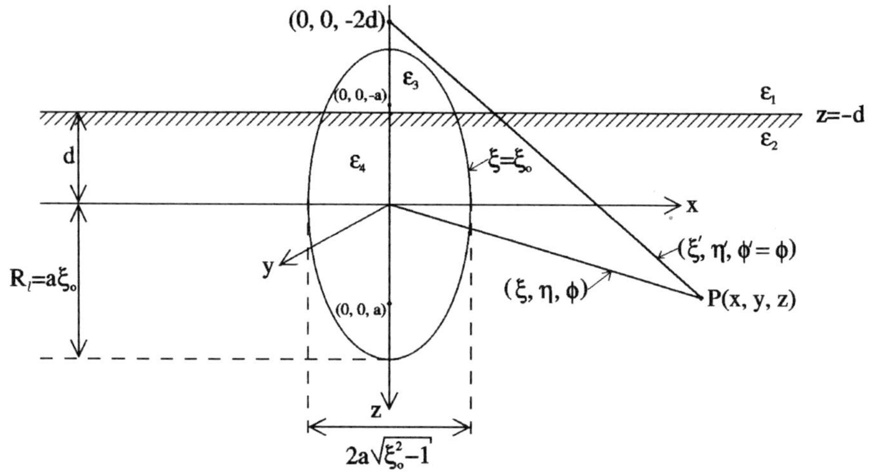

Figure 9.6 Prolate spheroidal cap on a substrate.

caps of different sizes and shapes, the polarizabilities in eq.(9.73) have to be replaced by their averages over the surface (cf. eq.(5.65). The invariants I_e, I_c, $I_{\delta,e}$ and I_τ are found by using eq.(5.66).

The main problem is therefore the solution of eq.(9.70)for such values of M, that the multipole coefficients A_{10}, A_{11}, A_{20} and A_{21}, necessary for the calculation of the polarizabilities, eq.(9.72), for the oblate spheroidal cap on the substrate, are obtained sufficiently accurate. The surface constitutive coefficients are then found, using eq.(9.73). For a further discussion the reader is referred to the last paragraph of section 9.2.

9.5 Prolate spheroidal caps on a substrate

Consider an prolate spheroidal cap with a frequency dependent dielectric constant $\epsilon_3(\omega)$ on a substrate with a dielectric constant $\epsilon_2(\omega)$ and surrounded by an ambient with dielectric constant $\epsilon_1(\omega)$, regions 3, 2 and 1 respectively; see fig. 9.6. As in the previous sections the dielectric constant in region 4 is replaced by $\epsilon_4(\omega)$. The axis of rotation of the spheroid is assumed to be normal to the surface of the planar substrate. The origin of the Cartesian coordinate system is chosen at the center of the spheroid. The z-axis is along the axis of rotation and the positive z-axis points into the substrate. The surface of the substrate is given by $z = -d$, where $0 \leq d \leq R_l$. Here R_l is half the length of the long axis of the spheroid.

Prolate spheroidal coordinates ξ, η and ϕ ($1 \leq \xi < \infty$, $-1 \leq \eta \leq 1$, $0 \leq \phi < 2\pi$) are again introduced, defined by eqs.(6.48)–(6.50). One can distinguish the following four interfaces:

$$\begin{aligned}
&\text{Interface between regions 1 and 3} \ : \quad \xi = \xi_0, \ \ -1 \leq \eta \leq -\frac{d}{a\xi_0} \\
&\text{Interface between regions 2 and 4} \ : \quad \xi = \xi_0, \ \ -\frac{d}{a\xi_0} \leq \eta \leq 1 \\
&\text{Interface between regions 3 and 4} \ : \quad \eta = -\frac{d}{a\xi}, \ \ 1 \leq \xi \leq \xi_0 \\
&\text{Interface between regions 1 and 2} \ : \quad \eta = -\frac{d}{a\xi}, \ \ \xi_0 \leq \xi < \infty
\end{aligned} \tag{9.74}$$

where $\xi_0 \geq 1$. Notice that this is, in fact, the same subdivision as in eq.(9.49). A positive truncation parameter is defined by

$$t_r \equiv \frac{d}{a\xi_0} = \frac{d}{R_l} \ \text{ with } \ 0 \leq t_r \leq 1 \tag{9.75}$$

Now $t_r = 1$ for an infinitely thin cap and $t_r = 0$ for a hemispheroid.

As in the previous section, image multipoles are now introduced at (0,0,$-2d$) to describe the fields. In the present case these will be prolate spheroidal multipoles, cf. section 6.5. The potentials in the regions 1,2,3 and 4 now become, cf. eq.(9.35):

$$\psi_1(\mathbf{r}) = (2\pi/3)^{1/2}[-2^{1/2}E_{0,z}\widetilde{X}_1^0(\xi,a)Y_1^0(\arccos\eta,\phi)$$

$$+(E_{0,x} - iE_{0,y})\widetilde{X}_1^1(\xi,a)Y_1^1(\arccos\eta,\phi) - (E_{0,x} + iE_{0,y})\widetilde{X}_1^{-1}(\xi,a)Y_1^{-1}(\arccos\eta,\phi)]$$
$$+\sum_{\ell m}{}' A_{\ell m}^t \widetilde{Z}_\ell^m(\xi,a)Y_\ell^m(\arccos\eta,\phi) \quad \text{for } \xi_0 \leq \xi < \infty, \ \eta \leq -d/a\xi,$$

$$\begin{aligned}
\psi_2(\mathbf{r}) \ &= \ \psi_2'' + c_1\widetilde{X}_1^0(\xi,a)Y_1^0(\arccos\eta,\phi) + c_2\widetilde{X}_1^1(\xi,a)Y_1^1(\arccos\eta,\phi) \\
&\quad +c_3\widetilde{X}_1^{-1}(\xi,a)Y_1^{-1}(\arccos\eta,\phi) + \sum_{\ell m}{}' A_{\ell m}\widetilde{Z}_\ell^m(\xi,a)Y_\ell^m(\arccos\eta,\phi) + \\
&\quad \sum_{\ell m}{}' A_{\ell m}^r \widetilde{Z}_\ell^m(\xi'(\xi,\eta),a)Y_\ell^m(\arccos\eta'(\xi,\eta),\phi) \ \text{ for } \ \xi_0 \leq \xi < \infty, \ \eta \geq -d/a\xi,
\end{aligned}$$

$$\psi_3(\mathbf{r}) = \psi_3'' + \sum_{\ell m}{}' B_{\ell m}^t \widetilde{X}_\ell^m(\xi,a)Y_\ell^m(\arccos\eta,\phi) \quad \text{for } \ 1 \leq \xi \leq \xi_0, \ \eta \leq -d/a\xi$$

$$\begin{aligned}
\psi_4(\mathbf{r}) \ &= \ \psi_4'' + \sum_{\ell m}{}' B_{\ell m}\widetilde{X}_\ell^m(\xi,a)Y_\ell^m(\arccos\eta,\phi) \\
&\quad + \sum_{\ell m}{}' B_{\ell m}^r \widetilde{X}_\ell^m(\xi'(\xi,\eta),a)Y_\ell^m(\arccos\eta'(\xi,\eta),\phi)
\end{aligned}$$

$$\text{for } 1 \leq \xi \leq \xi_0,\ \eta \geq -d/a\xi \tag{9.76}$$

The functions $\widetilde{X}_\ell^m$, $\widetilde{Z}_\ell^m$ and Y_ℓ^m are defined by eqs.(6.58), (6.59) and (6.5). Furthermore ξ' and η' are the first two prolate spheroidal coordinates of the point (x, y, z) in the shifted coordinate frame with origin at (0,0,$-2d$) (see fig.9.6). The functions $\xi'(\xi,\eta)$ and $\eta'(\xi,\eta)$ are found from eq.(6.50), now by noting that $x'^2 + y'^2 = x^2 + y^2$ and $z' = z + 2d$ for the coordinate transformation. One then obtains,cf. eq.(9.36),

$$\begin{aligned}
\xi'(\xi,\eta) &= \frac{1}{2}\sqrt{2}\xi\left\{1 + 4\frac{d^2}{a^2\xi^2} + 4\frac{d}{a\xi}\eta + \frac{\eta^2}{\xi^2}\right.\\
&\left.+\left[\left(1 + 4\frac{d^2}{a^2\xi^2} + 4\frac{d}{a\xi}\eta + \frac{\eta^2}{\xi^2}\right)^2 - 4\frac{1}{\xi^2}\left(2\frac{d}{a\xi} + \eta\right)^2\right]^{1/2}\right\}^{1/2}\\
\eta'(\xi,\eta) &= \sqrt{2}\left(\eta + 2\frac{d}{a\xi}\right)\left\{1 + 4\frac{d^2}{a^2\xi^2} + 4\frac{d}{a\xi}\eta + \frac{\eta^2}{\xi^2}\right.\\
&\left.+\left[\left(1 + 4\frac{d^2}{a^2\xi^2} + 4\frac{d}{a\xi}\eta + \frac{\eta^2}{\xi^2}\right)^2 - 4\frac{1}{\xi^2}\left(2\frac{d}{a\xi} + \eta\right)^2\right]^{1/2}\right\}^{-1/2}
\end{aligned} \tag{9.77}$$

If the boundary conditions $\psi_1 = \psi_2$ and $\epsilon_1\partial\psi_1/\partial z = \epsilon_2\partial\psi_2/\partial z$ are applied at the interface between the region 1 and 2, one finds, with the above equations and eq.(6.103), that eqs.(9.10), (9.58) and (9.59) are also valid for prolate spheroidal caps. In the same way one finds, by applying the boundary conditions $\psi_3 = \psi_4$ and $\epsilon_3\partial\psi_3/\partial z = \epsilon_4\partial\psi_4/\partial z$ at the interface between regions 3 and 4, that eqs.(9.60) and (9.61) are valid too. Next the boundary conditions, eqs.(9.62) and (9.63), have to be applied at the surface $\xi = \xi_0$ of the prolate spheroid. Upon substitution of the above potentials into these boundary conditions, using also eqs.(9.10), (9.58) and (9.59), one obtains:

$$\begin{aligned}
\sum_{\ell_1=|m|}^{\infty\prime} C_{\ell\ell_1}^m R_l^{-\ell_1-2} A_{\ell_1 m} + \sum_{\ell_1=|m|}^{\infty\prime} D_{\ell\ell_1}^m R_l^{\ell_1-1} B_{\ell_1 m} &= H_\ell^m \\
\sum_{\ell_1=|m|}^{\infty\prime} F_{\ell\ell_1}^m R_l^{-\ell_1-2} A_{\ell_1 m} + \sum_{\ell_1=|m|}^{\infty\prime} G_{\ell\ell_1}^m R_l^{\ell_1-1} B_{\ell_1 m} &= J_\ell^m
\end{aligned} \tag{9.78}$$

where $\ell = 0, 1, 2, 3, ...$ and $m = 0, \pm1, \pm2, \pm3, ..., \pm\ell$. The prime as superindex of the summation symbol means again, that the term $\ell_1 = 0$ should be omitted. Using the definition of the spherical harmonics and their normalization and orthogonality relations, eqs.(6.5) and (9.18), one finds that the matrix elements $C_{\ell\ell_1}^m$, $D_{\ell\ell_1}^m$, $F_{\ell\ell_1}^m$ and $G_{\ell\ell_1}^m$ are now given by

$$\begin{aligned}
C_{\ell\ell_1}^m &\equiv \left(\frac{2\epsilon_2}{\epsilon_2 + \epsilon_1}\right)\xi_0^{\ell_1+1}\widetilde{Z}_{\ell_1}^m(\xi_0)\delta_{\ell\ell_1}\\
&- \left(\frac{\epsilon_2 - \epsilon_1}{\epsilon_2 + \epsilon_1}\right)\zeta_{\ell\ell_1}^m \xi_0^{\ell_1+1}(-1)^{\ell+\ell_1}[\widetilde{Z}_{\ell_1}^m(\xi_0)Q_{\ell\ell_1}^m(t_r) - (-1)^{\ell_1+m}\widetilde{V}_{\ell\ell_1}^m(\xi_0, t_r)]
\end{aligned}$$

$$
\begin{aligned}
D^m_{\ell\ell_1} \equiv & -\left(\frac{2\epsilon_4}{\epsilon_4+\epsilon_3}\right)\xi_0^{-\ell_1}\widetilde{X}^m_{\ell_1}(\xi_0)\delta_{\ell\ell_1} \\
& +\left(\frac{\epsilon_4-\epsilon_3}{\epsilon_4+\epsilon_3}\right)\zeta^m_{\ell\ell_1}\xi_0^{-\ell_1}(-1)^{\ell+\ell_1}[\widetilde{X}^m_{\ell_1}(\xi_0)Q^m_{\ell\ell_1}(t_r)-(-1)^{\ell_1+m}\widetilde{W}^m_{\ell\ell_1}(\xi_0,t_r)]
\end{aligned}
$$

$$
\begin{aligned}
F^m_{\ell\ell_1} \equiv & \left(\frac{2\epsilon_2\epsilon_1}{\epsilon_2+\epsilon_1}\right)\xi_0^{\ell_1+2}\frac{d\widetilde{Z}^m_{\ell_1}(\xi_0)}{d\xi_0}\delta_{\ell\ell_1}+\epsilon_2\left(\frac{\epsilon_2-\epsilon_1}{\epsilon_2+\epsilon_1}\right)\zeta^m_{\ell\ell_1}\xi_0^{\ell_1+2}(-1)^{\ell+\ell_1} \\
& \times\left[\frac{d\widetilde{Z}^m_{\ell_1}(\xi_0)}{d\xi_0}Q^m_{\ell\ell_1}(t_r)+(-1)^{\ell_1+m}\frac{\partial\widetilde{V}^m_{\ell\ell_1}(\xi_0,t_r)}{\partial\xi_0}\right]
\end{aligned}
$$

$$
\begin{aligned}
G^m_{\ell\ell_1} \equiv & -\left(\frac{2\epsilon_4\epsilon_3}{\epsilon_4+\epsilon_3}\right)\xi_0^{-\ell_1+1}\frac{d\widetilde{X}^m_{\ell_1}(\xi_0)}{d\xi_0}\delta_{\ell\ell_1}-\epsilon_4\left(\frac{\epsilon_4-\epsilon_3}{\epsilon_4+\epsilon_3}\right)\zeta^m_{\ell\ell_1}\xi_0^{-\ell_1+1}(-1)^{\ell+\ell_1} \\
& \times\left[\frac{d\widetilde{X}^m_{\ell_1}(\xi_0)}{d\xi_0}Q^m_{\ell\ell_1}(t_r)+(-1)^{\ell_1+m}\frac{\partial\widetilde{W}^m_{\ell\ell_1}(\xi_0,t_r)}{\partial\xi_0}\right] \qquad (9.79)
\end{aligned}
$$

where the various functions are defined by eqs.(9.39)– (9.42), (8.14) and (8.15).

Just as for oblate spheroidal caps (see the previous section), the terms H^m_ℓ and J^m_ℓ at the right-hand sides of eq.(9.78), are found to be only unequal to zero, if $m = 0, 1$ or -1. Since terms with different m do not couple in this equation, as a consequence of the rotational symmetry of the system around the z-axis, only the case $m = 0, 1$ or -1 has to be considered. H^0_0 is again found to contain the constant ψ_0, from eqs.(9.60). Eq.(9.78a), for $\ell = 0$, can therefore be used to determine this unknown quantity ψ_0 in terms of the multipole coefficients. This results in

$$
\begin{aligned}
\psi_0 = & R_l\left(\frac{\epsilon_1}{\epsilon_2}-1\right)\left\{\frac{1}{\sqrt{3}}\zeta^0_{01}Q^0_{01}(t_r)-t_r\zeta^0_{00}Q^0_{00}(t_r)\right\}E_0\cos\theta_0 \\
& -\frac{1}{2\sqrt{\pi}}\sum_{\ell=1}^{\infty}A_{\ell 0}R_l^{-\ell-1}\left(\frac{\epsilon_2-\epsilon_1}{\epsilon_2+\epsilon_1}\right)\zeta^0_{0\ell}\xi_0^{\ell+1}[(-1)^\ell\widetilde{Z}^0_\ell(\xi_0)Q^0_{0\ell}(t_r)-\widetilde{V}^0_{0\ell}(\xi_0,t_r)] \\
& +\frac{1}{2\sqrt{\pi}}\sum_{\ell=1}^{\infty}B_{\ell 0}R_l^{\ell}\left(\frac{\epsilon_4-\epsilon_3}{\epsilon_4+\epsilon_3}\right)\zeta^0_{0\ell}\xi_0^{-\ell}[(-1)^\ell\widetilde{X}^0_\ell(\xi_0)Q^0_{0\ell}(t_r)-\widetilde{W}^0_{0\ell}(\xi_0,t_r)]
\end{aligned}
\qquad (9.80)
$$

Using the limiting formulae, for $\xi_0 \to \infty$, eqs.(9.134) and (9.135) of appendix A, it follows that this expression gives eq.(8.45) for spherical caps. Note that $R_l \to R$ in this limit.

For the calculation of the multipole coefficients, one can simply discard eq.(9.78), a as well as b, for $\ell = 0$. If free charge were present in the system this would not be the case. Eq.(9.78) can now be written as

$$
\begin{aligned}
\sum_{\ell_1=1}^{\infty}C^m_{\ell\ell_1}R_l^{-\ell_1-2}A_{\ell_1 m}+\sum_{\ell_1=1}^{\infty}D^m_{\ell\ell_1}R_l^{\ell_1-1}B_{\ell_1 m} &= H^m_\ell \\
\sum_{\ell_1=1}^{\infty}F^m_{\ell\ell_1}R_l^{-\ell_1-2}A_{\ell_1 m}+\sum_{\ell_1=1}^{\infty}G^m_{\ell\ell_1}R_l^{\ell_1-1}B_{\ell_1 m} &= J^m_\ell
\end{aligned}
$$

$$\text{for} \quad \ell = 1, 2, 3, \ldots.. \text{ and } m = 0, \pm 1 \tag{9.81}$$

It is found that

$$\begin{aligned} H_\ell^m &\equiv \sqrt{4\pi/3} E_0 \cos\theta_0 \delta_{m0} \{ \xi_0^{-1} \widetilde{X}_1^0(\xi_0) \delta_{\ell 1} + \left(\frac{\epsilon_2 - \epsilon_1}{\epsilon_2} \right) (-1)^{\ell+1} \\ &\times [\sqrt{3} t_r \zeta_{\ell 0}^0 Q_{\ell 0}^0(t_r) - \xi_0^{-1} \widetilde{X}_1^0(\xi_0) \zeta_{\ell 1}^0 Q_{\ell 1}^0(t_r)] \} + \sqrt{2\pi/3} E_0 \sin\theta_0 \delta_{\ell 1} \\ &\times [\exp(i\phi_0) \xi_0^{-1} \widetilde{X}_1^1(\xi_0) \delta_{m,-1} - \exp(-i\phi_0) \xi_0^{-1} \widetilde{X}_1^{-1}(\xi_0) \delta_{m1}] \end{aligned} \tag{9.82}$$

$$\begin{aligned} J_\ell^m &\equiv \sqrt{4\pi/3} E_0 \cos\theta_0 \epsilon_1 \frac{d\widetilde{X}_1^0(\xi_0)}{d\xi_0} \delta_{\ell 1} \delta_{m0} \\ &+ \sqrt{2\pi/3} E_0 \sin\theta_0 \left\{ \epsilon_1 \delta_{\ell 1} \left[\exp(i\phi_0) \frac{d\widetilde{X}_1^1(\xi_0)}{d\xi_0} \delta_{m,-1} - \exp(-i\phi_0) \frac{d\widetilde{X}_1^{-1}(\xi_0)}{d\xi_0} \delta_{m1} \right] \right. \\ &+ (\epsilon_2 - \epsilon_1)(-1)^{\ell+1} \left[\exp(i\phi_0) \frac{d\widetilde{X}_1^1(\xi_0)}{d\xi_0} \zeta_{\ell 1}^1 Q_{\ell 1}^1(t_r) \delta_{m,-1} \right. \\ &\left. \left. - \exp(-i\phi_0) \frac{d\widetilde{X}_1^{-1}(\xi_0)}{d\xi_0} \zeta_{\ell 1}^{-1} Q_{\ell 1}^{-1}(t_r) \delta_{m1} \right] \right\} \end{aligned} \tag{9.83}$$

If one applies the limiting formulae, (9.133)–(9.137) of appendix A, to the above expressions, (9.79), (9.12) and (9.83), one finds eqs.(8.43), (8.47) and (8.48), valid for the spherical cap, as should be the case in the limit $\xi_0 \to \infty$. The homogeneous linear equations (9.81) become equal to eq.(8.46) in this limit, since then $R_l \to R$.

Again one can prove, that the $m = -1$ equation (9.81) is superfluous. This equation may then be solved numerically by neglecting amplitudes $A_{\ell m}$ and $B_{\ell m}$ larger than a suitably chosen order M. One then solves:

$$\begin{aligned} \sum_{\ell_1=1}^{M} C_{\ell\ell_1}^m R_l^{-\ell_1-2} A_{\ell_1 m} + \sum_{\ell_1=1}^{M} D_{\ell\ell_1}^m R_l^{\ell_1-1} B_{\ell_1 m} &= H_\ell^m \\ \sum_{\ell_1=1}^{M} F_{\ell\ell_1}^m R_l^{-\ell_1-2} A_{\ell_1 m} + \sum_{\ell_1=1}^{M} G_{\ell\ell_1}^m R_l^{\ell_1-1} B_{\ell_1 m} &= J_\ell^m \\ \text{for } \ell &= 1, 2, 3, \ldots.. M; m = 0, 1 \end{aligned} \tag{9.84}$$

where H_ℓ^0 , J_ℓ^0 , H_ℓ^1 and J_ℓ^1 are given by

$$
\begin{aligned}
H_\ell^0 &\equiv \sqrt{4\pi/3}E_0\cos\theta_0\{\xi_0^{-1}\widetilde{X}_1^0(\xi_0)\delta_{\ell 1} \\
&\quad + \left(\frac{\epsilon_2-\epsilon_1}{\epsilon_2}\right)(-1)^{\ell+1}[\sqrt{3}t_r\zeta_{\ell 0}^0 Q_{\ell 0}^0(t_r) - \xi_0^{-1}\widetilde{X}_1^0(\xi_0)\zeta_{\ell 1}^0 Q_{\ell 1}^0(t_r)]\} \\
J_\ell^0 &\equiv \sqrt{4\pi/3}E_0\cos\theta_0\epsilon_1\frac{d\widetilde{X}_1^0(\xi_0)}{d\xi_0}\delta_{\ell 1} \\
H_\ell^1 &\equiv -\sqrt{2\pi/3}E_0\sin\theta_0\exp(-i\phi_0)\xi_0^{-1}\widetilde{X}_1^1(\xi_0)\delta_{\ell 1} \\
J_\ell^1 &\equiv -\sqrt{2\pi/3}E_0\sin\theta_0\exp(-i\phi_0)\frac{d\widetilde{X}_1^1(\xi_0)}{d\xi_0} \\
&\quad [\epsilon_1\delta_{\ell 1} + (\epsilon_2-\epsilon_1)(-1)^{\ell+1}\zeta_{\ell 1}^1 Q_{\ell 1}^1(t_r)]
\end{aligned}
\tag{9.85}
$$

and the matrix elements $C_{\ell\ell_1}^m$, $D_{\ell\ell_1}^m$, $F_{\ell\ell_1}^m$ and $G_{\ell\ell_1}^m$ by eq.(9.79).

The polarizabilities $\alpha_z(0)$, $\alpha_\|(0)$, $\alpha_z^{10}(0)$ and $\alpha_\|^{10}(0)$ of the prolate spheroidal caps on a substrate are then found again, using eq.(9.72), and the surface constitutive coefficients $\beta_e(-d)$, $\gamma_e(-d)$, $\delta_e(-d)$ and $\tau(-d)$ of a thin film of identical prolate spheroidal caps in the low density limit from eq.(9.73), of the previous section. For prolate spheroidal caps of different sizes and shapes, the polarizabilities in eq.(9.73) have to be replaced by their averages over the surface (cf. eq.(5.65)). The invariants I_e, I_c, $I_{\delta,e}$ and I_τ are found again by using eq.(5.66).

The main problem is therefore the solution of eq.(9.84) for a large enough value of M, that the multipole coefficients A_{10} , A_{11} , A_{20} and A_{21} , necessary for the calculation of the polarizabilities, eq.(9.72), for the prolate spheroidal cap on a substrate, are obtained sufficiently accurate. The surface constitutive coefficients are then found, using eq.(9.73). For a further discussion the reader is referred to the last paragraph of the previous section.

9.6 Spheroids and hemispheroids on a substrate

In this section two special cases will be considered. The first case is the spheroid, both oblate and prolate. This case has already been considered in chapter 6. It is instructive to see how the more general solution given in this chapter, reduces to the one given for the spheroid in chapter 6. The other case is the hemispheroid, both oblate and prolate. In this case the solution also simplifies considerably. It is therefore interesting and useful to give also this case some special attention.

In the case of an oblate spheroid, touching the substrate, the truncation parameter is one, $t_r = 1$. Just as in section 8.4, the region 4 in figs.9.3 and 9.4 disappears in this case, and the spheroidal multipole coefficients will therefore be independent of the value of ϵ_4. One can again conveniently put $\epsilon_4 = \epsilon_3$. For oblate spheroids the matrix elements $C^m_{\ell\ell_1}$, $D^m_{\ell\ell_1}$, $F^m_{\ell\ell_1}$ and $G^m_{\ell\ell_1}$, eq.(9.19), then become

$$\begin{aligned}
C^m_{\ell\ell_1} &= \left(\frac{2\epsilon_1}{\epsilon_1+\epsilon_2}\right)\xi_0^{\ell_1+1}Z^m_{\ell_1}(\xi_0)\delta_{\ell\ell_1} \\
&\quad -\left(\frac{\epsilon_1-\epsilon_2}{\epsilon_1+\epsilon_2}\right)\zeta^m_{\ell\ell_1}\xi_0^{\ell_1+1}[Z^m_{\ell_1}(\xi_0)Q^m_{\ell\ell_1}(1)-(-1)^{\ell_1+m}V^m_{\ell\ell_1}(\xi_0,1)] \\
D^m_{\ell\ell_1} &= -\xi_0^{-\ell_1}X^m_{\ell_1}(\xi_0)\delta_{\ell\ell_1} \\
F^m_{\ell\ell_1} &= \left(\frac{2\epsilon_1\epsilon_2}{\epsilon_1+\epsilon_2}\right)\xi_0^{\ell_1+2}\frac{dZ^m_{\ell_1}(\xi_0)}{d\xi_0}\delta_{\ell\ell_1} \\
&\quad +\epsilon_1\left(\frac{\epsilon_1-\epsilon_2}{\epsilon_1+\epsilon_2}\right)\zeta^m_{\ell\ell_1}\xi_0^{\ell_1+2}\left[\frac{dZ^m_{\ell_1}(\xi_0)}{d\xi_0}Q^m_{\ell\ell_1}(1)+(-1)^{\ell_1+m}\frac{\partial V^m_{\ell\ell_1}(\xi_0,1)}{\partial\xi_0}\right] \\
G^m_{\ell\ell_1} &= -\epsilon_3\xi_0^{-\ell_1+1}\frac{dX^m_{\ell_1}(\xi_0)}{d\xi_0}\delta_{\ell\ell_1}
\end{aligned} \tag{9.86}$$

where $\zeta^m_{\ell\ell_1}$ is given by eq.(8.14), $X^m_\ell(\xi)$ and $Z^m_\ell(\xi)$ by eqs.(9.20) and (9.21), and $Q^m_{\ell\ell_1}(1)$ and $V^m_{\ell\ell_1}(\xi,1)$ by eqs.(9.22) and (9.23) for $t_r=1$, respectively.

It will be obvious that eq.(8.58) holds again, i.e.

$$\zeta^m_{\ell\ell_1}Q^m_{\ell\ell_1}(1)=\delta_{\ell\ell_1} \tag{9.87}$$

Multiplication of the spheroidal multipole expansion, eq.(6.108), on both sides with the complex conjugate of the spherical harmonic $Y^{m'}_{\ell'}(\arccos\eta,\phi)$, and integration over η and ϕ, gives with eq.(9.18)

$$\begin{aligned}
&\int_{-1}^{1}d\eta\int_0^{2\pi}d\phi\; Z^m_\ell(\xi'(\xi,\eta),a)[Y^{m'}_{\ell'}(\arccos\eta,\phi)]^*Y^m_\ell(\arccos\eta'(\xi,\eta),\phi) \\
&= K^m_{\ell'\ell}(\xi_1)(2d)^{-\ell-\ell'-1}X^m_{\ell'}(\xi,a)\delta_{mm'}
\end{aligned} \tag{9.88}$$

Expressing the spherical harmonics in this equation by means of eq.(6.5) in terms of associated Legendre functions and using eqs.(8.14), (9.20), (9.21) and (9.23), one obtains the following results

$$\zeta^m_{\ell'\ell}V^m_{\ell'\ell}(\xi,1)=\left(\frac{a}{2d}\right)^{\ell'+\ell+1}K^m_{\ell'\ell}(\xi_1)X^m_{\ell'}(\xi) \tag{9.89}$$

and

$$\zeta^m_{\ell'\ell}\frac{\partial V^m_{\ell'\ell}(\xi,1)}{\partial\xi}=\left(\frac{a}{2d}\right)^{\ell'+\ell+1}K^m_{\ell'\ell}(\xi_1)\frac{dX^m_{\ell'}(\xi)}{d\xi} \tag{9.90}$$

Introducing for ℓ' and ℓ the new notation ℓ and ℓ_1, respectively, one finds for $\xi=\xi_0$, since $d=a\xi_1$ while furthermore $\xi_1=\xi_0$ for a spheroid which touches the substrate,

$$\zeta^m_{\ell\ell_1}V^m_{\ell\ell_1}(\xi_0,1)=\left(\frac{1}{2\xi_0}\right)^{\ell+\ell_1+1}K^m_{\ell\ell_1}(\xi_0)X^m_\ell(\xi_0) \tag{9.91}$$

and

$$\zeta^m_{\ell\ell_1}\frac{\partial V^m_{\ell\ell_1}(\xi_0,1)}{\partial\xi_0} = \left(\frac{1}{2\xi_0}\right)^{\ell+\ell_1+1} K^m_{\ell\ell_1}(\xi_0)\frac{dX^m_\ell(\xi_0)}{d\xi_0} \tag{9.92}$$

If the above formulae, eqs.(9.87), (9.91) and (9.92), are substituted into eq.(9.86), one obtains the following result for the matrix elements $C^m_{\ell\ell_1}$, $D^m_{\ell\ell_1}$, $F^m_{\ell\ell_1}$ and $G^m_{\ell\ell_1}$ in the case of an oblate spheroid on a substrate:

$$\begin{aligned}
C^m_{\ell\ell_1} &= \xi_0^{\ell+1}Z^m_\ell(\xi_0)\delta_{\ell\ell_1} + \left(\frac{\epsilon_1-\epsilon_2}{\epsilon_1+\epsilon_2}\right)\frac{(-1)^{\ell_1+m}}{2^{\ell+\ell_1+1}}K^m_{\ell\ell_1}(\xi_0)\xi_0^{-\ell}X^m_\ell(\xi_0)\\
D^m_{\ell\ell_1} &= -\xi_0^{-\ell}X^m_\ell(\xi_0)\delta_{\ell\ell_1}\\
F^m_{\ell\ell_1} &= \epsilon_1\xi_0^{\ell+2}\frac{dZ^m_\ell(\xi_0)}{d\xi_0}\delta_{\ell\ell_1} + \epsilon_1\left(\frac{\epsilon_1-\epsilon_2}{\epsilon_1+\epsilon_2}\right)\frac{(-1)^{\ell_1+m}}{2^{\ell+\ell_1+1}}K^m_{\ell\ell_1}(\xi_0)\xi_0^{-\ell+1}\frac{dX^m_\ell(\xi_0)}{d\xi_0}\\
G^m_{\ell\ell_1} &= -\epsilon_3\xi_0^{-\ell+1}\frac{dX^m_\ell(\xi_0)}{d\xi_0}\delta_{\ell\ell_1} \quad \text{for } \ell,\ell_1 = 1,2,3,.....
\end{aligned} \tag{9.93}$$

Furthermore it follows from eqs.(9.30) and (9.87), using the fact that $\ell \neq 0$, that

$$\begin{aligned}
H^0_\ell &= \sqrt{4\pi/3}E_0\cos\theta_0\xi_0^{-1}X^0_1(\xi_0)\delta_{\ell 1}\\
J^0_\ell &= \sqrt{4\pi/3}E_0\cos\theta_0\epsilon_1\frac{dX^0_1(\xi_0)}{d\xi_0}\delta_{\ell 1}\\
H^1_\ell &= -\sqrt{2\pi/3}E_0\sin\theta_0\exp(-i\phi_0)\xi_0^{-1}X^1_1(\xi_0)\delta_{\ell 1}\\
J^1_\ell &= -\sqrt{2\pi/3}E_0\sin\theta_0\exp(-i\phi_0)\frac{dX^1_1(\xi_0)}{d\xi_0}\epsilon_1\delta_{\ell 1} \quad \text{for } \ell = 1,2,3,.....
\end{aligned} \tag{9.94}$$

Using the above results, eqs.(9.93) and (9.94), in eq.(9.29), one can eliminate the spheroidal multipole coefficient $B_{\ell m}$, by multiplying both sides of the first equation (9.29) with the factor $\epsilon_3\xi_0^{-\ell+1}dX^m_\ell(\xi_0)/d\xi_0$ and those of the second with $\xi_0^{-\ell}X^m_\ell(\xi_0)$, and subsequently subtracting the resulting equations. The result can then be written in the following form for $m = 0$:

$$\begin{aligned}
&A_{\ell 0} + \left(\frac{\epsilon_1-\epsilon_2}{\epsilon_1+\epsilon_2}\right)(\epsilon_3-\epsilon_1)\xi_0^{-2\ell-1}R_s^{2\ell+1}\left[\epsilon_3\frac{Z^0_\ell(\xi_0)}{X^0_\ell(\xi_0)} - \epsilon_1\frac{dZ^0_\ell(\xi_0)/d\xi_0}{dX^0_\ell(\xi_0)/d\xi_0}\right]^{-1}\\
&\times\sum_{\ell_1=1}^{M}(-1)^{\ell_1}K^0_{\ell\ell_1}(\xi_0)(2R_s)^{-\ell-\ell_1-1}A_{\ell_1,0}\\
&= \sqrt{4\pi/3}E_0\cos\theta_0(\epsilon_3-\epsilon_1)\xi_0^{-3}R_s^3\left[\epsilon_3\frac{Z^0_1(\xi_0)}{X^0_1(\xi_0)} - \epsilon_1\frac{dZ^0_1(\xi_0)/d\xi_0}{dX^0_1(\xi_0)/d\xi_0}\right]^{-1}\delta_{\ell 1}\\
&\text{with } \ell = 1,2,3,.....,M
\end{aligned} \tag{9.95}$$

and for $m = 1$:

$$A_{\ell 1} - \left(\frac{\epsilon_1 - \epsilon_2}{\epsilon_1 + \epsilon_2}\right)(\epsilon_3 - \epsilon_1)\xi_0^{-2\ell-1} R_s^{2\ell+1} \left[\epsilon_3 \frac{Z_\ell^1(\xi_0)}{X_\ell^1(\xi_0)} - \epsilon_1 \frac{dZ_\ell^1(\xi_0)/d\xi_0}{dX_\ell^1(\xi_0)/d\xi_0}\right]^{-1}$$
$$\times \sum_{\ell_1=1}^{M} (-1)^{\ell_1} K_{\ell\ell_1}^1(\xi_0)(2R_s)^{-\ell-\ell_1-1} A_{\ell_1,1}$$
$$= -\sqrt{2\pi/3}E_0 \sin\theta_0 \exp(-i\phi_0)(\epsilon_3 - \epsilon_1)\xi_0^{-3} R_s^3 \left[\epsilon_3 \frac{Z_1^1(\xi_0)}{X_1^1(\xi_0)} - \epsilon_1 \frac{dZ_1^1(\xi_0)/d\xi_0}{dX_1^1(\xi_0)/d\xi_0}\right]^{-1} \delta_{\ell 1}$$
$$\text{for } \ell = 1, 2, 3, \ldots\ldots, M \tag{9.96}$$

where M is any natural number. Since ϵ_1, ϵ_2 and ϵ_3 were denoted by ϵ_a, ϵ_s and ϵ, respectively, in chapter 6, one finds, using eq.(6.22) for the multipole polarizabilities $\alpha_{\ell m}$ of the oblate spheroid in the infinite ambient and the definitions, eqs.(9.20) and (9.21), that eqs.(9.95) and (9.96) are identical with eqs.(6.130) and (6.131). Notice that the latter equations have been derived for the more general case, that the oblate spheroids are at a distance $d \geq R_s$ on or above the substrate, whereas eqs.(9.95) and (9.96) are only valid for oblate spheroids, touching the substrate ($d = R_s$).

For prolate spheroids on a substrate, linear inhomogeneous equations for the multipole coefficients $A_{\ell 0}$ and $A_{\ell 1}$ can be derived in an analogous way as above, starting from eqs.(9.47), (9.38) and (9.48). This set of linear equations can also be derived more quickly, by making the substitutions $\xi_0 \to -i\xi_0$ and $R_s \to R_l$ in eqs.(9.95) and (9.96). Using eqs.(6.159), (9.39) and (9.40),one obtains for $m = 0$:

$$A_{\ell 0} + \left(\frac{\epsilon_1 - \epsilon_2}{\epsilon_1 + \epsilon_2}\right)(\epsilon_3 - \epsilon_1)\xi_0^{-2\ell-1} R_l^{2\ell+1} \left(\epsilon_3 \frac{\widetilde{Z}_\ell^0(\xi_0)}{\widetilde{X}_\ell^0(\xi_0)} - \epsilon_1 \frac{d\widetilde{Z}_\ell^0(\xi_0)/d\xi_0}{d\widetilde{X}_\ell^0(\xi_0)/d\xi_0}\right)^{-1}$$
$$\times \sum_{\ell_1=1}^{M} (-1)^{\ell_1} \widetilde{K}_{\ell\ell_1}^0(\xi_0)(2R_l)^{-\ell-\ell_1-1} A_{\ell_1,0}$$
$$= \sqrt{4\pi/3}E_0 \cos\theta_0(\epsilon_3 - \epsilon_1)\xi_0^{-3} R_l^3 \left(\epsilon_3 \frac{\widetilde{Z}_1^0(\xi_0)}{\widetilde{X}_1^0(\xi_0)} - \epsilon_1 \frac{d\widetilde{Z}_1^0(\xi_0)/d\xi_0}{d\widetilde{X}_1^0(\xi_0)/d\xi_0}\right)^{-1} \delta_{\ell 1}$$
$$\text{with } \ell = 1, 2, 3, \ldots\ldots, M \tag{9.97}$$

and for $m = 1$:

$$A_{\ell 1} - \left(\frac{\epsilon_1 - \epsilon_2}{\epsilon_1 + \epsilon_2}\right)(\epsilon_3 - \epsilon_1)\xi_0^{-2\ell-1} R_l^{2\ell+1} \left(\epsilon_3 \frac{\widetilde{Z}_\ell^1(\xi_0)}{\widetilde{X}_\ell^1(\xi_0)} - \epsilon_1 \frac{d\widetilde{Z}_\ell^1(\xi_0)/d\xi_0}{d\widetilde{X}_\ell^1(\xi_0)/d\xi_0}\right)^{-1}$$
$$\times \sum_{\ell_1=1}^{M} (-1)^{\ell_1} \widetilde{K}_{\ell\ell_1}^1(\xi_0)(2R_l)^{-\ell-\ell_1-1} A_{\ell_1,1}$$
$$= -\sqrt{2\pi/3}E_0 \sin\theta_0 \exp(-i\phi_0)(\epsilon_3 - \epsilon_1)\xi_0^{-3} R_l^3 \left(\epsilon_3 \frac{\widetilde{Z}_1^1(\xi_0)}{\widetilde{X}_1^1(\xi_0)} - \epsilon_1 \frac{d\widetilde{Z}_1^1(\xi_0)/d\xi_0}{d\widetilde{X}_1^1(\xi_0)/d\xi_0}\right)^{-1} \delta_{\ell 1}$$
$$\text{for } \ell = 1, 2, 3, \ldots\ldots, M \tag{9.98}$$

where M is any natural number. Since ϵ_1, ϵ_2 and ϵ_3 were denoted by ϵ_a, ϵ_s and ϵ, respectively, in chapter 6, one finds, using eq.(6.68) for the multipole polarizabilities $\widetilde{\alpha}_{\ell m}$ of the prolate spheroid in the infinite ambient and the definitions, eqs.(9.39) and (9.40), that the above equations are identical with eqs.(6.173) and (6.174). Notice that the latter equations have been derived for the more general case, that the prolate spheroids are at a distance $d \geq R_l$ on or above the substrate, whereas eqs.(9.97) and (9.98) are only valid for prolate spheroids, touching the substrate ($d = R_l$).

As it should be the case, both eqs.(9.95) and (9.96), valid for oblate spheroids, and eqs.(9.97) and (9.98), valid for prolate spheroids, tend to eqs.(8.66) and (8.67) for spheres in the limit as $\xi_0 \to \infty$ and $a \to 0$, with $a\xi_0 = R_s = R_l \equiv R$. This can be proved by using the asymptotic relations, eqs.(9.130), (9.132), (9.135), (9.137), (6.109) and (6.158).

Next the case of a hemispheroidal island, touching the substrate with its flat side (normal to the axis of revolution), will be considered. This special case can be obtained for oblate hemispheroids by putting $t_r = 0$ and $\epsilon_4 = \epsilon_2$ in the general formulae for the matrix elements $C^m_{\ell\ell_1}$, $D^m_{\ell\ell_1}$, $F^m_{\ell\ell_1}$ and $G^m_{\ell\ell_1}$, and for H^m_ℓ and J^m_ℓ in section 9.2. It follows from eqs.(9.4), (9.5) and (9.2) that

$$\xi'(\xi_0, \eta) = \xi_0, \quad \eta'(\xi_0, \eta) = \eta \tag{9.99}$$

Using this result in eqs.(9.23) and (9.24), one finds with eq.(9.22) that

$$V^m_{\ell\ell_1}(\xi_0, 0) = Q^m_{\ell\ell_1}(0) Z^m_{\ell_1}(\xi_0) \tag{9.100}$$

and

$$W^m_{\ell\ell_1}(\xi_0, 0) = Q^m_{\ell\ell_1}(0) X^m_{\ell_1}(\xi_0) \tag{9.101}$$

Using these results in eq.(9.19), one finds, with $\epsilon_4 = \epsilon_2$, for oblate hemispheroids:

$$\begin{aligned}
C^m_{\ell\ell_1} &= \left(\frac{2\epsilon_1}{\epsilon_1 + \epsilon_2}\right) \xi_0^{\ell_1+1} Z^m_{\ell_1}(\xi_0) \delta_{\ell\ell_1} \\
&\quad - \left(\frac{\epsilon_1 - \epsilon_2}{\epsilon_1 + \epsilon_2}\right) \xi_0^{\ell_1+1} Z^m_{\ell_1}(\xi_0) [1 - (-1)^{\ell_1+m}] \zeta^m_{\ell\ell_1} Q^m_{\ell\ell_1}(0)
\end{aligned}$$

$$\begin{aligned}
D^m_{\ell\ell_1} &= -\left(\frac{2\epsilon_3}{\epsilon_2 + \epsilon_3}\right) \xi_0^{-\ell_1} X^m_{\ell_1}(\xi_0) \delta_{\ell\ell_1} \\
&\quad - \left(\frac{\epsilon_2 - \epsilon_3}{\epsilon_2 + \epsilon_3}\right) \xi_0^{-\ell_1} X^m_{\ell_1}(\xi_0) [1 - (-1)^{\ell_1+m}] \zeta^m_{\ell\ell_1} Q^m_{\ell\ell_1}(0)
\end{aligned}$$

$$\begin{aligned}
F^m_{\ell\ell_1} &= \left(\frac{2\epsilon_1\epsilon_2}{\epsilon_1 + \epsilon_2}\right) \xi_0^{\ell_1+2} \frac{dZ^m_{\ell_1}(\xi_0)}{d\xi_0} \delta_{\ell\ell_1} \\
&\quad + \epsilon_1 \left(\frac{\epsilon_1 - \epsilon_2}{\epsilon_1 + \epsilon_2}\right) \xi_0^{\ell_1+2} \frac{dZ^m_{\ell_1}(\xi_0)}{d\xi_0} [1 + (-1)^{\ell_1+m}] \zeta^m_{\ell\ell_1} Q^m_{\ell\ell_1}(0)
\end{aligned}$$

$$
\begin{aligned}
G_{\ell\ell_1}^m &= -\left(\frac{2\epsilon_2\epsilon_3}{\epsilon_2+\epsilon_3}\right)\xi_0^{-\ell_1+1}\frac{dX_{\ell_1}^m(\xi_0)}{d\xi_0}\delta_{\ell\ell_1} \\
&\quad +\epsilon_3\left(\frac{\epsilon_2-\epsilon_3}{\epsilon_2+\epsilon_3}\right)\xi_0^{-\ell_1+1}\frac{dX_{\ell_1}^m(\xi_0)}{d\xi_0}[1+(-1)^{\ell_1+m}]\zeta_{\ell\ell_1}^m Q_{\ell\ell_1}^m(0) \qquad (9.102)
\end{aligned}
$$

Furthermore it follows from eq.(9.30), that in this case

$$
\begin{aligned}
H_\ell^0 &= \sqrt{4\pi/3}E_0\cos\theta_0\xi_0^{-1}X_1^0(\xi_0)\left\{\frac{\epsilon_1}{\epsilon_2}\delta_{\ell 1}-\left(\frac{\epsilon_1-\epsilon_2}{\epsilon_2}\right)\zeta_{\ell 1}^0 Q_{\ell 1}^0(0)\right\} \\
J_\ell^0 &= \sqrt{4\pi/3}E_0\cos\theta_0\epsilon_1\frac{dX_1^0(\xi_0)}{d\xi_0}\delta_{\ell 1} \\
H_\ell^1 &= -\sqrt{2\pi/3}E_0\sin\theta_0\exp(-i\phi_0)\xi_0^{-1}X_1^1(\xi_0)\delta_{\ell 1} \\
J_\ell^1 &= -\sqrt{2\pi/3}E_0\sin\theta_0\exp(-i\phi_0)\frac{dX_1^1(\xi_0)}{d\xi_0}[\epsilon_2\delta_{\ell 1}+(\epsilon_1-\epsilon_2)\zeta_{\ell 1}^1 Q_{\ell 1}^1(0)]
\end{aligned} \qquad (9.103)
$$

In the above equations $\zeta_{\ell\ell_1}^m$ is given by eq.(8.14). The integrals

$$
Q_{\ell\ell_1}^m(0) \equiv \int_{-1}^{0} P_\ell^m(\eta)P_{\ell_1}^m(\eta)d\eta \qquad (9.104)
$$

can be evaluated explicitly. The result is given by eqs.(8.75) and (8.76).

In appendix B it is shown, that one can eliminate the multipole coefficients $B_{\ell m}$ by using eqs.(9.102), (9.103), (8.75) and (8.14) in eq.(9.29). For $m=0$ one then obtains:

$$
\begin{aligned}
&\left(-\epsilon_1(\epsilon_2+\epsilon_3)\xi_0^{\ell+2}\frac{dZ_\ell^0(\xi_0)}{d\xi_0}+\epsilon_3(\epsilon_1+\epsilon_2)\xi_0^{\ell+2}\frac{dX_\ell^0(\xi_0)}{d\xi_0}\frac{Z_\ell^0(\xi_0)}{X_\ell^0(\xi_0)}\right)R_s^{-\ell-2}A_{\ell 0} \\
&-2\epsilon_1(\epsilon_1-\epsilon_3)\sum_{\ell_1\ even}^{M}\zeta_{\ell\ell_1}^0 Q_{\ell\ell_1}^0(0)\xi_0^{\ell_1+2}\frac{dZ_{\ell_1}^0(\xi_0)}{d\xi_0}R_s^{-\ell_1-2}A_{\ell_1,0} \\
&= -\sqrt{\pi/3}(\epsilon_1+\epsilon_2)(\epsilon_2+\epsilon_3)\left[\frac{\epsilon_1}{\epsilon_2}-\frac{\epsilon_3(\epsilon_1+\epsilon_2)}{\epsilon_2(\epsilon_2+\epsilon_3)}\frac{\xi_0}{X_1^0(\xi_0)}\right]\frac{dX_1^0(\xi_0)}{d\xi_0}\delta_{\ell 1}E_0\cos\theta_0 \\
&\text{for } \ell \text{ odd} \qquad (9.105)
\end{aligned}
$$

and

$$
\begin{aligned}
&\epsilon_1\left((\epsilon_2+\epsilon_3)\xi_0^{\ell+1}Z_\ell^0(\xi_0)-(\epsilon_1+\epsilon_2)\xi_0^{\ell+1}X_\ell^0(\xi_0)\frac{dZ_\ell^0(\xi_0)/d\xi_0}{dX_\ell^0(\xi_0)/d\xi_0}\right)R_s^{-\ell-2}A_{\ell 0} \\
&+2\,\epsilon_2(\epsilon_3-\epsilon_1)\sum_{\ell_1\ odd}^{M}\zeta_{\ell\ell_1}^0 Q_{\ell\ell_1}^0(0)\xi_0^{\ell_1+1}Z_{\ell_1}^0(\xi_0)R_s^{-\ell_1-2}A_{\ell_1,0} \\
&= \sqrt{\pi/3}(\epsilon_1+\epsilon_2)(\epsilon_2+\epsilon_3)\left[\frac{(\epsilon_1+\epsilon_2)(\epsilon_2-\epsilon_3)}{\epsilon_2(\epsilon_2+\epsilon_3)}+\frac{(\epsilon_1-\epsilon_2)}{\epsilon_2}\frac{X_1^0(\xi_0)}{\xi_0}\right]\zeta_{\ell 1}^0 Q_{\ell 1}^0(0)E_0\cos\theta_0 \\
&\text{for } \ell \text{ even} \qquad (9.106)
\end{aligned}
$$

For $m = 1$ one obtains the following set of equations for the oblate spheroidal multipole coefficients $A_{\ell 1}$:

$$\epsilon_1 \left[(\epsilon_2+\epsilon_3)\xi_0^{\ell+1} Z_\ell^1(\xi_0) - (\epsilon_1+\epsilon_2)\xi_0^{\ell+1} X_\ell^1(\xi_0) \frac{dZ_\ell^1(\xi_0)/d\xi_0}{dX_\ell^1(\xi_0)/d\xi_0} \right] R_s^{-\ell-2} A_{\ell 1}$$
$$-2\epsilon_2\,(\epsilon_1-\epsilon_3) \sum_{\ell_1\ even}^{M} \zeta_{\ell\ell_1}^1 Q_{\ell\ell_1}^1(0)\xi_0^{\ell_1+1} Z_{\ell_1}^1(\xi_0) R_s^{-\ell_1-2} A_{\ell_1,1}$$
$$= \frac{1}{2}\sqrt{2\pi/3}\,(\epsilon_1-\epsilon_3)\,(\epsilon_1+\epsilon_2)\frac{X_1^1(\xi_0)}{\xi_0}\delta_{\ell 1} E_0 \sin\theta_0 \exp(-i\phi_0)$$
$$\text{for } \ell \text{ odd} \tag{9.107}$$

and

$$\left[\epsilon_1(\epsilon_2+\epsilon_3)\xi_0^{\ell+2}\frac{dZ_\ell^1(\xi_0)}{d\xi_0} - \epsilon_3(\epsilon_1+\epsilon_2)\xi_0^{\ell+2}\frac{dX_\ell^1(\xi_0)}{d\xi_0}\frac{Z_\ell^1(\xi_0)}{X_\ell^1(\xi_0)} \right] R_s^{-\ell-2} A_{\ell 1}$$
$$+2\,\epsilon_1\,(\epsilon_1-\epsilon_3) \sum_{\ell_1\ odd}^{M} \zeta_{\ell\ell_1}^1 Q_{\ell\ell_1}^1(0)\xi_0^{\ell_1+2}\frac{dZ_{\ell_1}^1(\xi_0)}{d\xi_0} R_s^{-\ell_1-2} A_{\ell_1,1}$$
$$= -\sqrt{2\pi/3}\,(\epsilon_1-\epsilon_3)\,(\epsilon_1+\epsilon_2)\frac{dX_1^1(\xi_0)}{d\xi_0}\zeta_{\ell 1}^1 Q_{\ell 1}^1(0) E_0 \sin\theta_0 \exp(-i\phi_0)$$
$$\text{for } \ell \text{ even} \tag{9.108}$$

(see also appendix B for proof).

In the above equations $Q_{\ell\ell_1}^m(0)$ for $m = 0, 1$ is again given by eq.(8.76) and $\zeta_{\ell\ell_1}^m$ by eq.(8.14). They are the generalizations for oblate hemispheroids of eqs.(8.77)–(8.80), which were valid for hemispheres. Notice that the latter equations follow from eqs.(9.105)–(9.108), if one uses the limits, eqs.(9.130) and (9.132), and the fact that $R_s \to R$, if $\xi \to \infty$.

For prolate hemispheroids, touching the substrate with their flat side (normal to the axis of revolution), linear inhomogeneous equations for the multipole coefficients $A_{\ell 0}$ and $A_{\ell 1}$ can be derived in a completely analogous way as above. As has been seen above, this set of linear equations can be obtained in a simple way, by making the substitutions $\xi_0 \to -i\xi_0$ and $R_s \to R_l$ in eqs.(9.105)–(9.108). Using eqs.(9.39) and (9.40), one then obtains for $m = 0$:

$$\left[-\epsilon_1(\epsilon_2+\epsilon_3)\xi_0^{\ell+2}\frac{d\widetilde{Z}_\ell^0(\xi_0)}{d\xi_0} + \epsilon_3(\epsilon_1+\epsilon_2)\xi_0^{\ell+2}\frac{d\widetilde{X}_\ell^0(\xi_0)}{d\xi_0}\frac{\widetilde{Z}_\ell^0(\xi_0)}{\widetilde{X}_\ell^0(\xi_0)} \right] R_l^{-\ell-2} A_{\ell 0}$$
$$-2\epsilon_1\,(\epsilon_1-\epsilon_3) \sum_{\ell_1\ even}^{M} \zeta_{\ell\ell_1}^0 Q_{\ell\ell_1}^0(0)\xi_0^{\ell_1+2}\frac{d\widetilde{Z}_{\ell_1}^0(\xi_0)}{d\xi_0} R_l^{-\ell_1-2} A_{\ell_1,0}$$
$$= -\sqrt{\pi/3}(\epsilon_1+\epsilon_2)(\epsilon_2+\epsilon_3)\left[\frac{\epsilon_1}{\epsilon_2} - \frac{\epsilon_3(\epsilon_1+\epsilon_2)}{\epsilon_2(\epsilon_2+\epsilon_3)}\frac{\xi_0}{\widetilde{X}_1^0(\xi_0)}\right]\frac{d\widetilde{X}_1^0(\xi_0)}{d\xi_0}\delta_{\ell 1} E_0 \cos\theta_0)$$
$$\text{for } \ell \text{ odd} \tag{9.109}$$

and

$$\epsilon_1\left[(\epsilon_2+\epsilon_3)\xi_0^{\ell+1}\widetilde{Z}_\ell^0(\xi_0)-(\epsilon_1+\epsilon_2)\xi_0^{\ell+1}\widetilde{X}_\ell^0(\xi_0)\frac{d\widetilde{Z}_\ell^0(\xi_0)/d\xi_0}{d\widetilde{X}_\ell^0(\xi_0)/d\xi_0}\right]R_l^{-\ell-2}A_{\ell 0}$$
$$+2\,\epsilon_2\,(\epsilon_3-\epsilon_1)\sum_{\ell_1\ odd}^{M}\zeta_{\ell\ell_1}^0 Q_{\ell\ell_1}^0(0)\xi_0^{\ell_1+1}\widetilde{Z}_{\ell_1}^0(\xi_0)R_l^{-\ell_1-2}A_{\ell_1,0}$$
$$=\ -\sqrt{\pi/3}(\epsilon_1+\epsilon_2)(\epsilon_2+\epsilon_3)\left[\frac{(\epsilon_1+\epsilon_2)(\epsilon_2-\epsilon_3)}{\epsilon_2(\epsilon_2+\epsilon_3)}+\frac{(\epsilon_1-\epsilon_2)}{\epsilon_2}\frac{\widetilde{X}_1^0(\xi_0)}{\xi_0}\right]$$
$$\times\zeta_{\ell 1}^0 Q_{\ell 1}^0(0)E_0\cos\theta_0 \qquad \text{for } \ell \text{ even} \tag{9.110}$$

and for $m=1$:

$$\epsilon_1\left[(\epsilon_2+\epsilon_3)\xi_0^{\ell+1}\widetilde{Z}_\ell^1(\xi_0)-(\epsilon_1+\epsilon_2)\xi_0^{\ell+1}\widetilde{X}_\ell^1(\xi_0)\frac{d\widetilde{Z}_\ell^1(\xi_0)/d\xi_0}{d\widetilde{X}_\ell^1(\xi_0)/d\xi_0}\right]R_l^{-\ell-2}A_{\ell 1}$$
$$-2\epsilon_2\,(\epsilon_1-\epsilon_3)\sum_{\ell_1\ even}^{M}\zeta_{\ell\ell_1}^1 Q_{\ell\ell_1}^1(0)\xi_0^{\ell_1+1}\widetilde{Z}_{\ell_1}^1(\xi_0)R_l^{-\ell_1-2}A_{\ell_1,1}$$
$$=\ \frac{1}{2}\sqrt{2\pi/3}\,(\epsilon_1-\epsilon_3)\,(\epsilon_1+\epsilon_2)\frac{\widetilde{X}_1^1(\xi_0)}{\xi_0}\delta_{\ell 1}E_0\sin\theta_0\exp(-i\phi_0)$$
$$\text{for } \ell \text{ odd} \tag{9.111}$$

and

$$\left[\epsilon_1(\epsilon_2+\epsilon_3)\xi_0^{\ell+2}\frac{d\widetilde{Z}_\ell^1(\xi_0)}{d\xi_0}-\epsilon_3(\epsilon_1+\epsilon_2)\xi_0^{\ell+2}\frac{d\widetilde{X}_\ell^1(\xi_0)}{d\xi_0}\frac{\widetilde{Z}_\ell^1(\xi_0)}{\widetilde{X}_\ell^1(\xi_0)}\right]R_l^{-\ell-2}A_{\ell 1}$$
$$+2\,\epsilon_1\,(\epsilon_1-\epsilon_3)\sum_{\ell_1\ odd}^{M}\zeta_{\ell\ell_1}^1 Q_{\ell\ell_1}^1(0)\xi_0^{\ell_1+2}\frac{d\widetilde{Z}_{\ell_1}^1(\xi_0)}{d\xi_0}R_l^{-\ell_1-2}A_{\ell_1,1}$$
$$=\ -\sqrt{2\pi/3}\,(\epsilon_1-\epsilon_3)\,(\epsilon_1+\epsilon_2)\frac{d\widetilde{X}_1^1(\xi_0)}{d\xi_0}\zeta_{\ell 1}^1 Q_{\ell 1}^1(0)E_0\sin\theta_0\exp(-i\phi_0)$$
$$\text{for } \ell \text{ even} \tag{9.112}$$

The matrices $Q_{\ell\ell_1}^m(0)$, for $m = 0, 1$, and $\zeta_{\ell\ell_1}^m$ are given by eqs.(8.76) and (8.14), respectively. Just as was the case for eqs.(9.105)– (9.108), the above equations tend to eqs.(8.77)–(8.80) in the limit as $\xi \to \infty$. This can immediately be shown by using eqs.(9.135) and (9.137) and using furthermore the fact that $R_l \to R$ in this limit.

The inhomogeneous linear set of equations (9.105)–(9.108) for oblate hemispheroids, as well as the set of equations (9.109)–(9.112) for prolate hemispheroids, can be solved on a computer (see the discussion at the end of sections 9.2 and 9.3). With the values for A_{10}, A_{20}, A_{11} and A_{20}, obtained in this way, one can then again calculate the polarizabilities, eq.(9.31), of a hemisphere on a substrate and the surface constitutive coefficients, eq.(9.32), of an island film of identical hemispheroids on a substrate in the low coverage limit. Finally the invariants I_e, I_c, $I_{\delta,e}$ and I_τ can be found by using eq.(5.66).

It is, of course, also possible to calculate the polarizabilities of hemispheroids using the equations in sections 9.4 and 9.5 for spheroidal caps in the limit $t_r \to 0$. Starting, in the case of oblate hemispheroids, from eqs.(9.70), (9.65) and (9.71), with $t_r = 0$ and $\epsilon_4 = \epsilon_2$, and using again eqs.(9.100), (9.101), (8.75) and (8.76), one can show, in an analogous way as in appendix B, that equations equivalent to eqs.(9.105)–(9.108) are valid. Replacing $\epsilon_2 A_{\ell m}$ by $\epsilon_1 A_{\ell m}$, for $\ell + m$ even in these equivalent equations one then obtains eqs.(9.105)–(9.108). This replacement is a consequence of the fact that the center of the oblate hemispheroid in the one case is above the substrate and in the other case below the substrate. Both sets of equations may alternatively be used if one combines them with the formulae for the polarizabilities and constitutive coefficients for the corresponding case. The resulting values for the constitutive coefficients are, of course, identical. See also the discussion at the end of section 8.4. In the case of prolate hemispheroids one comes to the same conclusion, starting from eqs.(9.84), (9.79) and (9.85).

9.7 Application: Truncated gold spheroids and caps on sapphire

In this section the theory, developed above, will be applied to the special case of gold islands on a flat sapphire surface. In particular, the dependence of the polarizabilities, the surface constitutive coefficients and the invariants on the shape, i.e. on the truncation parameter, t_r, and the axial ratio of the spheroid will be considered. The axial ratio of the spheroid is defined as the diameter of the spheroid along the substrate divided by the diameter along the symmetry axis normal to the substrate. This definition is not affected by the truncation parameter and is chosen to have two independent parameters, one for the shape of the spheroid and one as a measure of the truncation. One sometimes uses as axial ratio the diameter as seen from above divided by the height of the island. In this definition the shape of the spheroid and the amount of truncation are mixed. The dielectric constant of the island is a complex function of the wavelength, cf. table 5.1. The dielectric constant of sapphire is $\epsilon_2 = 3.13$ and of the ambient, vacuum, is $\epsilon_1 = 1$.

The following dimensionless amplitudes are introduced, cf. eqs.(8.106),

$$\begin{aligned}
\widehat{A}_{\ell 0} &\equiv A_{\ell 0} / \left[V (a\xi_0)^{\ell-1} \sqrt{4\pi/3} E_0 \cos\theta_0 \right] \\
\widehat{B}_{\ell 0} &\equiv B_{\ell 0} (a\xi_0)^{\ell+2} / \left[V \sqrt{4\pi/3} E_0 \cos\theta_0 \right] \\
\widehat{A}_{\ell 1} &\equiv -A_{\ell 1} / \left[V (a\xi_0)^{\ell-1} \sqrt{2\pi/3} E_0 \sin\theta_0 \exp(-i\phi_0) \right] \\
\widehat{B}_{\ell 1} &\equiv -B_{\ell 1} (a\xi_0)^{\ell+2} / \left[V \sqrt{2\pi/3} E_0 \sin\theta_0 \exp(-i\phi_0) \right]
\end{aligned} \tag{9.113}$$

where V is the volume of the truncated spheroid. In the case of a truncated oblate spheroid, see fig.9.3,

$$V = \frac{1}{3}\pi a^3 \xi_0 \left(1 + \xi_0^2\right) \left(2 + 3t_r - t_r^3\right) \tag{9.114}$$

and in the case of a truncated prolate spheroid, see fig.9.4,

$$V = \frac{1}{3}\pi a^3 \xi_0 \left(\xi_0^2 - 1\right) \left(2 + 3t_r - t_r^3\right) \tag{9.115}$$

Eq.(9.29) for truncated oblate spheroids, and eq.(9.47) for prolate spheroids, can then be written as

$$\begin{aligned}
\sum_{\ell_1=1}^{M} C^m_{\ell\ell_1}\widehat{A}_{\ell_1 m} + \sum_{\ell_1=1}^{M} D^m_{\ell\ell_1}\widehat{B}_{\ell_1 m} &= \widehat{H}^m_\ell \\
\sum_{\ell_1=1}^{M} F^m_{\ell\ell_1}\widehat{A}_{\ell_1 m} + \sum_{\ell_1=1}^{M} G^m_{\ell\ell_1}\widehat{B}_{\ell_1 m} &= \widehat{J}^m_\ell \\
\text{for } \ell &= 1,2,3,...,M \text{ and } m = 0,1
\end{aligned} \tag{9.116}$$

where

$$\begin{aligned}
\widehat{H}^0_\ell &\equiv \frac{3}{\pi}\left(2+3t_r-t_r^3\right)^{-1}\left\{\frac{\epsilon_1}{\epsilon_2}\xi_0^{-1}X^0_1(\xi_0)\delta_{\ell 1}\right. \\
&\quad \left. +\left(\frac{\epsilon_1-\epsilon_2}{\epsilon_2}\right)[\sqrt{3}t_r\zeta^0_{\ell 0}Q^0_{\ell 0}(t_r) - \xi_0^{-1}X^0_1(\xi_0)\zeta^0_{\ell 1}Q^0_{\ell 1}(t_r)]\right\} \\
\widehat{J}^0_\ell &\equiv \frac{3}{\pi}\left(2+3t_r-t_r^3\right)^{-1}\epsilon_1\frac{dX^0_1(\xi_0)}{d\xi_0}\delta_{\ell 1} \\
\widehat{H}^1_\ell &\equiv \frac{3}{\pi}\left(2+3t_r-t_r^3\right)^{-1}\xi_0^{-1}X^1_1(\xi_0)\delta_{\ell 1} \\
\widehat{J}^1_\ell &\equiv \frac{3}{\pi}\left(2+3t_r-t_r^3\right)^{-1}\frac{dX^1_1(\xi_0)}{d\xi_0}[\epsilon_2\delta_{\ell 1} + (\epsilon_1-\epsilon_2)\zeta^1_{\ell 1}Q^1_{\ell 1}(t_r)]
\end{aligned} \tag{9.117}$$

for truncated oblate spheroids, as follows using eq.(9.30). For truncated prolate spheroids it follows from eq.(9.48), that

$$\begin{aligned}
\widehat{H}^0_\ell &\equiv \frac{3}{\pi}\left(2+3t_r-t_r^3\right)^{-1}\left\{\frac{\epsilon_1}{\epsilon_2}\xi_0^{-1}\widetilde{X}^0_1(\xi_0)\delta_{\ell 1}\right. \\
&\quad \left. +\left(\frac{\epsilon_1-\epsilon_2}{\epsilon_2}\right)[\sqrt{3}t_r\zeta^0_{\ell 0}Q^0_{\ell 0}(t_r) - \xi_0^{-1}\widetilde{X}^0_1(\xi_0)\zeta^0_{\ell 1}Q^0_{\ell 1}(t_r)]\right\} \\
\widehat{J}^0_\ell &\equiv \frac{3}{\pi}\left(2+3t_r-t_r^3\right)^{-1}\epsilon_1\frac{d\widetilde{X}^0_1(\xi_0)}{d\xi_0}\delta_{\ell 1} \\
\widehat{H}^1_\ell &\equiv \frac{3}{\pi}\left(2+3t_r-t_r^3\right)^{-1}\xi_0^{-1}\widetilde{X}^1_1(\xi_0)\delta_{\ell 1} \\
\widehat{J}^1_\ell &\equiv \frac{3}{\pi}\left(2+3t_r-t_r^3\right)^{-1}\frac{d\widetilde{X}^1_1(\xi_0)}{d\xi_0}[\epsilon_2\delta_{\ell 1} + (\epsilon_1-\epsilon_2)\zeta^1_{\ell 1}Q^1_{\ell 1}(t_r)]
\end{aligned} \tag{9.118}$$

The matrix elements $C^m_{\ell\ell_1}$, $D^m_{\ell\ell_1}$, $F^m_{\ell\ell_1}$ and $G^m_{\ell\ell_1}(\ell,\ell_1 = 1,2,3,.....,M,\ m = 0,1)$ in eq.(9.116) are given by eq.(9.19) for truncated oblate spheroids and by eq.(9.38) for truncated prolate spheroids.

After solving the above linear set of equations, for a sufficiently large value of M, the dipole and quadrupole polarizabilities, in dimensionless form, are, using

$\epsilon_1 = 1$,

$$\begin{aligned}
\widehat{\alpha}_z &\equiv \alpha_z(0)/V = 4\pi\widehat{A}_{10} \\
\widehat{\alpha}_{\|} &\equiv \alpha_{\|}(0)/V = 4\pi\widehat{A}_{11} \\
\widehat{\alpha}_z^{10} &\equiv \alpha_z^{10}(0)/Va\xi_0 = 2\pi\widehat{A}_{20}\sqrt{5/3} \\
\widehat{\alpha}_{\|}^{10} &\equiv \alpha_{\|}^{10}(0)/Va\xi_0 = \frac{4\pi}{3}\widehat{A}_{21}\sqrt{5}
\end{aligned} \tag{9.119}$$

This equation follows from eqs.(9.31) and (9.113). The surface constitutive coefficients in dimensionless form, become, using $\epsilon_1 = 1$,

$$\begin{aligned}
\widehat{\gamma}_e &\equiv \frac{1}{\rho V}\gamma_e(d) = \widehat{\alpha}_{\|} \\
\widehat{\beta}_e &\equiv \frac{1}{\rho V}\beta_e(d) = \widehat{\alpha}_z \\
\widehat{\tau} &\equiv \frac{1}{\rho V a\xi_0}\tau(d) = t_r\widehat{\alpha}_{\|} - \widehat{\alpha}_{\|}^{10} \\
\widehat{\delta}_e &\equiv \frac{1}{\rho V a\xi_0}\delta_e(d) = t_r\widehat{\alpha}_z + t_r\widehat{\alpha}_{\|} - \widehat{\alpha}_z^{10} - \widehat{\alpha}_{\|}^{10}
\end{aligned} \tag{9.120}$$

This follows using eqs.(9.32), (9.119) and (9.2) for truncated oblate or (9.34) for truncated prolate spheroids. The invariants, eq.(5.66), become

$$\begin{array}{ll}
\widehat{I}_e \equiv I_e/\rho V = \widehat{\gamma}_e - \epsilon_2\widehat{\beta}_e & \widehat{I}_c \equiv (\epsilon_2 - 1)I_c/\rho V = \mathrm{Im}\,\widehat{\gamma}_e \\
\widehat{I}_\tau \equiv I_\tau/\rho V a\xi_0 = \widehat{\tau} & \widehat{I}_{\delta,e} \equiv I_{\delta,e}/\rho V a\xi_0 = \widehat{\delta}_e
\end{array} \tag{9.121}$$

Here it has been used that ϵ_2 is real.

In the case of a spheroidal cap eq.(9.113) can again be used, where V is now the volume of the spheroidal cap. In the case of an oblate spheroidal cap, see fig.9.5,

$$V = \frac{1}{3}\pi a^3\xi_0\left(1 + \xi_0^2\right)\left(2 - 3t_r + t_r^3\right) \tag{9.122}$$

and in the case of an prolate spheroidal cap, see fig.9.6,

$$V = \frac{1}{3}\pi a^3\xi_0\left(\xi_0^2 - 1\right)\left(2 - 3t_r + t_r^3\right) \tag{9.123}$$

The linear eq.(9.70) for oblate spheroidal caps, and (9.84) for prolate spheroidal caps, can again be written in the form (9.116), where

$$\begin{aligned}
\widehat{H}_\ell^0 &\equiv \frac{3}{\pi}\left(2 - 3t_r + t_r^3\right)^{-1}\Big\{\xi_0^{-1}X_1^0(\xi_0)\delta_{\ell 1} \\
&\quad + \left(\frac{\epsilon_2 - \epsilon_1}{\epsilon_2}\right)(-1)^{\ell+1}[\sqrt{3}t_r\zeta_{\ell 0}^0 Q_{\ell 0}^0(t_r) - \xi_0^{-1}X_1^0(\xi_0)\zeta_{\ell 1}^0 Q_{\ell 1}^0(t_r)]\Big\} \\
\widehat{J}_\ell^0 &\equiv \frac{3}{\pi}\left(2 - 3t_r + t_r^3\right)^{-1}\epsilon_1\frac{dX_1^0(\xi_0)}{d\xi_0}\delta_{\ell 1} \\
\widehat{H}_\ell^1 &\equiv \frac{3}{\pi}\left(2 - 3t_r + t_r^3\right)^{-1}\xi_0^{-1}X_1^1(\xi_0)\delta_{\ell 1} \\
\widehat{J}_\ell^1 &\equiv \frac{3}{\pi}\left(2 - 3t_r + t_r^3\right)^{-1}\frac{dX_1^1(\xi_0)}{d\xi_0}[\epsilon_1\delta_{\ell 1} + (\epsilon_2 - \epsilon_1)(-1)^{\ell+1}\zeta_{\ell 1}^1 Q_{\ell 1}^1(t_r)]
\end{aligned} \tag{9.124}$$

for oblate spheroidal caps, as follows using eq.(9.71). For prolate spheroidal caps it follows using eq.(9.85) that

$$\begin{aligned}
\widehat{H}_\ell^0 &\equiv \frac{3}{\pi}\left(2-3t_r+t_r^3\right)^{-1}\Big\{\xi_0^{-1}\widetilde{X}_1^0(\xi_0)\delta_{\ell 1} \\
&+\left(\frac{\epsilon_2-\epsilon_1}{\epsilon_2}\right)(-1)^{\ell+1}[\sqrt{3}t_r\zeta_{\ell 0}^{0}Q_{\ell 0}^0(t_r)-\xi_0^{-1}\widetilde{X}_1^0(\xi_0)\zeta_{\ell 1}^0Q_{\ell 1}^0(t_r)]\Big\} \\
\widehat{J}_\ell^0 &\equiv \frac{3}{\pi}\left(2-3t_r+t_r^3\right)^{-1}\epsilon_1\frac{d\widetilde{X}_1^0(\xi_0)}{d\xi_0}\delta_{\ell 1} \\
\widehat{H}_\ell^1 &\equiv \frac{3}{\pi}\left(2-3t_r+t_r^3\right)^{-1}\xi_0^{-1}\widetilde{X}_1^1(\xi_0)\delta_{\ell 1} \\
\widehat{J}_\ell^1 &\equiv \frac{3}{\pi}\left(2-3t_r+t_r^3\right)^{-1}\frac{d\widetilde{X}_1^1(\xi_0)}{d\xi_0}[\epsilon_1\delta_{\ell 1}+(\epsilon_2-\epsilon_1)(-1)^{\ell+1}\zeta_{\ell 1}^1Q_{\ell 1}^1(t_r)]
\end{aligned} \quad (9.125)$$

The matrix elements $C_{\ell\ell_1}^m$, $D_{\ell\ell_1}^m$, $F_{\ell\ell_1}^m$ and $G_{\ell\ell_1}^m$ $(\ell,\ell_1 = 1,2,3,.....,M,\ m = 0,1)$ in eq.(9.116) are given by eq.(9.65) for oblate spheroidal caps and by eq.(9.79) for prolate spheroidal caps.

After solving the above linear set of equations, eq.(9.116), for a sufficiently large value of M, the dipole and quadrupole polarizabilities, in dimensionless form, are, using $\epsilon_1 = 1$,

$$\begin{aligned}
\widehat{\alpha}_z &\equiv \alpha_z(0)/V = 4\pi\epsilon_2^2\widehat{A}_{10} \\
\widehat{\alpha}_\parallel &\equiv \alpha_\parallel(0)/V = 4\pi\epsilon_2\widehat{A}_{11} \\
\widehat{\alpha}_z^{10} &\equiv \alpha_z^{10}(0)/Va\xi_0 = 2\pi\epsilon_2^2\widehat{A}_{20}\sqrt{5/3} \\
\widehat{\alpha}_\parallel^{10} &\equiv \alpha_\parallel^{10}(0)/Va\xi_0 = \frac{4\pi}{3}\epsilon_2\widehat{A}_{21}\sqrt{5}
\end{aligned} \quad (9.126)$$

This equation follows from eqs.(9.72) and (9.113). The surface constitutive coefficients in dimensionless form, become, using $\epsilon_1 = 1$,

$$\begin{aligned}
\widehat{\gamma}_e &\equiv \frac{1}{\rho V}\gamma_e(-d) = \widehat{\alpha}_\parallel \\
\widehat{\beta}_e &\equiv \frac{1}{\rho V}\beta_e(-d) = \widehat{\alpha}_z/\epsilon_2^2 \\
\widehat{\tau} &\equiv \frac{1}{\rho V a\xi_0}\tau(-d) = -\left(t_r\widehat{\alpha}_\parallel+\widehat{\alpha}_\parallel^{10}\right)/\epsilon_2 \\
\widehat{\delta}_e &\equiv \frac{1}{\rho V a\xi_0}\delta_e(-d) = -\left(t_r\widehat{\alpha}_z+t_r\widehat{\alpha}_\parallel+\widehat{\alpha}_z^{10}+\widehat{\alpha}_\parallel^{10}\right)/\epsilon_2
\end{aligned} \quad (9.127)$$

This follows using eqs.(9.73), (9.126) and (9.50) for oblate or (9.75) for prolate spheroidal caps. The invariants are again given by eq.(9.121).

Using the interaction with the image to order $M = 16$ the dipole polarizabilities were calculated for oblate spheroids, with an axial ratio 4 as a function of the frequency. The resulting $\widehat{\beta}_e$ is plotted in figs. 9.7-9 for truncation parameters $t_r = 0.4, 0.0$ and -0.4. The minus sign of the truncation parameter will be used to indicate the case of a cap. In the actual calculations the formulae for the cap have

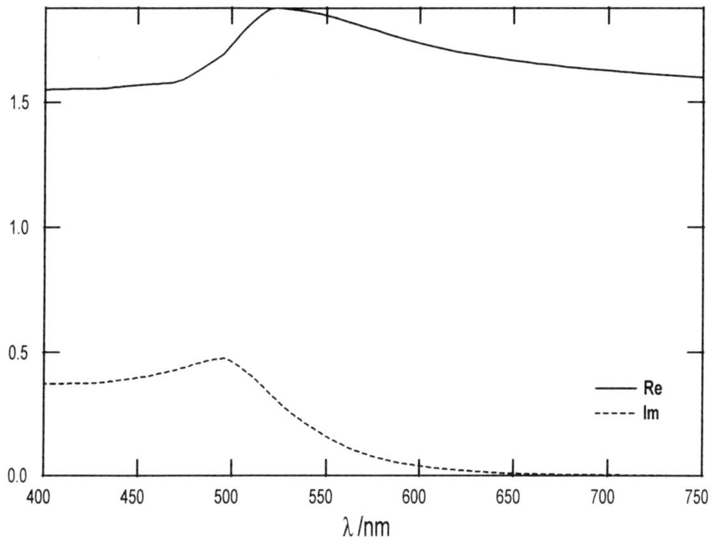

Figure 9.7 The real and imaginary parts of $\widehat{\beta}_e$ for a truncated oblate spheroid, with an axial ratio 4 and a truncation parameter 0.4, as a function of the wave length.

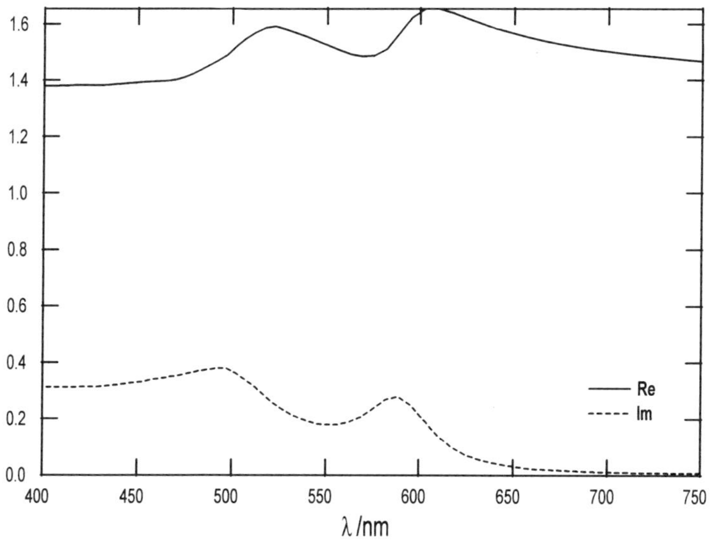

Figure 9.8 The real and imaginary parts of $\widehat{\beta}_e$ for a oblate hemispheroid, with an axial ratio 4 and a truncation parameter 0.0, as a function of the wave length.

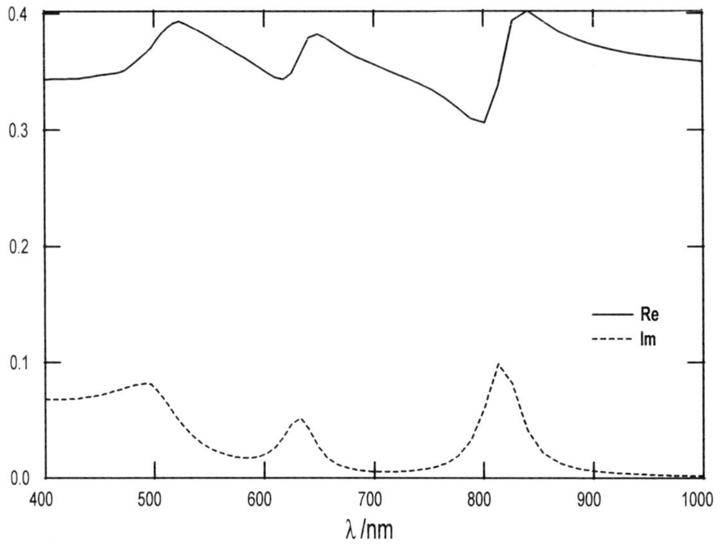

Figure 9.9 The real and imaginary parts of $\widehat{\beta}_e$ for an oblate spheroidal cap, with an axial ratio 4 and a truncation parameter -0.4, as a function of the wave length.

been used with the absolute value of this parameter. It is found for truncated oblate spheroids, oblate hemispheroids and oblate spheroidal caps, that the constitutive coefficient $\widehat{\beta}_e$, which is due to the dipole polarizability normal to the surface, is small. Its resonance structure is a very sensitive function of the truncation parameter. There is one resonance for the truncated spheroid and there are two for the hemispheroid and three for the cap in the optical domain.

The resulting $\widehat{\gamma}_e$ for an axial ratio of 4 is plotted in figs.9.10-12 for truncation parameters $t_r = 0.4, 0.0$ and -0.4. It is found for oblate spheroids that the constitutive coefficient $\widehat{\gamma}_e$, which is due to the dipole polarizability parallel to the surface, is orders of magnitude larger than the corresponding coefficient in the normal direction. The location and the size of the resonance is a very sensitive function of the truncation parameter. It shifts from the visible domain for a truncated spheroid to the far infrared for the cap. The size is largest for the hemispheroid.

Using the interaction with the image to order $M = 16$ the dipole polarizabilities were also calculated for prolate spheroids with an axial ratio 0.25 as a function of the wavelength. The resulting $\widehat{\beta}_e$ is plotted in figs.9.13-15 for truncation parameters $t_r = 0.4, 0.0$ and -0.4. It is found for prolate spheroids that the constitutive coefficient $\widehat{\beta}_e$, which is due to the dipole polarizability normal to the surface, is small. Its resonance structure is a sensitive function of the truncation parameter. There is a blue shift for a decreasing truncation parameter and an additional resonance appears for the cap in the optical domain.

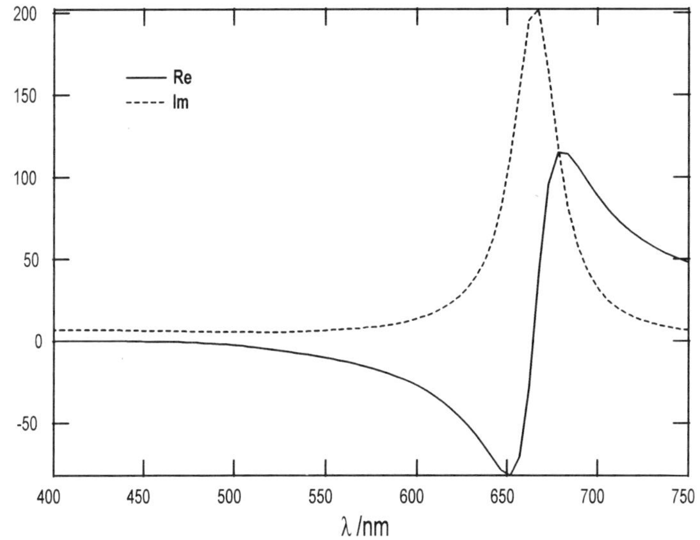

Figure 9.10 The real and imaginary parts of $\widehat{\gamma}_e$ for a truncated oblate spheroid, with an axial ratio 4 and a truncation parameter 0.4, as a function of the wave length.

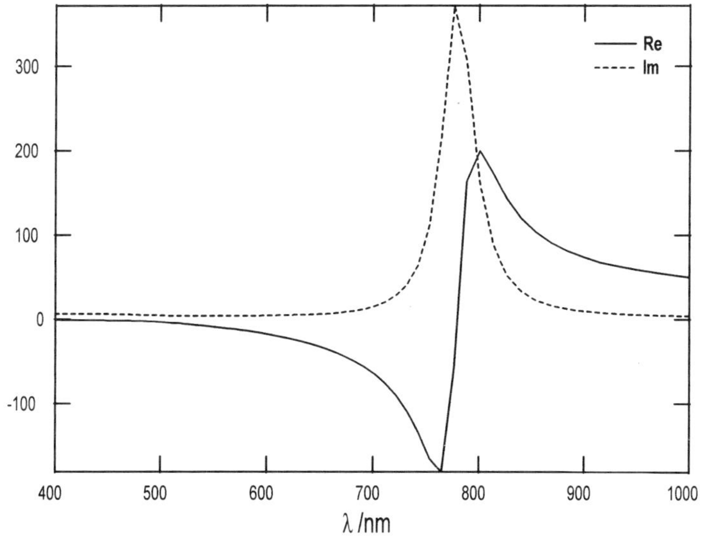

Figure 9.11 The real and imaginary parts of $\widehat{\gamma}_e$ for an oblate hemispheroid, with an axial ratio 4 and a truncation parameter 0.0, as a function of the wave length.

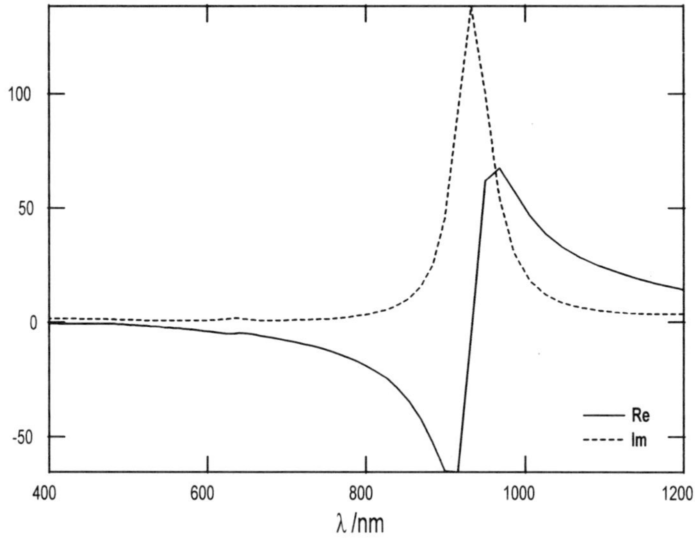

Figure 9.12 The real and imaginary parts of $\widehat{\gamma}_e$ for an oblate spheroidal cap, with an aial ratio 4 and a truncation parameter -0.4, as a function of the wave length.

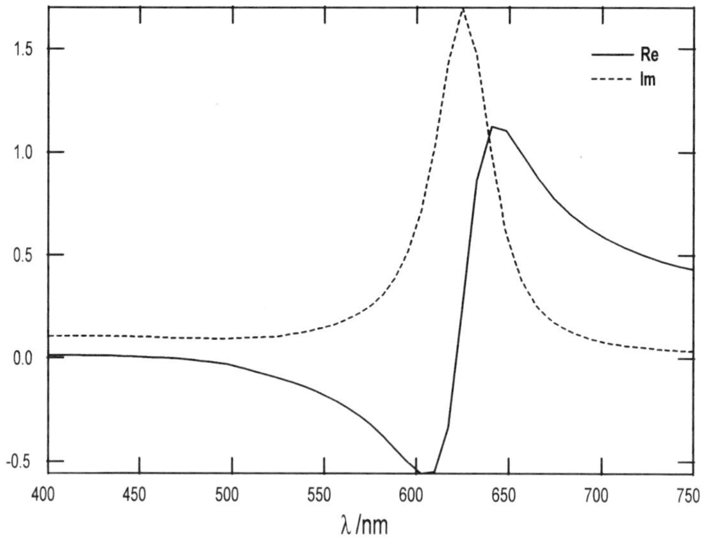

Figure 9.13 The real and imaginary parts of $\widehat{\beta}_e$ for a truncated prolate spheroid, with an axial ratio 0.25 and a truncation parameter 0.4, as a function of the wave length.

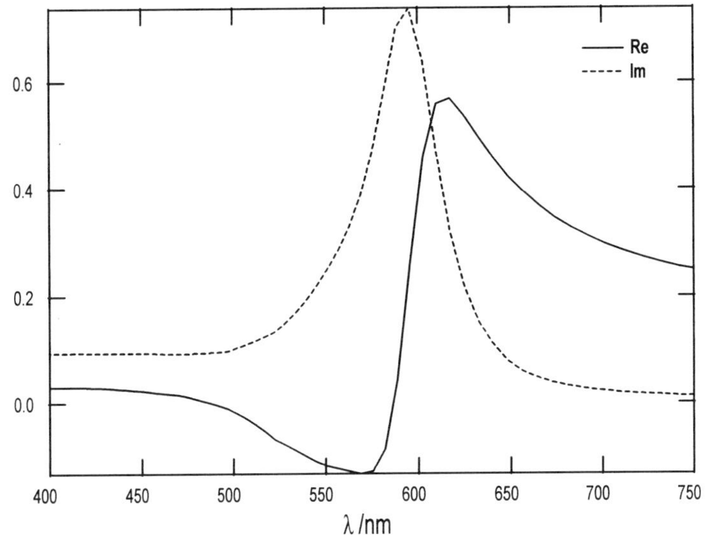

Figure 9.14 The real and imaginary parts of $\widehat{\beta}_e$ for a prolate hemispheroid, with an axial ratio 0.25 and a truncation parameter 0.0, as a function of the wave length.

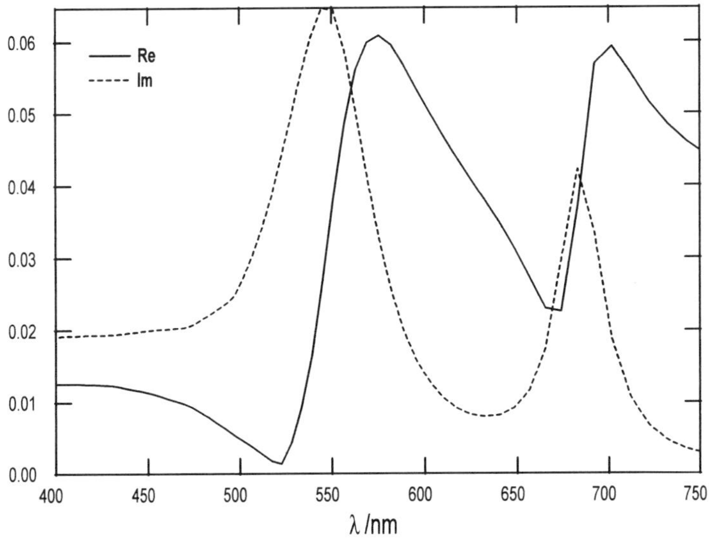

Figure 9.15 The real and imaginary parts of $\widehat{\beta}_e$ for a prolate spheroidal cap, with an axial ratio 0.25 and a truncation parameter -0.4, as a function of the wave length.

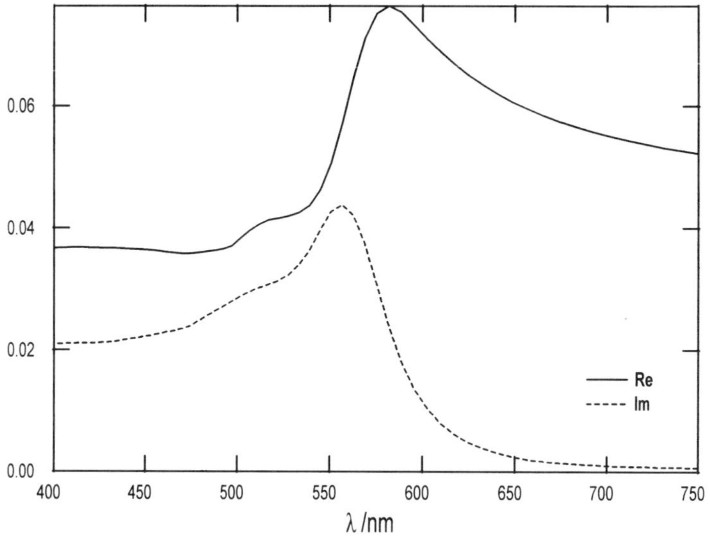

Figure 9.16 The real and imaginary parts of $\widehat{\gamma}_e$ for a truncated prolate spheroid, with an axial ratio 0.25 and a truncation parameter 0.4, as a function of the wave length.

The resulting $\widehat{\gamma}_e$ for an axial ratio of 0.25 is plotted in figs.9.16-18 for truncation parameters $t_r = 0.4, 0.0$ and -0.4. It is found for prolate spheroids that the constitutive coefficient $\widehat{\gamma}_e$, which is due to the dipole polarizability parallel to the surface, is small. For the truncated and the hemispheroid a small additional resonance appears.

One may overall conclude that the polarizability of oblate spheroids, for all values of the truncation parameter, along the surface is large. The normal polarizability of oblate spheroids and the normal and the parallel polarizabilities of prolate spheroids are, due to the interaction with the image, small. There is a very considerable dependence of all polarizabilities on the truncation parameter.

9.8 Appendix A

In the first part of this appendix it will be proved, that eq.(8.23), with eqs.(8.13), (8.24) and (8.25), for truncated spheres, follows from eq.(9.26), with eqs.(9.19), (9.27) and (9.28), for truncated oblate spheroids, in the limit as $\xi_0 \to \infty$. Similarly eq.(8.46), with eqs.(8.43), (8.47) and (8.48), for spherical caps, will be shown to follow from eqs.(9.67), with eqs.(9.65), (9.68) and (9.69), in this limit. In the second part, it will be shown, that the above mentioned equations for truncated spheres also follow from eq.(9.44), with eqs.(9.38), (9.45) and (9.46), for truncated prolate spheroids, in the limit $\xi_0 \to \infty$ and that the equations for the spherical caps also follow from eq.(9.81), with eqs.(9.79), (9.82) and (9.83), in the same limit.

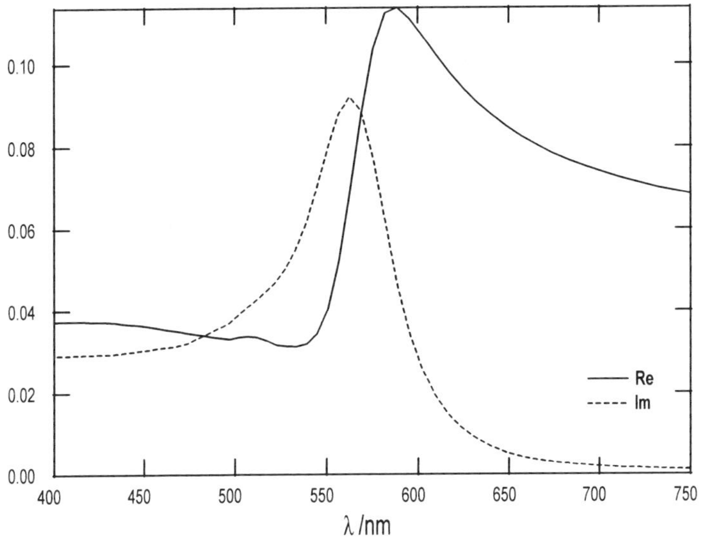

Figure 9.17 The real and imaginary parts of $\widehat{\gamma}_e$ for a prolate hemispheroid, with an axial ratio 0.25 and a truncation parameter 0.0, as a function of the wave length.

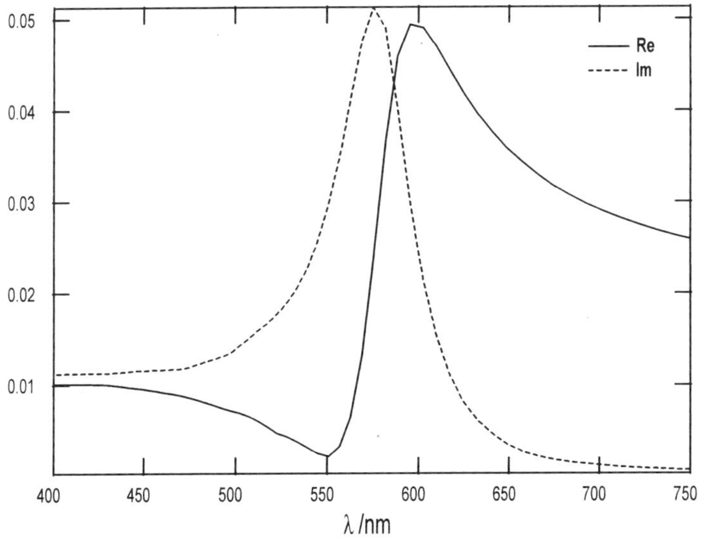

Figure 9.18 The real and imaginary parts of $\widehat{\gamma}_e$ for a prolate spheroidal cap, with an axial ratio 0.25 and a truncation parameter -0.4, as a function of the wave length.

From eqs.(9.23), (9.24), (9.4), (9.5), (9.20), (9.21) and the asymptotic relations, eqs.(6.17)–(6.20), one finds, in the limit as $\xi \to \infty$ (and $a \to 0$, with $a\xi = r$, $a\xi_0 = R$ and $a\xi_1 = d$), using also the definitions, eqs.(8.11), (8.16) and (8.20)

$$\begin{aligned} \lim_{\xi\to\infty} \xi^{\ell_1+1} V^m_{\ell\ell_1}(\xi, t_r) &= t^{\ell_1+1} S^m_{\ell\ell_1}(t, t_r) \\ \lim_{\xi\to\infty} \xi^{-\ell_1} W^m_{\ell\ell_1}(\xi, t_r) &= t^{-\ell_1} T^m_{\ell\ell_1}(t, t_r) \end{aligned} \tag{9.128}$$

where $S^m_{\ell\ell_1}(t, t_r)$ and $T^m_{\ell\ell_1}(t, t_r)$ are defined by eqs.(8.18) and (8.19). For $\xi = \xi_0$, which corresponds to $t = 1$, one furthermore finds with eq.(8.17) that

$$\begin{aligned} \lim_{\xi_0\to\infty} \xi_0^{\ell_1+1} V^m_{\ell\ell_1}(\xi_0, t_r) &= S^m_{\ell\ell_1}(t_r) \\ \lim_{\xi_0\to\infty} \xi_0^{-\ell_1} W^m_{\ell\ell_1}(\xi_0, t_r) &= T^m_{\ell\ell_1}(t_r) \end{aligned} \tag{9.129}$$

Since it follows from eqs.(6.17), (6.18), (9.20) and (9.21) that

$$\lim_{\xi\to\infty} \xi^{\ell+1} Z^m_\ell(\xi) = 1 \quad \text{and} \quad \lim_{\xi\to\infty} \xi^{-\ell} X^m_\ell(\xi) = 1 \tag{9.130}$$

it is found that the expressions, eq.(8.13), for the matrices $C^m_{\ell\ell_1}$ and $D^m_{\ell\ell_1}$ in the case of truncated spheres follow from the expressions, eq.(9.19), for truncated oblate spheroids, in the limit as $\xi_0 \to \infty$.

By differentiation of eqs.(9.128) and (9.130) with respect to ξ and using the asymptotic relation $t \simeq (a/R)\xi$, which follows from eqs.(6.19) and (8.20), one finds

$$\begin{aligned} \lim_{\xi\to\infty} \xi^{\ell_1+2} \frac{\partial V^m_{\ell\ell_1}(\xi, t_r)}{\partial \xi} &= t^{\ell_1+2} \frac{\partial S^m_{\ell\ell_1}(t, t_r)}{\partial t} \\ \lim_{\xi\to\infty} \xi^{-\ell_1+1} \frac{\partial W^m_{\ell\ell_1}(\xi, t_r)}{\partial \xi} &= t^{-\ell_1+1} \frac{\partial T^m_{\ell\ell_1}(t, t_r)}{\partial t} \end{aligned} \tag{9.131}$$

and

$$\lim_{\xi\to\infty} \xi^{\ell+2} \frac{dZ^m_\ell(\xi)}{d\xi} = -(\ell+1) \quad \text{and} \quad \lim_{\xi\to\infty} \xi^{-\ell+1} \frac{dX^m_\ell(\xi)}{d\xi} = \ell \tag{9.132}$$

With these relations for $\xi = \xi_0$ (and $t = 1$), it is now found that the expressions, eq.(8.13), for the matrices $F^m_{\ell\ell_1}$ and $G^m_{\ell\ell_1}$ in the case of truncated spheres also follow from the corresponding expressions, eq.(9.19), for truncated oblate spheroids in the limit as $\xi_0 \to \infty$.

By applying eqs.(9.130) and (9.132) to eqs.(9.27) and (9.28), one furthermore obtains the result that the expressions for H^m_ℓ and J^m_ℓ, valid for truncated oblate spheroids, tend to the corresponding expressions, eqs.(8.24) and (8.25), for truncated spheres in the limit as $\xi_0 \to \infty$. Since also $R_s \to R$ in this limit, it is found that eq.(8.23), with eqs.(8.13), (8.24) and (8.25), for truncated spheres, follows from eq.(9.26), with eqs.(9.19), (9.27) and (9.28), for truncated oblate spheroids, in the limit $\xi_0 \to \infty$. Similarly eq.(8.46), with eqs.(8.43), (8.47) and (8.48), for spherical caps, will be shown to follow from eqs.(9.67), with eqs.(9.65), (9.68) and (9.69), in this limit.

From eqs.(9.41), (9.42), (9.36), (9.39), (9.40) and the asymptotic relations, eqs.(6.62)– (6.65), one finds, in the limit as $\xi \to \infty$ (and $a \to 0$, with $a\xi = r$ and $a\xi_0 = R$), using also the definitions, eqs.(8.11), (8.16) and (8.20)

$$\begin{aligned} \lim_{\xi\to\infty} \xi^{\ell_1+1}\widetilde{V}^m_{\ell\ell_1}(\xi, t_r) &= t^{\ell_1+1} S^m_{\ell\ell_1}(t, t_r) \\ \lim_{\xi\to\infty} \xi^{-\ell_1}\widetilde{W}^m_{\ell\ell_1}(\xi, t_r) &= t^{-\ell_1} T^m_{\ell\ell_1}(t, t_r) \end{aligned} \tag{9.133}$$

where $S^m_{\ell\ell_1}(t, t_r)$ and $T^m_{\ell\ell_1}(t, t_r)$ are again defined by eqs.(8.18) and (8.19). For $\xi = \xi_0$, which corresponds to $t = 1$, one furthermore finds with eq.(8.17) that

$$\begin{aligned} \lim_{\xi_0\to\infty} \xi_0^{\ell_1+1}\widetilde{V}^m_{\ell\ell_1}(\xi_0, t_r) &= S^m_{\ell\ell_1}(t_r) \\ \lim_{\xi_0\to\infty} \xi_0^{-\ell_1}\widetilde{W}^m_{\ell\ell_1}(\xi_0, t_r) &= T^m_{\ell\ell_1}(t_r) \end{aligned} \tag{9.134}$$

Since it follows from eqs.(6.62), (6.63), (9.39) and (9.40) that

$$\lim_{\xi\to\infty} \xi^{\ell+1}\widetilde{Z}^m_\ell(\xi) = 1 \quad \text{and} \quad \lim_{\xi\to\infty} \xi^{-\ell}\widetilde{X}^m_\ell(\xi) = 1 \tag{9.135}$$

it is found that the expressions, eq.(8.13), for the matrices $C^m_{\ell\ell_1}$ and $D^m_{\ell\ell_1}$ in the case of truncated spheres, also follow from the expressions, eq.(9.38), for truncated prolate spheroids, in the limit as $\xi_0 \to \infty$.

Differentiating eqs.(9.133) and (9.135) with respect to ξ and using the asymptotic relation $t \simeq (a/R)\xi$, which now follows from eqs.(6.64) and (8.20), one finds

$$\begin{aligned} \lim_{\xi\to\infty} \xi^{\ell_1+2}\frac{\partial\widetilde{V}^m_{\ell\ell_1}(\xi, t_r)}{\partial\xi} &= t^{\ell_1+2}\frac{\partial S^m_{\ell\ell_1}(t, t_r)}{\partial t} \\ \lim_{\xi\to\infty} \xi^{-\ell_1+1}\frac{\partial\widetilde{W}^m_{\ell\ell_1}(\xi, t_r)}{\partial\xi} &= t^{-\ell_1+1}\frac{\partial T^m_{\ell\ell_1}(t, t_r)}{\partial t} \end{aligned} \tag{9.136}$$

and

$$\lim_{\xi\to\infty} \xi^{\ell+2}\frac{d\widetilde{Z}^m_\ell(\xi)}{d\xi} = -(\ell+1) \quad \text{and} \quad \lim_{\xi\to\infty} \xi^{-\ell+1}\frac{d\widetilde{X}^m_\ell(\xi)}{d\xi} = \ell \tag{9.137}$$

It will be obvious, that with these relations for $\xi = \xi_0$ (and $t = 1$), one obtains again the result, that the matrices $F^m_{\ell\ell_1}$ and $G^m_{\ell\ell_1}$, given by eq.(8.13) for truncated spheres, follow from the corresponding expressions, eq.(9.38), for truncated prolate spheroids in the limit as $\xi_0 \to \infty$.

Finally it is found, by applying eqs.(9.135) and (9.137) to eqs.(9.45) and (9.46), that the expressions for H^m_ℓ and J^m_ℓ, valid for truncated prolate spheroids, tend to the expressions, eqs.(8.24) and (8.25), for truncated spheres, in the limit as $\xi_0 \to \infty$. Since $R_l \to R$ in this limit, one obtains the result, that eq.(8.23), with eqs.(8.13), (8.24) and (8.25), for truncated spheres, follows from eq.(9.44), with eqs.(9.38), (9.45) and (9.46), for truncated prolate spheroids, in the limit as $\xi_0 \to \infty$. Similarly eq.(8.46), with eqs.(8.43), (8.47) and (8.48), for spherical caps, will be shown to follow from eqs.(9.81), with eqs.(9.79), (9.82) and (9.83), in this limit.

9.9 Appendix B

In this appendix eqs.(9.105)– (9.108) for the multipole coefficients $A_{\ell 0}$ and $A_{\ell 1}$ in the case of an oblate hemispheroid, touching with its flat side (normal to the axis of revolution) the surface of the substrate, will be derived. Using the results, eqs.(9.102) and (9.103), valid for oblate hemispheroids, in the general equation (9.29), one obtains the following set of equations for $m = 0$:

$$\begin{aligned}
&\left(\frac{2\epsilon_1}{\epsilon_1+\epsilon_2}\right)\xi_0^{\ell+1}Z_\ell^0(\xi_0)R_s^{-\ell-2}A_{\ell 0}\\
&-2\left(\frac{\epsilon_1-\epsilon_2}{\epsilon_1+\epsilon_2}\right)\sum_{\ell_1\ odd}^{M}\zeta_{\ell\ell_1}^{0}Q_{\ell\ell_1}^{0}(0)\xi_0^{\ell_1+1}Z_{\ell_1}^0(\xi_0)R_s^{-\ell_1-2}A_{\ell_1,0}\\
&-\left(\frac{2\epsilon_3}{\epsilon_2+\epsilon_3}\right)\xi_0^{-\ell}X_\ell^0(\xi_0)R_s^{\ell-1}B_{\ell 0}\\
&-2\left(\frac{\epsilon_2-\epsilon_3}{\epsilon_2+\epsilon_3}\right)\sum_{\ell_1\ odd}^{M}\zeta_{\ell\ell_1}^{0}Q_{\ell\ell_1}^{0}(0)\xi_0^{-\ell_1}X_{\ell_1}^0(\xi_0)R_s^{\ell_1-1}B_{\ell_1,0}\\
=\ &\sqrt{4\pi/3}\frac{X_1^0(\xi_0)}{\xi_0}\left[\frac{\epsilon_1}{\epsilon_2}\delta_{\ell 1}-\left(\frac{\epsilon_1-\epsilon_2}{\epsilon_2}\right)\zeta_{\ell 1}^0 Q_{\ell 1}^0(0)\right]E_0\cos\theta_0
\end{aligned}\tag{9.138}$$

and

$$\begin{aligned}
&\left(\frac{2\epsilon_1\epsilon_2}{\epsilon_1+\epsilon_2}\right)\xi_0^{\ell+2}\frac{dZ_\ell^0(\xi_0)}{d\xi_0}R_s^{-\ell-2}A_{\ell 0}\\
&+2\,\epsilon_1\left(\frac{\epsilon_1-\epsilon_2}{\epsilon_1+\epsilon_2}\right)\sum_{\ell_1\ even}^{M}\zeta_{\ell\ell_1}^{0}Q_{\ell\ell_1}^{0}(0)\xi_0^{\ell_1+2}\frac{dZ_{\ell_1}^0(\xi_0)}{d\xi_0}R_s^{-\ell_1-2}A_{\ell_1,0}\\
&-\left(\frac{2\epsilon_2\epsilon_3}{\epsilon_2+\epsilon_3}\right)\xi_0^{-\ell+1}\frac{dX_\ell^0(\xi_0)}{d\xi_0}R_s^{\ell-1}B_{\ell 0}\\
&+2\epsilon_3\left(\frac{\epsilon_2-\epsilon_3}{\epsilon_2+\epsilon_3}\right)\sum_{\ell_1\ even}^{M}\zeta_{\ell\ell_1}^{0}Q_{\ell\ell_1}^{0}(0)\xi_0^{-\ell_1+1}\frac{dX_{\ell_1}^0(\xi_0)}{d\xi_0}R_s^{\ell_1-1}B_{\ell_1,0}\\
=\ &\sqrt{4\pi/3}\epsilon_1\frac{dX_1^0(\xi_0)}{d\xi_0}\delta_{\ell 1}E_0\cos\theta_0
\end{aligned}\tag{9.139}$$

If one now applies eqs.(8.75) and (8.14) to eq.(9.138) for odd values of ℓ, and to eq.(9.139) for even values of ℓ, one finds the following relations:

$$\xi_0^{-\ell}X_\ell^0(\xi_0)R_s^{\ell-1}B_{\ell 0}=\xi_0^{\ell+1}Z_\ell^0(\xi_0)R_s^{-\ell-2}A_{\ell 0}-\sqrt{4\pi/3}\left(\frac{\epsilon_1+\epsilon_2}{2\epsilon_2}\right)\delta_{\ell 1}E_0\cos\theta_0$$

$$\text{for } \ell \text{ odd} \tag{9.140}$$

$$\xi_0^{-\ell+1}\frac{dX_\ell^0(\xi_0)}{d\xi_0}R_s^{\ell-1}B_{\ell 0}=\frac{\epsilon_1}{\epsilon_3}\xi_0^{\ell+2}\frac{dZ_\ell^0(\xi_0)}{d\xi_0}R_s^{-\ell-2}A_{\ell 0}\quad\text{for } \ell \text{ even} \tag{9.141}$$

With the help of these relations, one can eliminate the spheroidal multipole coefficients $B_{\ell 0}$ from eqs.(9.138) and (9.139). Applying eqs.(9.140) and (9.141) to eq.(9.139) for odd values of ℓ, one then obtains eq.(9.105). Application of eqs.(9.140) and (9.141) to eq.(9.138) for even values of ℓ gives eq.(9.106).

For $m = 1$ one obtains, by using eqs.(9.102) and (9.103) in eq.(9.29), the following set of equations:

$$\begin{aligned}
&\left(\frac{2\epsilon_1}{\epsilon_1+\epsilon_2}\right)\xi_0^{\ell+1}Z_\ell^1(\xi_0)R_s^{-\ell-2}A_{\ell 1}\\
&-2\left(\frac{\epsilon_1-\epsilon_2}{\epsilon_1+\epsilon_2}\right)\sum_{\ell_1\ even}^{M}\zeta_{\ell\ell_1}^1 Q_{\ell\ell_1}^1(0)\xi_0^{\ell_1+1}Z_{\ell_1}^1(\xi_0)R_s^{-\ell_1-2}A_{\ell_1,1}\\
&-\left(\frac{2\epsilon_3}{\epsilon_2+\epsilon_3}\right)\xi_0^{-\ell}X_\ell^1(\xi_0)R_s^{\ell-1}B_{\ell 1}\\
&-2\left(\frac{\epsilon_2-\epsilon_3}{\epsilon_2+\epsilon_3}\right)\sum_{\ell_1\ even}^{M}\zeta_{\ell\ell_1}^1 Q_{\ell\ell_1}^1(0)\xi_0^{-\ell_1}X_{\ell_1}^1(\xi_0)R_s^{\ell_1-1}B_{\ell_1,1}\\
= &-\sqrt{2\pi/3}\frac{X_1^1(\xi_0)}{\xi_0}\delta_{\ell 1}E_0\sin\theta_0\exp(-i\phi_0)
\end{aligned}\tag{9.142}$$

and

$$\begin{aligned}
&\left(\frac{2\epsilon_1\epsilon_2}{\epsilon_1+\epsilon_2}\right)\xi_0^{\ell+2}\frac{dZ_\ell^1(\xi_0)}{d\xi_0}R_s^{-\ell-2}A_{\ell 1}\\
&+2\epsilon_1\left(\frac{\epsilon_1-\epsilon_2}{\epsilon_1+\epsilon_2}\right)\sum_{\ell_1\ odd}^{M}\zeta_{\ell\ell_1}^1 Q_{\ell\ell_1}^1(0)\xi_0^{\ell_1+2}\frac{dZ_{\ell_1}^1(\xi_0)}{d\xi_0}R_s^{-\ell_1-2}A_{\ell_1,1}\\
&-\left(\frac{2\epsilon_2\epsilon_3}{\epsilon_2+\epsilon_3}\right)\xi_0^{-\ell+1}\frac{dX_\ell^1(\xi_0)}{d\xi_0}R_s^{\ell-1}B_{\ell 1}\\
&+2\epsilon_3\left(\frac{\epsilon_2-\epsilon_3}{\epsilon_2+\epsilon_3}\right)\sum_{\ell_1\ odd}^{M}\zeta_{\ell\ell_1}^1 Q_{\ell\ell_1}^1(0)\xi_0^{-\ell_1+1}\frac{dX_{\ell_1}^1(\xi_0)}{d\xi_0}R_s^{\ell_1-1}B_{\ell_1,1}\\
= &-\sqrt{2\pi/3}\frac{dX_1^1(\xi_0)}{d\xi_0}[\epsilon_2\delta_{\ell 1}+(\epsilon_1-\epsilon_2)\zeta_{\ell 1}^1 Q_{\ell 1}^1(0)]E_0\sin\theta_0\exp(-i\phi_0)
\end{aligned}\tag{9.143}$$

Applying eqs.(8.75) and (8.14) to eq.(9.143) for odd values of ℓ, and to eq.(9.142) for even values of ℓ, one obtains the relations

$$\begin{aligned}
&\xi_0^{-\ell+1}\frac{dX_\ell^1(\xi_0)}{d\xi_0}R_s^{\ell-1}B_{\ell 1}\\
= &\frac{\epsilon_1}{\epsilon_3}\xi_0^{\ell+2}\frac{dZ_\ell^1(\xi_0)}{d\xi_0}R_s^{-\ell-2}A_{\ell 1}+\frac{1}{2}\sqrt{2\pi/3}\frac{dX_1^1(\xi_0)}{d\xi_0}\left(\frac{\epsilon_1+\epsilon_2}{\epsilon_3}\right)\delta_{\ell 1}E_0\sin\theta_0\exp(-i\phi_0)\\
&\text{for } \ell \text{ odd}
\end{aligned}\tag{9.144}$$

and

$$\xi_0^{-\ell}X_\ell^1(\xi_0)R_s^{\ell-1}B_{\ell 1}=\xi_0^{\ell+1}Z_\ell^1(\xi_0)R_s^{-\ell-2}A_{\ell 1}\quad\text{for } \ell \text{ even}\tag{9.145}$$

Using these relations, one can now eliminate the multipole coefficients $B_{\ell 1}$ from eqs.(9.142) and (9.143). Applying eqs.(9.144) and (9.145) to eq.(9.142) for odd values of ℓ, one then obtains eq.(9.107), whereas application of these equations to eq.(9.143) for even values of ℓ, leads to eq.(9.108).

Chapter 10

FILMS OF TRUNCATED SPHERES OR SPHEROIDS FOR FINITE COVERAGE

10.1 Introduction

In this chapter the analysis, given in the two previous chapters for a low coverage, will be extended to a higher coverage. In this case the electromagnetic interaction between the different particles, lying on the substrate, has to be taken into account. As in those chapters these particles have the form of truncated spheres or spheroids (or of spherical or spheroidal caps). Their size is again much smaller than the wavelength of the incident light. The theory of the previous chapters is therefore applicable to calculate the polarizabilities of the truncated particles, interacting with the substrate. For truncated spheres the method of section 8.2 may be used, and for spherical caps that of section 8.3. For truncated spheroids one may apply the method of sections 9.2 and 9.3, and for spheroidal caps that of sections 9.4 and 9.5. In the first cases spherical multipole expansions of the fields are used and in the second cases spheroidal multipole expansions.

The electromagnetic interaction between the particles will be taken into account in an analogous way as in chapter 7, using spherical multipole expansions of the interaction fields. This is an approximation valid for truncated particles and caps, as long as the distance between the particles is not too short. The same technique as in chapter 7, using lattice sums for regular two-dimensional arrays or distribution integrals for random arrays, may therefore be applied in the present chapter. An equation of the type, eq.(7.10), can not be used here, since it contains the polarizability of the particles infinitely far from the substrate. With the method of the two previous chapters, one does not obtain this free polarizability of truncated particles or caps, but only the polarizability of this particle, interacting with the substrate. One should therefore use a generalization of eq.(7.10), which contains the latter kind of polarizability. Such type of equation, (7.60), has been derived in section 7.7 for spheroids on or above the substrate. This equation will be applied in section 10.2 to truncated spherical and spheroidal islands and, in a slightly modified form, in section 10.3 to spherical and spheroidal caps. In this way equations will be found for the amplitudes of the multipole fields, both for regular and for random two-dimensional arrays.

In section 10.4 the equations for the amplitudes will be solved in dipole approximation and in section 10.5 in quadrupole approximation. It should be noted that this approximation only refers to the mutual interaction between the particles,

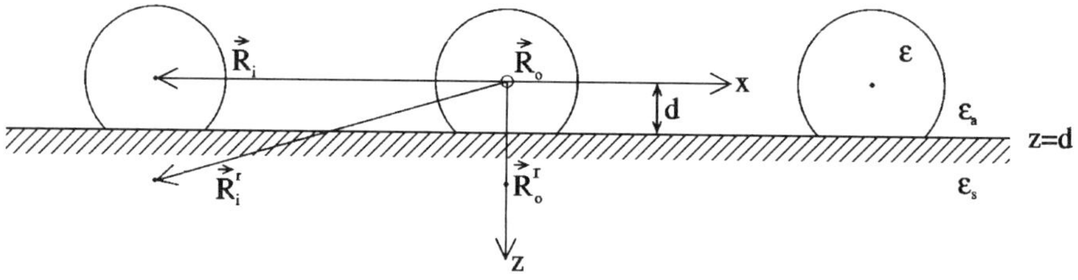

Figure 10.1 Two-dimensional array of identical truncated spheres on substrate.

and not to the interaction between particle and substrate, which has already been taken into account to arbitrary order in the polarizabilities. From the amplitudes, dipole and quadrupole polarizabilities can be calculated in the usual way for the truncated spheres or spheroids and for the spherical or spheroidal caps in regular or random two-dimensional arrays. In section 10.6 the theory is applied to gold islands on sapphire.

10.2 Two-dimensional arrays of truncated spherical or spheroidal islands

Consider first a two-dimensional array of identical truncated spherical islands of radius R, with centers at a distance d above a flat substrate ($0 \leq d \leq R$). This array may be either regular or random. The positions of these islands (centers of the spheres) are given by $\mathbf{R}_i = (\mathbf{R}_{i,\|}, 0)$ and the positions of their images by $\mathbf{R}_i^r = (\mathbf{R}_{i,\|}, 2d)$ (see fig.10.1). In the static approximation the incident field is independent of position and given in the ambient by

$$\mathbf{E}_0 = E_0(\sin\theta_0 \cos\phi_0, \sin\theta_0 \sin\phi_0, \cos\theta_0) \tag{10.1}$$

where θ_0 is the angle of the electric field with the z-axis and ϕ_0 the angle of its projection on the surface of the substrate with the x-axis.

In section 7.2 the case of an arbitrary two-dimensional array of identical spherical islands was treated and eq.(7.10) has been derived for the amplitudes $A_{\ell m,j}$ of the various multipole fields originating from the islands (j). An equation of this type can not be used for truncated spheres. The reason for this is the fact that eq.(7.10) contains the free polarizability of the sphere, which is the polarizability infinitely far from the substrate. With the method of chapter 8, one does not obtain the free polarizability of the truncated sphere, but rather the polarizability of this particle interacting with the substrate. One should therefore use an equation for the amplitudes, which contains the latter kind of polarizabilities. One could rewrite the equations for the spheres above or touching the substrate such that one first calculates the polarizability of a sphere together with its image, to a certain multipole order . Subsequently one may then take the interaction along the substrate between these sphere-image pairs to a multipole order $M_{parallel}$. In section 7.7 the equations were in fact rewritten in this way for oblate and prolate spheroids, on or above the substrate. This was not done for spheres. One may of course use the equations for spheroids in the limit that the axial ratio goes to one. In order to get converged results it was found that one must choose $M_{image} = 16$. For the interaction along the substrate the dipole approximation is enough until relatively high densities for spheres. For spheroids it was necessary to use spherical multipoles to describe the interaction along the substrate rather than spheroidal multipoles. This approximation is clearly inadequate if the islands approach each other to closely. This is an additional reason not to go beyond the dipole approximation for the interaction along the surface. Equation (7.60), derived for oblate and prolate spheroids, can now be used here equally well for identical truncated spheres on a substrate. One therefore has

$$\begin{aligned} A_{\ell m,j} &= -\sum_{\ell'=|m|}^{\infty\prime} \alpha_{\ell m,\ell' m}\{-E_0\sqrt{2\pi/3}\delta_{\ell' 1}[\cos\theta_0\delta_{m0}\sqrt{2} \\ &\quad +\sin\theta_0(\exp(i\phi_0)\delta_{m,-1} - \exp(-i\phi_0)\delta_{m1})] \\ &\quad +\sum_{\ell_2 m_2}{}^{\prime} H(\ell', m|\ell_2, m_2)\sum_{i\neq j} A_{\ell_2 m_2,i}[T_{\ell_1 m_1}(\mathbf{R}_i - \mathbf{R}_j) \\ &\quad +(-1)^{\ell_2+m_2}\left(\frac{\epsilon_a - \epsilon_s}{\epsilon_a + \epsilon_s}\right) T_{\ell_1 m_1}(\mathbf{R}_i^r - \mathbf{R}_j)]]\} \end{aligned} \tag{10.2}$$

where $\alpha_{\ell m,\ell' m'}$ are the multipole polarizabilities of the truncated sphere, interacting with the substrate. Furthermore $\ell_1 \equiv \ell' + \ell_2$ and $m_1 \equiv m_2 - m$. The matrices $H(\ell, m|\ell_2, m_2)$ and $T_{\ell_1 m_1}(\mathbf{R})$ are defined by eqs.(7.9) and (7.8). The dielectric constants of the ambient and the substrate by ϵ_a and ϵ_s.

In order to solve the above equation for the multipole amplitudes for truncated spheres, one should use the multipole polarizabilities obtained in section 8.2, for a low coverage. Solving eq.(8.29), for sufficiently large M, and using eqs.(8.31) and (5.67) gives the multipole polarizabilities $\alpha_{10,10}$, $\alpha_{11,11}$, $\alpha_{10,20}$, $\alpha_{20,10}$, $\alpha_{11,21}$ and $\alpha_{21,11}$. Higher multipole polarizabilities can, in principle, be calculated by considering non-homogeneous external fields. In appendix A of the present chapter this is done for the polarizabilities $\alpha_{20,20}$, $\alpha_{21,21}$ and $\alpha_{22,22}$, which are needed in this chapter.

Equation (10.2) will serve in the present chapter as the starting point, to incorporate the electro-magnetic interactions between the islands along the substrate. This is not only possible for truncated spheres, but also for truncated spheroids and, in fact, also for other particles, which possess rotational symmetry around an axis through their center and normal to the surface of the substrate. Notice that, as a consequence of this symmetry, $\alpha_{\ell m,\ell' m'} = \alpha_{\ell m,\ell' m}\delta_{mm'}$. For truncated spheroids one should use the polarizabilities, calculated with the method of sections 9.2 and 9.3 or appendix A in this chapter, see section 10.6 for details.

As has been discussed at the end of section 7.2, the solution of eq.(10.2), with an infinite number of coefficients $A_{\ell m,j}$, is only possible, if one makes some further assumptions, e.g. about the positions of the particles. If the islands are located on a regular array with only one truncated sphere or spheroid per unit cell

$$A_{\ell m,j} = A_{\ell m} \tag{10.3}$$

(cf. eq.(7.12)). (It is not difficult to generalize the analysis to the case of more particles per unit cell). Substituting the above relation into eq.(10.2), and using the definitions of lattice sums

$$S_{\ell m} \equiv L^{\ell+1} \sum_{i \neq j} T_{\ell m}(\mathbf{R}_i - \mathbf{R}_j) = L^{\ell+1} \sum_{i \neq 0} T_{\ell m}(\mathbf{R}_i) = \sum_{i \neq 0} (L/r)^{\ell+1} Y_\ell^m(\theta, \phi)|_{\mathbf{r}=\mathbf{R}_i} \tag{10.4}$$

(cf. eq.(7.14)) and

$$\widetilde{S}^r_{\ell m} \equiv L^{\ell+1} \sum_{i \neq j} T_{\ell m}(\mathbf{R}^r_i - \mathbf{R}_j) = L^{\ell+1} \sum_{i \neq 0} T_{\ell m}(\mathbf{R}^r_i) = \sum_{i \neq 0} (L/r)^{\ell+1} Y_\ell^m(\theta, \phi)|_{\mathbf{r}=\mathbf{R}^r_i} \tag{10.5}$$

(cf. eq.(7.63)) one obtains the following infinite set of linear inhomogeneous equations for the amplitudes $A_{\ell m}$:

$$\begin{aligned} A_{\ell m} &= -\sum_{\ell'=|m|}^{\infty\prime} \alpha_{\ell m,\ell' m}\{-E_0\sqrt{2\pi/3}\delta_{\ell' 1}[\cos\theta_0\delta_{m0}\sqrt{2} \\ &\quad + \sin\theta_0(\exp(i\phi_0)\delta_{m,-1} - \exp(-i\phi_0)\delta_{m1})] \\ &\quad + \sum_{\ell_2 m_2}^{\prime} H(\ell', m|\ell_2, m_2) A_{\ell_2 m_2} L^{-\ell_1-1}[S_{\ell_1 m_1} + (-1)^{\ell_2+m_2}\left(\frac{\epsilon_a-\epsilon_s}{\epsilon_a+\epsilon_s}\right)\widetilde{S}^r_{\ell_1 m_1}]\} \end{aligned} \tag{10.6}$$

which is of the same form as eq.(7.62). The solution of this equation for the amplitudes can be performed in an analogous way as that of eq.(7.16) described in section 7.3. One must again calculate the lattice sums, cf. appendix A of chapter 7. Notice that an important simplification is that for lattices with 2-, 4- and 6-fold symmetry $S_{\ell m} = \widetilde{S}^r_{\ell m} = 0$, if m is unequal to a multiple of 2, 4 or 6, respectively. As a consequence of the interaction with the other islands, there is no longer a symmetry for rotation around the normal on the substrate through the center of the truncated sphere of spheroid. This implies that the interaction along the surface, the last term in the above equation, leads to coupling of different values of m.

For truncated spheres and spheroids the solution of the above equation to arbitrary order in the multipole expansion is complicated by the need to first evaluate the polarizabilities $\alpha_{\ell m,\ell' m}$ to the appropriate order. In section 10.4 this will be done only to dipolar order in the interaction between the islands, and in section 10.5 to quadrupolar order. Higher order terms are not given.

Considering a random array of identical truncated spheres or spheroids, distributed homogeneously and isotropically on a flat substrate, and making the same statistical approximations as in section 7.4, one obtains from eq.(10.2) the following set of equations for the average amplitudes $< A_{\ell m} >=< A_{\ell m,j} >$:

$$\begin{aligned} < A_{\ell m} > &= -\sum_{\ell'=|m|}^{\infty\prime} \alpha_{\ell m,\ell' m}\{-E_0\sqrt{2\pi/3}\delta_{\ell' 1}[\cos\theta_0\delta_{m0}\sqrt{2} \\ &\quad + \sin\theta_0(\exp(i\phi_0)\delta_{m,-1} - \exp(-i\phi_0)\delta_{m1})] \\ + \sum_{\ell_2=|m|}^{\infty\prime} H(\ell', m|\ell_2, m) &< A_{\ell_2 m} > \rho^{(\ell_1+1)/2}[I_{\ell_1} + (-1)^{\ell_2+m}\left(\frac{\epsilon_a-\epsilon_s}{\epsilon_a+\epsilon_s}\right)\widetilde{I}^r_{\ell_1}]\} \end{aligned} \tag{10.7}$$

which has the same form as eq.(7.64). Here the "distribution integrals" I_ℓ and $\widetilde{I}_\ell^r$ are again given by eqs.(7.22) and (7.65) respectively, i.e. by

$$I_\ell \equiv \rho^{-(\ell-1)/2} \int d\mathbf{r}_\| g(\mathbf{r}_\|) T_{\ell 0}(\mathbf{r}_\|, 0) = 2\pi\rho^{-(\ell-1)/2} \int_0^\infty dr_\| g(r_\|) r_\|^{-\ell} P_\ell^0(0) \sqrt{(2\ell+1)/4\pi} \tag{10.8}$$

and

$$\begin{aligned}\widetilde{I}_\ell^r &\equiv \rho^{-(\ell-1)/2} \int d\mathbf{r}_\| g(\mathbf{r}_\|) T_{\ell 0}(\mathbf{r}_\|, 2d) \\ &= [2\pi\rho^{-(\ell-1)/2} \int_0^\infty dr_\| g(r_\|) r_\| (r_\|^2 + 4d^2)^{-(\ell+1)/2} \\ &\quad \times P_\ell^0(2d/(r_\|^2 + 4d^2)^{1/2})] \sqrt{(2\ell+1)/4\pi}\end{aligned} \tag{10.9}$$

where the pair correlation function $g(r_\|)$ of the islands is defined by eq.(7.21) and the Legendre function $P_\ell^0(x)$ by eq.(5.27). It follows from eq.(10.7) that different m's do not couple. As has been remarked in section 7.4, this is a consequence of the isotropy of the film and the statistical approximations made in the derivation of eq.(10.7). As a consequence

$$< A_{\ell m} > = < A_{\ell m,j} > = 0 \quad \text{for} \quad m \neq 0, \pm 1 \tag{10.10}$$

(cf. eq.(7.66)). Eq.(10.7) may therefore be written as the sum of separate contributions for $m = 0, \pm 1$. In this way one obtains for $m = 0$

$$\begin{aligned} < A_{\ell 0} > = \sum_{\ell'=1}^{\infty} \alpha_{\ell 0, \ell' 0} \{ E_0 \sqrt{4\pi/3} \cos\theta_0 \delta_{\ell' 1} \\ - \sum_{\ell_2=1}^{\infty} H(\ell', 0|\ell_2, 0) < A_{\ell_2 0} > \rho^{(\ell_1+1)/2} [I_{\ell_1} + (-1)^{\ell_2} \left(\frac{\epsilon_a - \epsilon_s}{\epsilon_a + \epsilon_s}\right) \widetilde{I}_{\ell_1}^r]\} \end{aligned} \tag{10.11}$$

and for $m = 1$

$$\begin{aligned} < A_{\ell 1} > = - \sum_{\ell'=1}^{\infty} \alpha_{\ell 1, \ell' 1} \{ E_0 \sqrt{2\pi/3} \sin\theta_0 \exp(-i\phi_0) \delta_{\ell' 1} \\ + \sum_{\ell_2=1}^{\infty} H(\ell', 1|\ell_2, 1) < A_{\ell_2 1} > \rho^{(\ell_1+1)/2} [I_{\ell_1} - (-1)^{\ell_2} \left(\frac{\epsilon_a - \epsilon_s}{\epsilon_a + \epsilon_s}\right) \widetilde{I}_{\ell_1}^r]\} \end{aligned} \tag{10.12}$$

where $\ell_1 \equiv \ell' + \ell_2$. The $m = -1$ equation is superfluous, as

$$< A_{\ell,-1} > \exp(-i\phi_0) = - < A_{\ell 1} > \exp(i\phi_0).$$

In sections 10.4 and 10.5 eqs.(10.11) and (10.12) will be solved in dipole and quadrupole approximation, respectively. This is sufficient, if the particles are not too densely packed on the substrate.

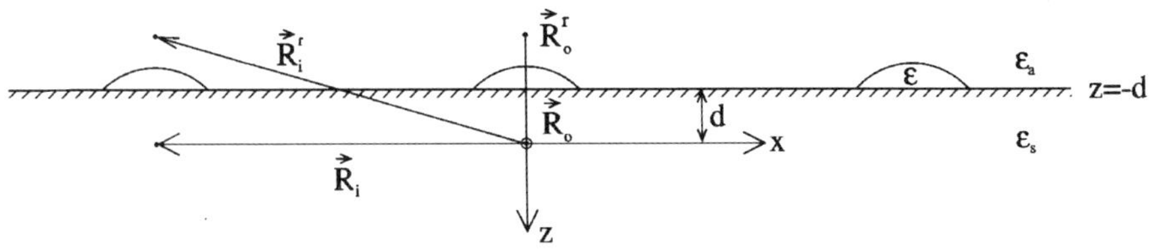

Figure 10.2 Two-dimensional array of identical spherical caps on substrate.

10.3 Two-dimensional arrays of spherical or spheroidal caps

Consider first a two-dimensional array of identical spherical caps with centers in $z = -d$ in the substrate at a distance d below the surface, $0 \leq d \leq R$ (see fig.10.2). The positions of these centers in the substrate are given by $\mathbf{R}_i = (\mathbf{R}_{i,\|}, 0)$, whereas the positions of their images, which are now in the ambient, are given by $\mathbf{R}_i^r = (\mathbf{R}_{i,\|}, -2d)$, (see fig.10.2). The incident field in the ambient is in the static approximation again given by eq.(10.1) and in the substrate by

$$\mathbf{E}_{0,s} = \mathbf{E}_0(\sin\theta_0\cos\phi_0, \sin\theta_0\sin\phi_0, \frac{\epsilon_a}{\epsilon_s}\cos\theta_0) \tag{10.13}$$

The question arises, which equation will replace eq.(10.2) of the previous section in the present case of spherical caps. This question can be answered in the following way. One first derives eq.(7.60), now for a two-dimensional array of oblate spheroids in the substrate, with centers in $\mathbf{R}_i = (\mathbf{R}_{i,\|}, 0)$, and images therefore in the ambient in $\mathbf{R}_i^r = (\mathbf{R}_{i,\|}, -2d)$. One then obtains, instead of eq.(7.60), the following equation for the amplitudes

$$\begin{aligned} A_{\ell m,j} &= -\sum_{\ell'=|m|}^{\infty\prime} \alpha_{\ell m,\ell' m}\{-E_0\sqrt{2\pi/3}\delta_{\ell' 1}[(\epsilon_a/\epsilon_s)\cos\theta_0\delta_{m0}\sqrt{2} \\ &\quad + \sin\theta_0(\exp(i\phi_0)\delta_{m,-1} - \exp(-i\phi_0)\delta_{m1})] \\ &\quad + \sum_{\ell_2 m_2}^{\prime} H(\ell', m|\ell_2, m_2)\sum_{i\neq j} A_{\ell_2 m_2,i}[T_{\ell_1 m_1}(\mathbf{R}_i - \mathbf{R}_j) \\ &\quad + (-1)^{\ell_2+m_2}\left(\frac{\epsilon_s-\epsilon_a}{\epsilon_s+\epsilon_a}\right) T_{\ell_1 m_1}(\mathbf{R}_i^r - \mathbf{R}_j)]\} \end{aligned} \tag{10.14}$$

These amplitudes now refer to the multipole fields in the substrate, with their center in $\mathbf{R}_j$. The z-component $E_0\cos\theta_0$ of the incident field in the ambient at the right-hand side of eq.(7.60) has been replaced by the z-component $(\epsilon_a/\epsilon_s)E_0\cos\theta_0$ of this field in the substrate in the above equation, whereas the parallel components are unchanged (see also eq.(10.13)). Furthermore the dielectric constants ϵ_a and ϵ_s of ambient and substrate have been interchanged in the last sum. The matrices $H(\ell,m|\ell_2,m_2)$ and $T_{\ell_1 m_1}(\mathbf{R})$ are given in eqs.(7.9) and (7.8), respectively. Furthermore $\ell_1 \equiv \ell' + \ell_2$, $m_1 \equiv m_2 - m$ and $\alpha_{\ell m,\ell' m}$ are the multipole polarizabilities of the oblate spheroids in the substrate.

This equation, which has been derived for a two-dimensional array of identical oblate spheroidal islands in the substrate, may now also be used in order to calculate the influence of the mutual interaction between the spherical caps. It gives the interaction between the multipoles located at the centers of these caps in the substrate (see fig.10.2). The lower multipole polarizabilities follow for spherical caps by solving eq.(8.49) together with eq.(8.51). The resulting multipole polarizabilities $\alpha_{\ell m,\ell' m}$, appearing in eq.(10.14), are then found using

$$\begin{aligned}
\alpha_{10,10} &= \alpha_z/4\pi\epsilon_s \\
\alpha_{11,11} &= \alpha_{1,-1;1,-1} = \alpha_{\parallel}/4\pi\epsilon_s \\
\alpha_{21,11} &= \frac{3}{5}\alpha_{11,21} = \alpha_{2,-1;1,-1} = \frac{3}{5}\alpha_{1,-1;2,-1} = 3\alpha_{\parallel}^{10}/4\pi\epsilon_s\sqrt{5} \\
\alpha_{20,10} &= \frac{3}{5}\alpha_{10,20} = \alpha_z^{10}/2\pi\epsilon_s\sqrt{5/3} \\
\alpha_{21,21} &= \alpha_{2,-1;2,-1} = 3(\alpha_{\parallel}^{11} + 2\alpha^{11})/4\pi\epsilon_s \\
\alpha_{20,20} &= (2\alpha_z^{11} + 3\alpha^{11})/2\pi\epsilon_s \\
\alpha_{22,22} &= \alpha_{2,-2;2,-2} = 3\alpha^{11}/2\pi\epsilon_s
\end{aligned} \tag{10.15}$$

which follows from eq.(5.67) by replacing ϵ_a by ϵ_s. This replacement is needed because the multipoles are now located in the substrate. Higher multipole polarizabilities can, in principle, also be calculated by extending the theory developed in section 8.3 to non-homogeneous external fields. In appendix B this is done for the multipole polarizabilities $\alpha_{20,20}$, $\alpha_{21,21}$ and $\alpha_{22,22}$.

Equation (10.14) will now serve as the starting point, to incorporate the electro-magnetic interactions between spherical caps along the substrate. This is not only possible for spherical caps, but also for spheroidal caps. For spheroidal caps one should use the polarizabilities, calculated with the method of sections 9.4 and 9.5, or appendix B in this chapter, to find the lower polarizabilities $\alpha_{\ell m,\ell' m}$ in eq.(10.14), see section 10.6 for details.

As has been discussed at the end of section 7.2, the solution of eq.(10.14), with an infinite number of coefficients $A_{\ell m,j}$, is only possible if one makes some further assumptions about, for instance, the locations of the islands. Two special cases are again considered. In the first case the caps are located on a regular two-dimensional array, with only one cap per unit cell, so that eq.(10.3) holds. Eq.(10.14) then becomes

$$\begin{aligned} A_{\ell m} &= -\sum_{\ell'=|m|}^{\infty\prime} \alpha_{\ell m,\ell' m}\{-E_0\sqrt{2\pi/3}\delta_{\ell' 1}[(\epsilon_a/\epsilon_s)\cos\theta_0\delta_{m0}\sqrt{2} \\ &+ \sin\theta_0(\exp(i\phi_0)\delta_{m,-1} - \exp(-i\phi_0)\delta_{m1})] \\ &+ \sum_{\ell_2 m_2}^{\prime} H(\ell', m|\ell_2, m_2)A_{\ell_2 m_2}L^{-\ell_1-1}[S_{\ell_1 m_1} + (-1)^{\ell_2+m_2}\left(\frac{\epsilon_s-\epsilon_a}{\epsilon_s+\epsilon_a}\right)\widetilde{S}^r_{\ell_1 m_1}]\} \end{aligned} \tag{10.16}$$

where the two-dimensional lattice sums $S_{\ell m}$ and $\widetilde{S}^r_{\ell m}$ are defined by eqs.(10.4) and (10.5).

The solution of this equation for the amplitudes can be performed in an analogous way as that of eq.(7.16) described in section 7.3. One must again calculate the lattice sums, cf. appendix A of chapter 7. Notice that, since now $\mathbf{R}^r_i = (\mathbf{R}_{i,\parallel}, -2d)$ in the lattice sums $\widetilde{S}^r_{\ell m}$, rather than $\mathbf{R}^r_i = (\mathbf{R}_{i,\parallel}, 2d)$, as was the case in chapter 7 and in the previous section, these lattice sums differ by a factor $(-1)^{\ell+m}$ from those given previously. An important simplification is again that for lattices with 2-, 4- and 6-fold symmetry $S_{\ell m} = \widetilde{S}^r_{\ell m} = 0$, if m is unequal to a multiple of 2, 4 or 6, respectively. As a consequence of the interaction with the other islands, there is no longer a symmetry for rotation around the normal on the substrate through the center of the truncated sphere of spheroid. This implies that the interaction along the surface, the last term in the above equation, leads to coupling of different values of m.

For spherical and spheroidal caps the solution of the above equation to arbitrary order in the multipole expansion is complicated by the need to first evaluate the polarizabilities $\alpha_{\ell m,\ell' m}$ to the appropriate order. In section 10.4 this will be done only to dipolar order in the interaction between the islands, and in section 10.5 to quadrupolar order. Higher order terms are not given.

In the second case one considers a random two-dimensional array of identical spherical or spheroidal caps, distributed homogeneously and isotropically on a flat substrate. Instead of eq.(10.7), one now obtains from eq.(10.14) the following set of equations for the average amplitudes

$$\begin{aligned} &< A_{\ell m} >= -\sum_{\ell'=|m|}^{\infty\prime} \alpha_{\ell m,\ell' m}\{-E_0\sqrt{2\pi/3}\delta_{\ell' 1}[(\epsilon_a/\epsilon_s)\cos\theta_0\delta_{m0}\sqrt{2} \\ &+ \sin\theta_0(\exp(i\phi_0)\delta_{m,-1} - \exp(-i\phi_0)\delta_{m1})] \\ &+ \sum_{\ell_2=|m|}^{\infty\prime} H(\ell', m|\ell_2, m) < A_{\ell_2 m} > \rho^{(\ell_1+1)/2}[I_{\ell_1} + (-1)^{\ell_2+m}\left(\frac{\epsilon_s-\epsilon_a}{\epsilon_s+\epsilon_a}\right)\widetilde{I}^r_{\ell_1}]\} \end{aligned} \tag{10.17}$$

where the "distribution integral" I_ℓ is given by eq.(10.8). The distribution integral

$\widetilde{I}^r_\ell$ is now given by

$$\begin{aligned}\widetilde{I}^r_\ell &\equiv \rho^{-(\ell-1)/2}\int d\mathbf{r}_{\|}g(\mathbf{r}_{\|})T_{\ell 0}(\mathbf{r}_{\|},-2d)\\ &= [2\pi\rho^{-(\ell-1)/2}\int_0^\infty dr_{\|}g(r_{\|})r_{\|}(r_{\|}^2+4d^2)^{-(\ell+1)/2}\\ &\quad \times P_\ell^0(-2d/(r_{\|}^2+4d^2)^{1/2})]\sqrt{(2\ell+1)/4\pi} \end{aligned} \qquad (10.18)$$

which differs by a factor by a factor $(-1)^\ell$ from the one given in eq.(10.9). It follows from eq.(10.17), that different m's do not couple, which is again a consequence of the isotropy of the film and the statistical approximations made in its derivation. One therefore finds (cf. eqs.(10.11) and (10.12)) for $m = 0$

$$\begin{aligned}< A_{\ell 0} >= \sum_{\ell'=1}^{\infty}\alpha_{\ell 0,\ell' 0}\{(\epsilon_a/\epsilon_s)E_0\sqrt{4\pi/3}\cos\theta_0\delta_{\ell' 1}\\ -\sum_{\ell_2=1}^{\infty}H(\ell',0|\ell_2,0) < A_{\ell_2 0} > \rho^{(\ell_1+1)/2}[I_{\ell_1}+(-1)^{\ell_2}\left(\frac{\epsilon_s-\epsilon_a}{\epsilon_s+\epsilon_a}\right)\widetilde{I}^r_{\ell_1}]\}\end{aligned} \qquad (10.19)$$

and for $m = 1$

$$\begin{aligned}< A_{\ell 1} >= -\sum_{\ell'=1}^{\infty}\alpha_{\ell 1,\ell' 1}\{E_0\sqrt{2\pi/3}\sin\theta_0\exp(-i\phi_0)\delta_{\ell' 1}\\ +\sum_{\ell_2=1}^{\infty}H(\ell',1|\ell_2,1) < A_{\ell_2 1} > \rho^{(\ell_1+1)/2}[I_{\ell_1}-(-1)^{\ell_2}\left(\frac{\epsilon_s-\epsilon_a}{\epsilon_s+\epsilon_a}\right)\widetilde{I}^r_{\ell_1}]\}\end{aligned} \qquad (10.20)$$

The $m = -1$ equation is superfluous, since

$$< A_{\ell,-1} > \exp(-i\phi_0) = - < A_{\ell 1} > \exp(i\phi_0).$$

In the next sections the above equations will be solved in dipole and in quadrupole approximation. This is sufficient if the islands are not too densely packed.

10.4 Dipole approximation

In this section first the case of truncated spheres or spheroids will be considered, and next that of spherical or spheroidal caps. In the dipole approximation all multipole moments of order $\ell \geq 2$ are neglected, so that eq.(10.6) becomes a finite set of linear equations for the amplitudes A_{10} , A_{11} and $A_{1,-1}$. In the regular array case one furthermore has for lattices with 4- or 6-fold symmetry (square or triangular lattice)

$$S_{2,-1} = S_{21} = \widetilde{S}^r_{2,-1} = \widetilde{S}^r_{21} = S_{2,-2} = S_{22} = \widetilde{S}^r_{2,-2} = \widetilde{S}^r_{22} = 0 \qquad (10.21)$$

(cf. eq.(7.28)). Substitution of all these zeros into eq.(10.6) and using eq.(7.9) leads to an expression in which $A_{\ell 0}$, $A_{\ell 1}$ and $A_{\ell,-1}(\ell = 1)$ are again decoupled and one finds:

$$\begin{aligned}
A_{10} &= \alpha_{10,10}\{E_0\sqrt{4\pi/3}\cos\theta_0 + 2A_{10}L^{-3}[S_{20} - \left(\frac{\epsilon_a - \epsilon_s}{\epsilon_a + \epsilon_s}\right)\widetilde{S}^r_{20}]\sqrt{4\pi/5}\} \\
A_{11} &= -\alpha_{11,11}\{E_0\sqrt{2\pi/3}\sin\theta_0\exp(-i\phi_0) + A_{11}L^{-3}[S_{20} + \left(\frac{\epsilon_a - \epsilon_s}{\epsilon_a + \epsilon_s}\right)\widetilde{S}^r_{20}]\sqrt{4\pi/5}\}
\end{aligned} \tag{10.22}$$

where again $A_{1,-1}\exp(-i\phi_0) = -A_{11}\exp(i\phi_0)$. It is crucial in this context to choose the x-axis along one of the lattice vectors. Notice that for lattices with 4- or 6-fold symmetry different m's do not couple. Solving these equations is straightforward. Using eq.(5.111), one finds for the dipole polarizabilities of the truncated spheres or spheroids on the substrate, as changed by the interaction with the other particles:

$$\begin{aligned}
\alpha_z(0) &= 4\pi\epsilon_a\alpha_{10,10}\{1 - 2\alpha_{10,10}L^{-3}[S_{20} - \left(\frac{\epsilon_a - \epsilon_s}{\epsilon_a + \epsilon_s}\right)\widetilde{S}^r_{20}]\sqrt{4\pi/5}\}^{-1} \\
\alpha_{\|}(0) &= 4\pi\epsilon_a\alpha_{11,11}\{1 + \alpha_{11,11}L^{-3}[S_{20} + \left(\frac{\epsilon_a - \epsilon_s}{\epsilon_a + \epsilon_s}\right)\widetilde{S}^r_{20}]\sqrt{4\pi/5}\}^{-1}
\end{aligned} \tag{10.23}$$

As indicated by the argument 0, these polarizabilities are for a dipole located in the center of the truncated sphere or spheroid. These equations have the same form as eq.(7.72) for spheroids. The polarizabilities $\alpha_{10,10}$ and $\alpha_{11,11}$ at the right-hand sides are now the polarizabilities of a truncated sphere or spheroid, interacting with the substrate. As has been remarked in section 10.2, these polarizabilities are obtained for truncated spheres and spheroids using the method of chapters 8 and 9 (see sections 8.2, 9.2 and 9.3). See also section 10.6 for further details. With eq.(10.23) one can then take into account, in first (dipolar) approximation, the interaction with the other islands in a regular array (with 4- or 6-fold symmetry). The lattice sums S_{20} and $\widetilde{S}^r_{20}$, are calculated using the technique described in appendix A of chapter 7.

In a similar way as above, one finds for random arrays in dipolar approximation from eqs.(10.11) and (10.12) expressions analogous to eq.(10.23), with S_{20} and $\widetilde{S}^r_{20}$ replaced by I_2 and $\widetilde{I}^r_2$, eqs.(10.8) and (10.9) respectively, and L^{-3} replaced by $\rho^{3/2}$.

The interfacial susceptibilities $\beta_e(d)$, $\gamma_e(d)$, $\delta_e(d)$ and $\tau(d)$ of a film of interacting identical truncated spherical or spheroidal islands can now be obtained from the expressions for the polarizabilities $\alpha_z(0)$ and $\alpha_{\|}(0)$, obtained above, by using the relations

$$\begin{aligned}
\beta_e(d) &= \rho\,\epsilon_a^{-2}\alpha_z(0) \\
\gamma_e(d) &= \rho\,\alpha_{\|}(0) \\
\delta_e(d) &= \rho\,d\,\epsilon_a^{-1}[\alpha_{\|}(0) + \alpha_z(0)] \\
\tau(d) &= \rho\,d\alpha_{\|}(0)
\end{aligned} \tag{10.24}$$

This has been proved at the end of section 7.5 (cf. eq.(7.39)). The argument d denotes the position of the dividing surface at $z = d$.

For a regular two-dimensional array (with 4- or 6-fold symmetry) of spherical or spheroidal caps on a substrate, one obtains from eqs.(10.16), (10.21) and (7.9), instead of eq.(10.22),

$$A_{10} = \alpha_{10,10}\{(\epsilon_a/\epsilon_s)E_0\sqrt{4\pi/3}\cos\theta_0 + 2A_{10}L^{-3}[S_{20} - \left(\frac{\epsilon_s - \epsilon_a}{\epsilon_s + \epsilon_a}\right)\widetilde{S}^r_{20}]\sqrt{4\pi/5}\}$$

$$A_{11} = -\alpha_{11,11}\{E_0\sqrt{2\pi/3}\sin\theta_0\exp(-i\phi_0) + A_{11}L^{-3}[S_{20} + \left(\frac{\epsilon_s - \epsilon_a}{\epsilon_s + \epsilon_a}\right)\widetilde{S}^r_{20}]\sqrt{4\pi/5}\} \tag{10.25}$$

where again $A_{1,-1}\exp(-i\phi_0) = -A_{11}\exp(i\phi_0)$. Notice that for lattices with 4- or 6-fold symmetry different m's do not couple. Solving these equations and using eq.(8.51) for spherical caps or eq.(9.72) for spheroidal caps, where $\epsilon_1 = \epsilon_a$ and $\epsilon_2 = \epsilon_s$, one finds, instead of eq.(10.23), for the polarizabilities of these interacting caps:

$$\begin{aligned} \alpha_z(0) &= 4\pi\epsilon_s\alpha_{10,10}\{1 - 2\alpha_{10,10}L^{-3}[S_{20} - \left(\frac{\epsilon_s - \epsilon_a}{\epsilon_s + \epsilon_a}\right)\widetilde{S}^r_{20}]\sqrt{4\pi/5}\}^{-1} \\ \alpha_{\|}(0) &= 4\pi\epsilon_s\alpha_{11,11}\{1 + \alpha_{11,11}L^{-3}[S_{20} + \left(\frac{\epsilon_s - \epsilon_a}{\epsilon_s + \epsilon_a}\right)\widetilde{S}^r_{20}]\sqrt{4\pi/5}\}^{-1} \end{aligned} \tag{10.26}$$

As indicated by the argument 0, these polarizabilities are for a dipole located in the center of the spherical or spheroidal cap, which is positioned in the substrate. The polarizabilities $\alpha_{10,10}$ and $\alpha_{11,11}$ at the right-hand sides of the above equations are polarizabilities of a cap, interacting with the substrate. As remarked in the previous section, these polarizabilities are obtained for spherical caps, using the method of section 8.3, and for spheroidal caps, using that of sections 9.4 and 9.5. With eq.(10.26) one then takes into account, in first (dipolar) approximation, the interaction with the other islands in a regular array (with 4- or 6-fold symmetry).

For a random array of caps S_{20} and $\widetilde{S}^r_{20}$ have to be replaced by I_2, eq.(10.8), and $\widetilde{I}^r_2$, eq.(10.18), respectively, and L^{-3} by $\rho^{3/2}$ in the above equation. Notice that, since $\ell = 2$ and $m = 0$, there is no sign difference between the lattice sums and distribution integrals for caps and truncated particles.

Finally the interfacial susceptibilities $\beta_e(-d)$, $\gamma_e(-d)$, $\delta_e(-d)$ and $\tau(-d)$ of a film of interacting spherical or spheroidal caps are obtained from the expressions, eq.(10.26), for the polarizabilities $\alpha_z(0)$ and $\alpha_{\|}(0)$, by using the relations

$$\begin{aligned} \beta_e(-d) &= \rho\,\epsilon_s^{-2}\alpha_z(0) \\ \gamma_e(-d) &= \rho\alpha_{\|}(0) \\ \delta_e(-d) &= -\rho d\epsilon_s^{-1}[\alpha_{\|}(0) + \alpha_z(0)] \\ \tau(-d) &= -\rho\, d\frac{\epsilon_a}{\epsilon_s}\alpha_{\|}(0) \end{aligned} \tag{10.27}$$

This follows from eq.(8.56) or (9.73), with $\epsilon_1 = \epsilon_a$, $\epsilon_2 = \epsilon_s$ and taking the quadrupole polarizabilities equal to zero. The argument $-d$ denotes the position of the dividing surface at $z = -d$.

10.5 Quadrupole approximation

Just as in the previous section, here first the case of truncated spheres or spheroids will be considered, and next the case of spherical or spheroidal caps. In the quadrupole approximation all multipole moments of order $\ell > 2$ are neglected, so that eq.(10.6) becomes a finite set of linear equations for $A_{1,-1}$, A_{10} , A_{11} , $A_{2,-2}$, $A_{2,-1}$, A_{20} , A_{21} and A_{22}. For the regular array case with a 4- or 6-fold symmetry (square and triangular lattices respectively)

$$S_{\ell m} = \widetilde{S}^{r}_{\ell m} = 0 \text{ for m} \neq \text{multiple of } 4 \text{ or } 6 \text{ respectively} \tag{10.28}$$

Substitution of all these zeros into the above mentioned finite set of equations leads, with the definition, eq.(7.9), of $H(\ell, m|\ell_2, m_2)$ to

$$\begin{aligned}
A_{10} &= \alpha_{10,10}\{2E_0 \cos\theta_0\sqrt{\pi/3} + 4\widetilde{S}^{-}_{20}A_{10}L^{-3}\sqrt{\pi/5} - 2\widetilde{S}^{+}_{30}A_{20}L^{-4}\sqrt{15\pi/7}\} \\
&\quad +\alpha_{10,20}\{6\widetilde{S}^{-}_{30}A_{10}L^{-4}\sqrt{3\pi/35} - 4\widetilde{S}^{+}_{40}A_{20}L^{-5}\sqrt{\pi}\} \\
A_{20} &= \alpha_{20,10}\{2E_0 \cos\theta_0\sqrt{\pi/3} + 4\widetilde{S}^{-}_{20}A_{10}L^{-3}\sqrt{\pi/5} - 2\widetilde{S}^{+}_{30}A_{20}L^{-4}\sqrt{15\pi/7}\} \\
&\quad +\alpha_{20,20}\{6\widetilde{S}^{-}_{30}A_{10}L^{-4}\sqrt{3\pi/35} - 4\widetilde{S}^{+}_{40}A_{20}L^{-5}\sqrt{\pi}\} \\
A_{11} &= \alpha_{11,11}\{-E_0 \sin\theta_0 e^{-i\phi_0}\sqrt{2\pi/3} - 2\widetilde{S}^{+}_{20}A_{11}L^{-3}\sqrt{\pi/5} + 2\widetilde{S}^{-}_{30}A_{21}L^{-4}\sqrt{5\pi/7}\} \\
&\quad +\alpha_{11,21}\{-6\widetilde{S}^{+}_{30}A_{11}L^{-4}\sqrt{\pi/35} + \frac{8}{3}\widetilde{S}^{-}_{40}A_{21}L^{-5}\sqrt{\pi}\} \\
A_{21} &= \alpha_{21,11}\{-E_0 \sin\theta_0 e^{-i\phi_0}\sqrt{2\pi/3} - 2\widetilde{S}^{+}_{20}A_{11}L^{-3}\sqrt{\pi/5} + 2\widetilde{S}^{-}_{30}A_{21}L^{-4}\sqrt{5\pi/7}\} \\
&\quad +\alpha_{21,21}\{-6\widetilde{S}^{+}_{30}A_{11}L^{-4}\sqrt{\pi/35} + \frac{8}{3}\widetilde{S}^{-}_{40}A_{21}L^{-5}\sqrt{\pi}\}
\end{aligned} \tag{10.29}$$

Here the abbreviations

$$\widetilde{S}^{\pm}_{\ell m} \equiv S_{\ell m} \pm \left(\frac{\epsilon_a - \epsilon_s}{\epsilon_a + \epsilon_s}\right)\widetilde{S}^{r}_{\ell m} \tag{10.30}$$

have been introduced (cf. eq.(7.80)). Notice that for lattices with 4- or 6-fold symmetry different m's do not couple. The equations for $A_{1,-1}$ and $A_{2,-1}$ are superfluous, since one can prove again, that $A_{\ell,-1}\exp(-i\phi_0) = -A_{\ell 1}\exp(i\phi_0)$, $(\ell = 1, 2)$. It is crucial in this context that the x-axis has been chosen along one of the lattice vectors. It is furthermore found that in quadrupole approximation

$$A_{22} = A_{2,-2} = 0 \tag{10.31}$$

(cf. eq.(7.43))

Solving eq.(10.29) and using eq.(5.111), one finds for the polarizabilities normal to the surface of the substrate

$$\begin{aligned}
\alpha_z(0) &= 4\pi\epsilon_a D_z^{-1}\{[1 + 4\alpha_{20,20}\widetilde{S}^{+}_{40}L^{-5}\sqrt{\pi}]\alpha_{10,10} \\
&\quad -4\alpha_{10,20}\alpha_{20,10}\widetilde{S}^{+}_{40}L^{-5}\sqrt{\pi}\} \\
\alpha_z^{10}(0) &= 2\pi\epsilon_a\sqrt{5/3}D_z^{-1}\{[1 - 6\alpha_{10,20}\widetilde{S}^{-}_{30}L^{-4}\sqrt{3\pi/35}]\alpha_{20,10} \\
&\quad +6\alpha_{20,20}\alpha_{10,10}\widetilde{S}^{-}_{30}L^{-4}\sqrt{3\pi/35}\}
\end{aligned} \tag{10.32}$$

where

$$
\begin{aligned}
D_z \equiv\ & [1 - 4\alpha_{10,10}\widetilde{S}_{20}^{-}L^{-3}\sqrt{\pi/5} - 6\alpha_{10,20}\widetilde{S}_{30}^{-}L^{-4}\sqrt{3\pi/35}] \\
& \times[1 + 2\alpha_{20,10}\widetilde{S}_{30}^{+}L^{-4}\sqrt{15\pi/7} + 4\alpha_{20,20}\widetilde{S}_{40}^{+}L^{-5}\sqrt{\pi}] \\
& +[4\alpha_{20,10}\widetilde{S}_{20}^{-}L^{-3}\sqrt{\pi/5} + 6\alpha_{20,20}\widetilde{S}_{30}^{-}L^{-4}\sqrt{3\pi/35}] \\
& \times[2\alpha_{10,10}\widetilde{S}_{30}^{+}L^{-4}\sqrt{15\pi/7} + 4\alpha_{10,20}\widetilde{S}_{40}^{+}L^{-5}\sqrt{\pi}]
\end{aligned}
\tag{10.33}
$$

and for the polarizabilities parallel to the surface of the substrate

$$
\begin{aligned}
\alpha_{\|}(0) =\ & 4\pi\epsilon_a D_{\|}^{-1}\{[1 - \frac{8}{3}\alpha_{21,21}\widetilde{S}_{40}^{-}L^{-5}\sqrt{\pi}]\alpha_{11,11} \\
& +\frac{8}{3}\alpha_{11,21}\alpha_{21,11}\widetilde{S}_{40}^{-}L^{-5}\sqrt{\pi}\} \\
\alpha_{\|}^{10}(0) =\ & \frac{4}{3}\pi\ \epsilon_a\sqrt{5}D_{\|}^{-1}\{[1 + 6\alpha_{11,21}\widetilde{S}_{30}^{+}L^{-4}\sqrt{\pi/35}]\alpha_{21,11} \\
& -6\alpha_{21,21}\alpha_{11,11}\widetilde{S}_{30}^{+}L^{-4}\sqrt{\pi/35}\}
\end{aligned}
\tag{10.34}
$$

where

$$
\begin{aligned}
D_{\|} \equiv\ & [1 + 2\alpha_{11,11}\widetilde{S}_{20}^{+}L^{-3}\sqrt{\pi/5} + 6\alpha_{11,21}\widetilde{S}_{30}^{+}L^{-4}\sqrt{\pi/35}] \\
& \times[1 - 2\alpha_{21,11}\widetilde{S}_{30}^{-}L^{-4}\sqrt{5\pi/7} - \frac{8}{3}\alpha_{21,21}\widetilde{S}_{40}^{-}L^{-5}\sqrt{\pi}] \\
& +[2\alpha_{21,11}\widetilde{S}_{20}^{+}L^{-3}\sqrt{\pi/5} + 6\alpha_{21,21}\widetilde{S}_{30}^{+}L^{-4}\sqrt{\pi/35}] \\
& \times[2\alpha_{11,11}\widetilde{S}_{30}^{-}L^{-4}\sqrt{5\pi/7} + \frac{8}{3}\alpha_{11,21}\widetilde{S}_{40}^{-}L^{-5}\sqrt{\pi}]
\end{aligned}
\tag{10.35}
$$

As indicated by the argument 0, these polarizabilities are for a dipole and a quadrupole located in the center of the truncated spheres or spheroids. The above equations have the same form as eqs.(7.81)–(7.84) for spheroids. The polarizabilities $\alpha_{\ell m,\ell' m}$ in the above equations are now for a truncated sphere or spheroid, interacting with the substrate. With the method, developed in the previous chapters, in which a homogeneous external field is used, one can only calculate $\alpha_{10,10}$, $\alpha_{11,11}$, $\alpha_{10,20}$, $\alpha_{20,10}$, $\alpha_{11,21}$ and $\alpha_{21,11}$. This follows from eqs.(5.67), (8.31) and (9.31). In order to be able to evaluate also the polarizabilities $\alpha_{20,20}$ and $\alpha_{21,21}$, the method of the two previous chapters has to be extended to the case of a non-homogeneous external electric field, which may be taken as a linear function in space. This problem is treated in appendix A of this chapter. The lattice sums $\widetilde{S}_{\ell 0}^{\pm}(\ell = 2, 3, 4)$ are calculated using the techniques described in appendix A of chapter 7.

With eqs.(10.32)– (10.35) one can take into account in quadrupolar approximation the interaction of a truncated sphere or spheroid with the other islands in a regular array with 4- or 6-fold symmetry. The interaction in a random array can be found, by solving eqs.(10.11) and (10.12), in the same (quadrupolar) approximation as eq.(10.6) has been solved above. The result is, that one simply has to replace the lattice sums $\widetilde{S}_{\ell 0}^{\pm}(\ell = 2, 3.4)$ by

$$
\widetilde{I}_{\ell}^{\pm} \equiv I_{\ell} \pm \left(\frac{\epsilon_a - \epsilon_s}{\epsilon_a + \epsilon_s}\right)\widetilde{I}_{\ell}^{r}
\tag{10.36}
$$

in the above expressions, eqs.(10.32)–(10.35), for the polarizabilities, and furthermore L by $\rho^{-1/2}$. Here I_ℓ and $\widetilde{I}_\ell^r$ are the distribution integrals, defined by eqs.(10.8) and (10.9) respectively.

The interfacial susceptibilities $\beta_e(d)$, $\gamma_e(d)$, $\delta_e(d)$ and $\tau(d)$ of a film of interacting identical truncated spherical or spheroidal islands can be obtained from the expressions for the polarizabilities $\alpha_z(0)$, $\alpha_z^{10}(0)$, $\alpha_\parallel(0)$ and $\alpha_\parallel^{10}(0)$, found above, by using the relations

$$\begin{aligned}
\beta_e(d) &= \rho\, \epsilon_a^{-2}\alpha_z(0) \\
\gamma_e(d) &= \rho\, \alpha_\parallel(0) \\
\delta_e(d) &= -\rho\, \epsilon_a^{-1}[\alpha_z^{10}(0) + \alpha_\parallel^{10}(0) - d\, \alpha_z(0) - d\, \alpha_\parallel(0)] \\
\tau(d) &= -\rho[\alpha_\parallel^{10}(0) - d\, \alpha_\parallel(0)]
\end{aligned} \tag{10.37}$$

This has been proven at the end of section 7.6 (cf. eq.(7.51)). The argument d denotes the position of the dividing surface at $z = d$.

For a regular two-dimensional array (with 4- or 6- fold symmetry) of spherical or spheroidal caps on a substrate, one obtains from eqs.(10.16), (10.28) and (7.9), instead of eqs. (10.29),

$$\begin{aligned}
A_{10} &= \alpha_{10,10}\{2(\epsilon_a/\epsilon_s)E_0\cos\theta_0\sqrt{\pi/3} + 4\widetilde{S}_{20}^{+}A_{10}L^{-3}\sqrt{\pi/5} - 2\widetilde{S}_{30}^{-}A_{20}L^{-4}\sqrt{15\pi/7}\} \\
&\quad +\alpha_{10,20}\{6\widetilde{S}_{30}^{+}A_{10}L^{-4}\sqrt{3\pi/35} - 4\widetilde{S}_{40}^{-}A_{20}L^{-5}\sqrt{\pi}\} \\
A_{20} &= \alpha_{20,10}\{2(\epsilon_a/\epsilon_s)E_0\cos\theta_0\sqrt{\pi/3} + 4\widetilde{S}_{20}^{+}A_{10}L^{-3}\sqrt{\pi/5} - 2\widetilde{S}_{30}^{-}A_{20}L^{-4}\sqrt{15\pi/7}\} \\
&\quad +\alpha_{20,20}\{6\widetilde{S}_{30}^{+}A_{10}L^{-4}\sqrt{3\pi/35} - 4\widetilde{S}_{40}^{-}A_{20}L^{-5}\sqrt{\pi}\} \\
A_{11} &= \alpha_{11,11}\{-E_0\sin\theta_0 e^{-i\phi_0}\sqrt{2\pi/3} - 2\widetilde{S}_{20}^{-}A_{11}L^{-3}\sqrt{\pi/5} + 2\widetilde{S}_{30}^{+}A_{21}L^{-4}\sqrt{5\pi/7}\} \\
&\quad +\alpha_{11,21}\{-6\widetilde{S}_{30}^{-}A_{11}L^{-4}\sqrt{\pi/35} + \frac{8}{3}\widetilde{S}_{40}^{+}A_{21}L^{-5}\sqrt{\pi}\} \\
A_{21} &= \alpha_{21,11}\{-E_0\sin\theta_0 e^{-i\phi_0}\sqrt{2\pi/3} - 2\widetilde{S}_{20}^{-}A_{11}L^{-3}\sqrt{\pi/5} + 2\widetilde{S}_{30}^{+}A_{21}L^{-4}\sqrt{5\pi/7}\} \\
&\quad +\alpha_{21,21}\{-6\widetilde{S}_{30}^{-}A_{11}L^{-4}\sqrt{\pi/35} + \frac{8}{3}\widetilde{S}_{40}^{+}A_{21}L^{-5}\sqrt{\pi}\}
\end{aligned} \tag{10.38}$$

where the lattice sums $\widetilde{S}_{\ell m}^{\pm}$ are given by eq.(10.30). Notice that for lattices with 4- or 6-fold symmetry different m's do not couple. Notice furthermore, that this equation may be obtained from eq.(10.29) by replacing at the right-hand sides the normal component $E_0\cos\theta_0$ of the external field in the ambient by its value $(\epsilon_a/\epsilon_s)E_0\cos\theta_0$ in the substrate, and by replacing $\widetilde{S}_{\ell m}^{\pm}$ by $\widetilde{S}_{\ell m}^{\mp}$, which is equivalent to interchanging the dielectric constants ϵ_a and ϵ_s, as follows from eq.(10.30). The equations for $A_{1,-1}$ and $A_{2,-1}$ are again superfluous, just as for truncated particles above. One can also prove, that in quadrupole approximation eq.(10.31) holds.

Solving eq. (10.38) and using eq. (8.51) for spherical caps, or eq. (9.72) for spheroidal caps, where $\epsilon_1 = \epsilon_a$ and $\epsilon_2 = \epsilon_s$, one finds, instead of eqs.(10.32)–(10.35), for the polarizabilities of the interacting caps, normal to the surface of the substrate:

$$
\begin{aligned}
\alpha_z(0) &= 4\pi\epsilon_s D_z^{-1}\{[1+4\alpha_{20,20}\widetilde{S}_{40}^{-}L^{-5}\sqrt{\pi}]\alpha_{10,10} \\
&\quad -4\alpha_{10,20}\alpha_{20,10}\widetilde{S}_{40}^{-}L^{-5}\sqrt{\pi}\} \\
\alpha_z^{10}(0) &= 2\pi\epsilon_s\sqrt{5/3}D_z^{-1}\{[1-6\alpha_{10,20}\widetilde{S}_{30}^{+}L^{-4}\sqrt{3\pi/35}]\alpha_{20,10} \\
&\quad +6\alpha_{20,20}\alpha_{10,10}\widetilde{S}_{30}^{+}L^{-4}\sqrt{3\pi/35}\}
\end{aligned}
\tag{10.39}
$$

where now

$$
\begin{aligned}
D_z \equiv\ & [1-4\alpha_{10,10}\widetilde{S}_{20}^{+}L^{-3}\sqrt{\pi/5}-6\alpha_{10,20}\widetilde{S}_{30}^{+}L^{-4}\sqrt{3\pi/35}] \\
& \times[1+2\alpha_{20,10}\widetilde{S}_{30}^{-}L^{-4}\sqrt{15\pi/7}+4\alpha_{20,20}\widetilde{S}_{40}^{-}L^{-5}\sqrt{\pi}] \\
& +[4\alpha_{20,10}\widetilde{S}_{20}^{+}L^{-3}\sqrt{\pi/5}+6\alpha_{20,20}\widetilde{S}_{30}^{+}L^{-4}\sqrt{3\pi/35}] \\
& \times[2\alpha_{10,10}\widetilde{S}_{30}^{-}L^{-4}\sqrt{15\pi/7}+4\alpha_{10,20}\widetilde{S}_{40}^{-}L^{-5}\sqrt{\pi}]
\end{aligned}
\tag{10.40}
$$

and for the polarizabilities parallel to the surface of the substrate

$$
\begin{aligned}
\alpha_\parallel(0) &= 4\pi\epsilon_s D_\parallel^{-1}\{[1-\frac{8}{3}\alpha_{21,21}\widetilde{S}_{40}^{+}L^{-5}\sqrt{\pi}]\alpha_{11,11} \\
&\quad +\frac{8}{3}\alpha_{11,21}\alpha_{21,11}\widetilde{S}_{40}^{+}L^{-5}\sqrt{\pi}\} \\
\alpha_\parallel^{10}(0) &= \frac{4}{3}\pi\ \epsilon_s\sqrt{5}D_\parallel^{-1}\{[1+6\alpha_{11,21}\widetilde{S}_{30}^{-}L^{-4}\sqrt{\pi/35}]\alpha_{21,11} \\
&\quad -6\alpha_{21,21}\alpha_{11,11}\widetilde{S}_{30}^{-}L^{-4}\sqrt{\pi/35}\}
\end{aligned}
\tag{10.41}
$$

where now

$$
\begin{aligned}
D_\parallel \equiv\ & [1+2\alpha_{11,11}\widetilde{S}_{20}^{-}L^{-3}\sqrt{\pi/5}+6\alpha_{11,21}\widetilde{S}_{30}^{-}L^{-4}\sqrt{\pi/35}] \\
& \times[1-2\alpha_{21,11}\widetilde{S}_{30}^{+}L^{-4}\sqrt{5\pi/7}-\frac{8}{3}\alpha_{21,21}\widetilde{S}_{40}^{+}L^{-5}\sqrt{\pi}] \\
& +[2\alpha_{21,11}\widetilde{S}_{20}^{-}L^{-3}\sqrt{\pi/5}+6\alpha_{21,21}\widetilde{S}_{30}^{-}L^{-4}\sqrt{\pi/35}] \\
& \times[2\alpha_{11,11}\widetilde{S}_{30}^{+}L^{-4}\sqrt{5\pi/7}+\frac{8}{3}\alpha_{11,21}\widetilde{S}_{40}^{+}L^{-5}\sqrt{\pi}]
\end{aligned}
\tag{10.42}
$$

As indicated by the argument 0, these polarizabilities are for a dipole and a quadrupole located in the centers of the spherical or the spheroidal cap, which are positioned in the substrate. The multipole polarizabilities $\alpha_{10,10}$, $\alpha_{10,20}$, $\alpha_{20,10}$, $\alpha_{20,20}$, $\alpha_{11,11}$, $\alpha_{11,21}$, $\alpha_{21,11}$ and $\alpha_{21,21}$, appearing at the right-hand sides of the above equations, are the polarizabilities of a cap, interacting with the substrate. Using the method of section 8.3 for spherical caps, or that of sections 9.4 and 9.5 in the case of spheroidal caps, one can only calculate $\alpha_{10,10}$, $\alpha_{11,11}$, $\alpha_{10,20}$, $\alpha_{20,10}$, $\alpha_{11,21}$ and $\alpha_{21,11}$. This follows from eqs.(10.15), (8.51) and (9.72). In order to be able to evaluate also the polarizabilities $\alpha_{20,20}$ and $\alpha_{21,21}$ for spherical and spheroidal caps on a substrate, the method of the previous chapters has to be extended to the case of a non-homogeneous external electric field, which may be taken as a linear function in space. This problem is treated in appendix B.

With eqs.(10.39)– (10.42) one can take into account in quadrupole approximation the interaction of a spherical or spheroidal cap with the other caps in a regular array with 4- or 6-fold symmetry. The interactions in a random array can be taken into account by replacing the lattice sums $\widetilde{S}^{\pm}_{\ell 0}(\ell = 2, 3, 4)$ in these equations by $\widetilde{I}^{\pm}_{\ell}$, cf. eq.(10.36); in this last equation $\widetilde{I}_{\ell}$ and $\widetilde{I}^{r}_{\ell}$ are defined by eqs.(10.8) and (10.18) respectively. Furthermore L should be replaced by $\rho^{-1/2}$.

Notice that, since $\widetilde{S}^{r}_{\ell m}$ differs by a factor $(-1)^{\ell+m}$ for truncated particles and caps on the same lattice (see section 10.3), it follows, with eq.(10.30), that $\widetilde{S}^{\pm}_{20}$ and $\widetilde{S}^{\pm}_{40}$ have the same value in the above equations for truncated particles and caps. For the same reason $\widetilde{S}^{\pm}_{30}$ in the case of truncated particles equals $\widetilde{S}^{\mp}_{30}$ in the case of caps. This is also true for the random array, where $\widetilde{I}^{\pm}_{2}$ and $\widetilde{I}^{\pm}_{4}$ are the same, whereas $\widetilde{I}^{\pm}_{3}$ for truncated particles equals $\widetilde{I}^{\mp}_{3}$ for caps.

Finally the interfacial susceptibilities $\beta_e(-d)$, $\gamma_e(-d)$, $\delta_e(-d)$ and $\tau(-d)$ of a film of interacting spherical or spheroidal caps are obtained from the expressions, eqs.(10.39)– (10.42) for the polarizabilities $\alpha_z(0)$, $\alpha^{10}_z(0)$, $\alpha_{\|}(0)$ and $\alpha^{10}_{\|}(0)$, by using the relations

$$\begin{aligned}
\beta_e(-d) &= \rho\, \epsilon_s^{-2}\alpha_z(0) \\
\gamma_e(-d) &= \rho\, \alpha_{\|}(0) \\
\delta_e(-d) &= -\rho\, \epsilon_s^{-1}[\alpha^{10}_z(0) + \alpha^{10}_{\|}(0) + d\, \alpha_z(0) + d\, \alpha_{\|}(0)] \\
\tau(-d) &= -\rho\frac{\epsilon_a}{\epsilon_s}[\alpha^{10}_{\|}(0) + d\, \alpha_{\|}(0)] \qquad (10.43)
\end{aligned}$$

This follows from eq.(8.44) or (9.73), with $\epsilon_1 = \epsilon_a$ and $\epsilon_2 = \epsilon_s$. The argument $-d$ denotes the position of the dividing surface at $z = -d$.

10.6 Application: Gold islands on sapphire

In order to demonstrate how the equations, derived above, can be used in practice, the special case of gold islands on sapphire will again be discussed. The influence of the interaction between the particles is considered first to dipolar and then quadrupolar order. The dielectric constant of the ambient is unity, $\epsilon_a = 1$. Sapphire has a small dispersion in the visible domain. An average value is used as dielectric constant of the substrate, $\epsilon_s = 3.13$. For the complex dielectric constant of gold a list of values is given in table 5.1, for the optical domain [40].

Consider first a regular array, with 4- or 6-fold symmetry, in which the interactions between the particles are taken into account until dipolar order. It follows from eq.(10.23), with $\epsilon_a = 1$, that for truncated spheres and spheroids the dipole polarizabilities, in dimensionless form, are given by

$$\begin{aligned}
\widehat{\alpha}_z &\equiv \frac{1}{V}\alpha_z(0) = 4\pi\widehat{\alpha}_{10,10}\{1 - 2\widehat{\alpha}_{10,10}\widehat{V}[S_{20} - \left(\frac{1-\epsilon_s}{1+\epsilon_s}\right)\widetilde{S}^{r}_{20}]\sqrt{4\pi/5}\}^{-1} \\
\widehat{\alpha}_{\|} &\equiv \frac{1}{V}\alpha_{\|}(0) = 4\pi\widehat{\alpha}_{11,11}\{1 + \widehat{\alpha}_{11,11}\widehat{V}[S_{20} + \left(\frac{1-\epsilon_s}{1+\epsilon_s}\right)\widetilde{S}^{r}_{20}]\sqrt{4\pi/5}\}^{-1} \qquad (10.44)
\end{aligned}$$

where

$$\widehat{\alpha}_{10,10} \equiv \frac{\alpha_{10,10}}{V}, \qquad \widehat{\alpha}_{11,11} \equiv \frac{\alpha_{11,11}}{V} \tag{10.45}$$

and

$$\widehat{V} \equiv \frac{V}{L^3} \tag{10.46}$$

For a truncated sphere with radius R and truncation parameter t_r it follows, with eq.(8.107), that

$$\widehat{V} = \frac{1}{3}\pi\left(\frac{R}{L}\right)^3 \left(2 + 3t_r - t_r^3\right) \tag{10.47}$$

For a truncated oblate spheroid, $0 \leq \xi_0$, one finds from eq.(9.114)

$$\widehat{V} = \frac{1}{3}\pi\left(\frac{a}{L}\right)^3 \xi_0\left(1 + \xi_0^2\right)\left(2 + 3t_r - t_r^3\right) \tag{10.48}$$

and in the case of a truncated prolate spheroid, $1 \leq \xi_0$, from eq.(9.115):

$$\widehat{V} = \frac{1}{3}\pi\left(\frac{a}{L}\right)^3 \xi_0\left(\xi_0^2 - 1\right)\left(2 + 3t_r - t_r^3\right) \tag{10.49}$$

See figs 9.2 and 9.3. The dimensionless dipole polarizabilities, eq.(10.45), follow from the dimensionless amplitudes, using eqs.(5.67) and (8.110) for a truncated sphere and eqs.(5.67) and (9.119) for a truncated spheroid:

$$\widehat{\alpha}_{10,10} = \widehat{A}_{10}, \qquad \widehat{\alpha}_{11,11} = \widehat{A}_{11} \tag{10.50}$$

These amplitudes are obtained for a truncated sphere by solving eq.(8.108), together with eqs.(8.109) and (8.13). For truncated spheroids one should solve eq.(9.116), together with eqs.(9.117) and (9.19) for oblate, and eqs.(9.118) and (9.38) for prolate spheroids . The lattice sums S_{20} and $\widetilde{S}_{20}^r$ are evaluated using the methods described in appendix A of chapter 7.

The surface constitutive coefficients, eq.(10.24), become in dimensionless form for truncated spheres with radius R and truncation parameter t_r

$$\begin{aligned}
\widehat{\gamma}_e &\equiv \frac{1}{\rho V}\gamma_e(d) = \widehat{\alpha}_\parallel \\
\widehat{\beta}_e &\equiv \frac{1}{\rho V}\beta_e(d) = \widehat{\alpha}_z \\
\widehat{\tau} &\equiv \frac{1}{\rho V R}\tau(d) = t_r\widehat{\alpha}_\parallel \\
\widehat{\delta}_e &\equiv \frac{1}{\rho V R}\delta_e(d) = t_r\left(\widehat{\alpha}_z + \widehat{\alpha}_\parallel\right)
\end{aligned} \tag{10.51}$$

for this case, where $\epsilon_a = 1$. The invariants, eq.(5.66), become for truncated spheres

$$\begin{aligned}\widehat{I}_e &\equiv \frac{1}{\rho V} I_e = \widehat{\gamma}_e - \epsilon_s \widehat{\beta}_e \ , \quad \widehat{I}_c \equiv \frac{1}{\rho V}(\epsilon_s - 1) I_c = \mathrm{Im}\,\widehat{\gamma}_e \\ \widehat{I}_\tau &\equiv \frac{1}{\rho V R} I_\tau = \widehat{\tau} \ , \qquad \widehat{I}_{\delta,e} \equiv \frac{1}{\rho V R} I_{\delta,e} = \widehat{\delta}_e \end{aligned} \tag{10.52}$$

where ϵ_s was taken real. For truncated spheroids one has to replace R by $a\xi_0$ in the above equations. All these quantities therefore follow in a simple way if the polarizabilities $\widehat{\alpha}_z$ and $\widehat{\alpha}_\parallel$, eq.(10.44), are known.

Next consider the case of a regular array, with 4- or 6-fold symmetry, of spherical or spheroidal caps. Now it follows from eq.(10.26) that, in dipolar approximation in the interaction between the particles,

$$\begin{aligned}\widehat{\alpha}_z &\equiv \frac{1}{V}\alpha_z(0) = 4\pi\epsilon_s \widehat{\alpha}_{10,10}\{1 - 2\widehat{\alpha}_{10,10}\widehat{V}[S_{20} + \left(\frac{1-\epsilon_s}{1+\epsilon_s}\right)\widetilde{S}^{\tau}_{20}]\sqrt{4\pi/5}\}^{-1} \\ \widehat{\alpha}_\parallel &\equiv \frac{1}{V}\alpha_\parallel(0) = 4\pi\epsilon_s \widehat{\alpha}_{11,11}\{1 + \widehat{\alpha}_{11,11}\widehat{V}[S_{20} - \left(\frac{1-\epsilon_s}{1+\epsilon_s}\right)\widetilde{S}^{\tau}_{20}]\sqrt{4\pi/5}\}^{-1} \end{aligned} \tag{10.53}$$

where $\widehat{\alpha}_{10,10}$, $\widehat{\alpha}_{11,11}$ and $\widehat{V}$ were defined above. For a spherical cap with radius R and truncation parameter t_r, it follows from eq.(8.113):

$$\widehat{V} = \frac{1}{3}\pi\left(\frac{R}{L}\right)^3 (2 - 3t_r + t_r^3) \tag{10.54}$$

For an oblate spheroidal cap, $0 \le \xi_0$, one finds from eq.(9.122)

$$\widehat{V} = \frac{1}{3}\pi\left(\frac{a}{L}\right)^3 \xi_0\left(1 + \xi_0^2\right)\left(2 - 3t_r + t_r^3\right) \tag{10.55}$$

and in the case of a prolate spheroidal cap, $1 \le \xi_0$, from eq.(9.123):

$$\widehat{V} = \frac{1}{3}\pi\left(\frac{a}{L}\right)^3 \xi_0\left(\xi_0^2 - 1\right)\left(2 - 3t_r + t_r^3\right) \tag{10.56}$$

See figs 9.5 and 9.6. The dimensionless dipole polarizabilities, eq.(10.45), on the right-hand side of eq.(10.53) follow from the dimensionless amplitudes, using eqs.(10.15) and (8.115) for a spherical cap and using eqs.(10.15) and (9.126) for a spheroidal cap:

$$\widehat{\alpha}_{10,10} = \epsilon_s \widehat{A}_{10}, \qquad \widehat{\alpha}_{11,11} = \widehat{A}_{11} \tag{10.57}$$

These dimensionless amplitudes follow for a spherical cap by solving eq.(8.108), together with eqs.(8.43) and (8.114). For an oblate spheroidal cap one should solve eq.(9.116), together with eqs.(9.65) and (9.124), and for a prolate spheroidal cap eqs.(9.116) with eqs.(9.79) and (9.125). The lattice sums S_{20} and $\widetilde{S}^{\tau}_{20}$ are evaluated using the methods described in appendix A of chapter 7. As was explained in section 10.4, these lattice sums are the same for truncated particles and caps on the same lattice.

The surface constitutive coefficients, eq.(10.27), become in dimensionless form for spherical caps, with radius R and truncation parameter t_r,

$$\begin{aligned}
\widehat{\gamma}_e &\equiv \frac{1}{\rho V}\gamma_e(-d) = \widehat{\alpha}_{\|} \\
\widehat{\beta}_e &\equiv \frac{1}{\rho V}\beta_e(-d) = \epsilon_s^{-2}\widehat{\alpha}_z \\
\widehat{\tau} &\equiv \frac{1}{\rho V R}\tau(-d) = -\epsilon_s^{-1}t_r\widehat{\alpha}_{\|} \\
\widehat{\delta}_e &\equiv \frac{1}{\rho V R}\delta_e(-d) = -\epsilon_s^{-1}t_r\left(\widehat{\alpha}_z + \widehat{\alpha}_{\|}\right)
\end{aligned} \qquad (10.58)$$

where $\epsilon_a = 1$. The invariants are again given by eq.(10.52) for real ϵ_s. For spheroidal caps one has to replace R by $a\xi_0$ in the above equations.

For random arrays the lattice sums S_{20} and $\widetilde{S}_{20}^r$, in eq.(10.44) for truncated particles and in eq.(10.53) for caps, have to be replaced by the distribution integrals I_2 and $\widetilde{I}_2^r$. These distribution integrals are given by eqs.(10.8) and (10.9) for truncated particles and by eqs.(10.8) and (10.18) for caps. Furthermore L^{-3} in eqs.(10.46)–(10.49) and (10.54)–(10.56), should be replaced by $\rho^{3/2}$.

Next the case of truncated spheres with a radius R is considered, in which the interaction between the particles along the substrate is taken into account to quadrupolar approximation. The dimensionless forms of eqs.(10.32) and (10.33), for the dipole and quadrupole polarizabilities normal to the surface of the substrate are given, with $\epsilon_a = 1$, by

$$\begin{aligned}
\widehat{\alpha}_z &\equiv \frac{\alpha_z(0)}{V} = 4\pi D_z^{-1}\{[1 + 4\widehat{V}\widehat{R}^2\widehat{\alpha}_{20,20}\widetilde{S}_{40}^{+}\sqrt{\pi}]\widehat{\alpha}_{10,10} \\
&\quad -4\widehat{V}\widehat{R}^2\widehat{\alpha}_{10,20}\widehat{\alpha}_{20,10}\widetilde{S}_{40}^{+}\sqrt{\pi}\} \\
\widehat{\alpha}_z^{10} &\equiv \frac{\alpha_z^{10}(0)}{VR} = 2\pi\sqrt{\frac{5}{3}}D_z^{-1}\{[1 - 6\widehat{V}\widehat{R}\widehat{\alpha}_{10,20}\widetilde{S}_{30}^{-}\sqrt{3\pi/35}]\widehat{\alpha}_{20,10} \\
&\quad +6\widehat{V}\widehat{R}\widehat{\alpha}_{20,20}\widehat{\alpha}_{10,10}\widetilde{S}_{30}^{-}\sqrt{3\pi/35}\}
\end{aligned} \qquad (10.59)$$

where

$$\begin{aligned}
D_z &\equiv [1 - 2\widehat{V}(2\widehat{\alpha}_{10,10}\widetilde{S}_{20}^{-}\sqrt{\pi/5} + 3\widehat{\alpha}_{10,20}\widetilde{S}_{30}^{-}\widehat{R}\sqrt{3\pi/35})] \\
&\quad \times[1 + 2\widehat{V}\widehat{R}(\widehat{\alpha}_{20,10}\widetilde{S}_{30}^{+}\sqrt{15\pi/7} + 2\widehat{\alpha}_{20,20}\widetilde{S}_{40}^{+}\widehat{R}\sqrt{\pi})] \\
&\quad +4\widehat{V}^2\widehat{R}[2\widehat{\alpha}_{20,10}\widetilde{S}_{20}^{-}\sqrt{\pi/5} + 3\widehat{\alpha}_{20,20}\widetilde{S}_{30}^{-}\widehat{R}\sqrt{3\pi/35}] \\
&\quad \times[\widehat{\alpha}_{10,10}\widetilde{S}_{30}^{+}\sqrt{15\pi/7} + 2\widehat{\alpha}_{10,20}\widetilde{S}_{40}^{+}\widehat{R}\sqrt{\pi}]
\end{aligned} \qquad (10.60)$$

The dimensionless forms of eqs.(10.34) and (10.35), for the dipole and quadrupole polarizabilities parallel to the surface of the substrate are given, with $\epsilon_a = 1$, by

$$\begin{aligned}\widehat{\alpha}_{\parallel} &\equiv \frac{\alpha_{\parallel}(0)}{V} = 4\pi D_{\parallel}^{-1}\{[1 - \frac{8}{3}\widehat{V}\widehat{R}^2\widehat{\alpha}_{21,21}\widetilde{S}_{40}^{-}\sqrt{\pi}]\widehat{\alpha}_{11,11} \\ &+\frac{8}{3}\widehat{V}\widehat{R}^2\widehat{\alpha}_{11,21}\widehat{\alpha}_{21,11}\widetilde{S}_{40}^{-}\sqrt{\pi}\} \\ \widehat{\alpha}_{\parallel}^{10} &\equiv \frac{\alpha_{\parallel}^{10}(0)}{VR} = \frac{4}{3}\pi\sqrt{5}D_{\parallel}^{-1}\{[1 + 6\widehat{V}\widehat{R}\widehat{\alpha}_{11,21}\widetilde{S}_{30}^{+}\sqrt{\pi/35}]\widehat{\alpha}_{21,11} \\ &-6\widehat{V}\widehat{R}\widehat{\alpha}_{21,21}\widehat{\alpha}_{11,11}\widetilde{S}_{30}^{+}\sqrt{\pi/35}\}\end{aligned} \tag{10.61}$$

where

$$\begin{aligned}D_{\parallel} &\equiv [1 + 2\widehat{V}(\widehat{\alpha}_{11,11}\widetilde{S}_{20}^{+}\sqrt{\pi/5} + 3\widehat{\alpha}_{11,21}\widetilde{S}_{30}^{+}\widehat{R}\sqrt{\pi/35})] \\ &\times[1 - 2\widehat{V}\widehat{R}(\widehat{\alpha}_{21,11}\widetilde{S}_{30}^{-}\sqrt{5\pi/7} + \frac{4}{3}\widehat{\alpha}_{21,21}\widetilde{S}_{40}^{-}\widehat{R}\sqrt{\pi})] \\ &+4\widehat{V}^2\widehat{R}[\widehat{\alpha}_{21,11}\widetilde{S}_{20}^{+}\sqrt{\pi/5} + 3\widehat{\alpha}_{21,21}\widetilde{S}_{30}^{+}\widehat{R}\sqrt{\pi/35}] \\ &\times[\widehat{\alpha}_{11,11}\widetilde{S}_{30}^{-}\sqrt{5\pi/7} + \frac{4}{3}\widehat{\alpha}_{11,21}\widetilde{S}_{40}^{-}\widehat{R}\sqrt{\pi}]\end{aligned} \tag{10.62}$$

In the above equations the abbreviations $\widetilde{S}_{\ell m}^{\pm}$, eq.(10.30), have been used, with $\epsilon_a = 1$. The following new dimensionless quadrupolar polarizabilities are defined

$$\begin{aligned}\widehat{\alpha}_{10,20} &\equiv \alpha_{10,20}/VR, \quad \widehat{\alpha}_{20,10} \equiv \alpha_{20,10}/VR \\ \widehat{\alpha}_{11,21} &\equiv \alpha_{11,21}/VR, \quad \widehat{\alpha}_{21,11} \equiv \alpha_{21,11}/VR\end{aligned} \tag{10.63}$$

and

$$\widehat{\alpha}_{20,20} \equiv \alpha_{20,20}/VR^2, \quad \widehat{\alpha}_{21,21} \equiv \alpha_{21,21}/VR^2 \tag{10.64}$$

where

$$\widehat{R} \equiv \frac{R}{L} \tag{10.65}$$

For truncated spheroids the above formulae also describe the case that the interaction along the substrate is taken into account to quadrupolar order, provided that one replaces R by $a\xi_0$. Furthermore V is in all cases the volume of the truncated particle. $\widehat{V}$ is defined by eq.(10.46) and given by eqs.(10.47)–(10.49) for the different cases.

First the calculation of the dipolar polarizabilities, eq.(10.45), and the first four quadrupolar polarizabilities, eq.(10.63), for the different shapes of the truncated particles will be discussed. The dimensionless dipole polarizabilities follow from the dimensionless amplitudes using eq.(10.50). The first four dimensionless quadrupole polarizabilities follow from the dimensionless amplitudes, using eqs.(5.67) and (8.110) for a truncated sphere and eqs.(5.67) and (9.119) for a truncated spheroid:

$$\widehat{\alpha}_{20,10} = \frac{3}{5}\widehat{\alpha}_{10,20} = \widehat{A}_{20}, \quad \widehat{\alpha}_{21,11} = \frac{3}{5}\widehat{\alpha}_{11,21} = \widehat{A}_{21} \tag{10.66}$$

The amplitudes $\widehat{A}_{10}$, $\widehat{A}_{11}$, $\widehat{A}_{20}$ and $\widehat{A}_{21}$ are obtained for a truncated sphere by solving eq.(8.108), together with eqs.(8.109) and (8.13). For truncated spheroids one should solve eq.(9.116), together with eqs.(9.117) and (9.19) for oblate, and eqs.(9.118) and (9.38) for prolate spheroids.

In order to calculate also the last two dimensionless quadrupole polarizabilities, eq.(10.64), one must use the method described in appendix A below. Now the dimensionless form of the equations found in this appendix will be given, as this has not been done before. The following dimensionless amplitudes are now needed for truncated spheres

$$
\begin{aligned}
\widehat{A}_{\ell 0} &\equiv -A_{\ell 0}/\left[VR^{\ell}\sqrt{\pi/5}\left(\phi_{xx}+\phi_{yy}\right)\right] \\
\widehat{B}_{\ell 0} &\equiv -B_{\ell 0}R^{\ell+1}/\left[V\sqrt{\pi/5}\left(\phi_{xx}+\phi_{yy}\right)\right] \\
\widehat{A}_{\ell 1} &\equiv -A_{\ell 1}/\left[VR^{\ell}\sqrt{2\pi/15}\left(\phi_{xy}-i\phi_{yz}\right)\right] \\
\widehat{B}_{\ell 1} &\equiv -B_{\ell 1}R^{\ell+1}/\left[V\sqrt{2\pi/15}\left(\phi_{xy}-i\phi_{yz}\right)\right]
\end{aligned}
\tag{10.67}
$$

where ϕ_{ij}, for $i, j = x, y, z$, are elements of the matrix which gives the gradient of the incident electric field in the ambient, cf. eq.(10.84). For truncated spheroids one may use the same definitions if one replaces R by $a\xi_0$. In all cases V is the volume of the truncated particle; which is given by eq.(8.107) for spheres and by eqs.(9.114) and (9.115) for oblate and prolate spheroids.

It follows, using eqs.(10.96), that

$$
\begin{aligned}
\sum_{\ell_1=1}^{M} C_{\ell\ell_1}^{m}\widehat{A}_{\ell_1 m}+\sum_{\ell_1=1}^{M} D_{\ell\ell_1}^{m}\widehat{B}_{\ell_1 m} &= \widehat{K}_{\ell}^{m} \\
\sum_{\ell_1=1}^{M} F_{\ell\ell_1}^{m}\widehat{A}_{\ell_1 m}+\sum_{\ell_1=1}^{M} G_{\ell\ell_1}^{m}\widehat{B}_{\ell_1 m} &= \widehat{L}_{\ell}^{m} \\
\text{for } \ell &= 1,2,3,...,M \text{ and } m=0,1
\end{aligned}
\tag{10.68}
$$

both for truncated spheres and spheroids. The matrix elements $C_{\ell\ell_1}^{m}$, $D_{\ell\ell_1}^{m}$, $F_{\ell\ell_1}^{m}$ and $G_{\ell\ell_1}^{m}$ are given for truncated spheres by eq.(8.13), for truncated oblate spheroids by eq.(9.19) and for truncated prolate spheroids by eq.(9.38). For truncated spheres it follows from eq.(10.89), that the right-hand side of the linear equations is given by

$$
\begin{aligned}
\widehat{K}_{\ell}^{0} &\equiv \frac{3}{\pi\left(2+3t_r-t_r^3\right)}\left\{\delta_{\ell 2}+2\sqrt{\frac{5}{3}}\left(\frac{1-\epsilon_s}{\epsilon_s}\right)t_r[\delta_{\ell 1}-\zeta_{\ell 1}^{0}Q_{\ell 1}^{0}(t_r)+\sqrt{3}t_r\zeta_{\ell 0}^{0}Q_{\ell 0}^{0}(t_r)]\right\} \\
\widehat{L}_{\ell}^{0} &\equiv \frac{6}{\pi\left(2+3t_r-t_r^3\right)}\left\{\epsilon_s\delta_{\ell 2}-(\epsilon_s-1)\left[\zeta_{\ell 2}^{0}Q_{\ell 2}^{0}(t_r)+\sqrt{\frac{5}{3}}t_r[\delta_{\ell 1}-\zeta_{\ell 1}^{0}Q_{\ell 1}^{0}(t_r)]\right]\right\} \\
\widehat{K}_{\ell}^{1} &\equiv \frac{3}{\pi\left(2+3t_r-t_r^3\right)}\left\{\frac{1}{\epsilon_s}\delta_{\ell 2}-\left(\frac{1-\epsilon_s}{\epsilon_s}\right)\left[\zeta_{\ell 2}^{1}Q_{\ell 2}^{1}(t_r)+\sqrt{5}t_r[\delta_{\ell 1}-\zeta_{\ell 1}^{1}Q_{\ell 1}^{1}(t_r)]\right]\right\} \\
\widehat{L}_{\ell}^{1} &\equiv \frac{3}{\pi\left(2+3t_r-t_r^3\right)}\left\{2\delta_{\ell 2}+\sqrt{5}(\epsilon_s-1)t_r[\delta_{\ell 1}-\zeta_{\ell 1}^{1}Q_{\ell 1}^{1}(t_r)]\right\}
\end{aligned}
\tag{10.69}
$$

The coefficients $\zeta^m_{\ell\ell_1}$ and $Q^m_{\ell\ell_1}$ are given by eqs.(8.14) and (8.15). For truncated oblate spheroids it follows from eq.(10.106), that these coefficients are given by

$$\begin{aligned}
\widehat{K}^0_\ell &\equiv \frac{3}{\pi\left(2+3t_r-t_r^3\right)}\frac{\xi_0^2}{1+\xi_0^2}\left\{\xi_0^{-2}X^0_2(\xi_0)\delta_{\ell 2}\right.\\
&\left.+2\sqrt{5/3}\left(\frac{1-\epsilon_s}{\epsilon_s}\right)t_r[\xi_0^{-1}X^0_1(\xi_0)\left(\delta_{\ell 1}-\zeta^0_{\ell 1}Q^0_{\ell 1}(t_r)\right)+\sqrt{3}t_r\zeta^0_{\ell 0}Q^0_{\ell 0}(t_r)]\right\}\\
\widehat{L}^0_\ell &\equiv \frac{3}{\pi\left(2+3t_r-t_r^3\right)}\frac{\xi_0^2}{1+\xi_0^2}\left\{\xi_0^{-1}\frac{dX^0_2(\xi_0)}{d\xi_0}\left[\epsilon_s\delta_{\ell 2}-\left(\epsilon_s-1\right)\zeta^0_{\ell 2}Q^0_{\ell 2}(t_r)\right]\right.\\
&\left.-2\sqrt{\frac{5}{3}}\left(\epsilon_s-1\right)t_r\frac{dX^0_1(\xi_0)}{d\xi_0}[\delta_{\ell 1}-\zeta^0_{\ell 1}Q^0_{\ell 1}(t_r)]\right\}
\end{aligned}$$

$$\begin{aligned}
\widehat{K}^1_\ell &\equiv \frac{3}{\pi\left(2+3t_r-t_r^3\right)}\frac{\xi_0^2}{1+\xi_0^2}\left\{\frac{1}{\epsilon_s}\xi_0^{-2}X^1_2(\xi_0)\delta_{\ell 2}\right.\\
&\left.-\left(\frac{1-\epsilon_s}{\epsilon_s}\right)\left[\xi_0^{-2}X^1_2(\xi_0)\zeta^1_{\ell 2}Q^1_{\ell 2}(t_r)+\sqrt{5}t_r\xi_0^{-1}X^1_1(\xi_0)[\delta_{\ell 1}-\zeta^1_{\ell 1}Q^1_{\ell 1}(t_r)]\right]\right\}\\
\widehat{L}^1_\ell &\equiv \frac{3}{\pi\left(2+3t_r-t_r^3\right)}\frac{\xi_0^2}{1+\xi_0^2}\left\{\xi_0^{-1}\frac{dX^1_2(\xi_0)}{d\xi_0}\delta_{\ell 2}\right.\\
&\left.+\sqrt{5}(\epsilon_s-1)t_r\frac{dX^1_1(\xi_0)}{d\xi_0}[\delta_{\ell 1}-\zeta^1_{\ell 1}Q^1_{\ell 1}(t_r)]\right\} \qquad (10.70)
\end{aligned}$$

The functions $X^m_\ell(\xi_0)$ are defined by eq.(9.20). For truncated prolate spheroids it follows from eq.(10.110), that these coefficients are given by

$$\begin{aligned}
\widehat{K}^0_\ell &\equiv \frac{3}{\pi\left(2+3t_r-t_r^3\right)}\frac{\xi_0^2}{\xi_0^2-1}\left\{\xi_0^{-2}\widetilde{X}^0_2(\xi_0)\delta_{\ell 2}\right.\\
&\left.+2\sqrt{5/3}\left(\frac{1-\epsilon_s}{\epsilon_s}\right)t_r[\xi_0^{-1}\widetilde{X}^0_1(\xi_0)\left(\delta_{\ell 1}-\zeta^0_{\ell 1}Q^0_{\ell 1}(t_r)\right)+\sqrt{3}t_r\zeta^0_{\ell 0}Q^0_{\ell 0}(t_r)]\right\}\\
\widehat{L}^0_\ell &\equiv \frac{3}{\pi\left(2+3t_r-t_r^3\right)}\frac{\xi_0^2}{\xi_0^2-1}\left\{\xi_0^{-1}\frac{d\widetilde{X}^0_2(\xi_0)}{d\xi_0}\left[\epsilon_s\delta_{\ell 2}-\left(\epsilon_s-1\right)\zeta^0_{\ell 2}Q^0_{\ell 2}(t_r)\right]\right.\\
&\left.-2\sqrt{\frac{5}{3}}\left(\epsilon_s-1\right)t_r\frac{d\widetilde{X}^0_1(\xi_0)}{d\xi_0}[\delta_{\ell 1}-\zeta^0_{\ell 1}Q^0_{\ell 1}(t_r)]\right\}
\end{aligned}$$

$$\begin{aligned}
\widehat{K}^1_\ell &\equiv \frac{3}{\pi\left(2+3t_r-t_r^3\right)}\frac{\xi_0^2}{\xi_0^2-1}\left\{\frac{1}{\epsilon_s}\xi_0^{-2}\widetilde{X}^1_2(\xi_0)\delta_{\ell 2}\right.\\
&\left.-\left(\frac{1-\epsilon_s}{\epsilon_s}\right)\left[\xi_0^{-2}\widetilde{X}^1_2(\xi_0)\zeta^1_{\ell 2}Q^1_{\ell 2}(t_r)+\sqrt{5}t_r\xi_0^{-1}\widetilde{X}^1_1(\xi_0)[\delta_{\ell 1}-\zeta^1_{\ell 1}Q^1_{\ell 1}(t_r)]\right]\right\}\\
\widehat{L}^1_\ell &\equiv \frac{3}{\pi\left(2+3t_r-t_r^3\right)}\frac{\xi_0^2}{\xi_0^2-1}\left\{\xi_0^{-1}\frac{d\widetilde{X}^1_2(\xi_0)}{d\xi_0}\delta_{\ell 2}\right.\\
&\left.+\sqrt{5}(\epsilon_s-1)t_r\frac{d\widetilde{X}^1_1(\xi_0)}{d\xi_0}[\delta_{\ell 1}-\zeta^1_{\ell 1}Q^1_{\ell 1}(t_r)]\right\} \qquad (10.71)
\end{aligned}$$

The functions $\widetilde{X}_\ell^m(\xi_0)$ are defined by eq.(9.39). Note that eqs.(10.70) and (10.71) for spheroids reduce to those for spheres, eq.(10.69), in the spherical limit, $\xi_0 \to \infty$. This can be proven, using asymptotic formulae given in appendix A in chapter 9.

Solving the above equations give the dimensionless quadrupole amplitudes which are identical to the following quadrupole polarizabilities

$$\widehat{\alpha}_{20,20} = \widehat{A}_{20}, \qquad \widehat{\alpha}_{21,21} = \widehat{A}_{21} \tag{10.72}$$

This follows from the definitions (10.64) and (10.67) together with eq.(10.101). These polarizabilities, together with those given above, must now be substituted into eqs. (10.59) and (10.61) to obtain the dipole and quadrupole polarizabilities including the interaction along the substrate. The lattice sums in the latter equations can be found using the method given in appendix A of chapter 7.

The surface constitutive coefficients, eq.(10.37), become in dimensionless form for truncated spheres with radius R and truncation parameter t_r

$$\begin{aligned}
\widehat{\gamma}_e &\equiv \frac{1}{\rho V}\gamma_e(d) = \widehat{\alpha}_{\|} \\
\widehat{\beta}_e &\equiv \frac{1}{\rho V}\beta_e(d) = \widehat{\alpha}_z \\
\widehat{\tau} &\equiv \frac{1}{\rho V R}\tau(d) = t_r\widehat{\alpha}_{\|} - \widehat{\alpha}_{\|}^{10} \\
\widehat{\delta}_e &\equiv \frac{1}{\rho V R}\delta_e(d) = t_r\left(\widehat{\alpha}_z + \widehat{\alpha}_{\|}\right) - \widehat{\alpha}_z^{10} - \widehat{\alpha}_{\|}^{10}
\end{aligned} \tag{10.73}$$

for this case, where $\epsilon_a = 1$. For truncated spheroids one has to replace R by $a\xi_0$ in the above equations. The dimensionless invariants are given by eq.(10.52) for truncated spheres, and for spheroids, with $a\xi_0$ instead of R. All these quantities therefore follow in a simple way if the polarizabilities $\widehat{\alpha}_z$, $\widehat{\alpha}_z^{10}$, $\widehat{\alpha}_{\|}$,and $\widehat{\alpha}_{\|}^{10}$, eqs.(10.59) and (10.61), are known.

Next the case of spherical caps with a radius R is considered, in which the interaction between the particles along the substrate is taken into account to quadrupolar approximation. The dimensionless forms of eqs.(10.39) and (10.40), for the dipole and quadrupole polarizabilities normal to the surface of the substrate are given, with $\epsilon_a = 1$, by

$$\begin{aligned}
\widehat{\alpha}_z &\equiv \frac{\alpha_z(0)}{V} = 4\pi\epsilon_s D_z^{-1}\{[1 + 4\widehat{V}\widehat{R}^2\widehat{\alpha}_{20,20}\widetilde{S}_{40}^{-}\sqrt{\pi}]\widehat{\alpha}_{10,10} \\
&\quad -4\widehat{V}\widehat{R}^2\widehat{\alpha}_{10,20}\widehat{\alpha}_{20,10}\widetilde{S}_{40}^{-}\sqrt{\pi}\} \\
\widehat{\alpha}_z^{10} &\equiv \frac{\alpha_z^{10}(0)}{VR} = 2\pi\epsilon_s\sqrt{\frac{5}{3}}D_z^{-1}\{[1 - 6\widehat{V}\widehat{R}\widehat{\alpha}_{10,20}\widetilde{S}_{30}^{+}\sqrt{3\pi/35}]\widehat{\alpha}_{20,10} \\
&\quad +6\widehat{V}\widehat{R}\widehat{\alpha}_{20,20}\widehat{\alpha}_{10,10}\widetilde{S}_{30}^{+}\sqrt{3\pi/35}\}
\end{aligned} \tag{10.74}$$

where

$$\begin{aligned} D_z \equiv\ & [1 - 2\widehat{V}(2\widehat{\alpha}_{10,10}\widetilde{S}_{20}^{+}\sqrt{\pi/5} + 3\widehat{\alpha}_{10,20}\widetilde{S}_{30}^{+}\widehat{R}\sqrt{3\pi/35})] \\ & \times[1 + 2\widehat{V}\widehat{R}(\widehat{\alpha}_{20,10}\widetilde{S}_{30}^{-}\sqrt{15\pi/7} + 2\widehat{\alpha}_{20,20}\widetilde{S}_{40}^{-}\widehat{R}\sqrt{\pi})] \\ & +4\widehat{V}^2\widehat{R}[2\widehat{\alpha}_{20,10}\widetilde{S}_{20}^{+}\sqrt{\pi/5} + 3\widehat{\alpha}_{20,20}\widetilde{S}_{30}^{+}\widehat{R}\sqrt{3\pi/35}] \\ & \times[\widehat{\alpha}_{10,10}\widetilde{S}_{30}^{-}\sqrt{15\pi/7} + 2\widehat{\alpha}_{10,20}\widetilde{S}_{40}^{-}\widehat{R}\sqrt{\pi}] \end{aligned} \tag{10.75}$$

The dimensionless forms of eqs.(10.41) and (10.42), for the dipole and quadrupole polarizabilities parallel to the surface of the substrate are given, with $\epsilon_a = 1$, by

$$\begin{aligned} \widehat{\alpha}_{\parallel} \equiv\ & \frac{\alpha_{\parallel}(0)}{V} = 4\pi\epsilon_s D_{\parallel}^{-1}\{[1 - \frac{8}{3}\widehat{V}\widehat{R}^2\widehat{\alpha}_{21,21}\widetilde{S}_{40}^{+}\sqrt{\pi}]\widehat{\alpha}_{11,11} \\ & +\frac{8}{3}\widehat{V}\widehat{R}^2\widehat{\alpha}_{11,21}\widehat{\alpha}_{21,11}\widetilde{S}_{40}^{+}\sqrt{\pi}\} \\ \widehat{\alpha}_{\parallel}^{10} \equiv\ & \frac{\alpha_{\parallel}^{10}(0)}{VR} = \frac{4}{3}\pi\epsilon_s\sqrt{5}D_{\parallel}^{-1}\{[1 + 6\widehat{V}\widehat{R}\widehat{\alpha}_{11,21}\widetilde{S}_{30}^{-}\sqrt{\pi/35}]\widehat{\alpha}_{21,11} \\ & -6\widehat{V}\widehat{R}\widehat{\alpha}_{21,21}\widehat{\alpha}_{11,11}\widetilde{S}_{30}^{-}\sqrt{\pi/35}\} \end{aligned} \tag{10.76}$$

where

$$\begin{aligned} D_{\parallel} \equiv\ & [1 + 2\widehat{V}(\widehat{\alpha}_{11,11}\widetilde{S}_{20}^{-}\sqrt{\pi/5} + 3\widehat{\alpha}_{11,21}\widetilde{S}_{30}^{-}\widehat{R}\sqrt{\pi/35})] \\ & \times[1 - 2\widehat{V}\widehat{R}(\widehat{\alpha}_{21,11}\widetilde{S}_{30}^{+}\sqrt{5\pi/7} + \frac{4}{3}\widehat{\alpha}_{21,21}\widetilde{S}_{40}^{+}\widehat{R}\sqrt{\pi})] \\ & +4\widehat{V}^2\widehat{R}[\widehat{\alpha}_{21,11}\widetilde{S}_{20}^{-}\sqrt{\pi/5} + 3\widehat{\alpha}_{21,21}\widetilde{S}_{30}^{-}\widehat{R}\sqrt{\pi/35}] \\ & \times[\widehat{\alpha}_{11,11}\widetilde{S}_{30}^{+}\sqrt{5\pi/7} + \frac{4}{3}\widehat{\alpha}_{11,21}\widetilde{S}_{40}^{+}\widehat{R}\sqrt{\pi}] \end{aligned} \tag{10.77}$$

In the above equations the abbreviations $\widetilde{S}_{\ell m}^{\pm}$, eq.(10.30), have been used, with $\epsilon_a = 1$. All the quantities on the right-hand side of these equations have been defined above. For spheroidal caps the above formulae also describe the case that the interaction along the substrate is taken into account to quadrupolar order, provided that one replaces R by $a\xi_0$. Furthermore V is in all cases the volume of the cap. $\widehat{V}$ is defined by eq.(10.46) and given by eqs.(10.54)–(10.56) for the different cases.

First the calculation of the dipolar polarizabilities, eq.(10.45), and the first four quadrupolar polarizabilities, eq.(10.63), for the different shapes of the caps will be discussed. The dimensionless dipole polarizabilities follow from the dimensionless amplitudes using eq.(10.57). The first four dimensionless quadrupole polarizabilities follow from the dimensionless amplitudes, using eqs.(10.15) and (8.115) for a spherical cap and eqs.(10.15) and (9.126) for a spheroidal cap:

$$\widehat{\alpha}_{20,10} = \frac{3}{5}\widehat{\alpha}_{10,20} = \epsilon_s\widehat{A}_{20}, \qquad \widehat{\alpha}_{21,11} = \frac{3}{5}\widehat{\alpha}_{11,21} = \widehat{A}_{21} \tag{10.78}$$

The amplitudes $\widehat{A}_{10}$, $\widehat{A}_{11}$, $\widehat{A}_{20}$ and $\widehat{A}_{21}$ are obtained for a spherical cap by solving eq.(8.108), together with eqs.(8.114)and (8.43). For spheroidal caps one should solve eq.(9.116), together with eqs.(9.124)and (9.65) for oblate, and eqs.(9.125) and (9.79) for prolate spheroidal caps.

In order to calculate also the last two dimensionless quadrupole polarizabilities, eq.(10.64), one must use the method described in appendix B below. Now the dimensionless form of the equations found in this appendix will be given, as this has not been done before. For spherical caps the dimensionless amplitudes are again given by eq.(10.67), where in this case ϕ_{ij}, $i, j = x, y, z$, are elements of the matrix which gives the gradient of the incident electric field in the substrate. For spheroidal caps the same formulae can be used with $a\xi_0$ instead of R. It follows, using eqs.(10.120), that eq.(10.68) for the dimensionless amplitudes is again valid, both for spherical and spheroidal caps. The matrix elements $C^m_{\ell\ell_1}$, $D^m_{\ell\ell_1}$, $F^m_{\ell\ell_1}$ and $G^m_{\ell\ell_1}$ are given for spherical caps by eq.(8.43), for oblate spheroidal caps by eq.(9.65) and for prolate spheroidal caps by eq.(9.79). For spherical caps it follows from eq.(10.117), that the right-hand side of the linear equations is given by

$$
\begin{aligned}
\widehat{K}^0_\ell &\equiv \frac{3}{\pi\left(2-3t_r+t_r^3\right)} \\
&\quad\times\left\{\delta_{\ell 2}+2\sqrt{\frac{5}{3}}\left(\epsilon_s-1\right)t_r\left(-1\right)^\ell\left[\delta_{\ell 1}-\zeta^0_{\ell 1}Q^0_{\ell 1}(t_r)+\sqrt{3}t_r\zeta^0_{\ell 0}Q^0_{\ell 0}(t_r)\right]\right\} \\
\widehat{L}^0_\ell &\equiv \frac{6}{\pi\left(2-3t_r+t_r^3\right)}\left\{\delta_{\ell 2}+\left(\epsilon_s-1\right)\left(-1\right)^\ell\left[\zeta^0_{\ell 2}Q^0_{\ell 2}(t_r)+\sqrt{\frac{5}{3}}t_r[\delta_{\ell 1}-\zeta^0_{\ell 1}Q^0_{\ell 1}(t_r)]\right]\right\} \\
\widehat{K}^1_\ell &\equiv \frac{3}{\pi\left(2-3t_r+t_r^3\right)}\left\{\epsilon_s\delta_{\ell 2}-\left(\epsilon_s-1\right)\left(-1\right)^\ell\left[\zeta^1_{\ell 2}Q^1_{\ell 2}(t_r)+\sqrt{5}t_r[\delta_{\ell 1}-\zeta^1_{\ell 1}Q^1_{\ell 1}(t_r)]\right]\right\} \\
\widehat{L}^1_\ell &\equiv \frac{3}{\pi\left(2-3t_r+t_r^3\right)}\left\{2\epsilon_s\delta_{\ell 2}-\sqrt{5}(\epsilon_s-1)t_r\left(-1\right)^\ell\left[\delta_{\ell 1}-\zeta^1_{\ell 1}Q^1_{\ell 1}(t_r)\right]\right\}
\end{aligned}
\tag{10.79}
$$

The coefficients $\zeta^m_{\ell\ell_1}$ and $Q^m_{\ell\ell_1}$ are given by eqs.(8.14) and (8.15). For oblate spheroidal caps it follows from eq.(10.127) that these coefficients are given by

$$
\begin{aligned}
\widehat{K}^0_\ell &\equiv \frac{3}{\pi\left(2-3t_r+t_r^3\right)}\frac{\xi_0^2}{1+\xi_0^2}\left\{\xi_0^{-2}X^0_2(\xi_0)\delta_{\ell 2}\right. \\
&\quad\left.+2\sqrt{5/3}\left(\epsilon_s-1\right)t_r\left(-1\right)^\ell\left[\xi_0^{-1}X^0_1(\xi_0)\left(\delta_{\ell 1}-\zeta^0_{\ell 1}Q^0_{\ell 1}(t_r)\right)+\sqrt{3}t_r\zeta^0_{\ell 0}Q^0_{\ell 0}(t_r)\right]\right\} \\
\widehat{L}^0_\ell &\equiv \frac{3}{\pi\left(2-3t_r+t_r^3\right)}\frac{\xi_0^2}{1+\xi_0^2}\left\{\xi_0^{-1}\frac{dX^0_2(\xi_0)}{d\xi_0}\left[\delta_{\ell 2}+\left(\epsilon_s-1\right)\left(-1\right)^\ell\zeta^0_{\ell 2}Q^0_{\ell 2}(t_r)\right]\right. \\
&\quad\left.+2\sqrt{\frac{5}{3}}\left(\epsilon_s-1\right)t_r\left(-1\right)^\ell\frac{dX^0_1(\xi_0)}{d\xi_0}\left[\delta_{\ell 1}-\zeta^0_{\ell 1}Q^0_{\ell 1}(t_r)\right]\right\}
\end{aligned}
$$

$$
\begin{aligned}
\widehat{K}_\ell^1 &\equiv \frac{3}{\pi\left(2-3t_r+t_r^3\right)}\frac{\xi_0^2}{1+\xi_0^2}\left\{\epsilon_s\xi_0^{-2}X_2^1(\xi_0)\delta_{\ell 2}\right.\\
&\quad\left.-\left(\epsilon_s-1\right)\left(-1\right)^\ell\left[\xi_0^{-2}X_2^1(\xi_0)\zeta_{\ell 2}^1Q_{\ell 2}^1(t_r)+\sqrt{5}t_r\xi_0^{-1}X_1^1(\xi_0)[\delta_{\ell 1}-\zeta_{\ell 1}^1Q_{\ell 1}^1(t_r)]\right]\right\}\\
\widehat{L}_\ell^1 &\equiv \frac{3}{\pi\left(2-3t_r+t_r^3\right)}\frac{\xi_0^2}{1+\xi_0^2}\left\{\epsilon_s\xi_0^{-1}\frac{dX_2^1(\xi_0)}{d\xi_0}\delta_{\ell 2}\right.\\
&\quad\left.-\sqrt{5}(\epsilon_s-1)t_r\left(-1\right)^\ell\frac{dX_1^1(\xi_0)}{d\xi_0}[\delta_{\ell 1}-\zeta_{\ell 1}^1Q_{\ell 1}^1(t_r)]\right\}
\end{aligned}
\tag{10.80}
$$

The functions $X_\ell^m(\xi_0)$ are defined by eq.(9.20). For prolate spheroidal caps it follows from eq.(10.130), that these coefficients are given by

$$
\begin{aligned}
\widehat{K}_\ell^0 &\equiv \frac{3}{\pi\left(2-3t_r+t_r^3\right)}\frac{\xi_0^2}{\xi_0^2-1}\left\{\xi_0^{-2}\widetilde{X}_2^0(\xi_0)\delta_{\ell 2}\right.\\
&\quad\left.+2\sqrt{5/3}\left(\epsilon_s-1\right)t_r\left(-1\right)^\ell\left[\xi_0^{-1}\widetilde{X}_1^0(\xi_0)\left(\delta_{\ell 1}-\zeta_{\ell 1}^0Q_{\ell 1}^0(t_r)\right)+\sqrt{3}t_r\zeta_{\ell 0}^0Q_{\ell 0}^0(t_r)\right]\right\}\\
\widehat{L}_\ell^0 &\equiv \frac{3}{\pi\left(2-3t_r+t_r^3\right)}\frac{\xi_0^2}{\xi_0^2-1}\left\{\xi_0^{-1}\frac{d\widetilde{X}_2^0(\xi_0)}{d\xi_0}\left[\delta_{\ell 2}+\left(\epsilon_s-1\right)\left(-1\right)^\ell\zeta_{\ell 2}^0Q_{\ell 2}^0(t_r)\right]\right.\\
&\quad\left.+2\sqrt{\frac{5}{3}}\left(\epsilon_s-1\right)t_r\left(-1\right)^\ell\frac{d\widetilde{X}_1^0(\xi_0)}{d\xi_0}[\delta_{\ell 1}-\zeta_{\ell 1}^0Q_{\ell 1}^0(t_r)]\right\}
\end{aligned}
$$

$$
\begin{aligned}
\widehat{K}_\ell^1 &\equiv \frac{3}{\pi\left(2-3t_r+t_r^3\right)}\frac{\xi_0^2}{\xi_0^2-1}\left\{\epsilon_s\xi_0^{-2}\widetilde{X}_2^1(\xi_0)\delta_{\ell 2}\right.\\
&\quad\left.-\left(\epsilon_s-1\right)\left(-1\right)^\ell\left[\xi_0^{-2}\widetilde{X}_2^1(\xi_0)\zeta_{\ell 2}^1Q_{\ell 2}^1(t_r)+\sqrt{5}t_r\xi_0^{-1}\widetilde{X}_1^1(\xi_0)[\delta_{\ell 1}-\zeta_{\ell 1}^1Q_{\ell 1}^1(t_r)]\right]\right\}\\
\widehat{L}_\ell^1 &\equiv \frac{3}{\pi\left(2-3t_r+t_r^3\right)}\frac{\xi_0^2}{\xi_0^2-1}\left\{\epsilon_s\xi_0^{-1}\frac{d\widetilde{X}_2^1(\xi_0)}{d\xi_0}\delta_{\ell 2}\right.\\
&\quad\left.-\sqrt{5}(\epsilon_s-1)t_r\left(-1\right)^\ell\frac{d\widetilde{X}_1^1(\xi_0)}{d\xi_0}[\delta_{\ell 1}-\zeta_{\ell 1}^1Q_{\ell 1}^1(t_r)]\right\}
\end{aligned}
\tag{10.81}
$$

The functions $\widetilde{X}_\ell^m(\xi_0)$ are defined by eq.(9.39). Note that eqs.(10.80) and (10.81) for spheroids reduce to those for spheres, eq.(10.79), in the spherical limit, $\xi_0 \to \infty$. This can be proven, using asymptotic formulae given in appendix A in chapter 9.

Substituting the above equations into eq.(10.68) gives, upon solution, the dimensionless quadrupole amplitudes which are identical to the following quadrupole polarizabilities

$$
\widehat{\alpha}_{20,20}=\widehat{A}_{20}, \qquad \widehat{\alpha}_{21,21}=\widehat{A}_{21}
\tag{10.82}
$$

This follows from the definitions (10.64) and (10.67) together with eq.(10.125). Note that, contrary to eq.(10.57), no ϵ_s appears in this equation. This is a consequence of the choice of the incident field in the calculation and the resulting manner in which the amplitude was made dimensionless, cf. appendix B for a more detailed explanation.

These polarizabilities, together with those given above, must now be substituted into eqs.(10.74) and (10.76) to obtain the dipole and quadrupole polarizabilities including the interaction along the substrate. The lattice sums in the latter equations can be found using the method given in appendix A of chapter 7. As mentioned at the end of the previous section, $\widetilde{S}^{r}_{\ell m}$ differs by a factor $(-1)^{\ell+m}$ for truncated particles and caps on the same lattice (see section 10.3). It follows, with eq.(10.30), that $\widetilde{S}^{\pm}_{20}$ and $\widetilde{S}^{\pm}_{40}$ have the same value in the above equations for truncated particles and caps. For the same reason $\widetilde{S}^{\pm}_{30}$ in the case of truncated particles equals $\widetilde{S}^{\mp}_{30}$ in the case of caps. This is also true for the random array, where $\widetilde{I}^{\pm}_{2}$ and $\widetilde{I}^{\pm}_{4}$ are the same, whereas $\widetilde{I}^{\pm}_{3}$ for truncated particles equals $\widetilde{I}^{\mp}_{3}$ for caps.

The surface constitutive coefficients, eq.(10.43), become in dimensionless form for spherical caps with radius R and truncation parameter t_r

$$\begin{aligned}
\widehat{\gamma}_e &\equiv \frac{1}{\rho V}\gamma_e(-d) = \widehat{\alpha}_{\|} \\
\widehat{\beta}_e &\equiv \frac{1}{\rho V}\beta_e(-d) = \epsilon_s^{-2}\widehat{\alpha}_z \\
\widehat{\tau} &\equiv \frac{1}{\rho V R}\tau(-d) = -\epsilon_s^{-1}\left(t_r\widehat{\alpha}_{\|} + \widehat{\alpha}_{\|}^{10}\right) \\
\widehat{\delta}_e &\equiv \frac{1}{\rho V R}\delta_e(-d) = -\epsilon_s^{-1}\left[t_r\left(\widehat{\alpha}_z + \widehat{\alpha}_{\|}\right) + \widehat{\alpha}_z^{10} + \widehat{\alpha}_{\|}^{10}\right]
\end{aligned} \tag{10.83}$$

for this case, where $\epsilon_a = 1$. For spheroidal caps one has to replace R by $a\xi_0$ in the above equations. The dimensionless invariants are given by eq.(10.52) for spherical caps, and for spheroidal caps, with $a\xi_0$ instead of R. All these quantities therefore follow in a simple way if the polarizabilities $\widehat{\alpha}_z$, $\widehat{\alpha}_z^{10}$, $\widehat{\alpha}_{\|}$,and $\widehat{\alpha}_{\|}^{10}$, eqs.(10.74) and (10.76), are known.

For random arrays the lattice sums $\widetilde{S}^{\pm}_{20}$, $\widetilde{S}^{\pm}_{30}$ and $\widetilde{S}^{\pm}_{40}$, in eqs.(10.59) and (10.61) for truncated particles and in eqs.(10.74) and (10.76) for caps, have to be replaced by the distribution integrals $\widetilde{I}^{r}_{2}$, $\widetilde{I}^{r}_{3}$ and $\widetilde{I}^{r}_{4}$. These distribution integrals are given by eqs.(10.8) and (10.9) for truncated particles and by eqs.(10.8) and (10.18) for caps. Furthermore L^{-3} in eqs.(10.46)–(10.49) and (10.54)–(10.56), should be replaced by $\rho^{3/2}$.

Using the interaction along the substrate to dipolar order and the interaction with the image to order $M = 16$ the dipole polarizabilities were calculated for oblate spheroids, with an axial ratio 4 and a number of truncation parameters, as a function of the frequency. The density was chosen such that the number of spheroids per square long axis, i.e. the diameter parallel to the surface of the substrate, is equal to 0.4. The spheroids were located on a square array. This implies that the close packing density is equal to 1 in these units. The resulting $\widehat{\beta}_e$ is plotted in figs.10.3-5 for truncation parameters $t_r = 0.4, 0.0$ and -0.4. The minus sign will be used to indicate the case of a cap; in the actual calculations the formulae for the cap have been used with the absolute value of this parameter. It is found that the constitutive coefficient β_e, which is due to the dipole polarizability normal to the surface, is not sensitive to the interaction along the substrate. The dipole polarizability normal

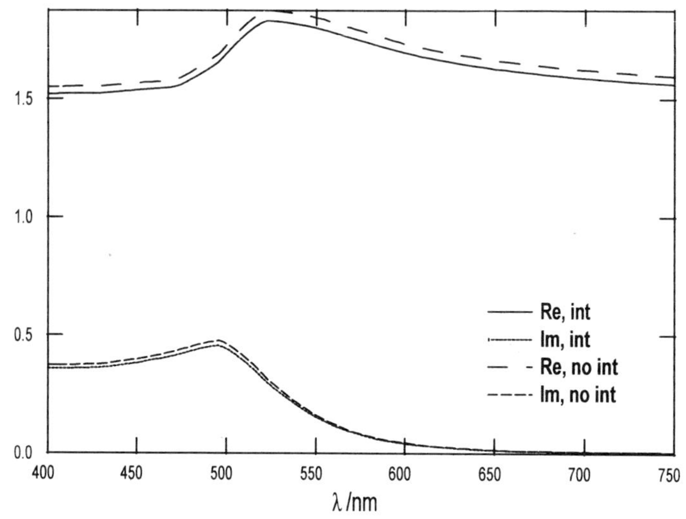

Figure 10.3 The real and imaginary parts of $\widehat{\beta}_e$ for a truncated oblate spheroid, with an axial ratio 4 and a truncation parameter 0.4, as a function of the wave length. Both the cases of interaction and of no interaction along the substrate are plotted.

to the surface is small. Its resonance structure is a very sensitive function of the truncation parameter. There is one resonance for the truncated spheroid and there are two for the hemispheroid and three for the cap in the optical domain.

The resulting $\widehat{\gamma}_e$ for an axial ratio of 4 is plotted in figs.10.6-8 for truncation parameters $t_r = 0.4, 0.0$ and -0.4. It is found that the constitutive coefficient $\widehat{\gamma}_e$, which is due to the dipole polarizability parallel to the surface, is not very sensitive to the interaction along the substrate. The interaction results in a small red shift of the resonance. The dipole polarizability parallel to the surface is large. The location and the size of the resonance is a very sensitive function of the truncation parameter. It shifts from the optical domain for a truncated spheroid to the far infrared for the cap. The size is largest for the hemispheroid.

Using the interaction along the substrate to dipolar order and the interaction with the image to order $M = 16$ the dipole polarizabilities were also calculated for prolate spheroids with an axial ratio 0.25 as a function of the frequency. The density was chosen such that the number of spheroids per square short axis, i.e. again the diameter parallel to the surface of the substrate, is equal to 0.4. One may again argue that the distance between the spheroids should be larger than their height divided by two. For $t_r = 0.4$ this implies that $\widehat{\rho}$ must be smaller than 0.5. For the hemisphere and the cap the condition is even less restrictive. The density chosen is 0.4 and therefore the results plotted below are expected to be reliable. The spheroids were located on a square array. This implies that the close packing density is equal to 1

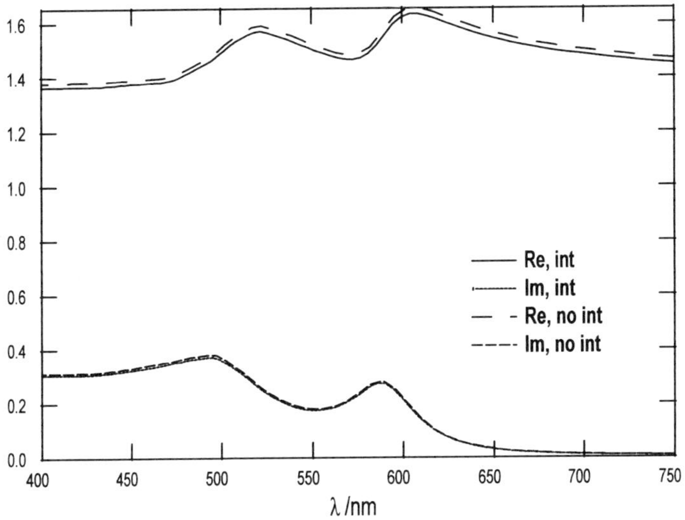

Figure 10.4 The real and imaginary parts of $\widehat{\beta}_e$ for a oblate hemispheroid, with an axial ratio 4 and a truncation parameter 0.0, as a function of the wave length. Both the cases of interaction and of no interaction along the substrate are plotted.

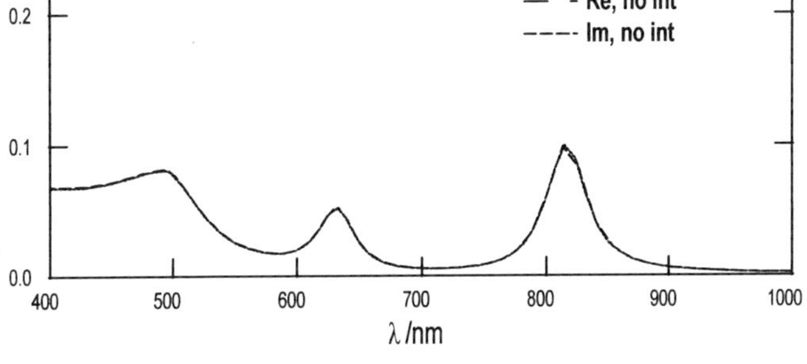

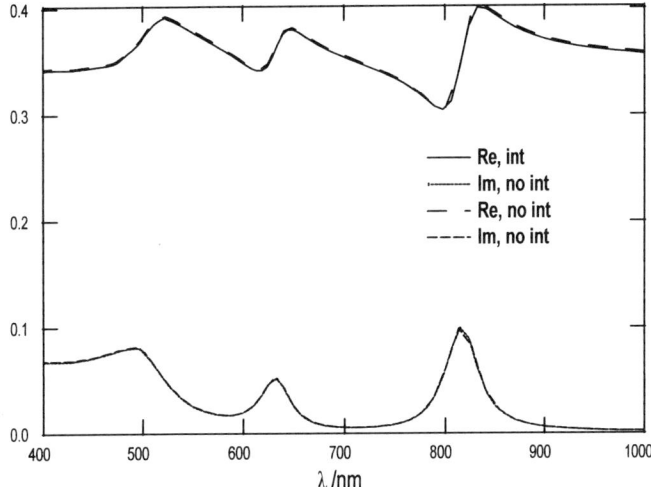

Figure 10.5 The real and imaginary parts of $\widehat{\beta}_e$ for a oblate spheroidal cap, with an axial ratio 4 and a truncation parameter -0.4, as a function of the wave length. Both the cases of interaction and of no interaction along the substrate are plotted.

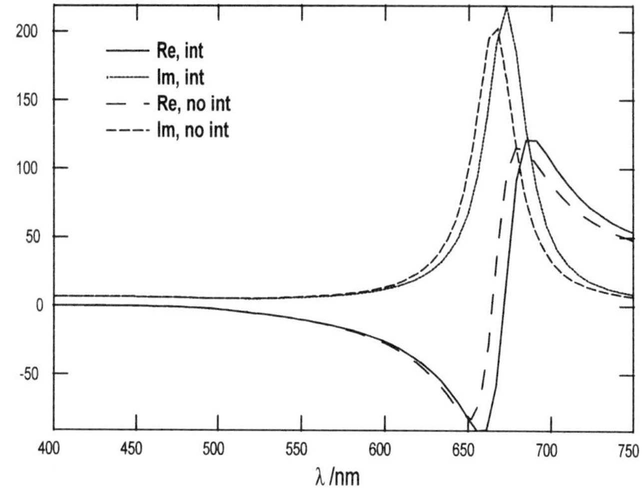

Figure 10.6 The real and imaginary parts of $\widehat{\gamma}_e$ for a truncated oblate spheroid, with an axial ratio 4 and a truncation parameter 0.4, as a function of the wave length. Both the cases of interaction and of no interaction along the substrate are plotted.

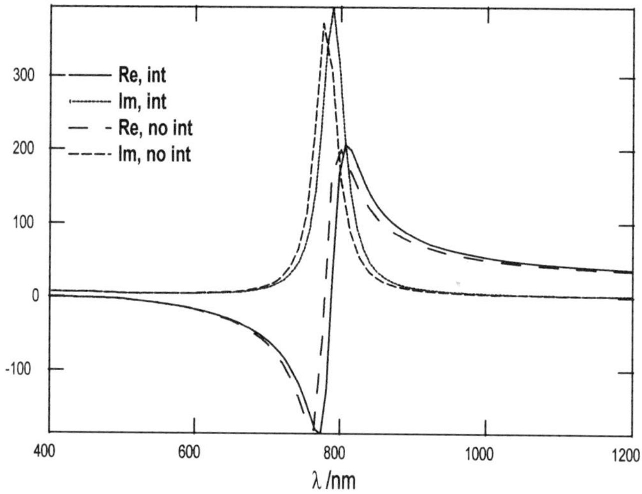

Figure 10.7 The real and imaginary parts of $\widehat{\gamma}_e$ for a oblate hemispheroid, with an axial ratio 4 and a truncation parameter 0.0, as a function of the wave length. Both the cases of interaction and of no interaction along the substrate are plotted.

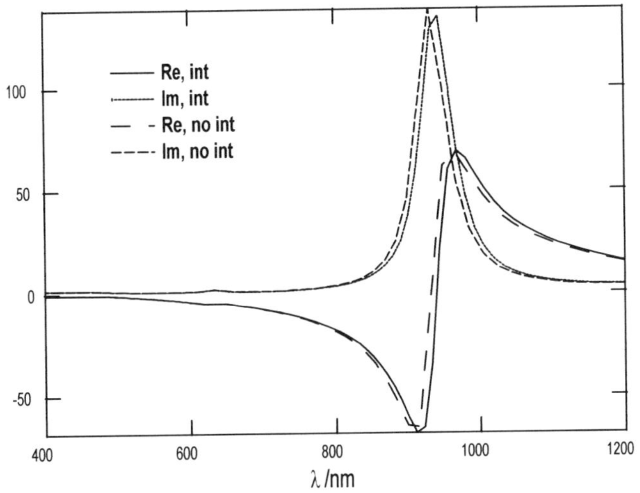

Figure 10.8 The real and imaginary parts of $\widehat{\gamma}_e$ for a oblate spheroidal cap, with an axial ratio 4 and a truncation parameter -0.4, as a function of the wave length. Both the cases of interaction and of no interaction along the substrate are plotted.

in these units. The resulting $\widehat{\beta}_e$ is plotted in figs.10.9-11 for truncation parameters $t_r = 0.4, 0.0$ and -0.4. It is found that the constitutive coefficient $\widehat{\beta}_e$, which is due to the dipole polarizability normal to the surface, for prolate spheroids is very sensitive to the interaction along the substrate. The location of the resonance shifts considerably to the blue for the truncated and the half spheroid in addition to reducing in size substantially. For the cap the blue shift is smaller but the size reduction is still considerable. In all cases the dipole polarizability normal to the surface is small. Its resonance structure is also a sensitive function of the truncation parameter. There is a blue shift for decreasing truncation parameter and an additional resonance appears for the cap in the optical domain.

The resulting $\widehat{\gamma}_e$ for an axial ratio of 0.25 is plotted in figs.10.12-14 for truncation parameters $t_r = 0.4, 0.0$ and -0.4. It is found that the constitutive coefficient $\widehat{\gamma}_e$, which is due to the dipole polarizability parallel to the surface, increases in size due to the interaction along the substrate. This increase is most pronounced for the truncated spheroid and not very substantial for the cap. The interaction does not effect the location of the main resonance. The dipole polarizability parallel to the surface is small. For the truncated and the hemispheroid a small additional resonance appears both in the presence and in the absence of interaction.

One may conclude that the polarizability along the surface of oblate spheroids, for all values of the truncation parameter, is large and not really dependent on the interaction along the substrate. The normal polarizability of oblate spheroids and the normal and the parallel polarizabilities of prolate spheroids are, due to the interaction

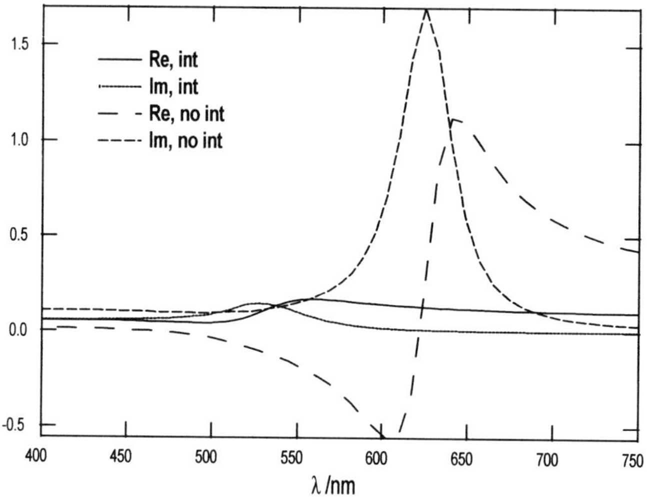

Figure 10.9 The real and imaginary parts of $\widehat{\beta}_e$ for a truncated prolate spheroid, with an axial ratio 0.25 and a truncation parameter 0.4, as a function of the wave length. Both the cases of interaction and of no interaction along the substrate are plotted.

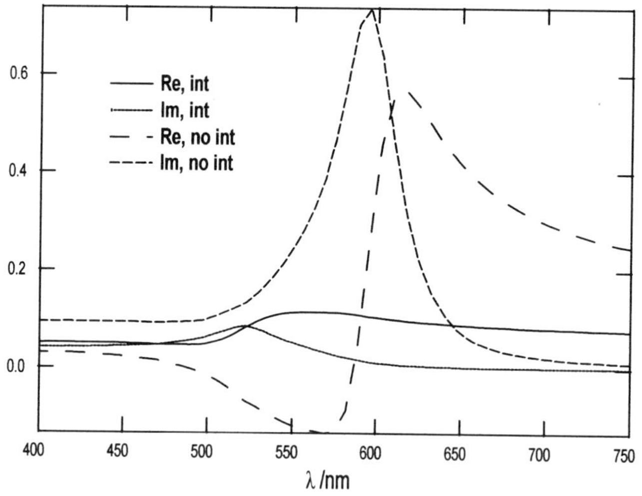

Figure 10.10 The real and imaginary parts of $\widehat{\beta}_e$ for a prolate hemispheroid, with an axial ratio 0.25 and a truncation parameter 0.0, as a function of the wave length. Both the cases of interaction and of no interaction along the substrate are plotted.

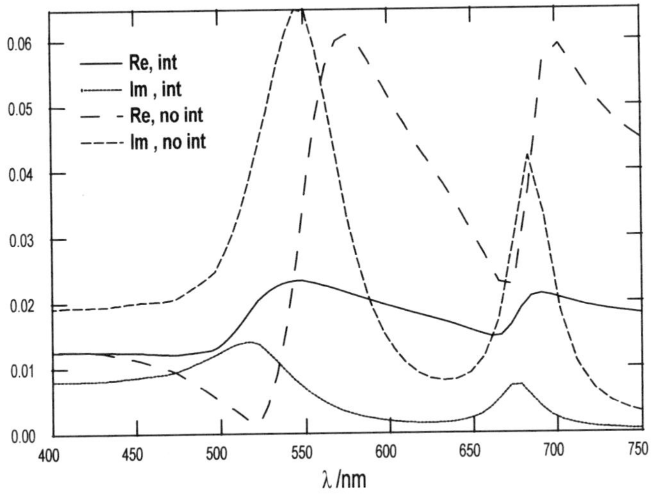

Figure 10.11 The real and imaginary parts of $\widehat{\beta}_e$ for a prolate spheroidal cap, with an axial ratio 0.25 and a truncation parameter -0.4, as a function of the wave length. Both the cases of interaction and of no interaction along the substrate are plotted.

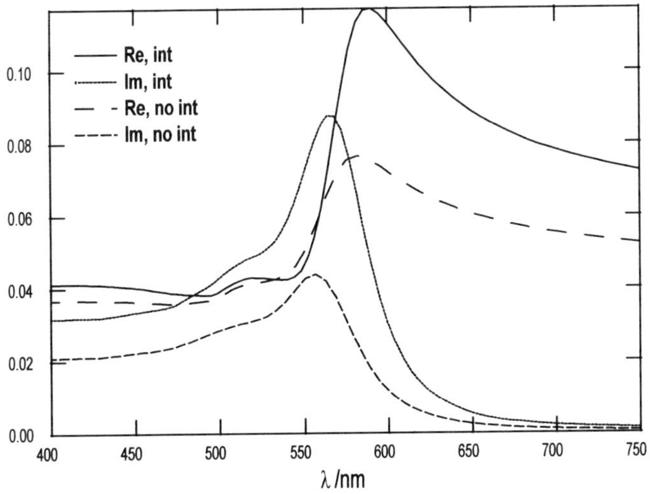

Figure 10.12 The real and imaginary parts of $\widehat{\gamma}_e$ for a truncated prolate spheroid, with an axial ratio 0.25 and a truncation parameter 0.4, as a function of the wave length. Both the cases of interaction and of no interaction along the substrate are plotted.

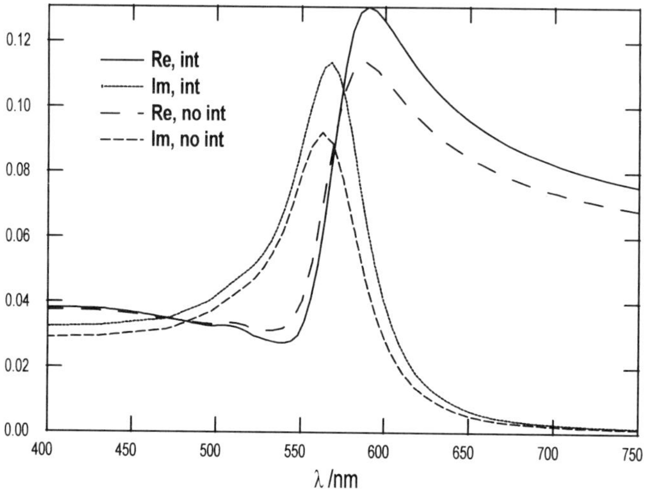

Figure 10.13 The real and imaginary parts of $\widehat{\gamma}_e$ for a prolate hemispheroid, with an axial ratio 0.25 and a truncation parameter 0.0, as a function of the wave length. Both the cases of interaction and of no interaction along the substrate are plotted.

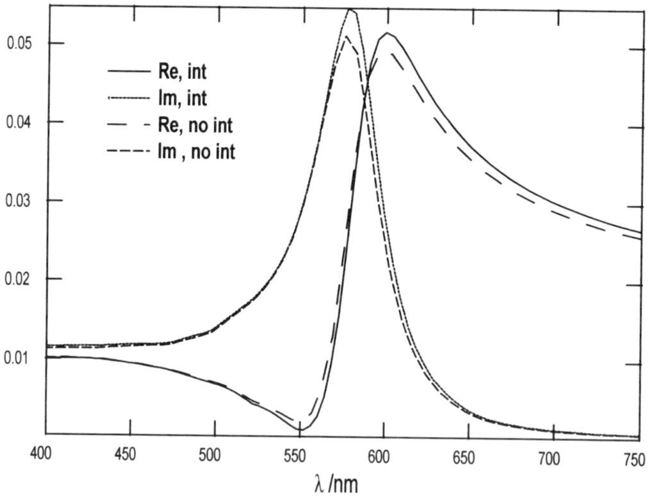

Figure 10.14 The real and imaginary parts of $\widehat{\gamma}_e$ for a prolate spheroidal cap, with an axial ratio 0.25 and a truncation parameter -0.4, as a function of the wave length. Both the cases of interaction and of no interaction along the substrate are plotted.

with the image and also due to the interaction along the substrate, small. There is a very considerable dependence of all polarizabilities on the truncation parameter.

10.7 Appendix A: Truncated particles

In this appendix it will be discussed, how the method for the calculation of the polarizabilities of a truncated sphere or spheroid on a substrate, developed in chapters 8 and 9 (sections 8.2, 9.2 and 9.3), must be changed, in order to be able to calculate the multipole polarizabilities $\alpha_{20,20}$, $\alpha_{21,21}$ and $\alpha_{22,22}$ for these particles. First the case of a truncated sphere will be considered (cf. section 8.2). Instead of the constant external electric field $\mathbf{E}_0$, eq.(8.3), a field will now be taken in the ambient, which is a linear function of space and therefore the gradient of a potential of the form

$$\psi_0(\mathbf{r}) = -\frac{1}{2}\boldsymbol{\phi} : \mathbf{rr} \equiv -\frac{1}{2}\sum_{i,j=1}^{3} \phi_{ij} r_i r_j \tag{10.84}$$

Here $r_1 \equiv x$, $r_2 \equiv y$, $r_3 \equiv z$ and ϕ_{ij} is a symmetric traceless matrix . The indices i and j denote the Cartesian components x, y and z. Introducing spherical coordinates r, θ, ϕ and eliminating $\phi_{zz}(= -\phi_{xx} - \phi_{yy})$, this potential can be written, in terms of the spherical harmonics, defined by eqs.(5.26) and (5.27), as

$$\begin{aligned}
\psi_0(\mathbf{r}) &= r^2\sqrt{2\pi/15}[\frac{1}{2}(\phi_{xx}+\phi_{yy})Y_2^0(\theta,\phi)\sqrt{6} - (\phi_{xz}+i\phi_{yz})Y_2^{-1}(\theta,\phi) \\
&\quad +(\phi_{xz}-i\phi_{yz})Y_2^1(\theta,\phi) - (\frac{1}{2}\phi_{xx}-\frac{1}{2}\phi_{yy}+i\phi_{xy})Y_2^{-2}(\theta,\phi) \\
&\quad -(\frac{1}{2}\phi_{xx}-\frac{1}{2}\phi_{yy}-i\phi_{xy})Y_2^2(\theta,\phi)]
\end{aligned} \tag{10.85}$$

This potential of the external field now has to replace the potential of the constant field $\mathbf{E}_0$ in the expression, eq.(8.2), for the potential in the region 1 (ambient), so that this becomes

$$\begin{aligned}
\psi_1(\mathbf{r}) &= r^2\sqrt{2\pi/15}[\frac{1}{2}(\phi_{xx}+\phi_{yy})Y_2^0(\theta,\phi)\sqrt{6} - (\phi_{xz}+i\phi_{yz})Y_2^{-1}(\theta,\phi) \\
&\quad +(\phi_{xz}-i\phi_{yz})Y_2^1(\theta,\phi) - (\frac{1}{2}\phi_{xx}-\frac{1}{2}\phi_{yy}+i\phi_{xy})Y_2^{-2}(\theta,\phi) \\
&\quad -(\frac{1}{2}\phi_{xx}-\frac{1}{2}\phi_{yy}-i\phi_{xy})Y_2^2(\theta,\phi)] \\
&\quad +\sum_{\ell m}{}' A_{\ell m} r^{-\ell-1} Y_\ell^m(\theta,\phi) + \sum_{\ell m}{}' A_{\ell m}^r \rho^{-\ell-1} Y_\ell^m(\theta^r,\phi^r)
\end{aligned} \tag{10.86}$$

For the meaning of the various symbols, the reader is referred to section 8.2. (See also fig.8.3).

The potential in region 2 (substrate), which was given by eq.(8.4), has to be replaced by

$$
\begin{aligned}
\psi_2(\mathbf{r}) \;=\; & \left(1-\frac{\epsilon_1}{\epsilon_2}\right)(\phi_{xx}+\phi_{yy})d^2 - r\sqrt{2\pi/3}\left(1-\frac{\epsilon_1}{\epsilon_2}\right)d[(\phi_{xx}+\phi_{yy})Y_1^0(\theta,\phi)\sqrt{2} \\
& +(\phi_{xz}+i\phi_{yz})Y_1^{-1}(\theta,\phi)-(\phi_{xz}-i\phi_{yz})Y_1^1(\theta,\phi)] \\
& +r^2\sqrt{2\pi/15}[\frac{1}{2}(\phi_{xx}+\phi_{yy})Y_2^0(\theta,\phi)\sqrt{6}-\frac{\epsilon_1}{\epsilon_2}(\phi_{xz}+i\phi_{yz})Y_2^{-1}(\theta,\phi) \\
& +\frac{\epsilon_1}{\epsilon_2}(\phi_{xz}-i\phi_{yz})Y_2^1(\theta,\phi)-(\frac{1}{2}\phi_{xx}-\frac{1}{2}\phi_{yy}+i\phi_{xy})Y_2^{-2}(\theta,\phi) \\
& -(\frac{1}{2}\phi_{xx}-\frac{1}{2}\phi_{yy}-i\phi_{xy})Y_2^2(\theta,\phi)]+\sum_{\ell m}{}' A_{\ell m}^t r^{-\ell-1}Y_\ell^m(\theta,\phi) \qquad (10.87)
\end{aligned}
$$

The potentials in the regions 3 (island) and 4 (substrate) are again given by eqs.(8.5) and (8.6), (see fig.8.3).

It can easily be checked, that the external potential, appearing at the right-hand sides of eqs.(10.86) and (10.87) (all terms except the sums $\Sigma'_{\ell m}$), satisfies the boundary conditions at the surface of the substrate. It furthermore follows from these conditions, that eqs.(8.7) and (8.8) hold again for the multipole coefficients $A_{\ell m}$ and $B_{\ell m}$.

Applying the boundary conditions, eqs.(8.9) and (8.10), at the surface of the sphere $r = R$, one obtains the infinite set of inhomogeneous linear equations (8.12), where the matrix elements $C^m_{\ell\ell_1}$, $D^m_{\ell\ell_1}$, $F^m_{\ell\ell_1}$ and $G^m_{\ell\ell_1}$ are again given by eq.(8.13), together with eqs.(8.14)– (8.20). The terms at the right-hand sides of eq.(8.12) are, however, given by expressions different from those found in section 8.2, and will be denoted here by K^m_ℓ and L^m_ℓ. They are now found to be unequal to zero, if $m = 0, \pm 1$ or ± 2. One therefore has to consider eq.(8.12) here not only for the values $m = 0, \pm 1$, but also for $m = \pm 2$. Furthermore one can prove again, that this equation has to be used only for $\ell = 1, 2, 3, ...$(see the discussion in section 8.2 below eq.(8.21)).

For $\ell = 1, 2, 3, ...$ and $m = 0, \pm 1$ one obtains

$$
\begin{aligned}
\sum_{\ell_1=1}^{\infty} C^m_{\ell\ell_1} R^{-\ell_1-2} A_{\ell_1 m} + \sum_{\ell_1=1}^{\infty} D^m_{\ell\ell_1} R^{\ell_1-1} B_{\ell_1 m} &= K^m_\ell \\
\sum_{\ell_1=1}^{\infty} F^m_{\ell\ell_1} R^{-\ell_1-2} A_{\ell_1 m} + \sum_{\ell_1=1}^{\infty} G^m_{\ell\ell_1} R^{\ell_1-1} B_{\ell_1 m} &= L^m_\ell \qquad (10.88)
\end{aligned}
$$

where

$$\begin{aligned}
K_\ell^0 &\equiv -\sqrt{\pi/5}(\phi_{xx}+\phi_{yy})R\{\delta_{\ell 2} \\
&\quad +2\sqrt{5/3}\left(\frac{\epsilon_1}{\epsilon_2}-1\right)t_r[\delta_{\ell 1}+\sqrt{3}t_r\zeta^0_{\ell 0}Q^0_{\ell 0}(t_r)-\zeta^0_{\ell 1}Q^0_{\ell 1}(t_r)]\} \\
K_\ell^{\pm 1} &\equiv \mp\sqrt{2\pi/15}(\phi_{xz}\mp i\phi_{yz})R\{\frac{\epsilon_1}{\epsilon_2}\delta_{\ell 2} \\
&\quad -\left(\frac{\epsilon_1}{\epsilon_2}-1\right)\left(\zeta^{\pm 1}_{\ell 2}Q^{\pm 1}_{\ell 2}(t_r)+\sqrt{5}t_r[\delta_{\ell 1}-\zeta^{\pm 1}_{\ell 1}Q^{\pm 1}_{\ell 1}(t_r)]\right)\} \\
L_\ell^0 &\equiv -2\sqrt{\pi/5}(\phi_{xx}+\phi_{yy})R\{\epsilon_2\delta_{\ell 2}-(\epsilon_2-\epsilon_1)\zeta^0_{\ell 2}Q^0_{\ell 2}(t_r) \\
&\quad -\sqrt{5/3}(\epsilon_2-\epsilon_1)t_r[\delta_{\ell 1}-\zeta^0_{\ell 1}Q^0_{\ell 1}(t_r)]\} \\
L_\ell^{\pm 1} &\equiv \mp\sqrt{2\pi/15}(\phi_{xz}\mp i\phi_{yz})R\{2\epsilon_1\delta_{\ell 2} \\
&\quad +\sqrt{5}(\epsilon_2-\epsilon_1)t_r[\delta_{\ell 1}-\zeta^{\pm 1}_{\ell 1}Q^{\pm 1}_{\ell 1}(t_r)]\}
\end{aligned} \tag{10.89}$$

For $\ell = 2, 3, 4, ...$ and $m = \pm 2$ one obtains

$$\begin{aligned}
\sum_{\ell_1=2}^{\infty} C^m_{\ell\ell_1}R^{-\ell_1-2}A_{\ell_1 m}+\sum_{\ell_1=2}^{\infty} D^m_{\ell\ell_1}R^{\ell_1-1}B_{\ell_1 m} &= K_\ell^m \\
\sum_{\ell_1=2}^{\infty} F^m_{\ell\ell_1}R^{-\ell_1-2}A_{\ell_1 m}+\sum_{\ell_1=2}^{\infty} G^m_{\ell\ell_1}R^{\ell_1-1}B_{\ell_1 m} &= L_\ell^m
\end{aligned} \tag{10.90}$$

where

$$\begin{aligned}
K_\ell^{\pm 2} &\equiv \sqrt{2\pi/15}R\left(\frac{1}{2}\phi_{xx}-\frac{1}{2}\phi_{yy}\mp i\,\phi_{xy}\right)\delta_{\ell 2} \\
L_\ell^{\pm 2} &\equiv 2\sqrt{2\pi/15}R\left(\frac{1}{2}\phi_{xx}-\frac{1}{2}\phi_{yy}\mp i\,\phi_{xy}\right) \\
&\quad \times\{\epsilon_2\delta_{\ell 2}-(\epsilon_2-\epsilon_1)\zeta^{\pm 2}_{\ell 2}Q^{\pm 2}_{\ell 2}(t_r)\}
\end{aligned} \tag{10.91}$$

The matrices $\zeta^m_{\ell\ell_1}$ and $Q^m_{\ell\ell_1}(t_r)$ are again given by eqs.(8.14) and (8.15), and the truncation parameter t_r by eq.(8.11).

In an analogous way as in section 8.2, one can now prove that eq.(10.88) for $m = -1$ and eq.(10.90) for $m = -2$ are redundant. First of all eq.(8.26) is valid. Furthermore one proves from the definition, eqs.(10.89) and (10.91), together with eqs.(8.14), (8.15) and (6.8), that

$$\begin{aligned}
(\phi_{xz}+i\,\phi_{yz})K_\ell^1 &= -(\phi_{xz}-i\,\phi_{yz})K_\ell^{-1} \\
(\phi_{xz}+i\,\phi_{yz})L_\ell^1 &= -(\phi_{xz}-i\,\phi_{yz})L_\ell^{-1}
\end{aligned} \tag{10.92}$$

and

$$\begin{aligned}
(\frac{1}{2}\phi_{xx}-\frac{1}{2}\phi_{yy}+i\,\phi_{xy})K_\ell^2 &= (\frac{1}{2}\phi_{xx}-\frac{1}{2}\phi_{yy}-i\,\phi_{xy})K_\ell^{-2} \\
(\frac{1}{2}\phi_{xx}-\frac{1}{2}\phi_{yy}+i\,\phi_{xy})L_\ell^2 &= (\frac{1}{2}\phi_{xx}-\frac{1}{2}\phi_{yy}-i\,\phi_{xy})L_\ell^{-2}
\end{aligned} \tag{10.93}$$

From eqs.(10.88), (8.26) and (10.92) one can then prove that

$$\begin{aligned}(\phi_{xz}+i\ \phi_{yz})A_{\ell 1} &= -(\phi_{xz}-i\ \phi_{yz})A_{\ell,-1}\\ (\phi_{xz}+i\ \phi_{yz})B_{\ell 1} &= -(\phi_{xz}-i\ \phi_{yz})B_{\ell,-1}\end{aligned} \tag{10.94}$$

whereas it follows from eqs.(10.90), (8.26) and (10.93) that

$$\begin{aligned}\left(\frac{1}{2}\phi_{xx}-\frac{1}{2}\phi_{yy}+i\ \phi_{xy}\right)A_{\ell 2} &= \left(\frac{1}{2}\phi_{xx}-\frac{1}{2}\phi_{yy}-i\ \phi_{xy}\right)A_{\ell,-2}\\ \left(\frac{1}{2}\phi_{xx}-\frac{1}{2}\phi_{yy}+i\ \phi_{xy}\right)B_{\ell 2} &= \left(\frac{1}{2}\phi_{xx}-\frac{1}{2}\phi_{yy}-i\ \phi_{xy}\right)B_{\ell,-2}\end{aligned} \tag{10.95}$$

Both the $m=-1$ equation (10.88) and the $m=-2$ equation (10.90) are therefore superfluous.

Equation (10.88) may be solved numerically, by neglecting the amplitudes $A_{\ell m}$ and $B_{\ell m}$ larger than a suitably chosen order. One then solves the finite set of inhomogeneous equations

$$\begin{aligned}\sum_{\ell_1=1}^{M} C^m_{\ell\ell_1}R^{-\ell_1-2}A_{\ell_1 m}+\sum_{\ell_1=1}^{M} D^m_{\ell\ell_1}R^{\ell_1-1}B_{\ell_1 m} &= K^m_\ell\\ \sum_{\ell_1=1}^{M} F^m_{\ell\ell_1}R^{-\ell_1-2}A_{\ell_1 m}+\sum_{\ell_1=1}^{M} G^m_{\ell\ell_1}R^{\ell_1-1}B_{\ell_1 m} &= L^m_\ell\\ \text{for }\ell &= 1,2,3,.....,M;\ m=0,1\end{aligned} \tag{10.96}$$

where K^0_ℓ, L^0_ℓ, K^1_ℓ and L^1_ℓ are given by eq.(10.89) and $C^m_{\ell\ell_1}$, $D^m_{\ell\ell_1}$, $F^m_{\ell\ell_1}$ and $G^m_{\ell\ell_1}$ by eq.(8.13). In this way the amplitudes A_{10} , A_{20} , A_{11} and A_{21} can be calculated with sufficient accuracy, by choosing M large enough. In a similar way one may solve eq.(10.90) numerically by solving the finite set of equations

$$\begin{aligned}\sum_{\ell_1=2}^{M} C^2_{\ell\ell_1}R^{-\ell_1-2}A_{\ell_1 2}+\sum_{\ell_1=2}^{M} D^2_{\ell\ell_1}R^{\ell_1-1}B_{\ell_1 2} &= K^2_\ell\\ \sum_{\ell_1=2}^{M} F^2_{\ell\ell_1}R^{-\ell_1-2}A_{\ell_1 2}+\sum_{\ell_1=2}^{M} G^2_{\ell\ell_1}R^{\ell_1-1}B_{\ell_1 2} &= L^2_\ell\\ \text{for }\ell &= 2,3,4,...,M\end{aligned} \tag{10.97}$$

where K^2_ℓ and L^2_ℓ are given by eq.(10.91). In this way the amplitude A_{22} may be calculated.

It follows from eqs.(10.96) and (10.97), together with eqs.(10.89) and (10.91), that A_{10} and A_{20} are proportional to $(\phi_{xx}+\phi_{yy})$, A_{11} and A_{21} to $(\phi_{xz}-i\ \phi_{yz})$ and A_{22} to $(\frac{1}{2}\phi_{xx}-\frac{1}{2}\phi_{yy}-i\ \phi_{xy})$, which quantities appear as factors of the spherical harmonics in the expression, eq.(10.85), for the potential of the external field in the ambient. If one writes this potential as

$$\psi_0(r)=r^2\sum_{m=-2}^{2} B^{(0)}_{2m}Y^m_2(\theta,\phi) \tag{10.98}$$

it is found that

$$
\begin{aligned}
B_{20}^{(0)} &= (\phi_{xx}+\phi_{yy})\sqrt{\pi/5}, \quad B_{2,\pm1}^{(0)} = \pm(\phi_{xz}\mp i\ \phi_{yz})\sqrt{2\pi/15} \\
B_{2,\pm2}^{(0)} &= -\left(\frac{1}{2}\phi_{xx}-\frac{1}{2}\phi_{yy}\mp i\ \phi_{xy}\right)\sqrt{2\pi/15}
\end{aligned} \tag{10.99}
$$

Therefore A_{10} and A_{20} are proportional to $B_{20}^{(0)}$, A_{11} and A_{21} to $B_{21}^{(0)}$ and A_{22} to $B_{22}^{(0)}$. The proportionality constants are, according to eq.(5.30), the multipole polarizabilities $\alpha_{\ell m,\ell' m'}$:

$$
\begin{aligned}
A_{10} &= -\alpha_{10,20}B_{20}^{(0)}, \quad A_{20} = -\alpha_{20,20}B_{20}^{(0)}, \quad A_{11} = -\alpha_{11,21}B_{21}^{(0)} \\
A_{21} &= -\alpha_{21,21}B_{21}^{(0)}, \quad A_{22} = -\alpha_{22,22}B_{22}^{(0)}
\end{aligned} \tag{10.100}
$$

The multipole polarizabilities $\alpha_{10,20}$ and $\alpha_{11,21}$ could also be found from eqs.(8.31) and (5.67), using the method of section 8.2. New are the multipole polarizabilities $\alpha_{20,20}$, $\alpha_{21,21}$ and $\alpha_{22,22}$. These quantities are therefore found by dividing the amplitudes A_{20}, A_{21} and A_{22}, obtained above, by $B_{20}^{(0)}$, $B_{21}^{(0)}$ and $B_{22}^{(0)}$ respectively or, more explicitly, from

$$
\begin{aligned}
\alpha_{20,20} &= -A_{20}/\left[(\phi_{xx}+\phi_{yy})\sqrt{\pi/5}\right], \quad \alpha_{21,21} = -A_{21}/\left[(\phi_{xz}-i\ \phi_{yz})\sqrt{2\pi/15}\right] \\
\alpha_{22,22} &= A_{22}/\left[\left(\frac{1}{2}\phi_{xx}-\frac{1}{2}\phi_{yy}-i\ \phi_{xy}\right)\sqrt{2\pi/15}\right]
\end{aligned} \tag{10.101}
$$

With eq.(5.67) one can then also calculate the quadrupole polarizabilities:

$$
\begin{aligned}
\alpha_z^{11}(0) &= -\frac{\pi\ \epsilon_1 A_{20}}{(\phi_{xx}+\phi_{yy})\sqrt{\pi/5}} - \frac{6\pi\ \epsilon_1 A_{22}}{(\phi_{xx}-\phi_{yy}-2i\phi_{xy})\sqrt{6\pi/5}} \\
\alpha_{\|}^{11}(0) &= -\frac{4\pi\ \epsilon_1 A_{21}}{(\phi_{xz}-i\phi_{yz})\sqrt{6\pi/5}} - \frac{8\pi\ \epsilon_1 A_{22}}{(\phi_{xx}-\phi_{yy}-2i\phi_{xy})\sqrt{6\pi/5}} \\
\alpha^{11}(0) &= \frac{4\pi\ \epsilon_1 A_{22}}{(\phi_{xx}-\phi_{yy}-2i\phi_{xy})\sqrt{6\pi/5}}
\end{aligned} \tag{10.102}
$$

Next the case of a truncated oblate spheroid on a substrate will be considered. Then the potential of the external field, eq.(10.84), has to be written in terms of oblate spheroidal coordinates ξ, η and ϕ, defined by eqs.(6.1) and (6.2) or by the inverse coordinate transformation, eq.(6.3), see section 6.2 for a further explanation. One obtains

$$
\begin{aligned}
\psi_0(\mathbf{r}) &= \sqrt{2\pi/15}[\frac{1}{2}(\phi_{xx}+\phi_{yy})X_2^0(\xi,a)Y_2^0(\arccos\eta,\phi)\sqrt{6}-(\phi_{xz}+i\ \phi_{yz}) \\
&\quad \times X_2^{-1}(\xi,a)Y_2^{-1}(\arccos\eta,\phi)+(\phi_{xz}-i\ \phi_{yz})X_2^1(\xi,a)Y_2^1(\arccos\eta,\phi) \\
&\quad -\left(\frac{1}{2}\phi_{xx}-\frac{1}{2}\phi_{yy}+i\ \phi_{xy}\right)X_2^{-2}(\xi,a)Y_2^{-2}(\arccos\eta,\phi) \\
&\quad -\left(\frac{1}{2}\phi_{xx}-\frac{1}{2}\phi_{yy}-i\ \phi_{xy}\right)X_2^{2}(\xi,a)Y_2^{2}(\arccos\eta,\phi)] \\
&\quad -\frac{1}{6}(\phi_{xx}+\phi_{yy})a^2
\end{aligned} \tag{10.103}
$$

For the meaning of the various symbols in the above equation and below, the reader is referred to section 6.2. The additive constant $-(1/6)(\phi_{xx}+\phi_{yy})a^2$ in the above potential will be omitted.

For the potential in region 1 (ambient) one now obtains

$$\begin{aligned}
\psi_1(\mathbf{r}) &= \sqrt{2\pi/15}[\frac{1}{2}(\phi_{xx}+\phi_{yy})X_2^0(\xi,a)Y_2^0(\arccos\eta,\phi)\sqrt{6}-(\phi_{xz}+i\ \phi_{yz})\\
&\quad \times X_2^{-1}(\xi,a)Y_2^{-1}(\arccos\eta,\phi)+(\phi_{xz}-i\ \phi_{yz})X_2^1(\xi,a)Y_2^1(\arccos\eta,\phi)\\
&\quad -\left(\frac{1}{2}\phi_{xx}-\frac{1}{2}\phi_{yy}+i\ \phi_{xy}\right)X_2^{-2}(\xi,a)Y_2^{-2}(\arccos\eta,\phi)\\
&\quad -\left(\frac{1}{2}\phi_{xx}-\frac{1}{2}\phi_{yy}-i\ \phi_{xy}\right)X_2^2(\xi,a)Y_2^2(\arccos\eta,\phi)]\\
&\quad +{\sum_{\ell m}}' A_{\ell m}Z_\ell^m(\xi,a)Y_\ell^m(\arccos\eta,\phi)\\
&\quad +{\sum_{\ell m}}' A_{\ell m}^r Z_\ell^m(\xi'(\xi,\eta),a)Y_\ell^m(\arccos\eta'(\xi,\eta),\phi)\\
&\quad \text{for } \xi_0\le\xi<\infty \text{ and } \eta\le\xi_1/\xi \qquad (10.104)
\end{aligned}$$

The potential in region 2 (substrate) becomes

$$\begin{aligned}
\psi_2(\mathbf{r}) &= \left(1-\frac{\epsilon_1}{\epsilon_2}\right)(\phi_{xx}+\phi_{yy})d^2-\sqrt{2\pi/3}\left(1-\frac{\epsilon_1}{\epsilon_2}\right)d[(\phi_{xx}+\phi_{yy})X_1^0(\xi,a)\\
&\quad \times Y_1^0(\arccos\eta,\phi)\sqrt{2}+(\phi_{xz}+i\phi_{yz})X_1^{-1}(\xi,a)Y_1^{-1}(\arccos\eta,\phi)\\
&\quad -(\phi_{xz}-i\phi_{yz})X_1^1(\xi,a)Y_1^1(\arccos\eta,\phi)]+\sqrt{2\pi/15}[\frac{1}{2}(\phi_{xx}+\phi_{yy})\\
&\quad \times X_2^0(\xi,a)Y_2^0(\arccos\eta,\phi)\sqrt{6}-\frac{\epsilon_1}{\epsilon_2}(\phi_{xz}+i\ \phi_{yz})X_2^{-1}(\xi,a)Y_2^{-1}(\arccos\eta,\phi)\\
&\quad +\frac{\epsilon_1}{\epsilon_2}(\phi_{xz}-i\ \phi_{yz})X_2^1(\xi,a)Y_2^1(\arccos\eta,\phi)-\left(\frac{1}{2}\phi_{xx}-\frac{1}{2}\phi_{yy}+i\ \phi_{xy}\right)\\
&\quad \times X_2^{-2}(\xi,a)Y_2^{-2}(\arccos\eta,\phi)-\left(\frac{1}{2}\phi_{xx}-\frac{1}{2}\phi_{yy}-i\ \phi_{xy}\right)\\
&\quad \times X_2^2(\xi,a)Y_2^2(\arccos\eta,\phi)]+{\sum_{\ell m}}' A_{\ell m}^t Z_\ell^m(\xi,a)Y_\ell^m(\arccos\eta,\phi)\\
&\quad \text{for } 0\le\xi\le\xi_0 \text{ and } \eta\ge\xi_1/\xi \qquad (10.105)
\end{aligned}$$

Finally the potential in the regions 3 (island) and 4 (substrate) are given by eqs.(9.8) and (9.9).

The external potential, appearing at the right-hand sides of eqs.(10.104) and (10.105) (all terms except the sums $\Sigma'_{\ell m}$), satisfies the boundary conditions at the surface of the substrate. It furthermore follows from these conditions that eqs.(9.12)–(9.14) are valid.

Applying the boundary conditions, eqs.(9.15) and (9.16), at the surface of the spheroid ($\xi = \xi_0$), one obtains again the infinite set of inhomogeneous linear equations (9.17), where the matrix elements $C^m_{\ell\ell_1}$, $D^m_{\ell\ell_1}$, $F^m_{\ell\ell_1}$ and $G^m_{\ell\ell_1}$ are given by eq.(9.19), together with eqs.(9.20)– (9.24). The terms at the right-hand sides of eq.(9.17) are now, however, given by different expressions as in section 9.2, and will be denoted again by the symbols K^m_ℓ and L^m_ℓ. Just as in the case of truncated spheres above, these quantities are only unequal to zero, if $m = 0, \pm 1$ or ± 2. One therefore has to consider eq.(9.17) only for these values of m. Furthermore one can take $\ell = 1, 2, 3, ...$just as above.

For $\ell = 1, 2, 3, ...$ and $m = 0, \pm 1$ one obtains eq.(10.88) in which R should be replaced by half the length of the short axis, $R_s = a\xi_0$ (see fig. 9.3 of chapter 9). The terms at the right-hand sides of the latter equation are given by

$$\begin{aligned}
K^0_\ell &\equiv -\sqrt{\pi/5}(\phi_{xx}+\phi_{yy})a\xi_0\{\xi_0^{-2}X^0_2(\xi_0)\delta_{\ell 2} \\
&\quad +2\sqrt{\frac{5}{3}}\left(\frac{\epsilon_1}{\epsilon_2}-1\right)t_r\left(\xi_0^{-1}X^0_1(\xi_0)[\delta_{\ell 1}-\zeta^0_{\ell 1}Q^0_{\ell 1}(t_r)]+\sqrt{3}t_r\zeta^0_{\ell 0}Q^0_{\ell 0}(t_r)\right)\} \\
K^{\pm 1}_\ell &\equiv \mp\sqrt{2\pi/15}(\phi_{xz}\mp i\phi_{yz})a\xi_0[\frac{\epsilon_1}{\epsilon_2}\xi_0^{-2}X^{\pm 1}_2(\xi_0)\delta_{\ell 2}-\left(\frac{\epsilon_1}{\epsilon_2}-1\right)\{\xi_0^{-2}X^{\pm 1}_2(\xi_0) \\
&\quad \times\zeta^{\pm 1}_{\ell 2}Q^{\pm 1}_{\ell 2}(t_r)+\sqrt{5}t_r\xi_0^{-1}X^{\pm 1}_1(\xi_0)[\delta_{\ell 1}-\zeta^{\pm 1}_{\ell 1}Q^{\pm 1}_{\ell 1}(t_r)]\}] \\
L^0_\ell &\equiv -\sqrt{\pi/5}(\phi_{xx}+\phi_{yy})a\xi_0\{\xi_0^{-1}\frac{dX^0_2(\xi_0)}{d\xi_0}[\epsilon_2\delta_{\ell 2}-(\epsilon_2-\epsilon_1)\zeta^0_{\ell 2}Q^0_{\ell 2}(t_r)] \\
&\quad -2\sqrt{5/3}(\epsilon_2-\epsilon_1)t_r\frac{dX^0_1(\xi_0)}{d\xi_0}[\delta_{\ell 1}-\zeta^0_{\ell 1}Q^0_{\ell 1}(t_r)]\} \\
L^{\pm 1}_\ell &\equiv \mp\sqrt{2\pi/15}(\phi_{xz}\mp i\phi_{yz})a\xi_0\{\epsilon_1\xi_0^{-1}\frac{dX^{\pm 1}_2(\xi_0)}{d\xi_0}\delta_{\ell 2} \\
&\quad +\sqrt{5}(\epsilon_2-\epsilon_1)t_r\frac{dX^{\pm 1}_1(\xi_0)}{d\xi_0}[\delta_{\ell 1}-\zeta^{\pm 1}_{\ell 1}Q^{\pm 1}_{\ell 1}(t_r)]\}
\end{aligned} \tag{10.106}$$

where the functions $X^m_\ell(\xi_0)$ are defined by eq.(9.20), the matrices $\zeta^m_{\ell\ell_1}$ and $Q^m_{\ell\ell_1}(t_r)$ by eqs.(8.14) and (8.15), and the truncation parameter t_r by eq.(9.2). For $\ell = 2, 3, 4, ...$ and $m = \pm 2$ one obtains eq.(10.90)) in which R should be replaced by half the length of the short axis, $R_s = a\xi_0$ (see fig. 9.3 of chapter 9). The terms at the right-hand sides of the latter equation are given by

$$\begin{aligned}
K^{\pm 2}_\ell &\equiv \sqrt{2\pi/15}a\xi_0\left(\frac{1}{2}\phi_{xx}-\frac{1}{2}\phi_{yy}\mp i\ \phi_{xy}\right)\xi_0^{-2}X^{\pm 2}_2(\xi_0)\delta_{\ell 2} \\
L^{\pm 2}_\ell &\equiv \sqrt{2\pi/15}a\xi_0\left(\frac{1}{2}\phi_{xx}-\frac{1}{2}\phi_{yy}\mp i\ \phi_{xy}\right)\xi_0^{-1}\frac{dX^{\pm 2}_2(\xi_0)}{d\xi_0} \\
&\quad \times[\epsilon_2\delta_{\ell 2}-(\epsilon_2-\epsilon_1)\zeta^{\pm 2}_{\ell 2}Q^{\pm 2}_{\ell 2}(t_r)]
\end{aligned} \tag{10.107}$$

Notice that the above expressions, eqs.(10.106) and (10.107), approach to the expressions, eqs.(10.89) and (10.91), for truncated spheres in the spherical limit, $\xi_0 \to \infty$ with $a\xi_0 = R$. This follows with eqs.(9.130) and (9.132).

In an analogous way as above for truncated spheres, one can prove that eq.(10.88) for $m = -1$ and eq.(10.90) for $m = -2$, with $a\xi_0$ instead of R, are also redundant for truncated oblate spheroids.

Equation (10.88), with $a\xi_0$ instead of R, may again be solved numerically by neglecting the amplitudes $A_{\ell m}$ and $B_{\ell m}$ larger than a sufficiently high order M. One solves eq.(10.96), with $a\xi_0$ instead of R, where $C^m_{\ell\ell_1}$, $D^m_{\ell\ell_1}$, $F^m_{\ell\ell_1}$ and $G^m_{\ell\ell_1}$ are given by eq.(9.19), and K^m_ℓ and L^m_ℓ by eq.(10.106). In this way the amplitudes A_{10}, A_{20}, A_{11} and A_{21} can be calculated for truncated oblate spheroids. In a similar way A_{22} can be evaluated by solving eq.(10.90), with $a\xi_0$ instead of R, numerically. One then has to solve eq.(10.97), with $a\xi_0$ instead of R, where K^m_ℓ and L^m_ℓ are given by eq.(10.107).

Writing the external potential, eq.(10.103) as

$$\psi_0(\mathbf{r}) = \sum_{m=-2}^{2} B^{(0)}_{2m} X^m_2(\xi, a) Y^m_2(\arccos\eta,\ \phi) \tag{10.108}$$

one obtains the same expressions, eq.(10.99), for the amplitudes $B^{(0)}_{2m}$. Since the amplitudes A_{10} and A_{20} are proportional to $B^{(0)}_{20}$, A_{11} and A_{21} to $B^{(0)}_{21}$, and A_{22} to $B^{(0)}_{22}$, eq. (10.100) holds again, defining now the oblate spheroidal multipole polarizabilities $\alpha_{10,20}$, $\alpha_{20,20}$, $\alpha_{11,21}$, $\alpha_{21,21}$ and $\alpha_{22,22}$. Since $\alpha_{10,20}$ and $\alpha_{11,21}$ can also be found from eqs.(9.31) and (5.67), the only new polarizabilities here are $\alpha_{20,20}$, $\alpha_{21,21}$ and $\alpha_{22,22}$, which can be obtained, using eq.(10.101). With eq.(10.102) one can then also calculate the quadrupole polarizabilities $\alpha^{11}_z(0)$, $\alpha^{11}_\parallel(0)$ and $\alpha^{11}(0)$ for a truncated oblate spheroid on a substrate. Notice, that it is again possible to calculate these quadrupole polarizabilities, using the same equation, as has been used for truncated spheres, since one has neglected octupole and higher order polarizabilities (see the discussion in chapters 6 and 9 on this point).

Finally the case of a truncated prolate spheroid will be considered. In section 6.3 prolate spheroidal coordinates ξ, η and ϕ were introduced by means of eqs.(6.48) and (6.49). In terms of these coordinates, the external potential $\psi_0(\mathbf{r})$, eq.(10.84), becomes

$$\begin{aligned}\psi_0(\mathbf{r}) &= \sqrt{2\pi/15}[\frac{1}{2}(\phi_{xx}+\phi_{yy})\widetilde{X}^0_2(\xi,a)Y^0_2(\arccos\eta,\phi)\sqrt{6} - (\phi_{xz}+i\ \phi_{yz}) \\ &\quad \times\widetilde{X}^{-1}_2(\xi,a)Y^{-1}_2(\arccos\eta,\phi) + (\phi_{xz}-i\ \phi_{yz})\widetilde{X}^1_2(\xi,a)Y^1_2(\arccos\eta,\phi) \\ &\quad - \left(\frac{1}{2}\phi_{xx}-\frac{1}{2}\phi_{yy}+i\ \phi_{xy}\right)\widetilde{X}^{-2}_2(\xi,a)Y^{-2}_2(\arccos\eta,\phi) \\ &\quad - \left(\frac{1}{2}\phi_{xx}-\frac{1}{2}\phi_{yy}-i\ \phi_{xy}\right)\widetilde{X}^2_2(\xi,a)Y^2_2(\arccos\eta,\phi)]\end{aligned} \tag{10.109}$$

For the definition of the various symbols in this equation the reader is referred to section 6.3. The additive constant has been omitted.

Notice that the above equation may be obtained from the corresponding expression, eq.(10.103), valid for oblate spheroidal coordinates, simply by replacing the functions $X^m_\ell(\xi,a)$ by $\widetilde{X}^m_\ell(\xi,a)$, and using prolate spheroidal coordinates, instead of oblate spheroidal coordinates. The same is true for the potential in the regions 1, 2, 3, and 4 (see fig. 9.4 of chapter 9): the functions $\psi_1(\mathbf{r})$ and $\psi_2(\mathbf{r})$ can be obtained from the expressions (10.104) and (10.105), by replacing $X^m_\ell(\xi,a)$ by $\widetilde{X}^m_\ell(\xi,a)$, eq.(6.58), $Z^m_\ell(\xi,a)$ by $\widetilde{Z}^m_\ell(\xi,a)$, eq.(6.59), and using prolate spheroidal coordinates. The functions $\xi'(\xi,\eta)$ and $\eta'(\xi,\eta)$ are then defined by eq.(9.36). The functions $\psi_3(\mathbf{r})$ and $\psi_4(\mathbf{r})$

are again given by eq.(9.35). The derivation of these results runs along the same lines as in section 9.3, and will not be given here.

Applying the boundary conditions at the surface of the substrate, as well as the boundary conditions, eqs.(9.15) and (9.16), at the surface of the prolate spheroid, one obtains the infinite set of inhomogeneous linear equations (9.37), where the matrix elements $C^m_{\ell\ell_1}$, $D^m_{\ell\ell_1}$, $F^m_{\ell\ell_1}$ and $G^m_{\ell\ell_1}$ are given by eq.(9.38), together with eqs.(9.39)–(9.42). The terms at the right-hand side of eq.(9.37) are, however, different, and will be denoted again by the symbols K^m_ℓ and L^m_ℓ . Just as in the case of a truncated oblate spheroid, these quantities are only different from zero, if $m = 0$, ±1 or ±2. Therefore equation (9.37) has to be considered only for these values of m. Furthermore it is sufficient to take $\ell = 1, 2, 3, ...$.

For $\ell = 1, 2, 3, ...$ and $m = 0$, ±1 one now obtains eq.(10.88) in which R should be replaced by half the length of the long axis, $R_l = a\xi_0$ (see fig. 9.4 of chapter 9). The terms at the right-hand sides of the latter equation are given by

$$\begin{aligned}
K^0_\ell &\equiv -\sqrt{\pi/5}(\phi_{xx}+\phi_{yy})a\xi_0\{\xi_0^{-2}\widetilde{X}^0_2(\xi_0)\delta_{\ell2} \\
&\quad +2\sqrt{5/3}\left(\frac{\epsilon_1}{\epsilon_2}-1\right)t_r\left(\xi_0^{-1}\widetilde{X}^0_1(\xi_0)[\delta_{\ell1}-\zeta^0_{\ell1}Q^0_{\ell1}(t_r)]+\sqrt{3}t_r\zeta^0_{\ell0}Q^0_{\ell0}(t_r)\right)\} \\
K^{\pm1}_\ell &\equiv \mp\sqrt{2\pi/15}(\phi_{xz}\mp i\phi_{yz})a\xi_0[\frac{\epsilon_1}{\epsilon_2}\xi_0^{-2}\widetilde{X}^{\pm1}_2(\xi_0)\delta_{\ell2}-\left(\frac{\epsilon_1}{\epsilon_2}-1\right)\{\xi_0^{-2}\widetilde{X}^{\pm1}_2(\xi_0) \\
&\quad \times\zeta^{\pm1}_{\ell2}Q^{\pm1}_{\ell2}(t_r)+\sqrt{5}t_r\xi_0^{-1}\widetilde{X}^{\pm1}_1(\xi_0)[\delta_{\ell1}-\zeta^{\pm1}_{\ell1}Q^{\pm1}_{\ell1}(t_r)]\}]
\end{aligned}$$

$$\begin{aligned}
L^0_\ell &\equiv -\sqrt{\pi/5}(\phi_{xx}+\phi_{yy})a\xi_0\{\xi_0^{-1}\frac{d\widetilde{X}^0_2(\xi_0)}{d\xi_0}[\epsilon_2\delta_{\ell2}-(\epsilon_2-\epsilon_1)\zeta^0_{\ell2}Q^0_{\ell2}(t_r)] \\
&\quad -2\sqrt{5/3}(\epsilon_2-\epsilon_1)t_r\frac{d\widetilde{X}^0_1(\xi_0)}{d\xi_0}[\delta_{\ell1}-\zeta^0_{\ell1}Q^0_{\ell1}(t_r)]\} \\
L^{\pm1}_\ell &\equiv \mp\sqrt{2\pi/15}(\phi_{xz}\mp i\phi_{yz})a\xi_0\{\epsilon_1\xi_0^{-1}\frac{d\widetilde{X}^{\pm1}_2(\xi_0)}{d\xi_0}\delta_{\ell2} \\
&\quad +\sqrt{5}(\epsilon_2-\epsilon_1)t_r\frac{d\widetilde{X}^{\pm1}_1(\xi_0)}{d\xi_0}[\delta_{\ell1}-\zeta^{\pm1}_{\ell1}Q^{\pm1}_{\ell1}(t_r)]\}
\end{aligned} \tag{10.110}$$

For $\ell = 2, 3, 4, ...$ and $m = \pm2$, one obtains eq.(10.90) in which R should be replaced by half the length of the long axis, $R_l = a\xi_0$ (see fig. 9.4 of chapter 9). The terms at the right-hand sides of the latter equation are given by

$$\begin{aligned}
K^{\pm2}_\ell &\equiv \sqrt{2\pi/15}a\xi_0\left(\frac{1}{2}\phi_{xx}-\frac{1}{2}\phi_{yy}\mp i\,\phi_{xy}\right)\xi_0^{-2}\widetilde{X}^{\pm2}_2(\xi_0)\delta_{\ell2} \\
L^{\pm2}_\ell &\equiv \sqrt{2\pi/15}a\xi_0\left(\frac{1}{2}\phi_{xx}-\frac{1}{2}\phi_{yy}\mp i\,\phi_{xy}\right)\xi_0^{-1}\frac{d\widetilde{X}^{\pm2}_2(\xi_0)}{d\xi_0} \\
&\quad \times[\epsilon_2\delta_{\ell2}-(\epsilon_2-\epsilon_1)\zeta^{\pm2}_{\ell2}Q^{\pm2}_{\ell2}(t_r)]
\end{aligned} \tag{10.111}$$

The above expressions, eqs.(10.110) and (10.111), may be obtained from the corresponding ones, eqs.(10.106) and (10.107), for truncated oblate spheroids, by replacing $X^m_\ell(\xi_0)$ by $\widetilde{X}^m_\ell(\xi_0)$. These expressions again tend to the corresponding ones, eqs.(10.89) and (10.91), for truncated spheres in the spherical limit, $\xi_0 \to \infty$ with $a\xi_0 = R$. This follows with eqs.(9.135) and (9.137).

In an analogous way as above for truncated spheres, one can prove that eq.(10.88) for $m = -1$ and eq.(10.90) for $m = -2$, with $a\xi_0$ instead of R, are also redundant for truncated prolate spheroids.

Equation (10.88), with $a\xi_0$ instead of R, may again be solved numerically by neglecting the amplitudes $A_{\ell m}$ and $B_{\ell m}$ larger than a sufficiently high order M. One solves eq.(10.96), with $a\xi_0$ instead of R, where $C^m_{\ell\ell_1}$, $D^m_{\ell\ell_1}$, $F^m_{\ell\ell_1}$ and $G^m_{\ell\ell_1}$ are given by eq.(9.38), and K^m_ℓ and L^m_ℓ by eq.(10.110). In this way the amplitudes A_{10}, A_{20}, A_{11} and A_{21} can be calculated for truncated prolate spheroids. In a similar way A_{22} can be evaluated by solving eq.(10.90), with $a\xi_0$ instead of R, numerically. One then has to solve eq.(10.97), with $a\xi_0$ instead of R, where K^m_ℓ and L^m_ℓ are given by eq.(10.111).

Writing the external potential, eq.(10.109), as

$$\psi_0(\mathbf{r}) = \sum_{m=-2}^{2} B^{(0)}_{2m} \widetilde{X}^m_2(\xi, a) Y^m_2(\arccos\eta, \phi) \tag{10.112}$$

one again obtains the expressions, eq.(10.99), for the amplitudes $B^{(0)}_{2m}$. Therefore eq. (10.100) is also valid in the present case of truncated prolate spheroids, and eq.(10.101) may be used to calculate the spheroidal multipole polarizabilities $\alpha_{20,20}$, $\alpha_{21,21}$ and $\alpha_{22,22}$ of these particles. With eq.(10.102) one then obtains the quadrupole polarizabilities $\alpha^{11}_z(0)$, $\alpha^{11}_\parallel(0)$ and $\alpha^{11}(0)$ for a truncated prolate spheroid on a substrate.

10.8 Appendix B: Caps

In this appendix it will be discussed, how the method for the calculation of the polarizabilities of a spherical or spheroidal cap on a substrate, developed in chapters 8 and 9 (sections 8.3, 9.4 and 9.5), must be changed, in order to be able to calculate the multipole polarizabilities $\alpha_{20,20}$, $\alpha_{21,21}$ and $\alpha_{22,22}$ for these particles. First the case of a spherical cap will be considered (see section 8.3 and fig. 8.4).

It is convenient to choose the potential of the incident external field such that it is given by eq.(10.85), now in the substrate rather than in the ambient, as was done in the previous section. It then follows from the boundary conditions at the surface of the substrate that the external potential in the ambient is given by

$$\begin{aligned}
\psi_0(\mathbf{r}) \;=\;& \left(1-\frac{\epsilon_2}{\epsilon_1}\right)(\phi_{xx}+\phi_{yy})d^2 + r\sqrt{\frac{2\pi}{3}}\left(1-\frac{\epsilon_2}{\epsilon_1}\right)d \\
&\times\left[(\phi_{xx}+\phi_{yy})Y^0_1(\theta,\phi)\sqrt{2}+(\phi_{xz}+i\phi_{yz})Y^{-1}_1(\theta,\phi)-(\phi_{xz}-i\phi_{yz})Y^1_1(\theta,\phi)\right] \\
&+r^2\sqrt{\frac{2\pi}{15}}[\frac{1}{2}(\phi_{xx}+\phi_{yy})Y^0_2(\theta,\phi)\sqrt{6}-\frac{\epsilon_2}{\epsilon_1}(\phi_{xz}+i\phi_{yz})Y^{-1}_2(\theta,\phi) \\
&+\frac{\epsilon_2}{\epsilon_1}(\phi_{xz}-i\phi_{yz})Y^1_2(\theta,\phi)-(\frac{1}{2}\phi_{xx}-\frac{1}{2}\phi_{yy}+i\phi_{xy})Y^{-2}_2(\theta,\phi) \\
&-(\frac{1}{2}\phi_{xx}-\frac{1}{2}\phi_{yy}-i\phi_{xy})Y^2_2(\theta,\phi)]
\end{aligned} \tag{10.113}$$

See for the meaning of the various symbols the previous appendix.

The potential in the ambient, region 1, then becomes

$$\begin{aligned}\psi_1(\mathbf{r}) &= \left(1-\frac{\epsilon_2}{\epsilon_1}\right)(\phi_{xx}+\phi_{yy})d^2 + r\sqrt{\frac{2\pi}{3}}\left(1-\frac{\epsilon_2}{\epsilon_1}\right)d \\ &\times\left[(\phi_{xx}+\phi_{yy})Y_1^0(\theta,\phi)\sqrt{2}+(\phi_{xz}+i\phi_{yz})Y_1^{-1}(\theta,\phi)-(\phi_{xz}-i\phi_{yz})Y_1^1(\theta,\phi)\right] \\ &+r^2\sqrt{\frac{2\pi}{15}}[\frac{1}{2}(\phi_{xx}+\phi_{yy})Y_2^0(\theta,\phi)\sqrt{6}-\frac{\epsilon_2}{\epsilon_1}(\phi_{xz}+i\phi_{yz})Y_2^{-1}(\theta,\phi) \\ &+\frac{\epsilon_2}{\epsilon_1}(\phi_{xz}-i\phi_{yz})Y_2^1(\theta,\phi)-(\frac{1}{2}\phi_{xx}-\frac{1}{2}\phi_{yy}+i\phi_{xy})Y_2^{-2}(\theta,\phi) \\ &-(\frac{1}{2}\phi_{xx}-\frac{1}{2}\phi_{yy}-i\phi_{xy})Y_2^2(\theta,\phi)]+\sum_{\ell m}{}' A^t_{\ell m}r^{-\ell-1}Y_\ell^m(\theta,\phi) \qquad (10.114)\end{aligned}$$

cf. eq.(8.33) and the text below that equation. The potential in the substrate, region 2, is now given by

$$\begin{aligned}\psi_2(\mathbf{r}) &= r^2\sqrt{\frac{2\pi}{15}}[\frac{1}{2}(\phi_{xx}+\phi_{yy})Y_2^0(\theta,\phi)\sqrt{6}-(\phi_{xz}+i\phi_{yz})Y_2^{-1}(\theta,\phi) \\ &+(\phi_{xz}-i\phi_{yz})Y_2^1(\theta,\phi)-(\frac{1}{2}\phi_{xx}-\frac{1}{2}\phi_{yy}+i\phi_{xy})Y_2^{-2}(\theta,\phi) \\ &-(\frac{1}{2}\phi_{xx}-\frac{1}{2}\phi_{yy}-i\phi_{xy})Y_2^2(\theta,\phi)] \\ &+\sum_{\ell m}{}' A_{\ell m}r^{-\ell-1}Y_\ell^m(\theta,\phi)+\sum_{\ell m}{}' A^r_{\ell m}\rho^{-\ell-1}Y_\ell^m(\theta,\phi) \qquad (10.115)\end{aligned}$$

The potentials in the regions 3, island, and 4, substrate, are again given by eqs.(8.35) and (8.36). It follows from the boundary conditions at the surface of the substrate that eqs.(8.37) and (8.38) hold again for the multipole coefficients $A_{\ell m}$ an $B_{\ell m}$.

Applying the boundary conditions, eqs.(8.39) and (8.40), at the surface of the sphere $r = R$, one obtains the infinite set of inhomogeneous linear equations (8.42), where the matrix elements $C^m_{\ell\ell_1}$, $D^m_{\ell\ell_1}$, $F^m_{\ell\ell_1}$ and $G^m_{\ell\ell_1}$ are again given by eq.(8.43), together with eqs.(8.14)– (8.20). The terms at the right-hand sides of eq.(8.42) are given by expressions different from those found in section 8.3, however, and will be denoted here by K^m_ℓ and L^m_ℓ. They are now found to be unequal to zero, if $m = 0, \pm 1$ or ± 2. One therefore has to consider eq.(8.42) here not only for the values $m = 0, \pm 1$, but also for $m = \pm 2$. Furthermore one can prove again, that this equation has to be used only for $\ell = 1, 2, 3, \ldots..$(see the discussion in section 8.3 below eq.(8.44)).

For $\ell = 1, 2, 3, \ldots.$ and $m = 0, \pm 1$ one obtains

$$\begin{aligned}\sum_{\ell_1=1}^{\infty} C^m_{\ell\ell_1}R^{-\ell_1-2}A_{\ell_1 m}+\sum_{\ell_1=1}^{\infty} D^m_{\ell\ell_1}R^{\ell_1-1}B_{\ell_1 m} &= K^m_\ell \\ \sum_{\ell_1=1}^{\infty} F^m_{\ell\ell_1}R^{-\ell_1-2}A_{\ell_1 m}+\sum_{\ell_1=1}^{\infty} G^m_{\ell\ell_1}R^{\ell_1-1}B_{\ell_1 m} &= L^m_\ell \qquad (10.116)\end{aligned}$$

with

$$
\begin{aligned}
K_\ell^0 &\equiv -\sqrt{\pi/5}(\phi_{xx}+\phi_{yy})R\{\delta_{\ell 2} \\
&+2\sqrt{5/3}\left(\frac{\epsilon_2}{\epsilon_1}-1\right)t_r(-1)^\ell[\delta_{\ell 1}+\sqrt{3}t_r\zeta^0_{\ell 0}Q^0_{\ell 0}(t_r)-\zeta^0_{\ell 1}Q^0_{\ell 1}(t_r)]\} \\
K_\ell^{\pm 1} &\equiv \mp\sqrt{2\pi/15}(\phi_{xz}\mp i\phi_{yz})R\{\frac{\epsilon_2}{\epsilon_1}\delta_{\ell 2} \\
&-\left(\frac{\epsilon_2}{\epsilon_1}-1\right)(-1)^\ell\left(\zeta^{\pm 1}_{\ell 2}Q^{\pm 1}_{\ell 2}(t_r)+\sqrt{5}t_r[\delta_{\ell 1}-\zeta^{\pm 1}_{\ell 1}Q^{\pm 1}_{\ell 1}(t_r)]\right)\}
\end{aligned}
$$

$$
\begin{aligned}
L_\ell^0 &\equiv -2\sqrt{\pi/5}(\phi_{xx}+\phi_{yy})R\{\epsilon_1\delta_{\ell 2} \\
&-(\epsilon_1-\epsilon_2)(-1)^\ell\left(\zeta^0_{\ell 2}Q^0_{\ell 2}(t_r)+\sqrt{5/3}t_r[\delta_{\ell 1}-\zeta^0_{\ell 1}Q^0_{\ell 1}(t_r)]\right)\} \\
L_\ell^{\pm 1} &\equiv \mp\sqrt{2\pi/15}(\phi_{xz}\mp i\phi_{yz})R\{2\epsilon_2\delta_{\ell 2} \\
&+\sqrt{5}(\epsilon_1-\epsilon_2)t_r(-1)^\ell[\delta_{\ell 1}-\zeta^{\pm 1}_{\ell 1}Q^{\pm 1}_{\ell 1}(t_r)]\}
\end{aligned}
\tag{10.117}
$$

For $\ell = 2, 3, 4, ...$ and $m = \pm 2$ one obtains

$$
\begin{aligned}
\sum_{\ell_1=2}^{\infty} C^m_{\ell\ell_1}R^{-\ell_1-2}A_{\ell_1 m}+\sum_{\ell_1=2}^{\infty} D^m_{\ell\ell_1}R^{\ell_1-1}B_{\ell_1 m} &= K^m_\ell \\
\sum_{\ell_1=2}^{\infty} F^m_{\ell\ell_1}R^{-\ell_1-2}A_{\ell_1 m}+\sum_{\ell_1=2}^{\infty} G^m_{\ell\ell_1}R^{\ell_1-1}B_{\ell_1 m} &= L^m_\ell
\end{aligned}
\tag{10.118}
$$

with

$$
\begin{aligned}
K_\ell^{\pm 2} &\equiv \sqrt{2\pi/15}R\left(\frac{1}{2}\phi_{xx}-\frac{1}{2}\phi_{yy}\mp i\,\phi_{xy}\right)\delta_{\ell 2} \\
L_\ell^{\pm 2} &\equiv 2\sqrt{2\pi/15}R\left(\frac{1}{2}\phi_{xx}-\frac{1}{2}\phi_{yy}\mp i\,\phi_{xy}\right)\{\epsilon_1\delta_{\ell 2} \\
&-(\epsilon_1-\epsilon_2)(-1)^\ell\zeta^{\pm 2}_{\ell 2}Q^{\pm 2}_{\ell 2}(t_r)\}
\end{aligned}
\tag{10.119}
$$

One proves, in a completely analogous way as in the previous appendix, that eq.(10.116) for $m = -1$ and eq.(10.118) for $m = -2$ are redundant. The proof will not be given here.

Equation (10.116) may now be solved numerically by neglecting the amplitudes $A_{\ell m}$ and $B_{\ell m}$ larger than a suitably chosen order. One then solves the finite set of inhomogeneous equations

$$
\begin{aligned}
\sum_{\ell_1=1}^{M} C^m_{\ell\ell_1}R^{-\ell_1-2}A_{\ell_1 m}+\sum_{\ell_1=1}^{M} D^m_{\ell\ell_1}R^{\ell_1-1}B_{\ell_1 m} &= K^m_\ell \\
\sum_{\ell_1=1}^{M} F^m_{\ell\ell_1}R^{-\ell_1-2}A_{\ell_1 m}+\sum_{\ell_1=1}^{M} G^m_{\ell\ell_1}R^{\ell_1-1}B_{\ell_1 m} &= L^m_\ell
\end{aligned}
$$

$$\text{for } \ell = 1, 2, 3, ..., M \text{ and } m = 0, 1 \tag{10.120}$$

where the matrix elements $C^m_{\ell\ell_1}$, $D^m_{\ell\ell_1}$, $F^m_{\ell\ell_1}$ and $G^m_{\ell\ell_1}$ are given by eq.(8.43) and K^0_ℓ , K^1_ℓ , L^0_ℓ and L^1_ℓ by eqs.(10.117) and (10.119). In this way the amplitudes A_{10} , A_{20} , A_{11} and A_{21} can be calculated with sufficient accuracy, by choosing M large enough. In a similar manner one may solve eq.(10.118) numerically by solving the finite set of equations

$$\begin{aligned}
\sum_{\ell_1=2}^{M} C^2_{\ell\ell_1} R^{-\ell_1-2} A_{\ell_1 2} + \sum_{\ell_1=2}^{M} D^2_{\ell\ell_1} R^{\ell_1-1} B_{\ell_1 2} &= K^2_\ell \\
\sum_{\ell_1=2}^{M} F^2_{\ell\ell_1} R^{-\ell_1-2} A_{\ell_1 2} + \sum_{\ell_1=2}^{M} G^2_{\ell\ell_1} R^{\ell_1-1} B_{\ell_1 2} &= L^2_\ell
\end{aligned}$$

$$\text{for } \ell = 2, 3, ..., M \tag{10.121}$$

where the matrix elements $C^2_{\ell\ell_1}$, $D^2_{\ell\ell_1}$, $F^2_{\ell\ell_1}$ and $G^2_{\ell\ell_1}$ are again given by eq.(8.43) for $m = 2$ and K^2_ℓ and L^2_ℓ by eq.(10.119). With these equations the amplitude A_{22} may be calculated.

The amplitudes A_{10} , A_{20} , A_{11} , A_{21} and A_{22}, found above, refer to multipole fields in the substrate. It follows from eqs.(10.120) and (10.121), together with eqs.(10.117) and (10.119), that A_{10} and A_{20} are proportional to $(\phi_{xx} + \phi_{yy})$, A_{11} and A_{21} to $(\phi_{xz} - i\phi_{yz})$ and A_{22} to $(\frac{1}{2}\phi_{xx} - \frac{1}{2}\phi_{yy} - i\phi_{xy})$, which quantities appear as factors of the spherical harmonics in the expression for the external field potential in the substrate. If one writes this potential as

$$\psi_{0,s}(\mathbf{r}) = r^2 \sum_{m=-2}^{2} B^{(0)}_{2m} Y^m_2(\theta, \phi) \tag{10.122}$$

it is found that

$$\begin{aligned}
B^{(0)}_{20} &= (\phi_{xx} + \phi_{yy})\sqrt{\pi/5}, \quad B^{(0)}_{2,\pm1} = \pm(\phi_{xz} \mp i\phi_{yz})\sqrt{2\pi/15} \\
B^{(0)}_{2,\pm2} &= -\left(\frac{1}{2}\phi_{xx} - \frac{1}{2}\phi_{yy} \mp i\ \phi_{xy}\right)\sqrt{2\pi/15}
\end{aligned} \tag{10.123}$$

Therefore A_{10} and A_{20} are proportional to $B^{(0)}_{20}$, A_{11} and A_{21} to $B^{(0)}_{21}$ and A_{22} to $B^{(0)}_{22}$. The proportionality constants are, according to eq.(5.30), the multipole polarizabilities $\alpha_{\ell m,\ell' m'}$:

$$\begin{aligned}
A_{10} &= -\alpha_{10,20} B^{(0)}_{20}, \quad A_{20} = -\alpha_{20,20} B^{(0)}_{20} \\
A_{11} &= -\alpha_{11,21} B^{(0)}_{21}, \quad A_{21} = -\alpha_{21,21} B^{(0)}_{21} \\
A_{22} &= -\alpha_{22,22} B^{(0)}_{22}
\end{aligned} \tag{10.124}$$

(cf. eq.(10.100)). The multipole polarizabilities $\alpha_{10,20}$ and $\alpha_{11,21}$ can also be found from eqs.(8.51) and (10.15), using the method of section 8.3. New are the multipole polarizabilities $\alpha_{20,20}$, $\alpha_{21,21}$ and $\alpha_{22,22}$. These quantities are therefore found by dividing the amplitudes A_{20} , A_{21} and A_{22} of multipole fields in the substrate respectively by the amplitudes $B_{20}^{(0)}$, $B_{21}^{(0)}$ and $B_{22}^{(0)}$ of the potential of the external field in the substrate, so that

$$\begin{aligned}\alpha_{20,20} &= -A_{20}/\left[(\phi_{xx}+\phi_{yy})\sqrt{\pi/5}\right], \quad \alpha_{21,21} = -A_{21}/\left[(\phi_{xz}-i\,\phi_{yz})\sqrt{2\pi/15}\right] \\ \alpha_{22,22} &= A_{22}/\left[\left(\frac{1}{2}\phi_{xx}-\frac{1}{2}\phi_{yy}-i\,\phi_{xy}\right)\sqrt{2\pi/15}\right] \qquad (10.125)\end{aligned}$$

(cf. eq.(10.101)). With eq.(10.15) one can then also calculate the quadrupole polarizabilities $\alpha_z^{11}(0)$, $\alpha_{\parallel}^{11}(0)$ and $\alpha^{11}(0)$ of a spherical cap on a substrate. One finds (cf. eq.(10.102))

$$\begin{aligned}\alpha_z^{11}(0) &= -\frac{\pi\,\epsilon_2 A_{20}}{(\phi_{xx}+\phi_{yy})\sqrt{\pi/5}} - \frac{6\pi\,\epsilon_2 A_{22}}{(\phi_{xx}-\phi_{yy}-2i\phi_{xy})\sqrt{6\pi/5}} \\ \alpha_{\parallel}^{11}(0) &= -\frac{4\pi\,\epsilon_2 A_{21}}{(\phi_{xz}-i\phi_{yz})\sqrt{6\pi/5}} - \frac{8\pi\,\epsilon_2 A_{22}}{(\phi_{xx}-\phi_{yy}-2i\phi_{xy})\sqrt{6\pi/5}} \\ \alpha^{11}(0) &= \frac{4\pi\,\epsilon_2 A_{22}}{(\phi_{xx}-\phi_{yy}-2i\phi_{xy})\sqrt{6\pi/5}} \qquad (10.126)\end{aligned}$$

Next the case of a spheroidal cap on a substrate will be considered. In a completely analogous way as has been done above for spherical caps, one may now derive the various equations for spheroidal caps starting from the expressions of the potentials in the regions 1- 4, see figs.9.5 and 9.6, in terms of spheroidal coordinates. This derivation will not be given here. In the following only the results will be given.

For an oblate spheroidal cap one obtains again eq.(10.120), with R replaced by half the short axis $R_s = a\xi_0$, where the matrix elements $C_{\ell\ell_1}^m$, $D_{\ell\ell_1}^m$, $F_{\ell\ell_1}^m$ and $G_{\ell\ell_1}^m$ are now given by eq.(9.65), and K_ℓ^0 , K_ℓ^1 , L_ℓ^0 and L_ℓ^1 by

$$\begin{aligned}K_\ell^0 &\equiv -\sqrt{\frac{\pi}{5}}(\phi_{xx}+\phi_{yy})a\xi_0\{\xi_0^{-2}X_2^0(\xi_0)\delta_{\ell 2} + 2\sqrt{\frac{5}{3}}\left(\frac{\epsilon_2}{\epsilon_1}-1\right)t_r(-1)^\ell \\ &\quad \times\left(\sqrt{3}t_r\zeta_{\ell 0}^0 Q_{\ell 0}^0(t_r) + \xi_0^{-1}X_1^0(\xi_0)[\delta_{\ell 1} - \zeta_{\ell 1}^0 Q_{\ell 1}^0(t_r)]\right)\} \\ K_\ell^1 &\equiv -\sqrt{\frac{2\pi}{15}}(\phi_{xz}-i\phi_{yz})a\xi_0 \\ &\quad \{\frac{\epsilon_2}{\epsilon_1}\xi_0^{-2}X_2^1(\xi_0)\delta_{\ell 2} - \left(\frac{\epsilon_2}{\epsilon_1}-1\right)(-1)^\ell(\xi_0^{-2}X_2^1(\xi_0)\zeta_{\ell 2}^1 Q_{\ell 2}^1(t_r) \\ &\quad +\sqrt{5}t_r\xi_0^{-1}X_1^1(\xi_0)[\delta_{\ell 1} - \zeta_{\ell 1}^1 Q_{\ell 1}^1(t_r)])\}\end{aligned}$$

$$
\begin{aligned}
L_\ell^0 &\equiv -\sqrt{\frac{\pi}{5}}(\phi_{xx}+\phi_{yy})a\xi_0\{\xi_0^{-1}\frac{dX_2^0(\xi_0)}{d\xi_0}[\epsilon_1\delta_{\ell 2}-(\epsilon_1-\epsilon_2)(-1)^\ell\zeta_{\ell 2}^0 Q_{\ell 2}^0(t_r)] \\
&\quad -2\sqrt{\frac{5}{3}}(\epsilon_1-\epsilon_2)t_r(-1)^\ell\frac{dX_1^0(\xi_0)}{d\xi_0}[\delta_{\ell 1}-\zeta_{\ell 1}^0 Q_{\ell 1}^0(t_r)]\} \\
L_\ell^1 &\equiv -\sqrt{\frac{2\pi}{15}}(\phi_{xz}-i\phi_{yz})a\xi_0\{\epsilon_2\xi_0^{-1}\frac{dX_2^1(\xi_0)}{d\xi_0}\delta_{\ell 2} \\
&\quad +\sqrt{5}(\epsilon_1-\epsilon_2)t_r(-1)^\ell\frac{dX_1^1(\xi_0)}{d\xi_0}[\delta_{\ell 1}-\zeta_{\ell 1}^1 Q_{\ell 1}^1(t_r)]\}
\end{aligned}
\qquad (10.127)
$$

Here the functions $X_\ell^m(\xi_0)$ are defined by eq.(9.20), the matrices $\zeta_{\ell\ell_1}^m$ and $Q_{\ell\ell_1}^m(t_r)$ by eqs.(8.14) and (8.15), and the truncation parameter t_r by eq.(9.50).

With eq.(10.120), with R replaced by $a\xi_0$, and eqs.(10.127) and (9.65) the amplitudes A_{10} , A_{20}, A_{11} and A_{21} can be calculated numerically for an oblate spheroidal cap on a substrate. In a similar way A_{22} can be found by solving eq.(10.121), with R replaced by $a\xi_0$, where the matrix elements $C_{\ell\ell_1}^2$, $D_{\ell\ell_1}^2$, $F_{\ell\ell_1}^2$ and $G_{\ell\ell_1}^2$ are given by eq.(9.65) for $m=2$, and K_ℓ^2 and L_ℓ^2 by

$$
\begin{aligned}
K_\ell^2 &\equiv \sqrt{\frac{2\pi}{15}}a\xi_0\left(\frac{1}{2}\phi_{xx}-\frac{1}{2}\phi_{yy}-i\ \phi_{xy}\right)\xi_0^{-2}X_2^2(\xi_0)\delta_{\ell 2} \\
L_\ell^2 &\equiv \sqrt{\frac{2\pi}{15}}a\xi_0\left(\frac{1}{2}\phi_{xx}-\frac{1}{2}\phi_{yy}-i\ \phi_{xy}\right)\xi_0^{-1}\frac{dX_2^2(\xi_0)}{d\xi_0} \\
&\quad \times[\epsilon_1\delta_{\ell 2}-(\epsilon_1-\epsilon_2)(-1)^\ell\zeta_{\ell 2}^2 Q_{\ell 2}^2(t_r)]
\end{aligned}
\qquad (10.128)
$$

The amplitudes found in this way are amplitudes of spheroidal multipole fields in the substrate, originating in the center of the cap. The potential of the external field in the substrate is

$$
\psi_{0,s}(\mathbf{r}) = \sum_{m=-2}^{2} B_{2m}^{(0)}X_2^m(\xi,a)Y_2^m(\arccos\eta,\phi) \qquad (10.129)
$$

where $B_{2m}^{(0)}(m=0,\pm1,\pm2)$ are again given by eq.(10.123). The oblate spheroidal multipole polarizabilities $\alpha_{20,20}$, $\alpha_{21,21}$ and $\alpha_{22,22}$ of the cap on the substrate may therefore be obtained using the same formula, eq.(10.125), as has been used above for a spherical cap. The quadrupole polarizabilities $\alpha_z^{11}(0)$, $\alpha_\parallel^{11}(0)$ and $\alpha^{11}(0)$ of an oblate spheroidal cap are again given by eq.(10.126).

Notice that the above equations for oblate spheroidal caps become equal to the corresponding equations for spherical caps in the limit as $\xi_0\to\infty$ and $a\xi_0=R$. This follows with eqs.(9.130) and (9.132).

For an prolate spheroidal cap one obtains again eq.(10.120), with R replaced by half the long axis $R_l=a\xi_0$, where the matrix elements $C_{\ell\ell_1}^m$, $D_{\ell\ell_1}^m$, $F_{\ell\ell_1}^m$ and $G_{\ell\ell_1}^m$

are now given by eq.(9.79), and K_ℓ^0 , K_ℓ^1 , L_ℓ^0 and L_ℓ^1 by

$$
\begin{aligned}
K_\ell^0 &\equiv -\sqrt{\frac{\pi}{5}}(\phi_{xx}+\phi_{yy})a\xi_0\{\xi_0^{-2}\widetilde{X}_2^0(\xi_0)\delta_{\ell 2}+2\sqrt{\frac{5}{3}}\left(\frac{\epsilon_2}{\epsilon_1}-1\right)t_r(-1)^\ell \\
&\quad \times\left(\sqrt{3}t_r\zeta_{\ell 0}^0 Q_{\ell 0}^0(t_r)+\xi_0^{-1}\widetilde{X}_1^0(\xi_0)[\delta_{\ell 1}-\zeta_{\ell 1}^0 Q_{\ell 1}^0(t_r)]\right)\} \\
K_\ell^1 &\equiv -\sqrt{\frac{2\pi}{15}}(\phi_{xz}-i\phi_{yz})a\xi_0 \\
&\quad \{\frac{\epsilon_2}{\epsilon_1}\xi_0^{-2}\widetilde{X}_2^1(\xi_0)\delta_{\ell 2}-\left(\frac{\epsilon_2}{\epsilon_1}-1\right)(-1)^\ell(\xi_0^{-2}\widetilde{X}_2^1(\xi_0)\zeta_{\ell 2}^1 Q_{\ell 2}^1(t_r) \\
&\quad +\sqrt{5}t_r\xi_0^{-1}\widetilde{X}_1^1(\xi_0)[\delta_{\ell 1}-\zeta_{\ell 1}^1 Q_{\ell 1}^1(t_r)])\} \\
L_\ell^0 &\equiv -\sqrt{\frac{\pi}{5}}(\phi_{xx}+\phi_{yy})a\xi_0\{\xi_0^{-1}\frac{d\widetilde{X}_2^0(\xi_0)}{d\xi_0}[\epsilon_1\delta_{\ell 2}-(\epsilon_1-\epsilon_2)(-1)^\ell\zeta_{\ell 2}^0 Q_{\ell 2}^0(t_r)] \\
&\quad -2\sqrt{\frac{5}{3}}(\epsilon_1-\epsilon_2)t_r(-1)^\ell\frac{d\widetilde{X}_1^0(\xi_0)}{d\xi_0}[\delta_{\ell 1}-\zeta_{\ell 1}^0 Q_{\ell 1}^0(t_r)]\} \\
L_\ell^1 &\equiv -\sqrt{\frac{2\pi}{15}}(\phi_{xz}-i\phi_{yz})a\xi_0\{\epsilon_2\xi_0^{-1}\frac{d\widetilde{X}_2^1(\xi_0)}{d\xi_0}\delta_{\ell 2} \\
&\quad +\sqrt{5}(\epsilon_1-\epsilon_2)t_r(-1)^\ell\frac{d\widetilde{X}_1^1(\xi_0)}{d\xi_0}[\delta_{\ell 1}-\zeta_{\ell 1}^1 Q_{\ell 1}^1(t_r)]\} \qquad (10.130)
\end{aligned}
$$

It will be clear from the derivation of this case in the previous appendix, that the results for a prolate spheroidal cap may be obtained from the above equations for an oblate spheroidal cap, simply by replacing $X_\ell^m(\xi_0)$ by $\widetilde{X}_\ell^m(\xi_0)$, which functions are defined by eq.(9.39). The matrices $\zeta_{\ell\ell_1}^m$ and $Q_{\ell\ell_1}^m(t_r)$ are defined by eqs.(8.14) and (8.15), and the truncation parameter t_r by eq.(9.75).

With eq.(10.120), with R replaced by $a\xi_0$, and eqs.(10.130) and (9.79) the amplitudes A_{10} , A_{20}, A_{11} and A_{21} can be calculated numerically for a prolate spheroidal cap on a substrate. In a similar way A_{22} can be found by solving eq.(10.121), with R replaced by $a\xi_0$, where the matrix elements $C_{\ell\ell_1}^2$, $D_{\ell\ell_1}^2$, $F_{\ell\ell_1}^2$ and $G_{\ell\ell_1}^2$ are given by eq.(9.79) for $m=2$, and K_ℓ^2 and L_ℓ^2 by

$$
\begin{aligned}
K_\ell^2 &\equiv \sqrt{\frac{2\pi}{15}}a\xi_0\left(\frac{1}{2}\phi_{xx}-\frac{1}{2}\phi_{yy}-i\ \phi_{xy}\right)\xi_0^{-2}\widetilde{X}_2^2(\xi_0)\delta_{\ell 2} \\
L_\ell^2 &\equiv \sqrt{\frac{2\pi}{15}}a\xi_0\left(\frac{1}{2}\phi_{xx}-\frac{1}{2}\phi_{yy}-i\ \phi_{xy}\right)\xi_0^{-1}\frac{d\widetilde{X}_2^2(\xi_0)}{d\xi_0} \\
&\quad \times[\epsilon_1\delta_{\ell 2}-(\epsilon_1-\epsilon_2)(-1)^\ell\zeta_{\ell 2}^2 Q_{\ell 2}^2(t_r)] \qquad (10.131)
\end{aligned}
$$

The prolate spheroidal multipole polarizabilities $\alpha_{20,20}$, $\alpha_{21,21}$ and $\alpha_{22,22}$ of the cap on the substrate may therefore be obtained using the same formula, eq.(10.125), as has been used above for a spherical cap. The quadrupole polarizabilities $\alpha_z^{11}(0)$, $\alpha_\parallel^{11}(0)$ and $\alpha^{11}(0)$ of an prolate spheroidal cap are again given by eq.(10.126).

Notice that the above equations for prolate spheroidal caps become equal to the corresponding equations for spherical caps in the limit as $\xi_0\to\infty$ and $a\xi_0=R$. This follows with eqs.(9.135) and (9.137).

Chapter 11
STRATIFIED LAYERS

11.1 Introduction

A planar stratified layer between two bulk media is a region, in which the dielectric constant and possibly also the magnetic permeability depend only on the coordinate normal to the layer, which is in this book always the z-coordinate. Most generally the system is characterized by the frequency dependent dielectric constants, $\epsilon_{\|}(z,\omega)$ and $\epsilon_z(z,\omega)$, and magnetic permeabilities, $\mu_{\|}(z,\omega)$ and $\mu_z(z,\omega)$, parallel and orthogonal to the layer, respectively. Away from the layer the dielectric constants and magnetic permeabilities approach their direction independent bulk values, ϵ^+ and μ^+ for $z \to \infty$ and ϵ^- and μ^- for $z \to -\infty$. In most of the literature on stratified media one considers non-magnetic media, $\mu_{\|}(z,\omega) = \mu_z(z,\omega) = 1$. Furthermore one usually takes a scalar dielectric tensor, i.e. $\epsilon_{\|}(z,\omega) = \epsilon_z(z,\omega) = \epsilon(z,\omega)$. An extensive discussion of this case is given for instance in the books by Born and Wolf[14] and by Lekner[11]. It is not the aim of this chapter to repeat this work in great detail. It is well understood and it does not need much further discussion. The main purpose of this chapter is to show that the description of surfaces, presented in this book, reproduces these results in a rather simple and direct manner. Given the relative ease of using the interfacial constitutive coefficients for this purpose an extension of the current formulae is given to systems which are magnetic and which have tensorial dielectric and magnetic permeabilities. The thickness of the layer is assumed to be small compared to the wavelength of the incident light. For the special case that the system is non-magnetic and the dielectric permeability is scalar the results are then compared to the results of the usual analysis. In particular the expressions containing invariants are compared to similar expressions given by Lekner. Explicit relations between the invariants he uses and those used in this book are given.

11.2 Constitutive coefficients

The first step to obtain the constitutive coefficients is to subdivide the stratified layer in thin sublayers. Such a sublayer between, for instance, z and $z + dz$ may be described to linear order in dz with the sublayer constitutive coefficients:

$$\begin{aligned}
\gamma_e(z) &= [\epsilon_{\|}(z) - \epsilon_0(z)]dz, \quad \beta_e(z) = -\{[\epsilon_z(z)]^{-1} - [\epsilon_0(z)]^{-1}\}dz \\
\gamma_m(z) &= [\mu_{\|}(z) - \mu_0(z)]dz, \quad \beta_m(z) = -\{[\mu_z(z)]^{-1} - [\mu_0(z)]^{-1}\}dz \\
\delta_e(z) &= 0, \quad \delta_m(z) = 0, \quad \tau(z) = 0
\end{aligned} \tag{11.1}$$

Here the extrapolated bulk dielectric and magnetic susceptibilities are defined by

$$\epsilon_0(z) \equiv \epsilon^- \theta(-z) + \epsilon^+ \theta(z) \quad \text{and} \quad \mu_0(z) \equiv \mu^- \theta(-z) + \mu^+ \theta(z) \tag{11.2}$$

θ is the Heaviside function, $\theta(s) \equiv 1$ if $s > 0$ and $\theta(s) \equiv 0$ if $s < 0$. It should be noticed that a function of a stepfunction is again a stepfunction. Thus one has

$$\begin{aligned} [\epsilon_0(z)]^{-1} &\equiv (\epsilon^-)^{-1}\theta(-z) + (\epsilon^+)^{-1}\theta(z) \\ [\mu_0(z)]^{-1} &\equiv (\mu^-)^{-1}\theta(-z) + (\mu^+)^{-1}\theta(z) \end{aligned} \tag{11.3}$$

for the extrapolated inverse bulk susceptibilities.

The second step is to shift all the layers towards $z = 0$. The layer which is first located between z and $z + dz$ then ends up in $z = 0$ and the appropriate constitutive coefficients of the shifted layer are given by eq.(3.52) and are

$$\begin{aligned} \gamma_e(0; z) &= \gamma_e(z), \quad \beta_e(0; z) = \beta_e(z) \\ \gamma_m(0; z) &= \gamma_m(z), \quad \beta_m(0; z) = \beta_m(z) \\ \delta_e(0; z) &= -z[\gamma_e(z)/\ \epsilon_0(z) + \beta_e(z)\epsilon_0(z)] \\ \delta_m(0; z) &= -z[\gamma_m(z)/\ \mu_0(z) + \beta_m(z)\mu_0(z)] \\ \tau(0; z) &= -z[\gamma_e(z)\mu_0(z) - \gamma_m(z)\epsilon_0(z)] \end{aligned} \tag{11.4}$$

In the argument the 0 gives the location of the shifted layer and the z indicates where this contribution originated. This last piece of information is needed to know the ordering of these contributions along the z-axis.

The last step is to add all the contributions together, using eq.(3.62). This results, using also eq.(11.4) with eq.(11.1), in

$$\begin{aligned} \gamma_e(0) &= \int_{-\infty}^{\infty} dz[\epsilon_\parallel(z) - \epsilon_0(z)], \quad \beta_e(0) = -\int_{-\infty}^{\infty} dz\{[\epsilon_z(z)]^{-1} - [\epsilon_0(z)]^{-1}\} \\ \gamma_m(0) &= \int_{-\infty}^{\infty} dz[\mu_\parallel(z) - \mu_0(z)], \quad \beta_m(0) = -\int_{-\infty}^{\infty} dz\{[\mu_z(z)]^{-1} - [\mu_0(z)]^{-1}\} \end{aligned}$$

$$\begin{aligned} \delta_e(0) &= -\int_{-\infty}^{\infty} dz\ z[\epsilon_\parallel(z) - \epsilon_0(z)]/\ \epsilon_0(z) + \int_{-\infty}^{\infty} dz\ z\{[\epsilon_z(z)]^{-1} - [\epsilon_0(z)]^{-1}\}\epsilon_0(z) \\ &+\frac{1}{2}\int_{-\infty}^{\infty} dz \int_{-\infty}^{\infty} dz' \operatorname{sign}(z' - z)[\epsilon_\parallel(z) - \epsilon_0(z)]\{[\epsilon_z(z')]^{-1} - [\epsilon_0(z')]^{-1}\} \\ \delta_m(0) &= -\int_{-\infty}^{\infty} dz\ z[\mu_\parallel(z) - \mu_0(z)]/\ \mu_0(z) + \int_{-\infty}^{\infty} dz\ z\{[\mu_z(z)]^{-1} - [\mu_0(z)]^{-1}\}\mu_0(z) \\ &+\frac{1}{2}\int_{-\infty}^{\infty} dz \int_{-\infty}^{\infty} dz' \operatorname{sign}(z' - z)[\mu_\parallel(z) - \mu_0(z)]\{[\mu_z(z')]^{-1} - [\mu_0(z')]^{-1}\} \\ \tau(0) &= -\int_{-\infty}^{\infty} dz\ z[\epsilon_\parallel(z) - \epsilon_0(z)]\mu_0(z) + \int_{-\infty}^{\infty} dz\ z[\mu_\parallel(z) - \mu_0(z)]\epsilon_0(z) \\ &+\frac{1}{2}\int_{-\infty}^{\infty} dz \int_{-\infty}^{\infty} dz' \operatorname{sign}(z' - z)[\epsilon_\parallel(z) - \epsilon_0(z)][\mu_\parallel(z') - \mu_0(z')] \end{aligned} \tag{11.5}$$

As is indicated by the 0 in the argument all these constitutive coefficients have been given with the $x - y$ plane as the dividing surface. It will be clear that the $x - y$ plane should be chosen in or very close to the stratified layer.

The constitutive coefficients given in eq.(11.5) fully describe the optical properties of the stratified layer under consideration. One simply substitutes them into the various reflection and transmission amplitudes, given in sections 4.2 and 4.3. In order to compare the resulting expressions for the measurable optical properties like reflectance, transmittance and the ellipsometric coefficients with those obtained, using the more conventional methods, one simply substitutes the above expressions into the appropriate formulae in the subsequent sections in chapter 4. For the special case of a non-magnetic system with $\epsilon_{\|}(z) = \epsilon_z(z)$ one then indeed recovers the usual results. As all the measurable optical properties have been given in terms of combinations of these constitutive coefficients, which are independent of the choice of the dividing surface, it is more convenient to first consider these invariant combinations. Such invariant combinations have also been introduced by Lekner [11] for this special case and now also follow, using the list of invariant combinations given in section 3.9.

11.3 Invariants

The invariants are obtained by substitution of eq.(11.5) into eqs.(3.53) – (3.55). This results in the following expressions for the linear invariants

$$\begin{aligned} I_e &= \gamma_e(0) - \epsilon^-\epsilon^+\beta_e(0) = \int_{-\infty}^{\infty} dz\left[\epsilon_{\|}(z) - \epsilon_0(z) + \epsilon^-\epsilon^+\{[\epsilon_z(z)]^{-1} - [\epsilon_0(z)]^{-1}\}\right] \\ &= \int_{-\infty}^{\infty} dz\left[\epsilon_{\|}(z) - \epsilon^- - \epsilon^+ + \epsilon^-\epsilon^+[\epsilon_z(z)]^{-1}\right] \end{aligned} \tag{11.6}$$

and similarly

$$\begin{aligned} I_m &= \gamma_m(0) - \mu^-\mu^+\beta_m(0) = \int_{-\infty}^{\infty} dz\left[\mu_{\|}(z) - \mu^- - \mu^+ + \mu^-\mu^+[\mu_z(z)]^{-1}\right] \\ I_{em} &= (\mu^+ - \mu^-)\gamma_e(0) - (\epsilon^+ - \epsilon^-)\gamma_m(0) \\ &= \int_{-\infty}^{\infty} dz[(\mu^+ - \mu^-)\epsilon_{\|} - (\epsilon^+ - \epsilon^-)\mu_{\|} - \mu^+\epsilon^- + \mu^-\epsilon^+] \\ I_c &= \mathrm{Im}\,\frac{(\mu^+ + \mu^-)\gamma_e(0) + (\epsilon^+ + \epsilon^-)\gamma_m(0)}{2\,(\mu^+\epsilon^+ - \mu^-\epsilon^-)} \\ &= \mathrm{Im}\int_{-\infty}^{\infty} dz\frac{(\mu^+ + \mu^-)\left(\epsilon_{\|}(z) - \epsilon_0(z)\right) + (\epsilon^+ + \epsilon^-)\left(\mu_{\|}(z) - \mu_0(z)\right)}{2\,(\mu^+\epsilon^+ - \mu^-\epsilon^-)} \\ &= \mathrm{Im}\int_{-\infty}^{\infty} dz\frac{(\mu^+ + \mu^-)\epsilon_{\|}(z) + (\epsilon^+ + \epsilon^-)\mu_{\|}(z) - (\mu^+ + \mu^-)(\epsilon^+ + \epsilon^-)}{2\,(\mu^+\epsilon^+ - \mu^-\epsilon^-)} \end{aligned} \tag{11.7}$$

The three complex and the one real linear invariants no longer contain the stepfunctions and are therefore manifestly invariant.

The three complex second order invariants are

$$
\begin{aligned}
I_{\delta,e} &= \delta_e(0) - \frac{1}{2}(\epsilon^- + \epsilon^+)(\epsilon^+ - \epsilon^-)^{-1}\gamma_e(0)\beta_e(0) \\
&= -\int_{-\infty}^{\infty} dz\ z\left[[\epsilon_\parallel(z) - \epsilon_0(z)]/\ \epsilon_0(z) - \{[\epsilon_z(z)]^{-1} - [\epsilon_0(z)]^{-1}\}\epsilon_0(z)\right] \\
&\quad +(\epsilon^+ - \epsilon^-)^{-1}\int_{-\infty}^{\infty} dz \int_{-\infty}^{\infty} dz'\epsilon_0(z'-z)[\epsilon_\parallel(z) - \epsilon_0(z)]\{[\epsilon_z(z')]^{-1} - [\epsilon_0(z')]^{-1}\}
\end{aligned}
$$

$$
\begin{aligned}
I_{\delta,m} &= \delta_m(0) - \frac{1}{2}(\mu^- + \mu^+)(\mu^+ - \mu^-)^{-1}\gamma_m(0)\beta_m(0) \\
&= -\int_{-\infty}^{\infty} dz\ z\left[[\mu_\parallel(z) - \mu_0(z)]/\ \mu_0(z) - \{[\mu_z(z)]^{-1} - [\mu_0(z)]^{-1}\}\mu_0(z)\right] + \\
&\quad (\mu^+ - \mu^-)^{-1}\int_{-\infty}^{\infty} dz \int_{-\infty}^{\infty} dz'\mu_0(z'-z)[\mu_\parallel(z) - \mu_0(z)]\{[\mu_z(z')]^{-1} - [\mu_0(z')]^{-1}\}
\end{aligned}
$$

$$
\begin{aligned}
I_\tau &= \tau(0) - \frac{1}{4}(\mu^- + \mu^+)(\epsilon^+ - \epsilon^-)^{-1}\gamma_e^2(0) + \frac{1}{4}(\epsilon^- + \epsilon^+)(\mu^+ - \mu^-)^{-1}\gamma_m^2(0) \\
&= -\int_{-\infty}^{\infty} dz\ z[\epsilon_\parallel(z) - \epsilon_0(z)]\mu_0(z) + \int_{-\infty}^{\infty} dz\ z[\mu_\parallel(z) - \mu_0(z)]\epsilon_0(z) \\
&\quad +\frac{1}{2}\int_{-\infty}^{\infty} dz \int_{-\infty}^{\infty} dz'\,\mathrm{sign}(z'-z)[\epsilon_\parallel(z) - \epsilon_0(z)][\mu_\parallel(z') - \mu_0(z')] \\
&\quad -\frac{1}{4}(\mu^- + \mu^+)(\epsilon^+ - \epsilon^-)^{-1}\{\int_{-\infty}^{\infty} dz[\epsilon_\parallel(z) - \epsilon_0(z)]\}^2 \\
&\quad +\frac{1}{4}(\epsilon^- + \epsilon^+)(\mu^+ - \mu^-)^{-1}\{\int_{-\infty}^{\infty} dz[\mu_\parallel(z) - \mu_0(z)]\}^2
\end{aligned}
\tag{11.8}
$$

These expressions are not manifestly invariant. Below expressions will be derived, which are manifestly invariant and which may conveniently be compared with analogous formulae given by Lekner[11]. For the actual calculation, given the profile, it is usually simpler to choose a convenient location of the dividing surface and to use the above formulae.

In order to derive the manifestly invariant form an identity will be used. Consider two functions $f(z)$ and $g(z)$, which have the limiting values f^- and g^- for $z \to -\infty$ and f^+ and g^+ for $z \to \infty$. Introduce the stepfunctions $f_0(z)$ and $g_0(z)$ in the usual way as well as an additional stepfunction $h_0(z)$. The required identity is given by

$$
\begin{aligned}
&\int_{-\infty}^{\infty} dz \int_{-\infty}^{\infty} dz' h_0(z'-z)[f(z) - f_0(z)][g(z') - g_0(z')] \\
&= \int_{-\infty}^{\infty} dz \int_{-\infty}^{\infty} dz' h_0(z'-z)[f(z) - f_0(z-z')][g(z') - g_0(z'-z)] \\
&\quad -(g^+ - g^-)\int_{-\infty}^{\infty} dz\ z\ h_0(-z)[f(z) - f_0(z)] \\
&\quad -(f^+ - f^-)\int_{-\infty}^{\infty} dz\ z\ h_0(z)[g(z) - g_0(z)]
\end{aligned}
\tag{11.9}
$$

This identity may now be used to rewrite the double integral contributions to the second order invariants. This results in

$$\begin{aligned} I_{\delta,e} &= (\epsilon^+ - \epsilon^-)^{-1} \int_{-\infty}^{\infty} dz \int_{-\infty}^{\infty} dz' \epsilon_0(z'-z)[\epsilon_{\|}(z) - \epsilon_0(z-z')] \\ &\quad \times \{[\epsilon_z(z')]^{-1} - [\epsilon_0(z'-z)]^{-1}\} \\ I_{\delta,m} &= (\mu^+ - \mu^-)^{-1} \int_{-\infty}^{\infty} dz \int_{-\infty}^{\infty} dz' \mu_0(z'-z)[\mu_{\|}(z) - \mu_0(z-z')] \\ &\quad \times \{[\mu_z(z')]^{-1} - [\mu_0(z'-z)]^{-1}\} \end{aligned}$$

$$\begin{aligned} I_\tau &= \frac{1}{4} \int_{-\infty}^{\infty} dz \int_{-\infty}^{\infty} dz' \{2\,\mathrm{sign}(z'-z)[\epsilon_{\|}(z) - \epsilon_0(z-z')][\mu_{\|}(z') - \mu_0(z'-z)] \\ &\quad -(\mu^- + \mu^+)(\epsilon^+ - \epsilon^-)^{-1}[\epsilon_{\|}(z) - \epsilon_0(z-z')][\epsilon_{\|}(z') - \epsilon_0(z'-z)] \\ &\quad +(\epsilon^- + \epsilon^+)(\mu^+ - \mu^-)^{-1}[\mu_{\|}(z) - \mu_0(z-z')][\mu_{\|}(z') - \mu_0(z'-z)]\} \end{aligned} \quad (11.10)$$

The above expressions for the second order invariants are now all manifestly invariant. This may be seen by shifting the origin of the z-axis over a distance d. This results in formulae in which the profiles $\epsilon_{\|}(z)$, $\epsilon_z(z)$, $\mu_{\|}(z)$ and $\mu_z(z)$ are replaced by $\epsilon_{\|}(z+d)$, $\epsilon_z(z+d)$, $\mu_{\|}(z+d)$ and $\mu_z(z+d)$, whereas all functions which depend on $z-z'$ remain the same. These are precisely the formulae one obtains by shifting the position of the profiles relative to dividing surface over a distance $-d$. The above expressions also have a rather shorter and more appealing form.

11.4 Non-magnetic stratified layers

If the system is non-magnetic one has $\mu_{\|}(z) = \mu_z(z) = 1$. Substituting this into the constitutive coefficients, eq.(11.5) gives

$$\begin{aligned} \gamma_e(0) &= \int_{-\infty}^{\infty} dz[\epsilon_{\|}(z) - \epsilon_0(z)] \\ \beta_e(0) &= -\int_{-\infty}^{\infty} dz\{[\epsilon_z(z)]^{-1} - [\epsilon_0(z)]^{-1}\} \\ \delta_e(0) &= -\int_{-\infty}^{\infty} dz\; z[\epsilon_{\|}(z) - \epsilon_0(z)]/\,\epsilon_0(z) + \int_{-\infty}^{\infty} dz\; z\{[\epsilon_z(z)]^{-1} - [\epsilon_0(z)]^{-1}\}\epsilon_0(z) \\ &\quad + \frac{1}{2}\int_{-\infty}^{\infty} dz \int_{-\infty}^{\infty} dz'\,\mathrm{sign}(z'-z)[\epsilon_{\|}(z) - \epsilon_0(z)]\{[\epsilon_z(z')]^{-1} - [\epsilon_0(z')]^{-1}\} \\ \tau(0) &= -\int_{-\infty}^{\infty} dz\; z[\epsilon_{\|}(z) - \epsilon_0(z)] \end{aligned} \quad (11.11)$$

The magnetic coefficients γ_m, β_m and δ_m are now all equal to zero.

Similarly one obtains for the invariants from eq.(11.6)–(11.8)

$$\begin{aligned}
I_e &= \int_{-\infty}^{\infty} dz\,\left[\epsilon_{\|}(z) - \epsilon^- - \epsilon^+ + \epsilon^-\epsilon^+[\epsilon_z(z)]^{-1}\right] \\
I_c &= \mathrm{Im}\int_{-\infty}^{\infty} dz\,\left[\epsilon_{\|}(z) - \epsilon_0(z)\right] / \left(\epsilon^+ - \epsilon^-\right) \\
&= \mathrm{Im}\int_{-\infty}^{\infty} dz\left[\epsilon_{\|}(z) - \frac{1}{2}\left(\epsilon^+ + \epsilon^-\right)\right] / \left(\epsilon^+ - \epsilon^-\right) \\
I_{\delta,e} &= -\int_{-\infty}^{\infty} dz\; z\left[[\epsilon_{\|}(z) - \epsilon_0(z)]/\,\epsilon_0(z) - \{[\epsilon_z(z)]^{-1} - [\epsilon_0(z)]^{-1}\}\epsilon_0(z)\right] \\
&\quad +(\epsilon^+ - \epsilon^-)^{-1}\int_{-\infty}^{\infty} dz\int_{-\infty}^{\infty} dz'\epsilon_0(z'-z)[\epsilon_{\|}(z) - \epsilon_0(z)]\{[\epsilon_z(z')]^{-1} - [\epsilon_0(z')]^{-1}\} \\
I_\tau &= -\int_{-\infty}^{\infty} dz\; z[\epsilon_{\|}(z) - \epsilon_0(z)] - \frac{1}{2}(\epsilon^+ - \epsilon^-)^{-1}\{\int_{-\infty}^{\infty} dz[\epsilon_{\|}(z) - \epsilon_0(z)]\}^2 \qquad (11.12)
\end{aligned}$$

In the manifestly invariant form, given in eq.(11.10), the second order invariants become

$$\begin{aligned}
I_{\delta,e} &= (\epsilon^+ - \epsilon^-)^{-1}\int_{-\infty}^{\infty} dz\int_{-\infty}^{\infty} dz'\epsilon_0(z'-z)[\epsilon_{\|}(z) - \epsilon_0(z-z')] \\
&\quad \times\{[\epsilon_z(z')]^{-1} - [\epsilon_0(z'-z)]^{-1}\} \\
I_\tau &= -\frac{1}{2}(\epsilon^+ - \epsilon^-)^{-1}\int_{-\infty}^{\infty} dz\int_{-\infty}^{\infty} dz'[\epsilon_{\|}(z) - \epsilon_0(z-z')] \\
&\quad \times[\epsilon_{\|}(z') - \epsilon_0(z'-z)] \qquad (11.13)
\end{aligned}$$

which is again a more elegant form.

If one further restricts the system by taking the dielectric tensor scalar, $\epsilon_{\|}(z) = \epsilon_z(z) = \epsilon(z)$, one finds

$$\begin{aligned}
I_e &= \int_{-\infty}^{\infty} dz[\epsilon(z) - \epsilon^-][\epsilon(z) - \epsilon^+]/\,\epsilon(z) \\
I_c &= \mathrm{Im}\int_{-\infty}^{\infty} dz\,[\epsilon(z) - \epsilon_0(z)] / \left(\epsilon^+ - \epsilon^-\right) \\
&= \mathrm{Im}\int_{-\infty}^{\infty} dz\left[\epsilon(z) - \frac{1}{2}\left(\epsilon^+ + \epsilon^-\right)\right] / \left(\epsilon^+ - \epsilon^-\right) \\
I_{\delta,e} &= (\epsilon^+ - \epsilon^-)^{-1}\int_{-\infty}^{\infty} dz\int_{-\infty}^{\infty} dz'\epsilon_0(z'-z)[\epsilon(z) - \epsilon_0(z-z')] \\
&\quad \times\{[\epsilon(z')]^{-1} - [\epsilon_0(z'-z)]^{-1}\} \\
I_\tau &= -\frac{1}{2}(\epsilon^+ - \epsilon^-)^{-1}\int_{-\infty}^{\infty} dz\int_{-\infty}^{\infty} dz'[\epsilon(z) - \epsilon_0(z-z')][\epsilon(z') - \epsilon_0(z'-z)] \\
&\qquad (11.14)
\end{aligned}$$

Lekner also finds the complex invariants. In particular one may identify I_e with Lekner's $-\Im_1 = \lambda_1 - \Lambda_1$,cf. eq.(37) on page 68 in [11] and I_τ with Lekner's i_2

divided by $2(\epsilon^+ - \epsilon^-)$, cf. eq.(33) on page 67 in [11]. Regarding the invariant $I_{\delta,e}$ there is a confusing typographical error in Lekner's definition of the invariant j_2, cf. eq.(39) on page 68 in [11]; a factor 2 is missing on the right hand side of his definition. One may verify this, for instance by comparing with eq.(54) on page 71 of his book. If the factor 2 is included all the subsequent formulae are correct. Upon inclusion of this factor of 2 the invariant $I_{\delta,e}$ may be identified with Lekner's j_2 divided by $2(\epsilon^+ - \epsilon^-)$. In Lekner's work invariants of arbitrary order in the thickness of the layer are constructed. For thin stratified layers these are not of much practical use. For adsorbing stratified layers he makes only a very limited use of the invariants. He mentions that those already constructed become complex, but he does not introduce I_c. He gives a convenient list of the values of the complex invariants for a number of different profiles. It should be noted that Lekner defines r_p with a different sign. As a consequence his expressions for the ellipsometric coefficient also have a different sign.

11.5 Conclusions

The subject of stratified layers will, in view of the extensive coverage in the literature, not be further discussed. The generalizations to, for instance, magnetic systems given in this section can easily be used along the same lines as the non-magnetic case. The usual results like, for instance, the Drude formula and the shift of the Brewster angle are found directly by substitution of I_e into eqs.(4.99) and (4.100), respectively. In a similar way one finds the formulae for the reflectances and transmittances. The extension to the magnetic case is directly found by substituting the invariants, including the magnetic contributions given above.

Chapter 12

THE WAVE EQUATION AND ITS GENERAL SOLUTION

12.1 Introduction

In the analysis in most of the previous chapters it was sufficient to solve the Laplace equation for the potential. Only in the chapter on reflection and transmission the wave equation was given for the bulk regions and its solution in terms of plane waves was used. In order to describe the properties of rough surfaces a more systematic analysis of the solution of the Maxwell equations including the effects on the interface is needed. In this chapter such an analysis will be given. As a first step it will be discussed how to obtain a wave equation valid not only in the bulk, but also at the surface. This extension to the surface poses a special problem, due to the singular nature of the normal components of the electric and magnetic fields at the surface. If one would try to construct the appropriate Green function to write the general solution in integral form, as one usually does, the Green function would have similar singularities. This would eliminate much of its usefulness. The way out of this dilemma is to use the solution of the wave equations for the normal components of the electric displacement field and the magnetic induction, which do not have such singular behavior at the interface, rather than for the normal components of the electric and magnetic fields. This results in a modification of one of the integral operators, which give the non-singular fields in terms of the polarization and magnetization densities, such that they no longer contain any singularities at the surface. These expressions are then used for further analysis.

12.2 The wave equations

In chapter 3 the Maxwell equations (3.6) were given and the consequences of the fundamental assumption made in this book, that these equations remain valid also if the fields contain singularities, and the consequences of this for the nature of the boundary conditions were discussed. If one absorbs currents in the electric displacement field, as discussed in section 3.5 the Maxwell equations become

$$\begin{aligned} \mathrm{rot}\,\mathbf{E}(\mathbf{r},\omega) &= i\frac{\omega}{c}\mathbf{B}(\mathbf{r},\omega), \qquad \mathrm{div}\,\mathbf{D}(\mathbf{r},\omega) = 0 \\ \mathrm{rot}\,\mathbf{H}(\mathbf{r},\omega) &= -i\frac{\omega}{c}\mathbf{D}(\mathbf{r},\omega), \quad \mathrm{div}\,\mathbf{B}(\mathbf{r},\omega) = 0 \end{aligned} \tag{12.1}$$

The wave equation for the electric field is now obtained, by taking the rotation of the first Maxwell equation and using the third, as well as the relations $\mathbf{B} = \mathbf{H} + \mathbf{M}$ and

$\mathbf{D} = \mathbf{E} + \mathbf{P}$. This then yields

$$\text{rot rot}\, \mathbf{E}(\mathbf{r}, \omega) - (\frac{\omega}{c})^2 \mathbf{E}(\mathbf{r}, \omega) = (\frac{\omega}{c})^2 \mathbf{P}(\mathbf{r}, \omega) + i\frac{\omega}{c}\, \text{rot}\, \mathbf{M}(\mathbf{r}, \omega) \tag{12.2}$$

The wave equation for the magnetic field is similarly obtained by taking the rotation of the third Maxwell equation and one obtains

$$\text{rot rot}\, \mathbf{H}(\mathbf{r}, \omega) - (\frac{\omega}{c})^2 \mathbf{H}(\mathbf{r}, \omega) = (\frac{\omega}{c})^2 \mathbf{M}(\mathbf{r}, \omega) - i\frac{\omega}{c}\, \text{rot}\, \mathbf{P}(\mathbf{r}, \omega) \tag{12.3}$$

In the problem under consideration the system is translationally invariant in the x- and the y-direction, but not in the z-direction. In order to construct solutions of these equations, it is therefore most appropriate to Fourier transform the above equation with respect to x and y, but not with respect to z. One then obtains, using rot rot $=$ grad div $-\Delta$,

$$\begin{aligned} &[k_x^2 + k_y^2 - \frac{\partial^2}{\partial z^2} - \begin{pmatrix} k_x k_x & k_x k_y & -ik_x\frac{\partial}{\partial z} \\ k_y k_x & k_y k_y & -ik_y\frac{\partial}{\partial z} \\ -ik_x\frac{\partial}{\partial z} & -ik_y\frac{\partial}{\partial z} & -\frac{\partial^2}{\partial z^2} \end{pmatrix} - (\frac{\omega}{c})^2].\mathbf{E}(k_x, k_y, z, \omega) \\ = \; &(\frac{\omega}{c})^2 \mathbf{P}(k_x, k_y, z, \omega) + i\frac{\omega}{c} \begin{pmatrix} 0 & -\frac{\partial}{\partial z} & ik_y \\ \frac{\partial}{\partial z} & 0 & -ik_x \\ -ik_y & ik_x & 0 \end{pmatrix} .\mathbf{M}(k_x, k_y, z, \omega) \end{aligned} \tag{12.4}$$

For ease of notation the unit tensor, multiplying a scalar, when adding this scalar to a tensor, will not usually be indicated explicitly. The analogous equation for the magnetic field becomes

$$\begin{aligned} &[k_x^2 + k_y^2 - \frac{\partial^2}{\partial z^2} - \begin{pmatrix} k_x k_x & k_x k_y & -ik_x\frac{\partial}{\partial z} \\ k_y k_x & k_y k_y & -ik_y\frac{\partial}{\partial z} \\ -ik_x\frac{\partial}{\partial z} & -ik_y\frac{\partial}{\partial z} & -\frac{\partial^2}{\partial z^2} \end{pmatrix} - (\frac{\omega}{c})^2].\mathbf{H}(k_x, k_y, z, \omega) \\ = \; &(\frac{\omega}{c})^2 \mathbf{M}(k_x, k_y, z, \omega) - i\frac{\omega}{c} \begin{pmatrix} 0 & -\frac{\partial}{\partial z} & ik_y \\ \frac{\partial}{\partial z} & 0 & -ik_x \\ -ik_y & ik_x & 0 \end{pmatrix} .\mathbf{P}(k_x, k_y, z, \omega) \end{aligned} \tag{12.5}$$

As is to be expected on the basis of the symmetry of the Maxwell equations the above two wave equations are very similar.

In order to construct the solutions needed, it is now convenient to write the polarization and the magnetization densities as sums of induced contributions and externally controlled contributions

$$\mathbf{P} = \mathbf{P}_{ind} + \mathbf{P}_{ext}, \quad \mathbf{M} = \mathbf{M}_{ind} + \mathbf{M}_{ext} \tag{12.6}$$

The induced, as well as the externally controlled contributions, may both have singular contributions on the $z = 0$ surface separating the $+$ and the $-$ phases. Before addressing the complications in the construction of the solution due to these singular terms, it is convenient to solve first the wave equations for the simple case of a homogeneous one phase medium, where such singular contributions do not occur. In that case one has

$$\mathbf{P}_{ind}(k_x,k_y,z,\omega) = [\epsilon(\omega)-1]\mathbf{E}(k_x,k_y,z,\omega) \tag{12.7}$$
$$\mathbf{M}_{ind}(k_x,k_y,z,\omega) = [\mu(\omega)-1][\mathbf{H}(k_x,k_y,z,\omega) \tag{12.8}$$

where $\epsilon(\omega)$ is the frequency dependent dielectric constant and $\mu(\omega)$ the frequency dependent magnetic permeability in the homogeneous bulk region. Substituting the above expressions for the polarization and magnetization densities into the wave equation (12.4), and using the third Maxwell equation (12.1) and the relation $\mathbf{D} = \mathbf{E}+\mathbf{P}$, one obtains for the electric field

$$[k_x^2+k_y^2-\frac{\partial^2}{\partial z^2}-\begin{pmatrix} k_xk_x & k_xk_y & -ik_x\frac{\partial}{\partial z} \\ k_yk_x & k_yk_y & -ik_y\frac{\partial}{\partial z} \\ -ik_x\frac{\partial}{\partial z} & -ik_y\frac{\partial}{\partial z} & -\frac{\partial^2}{\partial z^2}\end{pmatrix}-(\frac{\omega}{c})^2n^2(\omega)].\mathbf{E}(k_x,k_y,z,\omega)$$
$$= (\frac{\omega}{c})^2\mu(\omega)\mathbf{P}_{ext}(k_x,k_y,z,\omega)+i\frac{\omega}{c}\begin{pmatrix} 0 & -\frac{\partial}{\partial z} & ik_y \\ \frac{\partial}{\partial z} & 0 & -ik_x \\ -ik_y & ik_x & 0\end{pmatrix}.\mathbf{M}_{ext}(k_x,k_y,z,\omega) \tag{12.9}$$

where the complex refractive index is defined by $n^2(\omega) \equiv \epsilon(\omega)\mu(\omega)$. The analogous equation for the magnetic field becomes

$$[k_x^2+k_y^2-\frac{\partial^2}{\partial z^2}-\begin{pmatrix} k_xk_x & k_xk_y & -ik_x\frac{\partial}{\partial z} \\ k_yk_x & k_yk_y & -ik_y\frac{\partial}{\partial z} \\ -ik_x\frac{\partial}{\partial z} & -ik_y\frac{\partial}{\partial z} & -\frac{\partial^2}{\partial z^2}\end{pmatrix}-(\frac{\omega}{c})^2n^2(\omega)].\mathbf{H}(k_x,k_y,z,\omega)$$
$$= (\frac{\omega}{c})^2\epsilon(\omega)\mathbf{M}_{ext}(k_x,k_y,z,\omega)-i\frac{\omega}{c}\begin{pmatrix} 0 & -\frac{\partial}{\partial z} & ik_y \\ \frac{\partial}{\partial z} & 0 & -ik_x \\ -ik_y & ik_x & 0\end{pmatrix}.\mathbf{P}_{ext}(k_x,k_y,z,\omega) \tag{12.10}$$

In the next section the solutions will first be given for the wave equations in the above homogeneous case. Subsequently the solutions for a two phase medium, where the dielectric constant and the magnetic permeability in the + and in the − region are different, but still in the absence of singular contributions at the $z = 0$ surface, will be constructed, using the usual boundary conditions for this case. In the subsequent section the contributions due to the singularities at the $z = 0$ surface are then finally added, and the influence of the surface on the behavior of the fields is discussed.

12.3 The solution of the wave equations

In order to construct the general solution of the wave equation as an integral over the source density one first constructs the solution for an arbitrary externally controlled point source

$$\mathbf{P}_{ext}(\mathbf{r},\omega) = \mathbf{P}_{ext}(\omega)\delta(\mathbf{r}-\mathbf{r}_0), \quad \mathbf{M}_{ext}(\mathbf{r},\omega) = \mathbf{M}_{ext}(\omega)\delta(\mathbf{r}-\mathbf{r}_0) \tag{12.11}$$

In the (k_x, k_y, z, ω) representation this source becomes

$$
\begin{aligned}
\mathbf{P}_{ext}(k_x, k_y, z, \omega) &= \mathbf{P}_{ext}(\omega)\delta(z - z_0)\exp[-i(k_x x_0 + k_y y_0)] \\
\mathbf{M}_{ext}(k_x, k_y, z, \omega) &= \mathbf{M}_{ext}(\omega)\delta(z - z_0)\exp[-i(k_x x_0 + k_y y_0)] \qquad (12.12)
\end{aligned}
$$

In the case of a homogeneous medium one must therefore solve the following equation for the electric field

$$
\begin{aligned}
&[k_x^2 + k_y^2 - \frac{\partial^2}{\partial z^2} - \begin{pmatrix} k_x k_x & k_x k_y & -ik_x\frac{\partial}{\partial z} \\ k_y k_x & k_y k_y & -ik_y\frac{\partial}{\partial z} \\ -ik_x\frac{\partial}{\partial z} & -ik_y\frac{\partial}{\partial z} & -\frac{\partial^2}{\partial z^2} \end{pmatrix} - (\frac{\omega}{c})^2 n^2(\omega)].\mathbf{E}(k_x, k_y, z, \omega) \\
= \; &[(\frac{\omega}{c})^2\mu(\omega)\mathbf{P}_{ext}(\omega) + i\frac{\omega}{c}\begin{pmatrix} 0 & -\frac{\partial}{\partial z} & ik_y \\ \frac{\partial}{\partial z} & 0 & -ik_x \\ -ik_y & ik_x & 0 \end{pmatrix}.\mathbf{M}_{ext}(\omega)] \\
&\times\delta(z - z_0)\exp[-i(k_x x_0 + k_y y_0)] \qquad (12.13)
\end{aligned}
$$

which is obtained by substituting the above expression for the source into eq.(12.9). Similarly one finds from eq.(12.10)

$$
\begin{aligned}
&[k_x^2 + k_y^2 - \frac{\partial^2}{\partial z^2} - \begin{pmatrix} k_x k_x & k_x k_y & -ik_x\frac{\partial}{\partial z} \\ k_y k_x & k_y k_y & -ik_y\frac{\partial}{\partial z} \\ -ik_x\frac{\partial}{\partial z} & -ik_y\frac{\partial}{\partial z} & -\frac{\partial^2}{\partial z^2} \end{pmatrix} - (\frac{\omega}{c})^2 n^2(\omega)].\mathbf{H}(k_x, k_y, z, \omega) \\
= \; &[(\frac{\omega}{c})^2\epsilon(\omega)\mathbf{M}_{ext}(\omega) - i\frac{\omega}{c}\begin{pmatrix} 0 & -\frac{\partial}{\partial z} & ik_y \\ \frac{\partial}{\partial z} & 0 & -ik_x \\ -ik_y & ik_x & 0 \end{pmatrix}.\mathbf{P}_{ext}(\omega)] \\
&\times\delta(z - z_0)\exp[-i(k_x x_0 + k_y y_0)] \qquad (12.14)
\end{aligned}
$$

for the magnetic field.

For the homogeneous case under consideration the system is also translationally invariant in the z-direction. For that case the solution is most easily obtained by also Fourier transforming the z-coordinate. The wave equation for the electric field then reduces to

$$
[k^2 - \mathbf{kk} - (\frac{\omega}{c})^2 n^2(\omega)].\mathbf{E}(\mathbf{k}, \omega) = [(\frac{\omega}{c})^2\mu(\omega)\mathbf{P}_{ext}(\omega) - \frac{\omega}{c}\mathbf{k} \times \mathbf{M}_{ext}(\omega)]\exp(-i\mathbf{k}.\mathbf{r}_0) \qquad (12.15)
$$

and the wave equation for the magnetic field reduces to

$$
[k^2 - \mathbf{kk} - (\frac{\omega}{c})^2 n^2(\omega)].\mathbf{H}(\mathbf{k}, \omega) = [(\frac{\omega}{c})^2\epsilon(\omega)\mathbf{M}_{ext}(\omega) + \frac{\omega}{c}\mathbf{k} \times \mathbf{P}_{ext}(\omega)]\exp(-i\mathbf{k}.\mathbf{r}_0) \qquad (12.16)
$$

The solutions of these equations are given, as may be verified by substitution, by

$$
\begin{aligned}
\mathbf{E}(\mathbf{k}, \omega) &= [-\mu(\omega)\mathbf{F}(\mathbf{k}, \omega).\mathbf{P}_{ext}(\omega) + \mathbf{L}(\mathbf{k}, \omega).\mathbf{M}_{ext}(\omega)]\exp(-i\mathbf{k}.\mathbf{r}_0) \qquad (12.17) \\
\mathbf{H}(\mathbf{k}, \omega) &= [-\epsilon(\omega)\mathbf{F}(\mathbf{k}, \omega).\mathbf{M}_{ext}(\omega) - \mathbf{L}(\mathbf{k}, \omega).\mathbf{P}_{ext}(\omega)]\exp(-i\mathbf{k}.\mathbf{r}_0) \qquad (12.18)
\end{aligned}
$$

where the retarded "dipole propagator" is given by

$$\mathbf{F}(\mathbf{k},\omega) \equiv n^{-2}(\omega)\frac{\mathbf{kk}-(\frac{\omega}{c})^2 n^2(\omega)}{k^2-(\frac{\omega+i0}{c})^2 n^2(\omega)} \tag{12.19}$$

where $i0$ is an infinitesimally small positive imaginary number. Furthermore the other propagator is given by

$$\begin{aligned}\mathbf{L}(\mathbf{k},\omega) &\equiv \frac{c}{\omega}\mathbf{F}(\mathbf{k},\omega).\begin{pmatrix} 0 & -k_z & k_y \\ k_z & 0 & -k_x \\ -k_y & k_x & 0 \end{pmatrix} \\ &= -\frac{\frac{\omega}{c}}{k^2-(\frac{\omega+i0}{c})^2 n^2(\omega)}\begin{pmatrix} 0 & -k_z & k_y \\ k_z & 0 & -k_x \\ -k_y & k_x & 0 \end{pmatrix}\end{aligned} \tag{12.20}$$

Inverse Fourier transformation of the retarded dipole propagator, eq.(12.19), with respect to k_z gives, cf. appendix A,

$$\begin{aligned}\mathbf{F}(k_x,k_y,z,\omega) &= \frac{i}{2}[k_\perp n^2(\omega)]^{-1}[\mathbf{k}_\parallel\mathbf{k}_\parallel - k_\parallel^2\hat{\mathbf{z}}\hat{\mathbf{z}} - (\frac{\omega}{c})^2 n^2(\omega)(1-\hat{\mathbf{z}}\hat{\mathbf{z}})]\exp(ik_\perp|z|) \\ &+\frac{i}{2}\operatorname{sign}(z)n^{-2}(\omega)(\mathbf{k}_\parallel\hat{\mathbf{z}}+\hat{\mathbf{z}}\mathbf{k}_\parallel)\exp(ik_\perp|z|)+\delta(z)n^{-2}(\omega)\hat{\mathbf{z}}\hat{\mathbf{z}}\end{aligned} \tag{12.21}$$

where $\hat{\mathbf{z}}$ is the unit vector normal to the surface and $\mathbf{k}_\parallel \equiv (k_x, k_y, 0)$ the projection of the wave vector on the surface. Furthermore

$$k_\perp(k_x,k_y,\omega) \equiv \sqrt{[\frac{\omega}{c}n(\omega)]^2 - k_x^2 - k_y^2} \quad \text{with} \quad \operatorname{Im} k_\perp \geq 0 \tag{12.22}$$

It should be emphasized that $k_\perp$ is not a variable, but a function of k_x, k_y and ω. Furthermore it is good to realize, that $k_\perp$ can have an imaginary part, because of two reasons. One is the fact, that the refractive index may have an imaginary part, which is the case in an absorbing medium. The other is, when $k_\parallel$ is larger than $n\omega/c$, which is related to the phenomenon of total reflection. The expression for the $z-$dependent dipole propagator, eq.(12.21), contains three contributions with different behavior at $z = 0$. The first is continuous, the second discontinuous and the third singular at the $z = 0$ plane. Before elaborating on the singular contribution, also the inverse Fourier transformed expression for the other propagator, eq.(12.20), with respect to k_z is given, cf. appendix A,

$$\mathbf{L}(k_x,k_y,z,\omega) = -\frac{i}{2}\frac{\omega}{c}\exp(ik_\perp|z|)\{k_\perp^{-1}\begin{pmatrix} 0 & 0 & k_y \\ 0 & 0 & -k_x \\ -k_y & k_x & 0 \end{pmatrix} + \operatorname{sign}(z)\begin{pmatrix} 0 & -1 & 0 \\ 1 & 0 & 0 \\ 0 & 0 & 0 \end{pmatrix}\} \tag{12.23}$$

Also this propagator has a continuous and a discontinuous contribution, but in contrast with the retarded dipole propagator, there is no singular contribution.

Using the above expressions for the z-dependent propagators in a homogeneous medium one may calculate the electric and magnetic fields due to the point source, using the inverse Fourier transforms of eqs.(12.17) and (12.18) with respect to k_z. This results in

$$\begin{aligned} \mathbf{E}(\mathbf{k}_{\|}, z, \omega) &= [-\mu(\omega)\mathbf{F}(\mathbf{k}_{\|}, z - z_0, \omega).\mathbf{P}_{ext}(\omega) + \mathbf{L}(\mathbf{k}_{\|}, z - z_0, \omega).\mathbf{M}_{ext}(\omega)] \\ &\quad \times \exp(-i\mathbf{k}_{\|}.\mathbf{r}_{0,\|}) \qquad (12.24) \\ \mathbf{H}(\mathbf{k}_{\|}, z, \omega) &= [-\epsilon(\omega)\mathbf{F}(\mathbf{k}_{\|}, z - z_0, \omega).\mathbf{M}_{ext}(\omega) - \mathbf{L}(\mathbf{k}_{\|}, z - z_0, \omega).\mathbf{P}_{ext}(\omega)] \\ &\quad \times \exp(-i\mathbf{k}_{\|}.\mathbf{r}_{0,\|}) \qquad (12.25) \end{aligned}$$

As is apparent from eq.(12.21), these fields are singular in the $z = z_0$ plane, where the source is located, due to the first term on the right-hand side of the above two equations.

In order to eliminate these singularities it is now convenient to calculate the so-called non-singular fields defined by

$$\begin{aligned} \mathbf{N}_e(\mathbf{k}_{\|}, z, \omega) &\equiv \mathbf{E}(\mathbf{k}_{\|}, z, \omega) + \hat{\mathbf{z}}\hat{\mathbf{z}}.\mathbf{P}(\mathbf{k}_{\|}, z, \omega) = (E_x, E_y, D_z)(\mathbf{k}_{\|}, z, \omega) \quad (12.26) \\ \mathbf{N}_m(\mathbf{k}_{\|}, z, \omega) &\equiv \mathbf{H}(\mathbf{k}_{\|}, z, \omega) + \hat{\mathbf{z}}\hat{\mathbf{z}}.\mathbf{M}(\mathbf{k}_{\|}, z, \omega) = (H_x, H_y, B_z)(\mathbf{k}_{\|}, z, \omega) \quad (12.27) \end{aligned}$$

As $\mathbf{P}$ and $\mathbf{M}$ are not simply equal to $\mathbf{P}_{ext}$ and $\mathbf{M}_{ext}$ one must go back to the wave equations (12.2) and (12.3) and replace the electric and magnetic fields by the non-singular field. This results in

$$\begin{aligned} [\mathrm{rot}\,\mathrm{rot} - (\frac{\omega}{c})^2][\mathbf{N}_e(\mathbf{r}, \omega) - \hat{\mathbf{z}}\hat{\mathbf{z}}.\mathbf{P}(\mathbf{r}, \omega)] &= (\frac{\omega}{c})^2\mathbf{P}(\mathbf{r}, \omega) + i\frac{\omega}{c}\,\mathrm{rot}\,\mathbf{M}(\mathbf{r}, \omega) \\ &\qquad (12.28) \\ [\mathrm{rot}\,\mathrm{rot} - (\frac{\omega}{c})^2][\mathbf{N}_m(\mathbf{r}, \omega) - \hat{\mathbf{z}}\hat{\mathbf{z}}.\mathbf{M}(\mathbf{r}, \omega)] &= (\frac{\omega}{c})^2\mathbf{M}(\mathbf{r}, \omega) - i\frac{\omega}{c}\,\mathrm{rot}\,\mathbf{P}(\mathbf{r}, \omega) \\ &\qquad (12.29) \end{aligned}$$

The polarization and magnetization densities in these equations are the sums of the externally controlled point sources and the hereby induced polarization and magnetization densities. Using eqs.(12.6)– (12.8), (12.26) and (12.27), these densities may be written as

$$\begin{aligned} \mathbf{P}(\mathbf{r}, \omega) &= [\epsilon(\omega) - 1]\mathbf{E}(\mathbf{r}, \omega) + \mathbf{P}_{ext}(\mathbf{r}, \omega) \\ &= [\epsilon(\omega) - 1]\mathbf{N}_e(\mathbf{r}, \omega) - [\epsilon(\omega) - 1]\hat{\mathbf{z}}\hat{\mathbf{z}}.\mathbf{P}(\mathbf{r}, \omega) + \mathbf{P}_{ext}(\mathbf{r}, \omega) \quad (12.30) \\ \mathbf{M}(\mathbf{r}, \omega) &= [\mu(\omega) - 1]\mathbf{H}(\mathbf{r}, \omega) + \mathbf{M}_{ext}(\mathbf{r}, \omega) \\ &= [\mu(\omega) - 1]\mathbf{N}_m(\mathbf{r}, \omega) - [\mu(\omega) - 1]\hat{\mathbf{z}}\hat{\mathbf{z}}.\mathbf{M}(\mathbf{r}, \omega) + \mathbf{M}_{ext}(\mathbf{r}, \omega) \quad (12.31) \end{aligned}$$

Solving **P** and **M** from these equations, one obtains

$$\mathbf{P}(\mathbf{r},\omega) = (\epsilon-1)\begin{pmatrix} 1 & 0 & 0 \\ 0 & 1 & 0 \\ 0 & 0 & \epsilon^{-1} \end{pmatrix}.\mathbf{N}_e(\mathbf{r},\omega) + \begin{pmatrix} 1 & 0 & 0 \\ 0 & 1 & 0 \\ 0 & 0 & \epsilon^{-1} \end{pmatrix}.\mathbf{P}_{ext}(\mathbf{r},\omega) \tag{12.32}$$

$$\mathbf{M}(\mathbf{r},\omega) = (\mu-1)\begin{pmatrix} 1 & 0 & 0 \\ 0 & 1 & 0 \\ 0 & 0 & \mu^{-1} \end{pmatrix}.\mathbf{N}_m(\mathbf{r},\omega) + \begin{pmatrix} 1 & 0 & 0 \\ 0 & 1 & 0 \\ 0 & 0 & \mu^{-1} \end{pmatrix}.\mathbf{M}_{ext}(\mathbf{r},\omega) \tag{12.33}$$

In order to simplify the expressions, the frequency dependence of ϵ and μ is no longer explicitly indicated. For the electric displacement field $\mathbf{D} = \mathbf{E} + \mathbf{P}$ and the magnetic induction $\mathbf{B} = \mathbf{H} + \mathbf{M}$ one obtains, using eqs.(12.26), (12.27), (12.32) and (12.33), the analogous expressions

$$\mathbf{D}(\mathbf{r},\omega) = \begin{pmatrix} \epsilon & 0 & 0 \\ 0 & \epsilon & 0 \\ 0 & 0 & 1 \end{pmatrix}.\mathbf{N}_e(\mathbf{r},\omega) + \begin{pmatrix} 1 & 0 & 0 \\ 0 & 1 & 0 \\ 0 & 0 & 0 \end{pmatrix}.\mathbf{P}_{ext}(\mathbf{r},\omega) \tag{12.34}$$

$$\mathbf{B}(\mathbf{r},\omega) = \begin{pmatrix} \mu & 0 & 0 \\ 0 & \mu & 0 \\ 0 & 0 & 1 \end{pmatrix}.\mathbf{N}_m(\mathbf{r},\omega) + \begin{pmatrix} 1 & 0 & 0 \\ 0 & 1 & 0 \\ 0 & 0 & 0 \end{pmatrix}.\mathbf{M}_{ext}(\mathbf{r},\omega) \tag{12.35}$$

Substituting eqs.(12.31) and (12.32) into eq.(12.28), and using furthermore the third Maxwell equation (12.1) and eq.(12.34), one obtains

$$\begin{aligned}
&[\mathrm{rot\,rot} - (\frac{\omega}{c})^2][\begin{pmatrix} 1 & 0 & 0 \\ 0 & 1 & 0 \\ 0 & 0 & \epsilon^{-1} \end{pmatrix}.\mathbf{N}_e(\mathbf{r},\omega) - \epsilon^{-1}\hat{\mathbf{z}}\hat{\mathbf{z}}.\mathbf{P}_{ext}(\mathbf{r},\omega)] \\
&= (\frac{\omega}{c})^2[(n^2-1)\begin{pmatrix} 1 & 0 & 0 \\ 0 & 1 & 0 \\ 0 & 0 & \epsilon^{-1} \end{pmatrix}.\mathbf{N}_e(\mathbf{r},\omega) + \mu\begin{pmatrix} 1 & 0 & 0 \\ 0 & 1 & 0 \\ 0 & 0 & n^{-2} \end{pmatrix}.\mathbf{P}_{ext}(\mathbf{r},\omega)] \\
&\quad + i\frac{\omega}{c}\,\mathrm{rot}\,\mathbf{M}_{ext}(\mathbf{r},\omega)
\end{aligned} \tag{12.36}$$

Similarly one obtains, by substituting eqs.(12.30) and (12.33) into eq.(12.29), and using the first Maxwell equation (12.1) and eq.(12.35)

$$\begin{aligned}
&[\mathrm{rot\,rot} - (\frac{\omega}{c})^2][\begin{pmatrix} 1 & 0 & 0 \\ 0 & 1 & 0 \\ 0 & 0 & \mu^{-1} \end{pmatrix}.\mathbf{N}_m(\mathbf{r},\omega) - \mu^{-1}\hat{\mathbf{z}}\hat{\mathbf{z}}.\mathbf{M}_{ext}(\mathbf{r},\omega)] \\
&= (\frac{\omega}{c})^2[(n^2-1)\begin{pmatrix} 1 & 0 & 0 \\ 0 & 1 & 0 \\ 0 & 0 & \mu^{-1} \end{pmatrix}.\mathbf{N}_m(\mathbf{r},\omega) + \epsilon\begin{pmatrix} 1 & 0 & 0 \\ 0 & 1 & 0 \\ 0 & 0 & n^{-2} \end{pmatrix}.\mathbf{M}_{ext}(\mathbf{r},\omega)] \\
&\quad - i\frac{\omega}{c}\,\mathrm{rot}\,\mathbf{P}_{ext}(\mathbf{r},\omega)
\end{aligned} \tag{12.37}$$

The last two equations may now be simplified by bringing the terms with the non-singular fields and part of the terms with the external polarization and magnetization densities to the left. This leads to

$$\begin{aligned}
&[\text{rot}\,\text{rot} - (\frac{\omega}{c})^2 n^2][\begin{pmatrix} 1 & 0 & 0 \\ 0 & 1 & 0 \\ 0 & 0 & \epsilon^{-1} \end{pmatrix}.\mathbf{N}_e(\mathbf{r},\omega) - \epsilon^{-1}\hat{\mathbf{z}}\hat{\mathbf{z}}.\mathbf{P}_{ext}(\mathbf{r},\omega)] \\
&= (\frac{\omega}{c})^2 \mu \mathbf{P}_{ext}(\mathbf{r},\omega) + i\frac{\omega}{c}\,\text{rot}\,\mathbf{M}_{ext}(\mathbf{r},\omega) \qquad (12.38)
\end{aligned}$$

$$\begin{aligned}
&[\text{rot}\,\text{rot} - (\frac{\omega}{c})^2 n^2][\begin{pmatrix} 1 & 0 & 0 \\ 0 & 1 & 0 \\ 0 & 0 & \mu^{-1} \end{pmatrix}.\mathbf{N}_m(\mathbf{r},\omega) - \mu^{-1}\hat{\mathbf{z}}\hat{\mathbf{z}}.\mathbf{M}_{ext}(\mathbf{r},\omega)] \\
&= (\frac{\omega}{c})^2 \epsilon \mathbf{M}_{ext}(\mathbf{r},\omega) - i\frac{\omega}{c}\,\text{rot}\,\mathbf{P}_{ext}(\mathbf{r},\omega) \qquad (12.39)
\end{aligned}$$

These equations have, upon transformation to the (k_x, k_y, z, ω)-representation, in fact a form similar to the original equations (12.13) and (12.14). The only differences are that the electric and magnetic fields on the left-hand sides have been replaced by the terms appearing now between square brackets. As solutions one may therefore use the forms given in eqs.(12.24) and (12.25)

$$\begin{aligned}
&\begin{pmatrix} 1 & 0 & 0 \\ 0 & 1 & 0 \\ 0 & 0 & \epsilon^{-1} \end{pmatrix}.\mathbf{N}_e(\mathbf{k}_{\|}, z, \omega) - \epsilon^{-1}\hat{\mathbf{z}}\hat{\mathbf{z}}.\mathbf{P}_{ext}(\omega)\exp(-i\mathbf{k}_{\|}.\mathbf{r}_{0,\|})\delta(z - z_0) \\
&= [-\mu\mathbf{F}(\mathbf{k}_{\|}, z - z_0, \omega).\mathbf{P}_{ext}(\omega) + \mathbf{L}(\mathbf{k}_{\|}, z - z_0, \omega).\mathbf{M}_{ext}(\omega)]\exp(-i\mathbf{k}_{\|}.\mathbf{r}_{0,\|}) \\
&\qquad (12.40)
\end{aligned}$$

$$\begin{aligned}
&\begin{pmatrix} 1 & 0 & 0 \\ 0 & 1 & 0 \\ 0 & 0 & \mu^{-1} \end{pmatrix}.\mathbf{N}_m(\mathbf{k}_{\|}, z, \omega) - \mu^{-1}\hat{\mathbf{z}}\hat{\mathbf{z}}.\mathbf{M}_{ext}(\omega)\exp(-i\mathbf{k}_{\|}.\mathbf{r}_{0,\|})\delta(z - z_0) \\
&= [-\epsilon\mathbf{F}(\mathbf{k}_{\|}, z - z_0, \omega).\mathbf{M}_{ext}(\omega) - \mathbf{L}(\mathbf{k}_{\|}, z - z_0, \omega).\mathbf{P}_{ext}(\omega)]\exp(-i\mathbf{k}_{\|}.\mathbf{r}_{0,\|}) \\
&\qquad (12.41)
\end{aligned}$$

where eq.(12.12) was used for the externally controlled polarization and magnetization densities in this representation.

One may now bring the terms containing the polarization and magnetization densities to the right and it is then found from the above equations, that the non-singular fields $\mathbf{N}_e$ and $\mathbf{N}_m$ can be written in the forms

$$\begin{aligned}\mathbf{N}_e(\mathbf{k}_\parallel, z, \omega) &= [-\begin{pmatrix} \mu & 0 & 0 \\ 0 & \mu & 0 \\ 0 & 0 & n^2 \end{pmatrix}.\mathbf{G}(\mathbf{k}_\parallel, z - z_0, \omega).\mathbf{P}_{ext}(\omega) \\ &+ \begin{pmatrix} 1 & 0 & 0 \\ 0 & 1 & 0 \\ 0 & 0 & \epsilon \end{pmatrix}.\mathbf{L}(\mathbf{k}_\parallel, z - z_0, \omega).\mathbf{M}_{ext}(\omega)] \exp(-i\mathbf{k}_\parallel.\mathbf{r}_{0,\parallel})\end{aligned} \qquad (12.42)$$

$$\begin{aligned}\mathbf{N}_m(\mathbf{k}_\parallel, z, \omega) &= [-\begin{pmatrix} \epsilon & 0 & 0 \\ 0 & \epsilon & 0 \\ 0 & 0 & n^2 \end{pmatrix}.\mathbf{G}(\mathbf{k}_\parallel, z - z_0, \omega).\mathbf{M}_{ext}(\omega) \\ &- \begin{pmatrix} 1 & 0 & 0 \\ 0 & 1 & 0 \\ 0 & 0 & \mu \end{pmatrix}.\mathbf{L}(\mathbf{k}_\parallel, z - z_0, \omega).\mathbf{P}_{ext}(\omega)] \exp(-i\mathbf{k}_\parallel.\mathbf{r}_{0,\parallel})\end{aligned} \qquad (12.43)$$

where the propagator $\mathbf{G}(\mathbf{k}_\parallel, z, \omega)$ is equal to the propagator $\mathbf{F}(\mathbf{k}_\parallel, z, \omega)$, eq.(12.21), without the singular term:

$$\begin{aligned}\mathbf{G}(k_x, k_y, z, \omega) &\equiv \mathbf{F}(k_x, k_y, z, \omega) - n^{-2}(\omega)\hat{\mathbf{z}}\hat{\mathbf{z}}\delta(z) \\ &= \frac{i}{2}\exp(ik_\perp|z|)\{[k_\perp n^2(\omega)]^{-1}[\mathbf{k}_\parallel\mathbf{k}_\parallel - k_\parallel^2\hat{\mathbf{z}}\hat{\mathbf{z}} - (\frac{\omega}{c})^2 n^2(\omega)(1 - \hat{\mathbf{z}}\hat{\mathbf{z}})] \\ &+ \text{sign}(z) n^{-2}(\omega)(\mathbf{k}_\parallel\hat{\mathbf{z}} + \hat{\mathbf{z}}\mathbf{k}_\parallel)\}\end{aligned} \qquad (12.44)$$

The great advantage of this propagator is that it is no longer singular in the plane where the source is located, just as this is the case for the propagator $\mathbf{L}(k_x, k_y, z, \omega)$. As their names indicate, the non-singular fields $\mathbf{N}_e$ and $\mathbf{N}_m$ have therefore indeed no singularities in this plane.

For the following it will be necessary to express the non-singular fields in terms of the surface polarization and magnetization densities $\mathbf{P}^s(\mathbf{k}_\parallel, \omega)$ and $\mathbf{M}^s(\mathbf{k}_\parallel, \omega)$ in the $z = z_0$ plane, rather than in terms of the external surface polarization and magnetization densities $\mathbf{P}_{ext}(\omega)\exp(-i\mathbf{k}_\parallel.\mathbf{r}_{0,\parallel})$ and $\mathbf{M}_{ext}(\omega)\exp(-i\mathbf{k}_\parallel.\mathbf{r}_{0,\parallel})$ in this plane. For the definition of surface polarization and magnetization densities $\mathbf{P}^s(\mathbf{k}_\parallel, \omega)$ and $\mathbf{M}^s(\mathbf{k}_\parallel, \omega)$ one has to go back to eq.(2.43). Using eq.(12.34) in $(\mathbf{k}_\parallel, z)$-representation, eqs.(12.12), (12.24) and (12.21), and considering only the singular terms in these equations, one obtains:

$$\begin{aligned}\mathbf{D}_\parallel^s(\mathbf{k}_\parallel, \omega) &= \mathbf{P}_{ext,\parallel}(\omega)\exp(-i\mathbf{k}_\parallel.\mathbf{r}_{0,\parallel}) \\ E_z^s(\mathbf{k}_\parallel, \omega) &= -\epsilon^{-1}P_{ext,z}(\omega)\exp(-\mathbf{k}_\parallel.\mathbf{r}_{0,\parallel})\end{aligned} \qquad (12.45)$$

An analogous relation is found for $\mathbf{B}_\parallel^s$ and H_z^s, with μ instead of ϵ, if one uses eqs.(12.35), (12.12), (12.25) and (12.21). It then follows with eq.(2.43), that

$$\mathbf{P}^s(\mathbf{k}_\parallel,\omega) = \begin{pmatrix} 1 & 0 & 0 \\ 0 & 1 & 0 \\ 0 & 0 & \epsilon^{-1} \end{pmatrix} .\mathbf{P}_{ext}(\omega)\exp(-i\mathbf{k}_\parallel.\mathbf{r}_{0,\parallel})$$
$$\mathbf{M}^s(\mathbf{k}_\parallel,\omega) = \begin{pmatrix} 1 & 0 & 0 \\ 0 & 1 & 0 \\ 0 & 0 & \mu^{-1} \end{pmatrix} .\mathbf{M}_{ext}(\omega)\exp(-i\mathbf{k}_\parallel.\mathbf{r}_{0,\parallel}) \quad (12.46)$$

From eqs.(12.42)– (12.44), (12.46) and (12.23) one then obtains the following results:

$$\mathbf{N}_e(\mathbf{k}_\parallel,z,\omega) = -\mathbf{K}_{ee}(\mathbf{k}_\parallel,z-z_0,\omega).\mathbf{P}^s(\mathbf{k}_\parallel,\omega) + \mathbf{K}_{em}(\mathbf{k}_\parallel,z-z_0,\omega).\mathbf{M}^s(\mathbf{k}_\parallel,\omega) \quad (12.47)$$
$$\mathbf{N}_m(\mathbf{k}_\parallel,z,\omega) = -\mathbf{K}_{me}(\mathbf{k}_\parallel,z-z_0,\omega).\mathbf{P}^s(\mathbf{k}_\parallel,\omega) - \mathbf{K}_{mm}(\mathbf{k}_\parallel,z-z_0,\omega).\mathbf{M}^s(\mathbf{k}_\parallel,\omega) \quad (12.48)$$

where the propagators

$$\mathbf{K}_{ee}(\mathbf{k}_\parallel,z,\omega) \equiv \mu \begin{pmatrix} 1 & 0 & 0 \\ 0 & 1 & 0 \\ 0 & 0 & \epsilon \end{pmatrix} .\mathbf{G}(\mathbf{k}_\parallel,z,\omega). \begin{pmatrix} 1 & 0 & 0 \\ 0 & 1 & 0 \\ 0 & 0 & \epsilon \end{pmatrix}$$
$$= \frac{i}{2}\exp(ik_\perp z\,\mathrm{sign}(z))\{[k_\perp\epsilon(\omega)]^{-1}[\mathbf{k}_\parallel\mathbf{k}_\parallel - \epsilon^2(\omega)k_\parallel^2\hat{\mathbf{z}}\hat{\mathbf{z}}$$
$$-(\frac{\omega}{c})^2 n^2(\omega)(1-\hat{\mathbf{z}}\hat{\mathbf{z}})] + \mathrm{sign}(z)(\mathbf{k}_\parallel\hat{\mathbf{z}} + \hat{\mathbf{z}}\mathbf{k}_\parallel)\} \quad (12.49)$$

$$\mathbf{K}_{me}(\mathbf{k}_\parallel,z,\omega) \equiv \begin{pmatrix} 1 & 0 & 0 \\ 0 & 1 & 0 \\ 0 & 0 & \mu \end{pmatrix} .\mathbf{L}(\mathbf{k}_\parallel,z,\omega). \begin{pmatrix} 1 & 0 & 0 \\ 0 & 1 & 0 \\ 0 & 0 & \epsilon \end{pmatrix}$$
$$= -\frac{i}{2}\frac{\omega}{c}\exp(ik_\perp z\,\mathrm{sign}(z))\{k_\perp^{-1} \begin{pmatrix} 0 & 0 & \epsilon k_y \\ 0 & 0 & -\epsilon k_x \\ -\mu k_y & \mu k_x & 0 \end{pmatrix}$$
$$+\mathrm{sign}(z) \begin{pmatrix} 0 & -1 & 0 \\ 1 & 0 & 0 \\ 0 & 0 & 0 \end{pmatrix} \} \quad (12.50)$$

$$\mathbf{K}_{mm}(\mathbf{k}_\parallel,z,\omega) \equiv \epsilon \begin{pmatrix} 1 & 0 & 0 \\ 0 & 1 & 0 \\ 0 & 0 & \mu \end{pmatrix} .\mathbf{G}(\mathbf{k}_\parallel,z,\omega). \begin{pmatrix} 1 & 0 & 0 \\ 0 & 1 & 0 \\ 0 & 0 & \mu \end{pmatrix}$$
$$= \frac{i}{2}\exp(ik_\perp z\,\mathrm{sign}(z))\{[k_\perp\mu(\omega)]^{-1}[\mathbf{k}_\parallel\mathbf{k}_\parallel - \mu^2(\omega)k_\parallel^2\hat{\mathbf{z}}\hat{\mathbf{z}}$$
$$-(\frac{\omega}{c})^2 n^2(\omega)(1-\hat{\mathbf{z}}\hat{\mathbf{z}})] + \mathrm{sign}(z)(\mathbf{k}_\parallel\hat{\mathbf{z}} + \hat{\mathbf{z}}\mathbf{k}_\parallel)\} \quad (12.51)$$

$$\mathbf{K}_{em}(\mathbf{k}_\|, z, \omega) \equiv \begin{pmatrix} 1 & 0 & 0 \\ 0 & 1 & 0 \\ 0 & 0 & \epsilon \end{pmatrix} .\mathbf{L}(\mathbf{k}_\|, z, \omega). \begin{pmatrix} 1 & 0 & 0 \\ 0 & 1 & 0 \\ 0 & 0 & \mu \end{pmatrix}$$
$$= -\frac{i}{2}\frac{\omega}{c}\exp(ik_\perp z\,\mathrm{sign}(z))\{k_\perp^{-1}\begin{pmatrix} 0 & 0 & \mu k_y \\ 0 & 0 & -\mu k_x \\ -\epsilon k_y & \epsilon k_x & 0 \end{pmatrix}$$
$$+\mathrm{sign}(z)\begin{pmatrix} 0 & -1 & 0 \\ 1 & 0 & 0 \\ 0 & 0 & 0 \end{pmatrix}\} \tag{12.52}$$

have been introduced.

The next step in the analysis is to construct these propagators in the case that $\epsilon^- \neq \epsilon^+$ and (or) $\mu^- \neq \mu^+$. For this purpose the image dipole method may be used. As this method is rather well-known, it will not be discussed here in detail. In this more general case the propagators will depend on both the position of the source and the point of observation and in particular also on whether these points are in the same half space or in a different one. The solutions of the wave equations are given by

$$\mathbf{K}_{ee}(\mathbf{k}_\|, z, \omega|z') = \sum_{\nu,\nu'} \theta(\nu z)\theta(\nu' z')\{[\mathbf{K}^\nu_{ee}(\mathbf{k}_\|, z - z', \omega) + \mathbf{K}^\nu_{ee}(\mathbf{k}_\|, z + z', \omega).\mathbf{R}^\nu_e]\delta_{\nu,\nu'}$$
$$+\mathbf{K}^\nu_{ee}(\mathbf{k}_\|, z - \frac{k_\perp^{\nu'}}{k_\perp^\nu} z', \omega).\mathbf{T}^{\nu'}_e \delta_{\nu,-\nu'}\} \tag{12.53}$$

$$\mathbf{K}_{me}(\mathbf{k}_\|, z, \omega|z') = \sum_{\nu,\nu'} \theta(\nu z)\theta(\nu' z')\{[\mathbf{K}^\nu_{me}(\mathbf{k}_\|, z - z', \omega) + \mathbf{K}^\nu_{me}(\mathbf{k}_\|, z + z', \omega).\mathbf{R}^\nu_e]\delta_{\nu,\nu'}$$
$$+\mathbf{K}^\nu_{me}(\mathbf{k}_\|, z - \frac{k_\perp^{\nu'}}{k_\perp^\nu} z', \omega).\mathbf{T}^{\nu'}_e \delta_{\nu,-\nu'}\} \tag{12.54}$$

$$\mathbf{K}_{mm}(\mathbf{k}_\|, z, \omega|z') = \sum_{\nu,\nu'} \theta(\nu z)\theta(\nu' z')\{[\mathbf{K}^\nu_{mm}(\mathbf{k}_\|, z - z', \omega) + \mathbf{K}^\nu_{mm}(\mathbf{k}_\|, z + z', \omega).\mathbf{R}^\nu_m]\delta_{\nu,\nu'}$$
$$+\mathbf{K}^\nu_{mm}(\mathbf{k}_\|, z - \frac{k_\perp^{\nu'}}{k_\perp^\nu} z', \omega).\mathbf{T}^{\nu'}_m \delta_{\nu,-\nu'}\} \tag{12.55}$$

$$\mathbf{K}_{em}(\mathbf{k}_\|, z, \omega|z') = \sum_{\nu,\nu'} \theta(\nu z)\theta(\nu' z')\{[\mathbf{K}^\nu_{em}(\mathbf{k}_\|, z - z', \omega) + \mathbf{K}^\nu_{em}(\mathbf{k}_\|, z + z', \omega).\mathbf{R}^\nu_m]\delta_{\nu,\nu'}$$
$$+\mathbf{K}^\nu_{em}(\mathbf{k}_\|, z - \frac{k_\perp^{\nu'}}{k_\perp^\nu} z', \omega).\mathbf{T}^{\nu'}_m \delta_{\nu,-\nu'}\} \tag{12.56}$$

In these expressions ν and ν' can be $-$ and $+$ indicating respectively the half space in which the observer and the source are located. The superscript ν of the propagators $\mathbf{K}^\nu_{ee}(\mathbf{k}_\|, z, \omega)$, etc. indicates that one has to take $\epsilon = \epsilon^\nu$ and $\mu = \mu^\nu$ in the expressions

(12.49)– (12.52). Furthermore $\delta_{-,-} = \delta_{+,+} = 1$ and $\delta_{-,+} = \delta_{+,-} = 0$. In the above expressions the transmitted contribution contains $(k_\perp^\nu / \, k_\perp^{\nu'})z'$ as the position of the virtual source. If this position is real valued the substitution of the expressions given in eqs.(12.49)– (12.52) is straightforward. If the virtual position has an imaginary part one must use sign(z) rather than sign$(z - (k_\perp^\nu / \, k_\perp^{\nu'})z')$. Notice that for a real valued virtual position of the source in the transmitted contribution these signs are the same, which follows using the fact that $\nu = -\nu'$.

The three terms between curly brackets at the right of eqs.(12.53)– (12.56) have a simple physical interpretation. The first one represents a signal, which is emitted in the same half space and goes directly from z' to z. The second one represents a signal, which is reflected at the interface before reaching z. The last one represents a signal, which goes from one half space to the other and is therefore transmitted by the interface. The tensors $\mathbf{R}_e^\nu$, $\mathbf{T}_e^\nu$, $\mathbf{R}_m^\nu$ and $\mathbf{T}_m^\nu$ will be called Fresnel reflection and transmission tensors. Upon application of the above propagators on the surface polarization and magnetization densities $\mathbf{P}^s(\mathbf{k}_\|, \omega)$ and $\mathbf{M}^s(\mathbf{k}_\|, \omega)$, eq.(12.46), of the external dipole point source, eq.(12.11), one obtains, in a completely analogous way as above in eqs.(12.47) and (12.48), the non-singular fields $\mathbf{N}_e$ and $\mathbf{N}_m$:

$$\mathbf{N}_e(\mathbf{k}_\|, z, \omega) = -\mathbf{K}_{ee}(\mathbf{k}_\|, z, \omega|z_0).\mathbf{P}^s(\mathbf{k}_\|, \omega) + \mathbf{K}_{em}(\mathbf{k}_\|, z, \omega|z_0).\mathbf{M}^s(\mathbf{k}_\|, \omega) \quad (12.57)$$

$$\mathbf{N}_m(\mathbf{k}_\|, z, \omega) = -\mathbf{K}_{me}(\mathbf{k}_\|, z, \omega|z_0).\mathbf{P}^s(\mathbf{k}_\|, \omega) - \mathbf{K}_{mm}(\mathbf{k}_\|, z, \omega|z_0).\mathbf{M}^s(\mathbf{k}_\|, \omega) \quad (12.58)$$

The Fresnel reflection and transmission matrices may now be found from the boundary conditions of these fields, i.e. from their continuity at $z = 0$. One obtains:

$$\begin{aligned}
\mathbf{R}_e^\nu &= -\mathbf{R}_e^{-\nu} = \nu \left[\frac{\mu^- k_\perp^+ - \mu^+ k_\perp^-}{\mu^- k_\perp^+ + \mu^+ k_\perp^-}(1 - \hat{\mathbf{z}}\hat{\mathbf{z}} - \hat{\mathbf{k}}_\|\hat{\mathbf{k}}_\|) + \frac{\epsilon^- k_\perp^+ - \epsilon^+ k_\perp^-}{\epsilon^- k_\perp^+ + \epsilon^+ k_\perp^-}(\hat{\mathbf{z}}\hat{\mathbf{z}} - \hat{\mathbf{k}}_\|\hat{\mathbf{k}}_\|) \right] \\
\mathbf{T}_e^\nu &= 2 \left[\frac{\mu^\nu k_\perp^{-\nu}}{\mu^- k_\perp^+ + \mu^+ k_\perp^-}(1 - \hat{\mathbf{z}}\hat{\mathbf{z}} - \hat{\mathbf{k}}_\|\hat{\mathbf{k}}_\|) + \frac{1}{\epsilon^- k_\perp^+ + \epsilon^+ k_\perp^-}(\epsilon^\nu k_\perp^{-\nu}\hat{\mathbf{z}}\hat{\mathbf{z}} + \epsilon^{-\nu} k_\perp^\nu \hat{\mathbf{k}}_\|\hat{\mathbf{k}}_\|) \right]
\end{aligned} \quad (12.59)$$

$$\begin{aligned}
\mathbf{R}_m^\nu &= -\mathbf{R}_m^{-\nu} = \nu \left[\frac{\epsilon^- k_\perp^+ - \epsilon^+ k_\perp^-}{\epsilon^- k_\perp^+ + \epsilon^+ k_\perp^-}(1 - \hat{\mathbf{z}}\hat{\mathbf{z}} - \hat{\mathbf{k}}_\|\hat{\mathbf{k}}_\|) + \frac{\mu^- k_\perp^+ - \mu^+ k_\perp^-}{\mu^- k_\perp^+ + \mu^+ k_\perp^-}(\hat{\mathbf{z}}\hat{\mathbf{z}} - \hat{\mathbf{k}}_\|\hat{\mathbf{k}}_\|) \right] \\
\mathbf{T}_m^\nu &= 2 \left[\frac{\epsilon^\nu k_\perp^{-\nu}}{\epsilon^- k_\perp^+ + \epsilon^+ k_\perp^-}(1 - \hat{\mathbf{z}}\hat{\mathbf{z}} - \hat{\mathbf{k}}_\|\hat{\mathbf{k}}_\|) + \frac{1}{\mu^- k_\perp^+ + \mu^+ k_\perp^-}(\mu^\nu k_\perp^{-\nu}\hat{\mathbf{z}}\hat{\mathbf{z}} + \mu^{-\nu} k_\perp^\nu \hat{\mathbf{k}}_\|\hat{\mathbf{k}}_\|) \right]
\end{aligned} \quad (12.60)$$

where the unit vector $\hat{\mathbf{k}}_\| \equiv \mathbf{k}_\| / k_\|$. The reason to call these the Fresnel reflection and transmission matrices is the fact that, if one writes the Fresnel reflection and transmission amplitudes given in eqs.(4.34) and (4.43) in terms of $k_\perp^-$ and $k_\perp^+$, one obtains exactly the prefactors in the above matrices, taking into consideration, that one has taken $\mu^\nu = 1$ in those equations. This is, of course, not an accident as the above propagators also describe reflection and transmission at a similar interface. These matrices furthermore satisfy the relations

$$\mathbf{R}_e^\nu + \mathbf{T}_e^\nu = 1, \quad \mathbf{R}_m^\nu + \mathbf{T}_m^\nu = 1 \quad (12.61)$$

These conditions are also a direct consequence of the boundary conditions, but may also easily be verified by substitution of eqs.(12.59) and (12.60).

It is very important to stress that by construction, i.e. by the boundary conditions, the propagators are, given a location of the source away from the interface, continuous at the $z = 0$ location of the observer. Using the general symmetry of the solution of the Maxwell equations for the interchange of the source and the observer, it follows that, for a given position of the observer away from the $z' = 0$ interface, the field will also be independent from the precise position of the source. Whether one places the source just below the $z' = 0$ surface or just above, the field is the same. Another symmetry which is apparent from the above equations is the one for the interchange of the electric and the magnetic fields.

12.4 The fields due to the surface polarization and magnetization densities

The fields due to the polarization and magnetization densities $\mathbf{P}^s(\mathbf{k}_\|, \omega)$ and $\mathbf{M}^s(\mathbf{k}_\|, \omega)$ at the interface are now simply given by

$$\mathbf{N}_e(\mathbf{k}_\|, z, \omega) = -\mathbf{K}_{ee}(\mathbf{k}_\|, z, \omega|z' = 0).\mathbf{P}^s(\mathbf{k}_\|, \omega) + \mathbf{K}_{em}(\mathbf{k}_\|, z, \omega|z' = 0).\mathbf{M}^s(\mathbf{k}_\|, \omega)$$

$$\mathbf{N}_m(\mathbf{k}_\|, z, \omega) = -\mathbf{K}_{me}(\mathbf{k}_\|, z, \omega|z' = 0).\mathbf{P}^s(\mathbf{k}_\|, \omega) - \mathbf{K}_{mm}(\mathbf{k}_\|, z, \omega|z' = 0).\mathbf{M}^s(\mathbf{k}_\|, \omega) \qquad (12.62)$$

as follows from eqs.(12.57) and (12.58). The propagators in these equations have the property that they are discontinuous, if the order of z and z' is interchanged. As a consequence the non-singular fields due to the surface polarization and magnetization become discontinuous at the interface. Another important property of the propagators is that, for a given non-zero value of the position of the observer, the propagators are a continuous function of the position of the source also in $z' = 0$. This may be verified by substitution of $z' = 0$ into eqs.(12.53)– (12.56) and using the explicit form of the Fresnel reflection and transmission matrices given in eqs.(12.59) and (12.60). As a consequence it does not matter whether the polarization and magnetization densities are placed just above or just below the surface. This is one of the important advantages of using the non-singular field. The location of the source just above or just below the surface has in fact in the past led to confusion and incorrect predictions [cf. section 14.3]. In the present formulation such confusion can simply not arise.

Since the propagators with a source on the surface play such an important role, and in view of the fact that because of their continuous nature for $z' = 0$, it is useful to write them also down explicitly, using their continuity to simplify them.

One then finds

$$
\begin{aligned}
\mathbf{K}_{ee}(\mathbf{k}_{\|}, z, \omega|z'=0) &= \sum_{\nu} \theta(\nu z)\mathbf{K}^{\nu}_{ee}(\mathbf{k}_{\|}, z, \omega).(\mathbf{1}+\mathbf{R}^{\nu}_{e}) \\
&= \sum_{\nu} \theta(\nu z)\mathbf{K}^{\nu}_{ee}(\mathbf{k}_{\|}, z, \omega).\mathbf{T}^{-\nu}_{e} \\
\mathbf{K}_{me}(\mathbf{k}_{\|}, z, \omega|z'=0) &= \sum_{\nu} \theta(\nu z)\mathbf{K}^{\nu}_{me}(\mathbf{k}_{\|}, z, \omega).(\mathbf{1}+\mathbf{R}^{\nu}_{e}) \\
&= \sum_{\nu} \theta(\nu z)\mathbf{K}^{\nu}_{me}(\mathbf{k}_{\|}, z, \omega).\mathbf{T}^{-\nu}_{e} \\
\mathbf{K}_{mm}(\mathbf{k}_{\|}, z, \omega|z'=0) &= \sum_{\nu} \theta(\nu z)\mathbf{K}^{\nu}_{mm}(\mathbf{k}_{\|}, z, \omega).(\mathbf{1}+\mathbf{R}^{\nu}_{m}) \\
&= \sum_{\nu} \theta(\nu z)\mathbf{K}^{\nu}_{mm}(\mathbf{k}_{\|}, z, \omega).\mathbf{T}^{-\nu}_{m} \\
\mathbf{K}_{em}(\mathbf{k}_{\|}, z, \omega|z'=0) &= \sum_{\nu} \theta(\nu z)\mathbf{K}^{\nu}_{em}(\mathbf{k}_{\|}, z, \omega).(\mathbf{1}+\mathbf{R}^{\nu}_{m}) \\
&= \sum_{\nu} \theta(\nu z)\mathbf{K}^{\nu}_{em}(\mathbf{k}_{\|}, z, \omega).\mathbf{T}^{-\nu}_{m} \qquad (12.63)
\end{aligned}
$$

Substituting these expressions into eq.(12.62) for the fields then gives the fields due to sources on the surface.

12.5 Dipole-dipole interaction along the surface

An important phenomenon in the description of an excess dipole distribution along an interface are electromagnetic interactions between the dipoles. For this purpose one needs the field due to this dipole distribution, eq.(12.62), evaluated at the interface. In view of the fact that this field is discontinuous in $z = 0$, one obtains a field which depends on $\nu \equiv \text{sign}(z - z')$,

$$
\begin{aligned}
\mathbf{N}^{\nu}_{e}(\mathbf{k}_{\|}, \omega) &= -\mathbf{K}^{\nu}_{ee}(\mathbf{k}_{\|}, \omega).\mathbf{P}^{s}(\mathbf{k}_{\|}, \omega) + \mathbf{K}^{\nu}_{em}(\mathbf{k}_{\|}, \omega).\mathbf{M}^{s}(\mathbf{k}_{\|}, \omega) \\
\mathbf{N}^{\nu}_{m}(\mathbf{k}_{\|}, \omega) &= -\mathbf{K}^{\nu}_{me}(\mathbf{k}_{\|}, \omega).\mathbf{P}^{s}(\mathbf{k}_{\|}, \omega) - \mathbf{K}^{\nu}_{mm}(\mathbf{k}_{\|}, \omega).\mathbf{M}^{s}(\mathbf{k}_{\|}, \omega) \qquad (12.64)
\end{aligned}
$$

The propagators for the dipole field along the surface are found, using also eqs.(12.49)–(12.52), to be equal to

$$
\begin{aligned}
\mathbf{K}^{\nu}_{ee}(\mathbf{k}_{\|}, \omega) &= \frac{i}{2}\{[k^{\nu}_{\perp}\epsilon^{\nu}(\omega)]^{-1}[\mathbf{k}_{\|}\mathbf{k}_{\|} - (\epsilon^{\nu}(\omega))^{2}k_{\|}^{2}\hat{\mathbf{z}}\hat{\mathbf{z}} - (\frac{\omega}{c})^{2}(n^{\nu}(\omega))^{2}(1-\hat{\mathbf{z}}\hat{\mathbf{z}})] \\
&\quad +\nu(\mathbf{k}_{\|}\hat{\mathbf{z}} + \hat{\mathbf{z}}\mathbf{k}_{\|})\} \\
\mathbf{K}^{\nu}_{me}(\mathbf{k}_{\|}, \omega) &= -\frac{i}{2}\frac{\omega}{c}\{(k^{\nu}_{\perp})^{-1}\begin{pmatrix} 0 & 0 & \epsilon^{\nu}k_{y} \\ 0 & 0 & -\epsilon^{\nu}k_{x} \\ -\mu^{\nu}k_{y} & \mu^{\nu}k_{x} & 0 \end{pmatrix} + \nu\begin{pmatrix} 0 & -1 & 0 \\ 1 & 0 & 0 \\ 0 & 0 & 0 \end{pmatrix}\}
\end{aligned}
$$

$$
\begin{aligned}
\mathbf{K}^{\nu}_{mm}(\mathbf{k}_{\|},\omega) &= \frac{i}{2}\{[k^{\nu}_{\perp}\mu^{\nu}(\omega)]^{-1}[\mathbf{k}_{\|}\mathbf{k}_{\|} - (\mu^{\nu}(\omega))^2 k_{\|}^2\hat{\mathbf{z}}\hat{\mathbf{z}} - (\frac{\omega}{c})^2(n^{\nu}(\omega))^2(1-\hat{\mathbf{z}}\hat{\mathbf{z}})] \\
&\quad +\nu(\mathbf{k}_{\|}\hat{\mathbf{z}} + \hat{\mathbf{z}}\mathbf{k}_{\|})\} \\
\mathbf{K}^{\nu}_{em}(\mathbf{k}_{\|},\omega) &= -\frac{i}{2}\frac{\omega}{c}\{(k^{\nu}_{\perp})^{-1}\begin{pmatrix} 0 & 0 & \mu^{\nu}k_y \\ 0 & 0 & -\mu^{\nu}k_x \\ -\epsilon^{\nu}k_y & \epsilon^{\nu}k_x & 0 \end{pmatrix} + \nu\begin{pmatrix} 0 & -1 & 0 \\ 1 & 0 & 0 \\ 0 & 0 & 0 \end{pmatrix}\}
\end{aligned}
\tag{12.65}
$$

The discontinuous nature of the field on the surface has the somewhat disturbing consequence that in order to know the interaction between two dipoles (electric or magnetic), both located in the interfacial region and contributing to the surface polarization and magnetization densities, one must know $\nu = \text{sign}(z-z')$. In the definition of the surface polarization and magnetization densities this information is lost. The origin of this problem is related to the discussion in chapter 3 about the constitutive relations. The question then addressed was whether the excess surface polarization and magnetization should be expressed in the values of the fields on both sides of the surface or not. It was then found that one could, without any loss of generality, express the surface polarization and magnetization in terms of the average of these extrapolated fields. Thus one finds for instance in chapter 14 on rough surfaces that only the average fields appear in the analysis.

The interaction between dipoles on the surface can therefore be described using the average fields, which, using the above expressions, is found to be given by

$$
\begin{aligned}
\mathbf{N}_{e,+}(\mathbf{k}_{\|},\omega) &= -\mathbf{K}_{ee,+}(\mathbf{k}_{\|},\omega).\mathbf{P}^s(\mathbf{k}_{\|},\omega) + \mathbf{K}_{em,+}(\mathbf{k}_{\|},\omega)\mathbf{M}^s(\mathbf{k}_{\|},\omega) \\
\mathbf{N}_{m,+}(\mathbf{k}_{\|},\omega) &= -\mathbf{K}_{me,+}(\mathbf{k}_{\|},\omega).\mathbf{P}^s(\mathbf{k}_{\|},\omega) - \mathbf{K}_{mm,+}(\mathbf{k}_{\|},\omega).\mathbf{M}^s(\mathbf{k}_{\|},\omega)
\end{aligned}
\tag{12.66}
$$

where the propagators for the average dipole field along the surface are found to be equal to

$$
\begin{aligned}
\mathbf{K}_{ee,+}(\mathbf{k}_{\|},\omega) &= \frac{i}{4}\sum_{\nu}[k^{\nu}_{\perp}\epsilon^{\nu}(\omega)]^{-1}[\mathbf{k}_{\|}\mathbf{k}_{\|} - (\epsilon^{\nu}(\omega))^2 k_{\|}^2\hat{\mathbf{z}}\hat{\mathbf{z}} - (\frac{\omega}{c})^2(n^{\nu}(\omega))^2(1-\hat{\mathbf{z}}\hat{\mathbf{z}})] \\
\mathbf{K}_{me,+}(\mathbf{k}_{\|},\omega) &= -\frac{i}{4}\frac{\omega}{c}\sum_{\nu}(k^{\nu}_{\perp})^{-1}\begin{pmatrix} 0 & 0 & \epsilon^{\nu}k_y \\ 0 & 0 & -\epsilon^{\nu}k_x \\ -\mu^{\nu}k_y & \mu^{\nu}k_x & 0 \end{pmatrix} \\
\mathbf{K}_{mm,+}(\mathbf{k}_{\|},\omega) &= \frac{i}{4}\sum_{\nu}[k^{\nu}_{\perp}\mu^{\nu}(\omega)]^{-1}[\mathbf{k}_{\|}\mathbf{k}_{\|} - (\mu^{\nu}(\omega))^2 k_{\|}^2\hat{\mathbf{z}}\hat{\mathbf{z}} - (\frac{\omega}{c})^2(n^{\nu}(\omega))^2(1-\hat{\mathbf{z}}\hat{\mathbf{z}})] \\
\mathbf{K}_{em,+}(\mathbf{k}_{\|},\omega) &= -\frac{i}{4}\frac{\omega}{c}\sum_{\nu}(k^{\nu}_{\perp})^{-1}\begin{pmatrix} 0 & 0 & \mu^{\nu}k_y \\ 0 & 0 & -\mu^{\nu}k_x \\ -\epsilon^{\nu}k_y & \epsilon^{\nu}k_x & 0 \end{pmatrix}
\end{aligned}
\tag{12.67}
$$

These propagators give the average fields for arbitrary parallel wave vectors and frequencies.

In many applications one may neglect retardation. Setting $\omega = 0$ to obtain the relevant formulae for this case one finds, using eq.(12.22), the following propagators

$$\begin{aligned}
\mathbf{K}_{ee,+}(\mathbf{k}_{\|},\omega &= 0) = \frac{1}{4}\sum_{\nu} k_{\|}(\epsilon^{\nu})^{-1}[\hat{\mathbf{k}}_{\|}\hat{\mathbf{k}}_{\|} - (\epsilon^{\nu})^{2}\hat{\mathbf{z}}\hat{\mathbf{z}}] \\
\mathbf{K}_{me,+}(\mathbf{k}_{\|},\omega &= 0) = 0 \\
\mathbf{K}_{mm,+}(\mathbf{k}_{\|},\omega &= 0) = \frac{1}{4}\sum_{\nu} k_{\|}(\mu^{\nu})^{-1}[\hat{\mathbf{k}}_{\|}\hat{\mathbf{k}}_{\|} - (\mu^{\nu})^{2}\hat{\mathbf{z}}\hat{\mathbf{z}}] \\
\mathbf{K}_{em,+}(\mathbf{k}_{\|},\omega &= 0) = 0 \qquad (12.68)
\end{aligned}$$

This are the expressions used in chapter 14 on rough surfaces.

12.6 Appendix A

The integral to be calculated is

$$\begin{aligned}
\mathbf{F}(k_x,k_y,z,\omega) &= [2\pi\ n^2(\omega)]^{-1}\int dk_z \exp(ik_z z)\frac{\mathbf{k}\mathbf{k} - (\frac{\omega}{c})^2 n^2(\omega)}{k^2 - (\frac{\omega+i0}{c})^2 n^2(\omega)} \\
&= [2\pi\ n^2(\omega)]^{-1}[\begin{pmatrix} k_x k_x & k_x k_y & -ik_x\frac{\partial}{\partial z} \\ k_y k_x & k_y k_y & -ik_y\frac{\partial}{\partial z} \\ -ik_x\frac{\partial}{\partial z} & -ik_y\frac{\partial}{\partial z} & -\frac{\partial^2}{\partial z^2}\end{pmatrix} - (\frac{\omega}{c})^2 n^2(\omega)] \\
&\quad \int dk_z \exp(ik_z z)[k^2 - (\frac{\omega+i0}{c})^2 n^2(\omega)]^{-1} \qquad (12.69)
\end{aligned}$$

The poles in the denominator of the integrand are in $\pm k_{\perp}$ where

$$k_{\perp} \equiv \sqrt{[\frac{\omega}{c}n(\omega)]^2 - k_x^2 - k_y^2} \quad \text{with} \quad \operatorname{Im} k_{\perp} \geq 0 \qquad (12.70)$$

For $z > 0$ the integral must be closed in the upper half plane and for $z < 0$ in the lower half plane. This results in

$$\begin{aligned}
\mathbf{F}(k_x,k_y,z,\omega) &= \frac{i}{2}[k_{\perp}n^2(\omega)]^{-1}[\begin{pmatrix} k_x k_x & k_x k_y & -ik_x\frac{\partial}{\partial z} \\ k_y k_x & k_y k_y & -ik_y\frac{\partial}{\partial z} \\ -ik_x\frac{\partial}{\partial z} & -ik_y\frac{\partial}{\partial z} & -\frac{\partial^2}{\partial z^2}\end{pmatrix} \\
&\quad -(\frac{\omega}{c})^2 n^2(\omega)]\exp(ik_{\perp}|z|) \qquad (12.71)
\end{aligned}$$

If one now carries out the differentiations one obtains the result in eq.(12.21).

The propagator $\mathbf{L}(k_x,k_y,z,\omega)$ is obtained by inverse Fourier transformation of the expression, eq.(12.20), with respect to k_z, which gives

$$\begin{aligned}
\mathbf{L}(k_x,k_y,z,\omega) &= -\frac{\omega}{2\pi c}\begin{pmatrix} 0 & i\frac{\partial}{\partial z} & k_y \\ -i\frac{\partial}{\partial z} & 0 & -k_x \\ -k_y & k_x & 0\end{pmatrix}\int dk_z \frac{\exp(ik_z z)}{[k^2 - (\frac{\omega+i0}{c})^2 n^2(\omega)]} \\
&= -\frac{i}{2}\frac{\omega}{c}k_{\perp}^{-1}\begin{pmatrix} 0 & i\frac{\partial}{\partial z} & k_y \\ -i\frac{\partial}{\partial z} & 0 & -k_x \\ -k_y & k_x & 0\end{pmatrix}\exp(ik_{\perp}|z|) \qquad (12.72)
\end{aligned}$$

If one carries out the differentiation with respect to z, one obtains the result eq.(12.23).

Chapter 13

GENERAL LINEAR RESPONSE THEORY FOR SURFACES

13.1 Introduction

For many aspects of the analysis symmetry and other properties of the constitutive coefficients have been used, without giving a thorough discussion of the background of these relations. It is the aim of this chapter to do so. To first order in the non-locality of the response the constitutive relations were introduced in the third chapter, cf. eqs.(3.42)– (3.45). In the $(\mathbf{k}_\|, \omega)$ representation these constitutive equations become

$$\begin{aligned}
\mathbf{P}_\|^s(\mathbf{k}_\|,\omega) &= \gamma_e(\omega)\mathbf{E}_{\|,+}(\mathbf{k}_\|,\omega) - i\delta_e(\omega)\mathbf{k}_\| D_{z,+}(\mathbf{k}_\|,\omega) + i\frac{\omega}{c}\tau(\omega)\hat{\mathbf{z}}\times\mathbf{H}_{\|,+}(\mathbf{k}_\|,\omega)\\
P_z^s(\mathbf{k}_\|,\omega) &= \beta_e(\omega)D_{z,+}(\mathbf{k}_\|,\omega) + i\,\delta_e(\omega)\mathbf{k}_\|.\mathbf{E}_{\|,+}(\mathbf{k}_\|,\omega)\\
\mathbf{M}_\|^s(\mathbf{k}_\|,\omega) &= \gamma_m(\omega)\mathbf{H}_{\|,+}(\mathbf{k}_\|,\omega) - i\delta_m(\omega)\mathbf{k}_\| B_{z,+}(\mathbf{k}_\|,\omega) + i\frac{\omega}{c}\tau(\omega)\hat{\mathbf{z}}\times\mathbf{E}_{\|,+}(\mathbf{k}_\|,\omega)\\
M_z^s(\mathbf{k}_\|,\omega) &= \beta_m(\omega)B_{z,+}(\mathbf{k}_\|,\omega) + i\,\delta_m(\omega)\mathbf{k}_\|.\mathbf{H}_{\|,+}(\mathbf{k}_\|,\omega)
\end{aligned} \tag{13.1}$$

As discussed in chapter 3 this form of the constitutive relations was chosen such, that it agrees with the various symmetry relations. In order to show that this is indeed the case, and to discuss the general formulation of such relations, it is necessary and convenient to write the constitutive relations in a more general form.

For this purpose it is most appropriate to define a 6-dimensional field, which is a combination of the electric and the magnetic fields:

$$\underline{\mathbf{N}}(\mathbf{r},t) \equiv (\mathbf{N}_e(\mathbf{r},t), \mathbf{N}_m(\mathbf{r},t) = (E_x(\mathbf{r},t), E_y(\mathbf{r},t), D_z(\mathbf{r},t), H_x(\mathbf{r},t), H_y(\mathbf{r},t), B_z(\mathbf{r},t)) \tag{13.2}$$

Similarly a 6-dimensional polarization/magnetization density is introduced

$$\underline{\mathbf{P}}(\mathbf{r},t) \equiv (\mathbf{P}(\mathbf{r},t), \mathbf{M}(\mathbf{r},t)) \tag{13.3}$$

It should be noted that the $\underline{\mathbf{N}}$-field is in general discontinuous at the $z = 0$ surface, while the polarization/magnetization field $\underline{\mathbf{P}}$ also contains a singular contribution at this surface. The general form of the linear relation between these fields is given by

$$\underline{\mathbf{P}}(\mathbf{r},t) = \int d\mathbf{r}' \int dt' \underline{\underline{\boldsymbol{\xi}}}(\mathbf{r},t|\mathbf{r}',t').\underline{\mathbf{N}}(\mathbf{r}',t') \tag{13.4}$$

The response function, which now is a 6×6 tensor, is the sum of bulk and surface contributions:

$$\underline{\underline{\boldsymbol{\xi}}}(\mathbf{r},t|\mathbf{r}',t') = \underline{\underline{\boldsymbol{\xi}}}^{-}(\mathbf{r},t|\mathbf{r}',t')\theta(-z)\theta(-z') + \underline{\underline{\boldsymbol{\xi}}}^{s}(\mathbf{r},t|\mathbf{r}',t')\delta(z)\delta(z') + \underline{\underline{\boldsymbol{\xi}}}^{+}(\mathbf{r},t|\mathbf{r}',t')\theta(z)\theta(z') \tag{13.5}$$

In using this expression it should be noted that the δ-functions should be used in the following manner

$$\delta(z) \equiv \frac{1}{2}[\delta(z-0) + \delta(z+0)] \tag{13.6}$$

where 0 is an infinitesimally small positive number. This definition assures that the response of the surface couples to the average of the asymptotic values of the bulk fields on both sides of the interface. It also distributes the polarization and magnetization in equal amounts on both sides of the $z = 0$ surface. As explained in the previous chapter the fields due to the surface polarization and magnetization do not depend on this location. Thus one may simply shift the contribution on one side of the $z = 0$ surface to the other side and vice versa and add them together. The interfacial constitutive coefficients may be divided in equal parts and after the above procedure added again to give the original values. Cf. in this respect eq.(3.62) for the addition of adjacent films.

If the electric and magnetic fields are real the resulting polarization and magnetization fields are also real. Consequently the response function $\underline{\underline{\boldsymbol{\xi}}}(\mathbf{r},t|\mathbf{r}',t')$ is also real. Below it will be discussed how this reality condition restricts the behavior of the resulting complex wave vector and frequency dependent response function.

The systems considered in this book are assumed to be stationary. This implies that the response function depends only on $t-t'$ and not on both t and t'. Eq.(13.5) then reduces to

$$\begin{aligned}\underline{\underline{\boldsymbol{\xi}}}(\mathbf{r},t-t'|\mathbf{r}') &= \underline{\underline{\boldsymbol{\xi}}}^{-}(\mathbf{r},t-t'|\mathbf{r}')\theta(-z)\theta(-z') + \underline{\underline{\boldsymbol{\xi}}}^{s}(\mathbf{r},t-t'|\mathbf{r}')\delta(z)\delta(z') \\ &\quad + \underline{\underline{\boldsymbol{\xi}}}^{+}(\mathbf{r},t-t'|\mathbf{r}')\theta(z)\theta(z')\end{aligned} \tag{13.7}$$

Fourier transforming this equation with respect to the time one obtains

$$\underline{\underline{\boldsymbol{\xi}}}(\mathbf{r},\omega|\mathbf{r}') = \underline{\underline{\boldsymbol{\xi}}}^{-}(\mathbf{r},\omega|\mathbf{r}')\theta(-z)\theta(-z') + \underline{\underline{\boldsymbol{\xi}}}^{s}(\mathbf{r},\omega|\mathbf{r}')\delta(z)\delta(z') + \underline{\underline{\boldsymbol{\xi}}}^{+}(\mathbf{r},\omega|\mathbf{r}')\theta(z)\theta(z') \tag{13.8}$$

as a function of the frequency.

In the bulk regions the response function used in this book has the form

$$\underline{\underline{\boldsymbol{\xi}}}^{\pm}(\mathbf{r},\omega|\mathbf{r}') = \begin{pmatrix} \epsilon^{\pm}-1 & 0 & 0 & 0 & 0 & 0 \\ 0 & \epsilon^{\pm}-1 & 0 & 0 & 0 & 0 \\ 0 & 0 & 1-1/\epsilon^{\pm} & 0 & 0 & 0 \\ 0 & 0 & 0 & \mu^{\pm}-1 & 0 & 0 \\ 0 & 0 & 0 & 0 & \mu^{\pm}-1 & 0 \\ 0 & 0 & 0 & 0 & 0 & 1-1/\mu^{\pm} \end{pmatrix} \delta(\mathbf{r}-\mathbf{r}') \tag{13.9}$$

where ϵ^-, ϵ^+, μ^- and μ^+ are all functions of the frequency. The response function for the surface is given by

$$\underline{\underline{\boldsymbol{\xi}}}^{s}(\mathbf{r}_{\|},\omega|\mathbf{r}'_{\|}) = \begin{pmatrix} \gamma_e & 0 & -\delta_e\partial_x & 0 & -i\omega\tau/c & 0 \\ 0 & \gamma_e & -\delta_e\partial_y & i\omega\tau/c & 0 & 0 \\ \delta_e\partial_x & \delta_e\partial_y & \beta_e & 0 & 0 & 0 \\ 0 & -i\omega\tau/c & 0 & \gamma_m & 0 & -\delta_m\partial_x \\ i\omega\tau/c & 0 & 0 & 0 & \gamma_m & -\delta_m\partial_y \\ 0 & 0 & 0 & \delta_m\partial_x & \delta_m\partial_y & \beta_m \end{pmatrix} \delta(\mathbf{r}_{\|}-\mathbf{r}'_{\|}) \tag{13.10}$$

In this expression γ_e, β_e, δ_e, τ, γ_m, β_m and δ_m are in general functions of the frequency. Furthermore the short-hand notation $\partial_x \equiv \partial/\partial x$ and $\partial_y \equiv \partial/\partial y$ was used.

13.2 Green functions

In order to derive the appropriate symmetry relations for the constitutive matrix $\xi_{ij}^{\alpha\beta}$, where $i,j = x,y,z$ and $\alpha,\beta = e,m$, the so-called source observer symmetry will be used. Writing the solution for the field of an arbitrary polarization and magnetization distribution using Green functions, one has

$$\underline{\mathbf{N}}(\mathbf{r},t) = \int d\mathbf{r}' \int dt' \underline{\underline{\mathbf{G}}}_0(\mathbf{r},t|\mathbf{r}',t').\underline{\mathbf{P}}(\mathbf{r}',t') \tag{13.11}$$

The Fourier transformed Green function in this equation is the one for vacuum and can be found explicitly using eqs.(12.17)–(12.20) with $\epsilon = \mu = 1$. As this explicit expression will not be needed, it will not be given here. Source observer symmetry, [12], now implies that

$$G_{0,ij}^{\alpha\beta}(\mathbf{r},t|\mathbf{r}',t') = (2\ \delta_{\alpha\beta}-1)G_{0,ji}^{\beta\alpha}(\mathbf{r}',t|\mathbf{r},t') \tag{13.12}$$

Using stationarity, this gives as a function of the frequency

$$G_{0,ij}^{\alpha\beta}(\mathbf{r},\omega|\mathbf{r}') = (2\ \delta_{\alpha\beta}-1)G_{0,ji}^{\beta\alpha}(\mathbf{r}',\omega|\mathbf{r}) \tag{13.13}$$

Fourier transformation with respect to space, cf. eq. (5.20), then gives

$$\begin{aligned} G_{0,ij}^{\alpha\beta}(\mathbf{k},\omega|\mathbf{k}') &\equiv \int d\mathbf{r} \int d\mathbf{r}' e^{-i\mathbf{k}.\mathbf{r}} G_{0,ij}^{\alpha\beta}(\mathbf{r},\omega|\mathbf{r}') e^{i\mathbf{k}'.\mathbf{r}'} \\ &= (2\ \delta_{\alpha\beta}-1) G_{0,ji}^{\beta\alpha}(-\mathbf{k}',\omega|-\mathbf{k})) \end{aligned} \tag{13.14}$$

In the previous section only induced polarization/magnetization distributions were considered. The above expressions are valid also if one uses the total polarization and magnetization densities as sources. To distinguish contributions which are induced and externally controlled, the polarization/magnetization density is written as

$$\underline{\mathbf{P}}(\mathbf{r},t)=\underline{\mathbf{P}}_{ind}(\mathbf{r},t)+\underline{\mathbf{P}}_{ext}(\mathbf{r},t)=\int d\mathbf{r}'\int dt'\underline{\underline{\boldsymbol{\xi}}}(\mathbf{r},t|\mathbf{r}',t').\underline{\mathbf{N}}(\mathbf{r}',t')+\underline{\mathbf{P}}_{ext}(\mathbf{r},t) \tag{13.15}$$

In a more formal notation this may be written in the simple form

$$\underline{\mathbf{P}}=\underline{\mathbf{P}}_{ind}+\underline{\mathbf{P}}_{ext}=\underline{\underline{\boldsymbol{\xi}}}.\underline{\mathbf{N}}+\underline{\mathbf{P}}_{ext} \tag{13.16}$$

One may also express the fields in terms of the externally controlled polarization and magnetization densities alone:

$$\underline{\mathbf{N}}(\mathbf{r},t)=\int d\mathbf{r}'\int dt'\underline{\underline{\mathbf{G}}}(\mathbf{r},t|\mathbf{r}',t').\underline{\mathbf{P}}_{ext}(\mathbf{r}',t') \tag{13.17}$$

Because of the source observer symmetry the Green function in the medium satisfies the same symmetry relations as the Green function in vacuum. Therefore one has

$$G_{ij}^{\alpha\beta}(\mathbf{r},t|\mathbf{r}',t')=(2\ \delta_{\alpha\beta}-1)G_{ji}^{\beta\alpha}(\mathbf{r}',t|\mathbf{r},t') \tag{13.18}$$

Using stationarity, this gives as a function of the frequency

$$G_{ij}^{\alpha\beta}(\mathbf{r},\omega|\mathbf{r}')=(2\ \delta_{\alpha\beta}-1)G_{ji}^{\beta\alpha}(\mathbf{r}',\omega|\mathbf{r}) \tag{13.19}$$

Fourier transformation with respect to space furthermore gives

$$G_{ij}^{\alpha\beta}(\mathbf{k},\omega|\mathbf{k}')=(2\ \delta_{\alpha\beta}-1)G_{ji}^{\beta\alpha}(-\mathbf{k}',\omega|-\mathbf{k}) \tag{13.20}$$

In order to derive the symmetry relations for the constitutive coefficients, it is convenient to write eqs.(13.11) and (13.17) in the more formal form

$$\underline{\mathbf{N}}=\underline{\underline{\mathbf{G}}}_0.\underline{\mathbf{P}},\quad \underline{\mathbf{N}}=\underline{\underline{\mathbf{G}}}.\underline{\mathbf{P}}_{ext} \tag{13.21}$$

Inverting these equations one obtains

$$\underline{\mathbf{P}}=\underline{\underline{\mathbf{G}}}_0^{-1}.\underline{\mathbf{N}},\quad \underline{\mathbf{P}}_{ext}=\underline{\underline{\mathbf{G}}}^{-1}.\underline{\mathbf{N}} \tag{13.22}$$

Substitution into eq.(13.16) then leads to the identity

$$\underline{\underline{\boldsymbol{\xi}}}=\underline{\underline{\mathbf{G}}}_0^{-1}-\underline{\underline{\mathbf{G}}}^{-1} \tag{13.23}$$

In view of the fact that the inverse of the Green functions satisfy the same symmetry relations as the Green functions, it follows that the constitutive matrix also satisfies these symmetry relations.

Thus it follows that

$$\xi_{ij}^{\alpha\beta}(\mathbf{r},t|\mathbf{r}',t') = (2\ \delta_{\alpha\beta} - 1)\xi_{ji}^{\beta\alpha}(\mathbf{r}',t|\mathbf{r},t') \tag{13.24}$$

Using stationarity, this gives as a function of the frequency

$$\xi_{ij}^{\alpha\beta}(\mathbf{r},\omega|\mathbf{r}') = (2\ \delta_{\alpha\beta} - 1)\xi_{ji}^{\beta\alpha}(\mathbf{r}',\omega|\mathbf{r}) \tag{13.25}$$

Fourier transformation with respect to space furthermore gives

$$\xi_{ij}^{\alpha\beta}(\mathbf{k},\omega|\mathbf{k}') = (2\ \delta_{\alpha\beta} - 1)\xi_{ji}^{\beta\alpha}(-\mathbf{k}',\omega| - \mathbf{k}) \tag{13.26}$$

Comparing the general form of the symmetry relation given in eq.(13.25) with the form of the constitutive matrix used in this book and given explicitly in eq.(13.8) together with eqs.(13.9) and (13.10) it is clear that this form has been chosen in agreement with this general source observer symmetry [12]. In particular it follows that there is only one τ and not two coefficients τ_e and τ_m. This is a point, which is not so clear without explaining it in the general context given in this chapter.

The systems considered in this book are usually translationally invariant along the surface. Thus one has

$$\xi_{ij}^{\alpha\beta}(\mathbf{r},\omega|\mathbf{r}') = \xi_{ij}^{\alpha\beta}(\mathbf{r}_\parallel - \mathbf{r}'_\parallel, z, \omega|z') \tag{13.27}$$

It is then often convenient to Fourier transform along the surface but not in the direction normal to the surface. The appropriate symmetry relations then become

$$\xi_{ij}^{\alpha\beta}(\mathbf{k}_\parallel, z, \omega|z') = (2\ \delta_{\alpha\beta} - 1)\xi_{ji}^{\beta\alpha}(-\mathbf{k}_\parallel, z', \omega|z) \tag{13.28}$$

For the contribution due to the surface this reduces to

$$\xi_{ij}^{s,\alpha\beta}(\mathbf{k}_\parallel,\omega) = (2\ \delta_{\alpha\beta} - 1)\xi_{ji}^{s,\beta\alpha}(-\mathbf{k}_\parallel,\omega) \tag{13.29}$$

The surface contribution given in eq.(13.10) has the property of translational invariance along the surface and becomes, as a function of wave vector and frequency,

$$\underline{\underline{\boldsymbol{\xi}}}^s(\mathbf{k}_\parallel,\omega) = \begin{pmatrix} \gamma_e & 0 & -i\delta_e k_x & 0 & -i\omega\tau/c & 0 \\ 0 & \gamma_e & -i\delta_e k_y & i\omega\tau/c & 0 & 0 \\ i\delta_e k_x & i\delta_e k_y & \beta_e & 0 & 0 & 0 \\ 0 & -i\omega\tau/c & 0 & \gamma_m & 0 & -i\delta_m k_x \\ i\omega\tau/c & 0 & 0 & 0 & \gamma_m & -i\delta_m k_y \\ 0 & 0 & 0 & i\delta_m k_x & i\delta_m k_y & \beta_m \end{pmatrix} \tag{13.30}$$

This expression clearly has the symmetry derived above and given in eq.(13.29). Up to linear order in wave vector and frequency it is the most general expression, which satisfies this symmetry condition. This proves that the form of the constitutive equations given in eq.(13.1) is the most general form possible, consistent with the source observer symmetry. The general form of the symmetry relation given above, shows the way how to generalize this form to include, for instance, terms of a higher power in the wave vector and frequency.

Chapter 14
SURFACE ROUGHNESS

14.1 Introduction

In this chapter formulae for the susceptibilities of a so-called rough surface will be derived. Before this is done it is useful to describe the characteristics of a rough surface. First some examples are given. The most simple one is a surface of a solid substrate which is not flat. Due to the manner in which a surface is prepared it may have little holes or bumps which affect the optical properties of the surface. Usually these holes or bumps are stationary and have a shape which is rather irregular and cannot be modeled in terms of truncated spheroidal islands. A new method is needed. A non-stationary example are capillary waves. Such waves occur at a fluid-fluid interface due to thermal agitation and lead to a, on an optical time scale, slowly varying structure of the interface, very similar to the structure of the above mentioned rough solid substrate surface. A third example is a metal surface with a layer of oxide. In that case the metal surface resembles the above mentioned rough surface of a solid substrate while on top of this metal surface there are pockets of metal oxide.

All of these surfaces have a complicated structure, which usually occupies a layer with a thickness small compared to the wavelength of light. Though not mentioned yet, also the magnetic properties of the materials involved may be different from one phase to the next phase. As a consequence not only the electric but also the magnetic surface susceptibilities may be different from zero. In order to develop a theoretical description of these systems an expansion will be used in the thickness of this interface to second order. Such a development is most appropriate if the structural features along the interface are characterized by a length scale larger than the thickness. The analysis is quite general. Any distribution of material within the just formulated limitations may be described in this manner. The thin films of (truncated) spheroids and spheres discussed in the previous chapters are clearly not describable with the techniques developed in this chapter. The fact that they have already been analyzed in a much more rigorous fashion makes their analysis, using a roughness model, unnecessary. Caps, however, may also be described using the techniques developed in this chapter, this in particular for the case of thin caps. This case will, for comparison of the methods, be treated in some detail in the last section of this chapter.

The derivation given in this chapter, though based on earlier work of the authors and collaborators [[57], [58], [59], [60], [61], [62], [63], [64]], is new, more general and considerably simplified, using the insight gained in this earlier work.

Owing to the fact that the expansion is to second order in the thickness, the expressions for the interfacial susceptibilities will have contributions due to correlations in the distribution of the material along the surface. Thus one finds, that both the average square height of the roughness and the height-height correlation function at different positions contribute.

As a first step in the analysis one uses the formulae found for the stratified surface from chapter 11. For this purpose one simply divides the whole interface in thin layers, as was similarly done in chapter 11, and one proceeds to replace the whole stack of layers by interfacial susceptibilities. In this way one obtains the following, in this case $x-$ and $y-$ (and $\omega-$) dependent, susceptibilities,

$$\begin{aligned}
\gamma_e(x,y) &= \int_{-\infty}^{\infty} dz[\epsilon_{\parallel}(\mathbf{r}) - \epsilon_0(\mathbf{r})] \\
\beta_e(x,y) &= -\int_{-\infty}^{\infty} dz\{[\epsilon_z(\mathbf{r})]^{-1} - [\epsilon_0(\mathbf{r})]^{-1}\} \\
\gamma_m(x,y) &= \int_{-\infty}^{\infty} dz[\mu_{\parallel}(\mathbf{r}) - \mu_0(\mathbf{r})] \\
\beta_m(x,y) &= -\int_{-\infty}^{\infty} dz\{[\mu_z(\mathbf{r})]^{-1} - [\mu_0(\mathbf{r})]^{-1}\}
\end{aligned}$$

$$\begin{aligned}
\delta_e(x,y) &= -\int_{-\infty}^{\infty} dz\ z[\epsilon_{\parallel}(\mathbf{r}) - \epsilon_0(\mathbf{r})]/\ \epsilon_0(\mathbf{r}) + \int_{-\infty}^{\infty} dz\ z\{[\epsilon_z(\mathbf{r})]^{-1} - [\epsilon_0(\mathbf{r})]^{-1}\}\epsilon_0(\mathbf{r}) \\
&\quad + \frac{1}{2}\int_{-\infty}^{\infty} dz \int_{-\infty}^{\infty} dz' \operatorname{sign}(z'-z)[\epsilon_{\parallel}(\mathbf{r}) - \epsilon_0(\mathbf{r})]\{[\epsilon_z(\mathbf{r}')]^{-1} - [\epsilon_0(\mathbf{r}')]^{-1}\}
\end{aligned}$$

$$\begin{aligned}
\delta_m(x,y) &= -\int_{-\infty}^{\infty} dz\ z[\mu_{\parallel}(\mathbf{r}) - \mu_0(\mathbf{r})]/\ \mu_0(\mathbf{r}) + \int_{-\infty}^{\infty} dz\ z\{[\mu_z(\mathbf{r})]^{-1} - [\mu_0(\mathbf{r})]^{-1}\}\mu_0(\mathbf{r}) \\
&\quad + \frac{1}{2}\int_{-\infty}^{\infty} dz \int_{-\infty}^{\infty} dz' \operatorname{sign}(z'-z)[\mu_{\parallel}(\mathbf{r}) - \mu_0(\mathbf{r})]\{[\mu_z(\mathbf{r}')]^{-1} - [\mu_0(\mathbf{r}')]^{-1}\}
\end{aligned}$$

$$\begin{aligned}
\tau(x,y) &= -\int_{-\infty}^{\infty} dz\ z[\epsilon_{\parallel}(\mathbf{r}) - \epsilon_0(\mathbf{r})]\mu_0(\mathbf{r}) + \int_{-\infty}^{\infty} dz\ z[\mu_{\parallel}(\mathbf{r}) - \mu_0(\mathbf{r})]\epsilon_0(\mathbf{r}) \\
&\quad + \frac{1}{2}\int_{-\infty}^{\infty} dz \int_{-\infty}^{\infty} dz' \operatorname{sign}(z'-z)[\epsilon_{\parallel}(\mathbf{r}) - \epsilon_0(\mathbf{r})][\mu_{\parallel}(\mathbf{r}') - \mu_0(\mathbf{r}')] \quad (14.1)
\end{aligned}$$

All these constitutive coefficients have been given with the $x-y$ plane as the dividing surface. It will be clear that the $x-y$ plane should be chosen in, or very close to, the rough interfacial layer. If the dielectric constant and the magnetic permeability are independent of x, y the above formulae reduce to the formulae, (11.5), given in chapter 11 on stratified media.

The above constitutive coefficients vary as a function of the position along the surface. In order to obtain the desired non-fluctuating constitutive coefficients it will be necessary to average the solution of the wave equation over the distribution of material along the surface. This will be done in the next section. In subsequent sections the resulting formulae will be applied to a number of examples.

14.2 General theory

The analysis in this section shows, how one obtains the expressions for the effective susceptibilities in terms of the fluctuating susceptibilities, given in the preceding section. The final expressions are given at the end of this section in eq.(14.16). These expressions are the sum of straight averages of the fluctuating coefficients and contributions due to correlations of these coefficients along the surface. One may skip most of this section and directly go to these expressions, which can be used in combination with the definitions of the correlation functions given in eq.(14.11), together with eq.(14.15).

The variations of the constitutive coefficients, given above, along the substrate will result in corresponding fluctuations in the value of the interfacial polarization and magnetization densities. Using eq.(3.42) through (3.45) the fluctuating interfacial polarization density, $\mathbf{p}^s$, and magnetization density, $\mathbf{m}^s$, are given by

$$\begin{aligned}
\mathbf{p}_{\|}^s(\mathbf{r}_{\|},\omega) &= \gamma_e(\mathbf{r}_{\|},\omega)\mathbf{e}_{\|,+}(\mathbf{r}_{\|},\omega) - \delta_e(\mathbf{r}_{\|},\omega)\nabla_{\|}d_{z,+}(\mathbf{r}_{\|},\omega) + i\frac{\omega}{c}\tau(\mathbf{r}_{\|},\omega)\hat{\mathbf{z}}\times\mathbf{h}_{\|,+}(\mathbf{r}_{\|},\omega)\\
p_z^s(\mathbf{r}_{\|},\omega) &= \beta_e(\mathbf{r}_{\|},\omega)d_{z,+}(\mathbf{r}_{\|},\omega) + \delta_e(\mathbf{r}_{\|},\omega)\nabla_{\|}.\mathbf{e}_{\|,+}(\mathbf{r}_{\|},\omega)\\
\mathbf{m}_{\|}^s(\mathbf{r}_{\|},\omega) &= \gamma_m(\mathbf{r}_{\|},\omega)\mathbf{h}_{\|,+}(\mathbf{r}_{\|},\omega) - \delta_m(\mathbf{r}_{\|},\omega)\nabla_{\|}b_{z,+}(\mathbf{r}_{\|},\omega) + i\frac{\omega}{c}\tau(\mathbf{r}_{\|},\omega)\hat{\mathbf{z}}\times\mathbf{e}_{\|,+}(\mathbf{r}_{\|},\omega)\\
m_z^s(\mathbf{r}_{\|},\omega) &= \beta_m(\mathbf{r}_{\|},\omega)b_{z,+}(\mathbf{r}_{\|},\omega) + \delta_m(\mathbf{r}_{\|},\omega)\nabla_{\|}.\mathbf{h}_{\|,+}(\mathbf{r}_{\|},\omega) \qquad (14.2)
\end{aligned}$$

Due to the fluctuations of the interfacial polarization and magnetization densities the fields also fluctuate. These fluctuating fields have also been indicated by small letters.

The macroscopic polarization and magnetization densities and fields are given by the average of the fluctuating polarization and magnetization densities and fields. Averaging the above equation for the fluctuating interfacial polarization and magnetization densities one obtains

$$\begin{aligned}
\mathbf{P}_{\|}^s(\mathbf{r}_{\|},\omega) &= <\mathbf{p}_{\|}^s(\mathbf{r}_{\|},\omega)>=<\gamma_e(\mathbf{r}_{\|},\omega)\mathbf{e}_{\|,+}(\mathbf{r}_{\|},\omega)> - <\delta_e(\mathbf{r}_{\|},\omega)\nabla_{\|}d_{z,+}(\mathbf{r}_{\|},\omega)>\\
&\quad +i\frac{\omega}{c}<\tau(\mathbf{r}_{\|},\omega)\hat{\mathbf{z}}\times\mathbf{h}_{\|,+}(\mathbf{r}_{\|},\omega)>\\
P_z^s(\mathbf{r}_{\|},\omega) &= <p_z^s(\mathbf{r}_{\|},\omega)>=<\beta_e(\mathbf{r}_{\|},\omega)d_{z,+}(\mathbf{r}_{\|},\omega)> + <\delta_e(\mathbf{r}_{\|},\omega)\nabla_{\|}.\mathbf{e}_{\|,+}(\mathbf{r}_{\|},\omega)>\\
\mathbf{M}_{\|}^s(\mathbf{r}_{\|},\omega) &= <\mathbf{m}_{\|}^s(\mathbf{r}_{\|},\omega)>=<\gamma_m(\mathbf{r}_{\|},\omega)\mathbf{h}_{\|,+}(\mathbf{r}_{\|},\omega)>\\
&- <\delta_m(\mathbf{r}_{\|},\omega)\nabla_{\|}b_{z,+}(\mathbf{r}_{\|},\omega)> +i\frac{\omega}{c}<\tau(\mathbf{r}_{\|},\omega)\hat{\mathbf{z}}\times\mathbf{e}_{\|,+}(\mathbf{r}_{\|},\omega)>\\
M_z^s(\mathbf{r}_{\|},\omega) &= <m_z^s(\mathbf{r}_{\|},\omega)>\\
&= <\beta_m(\mathbf{r}_{\|},\omega)b_{z,+}(\mathbf{r}_{\|},\omega)> + <\delta_m(\mathbf{r}_{\|},\omega)\nabla_{\|}.\mathbf{h}_{\|,+}(\mathbf{r}_{\|},\omega)>
\end{aligned} \qquad (14.3)$$

The average is indicated by $< ... >$. The average is an average over the distribution of the roughness along the surface. Depending on ones preference, one may either consider this to be a surface average or an average over an ensemble of equivalent surfaces. In the analysis in this book the distribution of the material along the surface will always be both homogeneous and isotropic. The right-hand sides of the above expressions show, that all the terms on that side in principle contain contributions due to correlations of the fluctuating susceptibilities with the fluctuating fields. It follows from the expressions for the fluctuating susceptibilities given in the first section, that $\gamma_e(\mathbf{r}_\|,\omega)$, $\beta_e(\mathbf{r}_\|,\omega)$, $\gamma_m(\mathbf{r}_\|,\omega)$ and $\beta_m(\mathbf{r}_\|,\omega)$ are linear in the thickness of the surface, while $\delta_e(\mathbf{r}_\|,\omega)$, $\delta_m(\mathbf{r}_\|,\omega)$ and $\tau(\mathbf{r}_\|,\omega)$ are quadratic.

If one now writes the contribution containing $\gamma_e(\mathbf{r}_\|,\omega)$ one may write

$$\begin{aligned}
&< \gamma_e(\mathbf{r}_\|,\omega)\mathbf{e}_{\|,+}(\mathbf{r}_\|,\omega) >=< \gamma_e(\mathbf{r}_\|,\omega) >< \mathbf{e}_{\|,+}(\mathbf{r}_\|,\omega) > \\
+ &< (\gamma_e(\mathbf{r}_\|,\omega)- < \gamma_e(\mathbf{r}_\|,\omega) >)(\mathbf{e}_{\|,+}(\mathbf{r}_\|,\omega)- < \mathbf{e}_{\|,+}(\mathbf{r}_\|,\omega) >) \\
= &< \gamma_e(\mathbf{r}_\|,\omega) > \mathbf{E}_{\|,+}(\mathbf{r}_\|,\omega) \\
+ &< (\gamma_e(\mathbf{r}_\|,\omega)- < \gamma_e(\mathbf{r}_\|,\omega) >)(\mathbf{e}_{\|,+}(\mathbf{r}_\|,\omega) - \mathbf{E}_{\|,+}(\mathbf{r}_\|,\omega)) >
\end{aligned} \tag{14.4}$$

The first term on the right-hand side contains no contributions due to correlations between the fluctuating susceptibility and the fluctuating field. It gives as contribution to the susceptibility for the average fields the straight average of the fluctuating susceptibility. This term is of the first order in the thickness of the interface. The second term on the right-hand side gives the contributions due to correlations between the fluctuating susceptibility and the fluctuating field. This term is of the second order in the thickness of the interface. The contributions containing $\beta_e(\mathbf{r}_\|,\omega)$, $\gamma_m(\mathbf{r}_\|,\omega)$ and $\beta_m(\mathbf{r}_\|,\omega)$ similarly can be written as the sum of a first order contribution, given by the straight average of the fluctuating susceptibility, and a second order contribution due to correlations of this fluctuating susceptibility with the corresponding field. The contributions containing $\delta_e(\mathbf{r}_\|,\omega)$, $\delta_m(\mathbf{r}_\|,\omega)$ and $\tau(\mathbf{r}_\|,\omega)$ similarly can be written as the sum of a second order contribution, given by the straight average of the fluctuating susceptibility, and a third order contribution due to correlations of this fluctuating susceptibility with the corresponding field.

In view of the fact that the analysis will only go to second order, third order contributions may be dropped an one obtains in this way

$$\begin{aligned}
\mathbf{P}^s_\|(\mathbf{r}_\|,\omega) &= < \gamma_e(\mathbf{r}_\|,\omega) > \mathbf{E}_{\|,+}(\mathbf{r}_\|,\omega) \\
&+ < (\gamma_e(\mathbf{r}_\|,\omega)- < \gamma_e(\mathbf{r}_\|,\omega) >)(\mathbf{e}_{\|,+}(\mathbf{r}_\|,\omega) - \mathbf{E}_{\|,+}(\mathbf{r}_\|,\omega)) > \\
&- < \delta_e(\mathbf{r}_\|,\omega) > \nabla_\| D_{z,+}(\mathbf{r}_\|,\omega) + i\frac{\omega}{c} < \tau(\mathbf{r}_\|,\omega) > \hat{\mathbf{z}} \times \mathbf{H}_{\|,+}(\mathbf{r}_\|,\omega) \\
P^s_z(\mathbf{r}_\|,\omega) &= < \beta_e(\mathbf{r}_\|,\omega) > D_{z,+}(\mathbf{r}_\|,\omega) \\
&+ < (\beta_e(\mathbf{r}_\|,\omega)- < \beta_e(\mathbf{r}_\|,\omega) >)(d_{z,+}(\mathbf{r}_\|,\omega) - D_{z,+}(\mathbf{r}_\|,\omega)) > \\
&+ < \delta_e(\mathbf{r}_\|,\omega) > \nabla_\|.\mathbf{E}_{\|,+}(\mathbf{r}_\|,\omega) \\
\mathbf{M}^s_\|(\mathbf{r}_\|,\omega) &= < \gamma_m(\mathbf{r}_\|,\omega) > \mathbf{H}_{\|,+}(\mathbf{r}_\|,\omega) \\
&+ < (\gamma_m(\mathbf{r}_\|,\omega)- < \gamma_m(\mathbf{r}_\|,\omega) >)(\mathbf{h}_{\|,+}(\mathbf{r}_\|,\omega) - \mathbf{H}_{\|,+}(\mathbf{r}_\|,\omega)) > \\
&- < \delta_m(\mathbf{r}_\|,\omega) > \nabla_\| B_{z,+}(\mathbf{r}_\|,\omega) + i\frac{\omega}{c} < \tau(\mathbf{r}_\|,\omega) > \hat{\mathbf{z}} \times \mathbf{E}_{\|,+}(\mathbf{r}_\|,\omega) \\
M^s_z(\mathbf{r}_\|,\omega) &= < \beta_m(\mathbf{r}_\|,\omega) > B_{z,+}(\mathbf{r}_\|,\omega) \\
&+ < (\beta_m(\mathbf{r}_\|,\omega)- < \beta_m(\mathbf{r}_\|,\omega) >)(b_{z,+}(\mathbf{r}_\|,\omega) - B_{z,+}(\mathbf{r}_\|,\omega)) > \\
&+ < \delta_m(\mathbf{r}_\|,\omega) > \nabla_\|.\mathbf{H}_{\|,+}(\mathbf{r}_\|,\omega)
\end{aligned} \tag{14.5}$$

In order to obtain explicit expressions for the contributions due to correlations, one needs expressions for the field fluctuations in terms of the fluctuations of the interfacial susceptibilities.

The first important observation is that, in order to obtain the susceptibilities to the second order in the thickness of the interface, it is sufficient to use expressions for the fluctuations of the fields to linear order in this thickness. This is a consequence of the fact that in eq.(14.5) the fluctuations of the fields appear always in combination with the fluctuation of the constitutive coefficients. To first order the sources of these fields are

$$\begin{aligned}
\mathbf{p}^s_\parallel(\mathbf{r}_\parallel,\omega)-\mathbf{P}^s_\parallel(\mathbf{r}_\parallel,\omega) &= [\gamma_e(\mathbf{r}_\parallel,\omega)-<\gamma_e(\mathbf{r}_\parallel,\omega)>]\mathbf{E}_{\parallel,+}(\mathbf{r}_\parallel,\omega)\\
p^s_z(\mathbf{r}_\parallel,\omega)-P^s_z(\mathbf{r}_\parallel,\omega) &= [\beta_e(\mathbf{r}_\parallel,\omega)-<\beta_e(\mathbf{r}_\parallel,\omega)>]D_{z,+}(\mathbf{r}_\parallel,\omega)\\
\mathbf{m}^s_\parallel(\mathbf{r}_\parallel,\omega)-\mathbf{M}^s_\parallel(\mathbf{r}_\parallel,\omega) &= [\gamma_m(\mathbf{r}_\parallel,\omega)-<\gamma_m(\mathbf{r}_\parallel,\omega)>]\mathbf{H}_{\parallel,+}(\mathbf{r}_\parallel,\omega)\\
m^s_z(\mathbf{r}_\parallel,\omega)-M^s_z(\mathbf{r}_\parallel,\omega) &= [\beta_m(\mathbf{r}_\parallel,\omega)-<\beta_m(\mathbf{r}_\parallel,\omega)>]B_{z,+}(\mathbf{r}_\parallel,\omega) \qquad (14.6)
\end{aligned}$$

as follows from eq.(14.2). The resulting contribution to the field is found, using eq.(12.66), to be given by

$$\begin{aligned}
\mathbf{n}_{e,+}(\mathbf{r}_\parallel,\omega)-\mathbf{N}_{e,+}(\mathbf{r}_\parallel,\omega) &= -\int d\mathbf{r}'_\parallel \mathbf{K}_{ee,+}(\mathbf{r}_\parallel-\mathbf{r}'_\parallel,\omega).[\mathbf{p}^s(\mathbf{r}'_\parallel,\omega)-\mathbf{P}^s(\mathbf{r}'_\parallel,\omega)]\\
&\quad +\int d\mathbf{r}'_\parallel \mathbf{K}_{em,+}(\mathbf{r}_\parallel-\mathbf{r}'_\parallel,\omega).[\mathbf{m}^s(\mathbf{r}'_\parallel,\omega)-\mathbf{M}^s(\mathbf{r}'_\parallel,\omega)]\\
\mathbf{n}_{m,+}(\mathbf{r}_\parallel,\omega)-\mathbf{N}_{m,+}(\mathbf{r}_\parallel,\omega) &= -\int d\mathbf{r}'_\parallel \mathbf{K}_{me,+}(\mathbf{r}_\parallel-\mathbf{r}'_\parallel,\omega).[\mathbf{p}^s(\mathbf{r}'_\parallel,\omega)-\mathbf{P}^s(\mathbf{r}'_\parallel,\omega)]\\
&\quad -\int d\mathbf{r}'_\parallel \mathbf{K}_{mm,+}(\mathbf{r}_\parallel-\mathbf{r}'_\parallel,\omega).[\mathbf{m}^s(\mathbf{r}'_\parallel,\omega)-\mathbf{M}^s(\mathbf{r}'_\parallel,\omega)]
\end{aligned} \qquad (14.7)$$

where the fluctuating non-singular fields defined by $\mathbf{n}_e \equiv (e_x, e_y, d_z)$ and $\mathbf{n}_m \equiv (h_x, h_y, b_z)$, cf. eqs.(12.26) and (12.27), have been used. Substitution of eq.(14.6) then results in

$$\begin{aligned}
&\mathbf{e}_{\parallel,+}(\mathbf{r}_\parallel,\omega)-\mathbf{E}_{\parallel,+}(\mathbf{r}_\parallel,\omega)\\
=\ & -(\mathbf{1}-\hat{\mathbf{z}}\hat{\mathbf{z}}).\int d\mathbf{r}'_\parallel \mathbf{K}_{ee,+}(\mathbf{r}_\parallel-\mathbf{r}'_\parallel,\omega).\{[\gamma_e(\mathbf{r}'_\parallel,\omega)-<\gamma_e(\mathbf{r}'_\parallel,\omega)>]\mathbf{E}_{\parallel,+}(\mathbf{r}'_\parallel,\omega)\\
& +[\beta_e(\mathbf{r}'_\parallel,\omega)-<\beta_e(\mathbf{r}'_\parallel,\omega)>]D_{z,+}(\mathbf{r}'_\parallel,\omega)\hat{\mathbf{z}}\}\\
& +(\mathbf{1}-\hat{\mathbf{z}}\hat{\mathbf{z}}).\int d\mathbf{r}'_\parallel \mathbf{K}_{em,+}(\mathbf{r}_\parallel-\mathbf{r}'_\parallel,\omega).\{[\gamma_m(\mathbf{r}'_\parallel,\omega)-<\gamma_m(\mathbf{r}'_\parallel,\omega)>]\mathbf{H}_{\parallel,+}(\mathbf{r}'_\parallel,\omega)\\
& +[\beta_m(\mathbf{r}'_\parallel,\omega)-<\beta_m(\mathbf{r}'_\parallel,\omega)>]B_{z,+}(\mathbf{r}'_\parallel,\omega)\hat{\mathbf{z}}\}
\end{aligned}$$

$$\begin{aligned}
&d_{z,+}(\mathbf{r}_\parallel,\omega)-D_{z,+}(\mathbf{r}_\parallel,\omega)\\
=\ & -\hat{\mathbf{z}}.\int d\mathbf{r}'_\parallel \mathbf{K}_{ee,+}(\mathbf{r}_\parallel-\mathbf{r}'_\parallel,\omega).\{[\gamma_e(\mathbf{r}'_\parallel,\omega)-<\gamma_e(\mathbf{r}'_\parallel,\omega)>]\mathbf{E}_{\parallel,+}(\mathbf{r}'_\parallel,\omega)\\
& +[\beta_e(\mathbf{r}'_\parallel,\omega)-<\beta_e(\mathbf{r}'_\parallel,\omega)>]D_{z,+}(\mathbf{r}'_\parallel,\omega)\hat{\mathbf{z}}\}\\
& +\hat{\mathbf{z}}.\int d\mathbf{r}'_\parallel \mathbf{K}_{em,+}(\mathbf{r}_\parallel-\mathbf{r}'_\parallel,\omega).\{[\gamma_m(\mathbf{r}'_\parallel,\omega)-<\gamma_m(\mathbf{r}'_\parallel,\omega)>]\mathbf{H}_{\parallel,+}(\mathbf{r}'_\parallel,\omega)\\
& +[\beta_m(\mathbf{r}'_\parallel,\omega)-<\beta_m(\mathbf{r}'_\parallel,\omega)>]B_{z,+}(\mathbf{r}'_\parallel,\omega)\hat{\mathbf{z}}\}
\end{aligned}$$

$$\begin{aligned}
&\mathbf{h}_{\|,+}(\mathbf{r}_\|,\omega) - \mathbf{H}_{\|,+}(\mathbf{r}_\|,\omega) \\
= \; & -(\mathbf{1}-\hat{\mathbf{z}}\hat{\mathbf{z}}).\int d\mathbf{r}'_\| \mathbf{K}_{me,+}(\mathbf{r}_\| - \mathbf{r}'_\|,\omega).\{[\gamma_e(\mathbf{r}'_\|,\omega) - <\gamma_e(\mathbf{r}'_\|,\omega)>]\mathbf{E}_{\|,+}(\mathbf{r}'_\|,\omega) \\
& +[\beta_e(\mathbf{r}'_\|,\omega) - <\beta_e(\mathbf{r}'_\|,\omega)>]D_{z,+}(\mathbf{r}'_\|,\omega)\hat{\mathbf{z}}\} \\
& -(\mathbf{1}-\hat{\mathbf{z}}\hat{\mathbf{z}}).\int d\mathbf{r}'_\| \mathbf{K}_{mm,+}(\mathbf{r}_\| - \mathbf{r}'_\|,\omega).\{[\gamma_m(\mathbf{r}'_\|,\omega) - <\gamma_m(\mathbf{r}'_\|,\omega)>]\mathbf{H}_{\|,+}(\mathbf{r}'_\|,\omega) \\
& +[\beta_m(\mathbf{r}'_\|,\omega) - <\beta_m(\mathbf{r}'_\|,\omega)>]B_{z,+}(\mathbf{r}'_\|,\omega)\hat{\mathbf{z}}\}
\end{aligned}$$

$$\begin{aligned}
&b_{z,+}(\mathbf{r}_\|,\omega) - B_{z,+}(\mathbf{r}_\|,\omega) \\
= \; & -\hat{\mathbf{z}}.\int d\mathbf{r} K_{me,+}(\mathbf{r}_\| - \mathbf{r}'_\|,\omega).\{[\gamma_e(\mathbf{r}'_\|,\omega) - <\gamma_e(\mathbf{r}'_\|,\omega)>]\mathbf{E}_{\|,+}(\mathbf{r}'_\|,\omega) \\
& +[\beta_e(\mathbf{r}'_\|,\omega) - <\beta_e(\mathbf{r}'_\|,\omega)>]D_{z,+}(\mathbf{r}'_\|,\omega)\hat{\mathbf{z}}\} \\
& -\hat{\mathbf{z}}.\int d\mathbf{r}'_\| \mathbf{K}_{mm,+}(\mathbf{r}_\| - \mathbf{r}'_\|,\omega).\{[\gamma_m(\mathbf{r}'_\|,\omega) - <\gamma_m(\mathbf{r}'_\|,\omega)>]\mathbf{H}_{\|,+}(\mathbf{r}'_\|,\omega) \\
& +[\beta_m(\mathbf{r}'_\|,\omega) - <\beta_m(\mathbf{r}'_\|,\omega)>]B_{z,+}(\mathbf{r}'_\|,\omega)\hat{\mathbf{z}}\} \qquad (14.8)
\end{aligned}$$

In the cases considered in this book, retardation of the propagator will be neglected. It simplifies the further analysis considerably. Using the above formulae retardation effects may be accounted for in a straightforward manner. In the non retarded case the above formulae simplify to, cf. eq.(12.68),

$$\begin{aligned}
&\mathbf{e}_{\|,+}(\mathbf{r}_\|,\omega) - \mathbf{E}_{\|,+}(\mathbf{r}_\|,\omega) \\
= \; & -(\mathbf{1}-\hat{\mathbf{z}}\hat{\mathbf{z}}).\int d\mathbf{r}'_\| \mathbf{K}_{ee,+}(\mathbf{r}_\| - \mathbf{r}'_\|,\omega=0)(\gamma_e(\mathbf{r}'_\|,\omega) - <\gamma_e(\mathbf{r}'_\|,\omega)>).\mathbf{E}_{\|,+}(\mathbf{r}'_\|,\omega) \\
&d_{z,+}(\mathbf{r}_\|,\omega) - D_{z,+}(\mathbf{r}_\|,\omega) \\
= \; & -\hat{\mathbf{z}}.\int d\mathbf{r}'_\| \mathbf{K}_{ee,+}(\mathbf{r}_\| - \mathbf{r}'_\|,\omega=0).\hat{\mathbf{z}}(\beta_e(\mathbf{r}'_\|,\omega) - <\beta_e(\mathbf{r}'_\|,\omega)>)D_{z,+}(\mathbf{r}'_\|,\omega) \\
&\mathbf{h}_{\|,+}(\mathbf{r}_\|,\omega) - \mathbf{H}_{\|,+}(\mathbf{r}_\|,\omega) \\
= \; & -(\mathbf{1}-\hat{\mathbf{z}}\hat{\mathbf{z}}).\int d\mathbf{r}'_\| \mathbf{K}_{mm,+}(\mathbf{r}_\| - \mathbf{r}'_\|,\omega)(\gamma_m(\mathbf{r}'_\|,\omega) - <\gamma_m(\mathbf{r}'_\|,\omega)>).\mathbf{H}_{\|,+}(\mathbf{r}'_\|,\omega) \\
&b_{z,+}(\mathbf{r}_\|,\omega) - B_{z,+}(\mathbf{r}_\|,\omega) \\
= \; & -\hat{\mathbf{z}}.\int d\mathbf{r}'_\| \mathbf{K}_{mm,+}(\mathbf{r}_\| - \mathbf{r}'_\|,\omega=0).\hat{\mathbf{z}}(\beta_m(\mathbf{r}'_\|,\omega) - <\beta_m(\mathbf{r}'_\|,\omega)>)B_{z,+}(\mathbf{r}'_\|,\omega) \\
& \qquad (14.9)
\end{aligned}$$

These equations for the fluctuations of the fields may now be substituted into eq.(14.5) to obtain the average surface polarization and magnetization densities:

$$\begin{aligned}
\mathbf{P}^s_\|(\mathbf{r}_\|,\omega) \; = \; & <\gamma_e(\mathbf{r}_\|,\omega)> \mathbf{E}_{\|,+}(\mathbf{r}_\|,\omega) \\
& -\int d\mathbf{r}'_\| <(\gamma_e(\mathbf{r}_\|,\omega) - <\gamma_e(\mathbf{r}_\|,\omega)>)(\gamma_e(\mathbf{r}'_\|,\omega) - <\gamma_e(\mathbf{r}'_\|,\omega)>)> \\
& \times(\mathbf{1}-\hat{\mathbf{z}}\hat{\mathbf{z}}).\mathbf{K}_{ee,+}(\mathbf{r}_\| - \mathbf{r}'_\|,\omega=0).\mathbf{E}_{\|,+}(\mathbf{r}'_\|,\omega) \\
& - <\delta_e(\mathbf{r}_\|,\omega)> \nabla_\| D_{z,+}(\mathbf{r}_\|,\omega) + i\frac{\omega}{c} <\tau(\mathbf{r}_\|,\omega)> \hat{\mathbf{z}} \times \mathbf{H}_{\|,+}(\mathbf{r}_\|,\omega)
\end{aligned}$$

$$\begin{aligned}
P_z^s(\mathbf{r}_\|,\omega) &= <\beta_e(\mathbf{r}_\|,\omega)> D_{z,+}(\mathbf{r}_\|,\omega)\\
&-\int d\mathbf{r}'_\| <(\beta_e(\mathbf{r}_\|,\omega)-<\beta_e(\mathbf{r}_\|,\omega)>)(\beta_e(\mathbf{r}'_\|,\omega)-<\beta_e(\mathbf{r}'_\|,\omega)>)>\\
&\times\hat{\mathbf{z}}.\mathbf{K}_{ee,+}(\mathbf{r}_\|-\mathbf{r}'_\|,\omega=0).\hat{\mathbf{z}}D_{z,+}(\mathbf{r}'_\|,\omega)+<\delta_e(\mathbf{r}_\|,\omega)>\nabla_\|.E_{\|,+}(\mathbf{r}_\|,\omega)
\end{aligned}$$

$$\begin{aligned}
\mathbf{M}_\|^s(\mathbf{r}_\|,\omega) &= <\gamma_m(\mathbf{r}_\|,\omega)> \mathbf{H}_{\|,+}(\mathbf{r}_\|,\omega)\\
&-\int d\mathbf{r}'_\| <(\gamma_m(\mathbf{r}_\|,\omega)-<\gamma_m(\mathbf{r}_\|,\omega)>)(\gamma_m(\mathbf{r}'_\|,\omega)-<\gamma_m(\mathbf{r}'_\|,\omega)>)>\\
&\times(\mathbf{1}-\hat{\mathbf{z}}\hat{\mathbf{z}}).\mathbf{K}_{mm,+}(\mathbf{r}_\|-\mathbf{r}'_\|,\omega).\mathbf{H}_{\|,+}(\mathbf{r}'_\|,\omega)\\
&-<\delta_m(\mathbf{r}_\|,\omega)>\nabla_\| B_{z,+}(\mathbf{r}_\|,\omega)+i\frac{\omega}{c}<\tau(\mathbf{r}_\|,\omega)>\hat{\mathbf{z}}\times\mathbf{E}_{\|,+}(\mathbf{r}_\|,\omega)
\end{aligned}$$

$$\begin{aligned}
M_z^s(\mathbf{r}_\|,\omega) &= <\beta_m(\mathbf{r}_\|,\omega)> B_{z,+}(\mathbf{r}_\|,\omega)\\
&-\int d\mathbf{r}'_\| <(\beta_m(\mathbf{r}_\|,\omega)-<\beta_m(\mathbf{r}_\|,\omega)>)(\beta_m(\mathbf{r}'_\|,\omega)-<\beta_m(\mathbf{r}'_\|,\omega)>)>\\
&\times\hat{\mathbf{z}}.\mathbf{K}_{mm,+}(\mathbf{r}_\|-\mathbf{r}'_\|,\omega=0).\hat{\mathbf{z}}B_{z,+}(\mathbf{r}'_\|,\omega)+<\delta_m(\mathbf{r}_\|,\omega)>\nabla_\|.\mathbf{H}_{\|,+}(\mathbf{r}_\|,\omega)
\end{aligned} \tag{14.10}$$

Note that the symbol $\times$ used at the beginning of a line is always a simple multiplication, whereas it is an external product for other locations. It is now convenient to introduce the following correlation functions for the fluctuating surface susceptibilities of an isotropic surface

$$\begin{aligned}
S_{ee}^{\gamma\gamma}(\mathbf{r}_\|-\mathbf{r}'_\|) &\equiv <(\gamma_e(\mathbf{r}_\|,\omega)-<\gamma_e(\mathbf{r}_\|,\omega)>)(\gamma_e(\mathbf{r}'_\|,\omega)-<\gamma_e(\mathbf{r}'_\|,\omega)>)>\\
S_{ee}^{\beta\beta}(\mathbf{r}_\|-\mathbf{r}'_\|) &\equiv <(\beta_e(\mathbf{r}_\|,\omega)-<\beta_e(\mathbf{r}_\|,\omega)>)(\beta_e(\mathbf{r}'_\|,\omega)-<\beta_e(\mathbf{r}'_\|,\omega)>)>\\
S_{mm}^{\gamma\gamma}(\mathbf{r}_\|-\mathbf{r}'_\|) &\equiv <(\gamma_m(\mathbf{r}_\|,\omega)-<\gamma_m(\mathbf{r}_\|,\omega)>)(\gamma_m(\mathbf{r}'_\|,\omega)-<\gamma_m(\mathbf{r}'_\|,\omega)>)>\\
S_{mm}^{\beta\beta}(\mathbf{r}_\|-\mathbf{r}'_\|) &\equiv <(\beta_m(\mathbf{r}_\|,\omega)-<\beta_m(\mathbf{r}_\|,\omega)>)(\beta_m(\mathbf{r}',\omega)-<\beta_m(\mathbf{r}'_\|,\omega)>)>
\end{aligned} \tag{14.11}$$

Due to the homogeneous nature of the surface the correlation functions depend only on $\mathbf{r}_\| - \mathbf{r}'_\|$ and not on both $\mathbf{r}_\|$ and $\mathbf{r}'_\|$. Furthermore they are even functions of $\mathbf{r}_\| - \mathbf{r}'_\|$. Substitution of these correlation functions into the above expressions for the polarization and magnetization densities of the surface then gives

$$\begin{aligned}
\mathbf{P}_\|^s(\mathbf{r}_\|,\omega) &= <\gamma_e(\mathbf{r}_\|,\omega)> \mathbf{E}_{\|,+}(\mathbf{r}_\|,\omega)\\
&-\int d\mathbf{r}'_\| S_{ee}^{\gamma\gamma}(\mathbf{r}_\|-\mathbf{r}'_\|)(\mathbf{1}-\hat{\mathbf{z}}\hat{\mathbf{z}}).\mathbf{K}_{ee,+}(\mathbf{r}_\|-\mathbf{r}'_\|,\omega=0).\mathbf{E}_{\|,+}(\mathbf{r}'_\|,\omega)\\
&-<\delta_e(\mathbf{r}_\|,\omega)>\nabla_\| D_{z,+}(\mathbf{r}_\|,\omega)+i\frac{\omega}{c}<\tau(\mathbf{r}_\|,\omega)>\hat{\mathbf{z}}\times\mathbf{H}_{\|,+}(\mathbf{r}_\|,\omega)\\
P_z^s(\mathbf{r}_\|,\omega) &= <\beta_e(\mathbf{r}_\|,\omega)> D_{z,+}(\mathbf{r}_\|,\omega)\\
&-\int d\mathbf{r}'_\| S_{ee}^{\beta\beta}(\mathbf{r}_\|-\mathbf{r}'_\|)\hat{\mathbf{z}}.\mathbf{K}_{ee,+}(\mathbf{r}_\|-\mathbf{r}'_\|,\omega=0).\hat{\mathbf{z}}D_{z,+}(\mathbf{r}'_\|,\omega)\\
&+<\delta_e(\mathbf{r}_\|,\omega)>\nabla_\|.\mathbf{E}_{\|,+}(\mathbf{r}_\|,\omega)
\end{aligned}$$

$$\begin{aligned}
\mathbf{M}^s_\parallel(\mathbf{r}_\parallel,\omega) &= <\gamma_m(\mathbf{r}_\parallel,\omega)> \mathbf{H}_{\parallel,+}(\mathbf{r}_\parallel,\omega) \\
&\quad -\int d\mathbf{r}'_\parallel S^{\gamma\gamma}_{mm}(\mathbf{r}_\parallel-\mathbf{r}'_\parallel)(\mathbf{1}-\hat{\mathbf{z}}\hat{\mathbf{z}}).\mathbf{K}_{mm,+}(\mathbf{r}_\parallel-\mathbf{r}'_\parallel,\omega=0).\mathbf{H}_{\parallel,+}(\mathbf{r}'_\parallel,\omega) \\
&\quad -<\delta_m(\mathbf{r}_\parallel,\omega)> \nabla_\parallel B_{z,+}(\mathbf{r}_\parallel,\omega)+i\frac{\omega}{c}<\tau(\mathbf{r}_\parallel,\omega)> \hat{\mathbf{z}}\times\mathbf{E}_{\parallel,+}(\mathbf{r}_\parallel,\omega) \\
M^s_z(\mathbf{r}_\parallel,\omega) &= <\beta_m(\mathbf{r}_\parallel,\omega)> B_{z,+}(\mathbf{r}_\parallel,\omega) \\
&\quad -\int d\mathbf{r}'_\parallel S^{\beta\beta}_{mm}(\mathbf{r}_\parallel-\mathbf{r}'_\parallel)\hat{\mathbf{z}}.\mathbf{K}_{mm,+}(\mathbf{r}_\parallel-\mathbf{r}'_\parallel,\omega=0).\hat{\mathbf{z}}B_{z,+}(\mathbf{r}'_\parallel,\omega) \\
&\quad +<\delta_m(\mathbf{r}_\parallel,\omega)> \nabla_\parallel.\mathbf{H}_{\parallel,+}(\mathbf{r}_\parallel,\omega) \qquad (14.12)
\end{aligned}$$

Using the homogeneous nature of the surface, averages like $<\gamma_e(\mathbf{r}_\parallel,\omega)>$ are independent of the position along the surface. One may therefore write

$$\begin{aligned}
&<\gamma_e(\mathbf{r}_\parallel,\omega)>=<\gamma_e>(\omega),\quad <\gamma_m(\mathbf{r}_\parallel,\omega)>=<\gamma_m>(\omega) \\
&<\beta_e(\mathbf{r}_\parallel,\omega)>=<\beta_e>(\omega),\quad <\beta_m(\mathbf{r}_\parallel,\omega)>=<\beta_m>(\omega) \\
&<\delta_e(\mathbf{r}_\parallel,\omega)>=<\delta_e>(\omega),\quad <\delta_m(\mathbf{r}_\parallel,\omega)>=<\delta_m>(\omega) \\
&<\tau(\mathbf{r}_\parallel,\omega)>=<\tau>(\omega) \qquad (14.13)
\end{aligned}$$

Upon Fourier transformation along the surface the above eqs.(14.12) reduce to

$$\begin{aligned}
\mathbf{P}^s_\parallel(\mathbf{k}_\parallel,\omega) &= [<\gamma_e>(\omega) \\
&\quad -(2\pi)^{-2}\int d\mathbf{k}'_\parallel S^{\gamma\gamma}_{ee}(\mathbf{k}_\parallel-\mathbf{k}'_\parallel)(\mathbf{1}-\hat{\mathbf{z}}\hat{\mathbf{z}}).\mathbf{K}_{ee,+}(\mathbf{k}'_\parallel,\omega=0)].\mathbf{E}_{\parallel,+}(\mathbf{k}_\parallel,\omega) \\
&\quad -i<\delta_e>(\omega)\,\mathbf{k}_\parallel D_{z,+}(\mathbf{k}_\parallel,\omega)+i\frac{\omega}{c}<\tau>(\omega)\,\hat{\mathbf{z}}\times\mathbf{H}_{\parallel,+}(\mathbf{k}_\parallel,\omega) \\
P^s_z(\mathbf{k}_\parallel,\omega) &= [<\beta_e>(\omega) \\
&\quad -(2\pi)^{-2}\int d\mathbf{k}'_\parallel S^{\beta\beta}_{ee}(\mathbf{k}_\parallel-\mathbf{k}'_\parallel)\hat{\mathbf{z}}.\mathbf{K}_{ee,+}(\mathbf{k}'_\parallel,\omega=0).\hat{\mathbf{z}}]D_{z,+}(\mathbf{k}_\parallel,\omega) \\
&\quad +i<\delta_e>(\omega)\,\mathbf{k}_\parallel.\mathbf{E}_{\parallel,+}(\mathbf{k}_\parallel,\omega)
\end{aligned}$$

$$\begin{aligned}
\mathbf{M}^s_\parallel(\mathbf{k}_\parallel,\omega) &= [<\gamma_m>(\omega) \\
&\quad -(2\pi)^{-2}\int d\mathbf{k}'_\parallel S^{\gamma\gamma}_{mm}(\mathbf{k}_\parallel-\mathbf{k}'_\parallel)(\mathbf{1}-\hat{\mathbf{z}}\hat{\mathbf{z}}).\mathbf{K}_{mm,+}(\mathbf{k}'_\parallel,\omega=0)].\mathbf{H}_{\parallel,+}(\mathbf{k}_\parallel,\omega) \\
&\quad -i<\delta_m>(\omega)\,\mathbf{k}_\parallel B_{z,+}(\mathbf{k}_\parallel,\omega)+i\frac{\omega}{c}<\tau>(\omega)\,\hat{\mathbf{z}}\times\mathbf{E}_{\parallel,+}(\mathbf{k}_\parallel,\omega) \\
M^s_z(\mathbf{k}_\parallel,\omega) &= [<\beta_m>(\omega) \\
&\quad -(2\pi)^{-2}\int d\mathbf{k}'_\parallel S^{\beta\beta}_{mm}(\mathbf{k}_\parallel-\mathbf{k}'_\parallel)\hat{\mathbf{z}}.\mathbf{K}_{mm,+}(\mathbf{k}'_\parallel,\omega=0).\hat{\mathbf{z}}]B_{z,+}(_\parallel,\omega) \\
&\quad +i<\delta_m>(\omega)\,\mathbf{k}_\parallel.\mathbf{H}_{\parallel,+}(\mathbf{k}_\parallel,\omega) \qquad (14.14)
\end{aligned}$$

Comparing these expressions with eqs.(3.42)– (3.45) one may identify the following wave vector and frequency dependent surface susceptibilities

$$
\begin{aligned}
\gamma_e(\mathbf{k}_\|,\omega) &= <\gamma_e>(\omega) - (2\pi)^{-2}\int d\mathbf{k}'_\| S^{\gamma\gamma}_{ee}(\mathbf{k}_\| - \mathbf{k}'_\|)(\mathbf{1} - \hat{\mathbf{z}}\hat{\mathbf{z}}).\mathbf{K}_{ee,+}(\mathbf{k}'_\|,\omega = 0)] \\
\beta_e(\mathbf{k}_\|,\omega) &= <\beta_e>(\omega) - (2\pi)^{-2}\int d\mathbf{k}'_\| S^{\beta\beta}_{ee}(\mathbf{k}_\| - \mathbf{k}'_\|)\hat{\mathbf{z}}.\mathbf{K}_{ee,+}(\mathbf{k}'_\|,\omega = 0).\hat{\mathbf{z}} \\
\delta_e(\mathbf{k}_\|,\omega) &= <\delta_e>(\omega) \\
\gamma_m(\mathbf{k}_\|,\omega) &= <\gamma_m>(\omega) - (2\pi)^{-2}\int d\mathbf{k}'_\| S^{\gamma\gamma}_{mm}(\mathbf{k}_\| - \mathbf{k}'_\|)(\mathbf{1} - \hat{\mathbf{z}}\hat{\mathbf{z}}).\mathbf{K}_{mm,+}(\mathbf{k}'_\|,\omega = 0) \\
\beta_m(\mathbf{k}_\|,\omega) &= <\beta_m>(\omega) - (2\pi)^{-2}\int d\mathbf{k}'_\| S^{\beta\beta}_{mm}(\mathbf{k}_\| - \mathbf{k}'_\|)\hat{\mathbf{z}}.\mathbf{K}_{mm,+}(\mathbf{k}'_\|,\omega = 0).\hat{\mathbf{z}} \\
\delta_m(\mathbf{k}_\|,\omega) &= <\delta_m>(\omega) \\
\tau(\mathbf{k}_\|,\omega) &= <\tau>(\omega) \qquad (14.15)
\end{aligned}
$$

The wave vector dependence of the effective susceptibilities is due to the correlations in the distribution of the material along the surface. It is found that only the linear coefficients γ_e, β_e, γ_m and β_m exhibit such a dependence to second order in the "thickness" of the interface. If the correlation length is small compared to the wavelength of the light one may neglect this wave vector dependence and this then gives, using the fact that the correlation functions are even functions of $\mathbf{k}_\| - \mathbf{k}'_\|$,

$$
\begin{aligned}
\gamma_e(\omega) &= <\gamma_e>(\omega) - \frac{1}{2}(2\pi)^{-2}\int d\mathbf{k}_\| S^{\gamma\gamma}_{ee}(\mathbf{k}_\|)(\mathbf{1} - \hat{\mathbf{z}}\hat{\mathbf{z}}) : \mathbf{K}_{ee,+}(\mathbf{k}_\|,\omega = 0)] \\
\beta_e(\omega) &= <\beta_e>(\omega) - (2\pi)^{-2}\int d\mathbf{k}_\| S^{\beta\beta}_{ee}(\mathbf{k}_\|)\hat{\mathbf{z}}.\mathbf{K}_{ee,+}(\mathbf{k}_\|,\omega = 0).\hat{\mathbf{z}} \\
\delta_e(\omega) &= <\delta_e>(\omega) \\
\gamma_m(\omega) &= <\gamma_m>(\omega) - \frac{1}{2}(2\pi)^{-2}\int d\mathbf{k}_\| S^{\gamma\gamma}_{mm}(\mathbf{k}_\|)(\mathbf{1} - \hat{\mathbf{z}}\hat{\mathbf{z}}) : \mathbf{K}_{mm,+}(\mathbf{k}_\|,\omega = 0) \\
\beta_m(\omega) &= <\beta_m>(\omega) - (2\pi)^{-2}\int d\mathbf{k}_\| S^{\beta\beta}_{mm}(\mathbf{k}_\|)\hat{\mathbf{z}}.\mathbf{K}_{mm,+}(\mathbf{k}_\|,\omega = 0).\hat{\mathbf{z}} \\
\delta_m(\omega) &= <\delta_m>(\omega) \\
\tau(\omega) &= <\tau>(\omega) \qquad (14.16)
\end{aligned}
$$

In the expressions for γ_e and γ_m the original expression, which in fact gave a 2×2 matrix and allows for a directional dependence, was replaced by a half time the trace. For $\mathbf{k}_\| = 0$ such a directional dependence is impossible for an isotropic surface. The frequency dependence in the above expressions is not related to the motion of the material on the surface. This was assumed to be slow and the resulting retardation effects were neglected. The still remaining frequency dependence is due to the frequency dependence of the constitutive coefficients of the various materials involved.

A final simplification is due to the isotropic nature of the surface. This implies that the correlation functions do not depend on the direction.

$$\begin{aligned}
S_{ee}^{\gamma\gamma}(\mathbf{r}_\parallel - \mathbf{r}'_\parallel) &= S_{ee}^{\gamma\gamma}(|\mathbf{r}_\parallel - \mathbf{r}'_\parallel|) \Rightarrow S_{ee}^{\gamma\gamma}(\mathbf{k}_\parallel) = S_{ee}^{\gamma\gamma}(k_\parallel) \\
S_{ee}^{\beta\beta}(\mathbf{r}_\parallel - \mathbf{r}'_\parallel) &= S_{ee}^{\beta\beta}(|\mathbf{r}_\parallel - \mathbf{r}'_\parallel|) \Rightarrow S_{ee}^{\beta\beta}(\mathbf{k}_\parallel) = S_{ee}^{\beta\beta}(k_\parallel) \\
S_{mm}^{\gamma\gamma}(\mathbf{r}_\parallel - \mathbf{r}'_\parallel) &= S_{mm}^{\gamma\gamma}(|\mathbf{r}_\parallel - \mathbf{r}'_\parallel|) \Rightarrow S_{mm}^{\gamma\gamma}(\mathbf{k}_\parallel) = S_{mm}^{\gamma\gamma}(k_\parallel) \\
S_{mm}^{\beta\beta}(\mathbf{r}_\parallel - \mathbf{r}'_\parallel) &= S_{mm}^{\beta\beta}(|\mathbf{r}_\parallel - \mathbf{r}'_\parallel|) \Rightarrow S_{mm}^{\beta\beta}(\mathbf{k}_\parallel) = S_{mm}^{\beta\beta}(k_\parallel)
\end{aligned} \qquad (14.17)$$

Using this property one may integrate in the expressions for γ_e, β_e, γ_m and β_m above over the directions of the wave vector. One then obtains, using eq.(12.68) for the propagators,

$$\begin{aligned}
\gamma_e(\omega) &= <\gamma_e>(\omega) - \frac{1}{16\pi}((\epsilon^-)^{-1} + (\epsilon^+)^{-1}) \int_0^\infty dk_\parallel S_{ee}^{\gamma\gamma}(k_\parallel) k_\parallel^2 \\
\beta_e(\omega) &= <\beta_e>(\omega) + \frac{1}{8\pi}(\epsilon^- + \epsilon^+) \int_0^\infty dk_\parallel S_{ee}^{\beta\beta}(k_\parallel) k_\parallel^2 \\
\delta_e(\omega) &= <\delta_e>(\omega) \\
\gamma_m(\omega) &= <\gamma_m>(\omega) - \frac{1}{16\pi}((\mu^-)^{-1} + (\mu^+)^{-1}) \int_0^\infty dk_\parallel S_{mm}^{\gamma\gamma}(k_\parallel) k_\parallel^2 \\
\beta_m(\omega) &= <\beta_m>(\omega) + \frac{1}{8\pi}(\mu^- + \mu^+) \int_0^\infty dk_\parallel S_{mm}^{\beta\beta}(k_\parallel) k_\parallel^2 \\
\delta_m(\omega) &= <\delta_m>(\omega) \\
\tau(\omega) &= <\tau>(\omega)
\end{aligned} \qquad (14.18)$$

As mentioned above the frequency dependence in these equation for the effective susceptibilities is due to the frequency dependence of the materials distributed along the interface. In the following sections these equations will be used to analyze various special cases of surface roughness.

14.3 Rough surfaces

Several theories of this phenomenon have appeared in the literature, [65], [66], [67], [68], [69], [70], [71]. They do not all give the same result. In most of these theories the equivalent current model was used in which the roughness of the surface is replaced by a fluctuating surface current. The method presented in this chapter may be seen as a similar model using an equivalent surface polarization. The time derivative of this equivalent polarization is then the equivalent surface current used in the other theories. The origin of the disagreement between the different theories arises about the question: whether to put the equivalent current just inside the substrate or just outside the substrate. The first analysis was given by Stern, [65], who expressed the scattered field in terms of the equivalent current. After noting that the field depends on the position of the current he resolves this point in a completely ad hoc manner by putting half in the substrate and half in the ambient. Because of this choice he predicts a resonance at the plasma resonance of the substrate, which would not occur if the current was placed in the ambient. Subsequently Juranek, [68], and independently Beaglehole and Hunderi, [69], [70], calculate the scattered field using a completely different technique and find the result Stern would have found if he placed the current in the ambient. In a paper by Kröger and Kretchmann, [71], the different approaches are compared. They show, using Maxwell's equations, that an inconsistency arises if the current is placed either in the substrate or in the ambient. It should be placed in an infinitesimally thin layer of vacuum between the substrate and the ambient. In this way they find a field which agrees with the one found by Juranek and by Beaglehole and Hunderi. In the treatment in this book this problem does not arise. As explained by Vlieger and Bedeaux [57] the result of Kröger and Kretschmann is confirmed.

In this section the theory discussed above will be applied to a rough surface. The nature of the roughness is such that one has two different phases with the dielectric constants ϵ^- and ϵ^+ separated by a rough surface. This surface is located in the neighborhood of the $x-y$ plane and its location is characterized by its x- and y-dependent distance $h(x,y)$ to this plane. It is assumed that there are no overhangs so that $h(x,y)$ is a single-valued function, which may be both positive and negative. The system is furthermore non-magnetic so that the magnetic permeability is everywhere equal to one.

The position dependent dielectric constant can be written in the following form

$$\epsilon(\mathbf{r}) = \epsilon^-\theta(h(x,y)-z) + \epsilon^+\theta(z-h(x,y)) \tag{14.19}$$

The extrapolated dielectric constant is given as usual by

$$\epsilon_0(\mathbf{r}) = \epsilon^-\theta(-z) + \epsilon^+\theta(z) \tag{14.20}$$

The location of the $x-y$ plane has been chosen such that the average height of the surface above the plane is zero

$$< h(x,y) >= 0 \tag{14.21}$$

Notice that the height is measured in the positive z-direction and can be both negative and positive.

The height dependent constitutive coefficients describing the rough surface are now obtained by substitution of the above equations into the formulae given in eq.(14.1). This results in the following expressions

$$\begin{aligned}
\gamma_e(x,y) &= \epsilon^+\epsilon^-\beta_e(x,y) = -(\epsilon^+-\epsilon^-)h(x,y) \\
\delta_e(x,y) &= \frac{1}{2}\frac{(\epsilon^+-\epsilon^-)(\epsilon^++\epsilon^-)}{\epsilon^+\epsilon^-}h^2(x,y) \\
\tau(x,y) &= \frac{1}{2}(\epsilon^+-\epsilon^-)h^2(x,y)
\end{aligned} \tag{14.22}$$

The magnetic constitutive coefficients, $\gamma_m(x,y)$, $\beta_m(x,y)$ and $\delta_m(x,y)$, are all zero and will further be neglected.

In order to use the fluctuating constitutive coefficients one needs the height-height correlation function which is defined by

$$S(\mathbf{r}_\parallel - \mathbf{r}'_\parallel) \equiv < h(\mathbf{r}_\parallel)h(\mathbf{r}'_\parallel) > \tag{14.23}$$

It will be assumed that the roughness is isotropic so that the height-height correlations do not depend on the direction

$$S(\mathbf{r}_{\parallel} - \mathbf{r}'_{\parallel}) = S(|\mathbf{r}_{\parallel} - \mathbf{r}'_{\parallel}|) \tag{14.24}$$

Using the above expressions for the fluctuating constitutive coefficients, one finds that their correlation functions are also isotropic and given by,

$$S^{\gamma\gamma}_{ee}(r_{\parallel}) = (\epsilon^+\epsilon^-)^2 S^{\beta\beta}_{ee}(r_{\parallel}) = (\epsilon^+ - \epsilon^-)^2 S(r_{\parallel}), \tag{14.25}$$

in terms of the height-height correlation function. The wave vector dependent Fourier transforms or, using an other word, structure factors satisfy the same relation

$$S^{\gamma\gamma}_{ee}(k_{\parallel}) = (\epsilon^+\epsilon^-)^2 S^{\beta\beta}_{ee}(k_{\parallel}) = (\epsilon^+ - \epsilon^-)^2 S(k_{\parallel}), \tag{14.26}$$

A property of some special interest in the analysis is the root mean square height:

$$t^2 \equiv < h^2(x,y) >= S(\mathbf{r}_{\parallel} = 0) = (2\pi)^{-2} \int d\mathbf{k}_{\parallel} S(k_{\parallel}) = (2\pi)^{-1} \int_0^{\infty} dk_{\parallel} k_{\parallel} S(k_{\parallel}) \tag{14.27}$$

t is a measure of the thickness of the rough surface.

Substitution of the above expressions into eq.(14.18) gives the following formulae for the effective constitutive coefficients of the surface

$$\begin{aligned} \gamma_e &= -\frac{1}{2}\epsilon^+\epsilon^-\beta_e = -(16\pi)^{-1}\frac{(\epsilon^+ + \epsilon^-)}{\epsilon^+\epsilon^-}(\epsilon^+ - \epsilon^-)^2 \int_0^{\infty} dk_{\parallel} k_{\parallel}^2 S(k_{\parallel}) \\ \tau &= \frac{\epsilon^+\epsilon^-}{\epsilon^+ + \epsilon^-}\delta_e = \frac{1}{2}(\epsilon^+ - \epsilon^-)t^2 = (4\pi)^{-1}(\epsilon^+ - \epsilon^-) \int_0^{\infty} dk_{\parallel} k_{\parallel} S(k_{\parallel}) \end{aligned} \tag{14.28}$$

The prefactor given in [64] for γ_e and β_e is incorrect. In view of the fact that the average height is zero the corresponding contributions from the averages of $\gamma_e(x,y)$ and $\beta_e(x,y)$ are zero. The only contributions to γ_e and β_e are from correlations in the height distribution along the surface. The quadrupolar coefficients δ_e and τ are not affected by correlations along the surface. They are found to be proportional to the mean square height.

In order to gain some insight in the contribution due to correlations, it will be assumed that the height-height correlation function is Gaussian

$$S(r_{\parallel}) = t^2 \exp(-r_{\parallel}^2/\xi^2) \tag{14.29}$$

ξ is the correlation length along the surface. For $r_{\parallel} = 0$ this expression reduces to a relation for the mean square height. Fourier transformation of this expression gives the structure factor

$$S(k_{\parallel}) = \pi\, t^2\xi^2 \exp(-\frac{1}{4}k_{\parallel}^2\xi^2) \tag{14.30}$$

Substitution into eq.(14.28) and integration gives the dipolar constitutive coefficients

$$\gamma_e = -\frac{1}{2}\epsilon^+\epsilon^-\beta_e = -\frac{1}{8}\sqrt{\pi}\frac{(\epsilon^+ + \epsilon^-)}{\epsilon^+\epsilon^-}(\epsilon^+ - \epsilon^-)^2 t^2/\,\xi \tag{14.31}$$

It is interesting to compare the importance of the dipolar coefficients with the quadrupolar coefficients. In expressions for e.g. reflectivities the dipolar coefficients will appear divided by the wavelength λ while the quadrupolar coefficients appear divided by the wavelength squared. The relative importance of the quadrupolar coefficients is therefore of relative order ξ/λ. It may therefore be concluded that for surfaces with a correlation length small compared to the wavelength of light the dipolar coefficients are sufficient. If on the other hand the correlation length is larger than the wavelength of light, the quadrupolar coefficient will become more important. For most rough surfaces it is to be expected that the correlation length will be small compared to the wavelength of light. It depends of course on the way the surface is formed. For liquid surfaces, treated in more detail in the next section, the correlation length is always much larger than the wavelength of light. Both regimes are therefore of practical importance.

14.4 Capillary waves

The surface between two fluids, which are in coexistence with each other, is not stationary. Due to thermal fluctuations so-called capillary waves are constantly excited and as a result the height distribution along the surface is constantly changing. The typical velocity is very small compared to the velocity of light. This implies that the reflection and transmission of light take place so fast, that the surface may be considered to be temporarily frozen. The frozen shape is of the same nature as the shape of the rough surfaces considered in the previous section. In fact the fluctuating dielectric constant is given by the same equation (14.19), in terms of the height. All the subsequent equations are similarly valid.

The aspect that is different, is the nature of the correlations. The probability of a given shape of the fluid interface is given by a Boltzmann distribution in terms of the free energy of the surface. The proper averaging is therefore done, using this Boltzmann distribution, rather than using the spacial averaging appropriate for solid surfaces. There is a lot of literature, see for instance ref. [72], on this subject and there is no need to discuss this matter in detail here. It is sufficient to use the structure factor found using this Boltzmann distribution

$$S(k_{\parallel}) = \frac{k_B T/g\delta\rho}{1 + k_{\parallel}^2 L_c^2} \tag{14.32}$$

where k_B is Boltzmann's constant, T the temperature, g the gravitational acceleration and $\delta\rho$ the absolute value of the difference of the mass densities of the two fluids. Furthermore L_c is the so-called capillary length which is given by

$$L_c \equiv \sqrt{\sigma/g\delta\rho} \tag{14.33}$$

where σ is the surface tension. This capillary length is the correlation length for the height distribution along the surface. For most fluid-fluid interfaces it is of the order of a fraction of a millimeter. As such the correlation length is much larger than the wavelength of light for the fluid-fluid interface. As discussed in the previous section, this results in a relatively large importance of the quadrupolar coefficients.

Substitution of the above expression for the correlation function leads to an ultraviolet, i.e. large wave vector, divergence of the integrals. The usual method to eliminate such a divergence is the introduction of a molecular cutoff k_m for large wave vectors. This is equivalent to using as structure factor

$$S(k_{\|}) = \frac{k_BT/g\delta\rho}{1 + k_{\|}^2 L_c^2}\theta(k_{\|} - k_m) \tag{14.34}$$

in the above integrals. The molecular cutoff is of the typical order of 2π divided by the diameter of the molecules. The introduction of a cutoff reflects the physical fact that on a molecular length scale a description in terms of capillary waves is no longer realistic.

The average thickness of the interface is found from the correlation function by integration over the wave vector, cf. eq.(14.27). This results in

$$t^2 = (2\pi)^{-1}\int_0^\infty dk_{\|}k_{\|}S(k_{\|}) = \frac{k_BT}{4\pi\sigma}\ln(1 + k_m^2 L_c^2) \tag{14.35}$$

This formula may be simplified using the fact that the capillary length is many orders of magnitude larger than the molecular diameter. This results in

$$t^2 = \frac{k_BT}{4\pi\sigma}\ln(k_m^2\sigma/g\delta\rho) \tag{14.36}$$

For interfaces between simple liquids this formula gives a value, which is a little bit larger than the molecular diameter. An interesting property of the thickness of a flat interface is, that it is logarithmically divergent when the gravitational acceleration goes to zero. For a curved surface there is no divergence. In the zero gravity limit one should then replace the capillary length in the above formula for t by the radius of curvature. For a flat surface the finite size of the surface will similarly eliminate the divergence of the thickness in this limit [72].

The quadrupolar constitutive coefficients are now found by substituting the above expression for the thickness of the surface into eq.(14.28)

$$\tau = \frac{\epsilon^+\epsilon^-}{\epsilon^+ + \epsilon^-}\delta_e = -\frac{1}{2}(\epsilon^+ - \epsilon^-)t^2 = -(\epsilon^+ - \epsilon^-)\frac{k_BT}{8\pi\sigma}\ln(k_m^2\sigma/g\delta\rho) \tag{14.37}$$

The dipolar constitutive coefficients are also found using the expression in eq.(14.28). Substituting the above correlation function, and using again that the capillary length is much larger than the molecular diameter, yields

$$\gamma_e = -\frac{1}{2}\epsilon^+\epsilon^-\beta_e = -\frac{(\epsilon^+ + \epsilon^-)}{\epsilon^+\epsilon^-}(\epsilon^+ - \epsilon^-)^2\frac{k_BT}{16\pi\sigma}k_m \tag{14.38}$$

In the previous section on rough surfaces it was concluded that, if the correlation length along the surface was larger than the wavelength of light, the quadrupolar coefficients became important. For the fluid-fluid interface this correlation length is given by the capillary length and is therefore considerably larger than the wavelength.

In view of the rather different nature of the correlation function the relative importance should be calculated again. The size of the quadrupolar coefficient divided by the wavelength squared is of order $(t/\lambda)^2$. The size of the dipolar coefficients divided by the wavelength is of the order $t^2 k_m/\lambda$. Comparing these estimates one sees that the relative importance of the quadrupolar terms is of the order $(k_m\lambda)^{-1}$ which is in the percent range. Of course the relative importance is furthermore affected by which property is measured. In that respect the estimates are only meant to give a rough indication of values. For any particular case it is better to calculate their actual values. In particular one will see important modifications of these estimates, if one approaches the critical point of binary mixing or of the liquid-vapor transition. In that case k_m becomes of the order of 2π divided by the critical correlation length. Since this critical correlation length diverges in the critical point the dipolar coefficients will approach zero and disappear relative to the quadrupolar coefficients.

There is an alternative method suggested by Meunier, [73], to avoid the ultraviolet divergence in the capillary wave model. This method is in particular of importance if surfactants are present, reducing the surface tension. In that case bending energies of the surface become significant. This leads to the following modification of the structure factor:

$$S(k_\parallel) = \frac{k_B T}{g\delta\rho + \sigma k_\parallel^2 + \kappa k_\parallel^4} \tag{14.39}$$

κ is the bending elasticity of the surface and the corresponding term in the denominator is due to the bending energy.

The mean square thickness of the interface now becomes, cf. eq.(14.27),

$$\begin{aligned} t^2 &= (2\pi)^{-1} \int_0^\infty dk_\parallel k_\parallel S(k_\parallel) \\ &= \frac{k_B T}{4\pi\sigma}(1 + 4\kappa/\sigma L_c^2)^{-1/2} \ln[(1 + (1 + 4\kappa/\sigma L_c^2)^{-1/2})/(1 - (1 + 4\kappa/\sigma L_c^2)^{-1/2})] \end{aligned} \tag{14.40}$$

This formula may be simplified using the fact, that the capillary length is many orders of magnitude larger than $\sqrt{\kappa/\sigma}$. This results in

$$t^2 = \frac{k_B T}{4\pi\sigma} \ln(\sigma L_c^2/\kappa) \tag{14.41}$$

Notice that in fact $\sqrt{\sigma/\kappa}$ now plays the same role as the molecular cutoff. The physical significance of this parameter is very different, however.

The quadrupolar constitutive coefficients are again found by substituting the above expression for the thickness of the surface into eq.(14.28)

$$\tau = \frac{\epsilon^+\epsilon^-}{\epsilon^+ + \epsilon^-}\delta_e = -\frac{1}{2}(\epsilon^+ - \epsilon^-)t^2 = -(\epsilon^+ - \epsilon^-)\frac{k_B T}{8\pi\sigma} \ln(\sigma L_c^2/\kappa) \tag{14.42}$$

The dipolar constitutive coefficients are also found using the expression in eq.(14.28). Substituting the above correlation function, and using again that the capillary length is much larger than the molecular diameter, yields

$$\gamma_e = -\frac{1}{2}\epsilon^+\epsilon^-\beta_e = -\frac{(\epsilon^+ + \epsilon^-)}{\epsilon^+\epsilon^-}(\epsilon^+ - \epsilon^-)^2\frac{k_B T}{32\sigma}\sqrt{\sigma/\kappa} \tag{14.43}$$

In this case the relative significance of the quadrupolar coefficients is increased, if one for instance lowers the surface tension by adding surfactant. In microemulsions these coefficients are therefore relatively important. The value of κ in microemulsions has been measured using ellipsometry by Meunier[73], who derived the expression for the ellipsometric coefficient, which follows from eq.(14.43), for the first time. The first to measure the ellipsometric coefficient of a one component liquid-vapor interface was Beaglehole [74].

As the introduction of the rigidity and the cutoff wave vector have a different physical reason, one may also introduce this molecular cutoff in the capillary wave model for a finite rigidity κ. Data analysis close to the critical point has used such an analysis. For the relevant expressions giving the constitutive coefficients, see Blokhuis et al. [64].

14.5 Self-affine surfaces

Terraced surfaces are rough surfaces, which are simultaneously flat and rough over the same range of lengths. If they have a scaling property, these surfaces have a well defined roughness exponent and they are called self-affine. Terraced surfaces can be produced by methods like, for instance, molecular beam epitaxy [75, 76], etching [77] and electrodeposition [78]. They are locally flat and rough by virtue of the steps between the terraces. The fluctuating dielectric constant is given by the same equation (14.19), in terms of the height. All the subsequent equations in section 14.3 are similarly valid. The aspect that is different, is the nature of the correlations and the resulting interfacial susceptibilities.

Using a truncated scaling assumption [79, 80], the height-height auto correlation function is given by

$$S(\mathbf{r}_\parallel) = S(r_\parallel) = t^2 \exp\left[-\left(\frac{r_\parallel}{\xi}\right)^{2H}\right] \tag{14.44}$$

where ξ is the correlation length along the surface. The scaling exponent varies in the interval $0 < H \leq 1$. This correlation function decays from the mean square height of the surface $S(0) \equiv t^2$ at $r_\parallel = 0$ to zero for large $r_\parallel$. The surface is isotropic. For distances small compared to the correlation length one has

$$S(r_\parallel) = t^2\left[1 - \left(\frac{r_\parallel}{\xi}\right)^{2H}\right] \tag{14.45}$$

In order to obtain the interfacial susceptibilities the Fourier transform of the height-height auto-correlation function is needed

$$
\begin{aligned}
S(k_{\|}) &= S(\mathbf{k}_{\|}) = \int d\mathbf{r}_{\|} e^{-i\mathbf{k}_{\|}\cdot\mathbf{r}_{\|}} S(\mathbf{r}_{\|}) \\
&= t^2 \int d\mathbf{r}_{\|} \exp\left[-i\mathbf{k}_{\|}.\mathbf{r}_{\|} - \left(\frac{r_{\|}}{\xi}\right)^{2H}\right] \\
&= t^2 \int_0^{\infty} dr_{\|} \int_0^{2\pi} d\varphi\ r_{\|} \exp\left[-ik_{\|}r_{\|}\cos\varphi - \left(\frac{r_{\|}}{\xi}\right)^{2H}\right] \\
&= 2\pi t^2 \int_0^{\infty} dr_{\|}\ r_{\|} \exp\left[-\left(\frac{r_{\|}}{\xi}\right)^{2H}\right] J_0(k_{\|}r_{\|}) \\
&= 2\pi \left(\frac{t}{k_{\|}}\right)^2 \int_0^{\infty} ds\ s \exp\left[-\left(\frac{s}{k_{\|}\xi}\right)^{2H}\right] J_0(s) \qquad (14.46)
\end{aligned}
$$

where the integration variable $s \equiv k_{\|}r_{\|}$ was introduced. Furthermore J_0 is a Bessel function [48]. For $0 < H < 0.5$ the above expression leads to an ultraviolet, i.e. large wave vector, divergence of the integrals. This is corrected by the introduction of a molecular cutoff k_m for large wave vectors. This is equivalent to using

$$
S(k_{\|}) = 2\pi\Theta\left(k_{\|} - k_m\right) \left(\frac{t}{k_{\|}}\right)^2 \int_0^{\infty} ds\ s \exp\left[-\left(\frac{s}{k_{\|}\xi}\right)^{2H}\right] J_0(s) \qquad (14.47)
$$

in these integrals. The correlation length will be large compared to molecular length scales. Because of this the parameter $k_m\xi$ will be much larger than one.

The effective constitutive coefficients of the surface are given by eq.(14.28). Substituting eq.(14.47) into this equation gives for γ_e and β_e

$$
\begin{aligned}
\gamma_e &= -\frac{1}{2}\epsilon^+\epsilon^-\beta_e = -(16\pi)^{-1}\frac{(\epsilon^+ + \epsilon^-)}{\epsilon^+\epsilon^-}(\epsilon^+ - \epsilon^-)^2 \int_0^{\infty} dk_{\|} k_{\|}^2 S(k_{\|}) \\
&= -\frac{t^2}{\xi}\frac{(\epsilon^+ + \epsilon^-)}{8\epsilon^+\epsilon^-}(\epsilon^+ - \epsilon^-)^2 \int_0^{k_m\xi} du \int_0^{\infty} ds\ s \exp\left[-\left(\frac{s}{u}\right)^{2H}\right] J_0(s) \\
&\equiv -\frac{t^2}{\xi}\frac{(\epsilon^+ + \epsilon^-)}{8\epsilon^+\epsilon^-}(\epsilon^+ - \epsilon^-)^2 I\left(H, k_m\xi\right) \qquad (14.48)
\end{aligned}
$$

where $u \equiv k_{\|}\xi$ was introduced. It is thus found that the dipolar coefficients γ_e and β_e are proportional to the mean square height squared divided by the correlation length, times an H and $k_m\xi$ dependent double integral. For $H = 1$ the double integral is equal to $\sqrt{\pi}$, substitution in eq.(14.48) reproduces eq.(14.31). The integral has an ultraviolet divergence and the molecular cutoff is needed. In fig.14.1 the numerical value of the double integral I is plotted as a function of H for $k_m\xi$ = 10, 25, 50, 75 and 100. It is seen in the figure that the integral depends strongly on $k_m\xi$ for $H \neq 1$. Measurements of the reflection coefficients will give information about the scaling exponent H given the value of $k_m\xi$.

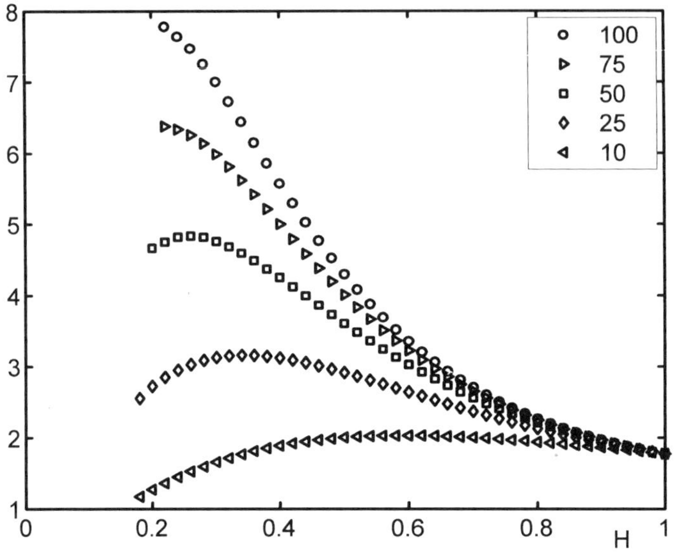

Figure 14.1 The integral $I(H, k_m\xi)$ as a function of H for $k_m\xi$ equal to 10, 25, 50, 75 and 100.

The quadrupolar coefficients τ and δ_e are given by

$$\tau = \frac{\epsilon^+\epsilon^-}{\epsilon^+ + \epsilon^-}\delta_e = \frac{1}{2}(\epsilon^+ - \epsilon^-)t^2 \tag{14.49}$$

The quadrupolar terms τ and δ_e are therefore proportional to the mean square height squared. They are independent of the correlation length, the molecular cutoff and the scaling exponent H. It is interesting to compare the importance of the dipolar coefficients with the quadrupolar coefficients. In expressions for e.g. reflectivities the dipolar coefficients will appear divided by the wavelength λ while the quadrupolar coefficients appear divided by the wavelength squared. The relative importance of the quadrupolar coefficients is therefore of relative order $\xi/\left(\lambda I\left(H, k_m\xi\right)\right)$. As the integral in figure 14.1 varies between roughly 1 and 8, it is to be expected that for surfaces with a correlation length small compared to the wavelength of light the dipolar coefficients are sufficient. If on the other hand the correlation length is comparable to the wavelength of light or larger, the quadrupolar coefficient will become more important. Larger values of the double integral make the quadrupolar coefficients even less important compared to the dipolar coefficient.

14.6 Intrinsic profile contributions

In the last three sections the roughness of the surface was assumed to be described by a sharp transition from one phase to another phase. If one averages over the roughness one may obtain an average dielectric constant profile. Due to contributions from correlations it was found that the expressions for the constitutive coefficients can not simply be obtained in terms of this average profile [64]. It is nevertheless not necessarily always realistic to use a transition which is sharp. On a length scale small compared to the correlation length along the surface, there is structure which is neglected in this manner. This structure is due to density and constitution fluctuations of the bulk phases close to the surface. The most appropriate manner to take this structure into account is to average over these fluctuations, keeping the height of the longer wavelength surface fluctuations constant. This leads to a stratified intrinsic dielectric constant profile, which is 'superimposed' on the height distribution:

$$\epsilon(\mathbf{r}) = \epsilon_{int}(z - h(\mathbf{r}_{\|})) \tag{14.50}$$

Correlation contributions to the surface constitutive coefficients, due to the short wavelength (bulk) fluctuations around the intrinsic profile, will be neglected. The amplitude of such fluctuations is usually to small to give a relevant contribution. In the bulk such contribution do not really contribute to the value of the dielectric constant either. The fluctuations of the profile due to long wavelength fluctuations are described in terms of the fluctuations of the height $h(\mathbf{r}_{\|})$ of the interface and they do give important contributions to the constitutive coefficients of the surface.

Away from the interface the intrinsic dielectric constant profile again approaches the values ϵ_0^+ and ϵ_0^-. The extrapolated dielectric constant is therefore again given by

$$\epsilon_0(\mathbf{r}) = \epsilon^-\theta(-z) + \epsilon^+\theta(z) \tag{14.51}$$

Subtracting eq.(14.51) from eq.(14.50) one may write

$$\begin{aligned}\epsilon(\mathbf{r}) - \epsilon_0(\mathbf{r}) &= \epsilon_{int}(z - h(\mathbf{r}_{\|})) - [\epsilon^-\theta(-z) + \epsilon^+\theta(z)] \\ &= \{[\epsilon^-\theta(h(\mathbf{r}_{\|}) - z) + \epsilon^+\theta(z - h(\mathbf{r}_{\|}))] - [\epsilon^-\theta(-z) + \epsilon^+\theta(z)]\} \\ &\quad + \{\epsilon_{int}(z - h(\mathbf{r}_{\|})) - [\epsilon^-\theta(h(\mathbf{r}_{\|}) - z) + \epsilon^+\theta(z - h(\mathbf{r}_{\|}))]\}\end{aligned} \tag{14.52}$$

If one now substitutes this expression into the expression for the x, y dependent susceptibility γ_e given by eq.(14.1) one finds

$$\begin{aligned}\gamma_e(\mathbf{r}_{\|}) &= \int_{-\infty}^{\infty} dz\{[\epsilon^-\theta(h(\mathbf{r}_{\|}) - z) + \epsilon^+\theta(z - h(\mathbf{r}_{\|}))] - [\epsilon^-\theta(-z) + \epsilon^+\theta(z)]\} \\ &\quad + \int_{-\infty}^{\infty} dz\{\epsilon_{int}(z - h(\mathbf{r}_{\|})) - [\epsilon^-\theta(h(\mathbf{r}_{\|}) - z) + \epsilon^+\theta(z - h(\mathbf{r}_{\|}))]\} \\ &= -(\epsilon^+ - \epsilon^-)h(\mathbf{r}_{\|}) + \int_{-\infty}^{\infty} dz\{\epsilon_{int}(z) - [\epsilon^-\theta(-z) + \epsilon^+\theta(z)]\}\end{aligned} \tag{14.53}$$

This last identity shows that the fluctuating surface susceptibility is now a sum of two contributions. The second one is due to the intrinsic profile and given by

$$\gamma_{int} = <\gamma_e(\mathbf{r}_{\|})> = \int_{-\infty}^{\infty} dz\{\epsilon_{int}(z) - [\epsilon^-\theta(-z) + \epsilon^+\theta(z)]\} \tag{14.54}$$

This contribution is identical to the expression one would directly obtain for a stratified layer with a dielectric constant profile given by $\epsilon_{int}(z)$ in the absence of any height fluctuations. The first contribution to the x, y dependent susceptibility,

$$\gamma_h(\mathbf{r}_\parallel) \equiv \gamma_e(\mathbf{r}_\parallel) - < \gamma_e(\mathbf{r}_\parallel) > = -(\epsilon^+ - \epsilon^-)h(\mathbf{r}_\parallel), \tag{14.55}$$

is simply the contribution due to the roughness used in the above three sections. Its average is zero and it therefore gives the contributions due to correlations calculated in these three sections. The intrinsic contribution does not fluctuate and gives a contribution equal to the one given in chapter 11 on stratified layers.

One may also analyze the other constitutive coefficients for the intrinsic profile in this way and one finds that they are all the sum of two contributions:

$$\begin{aligned} \gamma_e &= \gamma_{int} + \gamma_h, \quad \beta_e = \beta_{int} + \beta_h \\ \tau &= \tau_{int} + \tau_h, \quad \delta_e = \delta_{int} + \delta_h \end{aligned} \tag{14.56}$$

The intrinsic contributions are given in terms of the intrinsic profile by eq.(11.5). The contributions due to the roughness, indicated by a subindex h, are given by eqs.(14.28) and (14.31) for the rough surface, by eqs.(14.37) and (14.38) or by eqs.(14.42) and (14.43) for capillary waves and by eqs.(14.48) and (14.49) for self-affine surfaces. The additive nature of the contributions due to the intrinsic profile and the fluctuations of the height, is a very convenient feature in the analysis of experimental data.

14.7 Oxide layers

An important application of ellipsometry is the measurement of the thickness of thin layers of a different material on a substrate layers. In this way one may, for instance, obtain the thickness of an oxide layer on a metal substrate. This layer will in general not have a very homogeneous thickness. Not only will the substrate be rough, like the surfaces discussed in section 14.3, also the oxide layer will be very unevenly distributed. In the analysis it will be assumed that the oxide layer is bounded by two surfaces with respective height distributions $h^-(x, y)$ and $h^+(x, y)$. The metal has a frequency dependent dielectric constant ϵ^+ and is located in the region $z > h^+(x, y)$. The oxide has a frequency dependent dielectric constant ϵ and is located in the region $h^-(x, y) < z < h^+(x, y)$. It will be assumed that both height distributions are single-valued functions, which satisfy $h^-(x, y) \leq h^+(x, y)$. For parts of the surface where there is no oxide, both heights are taken to be the same. The ambient has a dielectric constant ϵ^- and fills the region $z < h^-(x, y)$.

The average position of the surface between the ambient and the oxide is used as the $x - y$ plane and one therefore has

$$< h^-(x, y) > = 0 \tag{14.57}$$

The average thickness of the oxide layer is given by d and one therefore similarly has

$$< h^+(x, y) > = d \tag{14.58}$$

One of the objectives of the measurement of the ellipsometric coefficient is in fact to obtain this thickness. If the layer was homogeneous this would be precisely what the usual expressions for the ellipsometric coefficient would give. As discussed in the previous sections there will be additional contributions due to height-height correlations. It is the aim of this section to give explicit expressions for these contributions.

The position dependent dielectric constant can be written as a function of both height distributions in the following way:

$$\epsilon(\mathbf{r}) = \epsilon^{-}\theta(h^{-}(\mathbf{r}_{\|}) - z) + \epsilon\, \theta(h^{+}(\mathbf{r}_{\|}) - z)\theta(z - h^{-}(\mathbf{r}_{\|})) + \epsilon^{+}\theta(z - h^{+}(\mathbf{r}_{\|})) \quad (14.59)$$

The extrapolated dielectric constant is given as usual by

$$\epsilon_0(\mathbf{r}) = \epsilon^{-}\theta(-z) + \epsilon^{+}\theta(z) \quad (14.60)$$

The height dependent constitutive coefficients describing the rough surface are now obtained by substitution of the above equations into the formulae given in eq.(14.1). This results in the following expressions

$$\begin{aligned}
\gamma_e(x,y) &= -(\epsilon - \epsilon^{-})h^{-}(x,y) - (\epsilon^{+} - \epsilon)h^{+}(x,y) \\
\beta_e(x,y) &= -\frac{\epsilon - \epsilon^{-}}{\epsilon\, \epsilon^{-}}h^{-}(x,y) - \frac{\epsilon^{+} - \epsilon}{\epsilon^{+}\epsilon}h^{+}(x,y) \\
\delta_e(x,y) &= \frac{1}{2}(\frac{\epsilon}{\epsilon^{-}} - \frac{\epsilon^{-}}{\epsilon})[h^{-}(x,y)]^2 + \frac{1}{2}(\frac{\epsilon^{+}}{\epsilon} - \frac{\epsilon}{\epsilon^{+}})[h^{+}(x,y)]^2 \\
&\quad + \frac{(\epsilon^{+} - \epsilon^{-})(\epsilon^{+} + \epsilon)(\epsilon - \epsilon^{-})}{\epsilon^{+}\epsilon\, \epsilon^{-}}h^{+}(x,y)h^{-}(x,y) \\
\tau(x,y) &= \frac{1}{2}(\epsilon - \epsilon^{-})[h^{-}(x,y)]^2 + \frac{1}{2}(\epsilon^{+} - \epsilon)[h^{+}(x,y)]^2
\end{aligned} \quad (14.61)$$

The magnetic constitutive coefficients, $\gamma_m(x,y)$, $\beta_m(x,y)$ and $\delta_m(x,y)$, are again all zero and will further be neglected.

Averaging the above expressions for the fluctuating interfacial susceptibilities gives the following values:

$$\begin{aligned}
&< \gamma_e(x,y) >= -(\epsilon^{+} - \epsilon)d \\
&< \beta_e(x,y) >= -\frac{\epsilon^{+} - \epsilon}{\epsilon^{+}\epsilon}d \\
\delta_e &= < \delta_e(x,y) >= \frac{1}{2}(\frac{\epsilon}{\epsilon^{-}} - \frac{\epsilon^{-}}{\epsilon})(t^{--})^2 + \frac{1}{2}(\frac{\epsilon^{+}}{\epsilon} - \frac{\epsilon}{\epsilon^{+}})[d^2 + (t^{++})^2] \\
&\quad + \frac{(\epsilon^{+} - \epsilon^{-})(\epsilon^{+} + \epsilon)(\epsilon - \epsilon^{-})}{\epsilon^{+}\epsilon\, \epsilon^{-}}(t^{-+})^2 \\
\tau &= < \tau(x,y) >= \frac{1}{2}(\epsilon - \epsilon^{-})(t^{--})^2 + \frac{1}{2}(\epsilon^{+} - \epsilon)[d^2 + (t^{++})^2]
\end{aligned} \quad (14.62)$$

Here root mean square displacements of both surfaces are defined by

$$(t^{--})^2 \equiv< (h^{-}(x,y))^2 >, \quad (t^{++})^2 \equiv< (h^{+}(x,y) - d)^2 > \quad (14.63)$$

while a cross correlation height is defined by

$$(t^{-+})^2 \equiv < h^-(x,y)(h^+(x,y) - d) > \tag{14.64}$$

The expressions for δ_e and τ are the final values needed. In order to obtain γ_e and β_e the contributions due to correlations must be added.

In order to use the fluctuating constitutive coefficients one needs the height-height correlation functions, which are defined by

$$S^{\nu\nu'}(\mathbf{r}_\parallel - \mathbf{r}'_\parallel) \equiv < (h^\nu(\mathbf{r}_\parallel) - d^\nu)(h^{\nu'}(\mathbf{r}'_\parallel) - d^{\nu'}) > \text{ with } \nu, \nu' = - \text{ or } + \tag{14.65}$$

Here $d^- \equiv 0$ and $d^+ \equiv d$. It will be assumed that the roughness is isotropic so that the height-height correlations do not depend on the direction

$$S^{\nu\nu'}(\mathbf{r}_\parallel - \mathbf{r}'_\parallel) = S^{\nu\nu'}(|\mathbf{r}_\parallel - \mathbf{r}'_\parallel|) \tag{14.66}$$

The root mean square displacements of the surfaces, as well as their cross correlation height, may be written in terms of the correlation functions in the following way

$$(t^{\nu\nu'})^2 = S^{\nu\nu'}(\mathbf{r}_\parallel = 0) = (2\pi)^{-2} \int d\mathbf{k}_\parallel S^{\nu\nu'}(k_\parallel) = (2\pi)^{-1} \int_0^\infty dk_\parallel k_\parallel S^{\nu\nu'}(k_\parallel) \tag{14.67}$$

Using the above expressions for the fluctuating constitutive coefficients one finds that their correlation functions are also isotropic and given by,

$$\begin{aligned} S_{ee}^{\gamma\gamma}(r_\parallel) &= \sum_{\nu,\nu'} \nu\nu'(\epsilon^\nu - \epsilon)(\epsilon^{\nu'} - \epsilon) S^{\nu\nu'}(r_\parallel) \\ S_{ee}^{\beta\beta}(r_\parallel) &= \sum_{\nu,\nu'} \nu\nu'(\frac{\epsilon^\nu - \epsilon}{\epsilon^\nu \epsilon})(\frac{\epsilon^{\nu'} - \epsilon}{\epsilon^{\nu'} \epsilon}) S^{\nu\nu'}(r_\parallel), \end{aligned} \tag{14.68}$$

in terms of the height-height correlation functions. The wave vector dependent Fourier transforms, or using an other word, structure factors satisfy the same relations

$$\begin{aligned} S_{ee}^{\gamma\gamma}(k_\parallel) &= \sum_{\nu,\nu'} \nu\nu'(\epsilon^\nu - \epsilon)(\epsilon^{\nu'} - \epsilon) S^{\nu\nu'}(k_\parallel) \\ S_{ee}^{\beta\beta}(k_\parallel) &= \sum_{\nu,\nu'} \nu\nu'(\frac{\epsilon^\nu - \epsilon}{\epsilon^\nu \epsilon})(\frac{\epsilon^{\nu'} - \epsilon}{\epsilon^{\nu'} \epsilon}) S^{\nu\nu'}(k_\parallel), \end{aligned} \tag{14.69}$$

in terms of the height-height structure factors.

Substitution of the above expressions into eq.(14.18) gives the following formulae for the effective constitutive coefficients of the surface

$$\begin{aligned} \gamma_e &= -(\epsilon^+ - \epsilon)d - (16\pi)^{-1}\frac{(\epsilon^+ + \epsilon^-)}{\epsilon^+\epsilon^-} \sum_{\nu,\nu'} \nu\nu'(\epsilon^\nu - \epsilon)(\epsilon^{\nu'} - \epsilon) \int_0^\infty dk_\parallel k_\parallel^2 S^{\nu\nu'}(k_\parallel) \\ \beta_e &= -\frac{\epsilon^+ - \epsilon}{\epsilon^+\epsilon} d + (8\pi)^{-1}(\epsilon^+ + \epsilon^-) \sum_{\nu,\nu'} \nu\nu'(\frac{\epsilon^\nu - \epsilon}{\epsilon^\nu \epsilon})(\frac{\epsilon^{\nu'} - \epsilon}{\epsilon^{\nu'} \epsilon}) \int_0^\infty dk_\parallel k_\parallel^2 S^{\nu\nu'}(k_\parallel) \end{aligned} \tag{14.70}$$

The expressions for δ_e and τ were already given above. The formulae for γ_e and β_e both have a contribution proportional to the average thickness of the oxide layer plus a contribution due to correlations in the height distributions along the surface. The quadrupolar coefficients δ_e and τ are not affected by correlations along the surface.

In order to gain some insight in the contribution due to correlations it will be assumed that the height-height correlation functions are Gaussian

$$S^{\nu\nu'}(r_{\|}) = (t^{\nu\nu'})^2 \exp[-r_{\|}^2/(\xi^{\nu\nu'})^2] \tag{14.71}$$

ξ^{--} and ξ^{++} are the correlation lengths of the two surfaces parallel to the surface while ξ^{-+} is a cross correlation length of the two surfaces. For $r_{\|} = 0$ this expression reduces to a relation for the mean square heights and the cross correlation height. Fourier transformation of this expression gives the structure factors

$$S^{\nu\nu'}(k_{\|}) = \pi(t^{\nu\nu'})^2(\xi^{\nu\nu'})^2 \exp[-\frac{1}{4}k_{\|}^2(\xi^{\nu\nu'})^2] \tag{14.72}$$

Substitution into eq.(14.70) and integration gives for the dipolar constitutive coefficients

$$\begin{aligned} \gamma_e &= -(\epsilon^+ - \epsilon)d - \sqrt{\pi}\frac{(\epsilon^+ + \epsilon^-)}{8\epsilon^+\epsilon^-}\sum_{\nu,\nu'}\nu\nu'(\epsilon^\nu - \epsilon)(\epsilon^{\nu'} - \epsilon)(t^{\nu\nu'})^2/\xi^{\nu\nu'} \\ \beta_e &= -\frac{\epsilon^+ - \epsilon}{\epsilon^+\epsilon}d + \frac{1}{8}\sqrt{\pi}(\epsilon^+ + \epsilon^-)\sum_{\nu,\nu'}\nu\nu'(\frac{\epsilon^\nu - \epsilon}{\epsilon^\nu\epsilon})(\frac{\epsilon^{\nu'} - \epsilon}{\epsilon^{\nu'}\epsilon})(t^{\nu\nu'})^2/\xi^{\nu\nu'} \end{aligned} \tag{14.73}$$

The interesting question now is, whether the correlation contribution is important compared with the average thickness of the oxide layer or not. If the oxide layer is thick the correlation contribution will generally be unimportant. For thin discontinuous layers of oxide the mean square heights will be of the same order of magnitude as the average thickness d. The correlation length will not be much larger. This implies that, depending of course on the contrast, i.e. the value of the dielectric constant differences, the contribution due to correlations can be of considerable importance. Neglecting these correlation contributions to the ellipsometric coefficient would lead to an incorrect interpretation of this coefficient in terms of an oxide layer thickness.

It is also interesting to compare the importance of the dipolar coefficients with the quadrupolar coefficients. In expressions for e.g. reflectivities the dipolar coefficients will appear divided by the wavelength λ, while the quadrupolar coefficients appear divided by the wavelength squared. The relative importance of the quadrupolar coefficients is therefore of relative order ξ/λ compared to the correlation contribution and $t^2/d\lambda$ compared to the average oxide layer contribution. Using the above estimates for the relative order of magnitude of the lengths involved, it may be concluded that the dipolar coefficients are the dominant contribution.

It should be noted that the Gaussian nature of the correlation functions represents surfaces covered in a relatively simple manner by a third material. There are much more complex possibilities to deposit material which leads to fractal structures of this material. Given the appropriate correlation functions one could apply the above theory also to these structures. This will not be done explicitly here.

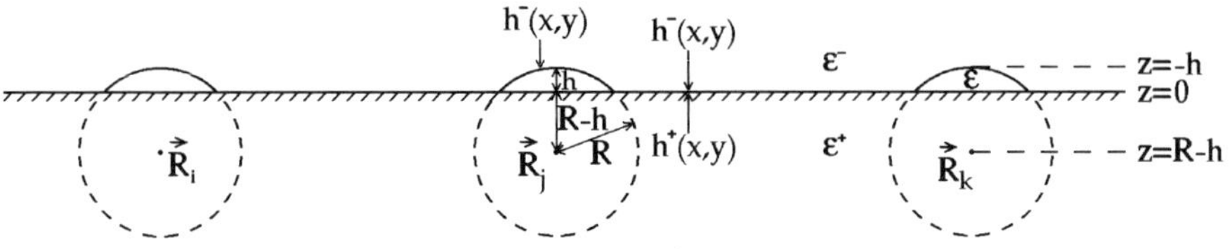

Figure 14.2 Thin spherical caps on a substrate.

14.8 Thin spherical caps on a substrate

In this last section, it will be shown how the theory of thin spherical caps, treated in chapter 8, section 5, can also be developed as a special case of the "oxide" layer, treated in the previous section. The analysis in this section is based on unpublished work by M.M. Wind. Consider for this purpose an isotropic array of identical thin spherical caps on the planar surface, $z = 0$, of the substrate. The dielectric constant of the substrate is equal to ϵ^+, of the ambient ϵ^- and of the cap ϵ, see fig. 14.2. This film will now be described as an, in this case discontinuous, oxide layer between the surface of the substrate, $h^+(x,y) = 0$, and the surface

$$h^-(x,y) = -\sum_i f\left(\mathbf{r}_\parallel - \mathbf{R}_{i,\parallel}\right) \tag{14.74}$$

Here $\mathbf{r}_\parallel = (x, y, 0)$ and $\mathbf{R}_{i,\parallel}$ are the positions of the centers of the interfaces between the caps and the surface of the substrate, see fig.14.2. The function f is given by

$$f(\mathbf{r}_\parallel) = f(r_\parallel) \equiv \begin{cases} \sqrt{R^2 - r_\parallel^2} - Rt_r & \text{for} \quad 0 \le r_\parallel \le R\sqrt{1-t_r^2} \\ 0 \quad \text{for} & R\sqrt{1-t_r^2} \le r_\parallel \end{cases} \tag{14.75}$$

where R is the radius of the sphere of which the cap is a part and the truncation parameter t_r is given by, see fig. 14.2,

$$t_r \equiv \frac{R-h}{R} = 1 - \frac{h}{R} \equiv 1 - h_r \tag{14.76}$$

Between the caps there is nothing on the surface so that both $h^-(x,y)$ and $h^+(x,y)$ are equal to zero between the caps.

For the surfaces defined above

$$\langle h^+(x,y) \rangle = 0 \tag{14.77}$$

and

$$\langle h^-(x,y) \rangle = -\rho V = -\pi \rho R^3 h_r^2 (1 - h_r/3) \equiv -t_w \tag{14.78}$$

where V, cf. eq.(8.103), is the volume of the cap, ρ the number of caps per unit of surface area and t_w the weight thickness. The latter quantity has also been discussed in the introduction, chapter 1, of this book.

The position dependent dielectric constant is given by eq.(14.59), which becomes in the present case, where $h^+(x,y) = 0$,

$$\epsilon(\mathbf{r}) = \epsilon^- \theta(h^-(\mathbf{r}_\parallel) - z) + \epsilon\, \theta(-z)\theta(z - h^-(\mathbf{r}_\parallel)) + \epsilon^+ \theta(z) \tag{14.79}$$

The extrapolated dielectric constant is given by eq.(14.60). The height dependent constitutive coefficients, describing the island film of thin spherical caps, are now obtained by substitution of eqs.(14.79) and (14.60) into eq.(14.1). This results in the following expressions

$$\begin{aligned}
\gamma_e(x,y) &= -(\epsilon - \epsilon^-)h^-(x,y) \\
\beta_e(x,y) &= -\frac{\epsilon - \epsilon^-}{\epsilon\, \epsilon^-} h^-(x,y) \\
\delta_e(x,y) &= \frac{1}{2}\left(\frac{\epsilon}{\epsilon^-} - \frac{\epsilon^-}{\epsilon}\right)[h^-(x,y)]^2 \\
\tau(x,y) &= \frac{1}{2}(\epsilon - \epsilon^-)[h^-(x,y)]^2
\end{aligned} \tag{14.80}$$

which also follows directly from Eq.(14.61).

Averaging the above expressions for the fluctuating interfacial susceptibilities gives the following values

$$\begin{aligned}
&< \gamma_e(x,y) >= (\epsilon - \epsilon^-)t_w \\
&< \beta_e(x,y) >= \frac{\epsilon - \epsilon^-}{\epsilon\epsilon^-} t_w \\
\delta_e &= < \delta_e(x,y) >= \frac{1}{2}\left(\frac{\epsilon}{\epsilon^-} - \frac{\epsilon^-}{\epsilon}\right)t^2 \\
\tau &= < \tau(x,y) >= \frac{1}{2}(\epsilon - \epsilon^-)t^2
\end{aligned} \tag{14.81}$$

where

$$t \equiv t^{--} = \sqrt{< (h^-(x,y) + t_w)^2 >} \tag{14.82}$$

is the root mean square height of the surface $z = h^-(x,y)$. The expressions for δ_e and τ are the final values needed. In order to obtain γ_e and β_e the contributions due to correlations must be added.

Since $h^+(x,y) = 0$ there is only one height-height correlation function, which is given by

$$S(\mathbf{r}_\parallel - \mathbf{r}'_\parallel) \equiv S^{--}(\mathbf{r}_\parallel - \mathbf{r}'_\parallel) =< (h^-(\mathbf{r}_\parallel) + t_w)(h^-(\mathbf{r}'_\parallel) + t_w) > \tag{14.83}$$

Since the distribution of the caps is isotropic the height-height correlation function does not depend on the direction, one has

$$S(\mathbf{r}_\parallel - \mathbf{r}'_\parallel) = S(|\mathbf{r}_\parallel - \mathbf{r}'_\parallel|) \tag{14.84}$$

The root mean square displacements of the surfaces may be written in terms of the correlation function in the following way

$$t^2 = S(\mathbf{r}_\parallel = 0) = (2\pi)^{-2}\int d\mathbf{k}_\parallel S(k_\parallel) = (2\pi)^{-1}\int_0^\infty dk_\parallel k_\parallel S(k_\parallel) \tag{14.85}$$

where the structure factor is defined by

$$S(k_\parallel) = \int d\mathbf{r}_\parallel S(r_\parallel)\exp\left(-i\mathbf{k}_\parallel.\mathbf{r}_\parallel\right) \tag{14.86}$$

Using the expressions for the fluctuating constitutive coefficients, one finds that their correlation functions are given by

$$\begin{aligned} S_{ee}^{\gamma\gamma}(r_\parallel) &= (\epsilon-\epsilon^-)^2 S(r_\parallel) \\ S_{ee}^{\beta\beta}(r_\parallel) &= (\frac{\epsilon-\epsilon^-}{\epsilon\epsilon^-})^2 S(r_\parallel) \end{aligned} \tag{14.87}$$

in terms of the height-height correlation function. Their Fourier transforms satisfy the same relations

$$\begin{aligned} S_{ee}^{\gamma\gamma}(k_\parallel) &= (\epsilon-\epsilon^-)^2 S(k_\parallel) \\ S_{ee}^{\beta\beta}(k_\parallel) &= (\frac{\epsilon-\epsilon^-}{\epsilon\epsilon^-})^2 S(k_\parallel), \end{aligned} \tag{14.88}$$

in terms of the structure factors.

Substitution of the above expressions into eq.(14.18) and using eqs.(14.81) and (14.88), gives for the zero frequency nonmagnetic terms

$$\begin{aligned} \gamma_e &= (\epsilon-\epsilon^-)t_w - \frac{1}{16\pi}((\epsilon^-)^{-1}+(\epsilon^+)^{-1})(\epsilon-\epsilon^-)^2\int_0^\infty dk_\parallel S(k_\parallel)k_\parallel^2 \\ \beta_e &= \frac{\epsilon-\epsilon^-}{\epsilon\epsilon^-}t_w + \frac{1}{8\pi}(\epsilon^-+\epsilon^+)(\frac{\epsilon-\epsilon^-}{\epsilon\epsilon^-})^2\int_0^\infty dk_\parallel S(k_\parallel)k_\parallel^2 \\ \delta_e &= \frac{1}{2}(\frac{\epsilon}{\epsilon^-}-\frac{\epsilon^-}{\epsilon})t^2 \\ \tau &= \frac{1}{2}(\epsilon-\epsilon^-)t^2 \end{aligned} \tag{14.89}$$

It remains to calculate the structure factor and the root mean square height. Substituting eq.(14.74) into eq.(14.83) the correlation function becomes

$$S(\mathbf{r}_\parallel - \mathbf{r}'_\parallel) = \sum_{i,j} < f\left(\mathbf{r}_\parallel - \mathbf{R}_{i,\parallel}\right) f\left(\mathbf{r}'_\parallel - \mathbf{R}_{j,\parallel}\right) > -t_w^2 \tag{14.90}$$

For $\mathbf{r}_\parallel = \mathbf{r}'_\parallel$ only terms with $i = j$ are unequal to zero so that

$$\begin{aligned} t^2 &= S(0) = \sum_i < f^2\left(\mathbf{r}_\parallel - \mathbf{R}_{i,\parallel}\right) > -t_w^2 = \rho\int f^2\left(\mathbf{r}_\parallel\right) d\mathbf{r}_\parallel - t_w^2 \\ &= 2\pi\rho\int_0^{R\sqrt{1-t_r^2}}\left(\sqrt{R^2-r_\parallel^2}-Rt_r\right)^2 r_\parallel dr_\parallel - t_w^2 \\ &= \frac{2}{3}\pi\rho R^4 h_r^3\left(1-\frac{1}{4}h_r\right) - \pi^2\rho^2 R^6 h_r^4\left(1-\frac{1}{3}h_r\right) \end{aligned} \tag{14.91}$$

where $h_r = 1 - t_r$. Furthermore eqs.(14.75) and (14.78) have been used. As the whole calculation of the surface susceptibilities above has been done neglecting terms of the third order in the height, it follows that

$$\delta_e = \tau = 0 \tag{14.92}$$

to this order for spherical caps

In order to calculate the structure factor it is convenient to introduce the pair correlation function for the centers of the caps

$$g(\mathbf{r}_\| - \mathbf{r}'_\|) \equiv \rho^{-2} \sum_{i \neq j} < \delta\left(\mathbf{r}_\| - \mathbf{R}_{i,\|}\right) \delta\left(\mathbf{r}'_\| - \mathbf{R}_{j,\|}\right) > \tag{14.93}$$

The correlation function then becomes

$$\begin{aligned} S(\mathbf{r}_\| - \mathbf{r}'_\|) &= \rho^2 \int d\mathbf{R}_\| \int d\mathbf{R}'_\| f\left(\mathbf{r}_\| - \mathbf{R}_\|\right) f\left(\mathbf{r}'_\| - \mathbf{R}'_\|\right) g\left(\mathbf{R}_\| - \mathbf{R}'_\|\right) \\ +\sum_i &< f\left(\mathbf{r}_\| - \mathbf{R}_{i,\|}\right) f\left(\mathbf{r}'_\| - \mathbf{R}_{i,\|}\right) > -t_w^2 \end{aligned} \tag{14.94}$$

The first and the last contribution are of the second order in the density ρ and are therefore negligible, in the low coverage limit to be considered, compared to the second term which is linear in ρ. To first order one therefore obtains

$$S(\mathbf{r}_\| - \mathbf{r}'_\|) = \rho \int d\mathbf{R}_\| f\left(\mathbf{r}_\| - \mathbf{R}_\|\right) f\left(\mathbf{r}'_\| - \mathbf{R}_\|\right) \tag{14.95}$$

In view of the fact that this is a convolution of the function f with itself, one has

$$S(\mathbf{k}_\|) = \rho f\left(\mathbf{k}_\|\right) f\left(-\mathbf{k}_\|\right) = \rho f^2\left(k_\|\right) \tag{14.96}$$

Here

$$f\left(\mathbf{k}_\|\right) = f\left(k_\|\right) \equiv \int d\mathbf{r}_\| f\left(r_\|\right) \exp\left(-i\mathbf{k}_\|.\mathbf{r}_\|\right) \tag{14.97}$$

Introducing plane polar coordinates, this integral can be written as

$$f\left(k_\|\right) \equiv \int_0^{R\sqrt{1-t_r^2}} dr_\| \int_{-\pi}^{\pi} d\phi \exp\left(-ik_\| r_\| \cos\phi\right) r_\| \left(\sqrt{R^2 - r_\|^2} - Rt_r\right) \tag{14.98}$$

Using the Bessel function $J_0\left(k_\| r_\|\right)$, cf. [48], and a new integration variable $s \equiv \left(r_\|/R\right)\left(1 - t_r^2\right)^{-1/2}$ one obtains

$$f\left(k_\|\right) \equiv 2\pi R^3\left(1 - t_r^2\right) \int_0^1 ds J_0\left(k_\| R s\sqrt{1 - t_r^2}\right) s\left[\sqrt{1 - s^2\left(1 - t_r^2\right)} - t_r\right] \tag{14.99}$$

Expanding the prefactor and the term between square brackets up to lowest order in terms of the small quantity $h_r = 1 - t_r$ this equation can be written as

$$f\left(k_\|\right) \equiv 4\pi R^3 h_r^2 \int_0^1 ds J_0\left(k_\| R s\sqrt{1 - t_r^2}\right) s\left(1 - s^2\right) \tag{14.100}$$

This integral can be evaluated rigorously, cf.[48] page 688, and gives

$$f\left(k_{\|}\right) \equiv 4\pi R h_r k_{\|}^{-2} J_2\left(k_{\|} R\sqrt{1-t_r^2}\right) \tag{14.101}$$

The structure factor is therefore to lowest order

$$S\left(k_{\|}\right) \equiv 16\pi^2 \rho R^2 h_r^2 k_{\|}^{-4}\left[J_2\left(k_{\|} R\sqrt{1-t_r^2}\right)\right]^2 \tag{14.102}$$

In order to obtain the linear susceptibilities one must evaluate the following integral to leading order

$$\begin{aligned}\int_0^\infty dk_{\|} S(k_{\|}) k_{\|}^2 &= 16\pi^2 \rho R^2 h_r^2 \int_0^\infty dk_{\|} k_{\|}^{-2}\left[J_2\left(k_{\|} R\sqrt{1-t_r^2}\right)\right]^2 \\ &= \frac{64\sqrt{2}}{15}\pi\rho R^3 h_r^{5/2}\end{aligned} \tag{14.103}$$

cf.[48] page 692 for the last equality. The resulting susceptibilities are, using eq.(14.78),

$$\begin{aligned}\gamma_e &= (\epsilon-\epsilon^-)\rho V\left[1-\frac{4\sqrt{2}}{15\pi}(\frac{\epsilon^-+\epsilon^+}{\epsilon^-\epsilon^+})(\epsilon-\epsilon^-)h_r^{1/2}+\bigcirc\left(h_r^{3/2}\right)\right] \\ \beta_e &= \frac{\epsilon-\epsilon^-}{\epsilon\epsilon^-}\rho V\left[1+\frac{8\sqrt{2}}{15\pi}(\epsilon^-+\epsilon^+)(\frac{\epsilon-\epsilon^-}{\epsilon\epsilon^-})h_r^{1/2}+\bigcirc\left(h_r^{3/2}\right)\right] \\ \delta_e &= \tau=\rho V\bigcirc\left(h_r\right)\end{aligned} \tag{14.104}$$

where $\bigcirc$ is Landau's order symbol.

It now interesting to compare this result with the results in chapter 8, section 5, for thin spherical caps. In that case one finds, cf. eq.(8.105),

$$\begin{aligned}\gamma_e &= (\epsilon-\epsilon^-)\rho V\left[1+\bigcirc\left(h_r^2\right)\right] \\ \beta_e &= \frac{\epsilon-\epsilon^-}{\epsilon\epsilon^-}\rho V\left[1+\bigcirc\left(h_r^2\right)\right] \\ \delta_e &= \tau=\rho V\bigcirc\left(h_r\right)\end{aligned} \tag{14.105}$$

One may conclude that the two approaches give the same leading term to second order in the height. Beyond this order the functional dependence on the height, in particular of the linear terms, has a completely different analytic structure, however. Intuitively the method in this chapter seems more appropriate. It is to be expected that the convergence of the expansion in spherical multipoles is unsatisfactory for a thin spherical cap. The fact that the asymptotic value is the same to the second order in the height is comforting.

Using the above formulae one may calculate the depolarization factors of a free thin spherical cap. Taking $\epsilon^- = \epsilon^+ = 1$ for this purpose, eq.(14.104) for the linear coefficients may be written in the form

$$\begin{aligned}\gamma_e &= \frac{(\epsilon-1)\rho V}{1+\left(8\sqrt{2}/15\pi\right)h_r^{1/2}(\epsilon-1)} \\ \beta_e &= \frac{(\epsilon-1)\rho V}{1+\left[1-\left(16\sqrt{2}/15\pi\right)h_r^{1/2}\right](\epsilon-1)}\end{aligned} \tag{14.106}$$

One may now identify the following depolarization factors

$$
\begin{aligned}
L_{\|} &= \frac{8\sqrt{2}}{15\pi} h_r^{1/2} \\
L_z &= 1 - \frac{16\sqrt{2}}{15\pi} h_r^{1/2}
\end{aligned}
\tag{14.107}
$$

Note that these depolarization factors satisfy the usual identity

$$L_z + 2L_{\|} = 1 \tag{14.108}$$

for free particles with cylindrical symmetry. In the limit that the cap becomes infinitely thin these factors approach the value

$$L_{\|} = 0 \quad \text{and} \quad L_z = 1 \tag{14.109}$$

This are the same values found for the infinitely thin oblate spheroid, eq.(6.146). One may in fact obtain these limiting value immediately using the definition of the electric and the displacement field in a medium, using thin slabs.

Chapter 15
REFLECTION OF A GYROTROPIC MEDIUM

15.1 Introduction

Until now the treatment has been restricted to isotropic surfaces and bulk phases. In this chapter an extension will be given to surfaces between gyrotropic media. In such media the left-handedness or right-handedness of their structure manifests itself by phenomena like optical rotation and circular dichroism. The description of the optical properties of such gyrotropic media is well established [81, 20]. There are two prevailing methods. In the first method, due to Condon [82], both the polarization and the magnetization contain a contribution due to the chiral structure of the medium. This description has the symmetry for the interchange of the electric with the magnetic fields, which was used throughout this book. See for instance sections 8 and 9 in chapter 3 for the relevance of this symmetry in the general context. In the second method one uses an alternative description in which the magnetization is adsorbed in the polarization [81, 20, 83]. Both methods are fully equivalent and as such the second method, which only needs a dielectric tensor, is usually preferred.

As one needs boundary conditions to interpret measurements on such systems, a proper formulation of these conditions is needed. It is found that if one uses the standard boundary conditions, which are appropriate in the absence of singular contributions to the polarization and magnetization at the surface, cf. eq.(3.30) with $\mathbf{P}'^s = \mathbf{M}^s = 0$, the above two methods lead to different reflection coefficients [84, 85, 86, 87, 88, 89]. Experimental results seem to indicate that the predictions of the symmetric method, due to Condon, are correct, while the predictions of the second method are incorrect. In this chapter it will be shown that the usual transformation from one to the other description, changes standard boundary conditions in one description to "non-standard" boundary conditions, like those given in eq.(3.30), in the other description. In the usual derivation of the equivalence of the two methods one only considers the transformation from one method to the other in the homogeneous bulk phases. As will be discussed below one should not only consider what the transformation implies in the bulk regions, but also what it implies for the surface. One then finds that singular contributions to the fields appear due to the transformation. When one accounts for these singular contributions of the fields in the boundary conditions, both methods give the same reflection coefficients.

On the basis of the experimental results in the work quoted above one could now conclude that, for the systems considered, the symmetric method together with the standard boundary condition is correct. In particular it is found that at normal incidence the differential reflection amplitude from a gyrotropic medium between left

(LCP) and right (RCP) circularly polarized light is zero. This implies that in the context of this method no singular contributions to the fields on the surface are needed to describe their experiments. It is noted that it was precisely to avoid the occurrence of such singular contributions, that Maxwell [90] used the displacement fields **D**, **B** in addition to the electric and magnetic fields **E**, **H**. As shall be verified below, one may equally well describe these experimental results using the second method, if one takes the singular contributions to the fields on the surface along and modifies the boundary conditions accordingly. This was also pointed out by Bungay et al. [91].

Following the general method outlined in this book singular contributions to the fields can also be taken along in the symmetric method. That such contributions are needed for certain systems is apparent from the work of Bungay et al. [91, 92], who measure a differential reflection amplitude at normal incidence from a gyrotropic medium, which is unequal to zero. They also introduce singular contributions to the polarization along the surface in the second method. The general form of the symmetry relations for surfaces between two gyrotropic media are found to be identical to those derived in chapter 13, cf. eq.(13.25). Contributions which break the symmetry are not needed to describe the various phenomena, like a finite differential reflection amplitude between left and right polarized light, as claimed by these authors.

The work in this chapter is based on a paper with M. Osipov [93], who drew the attention of the authors to the relevance of the methods in this book regarding the continuing debate about the validity of the two methods and to suggestions that the usual symmetry relations need to be violated in order to explain certain experimental results [91, 92].

Even though the reader of this book barely needs to be convinced in the last chapter that singular contributions to the fields at the surface play a role, the following sections are meant to convince the reader again of this fact. The reason to do this is that the role of these singularities has clearly been underestimated in the general discussion of the boundary conditions for gyrotropic media. In the following two sections the special case where the surface between the two media is a sharp transition from one medium to the other, is considered. Sharp is meant in the sense that no other material, or roughness or anything else modifies the boundary conditions. Even in this case it is found to be necessary to consider singular contributions to the fields in the second method. In the fourth section the non-standard boundary conditions, needed in the general case, are given. The origin of singular contributions to the fields in the first method will be discussed. Expressions for the constitutive coefficients, which give these singularities, are derived. In the fifth section the reflection of normally incident light by a gyrotropic medium is treated. It is shown that the non-standard boundary conditions, derived in the fourth section, predict that the nature of the polarization changes upon reflection. In general left and right polarized light are reflected as ellipsoidal polarized light. This is in agreement with the experimental evidence.

15.2 The relation between the two methods

The Maxwell equations for the electromagnetic fields as a function of the position and the frequency, cf. section 3.5, are

$$\text{rot }\mathbf{E} = \frac{i\omega}{c}\mathbf{B}\ , \qquad \text{div }\mathbf{D} = 0$$
$$\text{rot }\mathbf{H} = -\frac{i\omega}{c}\mathbf{D}\ , \qquad \text{div }\mathbf{B} = 0 \tag{15.1}$$

The polarization and magnetization are given by

$$\mathbf{P} = \mathbf{D} - \mathbf{E} \quad \text{and} \quad \mathbf{M} = \mathbf{B} - \mathbf{H} \tag{15.2}$$

Possible induced electric currents have been adsorbed in the polarization. The first method, followed throughout this book, uses all these fields. In the second method one transforms to a new displacement field by adsorbing the magnetization in the following manner

$$\mathbf{D}'' \equiv \mathbf{D} - \frac{c}{i\omega}\text{rot }\mathbf{M} \tag{15.3}$$

The Maxwell equations then become

$$\text{rot }\mathbf{E} = \frac{i\omega}{c}\mathbf{B}\ , \qquad \text{div }\mathbf{D}'' = 0$$
$$\text{rot }\mathbf{B} = -\frac{i\omega}{c}\mathbf{D}''\ , \qquad \text{div }\mathbf{B} = 0 \tag{15.4}$$

In the homogeneous phases all the fields, both before and after the transformation, are continuous and the transformation is straightforward. On the surface between two homogeneous materials, however, some of the fields are not continuous. In particular the magnetization is not continuous. Choosing the x, y-plane along the surface, the field $\mathbf{H}$ can be written in the form

$$\mathbf{H}(\mathbf{r},\omega) = \mathbf{H}^-(\mathbf{r},\omega)\Theta(-z) + \mathbf{H}^+(\mathbf{r},\omega)\Theta(z) \tag{15.5}$$

where $\mathbf{H}^-$ and $\mathbf{H}^+$ are continuous functions, determined by the constitutive relations in the two media. Now rot $\mathbf{H}$ can be expressed as

$$\begin{aligned}\text{rot}\,\mathbf{H}(\mathbf{r},\omega) &= \left[\text{rot }\mathbf{H}^-(\mathbf{r},\omega)\right]\Theta(-z) + \left[\text{rot }\mathbf{H}^+(\mathbf{r},\omega)\right]\Theta(z)\\ &\quad + \widehat{\mathbf{z}}\times\left[\mathbf{H}_\parallel^+(\mathbf{r},\omega) - \mathbf{H}_\parallel^-(\mathbf{r},\omega)\right]\ \delta(z)\end{aligned} \tag{15.6}$$

where $\widehat{\mathbf{z}}$ is a unit vector in the z-direction. The singular contribution on the right-hand side of eq.(15.6) disappears if one uses the standard boundary condition, $\mathbf{H}_\parallel^+ = \mathbf{H}_\parallel^-$. Such a singular contribution does not disappear, however, if one splits rot $\mathbf{H}$ into rot $\mathbf{B}$ and rot $\mathbf{M}$, and adsorbs rot $\mathbf{M}$ into the dielectric displacement field $\mathbf{D}''$. Writing the magnetization $\mathbf{M}$ in the same form as $\mathbf{H}$ in eq.(15.5) one obtains the following equation for $\mathbf{D}''$

$$\begin{aligned}\mathbf{D}''(\mathbf{r},\omega) &= \mathbf{D}(\mathbf{r},\omega) - \frac{c}{i\omega}\left[\text{rot }\mathbf{M}^-(\mathbf{r},\omega)\right]\Theta(-z) - \frac{c}{i\omega}\left[\text{rot }\mathbf{M}^+(\mathbf{r},\omega)\right]\Theta(z)\\ &\quad - \frac{c}{i\omega}\widehat{\mathbf{z}}\times\left[\mathbf{M}_\parallel^+(\mathbf{r},\omega) - \mathbf{M}_\parallel^-(\mathbf{r},\omega)\right]\ \delta(z)\end{aligned} \tag{15.7}$$

In this equation the singular term does not vanish because $\mathbf{M}_{\|}$ is discontinuous at the surface. At the same time it can be shown that the term rot $\mathbf{B}$ also contains a singular contribution. There is a good reason to use the field $\mathbf{H}$, because rot $\mathbf{H}$ does not contain singular contributions. Such contributions, which are present in rot $\mathbf{B}$ and rot $\mathbf{M}$, cancel against each other in rot $\mathbf{H}$. Maxwell, see ref.[90] vol.2, page 259, calls the term $\widehat{\mathbf{z}} \times \left[\mathbf{M}_{\|}^{+}(\mathbf{r},\omega) - \mathbf{M}_{\|}^{-}(\mathbf{r},\omega)\right] \delta(z)$ "free magnetism", which arises in his discussion because of a discontinuity in the magnetic permeability. He clearly states that such terms should be taken into account as sources in the calculation of the vector potential. He concludes that it is more convenient to calculate the magnetic and the induction fields on both sides of the surface separately and then to match them using the standard boundary conditions, or in other words to use the first method.

One concludes that the first method is consistent with the use of the standard boundary conditions, in the sense that no singular contributions to the fields appear directly in the Maxwell equations. At the same time it is clear from eq.(15.7) that one can not transform from the first to the second method without considering singular contributions to the $\mathbf{D}''$ field at the surface. Thus the second method is more singular than the first one, due to an extra derivative in rot $\mathbf{M}$ which is part of $\mathbf{D}''$. This eventually leads to non-standard boundary conditions, which should be used in the second method.

The standard boundary conditions for the $\mathbf{E}$, $\mathbf{D}$, $\mathbf{H}$, and $\mathbf{B}$ fields, used in the first method, are

$$\begin{aligned}
\mathbf{E}_{\|}^{+}(\mathbf{r}_{\|},0,\omega) - \mathbf{E}_{\|}^{-}(\mathbf{r}_{\|},0,\omega) &= 0 \\
D_{z}^{+}(\mathbf{r}_{\|},0,\omega) - D_{z}^{-}(\mathbf{r}_{\|},0,\omega) &= 0 \\
\mathbf{H}_{\|}^{+}(\mathbf{r}_{\|},0,\omega) - \mathbf{H}_{\|}^{-}(\mathbf{r}_{\|},0,\omega) &= 0 \\
B_{z}^{+}(\mathbf{r}_{\|},0,\omega) - B_{z}^{-}(\mathbf{r}_{\|},0,\omega) &= 0
\end{aligned} \tag{15.8}$$

Due to the singular contribution to the $\mathbf{D}_{\|}''$ field the boundary conditions for the second method must be modified into

$$\begin{aligned}
\mathbf{E}_{\|}^{+}(\mathbf{r}_{\|},0,\omega) - \mathbf{E}_{\|}^{-}(\mathbf{r}_{\|},0,\omega) &= 0 \\
D_{z}''^{+}(\mathbf{r}_{\|},0,\omega) - D_{z}''^{-}(\mathbf{r}_{\|},0,\omega) &= -\nabla_{\|}.\mathbf{D}_{\|}''^{s}(\mathbf{r}_{\|},\omega) \\
\mathbf{B}_{\|}^{+}(\mathbf{r}_{\|},0,\omega) - \mathbf{B}_{\|}^{-}(\mathbf{r}_{\|},0,\omega) &= \frac{i\omega}{c}\hat{\mathbf{z}} \times \mathbf{D}_{\|}''^{s}(\mathbf{r}_{\|},\omega) \\
B_{z}^{+}(\mathbf{r}_{\|},0,\omega) - B_{z}^{-}(\mathbf{r}_{\|},0,\omega) &= 0
\end{aligned} \tag{15.9}$$

where the amplitude of the singular displacement field $\mathbf{D}_{\|}''$ is given by

$$\mathbf{D}_{\|}''^{s}(\mathbf{r}_{\|},\omega) \equiv -\frac{c}{i\omega}\widehat{\mathbf{z}} \times \left[\mathbf{M}_{\|}^{+}(\mathbf{r}_{\|},0,\omega) - \mathbf{M}_{\|}^{-}(\mathbf{r}_{\|},0,\omega)\right] \tag{15.10}$$

Both sets of boundary conditions are equivalent. This one may verify by eliminating $\mathbf{D}''$ from eq.(15.9) using eq.(15.7). This results in the conditions given in eq.(15.8). It follows from the analysis given above that, when one uses standard boundary conditions for the first method, one should use non-standard boundary conditions for the second method. As has been clearly established in the literature cited above, the methods are not equivalent when one uses standard boundary conditions for both.

15.3 Constitutive equations

In order to further clarify this matter, the constitutive equations in both methods will be discussed in some more detail. In the first method the general form of the constitutive equations in a homogeneous medium is

$$\begin{aligned} \mathbf{D} &= \boldsymbol{\epsilon}.\mathbf{E}+\frac{i\omega}{c}\mathbf{g}^{em}.\mathbf{H} \\ \mathbf{B} &= -\frac{i\omega}{c}\mathbf{g}^{me}.\mathbf{E}+\boldsymbol{\mu}.\mathbf{H} \end{aligned} \tag{15.11}$$

Here $\boldsymbol{\epsilon}$ and $\boldsymbol{\mu}$ are the frequency dependent electric and magnetic susceptibility tensors, respectively, and $\mathbf{g}^{em}$ and $\mathbf{g}^{me}$ describe the frequency dependent gyrotropic effects. The gyrotropic tensors are always small. It can be shown, as will be clarified in the next section, that these tensors satisfy the general symmetry conditions which follow from time reversal invariance (source-observer symmetry):

$$\epsilon_{jk}(\omega)=\epsilon_{kj}(\omega) \quad , \quad g_{jk}^{em}(\omega)=g_{kj}^{me}(\omega) \quad , \quad \mu_{jk}(\omega)=\mu_{kj}(\omega) \tag{15.12}$$

For isotropic media, in which these tensors are constants, Schlagheck [94] proved that $g^{em}(\omega)=g^{me}(\omega)\equiv g(\omega)$. For the electric displacement field in the second method one obtains from the above constitutive equation to linear order in the gyrotropic tensors

$$\mathbf{D}''=\mathbf{D}-\frac{c}{i\omega}\text{rot } \mathbf{M}=\boldsymbol{\epsilon}.\mathbf{E}+\frac{i\omega}{c}\mathbf{g}^{em}.\boldsymbol{\mu}^{-1}.\mathbf{B}-\frac{c}{i\omega}\text{rot }\left[\left(\mathbf{1}-\boldsymbol{\mu}^{-1}\right).\mathbf{B}-\frac{i\omega}{c}\boldsymbol{\mu}^{-1}.\mathbf{g}^{me}.\mathbf{E}\right]$$

Eliminating $\mathbf{B}$ using the first Maxwell equation one finds

$$\mathbf{D}''=\boldsymbol{\epsilon}.\mathbf{E}+\mathbf{g}^{em}.\boldsymbol{\mu}^{-1}.\text{rot } \mathbf{E}+\text{rot }\left(\boldsymbol{\mu}^{-1}.\mathbf{g}^{me}.\mathbf{E}\right)+\left(\frac{c}{\omega}\right)^2\text{rot }\left[\left(\mathbf{1}-\boldsymbol{\mu}^{-1}\right).\text{rot } \mathbf{E}\right] \tag{15.13}$$

Considering the usual case, in which $\boldsymbol{\epsilon}$, $\mathbf{g}^{em}$, $\mathbf{g}^{me}$ and $\boldsymbol{\mu}$ are discontinuous on the $z=0$ surface, the standard boundary conditions imply that $\mathbf{E}_{\|}$ and $(\text{rot}\,\mathbf{E})_z$ are continuous while E_z and $(\text{rot}\,\mathbf{E})_{\|}$ are discontinuous at the surface. Writing $\mathbf{D}''$ in the form given in eq.(15.7) one obtains for the singular contribution

$$\begin{aligned} \mathbf{D}''^{s} &= \hat{\mathbf{z}}\times\left[\left(\boldsymbol{\mu}^{-1}.\mathbf{g}^{me}.\mathbf{E}\right)_{\|}^{+}-\left(\boldsymbol{\mu}^{-1}.\mathbf{g}^{me}.\mathbf{E}\right)_{\|}^{-}\right]_{z=0} \\ &+\left(\frac{c}{\omega}\right)^2\hat{\mathbf{z}}\times\left[\left(\left(\mathbf{1}-\boldsymbol{\mu}^{-1}\right).\text{rot } \mathbf{E}\right)_{\|}^{+}-\left(\left(\mathbf{1}-\boldsymbol{\mu}^{-1}\right).\text{rot } \mathbf{E}\right)_{\|}^{-}\right]_{z=0} \end{aligned} \tag{15.14}$$

This singular contribution is directed along the surface. For $\boldsymbol{\mu}=\mathbf{1}$ only the first term remains and this expression is precisely the one derived by Bungay et al. [92]. For an interface between two isotropic gyrotropic media, in which ϵ, $g^{em}=g^{me}=g$ and

μ are scalar, the expression reduces to

$$\mathbf{D}''^s = \left[\left(\frac{g}{\mu}\right)^+ - \left(\frac{g}{\mu}\right)^-\right]\widehat{\mathbf{z}} \times \mathbf{E}_\parallel$$
$$+\left(\frac{c}{\omega}\right)^2 \widehat{\mathbf{z}} \times \left[\left(\left(1-\frac{1}{\mu}\right)\text{rot }\mathbf{E}\right)^+_\parallel - \left(\left(1-\frac{1}{\mu}\right)\text{rot }\mathbf{E}\right)^-_\parallel\right]_{z=0}$$
$$= \left[\left(\frac{g}{\mu}\right)^+ - \left(\frac{g}{\mu}\right)^-\right]\widehat{\mathbf{z}} \times \mathbf{E}_\parallel - \left(\frac{c}{i\omega}\right)\left(\mu^+ - \mu^-\right)\widehat{\mathbf{z}} \times \mathbf{H}_\parallel \qquad (15.15)$$

where $\mathbf{E}_\parallel$ and $\mathbf{H}_\parallel$ are both taken at $z = 0$. In the discussions about what method to use to describe reflection between gyrotropic media one usually uses $\mu = 1$. The differences between the two methods are therefore a consequence of neglecting the first term by using the standard boundary conditions in the second method. As Bungay et al. also point out both methods predict the same if one takes this term along in the boundary condition. The second term, which was already found by Maxwell [90], shows that this difference between the two methods does manifest itself not only for surfaces between gyrotropic media. If $g = 0$ and μ is discontinuous, both methods, using standard boundary conditions, also predict different reflection coefficients. If one wants to use the standard boundary conditions, one should clearly use the first method.

15.4 General theory

In the second section it was found that singular contributions to the electric displacement field should be taken along in the second method, in order to make it equivalent to the first method. In the third section it was shown that this singular contribution is a consequence of possible discontinuities of $\mathbf{g}^{me}$ and $\boldsymbol{\mu}$ at the surface. As such this discontinuity does not really describe a physical property of the surface. The first method is in this respect more elegant. As has been discussed extensively in this book, singular contributions to the fields may also occur when one uses the first method. These are due to changes in the structure near the surface, roughness or the absorption of different materials. These are now characteristic properties of the surface. All fields then have the general form, cf. section 3.2,

$$\mathbf{E}(\mathbf{r},\omega) = \mathbf{E}^-(\mathbf{r},\omega)\Theta(-z) + \mathbf{E}^s(\mathbf{r}_\parallel,\omega)\delta(z) + \mathbf{E}^+(\mathbf{r},\omega)\Theta(z) \qquad (15.16)$$

The singular contributions modify the boundary conditions. Using the Maxwell equations for these position and frequency dependent fields, it follows, cf. eq.(3.30) together with eq.(3.27), that the non-standard boundary conditions in that case become

$$\mathbf{E}^+_\parallel(\mathbf{r}_\parallel,0,\omega) - \mathbf{E}^-_\parallel(\mathbf{r}_\parallel,0,\omega) = -\frac{i\omega}{c}\hat{\mathbf{z}} \times \mathbf{B}^s_\parallel(\mathbf{r}_\parallel,\omega) + \nabla_\parallel E^s_z(\mathbf{r}_\parallel,\omega)$$
$$D^+_z(\mathbf{r}_\parallel,0,\omega) - D^-_z(\mathbf{r}_\parallel,0,\omega) = -\nabla_\parallel.\mathbf{D}^s_\parallel(\mathbf{r}_\parallel,\omega)$$
$$\mathbf{H}^+_\parallel(\mathbf{r}_\parallel,0,\omega) - \mathbf{H}^-_\parallel(\mathbf{r}_\parallel,0,\omega) = \frac{i\omega}{c}\hat{\mathbf{z}} \times \mathbf{D}^s_\parallel(\mathbf{r}_\parallel,\omega) + \nabla_\parallel H^s_z(\mathbf{r},\omega)$$
$$B^+_z(\mathbf{r}_\parallel,0,\omega) - B^-_z(\mathbf{r}_\parallel,0,\omega) = -\nabla_\parallel.\mathbf{B}^s_\parallel(\mathbf{r}_\parallel,\omega) \qquad (15.17)$$

Furthermore it is found that, cf. eq.(3.26),

$$\mathbf{E}_{\parallel}^{s}(\mathbf{r}_{\parallel},\omega) = \mathbf{H}_{\parallel}^{s}(\mathbf{r}_{\parallel},\omega) = D_{z}^{s}(\mathbf{r}_{\parallel},\omega) = B_{z}^{s}(\mathbf{r},\omega) = 0 \tag{15.18}$$

This last relation implies that the amplitudes of the singular contributions to the polarization and the magnetization are defined by, cf. eq.(3.27),

$$\mathbf{P}^{s}(\mathbf{r}_{\parallel},\omega) \equiv \left(\mathbf{D}_{\parallel}^{s}(\mathbf{r}_{\parallel},\omega), -E_{z}^{s}(\mathbf{r}_{\parallel},\omega)\right) \quad \text{and} \quad \mathbf{M}^{s}(\mathbf{r}_{\parallel},\omega) \equiv \left(\mathbf{B}_{\parallel}^{s}(\mathbf{r}_{\parallel},\omega), -H_{z}^{s}(\mathbf{r}_{\parallel},\omega)\right) \tag{15.19}$$

Also in this case one may transform to the second method in which the magnetization is adsorbed in the electric displacement field using eq.(15.3). One now obtains

$$\begin{aligned}
\mathbf{D}''(\mathbf{r},\omega) &= \left[\mathbf{D}^{-}(\mathbf{r},\omega) - \frac{c}{i\omega}\text{rot } \mathbf{M}^{-}(\mathbf{r},\omega)\right]\Theta(-z) \\
&+ \left[\mathbf{D}^{+}(\mathbf{r},\omega) - \frac{c}{i\omega}\text{rot } \mathbf{M}^{+}(\mathbf{r},\omega)\right]\Theta(z) \\
&+ \left\{\mathbf{D}^{s}(\mathbf{r},\omega) - \frac{c}{i\omega}\widehat{\mathbf{z}} \times \left[\mathbf{M}_{\parallel}^{+}(\mathbf{r},\omega) - \mathbf{M}_{\parallel}^{-}(\mathbf{r},\omega)\right] - \frac{c}{i\omega}\text{rot } \mathbf{M}^{s}(\mathbf{r}_{\parallel},\omega)\right\}\delta(z) \\
&- \frac{c}{i\omega}\widehat{\mathbf{z}} \times \mathbf{M}^{s}(\mathbf{r}_{\parallel},\omega)\frac{d\delta(z)}{dz}
\end{aligned} \tag{15.20}$$

It follows that, when the singular contribution to the magnetization has a component along the surface, one needs to introduce in the second method a contribution to $\mathbf{D}''$ proportional to $d\delta(z)/dz$. Such a contribution will further complicate the boundary conditions. This is one of the reasons why this book systematically uses the first method, not only for surfaces of gyrotropic media but for all surfaces.

The standard boundary conditions imply that $\mathbf{E}_{\parallel}$, D_z, $\mathbf{H}_{\parallel}$ and B_z are continuous while $\mathbf{D}_{\parallel}$, E_z, $\mathbf{B}_{\parallel}$ and H_z are discontinuous at the surface. When one uses the non-standard boundary conditions (15.17), one finds that $\mathbf{E}_{\parallel}$, D_z, $\mathbf{H}_{\parallel}$ and B_z may have small discontinuities at the surface while $\mathbf{D}_{\parallel}$, E_z, $\mathbf{B}_{\parallel}$ and H_z not only have much larger discontinuities at the surface but are also singular. The difference between the standard boundary conditions and the general boundary conditions is determined by singular fields $\mathbf{B}^s, \mathbf{H}^s, \mathbf{E}^s$ and $\mathbf{D}^s$ at the surface. One notes that at this stage the system of Maxwell equations (15.1), together with the constitutive equations (15.11) in the bulk and the general boundary conditions (15.17), is still incomplete because one also needs constitutive equations for the singular surface polarization $\mathbf{P}^s$ and magnetization $\mathbf{M}^s$. These constitutive equations at the surface can not be written in the same form as eq.(15.11) in the bulk, because $\mathbf{D}$ and $\mathbf{B}$ in this equation are expressed in terms of the fields $\mathbf{E}$ and $\mathbf{H}$ which are generally discontinuous at the surface. Instead it is more convenient to use the non-singular fields introduced in chapter 12,

$$\begin{aligned}
\mathbf{N}_e &\equiv \left(\mathbf{E}_{\parallel}, D_z\right) = \mathbf{E}+\widehat{\mathbf{z}}\widehat{\mathbf{z}}.\mathbf{P} \\
\mathbf{N}_m &\equiv \left(\mathbf{H}_{\parallel}, B_z\right) = \mathbf{H}+\widehat{\mathbf{z}}\widehat{\mathbf{z}}.\mathbf{M}
\end{aligned} \tag{15.21}$$

which are composed of those components that are continuous for standard boundary conditions. At every point in the bulk the bulk polarization $\mathbf{P}$ and magnetization $\mathbf{M}$

can now be expressed in terms of $\mathbf{N}_e$ and $\mathbf{N}_m$. Expressing $\mathbf{P}$ and $\mathbf{M}$ in terms of these non-singular fields, using eqs.(15.2) and (15.11), one obtains after some algebra

$$\begin{aligned}\mathbf{P} &= \boldsymbol{\xi}^{ee}.\mathbf{N}_e + \boldsymbol{\xi}^{em}.\mathbf{N}_m \\ \mathbf{M} &= \boldsymbol{\xi}^{me}.\mathbf{N}_e + \boldsymbol{\xi}^{mm}.\mathbf{N}_m\end{aligned} \tag{15.22}$$

where the constitutive coefficients are related to the coefficients in eq.(15.11) by

$$\begin{aligned}\boldsymbol{\xi}^{ee} &= \left[(\boldsymbol{\epsilon}-\mathbf{1})^{-1}+\widehat{\mathbf{z}}\widehat{\mathbf{z}}\right]^{-1} \\ \boldsymbol{\xi}^{em} &= \frac{i\omega}{c}\left[\mathbf{1}+(\boldsymbol{\epsilon}-\mathbf{1}).\widehat{\mathbf{z}}\widehat{\mathbf{z}}\right]^{-1}.\mathbf{g}^{em}.\left[\mathbf{1}+\widehat{\mathbf{z}}\widehat{\mathbf{z}}.(\boldsymbol{\mu}-\mathbf{1})\right]^{-1} \\ \boldsymbol{\xi}^{me} &= -\frac{i\omega}{c}\left[\mathbf{1}+(\boldsymbol{\mu}-\mathbf{1}).\widehat{\mathbf{z}}\widehat{\mathbf{z}}\right]^{-1}.\mathbf{g}^{me}.\left[\mathbf{1}+\widehat{\mathbf{z}}\widehat{\mathbf{z}}.(\boldsymbol{\epsilon}-\mathbf{1})\right]^{-1} \\ \boldsymbol{\xi}^{mm} &= \left[(\boldsymbol{\mu}-\mathbf{1})^{-1}+\widehat{\mathbf{z}}\widehat{\mathbf{z}}\right]^{-1}\end{aligned} \tag{15.23}$$

to linear order in $\mathbf{g}^{em}$ and $\mathbf{g}^{me}$. From the symmetry relations for $\boldsymbol{\epsilon}$, $\mathbf{g}^{em}$ and $\mathbf{g}^{me}$ and $\boldsymbol{\mu}$, eq.(15.12), it follows that

$$\xi_{jk}^{ee}(\omega) = \xi_{kj}^{ee}(\omega) \quad , \quad \xi_{jk}^{em}(\omega) = -\xi_{kj}^{me}(\omega) \quad , \quad \xi_{jk}^{mm}(\omega) = \xi_{kj}^{mm}(\omega) \tag{15.24}$$

These symmetry relations are a special case of the general symmetry relations proven in chapter 13 and given in eq.(13.26).

Now the constitutive equations for the singular surface fields $\mathbf{P}^s$ and $\mathbf{M}^s$ can be derived from eqs.(15.22) and (15.23). Introduce for this purpose the excess polarization and magnetization in the boundary region, following the procedure discussed in section 2.6,

$$\begin{aligned}\mathbf{P}_{ex}(\mathbf{r},\omega) &= \left(\mathbf{D}_{\|}(\mathbf{r},\omega), -E_z(\mathbf{r},\omega)\right) \\ &\quad - \left(\mathbf{D}_{\|}^{+}(\mathbf{r},\omega), -E_z^{+}(\mathbf{r},\omega)\right)\Theta(z) - \left(\mathbf{D}_{\|}^{-}(\mathbf{r},\omega), -E_z^{-}(\mathbf{r},\omega)\right)\Theta(-z) \\ \mathbf{M}_{ex}(\mathbf{r},\omega) &= \left(\mathbf{B}_{\|}(\mathbf{r},\omega), -H_z(\mathbf{r},\omega)\right) \\ &\quad - \left(\mathbf{B}_{\|}^{+}(\mathbf{r},\omega), -H_z^{+}(\mathbf{r},\omega)\right)\Theta(z) - \left(\mathbf{B}_{\|}^{-}(\mathbf{r},\omega), -H_z^{-}(\mathbf{r},\omega)\right)\Theta(-z)\end{aligned} \tag{15.25}$$

These expressions can be written in the form

$$\begin{aligned}\mathbf{P}_{ex}(\mathbf{r},\omega) &\equiv \mathbf{P}(\mathbf{r},\omega) - \mathbf{P}^{+}(\mathbf{r},\omega)\Theta(z) - \mathbf{P}^{-}(\mathbf{r},\omega)\Theta(-z) \\ &\quad + (1-2\widehat{\mathbf{z}}\widehat{\mathbf{z}}).\left[\mathbf{N}_e(\mathbf{r},\omega) - \mathbf{N}_e^{+}(\mathbf{r},\omega)\Theta(z) - \mathbf{N}_e^{-}(\mathbf{r},\omega)\Theta(-z)\right] \\ \mathbf{M}_{ex}(\mathbf{r},\omega) &\equiv \mathbf{M}(\mathbf{r},\omega) - \mathbf{M}^{+}(\mathbf{r},\omega)\Theta(z) - \mathbf{M}^{-}(\mathbf{r},\omega)\Theta(-z) \\ &\quad + (1-2\widehat{\mathbf{z}}\widehat{\mathbf{z}}).\left[\mathbf{N}_m(\mathbf{r},\omega) - \mathbf{N}_m^{+}(\mathbf{r},\omega)\Theta(z) - \mathbf{N}_m^{-}(\mathbf{r},\omega)\Theta(-z)\right]\end{aligned} \tag{15.26}$$

$\mathbf{P}(\mathbf{r},\omega)$ and $\mathbf{M}(\mathbf{r},\omega)$ are the polarization and magnetization densities given by eqs. (15.22) and (15.23). They depend on the functions $\boldsymbol{\epsilon}(z,\omega)$, $\boldsymbol{\mu}(z,\omega)$, $\mathbf{g}^{em}(z,\omega)$ and $\mathbf{g}^{me}(z,\omega)$, which vary rapidly through the surface. The superscripts "+" and "−"

indicate the extrapolated values of the corresponding quantity in the bulk medium to the right, $z > 0$, and to the left, $z < 0$, from the surface, respectively. The extrapolated values of the polarization and magnetization depend on the constant tensors $\boldsymbol{\epsilon}^{+}$, $\boldsymbol{\mu}^{+}$, $\mathbf{g}^{em,+}$, $\mathbf{g}^{me,+}$ and $\boldsymbol{\epsilon}^{-}$, $\boldsymbol{\mu}^{-}$, $\mathbf{g}^{em,-}$, $\mathbf{g}^{me,-}$. The singular surface polarization and magnetization, cf. section 2.6, can be expressed as

$$\begin{aligned}\mathbf{P}^{s}(\mathbf{r}_{\|},\omega) &= \int_{-\infty}^{\infty} \mathbf{P}_{ex}(\mathbf{r},\omega)dz \\ \mathbf{M}^{s}(\mathbf{r}_{\|},\omega) &= \int_{-\infty}^{\infty} \mathbf{M}_{ex}(\mathbf{r},\omega)dz \end{aligned} \tag{15.27}$$

The fields $\mathbf{P}$, $\mathbf{P}^{+}$,$\mathbf{P}^{-}$ and $\mathbf{M}$, $\mathbf{M}^{+}$,$\mathbf{M}^{-}$ can be expressed in terms of the non-singular fields $\mathbf{N}_{e}$ and $\mathbf{N}_{m}$ using eqs.(15.22) and (15.23) and then can be substituted into eq.(15.27). It is assumed that the tensors $\boldsymbol{\epsilon}(z)$, $\boldsymbol{\mu}(z)$, $\mathbf{g}^{em}(z)$ and $\mathbf{g}^{me}(z)$ depend on z only in the narrow boundary region around $z = 0$. In this case the fields $\mathbf{N}_{e}$ and $\mathbf{N}_{m}$ can be treated as constants within this interval in the first approximation, see the paragraph under eq.(15.30) below. This implies that the terms with the prefactor $(1 - 2\widehat{\mathbf{z}\mathbf{z}})$ in eq.(15.26) do not contribute to the surface polarization and magnetization in this approximation*. The fields $\mathbf{N}_{e}$ and $\mathbf{N}_{m}$ can furthermore be taken out of the integrals in eq.(15.27). As a result one obtains the following constitutive equations for the surface fields $\mathbf{P}^{s}$ and $\mathbf{M}^{s}$

$$\begin{aligned}\mathbf{P}^{s} &= \boldsymbol{\xi}^{ee,s}.\mathbf{N}_{e,+} + \boldsymbol{\xi}^{em,s}.\mathbf{N}_{m,+} \\ \mathbf{M}^{s} &= \boldsymbol{\xi}^{me,s}.\mathbf{N}_{e,+} + \boldsymbol{\xi}^{mm,s}.\mathbf{N}_{m,+}\end{aligned} \tag{15.28}$$

The subscript + is explained in the paragraph below eq.(15.30). The constitutive coefficients are given by

$$\boldsymbol{\xi}^{\alpha\beta,s}(\omega) = \int_{-\infty}^{\infty} \left[\boldsymbol{\xi}^{\alpha\beta}(z,\omega) - \boldsymbol{\xi}^{\alpha\beta,+}(\omega)\Theta(z) - \boldsymbol{\xi}^{\alpha\beta,-}(\omega)\Theta(-z)\right] dz \tag{15.29}$$

where $\alpha, \beta = e$ or m. For the special case that the system is isotropic and non-gyrotropic this expression reproduces those given in eq.(11.5) for $\gamma_{e}, \beta_{e}, \gamma_{m}$ and β_{m} for stratified media. As the analysis in this chapter is restricted to contributions of linear order the second order terms in eq.(11.5) are not reproduced. Eq.(15.29) presents explicit expressions for the constitutive coefficients at the surface in terms of the position dependent material tensors $\boldsymbol{\epsilon}(z,\omega)$, $\boldsymbol{\mu}(z,\omega)$, $\mathbf{g}^{em}(z,\omega)$ and $\mathbf{g}^{me}(z,\omega)$, using eq.(15.23) for all values of z, and which are reduced to the corresponding bulk values for the two media in the limits $z \to \pm\infty$. The symmetry relations for the constitutive coefficients are

$$\xi_{jk}^{ee,s}(\omega) = \xi_{kj}^{ee,s}(\omega) \quad , \quad \xi_{jk}^{em,s}(\omega) = -\xi_{kj}^{me,s}(\omega) \quad , \quad \xi_{jk}^{mm,s}(\omega) = \xi_{kj}^{mm,s}(\omega) \tag{15.30}$$

*In ref. [93] this contribution was already neglected by choosing a different definition of the excess polarization and magnetization. This was done because the authors felt that the alternative definition was more transparant for people who have not read the earlier chapters of this book.

similar to those given in eq.(15.24). These symmetry relations are a special case of the general symmetry relations proven in chapter 13 and given for the surface in eq.(13.29).

One notes that the fields $\mathbf{N}_e$ and $\mathbf{N}_m$ are continuous across the surface only if the surface constitutive coefficients vanish, i.e. if the standard boundary conditions are correct. However, the constitutive coefficients for the surface are assumed to be small and small discontinuities in $\mathbf{N}_e$ and $\mathbf{N}_m$ may therefore be neglected to linear order in these coefficients. Corrections would be of relative order of the ratio between the thickness of the surface and the wavelength. In order to be precise it was specified in chapter 3 that the constitutive equations contain the averages of the fields $\mathbf{N}_e$ and $\mathbf{N}_m$ on both sides of the surface. This is indicated by a subscript + in equation (15.28). This choice enables one to perform a systematic expansion into powers of the ratio between the thickness of the surface and the wavelength. The reader is referred to section 3.6 for a discussion of this point. As discussed in section 3.7, the second order contribution in this expansion gives contributions of quadrupolar order. This will not be discussed here. To linear order one may replace this average by the extrapolated values on either side of the surface. This property will in fact be convenient in the following section where the reflection amplitudes are calculated.

Eq.(15.29) can be used to prove that at the ideal flat "mathematical" boundary between two gyrotropic media the first method, based on eq.(15.11), yields the standard boundary conditions without any singular surface terms. Indeed, in this ideal case the tensors $\boldsymbol{\xi}^{\alpha\beta}(z,\omega)$ can be written as

$$\boldsymbol{\xi}^{\alpha\beta}(z,\omega) = \boldsymbol{\xi}^{\alpha\beta,+}(\omega)\Theta(z) + \boldsymbol{\xi}^{\alpha\beta,-}(\omega)\Theta(-z) \tag{15.31}$$

and therefore all constitutive coefficients for the surface given in eq.(15.29) vanish. This means that there is no excess polarization and/or magnetization at such a surface, i.e. $\mathbf{P}^s = \mathbf{M}^s = 0$. One may therefore conclude that also $\mathbf{E}^s = \mathbf{D}^s = \mathbf{H}^s = \mathbf{B}^s = 0$ in eq.(15.17). As a consequence the general boundary conditions reduce to the standard ones. Thus one arrives at the important conclusion that in the first method the non-standard boundary conditions are obtained when the surface is not perfect, i.e. it is different from the "mathematical" boundary given above. Mathematically this is due to a z-dependance of the material tensors $\boldsymbol{\epsilon}(z,\omega)$, $\boldsymbol{\mu}(z,\omega)$, $\mathbf{g}^{em}(z,\omega)$ and $\mathbf{g}^{me}(z,\omega)$ close to the surface. As has been illustrated extensively in the earlier chapters, such dependence arises due to structural changes close to the surface, surface roughness or adsorbed materials, like oxides, for example.

As was explained in section 3.8, there is one more possible reason why the coefficients $\boldsymbol{\xi}^{\alpha\beta,s}$ arise and that is the choice of the location of the surface. If one shifts this location over a distance d in a positive direction, it follows from eq.(15.29) that the interfacial constitutive coefficients increase by $d\left(\boldsymbol{\xi}^{\alpha\beta,+} - \boldsymbol{\xi}^{\alpha\beta,-}\right)$. The proper choice of the location of an ideal surface, see eq.(15.31), is such that there is no such contribution. It should be emphasized that the reflection coefficients do not depend on this shift. As explained in section 3.9, the reflection coefficients depend only on combinations of these coefficients which are independent of such a shift. A further reason to mention this rather trivial contribution to the constitutive coefficients of the surface, is the transformation to $\mathbf{D}''$ in the second method. Eq.(15.20) shows that after shifting the location of the surface, one needs terms proportional to $\delta(z)$ and to $d\delta(z)/dz$ (times the appropriate elements of $d\left(\boldsymbol{\xi}^{\alpha\beta,+} - \boldsymbol{\xi}^{\alpha\beta,-}\right)$) in the second method in order to obtain an equivalent description. It is clear that the second method, though perfectly correct, leads to very unpleasant complications in the boundary conditions and in the constitutive equations. As the analysis makes clear the order of the singularity in the second method is always one higher than in the first method. This makes the second method more difficult to use, when one wants to address simple problems, like how to choose the location of the surface.

The above equations, which were first derived in [93], give a complete set of constitutive relations not only in the bulk but, more importantly, also for the surface. The constitutive coefficients for the surface found by these authors are new in the sense that they were not earlier given for surfaces between gyrotropic media. In the following section it will be worked out how these coefficients modify the reflection amplitudes for normally incident light for the special case that both media are isotropic and that only one of them is gyrotropic.

15.5 Reflection at normal incidence

Consider reflection of normally incident light by an isotropic gyrotropic medium. The reason to consider in particular this case is that both methods, using standard boundary conditions, lead to a fundamentally different result. The first method predicts that the reflection coefficients for left and right circularly polarized light are the same, while the second method predicts that they are different. As such this case is used as one of the standard examples. It will be investigated to what extent the presence of singular fields in the first method may give rise to a difference between these two reflection coefficients. The magnetic permeability is everywhere taken equal to one.

Consider linearly polarized light incident from the left, $z < 0$, through an ambient medium which is isotropic and non-gyrotropic. The dielectric constant is $\epsilon_a = n_a^2$. The incident fields are

$$\begin{aligned}\mathbf{E}^{inc} &= (1,0,0)\exp\left[i\left(\frac{\omega}{c}n_a z - \omega t\right)\right]\\ \mathbf{B}^{inc} &= n_a\,(0,1,0)\exp\left[i\left(\frac{\omega}{c}n_a z - \omega t\right)\right]\end{aligned} \tag{15.32}$$

The polarization of the incident light is chosen along the x-axis. The analysis for an incident polarization along the y-axis is analogous. The reflected fields are

$$\begin{aligned}\mathbf{E}^{refl} &= (r_x, r_y, 0)\exp\left[i\left(-\frac{\omega}{c}n_a z - \omega t\right)\right]\\ \mathbf{B}^{refl} &= -n_a\,(-r_y, r_x, 0)\exp\left[i\left(-\frac{\omega}{c}n_a z - \omega t\right)\right]\end{aligned} \tag{15.33}$$

The reason to give the $\mathbf{E}$ and the $\mathbf{B}$ fields is that they are normal to each other and to the wave vector, $(\omega n_a/c)\,\widehat{\mathbf{z}}$. This follows from the first Maxwell equation. In the ambient $\mathbf{D} = \epsilon_a\mathbf{E}$ and $\mathbf{H} = \mathbf{B}$. The isotropic substrate has a dielectric constant $\epsilon_s = n_s^2$ and a gyrotropic coefficient g_s. In the substrate one finds after some algebra the following dispersion relation for left (−) and right (+) polarized light

$$k_{\pm} = \frac{\omega}{c}\left(n_s \pm \frac{\omega}{c} g_s\right) \equiv \frac{\omega}{c} n_{\pm} \tag{15.34}$$

to linear order in g_s. Subscripts a and s are used rather than the superscripts − and + to indicate values in the regions for $z < 0$ and $z > 0$. One should not be confused between the subscripts − and +, which indicate left and right polarized light, and the same subscripts, which indicate the difference and the average of a discontinuous quantity at the surface. Below these subscripts are only used to indicate left and right polarized light. The corresponding transmitted electric and magnetic fields in the direction normal to the surface are

$$\begin{aligned}\mathbf{E}^{tr} &= t_{-}\,(1,-i,0)\exp\left[i\left(\frac{\omega}{c}n_{-}z-\omega t\right)\right]+t_{+}\,(1,i,0)\exp\left[i\left(\frac{\omega}{c}n_{+}z-\omega t\right)\right]\\ \mathbf{B}^{tr} &= t_{-}n_{-}\,(i,1,0)\exp\left[i\left(\frac{\omega}{c}n_{-}z-\omega t\right)\right]+t_{+}n_{+}\,(-i,1,0)\exp\left[i\left(\frac{\omega}{c}n_{+}z-\omega t\right)\right]\end{aligned} \tag{15.35}$$

The corresponding **D** and the **H** fields can be found using the constitutive relations (15.11).

As all the fields are parallel to the surface one can simplify the general boundary conditions (15.17), using also eq.(15.19), to

$$\begin{aligned}\mathbf{E}_{\|}^{+} - \mathbf{E}_{\|}^{-} &= -\frac{i\omega}{c}\hat{\mathbf{z}} \times \mathbf{M}_{\|}^{s}\\ \mathbf{H}_{\|}^{+} - \mathbf{H}_{\|}^{-} &= \frac{i\omega}{c}\hat{\mathbf{z}} \times \mathbf{P}_{\|}^{s}\end{aligned} \tag{15.36}$$

Because the magnetic permeability is equal to one in the bulk phases $\boldsymbol{\xi}_{\|\|}^{mm} = 0$ both in the ambient and in the substrate. From eq.(15.29) it follows that also $\boldsymbol{\xi}_{\|\|}^{mm,s} = 0$. In the ambient, which is non-gyrotropic, one has $\boldsymbol{\xi}_{\|\|}^{em} = \boldsymbol{\xi}_{\|\|}^{me} = 0$. It follows that $\mathbf{H}_{\|}^{-} = \mathbf{B}_{\|}^{-}$ in the ambient. The substrate is an isotropic gyrotropic medium and one has

$$\mathbf{H}_{\|}^{+} = \mathbf{B}_{\|}^{+} - \mathbf{M}_{\|}^{+} = \mathbf{B}_{\|}^{+} + \frac{i\omega}{c} g_s \mathbf{E}_{\|}^{+} \tag{15.37}$$

Define the two by two matrices

$$\boldsymbol{\xi}_{\|\|}^{ee,s} \equiv \boldsymbol{\gamma} \quad , \quad \boldsymbol{\xi}_{\|\|}^{em,s} \equiv \frac{i\omega}{c}\mathbf{g}_{\|\|}^{em,s} \quad \text{and} \quad \boldsymbol{\xi}_{\|\|}^{me,s} \equiv -\frac{i\omega}{c}\mathbf{g}_{\|\|}^{me,s} \tag{15.38}$$

for the general case that the surface is non-isotropic. For the $\boldsymbol{\gamma}$ matrix, which is Hermitian, there are two orthogonal principle axes. The x and the y-axis will be chosen along these axes. The $\boldsymbol{\gamma}$ matrix is then diagonal, with diagonal elements γ_x and γ_y. The symmetry relations for the gyrotropic matrices along the surface are

$$\begin{aligned}g_{xx}^{em,s} &= g_{xx}^{me,s} \equiv g_{xx}^{s} \quad , \quad g_{yy}^{em,s} = g_{yy}^{me,s} \equiv g_{yy}^{s}\\ g_{xy}^{em,s} &= g_{yx}^{me,s} \equiv g_{xy}^{s} \quad , \quad g_{yx}^{em,s} = g_{xy}^{me,s} \equiv g_{yx}^{s}\end{aligned} \tag{15.39}$$

Using eqs.(15.28), (15.21), (15.37), (15.38) and (15.39), the boundary conditions (15.36) can be written as

$$\begin{aligned}
E_x^+ - E_x^- &= \left(\frac{\omega}{c}\right)^2 \left(g_{xy}^s E_x^+ + g_{yy}^s E_y^+\right) \\
E_y^+ - E_y^- &= -\left(\frac{\omega}{c}\right)^2 \left(g_{xx}^s E_x^+ + g_{yx}^s E_y^+\right) \\
H_x^+ - H_x^- &= -\frac{i\omega}{c}\left[\gamma_y E_y^+ + \frac{i\omega}{c}\left(g_{yx}^s H_x^+ + g_{yy}^s H_y^+\right)\right] \\
H_y^+ - H_y^- &= \frac{i\omega}{c}\left[\gamma_x E_x^+ + \frac{i\omega}{c}\left(g_{xx}^s H_x^+ + g_{xy}^s H_y^+\right)\right]
\end{aligned} \tag{15.40}$$

Here it has been used that to linear order in the interfacial constitutive coefficients one may replace the averaged fields on the right-hand side of eq.(15.28) by their values in the substrate. Substituting the fields, given in eqs.(15.32), (15.33) and (15.35), into eq.(15.40) and using eq.(15.34), one finds for the reflection coefficients of incident light with a direction of polarization along the x-axis

$$\begin{aligned}
r_x &= \frac{n_a - n_s}{n_a + n_s} + 2\frac{i\omega}{c}\gamma_x \frac{n_a}{(n_a+n_s)^2} - 4\left(\frac{\omega}{c}\right)^2 g_{xy}^s \frac{n_a n_s}{(n_a+n_s)^2} \\
r_y &= 2\left(\frac{\omega}{c}\right)^2 \left(g_{xx}^s - g_{yy}^s\right) \frac{n_a n_s}{(n_a+n_s)^2}
\end{aligned} \tag{15.41}$$

One may do a similar analysis for incident light with a direction of polarization along the y-axis. This results in reflection amplitudes

$$\begin{aligned}
r_x' &= 2\left(\frac{\omega}{c}\right)^2 \left(g_{xx}^s - g_{yy}^s\right) \frac{n_a n_s}{(n_a+n_s)^2} \\
r_y' &= \frac{n_a - n_s}{n_a + n_s} + 2\frac{i\omega}{c}\gamma_y \frac{n_a}{(n_a+n_s)^2} + 4\left(\frac{\omega}{c}\right)^2 g_{yx}^s \frac{n_a n_s}{(n_a+n_s)^2}
\end{aligned} \tag{15.42}$$

It follows from these formulae that the polarization of the reflected light is the same as the polarization of the incident light, both linear and circular, if $\gamma_x = \gamma_y$, $g_{xx}^s = g_{yy}^s$ and $g_{xy}^s = -g_{yx}^s$. An isotropic surface is invariant for rotations around a normal to the surface. For such an isotropic surface these relations follow. If one compares with the constitutive matrix given in eq.(13.30), one finds that one has for that case an isotropic surface with $\gamma_x = \gamma_y = \gamma$, $g_{xx}^s = g_{yy}^s = 0$ and $g_{xy}^s = -g_{yx}^s = -\tau$. When the interfacial constitutive coefficients are zero the reflection coefficients reduce to the value given by Fresnel for the surface of a non-gyrotropic medium. The bulk gyrotropic tensor does not change the reflection coefficient for normal reflection. This is a property one generally agrees upon [84, 86, 92, 94, 95].

If the interface is not isotropic, there are three different reasons why the surface modifies the nature of the polarization of normally incident light upon reflection. The first is $\gamma_x \neq \gamma_y$, the second $g_{xx}^s \neq g_{yy}^s$ and the third $g_{xy}^s \neq -g_{yx}^s$. It is therefore in general not correct to say that the reflection coefficients for left and right polarized light are the same. The surface leads to differences between the reflection of left and right polarized light. The first reason is that the polarization of the surface is not rotationally invariant. The second reason is due to a similar asymmetry for the diagonal of the gyrotropic tensor. It is noted that a shift d in the location of the dividing surface would lead to a change in both these coefficients of dg_s. This does not modify the reflection amplitudes. The third reason for differences between the reflection of left and right polarized light are the off-diagonal coefficients of the gyrotropic tensor.

REFERENCES

1. J.C. Maxwell Garnett. *Phil. Trans. Roy. Soc. London, 203A*, page 385, 1904.
2. D.A.G. Bruggeman. *Ann.Physik (Leipz.), 24*, page 636, 1935.
3. D.A.G. Bruggeman. *Ann.Physik (Leipz.), 29*, page 160, 1937.
4. R. Landauer, eds. J.C. Garland and D.B. Tanner. *Proc. of the First Conf. on the Electr. and Opt. Prop. of Inhomogeneous Media, no. 40, p. 2.* AIP, New York, 1978.
5. L.K.H. van Beek. *Prog. Dielectr., 7*, page 69, 1967.
6. W.R. Tinga, W.A.G. Voss and D.F. Blossey. *J. Appl. Phys., 44*, page 3897, 1973.
7. C.J.F. Böttcher and P. Bordewijk. *Theory of Electric Polarization, vol. 2, p. 476.* Elsevier, Amsterdam, 1978.
8. C. Grosse and J.L. Greffe. *J. Chem. Phys., 67*, page 305, 1979.
9. G.A. Niklasson, C.G. Granqvist and O. Hunderi. *Applied Optics, 20*, page 26, 1981.
10. T. Yamaguchi, S. Yoshida and A. Kinbara. *Thin Solid Films, 21*, page 173, 1974.
11. J. Lekner. *Theory of Reflection.* Martinus Nijhoff Publishers, Dordrecht, The Netherlands, 1987.
12. H.A. Lorentz. *Proc. Akad. Wetensch. Amsterdam, 4*, page 176, 1895.
13. J.W. Gibbs. *Collected Works, 2 vols.* Dover, New York, 1961.
14. M. Born and E. Wolf. *Principles of Optics.* Pergamon Press, New York, 1959.
15. A.M. Albano, D. Bedeaux and J. Vlieger. *Physica A, 99*, page 293, 1979.
16. A.M. Albano, D. Bedeaux and J. Vlieger. *Physica A, 102*, page 105, 1980.
17. D.Bedeaux and J. Vlieger. *Physica A, 67*, page 55, 1973.
18. A.M. Albano and D. Bedeaux. *Physica A, 147*, page 407, 1987.
19. A. Stahl and H. Wolters. *Z. Physik, 255*, page 227, 1972.
20. V.M. Agranovich and V.L. Ginzburg. *Crystal Optics with Spacial Dispersion and Excitons.* Springer Verlag, Berlin, 1984.
21. D. Bedeaux and J. Vlieger. *Physica, 73*, page 287, 1974.
22. D. Bedeaux, G.J.M. Koper, E.A. van der Zeeuw, J. Vlieger and M.M. Wind. *Physica A, 207*, page 285, 1994.
23. M.T. Haarmans and D. Bedeaux. *Thin Solid Films, 258*, page 213, 1995.
24. E.K. Mann, L. Heinrich, J.C. Voegel and P. Schaaf. *J. Chem. Phys., 105*, page 6082, 1996.
25. E.K. Mann, L. Heinrich, M.J. Semmler, J.C. Voegel and P. Schaaf. *J. Chem. Phys.,*

108, page 7416, 1998.

26. J.D. Jackson. *Classical Electrodynamics.* John Wiley, New York, 1962.
27. R.M.A. Azzam and N.M. Bashara. *Ellipsometry and Polarized Light.* North Holland Publ. Co., Amsterdam, 1989.
28. M. Abramowitz and I.A. Stegun. *Handbook of Mathematical Functions.* National Bureau of Standards, Washington D.C., 1970.
29. R.S. Sennett and G.D. Scott. *J. Opt. Soc. Am., 40*, page 203, 1950.
30. S. Norrman, T. Anderson, C.G. Granqvist and O. Hunderi. *Phys. Rev. B, 18*, page 674, 1978.
31. T. Yamaguchi, S. Yoshida and A. Kinbara. *Thin Solid Films, 18*, page 63, 1973.
32. T. Yamaguchi. *J. Phys. Soc. Jpn., 17*, page 184, 1962.
33. A. Meessen. *J. Phys. (Paris), 33*, page 371, 1972.
34. H.G. Craighead and G.A. Niklasson. *Appl. Phys. Lett., 44*, page 1134, 1984.
35. G.A. Niklasson and H.G. Craighead. *Thin Solid Films, 125*, page 165, 1985.
36. M.E. Rose. *Elementary Theory of Angular Momentum.* Academic Press, New York, 1957.
37. P.M. Morse and H. Feshbach. *Methods of Theoretical Physics, vols. I and II.* McGraw-Hill, New York, 1953.
38. R. Ruppin. *Surface Science, 127*, page 108, 1983.
39. M.A. van Dijk, J.G.H. Joosten, Y.K. Levine and D. Bedeaux. *J. Phys. Chem., 93*, page 2506, 1989.
40. P.B. Johnson and R.W. Christy. *Phys. Rev. B, 6*, page 4370, 1972.
41. E.D. Palik, editor. *Handbook of Optical Constants of Solids.* Press, 1985.
42. P.A. Bobbert and J. Vlieger. *Physica A, 147*, page 115, 1987.
43. B.U. Felderhof. *Physica A, 130*, page 34, 1985.
44. V.V. Truong and G.D. Scott. *J. Opt. Soc Am., 66*, page 124, 1976.
45. V.V. Truong and G.D. Scott. *J. Opt. Soc Am., 67*, page 502, 1977.
46. D. Bedeaux and J. Vlieger. *Thin Solid Films, 102*, page 265, 1983.
47. J.A. Stratton. *Electromagnetic Theory.* McGraw-Hill, New York, 1941.
48. I.S. Gradshteyn and I.M. Ryzhik. *Tables of Integrals, Series and Products.* Academic Press, New York, 1965.
49. M.T. Haarmans and D. Bedeaux. *Thin Solid Films, 224*, page 117, 1993.
50. M.T. Haarmans and D. Bedeaux. *Physica A, 207*, page 340, 1994.
51. B.R.A. Nijboer and F.W. de Wette. *Physica, 23*, page 309, 1957.
52. M.M. Wind, J. Vlieger and D. Bedeaux. *Physica A, 141*, page 33, 1987.
53. M.M. Wind, P.A. Bobbert, J. Vlieger and D. Bedeaux. *Physica A, 143*, page 164, 1987.
54. D.W. Berreman. *Phys.Rev., 163*, page 855, 1967.
55. R. Chauvaux and A. Meessen. *Thin Solid Films, 62*, page 125, 1979.
56. I. Simonsen, R. Lazzari, J. Jupille and S. Roux. *Phys. Rev. B, 61*, 2000.
57. J. Vlieger and D. Bedeaux. *Physica A, 85*, page 389, 1976.

58. B.J.A. Zielinska, D. Bedeaux and J. Vlieger. *Physica A, 107*, page 91, 1981.
59. B.J.A. Zielinska, D. Bedeaux and J. Vlieger. *Physica A, 117*, page 28, 1983.
60. J. Vlieger and M.M. Wind. *J. de Physique, Colloques C, 10, nr. 44*, page 363, 1983.
61. M.M. Wind and J. Vlieger. *Physica A, 125*, page 75, 1984.
62. M.M. Wind and J. Vlieger. *Physica A, 131*, page 1, 1985.
63. D. Bedeaux, E.M. Blokhuis and J.W. Schmidt. *Int. Journ. of Thermophys., 11*, page 13, 1990.
64. E.M. Blokhuis and D. Bedeaux. *Physica A, 164*, page 515, 1990.
65. E.A. Stern. *Phys. Rev. Letters, 19*, page 1321, 1967.
66. E. Kretschmann. *Z. Physik, 227*, page 412, 1969.
67. E. Kretschmann and E. Kröger. *J. Opt. Soc. Am., 65*, page 150, 1974.
68. H.J. Juranek. *Z. Physik, 233*, page 324, 1970.
69. D. Beaglehole and O. Hunderi. *Phys. Rev. B, 2*, page 309, 1970.
70. D. Beaglehole and O. Hunderi. *Phys. Rev. B, 2*, page 321, 1970.
71. E. Kröger and E. Kretschmann. *Z. Physik, 237*, page 1, 1970.
72. D. Bedeaux. *Adv. Chem. Phys., 64*, page 47, 1986.
73. J. Meunier. *J. Physique, 48*, page 1819, 1987.
74. D. Beaglehole. *Physica B, 100*, page 163, 1980.
75. J. Vrijmoeth et al. *Phys. Rev. Lett., 72*, page 3843, 1994.
76. J.-K. Zuo and J.F. Wendelken. *Phys. Rev. Lett., 78*, page 2791, 1997.
77. J.J. Boland and J.H. Weaver. *Phys. Today, 51*, page 34, 1998.
78. K. Sieradzki, S.R. Brankovic and N. Dimitrov. *Science, 284*, page 138, 1999.
79. S. Gheorghiu and P. Pfeifer. *Phys. Rev. Lett., 85*, page 3894, 2000.
80. P. Pfeifer and S. Gheorghiu. *Int. J. Mod. Phys., 15*, page 3197, 2001.
81. L.D. Landau and E.M. Lifshitz. *Electrodynamics of Continuous Media.* Pergamon Press, Oxford, 1980.
82. E.U. Condon. *Rev. Mod. Phys.,9*, page 432, 1937.
83. M. Born. *Optik.* Springer Verlag, Heidelberg, 1972.
84. T. Takizawa. *J. Phys. Soc. Japan, 50*, page 3054, 1981.
85. M.P. Silverman. *Lett. Nuovo Cimento, 43*, page 378, 1985.
86. M.P. Silverman. *J. Opt. Soc. Am., A 3*, page 830, 1986.
87. M.P. Silverman and J. Badoz. *Opt. Comm., 74*, page 129, 1989.
88. A.Yu. Luk'yanov and M.A. Novikov. *Pis'ma Zh. Eksp. Teor. Fiz., 51*, page 673, 1990.
89. M.P. Silverman, J. Badoz and B. Briat. *Opt. Lett., 17*, page 886, 1992.
90. J.C. Maxwell. *A Treatise on Electricity and Magnetism.* Dover Publ. Inc., New York, 1954.
91. A.R. Bungay, Yu.P. Svirko and N.U. Zheludev. *Phys. Rev. Lett., 70*, page 3039, 1993.
92. A.R. Bungay, Yu.P. Svirko and N.U. Zheludev. *Phys. Rev., B 47*, page 16141, 1993.

93. M. Osipov, D. Bedeaux and J. Vlieger. *In preparation*, 2004.
94. U. Schlagheck. *Z. Physik, 258*, page 223, 1973.
95. M.P. Silverman and J. Badoz. *J. Opt. Soc. Am., A 7*, page 1163, 1990.

INDEX